T0269367

Book of Abstracts of the 64th Annual Meeting of the European Federation of Animal Science

EAAP - European Federation of Animal Science

The European Federation of Animal Science wishes to express its appreciation to the
Ministero delle Politiche Agricole e Forestali (Italy) and the
Associazione Italiana Allevatori (Italy)
for their valuable support of its activities.

Book of Abstracts of the 64th Annual Meeting of the European Federation of Animal Science

Nantes, France, 26-30 August, 2013

EAAP Scientific Committee:

H. Simianer
J.E. Lindberg
C. Fourichon
M. Vestergaard
A. Bernués
A. Kuipers
L. Bodin
P. Knap
N. Miraglia
G. Pollott

Proceedings publication and Abstract Submission System (OASES) by

Wageningen Academic Publishers
P.O. Box 220
6700 AE Wageningen
The Netherlands
www.WageningenAcademic.com
copyright@WageningenAcademic.com

ISBN: 978-90-8686-228-3
e-ISBN: 978-90-8686-782-0
DOI: 10.3920/978-90-8686-782-0

ISSN 1382-6077

First published, 2013

Welcome to EAAP 2013 in Nantes - France

On behalf of the French Organising Committee, I am pleased to invite you to attend the 64th Annual Meeting of the EAAP which will take place in Nantes from the 26th to the 30th of August, 2013. This will be the third time that France has hosted this meeting, with the last time being in Toulouse in 1990.

The main theme of the meeting will be "New challenges facing animal production for diversified territories, market demands and social expectations". The programme will cover all aspects of scientific achievements within animal production, including genetics, physiology, nutrition, management and health. You will have the opportunity to attend a number of selected oral presentations and study posters from a great number of scientists from Europe and world-wide. You will also take part in workshops and discussions of the latest and most relevant research in the field of Animal Science. You will see good examples of successful partnerships of international teams bringing scientists and stakeholders together. We will also pay attention to efficient and faster transfer of knowledge and life education of professionals in the livestock sector. This is a unique occasion for updating knowledge and acquiring new ideas, and we especially encourage young scientists and students to attend.

We are sure that all of you will have a productive meeting from a scientific point of view and that you will also enjoy the social events, landscape, culture and hospitality in France.

Henri Seegers

President of the French Organising Committee

National Organisers of the 64th EAAP Annual Meeting

French National Organising Committee

President
- **Henri Seegers**,
  INRA, president@nantes.inra.fr

Vice President
- **Joël Merceron**,
  Institut de l'Elevage, The Breeding Institute, joel.merceron@idele.fr

Executive Secretaries
- **Michel Bonneau**,
  INRA, michel.bonneau@rennes.inra.fr
- **Bernard Coudurier**,
  INRA, bernard.coudurier@tours.inra.fr

Members
- **Daniel Sauvant**,
  French Association for Animal Science, sauvant@agroparistech.fr
- **Dominique Tremblay**,
  Pays de Loire Regional Council, dominique.tremblay@paysdelaloire.fr
- **Alain Montembault**,
  TERRENA, amontembault@terrena.fr
- **Philippe Lecomte**,
  CIRAD, Agricultural Research for Development, philippe.lecomte@cirad.fr
- **Gilles Salvat**,
  ANSES, French Agency for Food, Health and Safety, gilles.salvat@anses.fr
- **Thierry Geslain**,
  CNIEL, The Milk House, tgeslain@cniel.com
- **Stéphane Radet**,
  SNIA, Union of Animal Nutrition Industries, s.radet@nutritionanimale.org
- **Valérie Bris**,
  COOP de France Animal Nutrition, valerie.bris@coopdefrance.coop
- **Bernard Fostier**,
  IFIP, The French Pork and Pig Institute, bernard.fostier@ifip.asso.fr
- **Françoise Clément**,
  IFCE, Institute for Horse & Horse Riding, francoise.clement@haras-nationaux.fr
- **Pierre-Louis Gastinel**,
  FGE, Genetic Improvement of Ruminants, pierre-louis.gastinel@france-genetique-elevage.fr

French Scientific Committee

President

- **Jean-Louis Peyraud**,
  INRA, jean-louis.peyraud@rennes.inra.fr

Vice Presidents

- **Catherine Disenhaus**,
  Agrocampus Ouest, catherine.disenhaus@agrocampus-ouest.fr
- **Philippe Lecomte**,
  CIRAD, philippe.lecomte@cirad.fr

Executive Secretaries

- **Michel Bonneau**,
  INRA, michel.bonneau@rennes.inra.fr
- **Bernard Coudurier**,
  INRA, bernard.coudurier@tours.inra.fr

Animal Genetics

- **Vincent Ducrocq**, INRA, vincent.ducrocq@jouy.inra.fr
- **Sophie Mattalia**, Institut de l'Elevage, sophie.mattalia@idele.fr

Animal Management and Health

- **Nathalie Bareille**, ONIRIS, nathalie.bareille@oniris-nantes.fr
- **Nicolas Rose**, ANSES, nicolas.rose@anses.fr

Animal Physiology

- **Xavier Druard**, INRA, xavier.druart@tours.inra.fr
- **Claire Ponsart**, UNCEIA, claire.ponsart@unceia.fr

Livestock Farming Systems

- **Muriel Tichit**, AgroParisTech, muriel.tichit@agroparistech.fr
- **Sophie Bertrand**, Institut de l'Elevage, sophie.bertrand@idele.fr

Animal Nutrition

- **Daniel Sauvant**, AgroParisTech, sauvant@agroparistech.fr
- **Alain Guyonvarc'h**, In Vivo, aguyonvarch@invivo-nsa.com

Cattle Production

- **Jean-François Hocquette**, INRA, jean-francois.hocquette@clermont.inra.fr
- **André Le Gall**, Institut de l'Elevage, andre.legall@idele.fr

Sheep & Goat Production

- **Loys Bodin**, INRA, loys.bodin@toulouse.inra.fr
- **Jean Michel Astruc**, Institut de l'Elevage, jean-michel.astruc@idele.fr

Pig Production

- **Jean-Yves Dourmad**, INRA, jean-yves.dourmad@rennes.inra.fr
- **Yvon Salaün**, IFIP, the French Pork and Pig Institute, yvon.salaun@ifip.asso.fr

Horse Production

- **William Martin-Rosset**, INRA, william.martin-rosset@clermont.inra.fr
- **Françoise Clément**, IFCE, francoise.clement@haras-nationaux.fr

EAAP Program Foundation

Aims

EAAP aims to bring to our annual meetings, speakers who can present the latest findings and views on developments in the various fields of science relevant to animal production and its allied industries. In order to sustain the quality of the scientific program that will continue to entice the broad interest in EAAP meetings we have created the 'EAAP Program Foundation'. This Foundation aims to support:

- Invited speakers with a high international profile by funding part or all of registration and travel costs.
- Delegates from less favoured areas by offering scholarships to attend EAAP meetings.
- Young scientists by providing prizes for best presentations.

The '**EAAP Program Foundation**' is an initiative of the Scientific committee (SC) of EAAP. The Foundation aims to stimulate the quality of the scientific program of the EAAP meetings and to ensure that the science meets societal needs. The Foundation Board of Trustees oversees these aims and seeks to recruit sponsors to support its activities.

Sponsorships

1. **Meeting sponsor – From 5000 euro**
   - acknowledgements in the final booklet with contact address and logo
   - one page allowance in the final booklet
   - advertising/information material inserted in the bags of delegates
   - advertising/information material on a stand display (at additional cost to be negotiated)
   - acknowledgement in the EAAP Newsletter with possibility of a page of publicity
   - possibility to add session and speaker support (at additional cost to be negotiated)
2. **Session sponsor – from 3000 to 5000 euro**
   - acknowledgements in the final booklet with contact address and logo
   - one page allowance in the final booklet
   - advertising/ information material in the delegate bag
   - slides at beginning of session to acknowledge support and recognition by session chair
   - acknowledgement in the EAAP Newsletter.
3. **Speaker sponsor - from 2000 euro (cost will be defined according to speakers country of origin)**
   - half page allowance in the final booklet
   - recognition by speaker of the support at session (at additional cost to be negotiated)
   - acknowledgement in the EAAP Newsletter
4. **Registration Sponsor - (equivalent to a full registration fee of the Annual Meeting)**
   - acknowledgements in the booklet with contact address and logo
   - advertising/information material in the delegate bag

The Association

EAAP (The European Federation of Animal Science) organises every year an international meeting which attracts between 900 and 1500 people. The main aims of EAAP are to promote, by means of active co-operation between its members and other relevant international and national organisations, the advancement of scientific research, sustainable development and systems of production; experimentation, application and extension; to improve the technical and economic conditions of the livestock sector; to promote the welfare of farm animals and the conservation of the rural environment; to control and optimise the use of natural resources in general and animal genetic resources in particular; to encourage the involvement of young scientists and technicians. More information on the organisation and its activities can be found at www.eaap.org.

Contact and further information

If you are interested to become a sponsor of the 'EAAP Program Foundation' or want to have further information, please contact the EAAP Secretariat (eaap@eaap.org, Phone +39 06 44202639).

Acknowledgements

European Federation of Animal Science (EAAP)

President: P. Chemineau
Secretary General: A. Rosati
Address: Via G.Tomassetti 3, A/I
I-00161 Rome, Italy
Phone: +39 06 4420 2639
Fax: +39 06 4426 6798
E-mail: eaap@eaap.org
Web: www.eaap.org

65th EAAP Annual meeting of the European Federation of Animal Science

25 – 28 August 2014, Copenhagen, Denmark

Organising Committee

President:
- Brian Bech Nielsen, Aarhus University – dean.scitech@au.dk

Vice-President:
- Birgit Nørrung, University of Copenhagen – birgit.noerrung@sund.ku.dk

Executive Secretaries:
- John Hermansen, Aarhus University – john.hermansen@agrsci.dk
- Vivi Hunnicke Nielsen, Aarhus University – vivih.nielsen@agrsci.dk

Conference website: www.eaap2014.org

Scientific Programme EAAP 2013

<table>
<tr><th>Monday 26 August
8.30 – 12.30</th><th>Monday 26 August
14.00 – 18.00</th><th>Tuesday 27 August
8.30 – 12.30</th><th>Tuesday 27 August
14.00 – 17.30</th></tr>
<tr><td colspan="2">Interbull workshop</td><td rowspan="8">Session 16
Advances in genomic analysis and prediction: 1
Chair: T. Meuwissen

Session 17
Dairy farming after 2015: sector, farm management and cow aspects
Chair: A. Kuipers

Session 18
The new equine economy: growth in new sectors and activities in the 21st Century
Chair: C. Vial/ R. Evans

Session 19
Milking physiology of sheep and goats
Chair: P. Marnet

Session 20a
Welfare and mortality aspects of sheep and goat production
Chair: J. Conington

Session 20b
Sheep and Goat production: Economics and marketing
Chair: M. Benoit

Session 21a
Multi-species database of feeding values of animal feed for warm countries
Chair: A. Guyonvarch

Session 21b
Use of knowledge in animal nutrition in specifications for 'label' and other higher quality production systems
Chair: F. Casabianca

Session 22a
Services provided by livestock farming systems: ecological, economic and social dimensions
Chair: J. Hermansen

Session 22b
LFS innovations for local/rural development
Chair: I. Casasus/B. Dedieu

Session 23a
Effects of high temperature on reproductive physiology
Chair: M. Driancourt

Session 23b
Physiological biomarkers of stress and production diseases
Chair: M. Vestergaard/S. Edwards</td><td rowspan="8">Session 24
Plenary session
and
Leroy Fellowship
Award lecture
Chair: Ph. Chemineau</td></tr>
<tr><td>Session 1
Genomic selection: impact on the organisation of the breeding sector
Chiar: V. Ducrocq</td><td>Session 10
How can farmer benefit from genomic information?
Chair: G. Thaller</td></tr>
<tr><td colspan="2">One day symposium – Session 2
Carcase and meat quality: from measurement to payment
Chair: L. Bunger</td></tr>
<tr><td colspan="2">Hot topic session – Reproduction Symposium – Session 3
New advances in biotechnology of equine reproduction
Chair: D. Varner/M. Magistrini</td></tr>
<tr><td>Session 4
Redesigning selection objectives to account for long-term challenges
Chair: P. Knap

Session 5a
New feed systems for ruminants
Chair: C. Thomas

Session 5b – Hot topic
Alternative and novel protein sources for livestock
Chair: L. Bailoni

Session 6a
Aquaculture nutrition in the context of animal nutrition: common issues and differences
Chair: S. Kaushik

Session 6b – Industry session
Automation in livestock farming
Chair: I. Halachmi</td><td>Session 11 – Genetic resources WG workshop
The role of imported genetics for sustainable breeding programmes
Chair: R. Baumung

Session 12
Beef production: environmental and economic aspects and impact
Chair: J. Hocquette

Session 13a
Sow nutrition to cope with increased reproductive potential
Chair: E. Knol

Session 13b
Enriching environment and improving comfort in intensive indoor systems
Chair: L. Boyle</td></tr>
<tr><td colspan="2">One-day Symposium – Session 7
Sustainable animal production in the tropics (SAPT2) and high constraint areas: multidisciplinary approaches
Chair: M. Boval/S. Oosting</td></tr>
<tr><td rowspan="2">Session 8
Social genetic effects: free communications
Chair: P. Bijma

Session 9
Animal health and welfare in loose housing and free range systems
Chair: G. Das</td><td>Session 14
Genetic control of adaptation
Chair: S. Bishop

Session 15a
Integrated management of animal health to reduce use of pharma
Chair: N. Bareille

Session 15b
Modulation of the immune defence system to cope with antibiotic use reduction
Chair: C. Lauridsen</td></tr>
<tr><td>18:00-21:00
Poster session and dinner</td></tr>
</table>

<table>
<tr><th>Wednesday 28 August
8.30 – 12.30</th><th>Wednesday 28 August
14.00 – 18.00</th><th>Thursday 29 August
8.30 – 12.30</th><th>Thursday 29 August
14.00 – 18.00</th></tr>
<tr>
<td rowspan="9">Free communications
followed by
Commission business meetings

Session 25
Molecular genetics: free communications
Chair: H. Simianer

Session 26a
Cattle debate: new production methods and society's response
Chair: B. Whitelaw/ M. Coffey

Session 26b
Cattle: free communications
Chair: A. Kuipers

Session 27
Horse: free communications
Chair: S. Janssens

Session 28
Sheep and Goats: free communications
Chair: L. Bodin

Session 29
Pigs: free communications
Chair: P. Knap

Session 30
Nutrition: free communications
Chair: J.-E. Lindberg

Session 31
LFS: free communications
Chair: K. Eilers

Session 32
Physiology: free communications
Chair: H. Quesnel

Session 33
Health and Welfare: free communications
Chair: C. Fourichon</td>
<td rowspan="9">Session 34
Advances in genomic analysis and prediction: 2
Chair: J. Bennewitz

Session 35
YoungTrain: Young scientist's session: challenges and aspects of food production
Chair: I. Stokovic/ M. Klopcic

Session 36
Behavioural evaluation of the working horse
Chair: M. Hausberger/ U. von Borstel

Session 37
Ethical aspects of breeding
Chair: G. Gandini

Session 38
Group housing of sows
Chair: A. Velarde

Session 39a
Industry session
Feed additives: Impact on health and performance in livestock
Chair: E. Auclair

Session 39b
Pigs: free communications
Chair: G. Bee

Session 40
Robust and resilient livestock farming systems in a changing world
Chair: S. Ingrand

Session 41
Cellular physiology of secretory processes
Chair: C. Knight

Session 42
Breeding in low input production systems
Chair: H. Simianer/V. Maurer</td>
<td colspan="2">One day symposium
Session 43
6th European Symposium on South American Camelids and 2nd European Meeting on Fibre Animals
Chair: D. Allain</td>
</tr>
<tr><td>Session 44
Improving fertility and udder health
Chair: M. Klopcic</td><td>Session 51a
Education in animal science for the future
Chair: C. Disenhaus

Session 51b
Well functioning dogs
Chair: K. Grandinson</td></tr>
<tr><td colspan="2">One day symposium: Feed efficiency</td></tr>
<tr><td>Session 45
Feed efficiency: physiological and genetic mechanisms
Chair: N. Scollan</td><td>Session 52
Feed efficiency in ruminants
Chair: P. Faverdin

Session 53
Feed efficiency in non-ruminants
Chair: P. Knap</td></tr>
<tr><td colspan="2">One day symposium: Genomic tools and technologies</td></tr>
<tr><td>Session 46
Tools and technologies for the exploitation of livestock genomes
Chair: T Roozen</td><td>Session 54
Recent results provided by genomic tools in sheep and goats
Chair: H. Jones</td></tr>
<tr><td colspan="2">Horse technical Tour
Organized by W. Martin-Rosset and Françoise Clement</td></tr>
<tr><td>Saumur:
exhibition and technical reports</td><td>Lion d'Angers:
exhibition and technical reports</td></tr>
<tr><td>Session 47
Gut health and immune response in mono-gastric animals
Chair: G. van Duinkerken

Session 48
Cattle production in a changing policy environment in Europe
Chair: M. Zehetmeier/ P. Sarzeaud

Session 49
Management and nutrition of entire male growing pigs (not boar taint)
Chair: G. Bee

Session 50
Evaluation of intrinsic and extrinsic qualities of cattle, dairy and meat products
Chair: K. Duhem</td><td>Session 55
Cattle genetics: free communications
Chair: S. Eaglen

Disease symposium
Session 56a
Emergent and re-emergent diseases of livestock
Chair: C. Belloc

Session 56b
Design and evaluation of control plans for infectious diseases with epidemiological models
Chair: C. Belloc

Session 57
Generating, handling and exploiting very large data for management and breeding
Chair: B. Guldbrandtsen

Session 58
Cattle: free communications
Chair: M. Coffey</td></tr>
</table>

Commission on Animal Genetics

Dr Simianer	President Germany	University of Goettingen hsimian@gwdg.de
Dr Baumung	Vice-President Italy	FAO Rome roswitha.baumung@fao.org
Dr Meuwissen	Vice-President Norway	Norwegian University of Life Sciences theo.meuwissen@umb.no
Dr Szyda	Vice-President Poland	Agricultural University of Wroclaw szyda@karnet.ar.wroc.pl
Dr Ibañez	Secretary Spain	IRTA noelia.ibanez@irta.es
Dr De Vries	Industry rep. Netherlands	CRV alfred.de.vries@crv4all.com

Commission on Animal Nutrition

Dr Lindberg	President Sweden	Swedish University of Agriculture jan-eric.lindberg@huv.slu.se
Dr Bailoni	Vice-President Italy	University of Padova lucia.bailoni@unipd.it
Dr Auclair	Vice President/ Industry rep. France	 LFA Lesaffre ea@lesaffre.fr
Dr Tsiplakou	Secretary Greece	Agricultural University of Athens eltsiplakou@aua.gr
Dr Van Duinkerken	Secretary Netherlands	Wageningen University gert.vanduinkerken@wur.nl

Commission on Health and Welfare

Dr Fourichon	President France	Oniris INRA christine.fourichon@oniris-nantes.fr
Dr Spoolder	Vice-President Netherlands	ASG-WUR hans.spoolder@wur.nl
Dr Krieter	Vice-President Germany	University Kiel jkrieter@tierzucht.uni-kiel.de
Mr Pearce	Vice-President/ Industry rep. United Kingdom	 Pfizer michael.c.pearce@pfizer.com
Dr. Boyle	Secretary Ireland	Teagasc laura.boyle@teagasc.ie
Dr Das	Secretary Germany	University of Goettingen gdas@gwdg.de

Commission on Animal Physiology

Dr Vestergaard	President	Aarhus University
	Denmark	mogens.vestergaard@agrsci.dk
Dr Kuran	Vice-President	Gaziosmanpasa University
	Turkey	mkuran@gop.edu.tr
Dr Driancourt	Vice president/	
	Industry rep.	Intervet
	France	marc-antoine.driancourt@sp.intervet.com
Dr Bruckmaier	Vice-President	University of Bern
	Switzereland	rupert.bruckmaier@physio.unibe.ch
Dr Quesnel	Secretary	INRA Saint Gilles
	France	helene.quesnel@rennes.inra.fr
Dr Scollan	Secretary	Institute of Biological, Environmental and rural sciences
	United Kingdom	ngs@aber.ac.uk

Commission on Livestock Farming Systems

Dr Bernués Jal	President	Norwegian University of Life Sciences
	Norway	alberto.bernues@umb.no
Dr Ingrand	Vice-President	INRA
	France	ingrand@clermond.inra.fr
Dr Tichit	Vice-President	INRA
	France	muriel.tichit@agroparistech.fr
Dr Eilers	Secretary	Schuttelaar & Partners
	Netherlands	keilers@schuttelaar.nl
Mrs Zehetmeier	Secretary	Institute Agricultural Economics and Farm Management
	Germany	monika.zehetmeier@tum.de

Commission on Cattle Production

Dr Kuipers	President	Wageningen UR
	Netherlands	abele.kuipers@wur.nl
Dr Thaller	Vice-President	Animal Breeding and Husbandry
	Germany	georg.thaller@tierzucht.uni-kiel.de
Dr Coffey	Vice president/	
	Industry rep.	SAC, Scotland
	United Kingdom	mike.coffey@sac.ac.uk
Dr Hocquette	Secretary	INRA
	France	jean-francois.hocquet@clermont.inra.fr
Dr Klopcic	Secretary/	
	Industry rep.	University of Ljublijana
	Slovenia	marija.klopcic@bf.uni-lj.si

Commission on Sheep and Goat Production

Dr Bodin	President	INRA-SAGA
	France	loys.bodin@toulouse.inra.fr
Dr Ringdorfer	Vice-President	LFZ Raumberg-Gumpenstein
	Austria	ferdinand.ringdorfer@raumberg-gumpenstein.at
Dr Conington	Vice-President	SAC
	United Kingdom	joanne.conington@sac.ac.uk
Dr Papachristoforou	Vice President	Agricultural Research Institute
	Cyprus	c.papachristoforou@cut.ac.cy
Dr Milerski	Secretary/	
	Industry rep.	Research Institute of Animal Science
	Czech Republic	m.milerski@seznam.cz
Dr Ugarte	Secretary/	
	Industry rep.	NEIKER-Tecnalia
	Spain	eugarte@neiker.net

Commission on Pig Production

Dr Knap	President/	
	Industry rep.	PIC International Group
	Germany	pieter.knap@genusplc.com
Dr Lauridsen	Vice-President	Aarhus University
	Denmark	charlotte.lauridsen@agrsci.dk
Dr Bee	Vice-President	Agroscope Liebefeld-Posieux ALP
	Switzerland	giuseppe.bee@alp.admin.ch
Dr Pescovicova	Secretary	Research Institute of Animal Production
	Slovak Republic	peskovic@vuzv.sk
Dr Velarde	Secretary	IRTA
	Spain	antonio.velarde@irta.es

Commission on Horse Production

Dr Miraglia	President	Molise University
	Italy	miraglia@unimol.it
Dr Burger	Vice president	Clinic Swiss National Stud
	Switzerland	dominique.burger@mbox.haras.admin.ch
Dr Janssen	Vice president	BIOSYST
	Belgium	steven.janssens@biw.kuleuven.be
Dr Lewczuk	Vice president	IGABPAS
	Poland	d.lewczuk@ighz.pl
Dr Saastamoinen	Vice president	MTT Agrifood Research Finland
	Finland	markku.saastamoinen@mtt.fi
Dr Palmer	Vice president/	
	Industry rep.	CRYOZOOTECH
	France	ericpalmer@cryozootech.com
Dr Holgersson	Secretary	Swedish University of Agriculture
	Sweden	anna-lena.holgersson@hipp.slu.se
Dr Hausberger	Secretary	CNRS University
	France	martine.hausberger@univ-rennes1.fr

Scientific programme

Session 01. Genomic selection: impact on the organisation of the breeding sector

Date: 26 August 2013; 8:30 – 12:30 hours
Chairperson: Ducrocq

Theatre Session 01

Session 02. Carcase and meat quality: from measurement to payment

Date: 26 August 2013; 8:30 – 18:00 hours
Chairperson: Bunger

Theatre Session 02

Poster Session 02 **Page**

Session 03. New advances in biotechnology of equine reproduction

Date: 26 August 2013; 8:30 – 18:00 hours
Chairperson: Magistrini / Varner

Theatre Session 03 **Page**

Poster Session 03 **Page**

Session 04. Redesigning selection objectives to account for long-term challenges

Date: 26 August 2013; 8:30 – 12:30 hours
Chairperson: Knap

Theatre Session 04 **Page**

Poster Session 04 **Page**

Session 05a. New feed unit systems for ruminants

Date: 26 August 2013; 8:30 – 10:15 hours
Chairperson: Thomas

Theatre Session 05a **Page**

Poster Session 05a **Page**

Session 05b. Alternative and novel protein sources for livestock

Date: 26 August 2013; 10:45 – 12:30 hours
Chairperson: Bailoni

Theatre Session 05b **Page**

Poster Session 05b **Page**

Session 06a. Aquaculture nutrition in the context of animal nutrition: common issues and differences

Date: 26 August 2013; 8:30 – 10:15 hours
Chairperson: Kaushik

Theatre Session 06a **Page**

Session 06b. Industry session: automation in livestock farming

Date: 26 August 2013; 10:45 – 12:30 hours
Chairperson: Halachmi

Theatre Session 06b — Page

Poster Session 06b — Page

Session 07. Sustainable animal production in the tropics and high constraint areas: multidisciplinary approaches

Date: 26 August 2013; 8:30 – 18:00 hours
Chairperson: Boval / Oosting

Theatre Session 07 — Page

Session 08. Social genetic effects / free communications

Date: 26 August 2013; 8:30 – 12:30 hours
Chairperson: Bijma

Session 09. Animal health and welfare in loose housing and free range systems

Date: 26 August 2013; 8:30 – 12:30 hours
Chairperson: Das

Theatre Session 09 — Page

Poster Session 09 **Page**

Session 10. How can farmer benefit from genomic information

Date: 26 August 2013; 14:00 – 18:00 hours
Chairperson: Thaller

Theatre Session 10 **Page**

Poster Session 10 **Page**

Session 11. The role of imported genetics for sustainable breeding programmes

Date: 26 August 2013; 14:00 – 18:00 hours
Chairperson: Baumung

Theatre Session 11 **Page**

Poster Session 11 **Page**

Session 12. Beef production: environmental and economic aspects and impact

Date: 26 August 2013; 14:00 – 18:00 hours
Chairperson: Hocquette

Theatre Session 12 **Page**

Poster Session 12 **Page**

Session 13a. Sow nutrition to cope with increased reproductive potential

Date: 26 August 2013; 14:00 – 15:45 hours
Chairperson: Knol

Theatre Session 13a — Page

Poster Session 13a — Page

Session 13b. Enriching environment and improving comfort in intensive indoor systems

Date: 26 August 2013; 16:15 – 18:00 hours
Chairperson: Boyle

Theatre Session 13b — Page

Poster Session 13b **Page**

Session 14. Genetic control of adaptation

Date: 26 August 2013; 14:00 – 18:00 hours
Chairperson: Bishop

Theatre Session 14 **Page**

Poster Session 14 **Page**

Session 15a. Integrated management of animal health to reduce usage of veterinary pharmaceuticals

Date: 26 August 2013; 14:00 – 15:45 hours
Chairperson: Bareille

Theatre Session 15a **Page**

Session 15b. Modulation of the immune defence system to cope with antibiotic use reduction

Date: 26 August 2013; 16:15 – 18:00 hours
Chairperson: Lauridsen

Session 16. Advances in genomic analysis and prediction

Date: 27 August 2013; 8:30 – 12:30 hours
Chairperson: Meuwissen

Poster Session 16 **Page**

Session 17. Dairy farming after 2015: sector, farm management and cow aspects

Date: 27 August 2013; 8:30 – 12:30 hours
Chairperson: Kuipers

Theatre Session 17 — Page

Poster Session 17 **Page**

Session 18. The new equine economy: growth in new sectors and activities in the 21st century

Date: 27 August 2013; 8:30 – 12:30 hours
Chairperson: Vial / Evans

Theatre Session 18 **Page**

Poster Session 18 **Page**

Session 19. Milking physiology, milkability and milk production and quality in sheep and goats

Date: 27 August 2013; 8:30 – 12:30 hours
Chairperson: Marnet

Theatre Session 19 **Page**

Poster Session 19 **Page**

Session 20a. Welfare and mortality aspects of sheep and goat production

Date: 27 August 2013; 8:30 – 10:15 hours
Chairperson: Conington

Theatre Session 20a **Page**

Poster Session 20a **Page**

Session 20b. Sheep and goat production: economics and marketing

Date: 27 August 2013; 10:45 – 12:30 hours
Chairperson: Benoit

Theatre Session 20b **Page**

Session 21a. Multi-species database of feeding values of animal feed for warm countries

Date: 26 August 2013; 8:30 – 10:15 hours
Chairperson: Guyonvarch

Theatre Session 21a **Page**

Session 21b. Use of knowledge in animal nutrition in specifications for 'label' and other higher quality production systems

Date: 27 August 2013; 10:45 – 12:30 hours
Chairperson: Casabianca

Theatre Session 21b — Page

Poster Session 21b — Page

Session 22a. Services provided by livestock farming systems: ecological, economic and social dimensions

Date: 27 August 2013; 8:30 – 10:15 hours
Chairperson: Hermansen

Theatre Session 22a — Page

Poster Session 22a **Page**

Session 22b. LFS innovations for local/rural development

Date: 27 August 2013; 10:45 – 12:30 hours
Chairperson: Casasus / Dedieu

Theatre Session 22b **Page**

Poster Session 22b **Page**

Session 23a. Effects of high temperature on reproductive physiology

Date: 27 August 2013; 8:30 – 10:15 hours
Chairperson: Driancourt

Theatre Session 23a — Page

Poster Session 23a — Page

Session 23b. Physiological biomarkers of stress and production diseases

Date: 27 August 2013; 10:45 – 12:30 hours
Chairperson: Vestergaard / Edwards

Theatre Session 23b — Page

Poster Session 23b

Session 24. Plenary session: new challenges facing animal production for diversified territories, market demands and social expectations

Date: 27 August 2013; 15:00 – 17:30 hours
Chairperson: Ph. Chemineau

Theatre Session 24

Session 25. Free communications – molecular genetics

Date: 28 August 2013; 8:30 – 11:30 hours
Chairperson: Simianer

Theatre Session 25

Poster Session 25 Page

Session 26a. Cattle debate: new production methods and society response

Date: 28 August 2013; 8:30 – 11:30 hours
Chairperson: Whitelaw / Coffey

Theatre Session 26a **Page**

Session 26b. Cattle production free communications

Date: 28 August 2013; 8:30 – 9:30 hours
Chairperson: Kuipers

Theatre Session 26b **Page**

Poster Session 26b **Page**

Session 27. Horse production free communications

Date: 28 August 2013; 9:30 – 11:30 hours
Chairperson: Janssens

Theatre Session 27 Page

Poster Session 27

Session 28. Sheep and goats production free communications

Date: 28 August 2013; 8:30 – 11:30 hours
Chairperson: Bodin

Theatre Session 28 — Page

Poster Session 28 — Page

Session 29. Pig production free communications

Date: 28 August 2013; 8:30 – 11:30 hours
Chairperson: Knap

Theatre Session 29 **Page**

Poster Session 29 **Page**

Session 30. Animal nutrition free communications

Date: 28 August 2013; 8:30 – 11:30 hours
Chairperson: Lindberg

Theatre Session 30 Page

Poster Session 30 Page

Session 31. Livestock farming systems free communications

Date: 28 August 2013; 8:30 – 11:30 hours
Chairperson: Eilers

Theatre Session 31 Page

Poster Session 31 Page

Session 32. Animal physiology free communications

Date: 28 August 2013; 8:30 – 11:30 hours
Chairperson: Quesnel

Theatre Session 32 — Page

Poster Session 32 — Page

Session 33. Animal health and welfare free communications

Date: 28 August 2013; 8:30 – 11:30 hours
Chairperson: Fourichon

Theatre Session 33 **Page**

Poster Session 33

Session 34. Advances in genomic analysis and prediction

Date: 28 August 2013; 14:00 – 18:00 hours
Chairperson: Bennewitz

Theatre Session 34

Session 35. Young train/young scientist session: challenges and aspects of food production

Date: 28 August 2013; 14:00 – 18:00 hours
Chairperson: Štokovic / Klopcic

Theatre Session 35 — Page

Poster Session 35

Session 36. Behavioural evaluation of the working horse

Date: 28 August 2013; 14:00 – 18:00 hours
Chairperson: Hausberger / König Von Borstel

Theatre Session 36

Poster Session 36

Session 37. Ethical aspects of breeding

Date: 28 August 2013; 14:00 – 18:00 hours
Chairperson: Gandini

Theatre Session 37 — Page

Poster Session 37 — Page

Session 38. Group housing of sows

Date: 28 August 2013; 14:00 – 18:00 hours
Chairperson: Velarde

Theatre Session 38 — Page

Session 39a. Industry session: feed additives; impact on health and performance in livestock

Date: 28 August 2013; 14:00 – 15:45 hours
Chairperson: Auclair

Theatre Session 39a — Page

Poster Session 39a — Page

Session 39b. Free communications in pig production

Date: 28 August 2013; 16:15 – 18:00 hours
Chairperson: Bee

Theatre Session 39b **Page**

Poster Session 39b **Page**

Session 40. Robust and resilient livestock farming systems in a changing world

Date: 28 August 2013; 14:00 – 18:00 hours
Chairperson: Ingrand

Theatre Session 40 Page

Poster Session 40 **Page**

Session 41. Cellular physiology of secretory processes

Date: 28 August 2013; 14:00 – 18:00 hours
Chairperson: Knight

Theatre Session 41 **Page**

Poster Session 41 **Page**

Session 42. Breeding and management in low input production systems

Date: 28 August 2013; 14:00 – 18:00 hours
Chairperson: Simianer / Maurer

Theatre Session 42 **Page**

Poster Session 42 **Page**

Session 43. South American camelids and fibre animals

Date: 29 August 2013; 8:30 – 18:00 hours
Chairperson: Allain

Theatre Session 43 **Page**

Poster Session 43 **Page**

Session 44. Improving fertility and udder health

Date: 29 August 2013; 8:30 – 12:30 hours
Chairperson: Klopcic

Theatre Session 44 Page

Poster Session 44 Page

Session 45. Feed efficiency: physiological and genetic mechanisms

Date: 29 August 2013; 8:30 – 12:30 hours
Chairperson: Scollan

Theatre Session 45 Page

Poster Session 45 Page

Session 46. Tools and technologies for the exploitation of livestock genomes

Date: 29 August 2013; 8:30 – 12:30 hours
Chairperson: Roozen

Theatre Session 46 — Page

Poster Session 46 — Page

Session 47. Gut health and immune response in mono-gastric animals

Date: 29 August 2013; 8:30 – 12:30 hours
Chairperson: Van Duinkerken

Theatre Session 47 **Page**

Poster Session 47 **Page**

Session 48. Cattle production in a changing policy environment in Europe

Date: 29 August 2013; 8:30 – 12:30 hours
Chairperson: Zehetmeier / Sarzeaud

Theatre Session 48 **Page**

Session 49. Management and nutrition of entire male growing pigs (not boar taint)

Date: 29 August 2013; 8:30 – 12:30 hours
Chairperson: Bee

Theatre Session 49

Session 50. Evaluation of intrinsic and extrinsic qualities of cattle, dairy and meat products

Date: 29 August 2013; 8:30 – 12:30 hours
Chairperson: Duhem

Theatre Session 50

Poster Session 50 **Page**

Session 51a. Education in animal science for the future

Date: 29 August 2013; 14:00 – 15:45 hours
Chairperson: Disenhaus

Theatre Session 51a **Page**

Poster Session 51a **Page**

Session 51b. Well-functioning dogs

Date: 29 August 2013; 16:15 – 18:00 hours
Chairperson: Strandberg

Theatre Session 51b **Page**

Session 52. Feed efficiency in ruminants

Date: 29 August 2013; 14:00 – 18:00 hours
Chairperson: Faverdin

Theatre Session 52 **Page**

Poster Session 52 **Page**

Session 53. Feed efficiency in non-ruminants

Date: 29 August 2013; 14:00 – 18:00 hours
Chairperson: Knap

Theatre Session 53 Page

Poster Session 53 Page

Session 54. Recent results provided by genomic tools in sheep and goats

Date: 29 August 2013; 14:00 – 18:00 hours
Chairperson: Jones

Theatre Session 54 — Page

Poster Session 54 — Page

Session 55. Animal genetics free communications

Date: 29 August 2013; 14:00 – 18:00 hours
Chairperson: Eaglen

Theatre Session 55 Page

Session 56a. Emergent and re-emergent diseases of livestock

Date: 29 August 2013; 14:00 – 15:45 hours
Chairperson: Belloc

Theatre Session 56a **Page**

Session 56b. Design and evaluation of control plans for infectious diseases with epidemiological models

Date: 29 August 2013; 16:15 – 18:00 hours
Chairperson: Belloc

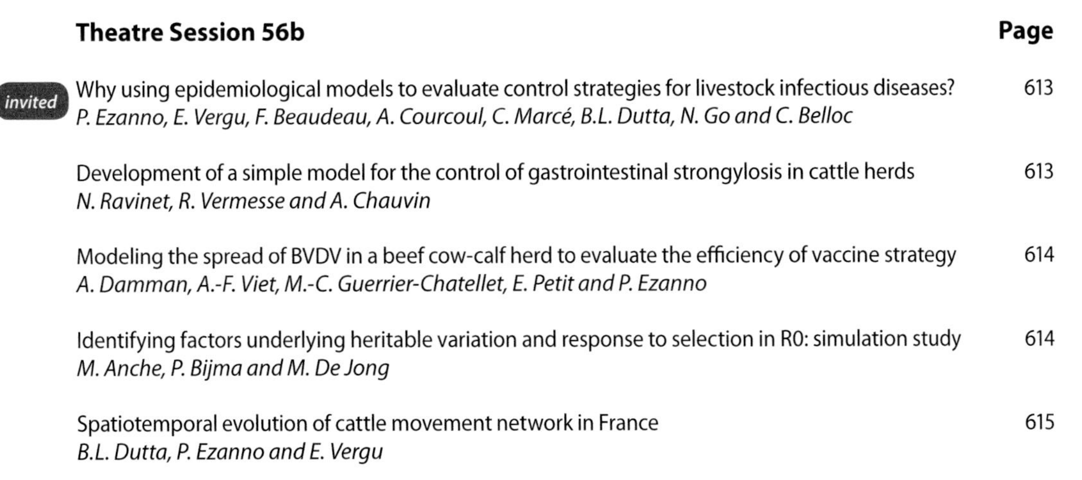

Theatre Session 56b **Page**

Session 57. Generating, handling and exploiting very large data for management and breeding

Date: 29 August 2013; 14:00 – 18:00 hours
Chairperson: Guldbrandtsen

Theatre Session 57 — Page

Poster Session 57 Page

Session 58. Cattle production free communications

Date: 29 August 2013; 14:00 – 18:00 hours
Chairperson: Coffey

Theatre Session 58 Page

Session 01 Theatre 1

Emerging markets, emerging strategies under the genomic era

J.L. Labatut
INRA, SAD, UMR AGIR, Chemin Borde Rouge, 31326 Castanet Tolosan, France; julie.labatut@toulouse.inra.fr

In the past 7 years, the organization of the dairy breeding industry in France has faced two deep changes: the development of genomic evaluations and the reform of 1966 Law of Breeding, leading to a suppression of breeding companies' territorial monopoly and a decrease in French government subsidies to the breeding industry. Until then, this industry was based on a strong public and cooperative regime of breeding, involving public institutes, breeders' cooperatives and associations in the development of collective tools and public information on animals. Breeds and breeding schemes were managed as common goods: there was no individual property right on breeds (collective property of farmers) nor on the breeding information on animals (public EBV's). This organization was highly efficient in terms of research capacities and genetic gain production. The development of genomic evaluations in a context of a liberalization of breeding markets, destabilizes this collective organization of knowledge production, allowing for the emergence of individual dynamics and increasing competition among participants. First, this contribution describes where and how this cooperative organization is destabilized. Several new forms of breeding products and services are identified on the market, as well as new actors and new roles on this market, with new research activities. Where does competition increase? With what risks and opportunities? Using several examples of dairy breeds, from local to national and international ones, this contribution seeks to identify various strategies and crises of cooperation among actors of the breeding industry. Several dimensions are analysed: relation between research and industry, relation among actors of the industry, evolving governance systems. Using examples from other types of industries, the benefit of new forms of inter-firm cooperation is considered. What type of joint platforms could be designed? In such collaborations, relationships are precarious, but the question of building collective identity is crucial.

Session 01 Theatre 2

What are the changes that GS generated for the EuroGenomics countries?

S. Borchersen
VikingGenetics, Ebeltoftvej 16, 8960 Randers SØ, Denmark; sobor@vikinggenetics.com

The introduction of Illumina BovineSNP50 BeadChip© in 2008 initiate fast and huge evolution of Genomic Selection in dairy cattle. Since then, many new platforms for genotyping have become available. Three main key factors of genomic selection have been stated: Size of reference population, quality of performance recording and efficiency of the methodology including chip technology. A crucial factor influencing the reliability of genomic predictions has shown to be the size of reference population. In this perspective four national (unions of) breeding organizations – UNCEIA (France), VikingGenetics (Finland, Sweden Denmark), DHV-VIT (Germany) and CRV (the Netherlands, Flanders) and their scientific partners in September 2009 agreed to merge their reference population with a contribution of 4,000 bulls from each party. The results showed that combined data led to significant increase in reliability (10%) of genomic prediction for bulls in all four parties. Since then CONAFE 2011, (Spain), Genomika Polska 2012 (Poland) and their respective research partners have joined EuroGenomics. The joined reference population now exceeds 25,000 reliable proven bulls. Further aim of the EuroGenomic project is to continue exchange of genotypes, supporting further developments of methods by strong research collaboration, use of next chip generation, EuroG10 k for large scale genotyping of females. Use of imputation methods that enable accurate prediction of animals genotyped with low density chips Future steps will be made to use all available genomic marker information including full genome sequence information as well as the potential in large scale genotyping of cows. Further benefit from genomic selection is the ability to provide genetic evaluations for low heritable traits and in future also including new traits as feed efficiency. EuroGenomics aims to facilitate implementation of new international standards for increased genetic benefit in the dairy farming industry.

The benefits of taking a national perspective in the organisation of breeding programs

A.R. Cromie
Irish Cattle Breeding Federation, Highfield House, Bandon, Co. Cork, N7 Cork, Ireland; acromie@icbf.com

Unlike most other cattle breeding countries, the breeding program in Ireland is owned by the industry, with AI companies, herdbooks, bull breeders and commercial farmers each playing an integral role in ensuring the rapid dissemination of genes from elite animals into the commercial population. The program is organised by Irish Cattle Breeding Federation (ICBF) under the brand name G€N€ IR€LAND. Within this organisational structure AI companies and private breeders can own bulls, with ICBF then responsible for the progeny testing of these bulls, including distribution of semen to targeted herds, collection of all relevant data and subsequent genetic evaluations. By taking a national perspective in the organisation of the breeding program, the Irish cattle breeding industry can more effectively manage: (1) the identification of elite animals (including provision of breeding program advice) through the use of the ICBF database; (2) the sharing of semen collection and progeny test resources across AI centres; and (3) the minimisation of risk to the breeding program, through maintenance of genetic diversity and monitoring traits of interest. In addition the potential of genomics is more fully utilised, with widespread genotyping of young animals by both AI companies and private breeders now prevalent. Indeed in 2013, we expect some 25,000 young males to be genotyped as part of the pedigree registration process, with all of these then available to the breeding program. As a consequence genomics is expected to have minimal impact on the organisation of breeding programs in Ireland, but certainly will have a huge impact on the level of genetic gain that we can generate for Irish dairy and beef farmers.

Session 01 Theatre 4

Impact of genomic selection on the Evolution breeding program

M. Philippe and T. De Bretagne
EVOLUTION, 69 rue de la Motte Brûlon, CS30621, 35706 Rennes Cedex 7, France; maelle.philippe@evolution-xy.fr

The implementation of genomic selection in dairy cattle has generated many changes in the design of breeding programs as well as in the organization of AI breeding companies. The acceptable reliability obtained on a large number of traits whatever their heritability, for males as well as for females, turns upside down the typical selection programs used for more than 50 years. The birth in 2013 of Evolution, a new French AI company merging the Amelis and Creavia cooperatives and marketing 5 million doses a year in France and abroad, is directly linked to this technological breakthrough. Its Holstein breeding program is now completely based on genomic evaluation for the selection of best candidates and involves more than 3,200 females and 3,000 males genotyped every year. Somewhat overlooked in the past, the female pathway becomes strategic with 720 elite cows selected as potential bull dams. More intense selection entails production of more progeny from the best females, especially heifers, through an increase in the number of ovum pick-ups and embryos transferred. The objective set by Evolution is to produce and transplant 8,500 embryos every year. To improve the management of genetic variability, more than 130 distinct sires of sons are mated with elite bulls dams, of which 90% are young bulls themselves. Selection intensity on bulls used AI – about 3.5% – is as crucial as before and aims at marketing 150 new young bulls every year. This ambitious program requires a strong partnership with the breeders, which guarantees both the development of their genetics and the economic sustainability of their herd. As an example, Evolution proposes stations for donor cows and a network of recipient cows available for the multiplication of elite cows. Genomic selection in dairy cattle has become a real race against the clock with more intense competition. On top of the acknowledged higher genetic gain permitted by genomic selection, new challenges lie ahead regarding research and inclusion of new traits.

How the Genomic era is shaping the dairy industry in BSW breed

E. Santus
ANARB, loc Ferlina 204, 37012 Bussolengo, Italy; enrico.santus@anarb.it

Genomic analysis, since their first introduction, have been considered a unique possibility to radically change the central paradigm of genetic improvement in dairy cattle. Instead of using field data in order to estimate the value of single individuals the final product of field data is rapidly switching to a set of robust and reliable prediction equations that can be used to estimate the values of single individuals regardless of the availability of phenotypes for specific animals. The challenge for small to medium sized populations has been clear since the beginning: in order to have reliable prediction equations the amount of data in each country was too small. That called for a new way of cooperating at international level. In 2009 the European federation of BSW breeders set up a project, later called InterGenomics, with the goal of setting up a unique international system of genomic evaluation for the breed. In 2012 the service began to be provided routinely by Interbull The participants in the projects decided also to share the pool of genotypes used for prediction equation estimation in order to allow more scientific groups to work on the same database with the aim of improving continuously the quality of the service. Since the introduction of genomic evaluations the industry reacted very fast. The number of test made on candidates bulls increased exponentially, the rules for the introduction of 'proven' bulls on the market changed almost everywhere in some months. The acceptance of genomic proven bulls without daughters has been cautious at the beginning but is rapidly improving after the first runs of consecutive proofs that showed how the system is reliable. The introduction of the service of genomic evaluation for cows, already active in most countries, is pushing the changes even further with implications in parentage verifications, herdbook rules and selection services provided to breeders.

Changes in the French Brown Swiss AI program related to the use of intergenomics breeding values

O. Bulot
BGS, 149 rue de Bercy, 75595 Paris cedex 12, France; olivier.bulot@acta.asso.fr

The Brown Swiss population in France is about 18,000 purebred milk recorded cows, representing 1% of dairy cows in the country and 4% of European Brown Swiss cows. The national organisation, Brune Génétique Services (BGS) includes Herd Book tasks and the management of a small AI program. BGS used to sample 8 to 10 young sires per year and release 2 or 3 new sires each year based on progeny proofs. Most French BS breeders use BGS genetics but also a large share of foreign semen (30% of AI) Since November 2011, the French Brown Swiss breeding scheme has been using genomic breeding values under the Intergenomics agreement. This international cooperation aimed at gathering the most complete reference population for the Brown Swiss breed and at delivering an international genomic evaluation for all genotyped sires. The traditional progeny test program was replaced by an intensive pre-selection of very young males on the field. Among 130 bull calves genotyped each year, the best 10% now enter AI centre to produce semen. Every year, a new group of 10 to 12 young sires is marketed with official genomic values published. Brown Swiss breeders show a lot of confidence in genomics and the fraction of AI done with young sires went up from 9% up to 19% during the 2nd semester of 2012. The forecast for 2013 is about 30-40% of total AI. The use of this technology is coupled with the development of the use of sexed semen. Though the national BS population is small in France compared to the main BS countries, the first results show that French BS bulls can compete with the best of the breed worldwide. Among the Top 100 Intergenomics sires for total net merit on the French scale, there are 9 French sires. This new tool allows BGS to widen and strengthen its genetic offer to the breeders and to drastically improve the competitiveness of the French BS genetics, in spite of its restricted base population incompatible with large progeny testing.

Session 01 Theatre 7

Impact of genomic selection on North American dairy cattle breeding organizations

J.P. Chesnais[1], G. Wiggans[2] and F. Miglior[3,4]

[1]Semex Alliance, Guelph, N1G 3Z2, Canada, [2]Animal Improvement Programs Laboratory, USDA, Beltsville, 20705-2350, USA, [3]Guelph Food Research Center, Agriculture and Agri-Food Canada, Guelph, N1G 5C9, Canada, [4]Canadian Dairy Network, Guelph, N1K 1E5, Canada; jpchesnais@rogers.com

Statistics were obtained from industry organizations in order to assess the impact of genomics on their operations. Genomic selection is expected to have a strong long-term impact on dairy cattle breeding structures in North America, but is relatively recent. Up to now, most organizations in the breeding sector have made internal changes to adapt to the new situation, but their roles have not changed dramatically. After April 2013, genomic evaluation of males will be open to all, which will increase competition for acquisition of top young males and for the marketing of their semen. More than 200,000 dairy males and females have now been genotyped with panels of various densities. The size of formal progeny-testing programs has decreased by a modest 22%, from 1,623 bulls entering such programs in 2012 compared to 2,079 in 2008. However, an increasing number of young genotyped bulls (678 in 2012 versus none in 2008) have entered AI studs outside of progeny-test programs, so that the total number of bulls entering AI has actually increased. Usage of bulls less than 4 years of age has increased from about 28% to 51%, but from 2011 to 2012 the increase was only 3% in Holsteins. The decrease in formal progeny-testing programs has not affected enrolment in milk recording programs. The number of classified cows has decreased by about 10% from 2008 to 2011, but has increased slightly from 2011 to 2012. Commercial producers are benefitting from faster rates of genetic change. Genotyping of heifers is progressively gaining ground. Genomics offers new opportunities for the selection of novel traits (health, feed efficiency, milk composition) but this requires large enough resource populations where these traits are recorded. It is not yet clear which industry organizations will step forward to create these.

Session 01 Theatre 8

Impact of genomics on international cooperation for dairy genetics

J.W. Dürr

SLU, Interbull Centre, Box 7023, 750 07 Uppsala, Sweden; joao.durr@slu.se

The ultimate goal of Interbull has been to facilitate international trade of dairy genetics through networking, harmonization, standardization and across-countries comparisons. This existing framework proved to be instrumental in the onset of genomic selection by providing the forum for development of methods and policies to incorporate the new technologies into the existing national and international evaluation systems. The early realization that the accuracy of genomic prediction equations is highly dependent on the size of the reference populations has amplified the importance of the Interbull MACE EBVs as source of phenotypic information on the foreign animals required as references. Genomics diversified the Interbull services portfolio: validation of national genomic EBVs, multi-country genomic evaluations of Brown Swiss populations (Intergenomics), international genomic evaluations of young bulls (GMACE) and international repository of genomic information. Few years after the bovine SNP panels started to be massively used, the scenario of the dairy breeding industry has already changed substantially and it is fated to continue moving towards a new paradigm. Exporting countries took the lead in adopting genomic technologies, formed consortia to share genotypes and polarized the market into two major competing blocks. The technological gap between exporters and importers widened rapidly, both due to the required investments and to the scarcer local expertise. Poor results from multi-breed genomic predictions have hindered genomic applications for smaller populations and the Holstein hegemony advances at a higher speed. Potential uses of genomics are limitless, but new actors with differentiated resources are more likely to take the lead and supply innovative options to dairy breeders in a similar fashion to what happens in the poultry and swine industries. Finally, data ownership became the centre of the debate and the control over the animal improvement process is shifting. Interbull is being transformed as well, but its unique role in harmonization is more needed than ever.

The impact of genomic selection on the South African dairy breeding sector

C.B. Banga and A. Maiwashe
ARC, P Bag X2, Irene 0062, South Africa; Cuthbert@arc.agric.za

Genomic selection is being adopted rapidly in breeding programmes worldwide. This paper discusses the influence that genomic selection has had on the South African dairy breeding sector. Although it is yet to be implemented, genomic selection has undoubtedly been the most topical subject in the South African breeding fraternity in recent years. Unfortunately, however, excitement about the new technology overwhelmed any efforts to educate industry about how genomic selection works and the extent to which benefits accrue to the various sectors. There was a wide-spread misconception that genomically tested means superior breeding value. Some farmers, gripped by fever of the new technology, spent large amounts of money on DNA tests for commercial marker panels. This was despite a lack of knowledge of how accurately these tests predict breeding values. Companies providing such tests quickly cashed in on this frenzy. Foreign AI companies also capitalised by pushing up sales of semen of genomically tested sires, without any regard to genotype by environment interaction. This unduly disadvantaged local sires and adversely affected the business of local AI companies. Breed societies, believing that implementation of genomic selection would put their breeds ahead of the pack, made frantic efforts to get service providers to provide genomic EBVs. Any efforts by research and academic institutions to respond to this need were however stymied by the large financial investments required to implement genomic selection. Realising the importance of a concerted approach to genomic selection, scientists and industry came together and formed a genomics consortium. The consortium is spearheading efforts to implement genomic selection programmes in the country, particularly the sourcing of funds. Government funding agencies are gradually buying into the technology and prospects of financial support are now bright. The co-opetition among industry players, brought about by the genomics consortium, is unprecedented in the South African livestock industry.

Genomic selection in dairy cattle: opportunities and challenges for Brazil

C.N. Costa[1], A.R. Caetano[2], J.A. Cobuci[3], G.G. Santos[1] and W.A. Arbex[1]
[1]Embrapa Gado de Leite, R&D, Rua Eugênio do Nascimento, 610, 36038-330 Juiz de Fora, MG, Brazil, [2]Embrapa Recursos Genéticos e Biotecnologia, R&D, Parque Estação Biológica – PqEB, Av. W5 Norte, 70770-917 Brasília, DF, Brazil, [3]Universidade Federal do Rio Grande do Sul, Animal Science, Av. Bento Goncalves, 7712 São José, 91540-000 Porto Alegre, RS, Brazil; claudio.napolis@embrapa.br

Most semen of Holstein cattle used by dairy farmers in Brazil is imported, but they have not relied on international bull rankings because Brazil is not an Interbull member. Joining Interbull is a major goal in order to evaluate imported bulls in Brazil. A common challenge with genomic selection is to obtain high reliabilities for genomic breeding values from larger reference populations. Building on north and south hemispheres collaboration, genomic selection could radically transform cattle genetic improvement in the tropics. Embrapa Dairy Cattle has played a key role in running genetic evaluations of dairy cattle in Brazil, in collaboration with breeder associations and the AI industry. Integration with international organizations is undertaken by a project involving collaboration with European scientists from key institutional partners. This initiative includes the development of a R&D agenda to align genetic evaluation procedures to facilitate affiliation to Interbull and design strategies to apply genomic selection in Holstein cattle in Brazil. The strategic objectives focus on capacity building in genetic and genomic evaluations; exchange of genotypes with partners and development of collaborative research by effective networking. The expected outcomes are the implementation of a sustainable breeding program of Holstein cattle by selection of local and international bulls evaluated in Brazilian environments and the integration of Brazil in the international set-up of knowledge transfer and scientific collaboration. We will illustrate these points through examples, advocating a network of excellence for the improvement of efficiency and competitiveness of dairy cattle breeding in Brazil.

Genomic selection in Kazakhstan: example of a country without existing selection program

C. Patry[1], S. Fritz[2], A. Baur[2], L. Genestout[3], I. Tretyakov[4], T. Echshzhanov[4] and V. Ducrocq[1]
[1]INRA, UMR1313, Domaine de Vilvert, 78352 Jouy-en-Josas, France, [2]UNCEIA, 149 rue de Bercy, 75012 Paris, France, [3]LABOGENA, Domaine de Vilvert, 78352 Jouy-en-Josas, France, [4]KazAgroInnovation, JSC, 26 rue Akzhol, 010000 Astana, Kazakhstan; clotilde.patry@jouy.inra.fr

In Kazakhstan, milk production is about 5 million tons per year, as in Denmark, a country 60 times smaller. Production has been growing at a yearly rate of 4.5% for 10 years, mainly due to an increase of the cattle population. While 90% of milk is still produced by household farms, about 3% is delivered by Agricultural Enterprises (AE) assumed to have better management and genetics. Kazakhstan is now strongly investing in breeding improvement, relying on imported animals and semen and, on the longer term, on national genomic selection (GS). A GS strategy does not only require the genotypes of bull candidates but also estimates of SNP effects under local conditions. However, neither corrected cow performances nor bull estimated breeding values to be used as phenotypes for genomic prediction are currently available in Kazakhstan. KazAgroInnovation, a research organization, undertook together with 4 large AE the creation of a female reference population of at least 10,000 genotyped and phenotyped Holstein cows. To encourage genotyping, direct genomic values (DGV) were made available for 1,000 cows with DNA in 2012. These DGV were computed using the French prediction equations, i.e. based on performances collected under European conditions. No pedigree information was considered. The consistency between raw performances for production and type and the animal DGV differs depending on the trait and herd. Such genomic breeding values can be temporarily used as a rough indicator for herd improvement. However, they should not prevent from intense efforts to improve the quality of identification and data collection. This is important to build long-term confidence not only in GS but also in further technological and commercial international collaborations.

Session 01 Theatre 12

Present and future of genomic selection at the commercial level

I. Misztal
University of Georgia, Athens, GA 30605, USA; ignacy@uga.edu

Experiences with genomic selection (GS) in the last couple years were interesting and somewhat unexpected. GS seems to be successful in dairy in populations with large number of genotyped high-accuracy bulls. Successes in smaller populations or with low-accuracy animals genotyped are smaller or none. Improvement in accuracies in some species (e.g. sheep) despite large number of genotypes is limited. Improvement of accuracies may be high for one group of animals and low for other groups. Adding external genotypes is not always helpful. The experiences suggest that: (1) LD for most traits is small; (2) gains with GS are mostly to better relationships; (3) gains from genotyping low-accuracy animals are small; (4) old genotypes are less useful; (5) genotypes from other populations are less useful and from other breeds are likely not useful. Current hope in GS is placed on detecting causative SNP in full-genome sequencing, however, the success of such scheme is based on strong assumptions that substitution effects of these SNP are constant, and epistasis across line/breeds/generations is negligible. As genotyping costs decrease, eventually millions of animals may be genotyped. Using more genotypes for predictions directly will greatly increase costs at low marginal gains (or even losses). Also emphasis on GS reduces attention to model refinement and correctly modelling environmental changes. Future GS needs to be simple, account for line and breed differences, and cannot limit modelling flexibility. It requires a breakthrough like the Henderson's algorithm to create A^{-1}, which changed the animal model evaluation from tough to simple.

Non-invasive measurement of body and carcass composition in livestock by CT, DXA, MRI, and US

A.M. Scholz[1], L. Bünger[2], J. Kongsro[3], U. Baulain[4] and A.D. Mitchell[5]
[1]LMU Munich, St. Hubertus-Str. 12, 85764 Oberschleissheim, Germany, [2]SRUC, Easter Bush, Midlothian, EH25 9RG, United Kingdom, [3]NORSVIN, P.O. Box 504, 2304 Hamar, Norway, [4]FLI, Hoeltystrasse 10, 31535 Neustadt-Mariensee, Germany, [5]USDA, BARC-East, Beltsville, MD 20705, USA; a.s@lmu.de

The ability to accurately measure body or carcass composition is an important application for performance testing, grading, and finally selection or payment of meat producing animals like for example beef cattle, sheep, swine or poultry, and also fish. Advances in especially non-invasive techniques are mainly based on the development of electronic and computer driven methods in order to provide objective phenotypic data. The preference for a specific technique depends on the target animal species or carcass – combined with technical and practical aspects like accuracy, reliability, cost, portability, speed, ease of use, safety, and for *in vivo* measurements the need for fixation or sedation. The techniques rely on specific device driven signals which interact with tissues in the body or carcass at the atomic or molecular level resulting in secondary or attenuated signals detected by the instruments and analysed quantitatively. The electromagnetic signal produced by the instrument may be in the form of sound waves (ultrasound – US), X-rays (e.g. Computer Tomography – CT, Dual Energy X-ray Absorptiometry – DXA), or radio frequency waves (Magnetic Resonance Imaging – MRI). The signals detected by these instruments are being processed to measure for example tissue depths, areas, volumes or distributions of fat, muscle (water, protein), and partly bone or bone mineral. Among the techniques, CT is the most accurate one followed by MRI and DXA, while alone US can be used for all sizes of farm animal species. CT, MRI and US can provide volume data, while alone DXA delivers immediate composition results without (2D) image manipulation. A combination of 'simple' US and 'high-end' CT or MRI may be used in a two-step performance testing and selection procedure.

Video image analysis (VIA) and value-based marketing of beef and lamb carcasses

C.R. Craigie[1], L. Bunger[2], R. Roehe[2], S.T. Morris[3], R.W. Purchas[3], D.W. Ross[2] and C.A. Maltin[1]
[1]Quality Meat Scotland, The Rural Centre, Ingliston, EH28 8NZ, United Kingdom, [2]Scotland's Rural College, Kings Buildings, Edinburgh, EH9 3JG, United Kingdom, [3]Massey University, College of Sciences, Private Bag 11-222, Palmerston North, 4442, New Zealand; ccraigie@qmscotland.co.uk

Value-based marketing (VBM) of beef and lamb carcasses requires accurate determination of saleable meat yield (SMY%) since this, along with meat quality characteristics, currently constitutes the main value component of the carcass. The present EUROP system comprises subjective carcass descriptions of conformation and fatness, but the relationship between these categories and SMY% is inconsistent. Basing carcass evaluation on SMY% should ensure the main value component of the carcass is reflected in the price. Video image analysis (VIA) systems are available that can predict with moderate to good accuracy the SMY% and also carcass conformation and fatness of beef and lamb carcasses before chilling. VIA systems do potentially have a role in VMB systems because they enable carcass evaluation to be based on SMY% without needing to weigh all pieces of saleable meat. VIA is automated, objective, and reproducible as it can be replicated across different processing sites. VIA carcass information also forms important inputs to farm management software and animal breeding programmes that enable producers to align production to meet market requirements. In order for producers and processors to have confidence in VIA for predicting SMY%, and accept VIA-predicted SMY% as a basis of payment, validation of VIA needs to include SMY% as a performance measure. There is still a need to develop robust prediction equations that can be applied at an industry level to identify carcases that have a higher yield of quality meat rather than to simply classify carcases according to the EUROP grid. There is also a need to refine the prediction of SMY% to a level of resolution that is affordable and of significant value to the industry, preferably at the individual cut level.

Session 02 Theatre 3

Computed tomography and automatic imaging systems for a value based marketing system in pig

G. Daumas[1], T. Donko[2], M. Font-I-Furnols[3], M. Gispert[3], M. Monziols[1] and E. Olsen[4]
[1]IFIP, BP 35104, 35651 Le Rheu Cedex, France, [2]Kaposvar University, Guba S. str. 40, 7400 Kaposvar, Hungary, [3]IRTA, Finca Camps i Armet, 17121 Monells, Spain, [4]DMRI, Maglegaardsvej 2, 4000 Roskilde, Denmark; gerard.daumas@ifip.asso.fr

Value-based payment and marketing systems in pig production in Europe are mainly based on carcass weight and lean meat percentage (LMP). The golden standard of LMP is still today dissection. Computed Tomography (CT) can be used but has to be calibrated against dissection. An international CT reference is therefore the next step, which is planned in 2014 within the COST Action FA 1102 (FAIM). Implementation of automatic on line imaging for grading is being spread. This communication reviews: (1) the CT procedures which have been nationally developed; (2) the accuracy of the automatic on line imaging systems which have been approved for pig classification; and (3) the payment grids in some EU countries. The root mean square error of prediction (RMSEP) of LMPdis from LMPct ranged between 0.3% and 1.4% LMP. The lowest error was obtained by muscle segmentation followed by mathematical morphology. Three automatic on line systems have been approved in a total of ten EU member states: one using ultrasounds and the others vision technology (VIA). The RMSEP ranged between 1.1% and 2.2% for ultrasounds and between 2.0% and 2.5% for VIA. Number of predictors can be very high. These systems differ also in reproducibility, cost, place, speed, all criteria which could be practically more important than accuracy. Higher payment is got within a specific carcass weight range with penalties for lighter and heavier carcasses. Range greatly differ between countries and can be local or linked with specific breeds. For LMP there is a base which is generally close to the population mean. Often there is a linear incentive above and below with different slopes. These incentives differ greatly between countries. Another payment system is based on a quality index calculated from the estimated composition of the main joints.

Session 02 Theatre 4

New lean meat formulas for progeny testing of intact boars: developed by using MRI and DXA

M. Bernau[1], P.V. Kremer[2], E. Tholen[3], S. Müller[4], E. Pappenberger[1], P. Gruen[5], E. Lauterbach[1] and A.M. Scholz[1]
[1]Ludwig-Maximilians-University Munich, St. Hubertusstrasse 12, 85764 Oberschleissheim, Germany, [2]University of Applied Sciences Weihenstephan-Triesdorf, Steingruberstr. 2, 91746 Weidenbach, Germany, [3]University Bonn, Endenicher Allee 15, 53115 Bonn, Germany, [4]Thueringer Landesanstalt fuer Landwirtschaft, Naumburger Strasse 98, 07743 Jena, Germany, [5]Landesanstalt für Schweinezucht, Seehoefer Strasse 50, 97944 Boxberg-Windischbuch, Germany; Maren.Bernau@lvg.vetmed.uni-muenchen.de

The objective was to evaluate formulas for progeny testing of boars instead of castrates or female pigs in order to replace the so called Bonner Formula 2004 for castrates of German Landrace and Large White including crossbreds. For the first step, 61 boars (Pietrain × crossbred sows) were fattened and analysed according to standard rules at three German performance test stations. A number of 20 right carcass halves was additionally dissected into lean meat and fat tissue providing the reference meat% at one station. All 61 left carcass halves were scanned by using magnetic resonance imaging (MRI) and dual energy X-ray absorptiometry (DXA). All DXA images were dissected into four body parts: shoulder/neck, back/belly, ham, and rear lower leg. The MR images were segmented semi-automatically into muscle and fat tissue for the regions: shoulder, caudal thorax with belly, loin, and ham. Resulting MRI and DXA traits were used in combination with the reference meat% to calculate new meat% formulas. The estimated meat% based on DXA and MRI was then used to calculate formulas based on the standard performance test traits. Finally, 33 boar carcasses from a fourth station served as independent sample. The new MRI/DXA 'boar' formulas are as accurate as the Bonner Formula (R^2=0.78-0.87; $\sqrt{MSE}$=0.58-0.76%). They contain, however, only 3 instead of 7 variables reducing the effort for progeny testing of intact boars. Additionally, MRI plus DXA or each technique alone could be used instead of dissection.

Main characteristics and evolution of the most valuable pig carcasses in Europe

M. Gispert[1], G. Daumas[2], J. Larsen[3] and M. Font I Furnols[4]
[1]IRTA, Product Quality, Sp Carcass quality, Finca Camps i Armet, 17121 Monells, Spain, [2]IFIP, BP 35104, 35651 Le Rheu, France, [3]Danish Classification Inspection, Axeltorv 3, 1609 Copenhagen, Denmark, [4]IRTA, Product Quality. Sp Carcass quality, Finca Camps i Armet s/n, 17121 Monells, Spain; marina.gispert@irta.es

The main criterion for pig carcass grading in the EU is carcass lean meat content that has to be estimated by objective methods. Payment system in some EU countries, that links carcass grading with their economic value, is a reward for producers who had been striving for to obtain leaner carcasses by means of breeding programs or productions systems. The aim of the present work is: (1) to review the different devices used to classify pig carcasses in Europe; (2) to describe the carcasses characteristics according to their lean meat content in Europe; (3) to describe payment strategies according to lean content in Denmark and France; and (4) to link lean content of the carcass with those of the main cuts and relate them with carcass value. A representative sample – at least 120 carcasses – of the country pig population should be described and dissected to obtain the carcass lean meat content to be used to calibrate the different classification devices. A lot of methods are approved in the EU (RMSEP<2.5%), either automatic or semi-automatic or manual. Reflectance is still the most common technology, but is invasive. Ultrasound and VIA are non-invasive methods which are in progress, especially because of automation. In EU, 83% of the carcasses are classified as S or E (lean content ≥55%) and the average carcass weight is 88.8 kg. Premiums and penalties for lean meat percentage and carcass weight show, for a bulk production, that the leanest carcasses are the most valuable from the economical point of view.

Altering the Carcase Plus Index has weakened its impact on lean meat yield %

F. Anderson[1,2], A. Williams[1,2], L. Pannier[1,2], D.W. Pethick[1,2] and G.E. Gardner[1,2]
[1]Australian Sheep CRC, University of New England, Armidale 2351, Australia, [2]Murdoch University, Veterinary and Life Sciences, South Street, Murdoch WA 6019, Australia; f.anderson@murdoch.edu.au

Increasing lean meat yield % (LMY%) and rapid growth are important profit drivers in the sheep supply chain. To optimise these traits, Australian prime lamb producers use the Carcase Plus Index (CP) for selecting sires. The index originally combined breeding values for weight, eye muscle depth and decreased fat depth, with weightings of 60:20:20 (old CP). Due to perceived gains in leanness in Terminal sires, and concern over reducing intramuscular fat levels in lamb meat, this index was altered to the current weightings of 65:30:5 (new CP). Given that selection for reduced fat depth results in an increase in LMY%, we hypothesised that the new CP will return less LMY% and therefore reduced carcase value compared to the old CP. Lamb carcases (n=1,800) from the Sheep CRC Information Nucleus Flock were collected from 6 research stations over 5 years. Carcases were scanned in 3 sections, fore, saddle and hind, using Computed Tomography (CT) to determine fat, lean and bone weights. Data was analysed using the allometric equation $y=ax^b$, fitted in its log-linearised form $\log y = \log a + b.\log x$. The impact on carcase value was determined for new and old CP indexes in a 23 kg carcase. Within the 76 unit range of Terminal sire CP index values, the old CP Index delivered 2.2, 1.2 and 0.5 percentage units more lean in the fore, saddle and hind sections ($P<0.01$) compared to the new CP index. Across this same range in CP index values the old CP increased the value of carcase lean by $9.27, compared to $6.49 for the new CP, equating to a value difference of $2.78 within a 23 kg carcase. Aligning with our hypothesis there was decreased gain in LMY% and carcase value using the new CP weightings. The cost in LMY% represents a substantial loss in daily profit for processors, especially when the potential improvement to intramuscular fat levels is not currently rewarded.

The use of robust regression models for *in vivo* prediction of lean meat proportion of lambs carcass

C. Xavier and V.A.P. Cadavez
Mountain Research Center (CIMO), ESA, Instituto Politécnico de Bragança, Animal Science, Campus de Santa Apolónia, Apartado 1172, 5301-855 Bragança, Portugal; vcadavez@ipb.pt

The aim of this study was to develop and evaluate robust regression models for *in vivo* prediction of lean meat proportion of lamb carcasses. Fifty lambs, 24 Suffolk and 26 Churro Galego Bragançano breeds, were used. The subcutaneous fat thickness (C1) was measured *in vivo*, between the 1st and 2nd lumbar vertebrae, using a real time ultrasound device (ALOKA SSD 500).The lambs were slaughtered, and their carcasses were cooled for 24 h. The left side of the carcasses was dissected, and the lean meat proportion (LMP) was calculated. Multiple regression models were fitted using ordinary least squares (OLS) and robust regression (RR) methods, and models fitting quality was evaluated by the following statistics: the root mean square error, the median absolute deviation, the mean absolute error, and the coefficient of determination (R^2). The RR model presented higher (0.358) coefficient of determination than the OLS model (0.308). The RR methods should be preferred to the OLS approach for predicting the LMP of lambs carcass.

Evaluating a new numerical approach for carcass lean meat yield determination

M. Font-I-Furnols[1,2], F.-A. Ouellet[3], H. Larochelle[3], M. Marcoux[1] and C. Pomar[1]
[1]Agriculture and Agri-Food Canada, Dairy and Swine Research and Development Centre, 2000 College Str., J1M 0C8 Sherbrooke, Canada, [2]IRTA, Product Quality, Finca Camps i Armet, 17121 Monells, Spain, [3]Université de Sherbrooke, Computer Science Department, 2500 Boul. de l'Université, J1K 2R1 Sherbrooke, Canada; maria.font@irta.cat

Carcass classification is based on carcass lean yield (CLY). Many pig carcass classification devices (semiautomatic probes) base their estimation of CLY on one or two fat and muscles thicknesses. Prediction equations are usually obtained from linear regression (LR). Gaussian process (GP) regression is a flexible alternative to LR. It allows expressing priors over the shape of the unknown predictive function that we are trying to uncover. In particular, using non-linear covariance functions (or kernel), GP regression can yield more powerful, non-linear predictors of the variable of interest (CLY in our case). The aim of the present work is to study the potential of GP regression to predict CLY, compared to the more common LR approach. For this purpose, 395 pig carcasses were used. Muscle and fat depth were measured from a digital image at 6 cm and parallel to the midline in the Longissimus Thoracis area at the level of 3-4th last ribs. Then, carcasses were dissected and CLY measured. For this experiment, the data were further randomly divided into a calibration (training) set (n=263) and a validation (evaluation) set (n=132). While LR obtained a prediction root mean square error (RMSEP) of 2.16% on the validation data set, GP regression based on a non-linear covariance function (Matern32) obtained a slightly lower RMSEP of 2.10%. This small improvement suggests that the relationship between muscle/fat depths and CLY could be better modelled using a non-linear predictor. Since muscle depths and fat depths variables explain only part of the variance observed in CLY, the inclusion of other predictor variables should be evaluated, in the hope of decreasing the prediction error further.

Consequences of different national ZP equations in EU to estimate lean meat content in pig carcasses

M. Font-I-Furnols[1], G. Daumas[2], M. Judas[3], M. Seynaeve[4], M. Gispert[1] and M. Candek-Potokar[5]
[1]IRTA, Product Quality, Finca Camps i Armet, 17121 Monells, Spain, [2]IFIP, BP 35104, 35601 Le Rheu Cedex, France, [3]MRI, Institute of Safety and Quality of Meat, E.-C.-Baumann-Straße 20, 95326 Kulmbach, Germany, [4]LANUPRO, Gent University, Proefhoevestraat 10, 9090 Melle, Belgium, [5]Agricultural Institute of Slovenia, Hacquetova ulica 17, 1000 Ljubljana, Slovenia; marina.gispert@irta.es

Pig carcass grading in the EU is based on the prediction of carcass lean meat percentage (LMP), which should be comparable between EU countries for market transparency. LMP prediction equations are determined for each country and classification method. Twelve EU countries use the so-called ZP method, based on one fat and one muscle thickness measure in the area of M. Gluteus Medius. Measurements can be taken with ruler, calliper or the specific device OptiScan-TP©. The aim of the present work was to compare different national ZP equations for LMP estimation. For this purpose, data from national dissection trials in Spain (n=132), Slovenia (n=121), France (n=250), Germany (n=308) and Belgium (n=140) have been pooled (n=951) to ensure large variability of carcasses. Prediction equations from the 12 EU countries which authorized ZP method were applied on the pooled data set. Results showed that average LMP varied between 57 and 62%, and consequently % of carcasses classified as S (≥60), E (≥55) or U (≥50) using the different equations varied between 20 and 74%, 23 and 57% and 2 and 21%, respectively. Important differences can be seen in LMP when different approved EU equations are used on the same data set. In each country, equations were developed on a representative sample of the national population. Differences in the results can be ascribed to the diversity of pig national populations across countries (sexes and genotypes) or methodological differences (different tools used to measure ZP, differences in the measurement due to the operator). Observed differences between approved ZP equations can be considered too high to be acceptable in view of an EU harmonization.

On-field and laboratory performance of electronic ear tag used for tracing pigs from farm to carcass

P. Grassi[1], G. Caja[1], J.H. Mocket[1], H. Xuriguera[2], S. González-Martín[2], J. Soler[3], M. Gispert[3] and J. Tibau[3]
[1]Universitat Autònoma de Barcelona, Animal & Food Sci., 08193 Bellaterra, Spain, [2]Universitat de Barcelona, DIOPMA, 08028 Barcelona, Spain, [3]Institut de Recerca i Tecnologia Agroalimentàries, IRTA, 17121 Monells, Spain; paulagrassi.ledo@gmail.com

A total of 1,540 crossbred pigs were used in 3 traceability experiments (fattening to 100 kg BW) using 3 types of button electronic ear tags (FDX-B, full duplex B; HDX, half-duplex). In Exp. 1, piglets (n=1,033) were tagged at birth with EF1 (FDX-B, 2.6 g) and slaughtered under commercial conditions. No EF1 losses were reported until weaning, but fattening losses were 6.3% and transportation and slaughtering losses were 44.9%, EF1 overall traceability resulting 52.8%. In Exp. 2, 133 weaned piglets were tagged with EH (HDX, 4.4 g). Fattening and slaughtering were done under experimental conditions. On-farm losses were 1.5 and 0.8% more were lost at slaughter, being EH overall traceability 97.7%. In Exp. 3, 374 weaned piglets were tagged with 453 ear tags: EF1 (n=151), EF2 (FDX-B, 4.2 g; n=140) and EH (n=162). Fattening was done in the farm Exp. 1 and slaughtering in 3 commercial slaughterhouses. By ear tag type (EF1, EF2 and EH), on-farm losses were 0, 1.9 and 0.9%, and failures 5.0, 5.6 and 0.9%, respectively. Additionally, 5.3, 5.0 and 1.8 ear tags were lost or failed during transportation, respectively. Slaughtering losses and failures were 28.4, 18.9 and 4.5%, resulting in 64.4, 71.2 and 92.1% overall traceability ($P<0.05$), respectively. Samples of ear tags were collected (n=10) to study their performances under laboratory conditions. Separation strength of EF1 from Exp. 1 was weaker ($P<0.05$) than the rest of ear tags, and reading distances varied (4 to 28 cm) according to type, reader and antenna orientation. In conclusion, EH ear tags were more efficient than EF1 and EF2 for tracing pigs under commercial conditions, but results depended on the device and the fattening and slaughtering conditions used.

Growth and carcase parameters of lambs sired by extreme muscle density rams at differing end points

E.M. Price[1], J.A. Roden[1], W. Haresign[1], J.A. Finch[2], G.E. Gardner[3] and N.D. Scollan[1]
[1]Aberytwyth University, Wales, SY23 3EB, United Kingdom, [2]Innovis Breeding Sheep Ltd, Aberystwyth, SY23 3HU, United Kingdom, [3]Murdoch University, Murdoch, WA 6150, Australia; emp15@aber.ac.uk

In vivo measures of carcase quality can aid selection. Muscle density, (measured by computer tomography) has strong negative genetic and phenotypic correlations with intramuscular fat. The aim of this study was to evaluate growth and carcase parameters of lambs, sired by five high and low muscle density Terminal sire rams, using three different covariates. Progeny (n=394) were recorded for live weight at: birth; 8 weeks, 16 weeks and immediately pre-slaughter. Ultrasonic measurements were made at the third lumbar vertebrae, pre-slaughter. Carcase records include carcase weights, EUROP classification scores for fatness and conformation. Data were analysed using GENSTAT 15, using a sire nested within muscle density ANOVA design model, fitting sex; dam age (slaughter batch for carcase traits); birth/rear type and MyoMax™ carrier status. The model was adjusted for the covariates, age at measurement (age); live weight at slaughter (Slwt); or slaughter fat grade (fat). Low muscle density sired lambs were heavier at 8 weeks (age P=0.022; Slwt P=0.043; fat P=0.04) and 16 weeks (age 0.014; Slwt P=0.05; fat P=0.036). However high muscle density sired lambs had increased ultrasonic muscle depth (age 0.019; Slwt P=0.017; and fat P=0.017), heavier hot carcase weight (age P=0.004; Slwt P<0.001; fat P=0.004), 48 h cold carcase weight (age P=0.027; Slwt P<0.001; fat P=0.013), higher killing out percentages (age, Slwt and fat P<0.001), and improved carcase conformation (age, Slwt and fat P<0.001). It is proposed that when compared at the same end point, high muscle density sired lambs have increased lean tissue which is expressed as higher carcase weight rather than live weight. Further data analysis is underway to investigate primal yields, waste data and eating quality.

Carcase and meat quality of lambs from different breeds and production systems

F. Ringdorfer
LFZ Raumberg-Gumpenstein, Sheep and goats, Raumberg 38, 8952 Irdning, Austria; ferdinand.ringdorfer@raumberg-gumpenstein.at

Lamb meat production under the hard production conditions in the alpine regions and concerning to the increasing price of concentrate is a special challenge. On the one hand the consumers prefer lamb meat from young lambs with high portion of meat and less content of fat. On the other hand they expect a natural production system. But there is a negative correlation between high carcase quality and natural production system in alpine regions. For marketing not only carcase quality is important, also meat quality can be an argument to buy a product. Characteristics of meat quality are tenderness, juiciness and flavour and also the content of valuable fatty acids. Carcase and meat parameters of 194 male and female lambs of 3 different production systems (alpine pasture A, semi intensive fattening indoor B and intensive fattening indoor C) where compared. Lambs of group A and B was mountain sheep breed and crossbreeds, lambs of group C was Merino land breed. Carcase dressing percentage was significant different between groups, 43.3, 49.8 and 47.8% (P<0.01)for A, B and C. Also kidney fad, expressed as percentage of cold carcass weight, was significant different, 1.26, 2.64 and 2.34 for group A, B and C. The c9, t11-18:2 isomer of conjugated linoleic acid (c9, t11-CLA) represents the main dietary CLA form with putative health benefits. Highest content was found in group A and B, the significant lowest in C, 0.97 and 0.90 vs. 0.59. The ratio of n-6:n-3 faddy acids was lowest in group A and B and significant higher in C, 2.6, 2.5 vs. 5.1. Some other meat quality parameters are cooking characteristics like drip losses and tenderness, measured as shearing force. Drip losses was 1.9, 2.3 and 2.7% for group A, B and C (P<0.01) and shear force was 4.0, 4.4 and 4.4 for group A, B and C but the differences was not significant (P=0.1).

Effects of castration age, protein level and lys/met on performance and carcass quality of steers

M.M. Campo[1], I.N. Prado[1], E. Muela[1], O. Catalan[2] and C. Sañudo[1]
[1]University of Zaragoza, Miguel Servet 177, 50013 Zaragoza, Spain, [2]INZAR, Julio García 42, 50018 Zaragoza, Spain; marimar@unizar.es

Castration is used as a tool for better animal handling, but also to create a differentiated product in terms of meat quality. Besides, changes in the feeding are used to improve beef production, although protein and amino acid levels are not fully optimised to reduce nitrogen in the environment. The aim of this study was to assess the effects of pre-pubertal castration, early (15 days) or late (5 months), dietary protein level (13 vs. 15% of DM) and the lys/met ratio (3.0 vs. 3.4) on animal performance and carcass characteristics of Friesian steers finished with diets containing high levels of concentrate. Sixty-four Friesian calves were used in a complete factorial design. Half of the calves were selected at random and castrated at 15 days of age. The rest of the animals were castrated at 5 months of age. Each of the castrated group was divided afterwards into 2 dietary protein levels and 2 lys/met groups, for a total of 8 animals/pen in the fattening period, which lasted 6 months. Steers were slaughtered at an average live weight of 414.6±38.6 kg. No significant differences neither interactions, were found on cold carcass weight, morphology or conformation and fatness scores between the treatments. The same genetics and the fact that all animals were relatively early castrated might explain this results. However, although castration did not show any effect on tissue composition, high protein in the diet (15%) showed higher muscle in the carcass (57.4 vs. 55.6%; $P<0.05$). A high lys/met of 3.4, also increased the percentage of muscle (57.2 vs. 55.8%; $P<0.001$). Thus, it is recommend using diets with a higher protein level (15%) and a higher lysine/methionine ratio (3.4), by reducing the methionine content, in order to obtain carcasses with higher muscle content. However, the price of feed and the subsequent potential return must be taken into consideration for such diets to be profitable.

Session 02 Theatre 14

Analytical comparison of online techniques to measure meat quality

R. Roehe[1], D.W. Ross[1], C.-A. Duthie[1], C.R. Craigie[1], M. Font-I-Furnols[2], M. Candek-Potokar[3], C.A. Maltin[4] and L. Bunger[1]
[1]Scotland's Rural College, Roslin Institute Building, EH259RG, United Kingdom, [2]IRTA, Finca Camps i Armet, 17121 Monells, Spain, [3]Agricultural Institute of Slovenia, Hacquetova ul. 17, 1000 Ljubljana, Slovenia, [4]Quality Meat Scotland, Edinburgh, EH28 8NZ, United Kingdom; rainer.roehe@sruc.ac.uk

Establishing and evaluating accurate, reliable and objective techniques for measuring or predicting eating and nutritional quality of meat is a key step for improving these traits in the industry. For meat eating quality, the sensory assessment of tenderness, juiciness and flavour, based on a trained taste panel, is the reference method of choice. Nutritional quality is largely related to fatty acid profiles for which increased polyunsaturated fatty acids, in particular the Omega-3 fatty acids, is associated with higher human health benefits. Online techniques for predicting meat quality in the abattoir have to be accurate and precise, easy and fast to operate, cost-effective with the ability to be applied as early as possible post-slaughter. Using these criteria, visual near infrared spectroscopy (VisNIR), hyperspectral imaging (HSI) and Raman spectroscopy (RS) were analytically compared. All techniques have been used to predict, besides numerous other attributes (e.g. cooking loss, colour, pH), eating and nutritional quality of meat and are suitable for online application in the abattoir. Under abattoir conditions, VisNIR predicted sensory characteristics at accuracies in the range of $R^2=0.21$ (juiciness) to 0.59 (flavour), fatty acid profiles at $R^2=0.16$ to 0.73, with Omega-3 fatty acids at $R^2=0.43$. HSI combines imaging and NIR technology and has the potential to improve the prediction of meat quality traits. RS is measuring more specifically the energy of molecular vibrations. The analytical comparison indicates that VisNIR is a robust and accurate technique to predict meat quality characteristics online in the abattoir which are otherwise difficult to determine. HSI has the potential to improve the accuracy of prediction under laboratory conditions but requires further development and testing under abattoir conditions.

Session 02 Theatre 15

Does selection for lean meat yield reduce the sensory scores of Australian lamb?

L. Pannier[1], G.E. Gardner[1], A.J. Ball[2] and D.W. Pethick[1]
[1]Murdoch University, 90 South Street, 6150 Murdoch, Australia, [2]University of New England, Armidale NSW, 2351 New South Wales, Australia; d.pethick@murdoch.edu.au

Sensory enjoyment is one of the key drivers that influences the consumer demand for lamb in Australia. One of the key factors that determines consumer satisfaction of lamb is intramuscular fat (IMF). Yet the challenge is to balance this against the industry aim of selecting for lean meat yield using Australian Sheep Breeding values for post-weaning eye muscle depth (PEMD) and subcutaneous fat depth at the c-site (PFAT), as these have been shown to decrease IMF levels. Hence, we hypothesised that selection for reduced PFAT and increased PEMD will reduce the sensory scores of lamb and that this relationship is driven through reduced IMF levels. Sensory scores were generated on both the longissimus thoracis et lumborum (loin) and semimembranosus (topside) muscle from 1,434 lambs. Five day aged grilled steaks were tasted by untrained consumers who scored (1-100 score) the samples for tenderness, juiciness, flavour, odour and overall liking. Increasing PEMD was associated with 5.3, 3.6 and 3.1 lower sensory scores for tenderness, overall liking and flavour for both the loin and topside samples. Decreasing PFAT was associated with a 3.1 score reduction for tenderness within the loin samples only. All sensory scores increased with higher IMF levels, most strongly for juiciness and flavour, however in this analysis variation in IMF levels did not appear to explain the impact of either PEMD or PFAT. This illustrates that the associations seen between PEMD and PFAT with the sensory scores are not solely driven through the phenotypic impact of IMF, in contrast to our initial hypothesis. Yet in support of our hypothesis, selection for more muscular and leaner animals did reduce the sensory score, confirming our growing concerns that selecting for lean meat yield would reduce consumer eating quality. This highlights the need for careful monitoring of selection programs to maintain the eating quality of lamb.

Session 02 Theatre 16

A feasibility study for the prediction of the technological quality of ham with NIR spectroscopy

A. Vautier, T. Lhommeau and G. Daumas
IFIP, Meat Quality and Safety, La motte au Vicomte, BP 35104, 35651 Le Rheu, France; antoine.vautier@ifip.asso.fr

The ability of visible and near infrared spectroscopy to predict the cooking yield was tested on industrial Jambon Cuit Supérieur process. The precision of NIRS calibration to predict PSE-Like zone classification was evaluated too. A set of 150 hams was processed following industrial standards and materials. Individual traceability was maintained from brine injection to slicing. One third of the population was selected as an external validation data set while the other part was used to construct PLS calibration models with cross validation. Spectra (350-1,800 nm) were collected using a surface probe and an insertion probe. Five muscle sites were measured on deboned hams with the surface probe (Gluteus Medius, Semimembranosus, Vastus Medialis, Biceps Femoris and Semitendinosus), whereas the intramuscular part of the Semimembranosus was measured with the insertion probe for the specific need of an early detection of PSE-like zones (bone-in hams). Ultimate pH measurement was applied on the NIRS sites and on the reference site of the Semimembranosus. Best external validation results for cooking yield were obtained with NIRS calibration on the Gluteus Medius cross section. Predicted/observed cooking yield correlation and predicting error (r=0.82, error=1.62) confirm the better precision of NIRS compare to the Semimembranosus ultimate pH based prediction (r=0.81, error=1.75). Nirs classification of hams according to the PSE-like zone defect showed a good level of precision for external validation with 84% of correct classification on deboned hams (surface probe) and 77% of correct classification on bone-in hams (insertion probe). External validation test confirms that visible and near infrared spectroscopy is an efficient alternative to pHu for cooking yield prediction. Developments are still needed for a PSE-Like zone classification based on NIRS and using an insertion probe to recover the precision level of the surface probe.

Visualization of marbling and prediction of intramuscular fat of pork loins with computed tomography

M. Font-I-Furnols[1], A. Brun[1], N. Tous[2,3] and M. Gispert[1]
[1]IRTA, Product Quality, Finca Camps i Armet, 17121 Monells, Spain, [2]IRTA, Monogastric Nutrition, Ctra. Reus-El Morell Km 3.8, 43120 Constantí, Spain, [3]URV, Laboratory of Toxicology and Environmental Health, St Llorenç 21, 43201 Reus, Spain; maria.font@irta.es

Intramuscular fat content (IMF) improves the acceptability of pork when tasted by consumers beside their preferences for meat with more/less marbling and it is a quality attribute in dry cured ham production. The non-destructive quantification of IMF and the visualization of marbling would allow classify meat and give it an added value. The aim of the work was to use computed tomography (CT) to visualize marbling and quantify IMF in pork. Loins (n=365) were CT scanned using 2 axial acquisition procedures, one that allows the visualization of marbling A120 (120 kV, 200 mA, 1 mm-thick and edge algorithm) and one used to quantify tissues in pig carcasses A140 (140 kV, 145 mA, 10 mm-thick and standard algorithm). Marbling was evaluated by one trained technician using NPPC scale, in fresh loin and in A120 images (n=227). IMF was determined with a near infrared transmittance device. The relative volume associated to each Hounsfield value (HU) was obtained for the longissimus region of each image and partial volume associated to an interval of 10-20HU values were obtained. To evaluate marbling the % of agreement between scores from fresh meat and from CT images was calculated. Results showed that NPPC scale used on CT images underestimates marbling comparing with fresh loins: 55% of the loins have higher marbling scores in fresh meat than in CT image. To estimate IMF linear regression was performed using partial volumes selected by stepwise procedure as predictors. The best IMF estimation was found combining data from both CT images (R^2=0.84; RMSEP=0.45%). It can be concluded that to predict IMF in pork loins non-destructively with CT, the combination of data from two images with different acquisition conditions produces the highest accuracy. Moreover, it is necessary to create a new marbling scale based en CT images for its evaluation.

Breeding for meat quality in Swiss beef cattle

H. Joerg[1], M. Scheeder[1], A. Burren[1], M. Schafroth[1], R. Tobler[1], U. Vogt[2], T. Aufdermauer[2] and C. Flury[1]
[1]Bern University of Applied Sciences, School of Agricultural, Forest and Food Sciences, Länggasse 85, 3052 Zollikofen, Switzerland, [2]Swiss Beef Cattle Mutterkuh Schweiz, Stapferstrasse 2 Postfach, 5201 Brugg, Switzerland; hannes.joerg@bfh.ch

In Switzerland beef meat quality is not routinely measured, so a classical breeding based on estimated breeding values is not feasible. Under these conditions, a study of a molecular basis for variations in meat tenderness may provide a solution to improve meat quality by developing genome-wide selection. Most Swiss beef cattle are crossbreeds between dairy and beef breeds, therefore we analysed across breed. The objectives of the study were to estimate phenotypic variation of meat quality traits in Swiss beef cattle. Beef samples were selected due to extreme values in carcass weights, lean and fat yield, respectively. Meat quality traits were analysed of longissimus dorsi and biceps femoris muscles. Measurements were performed at 21 days of postmortem aging for pH, colour, intra muscular fat content, cooking loss and Warner-Bratzler shear force. There were highly different values for red colour, the intra muscular fat content, the cooking loss and the Warner-Bratzler shear force. These results show that approximately 1000 animals will be necessary to estimate the SNP effects in the Swiss beef population. The mean phenotypes for Warner-Bratzler shear force are in Simmental and Angus breed of 3.74 and 4.36, respectively. The large difference of the means compared to the phenotypic standard deviations of approximately 0.6 require a minimum number of animals with Simmental and Angus sire to estimate variation due to the breed. In a further study we attempt to perform a genome-wide association analysis for colour, intra muscular fat content, cooking loss and Warner-Bratzler shear force in Swiss beef cattle having a beef cattle sire and a dairy cattle dam.

Reducing the pH of lamb carcasses will improve retail meat colour

H.B. Calnan[1,2], R.H. Jacob[1], D.W. Pethick[1,2] and G.E. Gardner[1,2]
[1]Cooperative Research Centre for Sheep Industry Innovation, University of New England, Armidale, 2351 NSW, Australia, [2]Murdoch University, Division of Veterinary Biology and Biomedical Science, Murdoch, 6150 WA, Australia; h.calnan@murdoch.edu.au

Meat colour data was collected from 4,953 lambs produced at 5 sites across Australia over a 5 year period (2007-2011) as part of the Sheep Cooperative Research Centre's information nucleus flock experiment. Longissimus muscle samples were collected 24 h post-mortem, vacuum packaged, aged for 5 days and then placed under simulated retail display conditions for 3 days. At the end of this period light reflectance of the meat surface was measured with a Hunterlab reflectometer and a ratio was calculated (630 nm/580 nm reflectance) to represent redness, with higher values redder and hence more desirable. These ratios were analysed using linear mixed effects models. The base model included fixed effects for site, year of birth, kill group, sire type and dam breed. In a second analysis pH measured 24 h post-mortem was included in the base model as a covariate. Of the dam breeds, Merino progeny had 0.19 units lower redness than those of Maternal dam breed. Similarly the Merino sire type produced lower redness values than Maternal or Terminal sired lambs, with 0.39 units difference seen between lambs of Merino and Terminal sire types. pH was negatively associated with meat redness, with a 0.92 unit decrease ($P<0.01$) in redness across the pH range of 5.4 to 6. With pH included in the model, the effect of sire type was not significant, suggesting that differences in post-mortem muscle pH between sire types may underpin the observed variation in retail meat colour. Our findings suggest that by reducing the pH of lamb loins we can improve the redness of the meat whilst on retail display.

Fatty acid profile from plasma and adipose tissues of lambs grazing two secondary successions

M.A. Gallardo[1], D. Dannenberger[2], J. Rivero[3] and R. Pulido[1]
[1]Universidad Austral de Chile, Animal Science, Graduated School. Faculty of Veterinary Science. Casilla 567, 5090000 Valdivia, Chile, [2]Leibniz-Institute for Farm Animal Biology, Department of Muscle Biology and Growth, Wilhelm-Stahl-Allee 2, 18196 Dummerstorf, Germany, [3]Universidad Austral de Chile, Animal Production Institute, Graduated School. Agriculture Faculty. Casilla 567, 5090000 Valdivia, Chile; mugallar@gmail.com

This experiment was carried out for evaluation and determination breed and pasture effects on fatty acid profiles in two lamb breeds. The experiment was conducted at the Butalcura Research Station (Chiloé-Chile), during October to December 2011. Eight Chilota and 6 Suffolk Down lambs, 2 months age, males, no twins, average live weight 14.36±2.37 kg and average body condition 2.71±0.27, were located to graze a 'calafatal', a typical secondary succession of Chiloé archipelago. Simultaneously, 8 Chilota lambs breed were located to graze a naturalized pasture of the Chiloé archipelago. Botanical and chemical composition of pastures and fatty acid profile analysis from pastures and different lambs tissues (intramuscular, subcutaneous and plasma) were performed. Under the conditions of this study, fatty acids profile from Longissimus dorsi muscle is affected by pasture type, but not lamb breed. Thus, Chilota lambs grazing calafatal showed higher PUFA contents (13.42±0.89 versus 10.34±0.83%), but not MUFA contents (40.89±0.40 versus 43.02±0.78%) nor $\omega 6/\omega 3$ ratios (3.00±0.12 versus 2.58±0.16) than Chilota grazing naturalized pasture. Otherwise, although fatty acids profile from back fat is notable affected by lamb breed and tail fat is affected by both, lamb breed (not as closely as back fat) and pasture type (not as closely as Longissimus dorsi muscle), both subcutaneous adipose tissues only reflected a fourth (2.85%, as mean value) of the PUFA content found in Longissimus dorsi muscle (12.01% as mean value).

Effect of dietary vitamin E on trans fatty acid profile of muscle and adipose tissues of indoor lamb

V. Berthelot[1], L. Broudiscou[2] and P. Schmidely[1]
[1]AgroParisTech, UMR 791 MoSAR, 16 rue Claude Bernard, 75005 Paris, France, [2]INRA, UMR 791 MoSAR, 16 rue Claude Bernard, 75005 Paris, France; valerie.berthelot@agroparistech.fr

Lambs fed high concentrate diets usually exhibit high proportions of C18:1 trans, especially C18:1 10t, in muscle and adipose tissues. Cardiovascular health risks have been associated with these fatty acids (FA). As it was shown in steers, a way to prevent the '11t to 10t shift' in the rumen of lambs could be the addition of vitamin E to high-concentrate diets. Thus the aim of the study was to investigate the effect of dietary vitamin E supplementation on trans FA profile of tissues of lambs fed high concentrate diets. Thirty male lambs were assigned to 3 experimental groups corresponding with three levels of supplementary vitamin E (45 (E0), 286 (E1) and 551 (E2) mg/kg DM) included in a barley-based diet. All lambs had *ad libitum* access to the diets from 29 to 46 kg BW. After slaughter, perirenal (PR), caudal (CA) adipose tissues and muscle (MU; extensior carpi radialis) were sampled to analyse their FA composition. GLM procedure was used for statistical analyses. Vitamin E supplementation did not modify lamb growth (377 g/d, $P>0.10$) and slaughter parameters. The proportion of ∑C18:1 trans was the highest in PR, intermediate in CA and the lowest in MU (7.78%, 6.71%, 2.58% of total FA respectively; $P<0.0001$). The proportion of ∑C18:1 trans in tissues was not affected by dietary treatments (5.74, 5.69 and 5.63 for E0, E1 and E2 respectively). Among trans isomers of C18:1, E2 lambs had a higher proportion of C18:1 10t and a lower proportion of C18:1 11t than E0 and E1 lambs (50% vs. 38% and 17% vs. 26% of ∑C18:1 trans for E2 vs. E0 and E1, for C18:1 10t and C18:1 11t, respectively; $P<0.0001$). Contrary to steers fed high barley diets, vitamin E supplementation did not decrease the C18:1 trans proportion tissues or improve its isomeric profile in lambs fed high concentrate diets. It even increased the C18:1 10t proportion which possibly lowered the lamb meat nutritional value.

Effect of the age at 2nd Improvac® vaccination on fatty acid composition in back fat of male pigs an

T. Sattler[1], F. Sauer[1] and F. Schmoll[2]
[1]University Leipzig, Large Animal Clinic for Internal Medicine, An den Tierkliniken 11, 04103 Leipzig, Germany, [2]AGES, Institute for Veterinary Disease Control, Robert-Koch-Gasse 17, 2340 Mödling, Austria; tasat@vetmed.uni-leipzig.de

Vaccination of male fatteners to avoid boar taint has been proven to be a practical and animal friendly alternative to surgical castration. Since meat quality is also an important aspect for consumers and meat processors, the objective of the study was to compare fatty acid composition in back fat of Improvac® vaccinated, surgical castrated and entire male fatteners and to evaluate, if the age at 2nd vaccination can influence the fatty acid composition which might be used to match consumer´s wishes. Experimental methods: A whole of 166 male fatteners were included in the study. Group 1 and 2 consisted each of 40 Improvac® vaccinated pigs, whereas Group 1 received 2nd vaccination at the age of 21 weeks (5-6 weeks before slaughter) and Group 2 respectively 18 weeks (8-9 weeks before slaughter). Group 3 included 41 surgical castrated pigs and group 4 consisted of 45 entire boars. The same amount of pigs was slaughtered at the age of 26 and 27 weeks, respectively. Fatty acid composition in back fat was measured with gas chromatography. Differences between groups were tested with Anova and Bonferroni as post-hoc test. Results and conclusions: Percentages of saturated and monounsaturated fatty acids in back fat of both groups of Improvac® vaccinated fatteners were significantly lower than in surgical castrated fatteners but significantly higher than in entire boars. Polyunsaturated fatty acids, on the other hand, were significantly higher in Improvac® vaccinated pigs than in castrates, but lower than in entire boars. No differences were seen between the two Improvac® groups. Results show that fatty acid composition differs between Improvac® vaccinated, surgical castrated and entire boars, whereat a difference of three weeks of age at 2nd vaccination with Improvac® and same age at slaughter in this study had no influence of fatty acid composition in back fat.

Sequencing of PRKAG3 gene revealed several novel amino acid substitutions in the Finnish Yorkshire

P. Uimari and A. Sironen
Agrifood Research Finland, MTT, Alimentum, 36100 Jokioinen, Finland; pekka.uimari@mtt.fi

PRKAG3 gene on porcine chromosome 15, also known as rendement napole, is known to be associated with meat quality traits. The best known amino acid substitution within PRKAG3 is R200Q (RN-allele) that is found only in the Hampshire breed and is associated with high glycogen content, low pH, and low technological yield. However, greater tenderness of meat was also found in animals carrying the RN-allele. Four other amino acid substitutions (T30N, G52S, L53P, and I199V) have also been reported for PRKAG3 in different breeds. A GWAS including over 400 Finnish Yorkshire AI-boars revealed a highly significant association between a SNP (ASGA0070625) in the vicinity of PRKAG3 and pH measured from the loin. Fine mapping of the region was done by sequencing 220 AI-boars with 3500×L Genetic Analyzer. Primer pairs for genomic sequence of PRKAG3 gene covering all exons and 520 bp of the promoter region were designed and DNA fragments were amplified with the gene specific primers. PCR fragments were sequenced in both directions with the same primers used in the amplification procedures. Sequencing of PRKAG3 revealed several novel SNPs on the promoter region at positions -388 bp (A/G), -298 bp (G/A), -234 bp (C/G), and -58 bp (A/G) and amino acid substitutions K36E, I41V, K131R, and P134L within exons 2, 3, 4 and 5, respectively. All new variations were in complete linkage disequilibrium with ASGA0070625 and the known alleles 30T, 53P and 200R were fixed in Finnish Yorkshire. For the known amino acid substitutions the best P-value for association with pH measured from the loin was found for I199Q (P-value=0.02). P-values for the novel amino acid substitutions were the same as for ASGA0070625 (P-value=$2.18E^{-10}$). Because of the complete linkage disequilibrium in the region it is currently impossible to say which of the novel amino acid substitutions are causative. The effect on meat quality may also be due to combination of several amino acid substitutions or variation in the promoter region.

Use of X-ray computed tomography for predicting an industrial lamb carcass lean meat yield

J. Normand[1] and M. Ferrand[2]
[1]Institut de l'Elevage (French Livestock Institute), Meat Quality Department, Agrapole, 23 rue Jean Baldassini, 69364 Lyon cedex 07, France, [2]Institut de l'Elevage, Biometrics Department, 149 rue de Bercy, 75595 Paris cedex 12, France; jerome.normand@idele.fr

Interbev Ovins, (the French ovine interprofession) is prospectively interested in the ability of the video image analysis system VIAscan® to estimate the lamb carcasses percentage of lean meat yield. To calibrate this system, it is necessary to bone a large number of carcasses following specifications closed to the anatomical dissection. But the dissection of carcasses into fat, muscle and bone is a destructive, time-consuming, and therefore a costly method. Recent studies on pig carcasses showed that X-ray computed tomography (CT), a non-invasive method, can replace dissection with the following advantages: repeatability, reproducibility and standardization of the measurement protocol. Therefore, the potential of CT to predict an industrial lean meat yield was investigated in lamb. Composition of 158 lamb carcasses representative of carcasses usually cut in the French sheep industry (conformation: R and O; fat score: 2, 3 and 4; carcass weight: 14-25 kg) was determined from CT and industrial cut, following specifications closed to the dissection. The results show that the CT overestimates the percentage of muscle (11.5 points on average) and fat (6.2 points) and underestimates the percentage of bone (-17.4 points) compared to industrial cut. However, the difference remains relatively constant between carcasses and the coefficients of determination (R^2) ranged from 0.8 to 0.9. CT prediction is presently not accurate enough for use by the sheep industry: in 61% of cases, the percentage of muscle evaluated with CT is outside a tolerance interval set at ±2% of the percentage of muscle measured from the cut. The determination of carcass composition by anatomical cut in an industrial context is probably the source of the prediction error because, in this context, the cut is somewhat difficult to standardize and not repeatable.

Session 02 — Poster 25

Effects of weaning age, period at fattening unit and slaughter age on lamb depot's composition

M.M. Campo[1], V. Resconi[1], A. Cònesa[2], E. Horcas[3] and C. Sañudo[1]
[1]University of Zaragoza, Miguel Servet 177, 50013 Zaragoza, Spain, [2]PGI Ternasco de Aragon, MercaZaragoza, 50014 Zaragoza, Spain, [3]Pastores Grupo Cooperativo, MercaZaragoza, 50014 Zaragoza, Spain; marimar@unizar.es

In Spain, traditional lamb production systems have evolved towards more intensive farming. The new farming scheme is divided into the breeding stage, carried out by traditional farmers, and the fattening stage, which takes place at cooperative fattening units (FU), in order to obtain a more homogeneous product with a quality brand. This implies differences in animal management prior to FU that provoke different lengths of the common fattening period at FU. The aim of this work was to assess differences in fatty acid (FA) composition of two fat depots [subcutaneous (SC) and kidney and knob channel (KKC)] due to different age at weaning (40 vs. 50 days old), period in FU (1 vs. 20-30 days) and age of slaughter (67 vs. 104 days old) in 85 lambs of Rasa Aragonesa breed. These depots do not reduce the value of the carcass after sampling. The effect of age of weaning was more obvious in KKC than in SC, where weaning at 50 days old increased the percentage of saturated FA (SFA) (56.0 vs. 54.3), especially due to an increase in C14:0 ($P<0.001$), and lower n-6/n-3 ratio due to a higher n-3 FA%, mainly α–linolenic acid (0.37 vs. 0.42), reflecting the grazing diet of their dams through the milk composition during a longer period of time. No significant effect of polyunsaturated FA or FA ratio was found in SC due to weaning. However, the longer the time at the fattening unit the higher the percentage of monounsaturated FA (MUFA), especially in SC (43.6 vs. 47.6%, 1 vs. 20-30 d). This was due to a more unsaturated diet in the CC in order to improve the quality of the meat. Increasing the age at slaughter decreased the SFA and increased the MUFA percentage in both depots, although it was more intense in SC (43.4 vs. 47.8%, 67 vs. 104 days old) than in KKC (38.6 vs. 40.1%). KKC reflects better the management prior to the FU, and SC is a good indicator of the last fattening phase.

Session 02 — Poster 26

Carcass quality studies of two commercial hybrid pig groups

D. Ribikauskiene and V. Razmaite
Institute of Animal Science of Lithuanian University of Health Sciences, Animal Breeding and Genetics, R. Zebenkos 12, Baisogala, Radviliskis distr., 82317, Lithuania; daiva@lgi.lt

The objective of the study was to determine the influence of different methods of carcass evaluation and genotype pigs on carcass quality. The study was carried out at the LUHS Institute of Animal Science and joint-stock company 'Utenos mesa'. Pigs were allocated into two groups of commercial hybrid pigs. Group 1: (German Large White × German Landrace) × Norvegian Landrace (GLW×GL)×NL, n=12), group 2: German Large White × German Landrace (GLW×GL, n=19). Three methods were used for the analysis of commercial hybrids, i.e. dissection, ruler (ZP) and FOM device. The chosen carcasses were dissected by the method of Walstra and Merkus. All analyses were performed in MINITAB 15. The lean meat content of the carcasses in both groups of pigs determined by dissection was by 0.8 and 0.1%, respectively, higher than that determined by FOM. The muscle depth in groups 1 and 2 between ribs 3-4 from the last rib determined by FOM was, respectively, 5.4 and 3.6 mm higher than that measured by the ruler at the same point. The lean meat content of (GLW×GL)×NL hybrid pigs when determined by dissection was 2.9% higher than that of GLW×GL hybrid pigs.

The use of artificial neural networks for predicting meat content in pig carcasses

M. Szyndler-Nędza[1], K. Bartocha[2], M. Maśko[2] and M. Tyra[1]
[1]National Research Institute of Animal Production, Department of Animal Genetics and Breeding, ul. Sarego 2, 31-047 Krakow, Poland, [2]National Research Institute of Animal Production, Department IT, ul. Sarego 2, 31-047 Krakow, Poland; magdalena.szyndler@izoo.krakow.pl

The Polish pig breeding program uses performance test results for animal evaluation and selection. This method is based on determining carcass meat content using ultrasonic measurements of backfat thickness and height of loin eye. Measurements are standardized for 110 kg of body mass and for the 180th day of life to compare between animals. For estimating carcass meat content a linear regression equation has been introduced in previous research achieving an error of 3.01%. The aim of the study was to evaluate Artificial Neural Networks (ANN) as an estimator for meat content in carcasses based on ultrasonic measurements on alive pigs. The study included 628 pigs of different breeds. These animals were evaluated in the Polish Pig Testing Station (SKURTCh) during the period from 2008 to 2012. On the day of slaughter, live ultrasonic measurements of backfat thickness were taken at points P2, P4 and of the longissimus muscle at the P4M point. Then, animals were slaughtered and after a 24-h cooling period their right sides were divided into primal cuts and subjected to a detailed dissection. Among other things, gathered records include ultrasonic measurements and real meat contents in carcasses. A two-layer ANN architecture was used to predict carcass meat content with a breed indicator, P2, P4 and P4M measurements as inputs. A total of 100 training cycles were conducted and in each the data set was split into subsets for training (70% of samples), validation (15%) and testing (15%). The architecture contained 7 hidden neurons and was trained using the Levenberg-Marquardt algorithm. After training, 20 top networks were chosen as a representative sample of high performing models. The ANN model achieved an average testing error of 2.1% (with 0.78 correlation). The work was funded by the National Science Centre N N311 082240.

Growth and carcass traits of purebred Ayrshire bulls and crossbred Ayrshire×beef-breed bulls

M. Pesonen[1], A. Huuskonen[1], M. Hyrkäs[2], H. Kämäräinen[3] and R. Kauppinen[3]
[1]MTT Agrifood Research Finland, Animal Production Research, Tutkimusasemantie 15, 92400 Ruukki, Finland, [2]MTT Agrifood Research Finland, Animal Production Research, Halolantie 31A, 71750 Maaninka, Finland, [3]Savonia University of Applied Sciences, P.O. Box 72, 74101 Iisalmi, Finland; maiju.pesonen@mtt.fi

Crossbreeding between dairy cows and beef-breed bulls is suggested to improve carcass production compared to pure dairy bulls. The objective of the study was to compare growth and carcass traits of purebred Finnish Ayrshire (Ay) and Ay×beef breed crossbred bulls. The data from four Finnish slaughterhouses included observations of 164,812 purebred Ay, 2,329 Ay×Aberdeen angus (Ay×Ab), 1,466 Ay×Blonde d'Aquitaine (Ay×Ba), 1,044 Ay×Charolais (Ay×Ch), 782 Ay×Hereford (Ay×Hf), 5,293 Ay×Limousin (Ay×Li) and 1,270 Ay×Simmental (Ay×Si) bulls. An estimated daily carcass gain was calculated by subtracting 16 kg birth carcass weight from the reported slaughter weight and dividing the result by age at slaughter. The statistical analyses were performed using the SAS Mixed procedure. Differences between the breeds were compared using a Dunnett's test so that purebred Ay was used as a control breed. The average slaughter age of the bulls was 587 days. Mean carcass weights for the Ay, Ay×Ab, Ay×Ba, Ay×Ch, Ay×Hf, Ay×Li and Ay×Si bulls were 330, 357, 374, 383, 356, 372 and 381 kg, and daily carcass gains were 532, 576, 616, 629, 580, 605 and 628 g/d, respectively. All crossbred groups differed significantly ($P<0.001$) from Ay bulls in both carcass weight and carcass gain. The EUROP conformation score of the Ay bulls was 4.7, on average, and improved 28, 57, 51, 23, 55 and 36% with Ay×Ab, Ay×Ba, Ay×Ch, Ay×Hf, Ay×Li and Ay×Si crossbreds, respectively, compared to pure Ay bulls ($P<0.001$). It can be concluded that crossbreeding with late maturing breeds largely improve carcass production compared to purebred Ay bulls. With early maturing breeds the improvements in gain and conformation score were intermediate compared to the late maturing crossbreds.

Effect of season and farmer to plenitude of carcass for Holstein steers

S. Yamamoto[1], S. Ito[2], S. Tsubosaka[2], T. Wakisaka[3], S. Okada[3] and K. Kuchida[1]
[1]Obihiro University of A &VM, Inada-cho, Obihiro, Hokkaido, 080-8555, Japan, [2]Tokachi Shimizu Food Service, Shimizu-cho, Hokkaido, 089-0103, Japan, [3]Japan Agricultural Cooperatives Tokachi Simizu-cho, Shimizu-cho, Hokkaido, 089-0198, Japan; s20251@st.obihiro.ac.jp

In Holstein fattening steers, it is difficult to evaluate meat quality differences between carcasses using Japanese beef grading system. So they require some new methods to quantify the variability objectively. 'Plenitude of Carcass' (PC) might solve above problem. It is calculated from image analysis traits and carcass grading traits and indicates meat volume and muscle/fat balance. The purpose of this study is to investigate yearly transition of PC and effect of season and farmers to PC. Carcass data were collected from Holstein steers shipped between August 2011 and December 2012. The number of records of was 1,307. The image analysis traits were calculated from the rib images of the subprimal cuts taken by an originally developed imaging device placed in a meat processing factory. This device always could take pictures under the same focal length and lighting conditions. Thus, we could accurately measure and compare the area and color of the meat. The average of PC from 2011-summer to 2012-winter by season were respectively 202.7, 188.7, 231.0, 289.2, 298.0, 253.9 and 262.4. 2012-spring and 2012-summer shows high PC (There are significant differences between other seasons $P<0.05$). Coefficient of variance about PC was 0.44 in 2011-summer, and tended to decrease in 2012-spring. Afterwards it remained between 0.22 and 0.27, and increased in 2012-winter. The averages of 5 farmers' PC were 271.9, 210.9, 240.4, 217.8 and 271.7, respectively. It was consistent with the view of the meat processing factory staff. Coefficient of variance of each farmer about PC was 0.32, 0.34, 0.30, 0.33 and 0.26. From results, season and farm affect PC. Therefore CV was suggested a possible beneficial effect of control dispersion.

Session 02 Poster 30

Rating pork meat tenderness using visible/near-infrared spectroscopy and artificial neural networks

J.M. Balage[1], S.L. Silva[1], A.C.O.L. Figueira[2] and M.N. Bonin[1]
[1]University of Sao Paulo, Faculty of Animal Science and Food Engeneering, Dept. Animal Science, Av. Duque de Caxias Norte, 225, 13635900 Pirassununga, SP, Brazil, [2]Instituto Superior de Engenharia, Universidade do Algarve, Engenharia Alimentar, Campus da Penha, 8005, 8005-139 Faro, Algarve, Portugal; sauloluz@usp.br

Visible/near-infrared spectroscopy (Vis/NIR) combined with artificial neural networks (ANN) was tested as a non-destructive method to classify pork loin in tenderness classes. Samples from Longissimus muscle at the 11th rib were removed from 134 pig carcasses and the Vis/NIR spectra were collected from 400 to 1,495 nm. Spectral resolution was set to five nm and five scans were averaged for every spectrum. Each sample was scanned two times at different locations throughout the sample. Tenderness was determined using the Warner-Bratzler shear force (WBSF) method. Tenderness classes were defined as tender (n=44), when $WBSF \leq 32 N/cm^2$, and tough (n=70), when $WBSF \geq 45 N/cm^2$. Intermediate values have been removed to reduce errors, thereby forming a group of 114 samples to be evaluated. Machine learning based on ANN was applied to build the classification model of tenderness, performed by the Weka Data Mining Software. The reflectance of 600 nm to 1495 nm wavelength range was considerate for the analyses. A principal component analysis (PCA) filter was applied in the spectra collected with the Vis/NIR. The classifier used a multilayer perceptron ANN with two hidden layers, one input (filtered spectra by PCA) and one output (WBSF). Data evaluation was done by cross validation in eight groups. The results indicate a tendency where tender meat shows higher reflectance values than tough meat. The Vis/NIR and ANN generated a classifier with 76% classification accuracy. Moreover, the classifier rated 90% of the tender samples correctly (recall). A classifier capable of grouping tender pork loins with 90% accuracy has great importance for the meat industry enabling the creation of a line with high quality standard products and high added value.

Determination of drip loss and intramuscular fat at different positions in the porcine carcass

U. Baulain[1], L. Frieden[2], F. Adam[3], M. Henning[1] and E. Tholen[2]
[1]Institute of Farm Animal Genetics, Friedrich-Loeffler-Institut, Höltystr. 10, 31535 Neustadt, Germany, [2]Institute of Animal Science, University of Bonn, Endenicher Allee 15, 53115 Bonn, Germany, [3]Chamber of Agriculture North Rhine-Westphalia, Nevinghoff 40, 48147 Münster, Germany; ulrich.baulain@fli.bund.de

A total of 203 carcasses of four genetically different pig breeds were investigated to evaluate the effect of different measuring positions on drip loss and intramuscular fat content. Both traits were measured at three positions in loin and ham, respectively, using the repeated EZ-DripLoss method and near-infrared spectroscopy. Moreover, meat quality was indirectly characterized by pH, meat colour and conductivity recorded 1, 2 and/or 24 h p.m. in loin at the 13th/14th rib and in ham. Regarding drip loss, significant differences between the recordings at single measuring positions within and between ham and loin muscle groups were detected. Correlations between different positions ranged from r=0.3 to 0.5. Out of these results, it can be concluded that it is not possible to accurately characterize drip loss of the entire carcass with one single measurement. Correlations between indirect traits and drip loss were most expressed at the loin position 13th/14th rib, ranging from 0.5 to 0.6. These results could be expected because of the identical measurement position of these traits. Relationship between indirect meat quality traits and drip loss recorded at remote positions were lower, ranging from 0.2 to 0.4. With respect to the practicability of drip loss detection under routine slaughterhouse conditions, a combination of drip loss measurements at the cranial and caudal ends of the loin plus conductivity (2 h p.m.) can be used to estimate carcass drip loss with acceptable accuracy (R^2=0.65). In contrast to drip loss, intramuscular fat of the entire carcass can be sufficiently estimated with one single measurement. Correlations between intramuscular fat records at different positions ranged from 0.7 to 0.8.

High MUFA in beefs lowers serum triacylglycerol and increases insulin sensitivity in SD rats

C. Choi[1], H. Kwon[1], K.H. Baik[1], K.H. Lee[1] and Y.W. Kim[2]
[1]School of Biotechnolgy, Department of Statistics, 280, Daehak-ro, 712-749, Korea, South, [2]Department of Physiology, College of Medicine, Yeungnam University, 170, Hyunchoong-ro, Daegu, 705-703, Korea, South; cbchoi@yu.ac.kr

Recent epidemiological data provide evidences that monounsaturated fatty acid (MUFA) increases HDL cholesterol, improves insulin sensitivity, and reduces LDL cholesterol. Korean cattle (Hanwoo) is a unique cattle breed with a high proportion [~30% for the highest intramuscular fat (IMF)] of fat and oleic acid (C18:1, ~50%) in M. longissimus dorsi comparing to other cattle breeds. The objective of the current study was to determine the effects of MUFA in beef on the health related indices of experimental animals. M. longissimus dorsi samples were taken from three individual beef cattle on the basis of five different IMF levels, B1, B2, B3, B4, and B5 (total 15 samples). Diets with beef (10% of the total amount) were fed to 6 week old Sprague-Dawley (SD) rats (3 rats for each beef sample, total 45 rats) for 4 weeks. Fat contents and fatty acid compositions in beef samples, lipid profiles, homeostatic model assessment-insulin resistance (HOMA-IR) in the blood of rats were analyzed by One-way ANOVA. Duncan's multiple test was used to determine significances among beef samples. Fat contents for B1 through B5 were 4.47, 8.36, 11.91, 13.23 and 23.48%, respectively. Percentages of stearic acid (C18:0) were significantly ($P<0.05$) higher in B1 (18.07%) and B2 (14.55%) whereas those of C18:1 was higher in B4 (47.19%) and B5 (47.70%). MUFA was also higher in B3, B4, and B5 (51.81, 53.81 and 54.37%, respectively) than B1 (43.86%) and B2 (49.83%). Blood triacylglycerol (TG) was 50.75, 36.88, 23.63, 16.89 and 12.44 mg/dl for B1, B2, B3, B4, and B5, respectively. HOMA-IR for B1 through B5 was 4.56, 2.96, 2.88, 2.96 and 2.87, respectively. In conclusion, the current results imply that the high proportion of MUFA in beef would be beneficial for the health of experimental animals in the aspect of blood TG and insulin sensitivity.

Genetic parameters of growth and beef quality traits in Nellore

J.B.S. Ferraz[1], M.N. Bonin[1], R.C. Gomes[2], F.M. Rezende[3] and J.P. Eler[1]
[1]Center of Animal Genetic Improvement, Biotechnology and Transgenesis, FZEA/University of Sao Paulo, Veterinary Medicine, Rua Duque de Caxias Norte, 225 USP/FZEA/ZMV/NAP-GMABT, 13635900 Pirassununga, SP, Brazil, [2]Embrapa Gado de Corte, Avenida Rádio Maia, 830, 79106-550 Campo Grande, MS, Brazil, [3]Federal University of Uberlandia, INGEB, Av. Getúlio Vargas, número 230, 38.700 128 Patos de Minas, MG, Brazil; jbferraz@usp.br

Genetic parameters for growth, carcass composition and beef quality traits were evaluated. Data 12,920 Nellore steers and bulls, grown in pastures and feed for finishing, between 90 and 120 days, were analyzed. Individual information on live ultrasound carcass measurements and frame were, also, were collected. From the total of animals, 2,048 were slaughtered and carcass, beef quality traits and video image analyses (VIA) of muscle Longissimus were obtained for estimation of beef quality attributes like color, marbling and pH. Evaluations with Near Infrared Technology (NIRS) were, also, performed to quantify tenderness and lipid in Longissimus muscle. Laboratory analysis of tenderness and total lipids were made to compare with VIA and NIRS data. Variance components, genetic and phenotypic parameters and genetic values for 26 traits were estimated by mixed models, under full animal model, using PEST/VCE 6.0, considering a relationship matrix of 42,028 animals and proper models. Estimate of heritability of rib-eye area, measured by ultrasound, was 0.32±0.04 and the genetic correlations with carcass measure were high. Heritabilty for ultrasound backfat was estimated in 0.46±0.05 and the genetic correlation with rump fat was very high, while only moderate with marbling and fat extracted from meat. Frame had a moderate heritability estimate (0.28±0.03) and its' genetic correlation with fat deposition traits was very low, although negative. Marbling and tenderness had low genetic correlation. This research provides important information to development of auxiliary tools for genetic improvement of growth, carcass and meat quality traits in Nellore herds.

Association of SCD, DGAT, LEP genes' polymorphism with meat quality of HF bulls fed different diets

J.M. Oprzadek, P. Urtnowski, A. Brzozowska, G. Sender and A. Pawlik
Institute of Genetics and Animal Breeding, Animal Science, Jastrzebiec, 05-552, Poland; j.oprzadek@ighz.pl

Fat in meat is the source of essential fatty acids which cannot be synthesized by an organism and must be provided in a diet. The aim of this study was to determine the relationship between the SCD, DGAT1and LEP genes' polymorphism and the composition of fatty acids (FA) in muscle tissue of HF bulls from different fattening systems. The DGAT1, SCD and LEP show significant activity in the metabolism of FA in animals. Previous studies conducted in dairy cattle showed a significant effect of DGAT1 and SDC on fat content in milk as well as relationship between LEP gene polymorphism with carcass fatness. In this work we focused on utilization genes' polymorphism as a markers for fat synthesis and deposition in the MLD and influence on meat quality. Polish Holstein-Friesian bulls (n=70), aged 15 months, were divided into 3 feeding groups. Animal were either grazed (n=30), fed TMR with supplementation of linseed (n=20) or fed TMR with linseed and selenium yeast (n=20), for 3 months. Afterwards, the animals were slaughtered and the carcasses were dissected into retail cuts. Samples of Musculus Longissimus Dorsi (MLD) were taken to determine the content of FA, from C:12 to C20:5, including conjugated linoleic acid (CLA) and trans FA (TFA). We found correlation between the polymorphisms of the DGAT1/EaeI, SCD/Fnu4HI, and LEP/Kpn2I genes and FA content in MLD. Differences were significant at $P \leq 0.05$ for myristic acid and unsaturated FA, including palmitooleic, cis-oleic, linoleic, CLA, eicosanoic, arachidonic and eicosapentaenoic acids.
Research was realized within the project 'BIOFOOD – innovative, functional products of animal origin' no. POIG.01.01.02-014-090/09 co-financed by the European Union from the European Regional Development Fund within the innovative Economy Operational Programme 2007-2013.

Effect of production system on carcass composition and the fatty acids profile in musle of HF bulls

J.M. Oprzadek, P. Urtnowski, E. Poławska, M. Gabryszuk, A. Brzozowska and G. Sender
Institute of Genetics and Animal Breeding, Animal Science, Jastrzebiec, 05-552, Poland; j.oprzadek@ighz.pl

The objective of the study was to determine the effect of different nutrition systems on tissue composition and meat quality. The experiment was conducted on 50 PHF bulls. The animals were divided into two groups: first group (30 animals) were grazed in the summer period and were given additional meadow hay, whereas the second group (20 animals) were fed maze silage, meadow hay and 2 kg/bull/day of concentrate. The analysis of the influence of nutrition on the slaughter performance showed that animals fed the maize silage had higher content of fat (by about 14%) and lower bone percentage (by about 6%) than animals kept on the pasture. The level of cholesterol in blood of bulls from the pasture group was 54.5±13.9 μg/ml, and was lower by about 39% from the level of the other group. Meat of animals of both groups was characterised by high content of protein (above 22%) and low content of fat (about 1%). Pasture nutrition increased PUFA in meat. The PUFA:SFA in beef from the pasture group was 0.6 and was higher than of the animals of maze silage group – 0.3. The main PUFA n-3 in beef meat were α-linolenic acid (18:3n-3), eicosapentaenoic acid (20:5n-3) and docosahexaenoic acid (22:5n-3). The profile of the fatty acids in longissimus dorsi muscle of bulls kept on the pasture was more appropriate from the dietetic point of view. Research was realized within the project 'BIOFOOD – innovative, functional products of animal origin' no. POIG.01.01.02-014-090/09 co-financed by the European Union from the European Regional Development Fund within the innovative Economy Operational Programme 2007-2013.

Influence of spruce needle extractives feeding on broiler chicken meat composition

I.I. Vitina[1], V. Krastina[1], A. Jemeljanovs[1], I.H. Konosonoka[1], S. Cerina[1], V. Sterna[1], U. Antone[1] and M. Daugavietis[2]
[1]Research Institute of Biotechnology and Veterinary Medicine Sigra, Instituta street 1, Sigulda 2150, Latvia, [2]Latvian State Forest Research Institute Silava, Rigas street 111, Salaspils, Latvia; biolab.sigra@lis.lv

The objective of the present study was to evaluate the influence of spruce needles' extractives on broiler chicken productivity and meat quality. The investigation was carried out with 300 cross ROSS 308 broiler chickens from 0 to 42 days age. Broiler chickens were randomly allocated in the trial and divided in three groups of 100 birds each. The birds in the control group (1st group) were fed with the commercial basic feed for cross ROSS 308 birds. Basic feed ration of the 2nd group chickens was supplemented with the spruce needle total extractives additive in doses of 0.04-0.05%, but basic feed ration of the 3rd group chickens was supplemented with the spruce needle neutral extractives additive that was obtained by extraction from the total extractives mass in doses of 0.04-0.05%. The main indices of poultry productivity and meat quality were recorded during the investigation. Results confirmed that by including the spruce needle total extractives and the spruce needle neutral extractives substance additives in broiler chicken diet, their live weight increased by 4.31-7.58% ($P<0.05$). The amount of monounsaturated and polyunsaturated fat, desirable for the human organism, was increased correspondingly by 3.10-2.59% and 1.13-1.15% in broiler chicken muscle tissue ($P<0.05$). At the same time the ω-6 fatty acids level in muscle tissue was not influenced by the spruce needle extractive additives to the broiler chicken diet. Research presentation is financed by the Project 'Raising awareness and fostering international cooperation of the Research Institute of Biotechnology and Veterinary Medicine 'Sigra', Agreement 2010/0197/2DP/2.1.1.1.0/10/APIA/VIAA/016.

Session 02 Poster 37

Effect of chilled storage and type of packaging on fatty acids composition of ostrich meat

J.O. Horbańczuk[1], E. Poławska[1], A. Jóźwik[1], N. Strzałkowska[1], M. Pierzchała[1], K. Raes[2] and S. De Smet[3]
[1]Institute of Genetics and Animal Breeding PAS, Postępu str. 36a Jastrzębiec, 05-552, Poland, [2]University College West-Flanders, Research Group EnBiChem, Department of Industrial Engineering and Technology, Graaf Karel deGoedelaan 5, 8500 Kortrijk, Belgium, [3]Ghent University, Department of Animal Production, Proefhoevestraat 10, 9090 Melle, Belgium; j.horbanczuk@ighz.pl

The aim of the study was to assess the changes in FA composition of chilled ostrich meat enriched (with linseed) as related to type of packaging i.e. vacuum vs. skin pack. The study was carried out on 40 ostriches. From the age of 5 months (ca. 40 kg BW), birds were randomly allocated into two groups: C (control) group fed on standard diet and L (linseed) group (4% suppl.). After slaughter (ca. 85 kg BW), meat samples were taken and pooled from the M. gastrocnemius pars interna (GM) muscle, packed in two different vacuum packing systems: foil bags (VAC) and on trays with 'skin' effect (SP), and stored for 14 days at 4C until FA analysis (GC-analysis). No significant differences in the SFA content in meat during storage in both types of packaging types occurred. A significant decrease ($P<0.05$) in the content of PUFA after 7 and 14 days of storage was observed in VAC packed meat as compared to fresh meat, especially when ostriches fed linseed diet. When skin packaging was used, no differences in the PUFA content during storage were found. Although, the PUFA (n6 and n3) content in VAC packed meat was lower, there were no differences in n6/n3 and PUFA/SFA ratios values during storage. In conclusion, in the skin pack treatment PUFA oxidation processes rate in ostrich meat during storage were lower as compared to vacuum packaging. Considering this, the former type of packaging can be recommended for ostrich meat industry [$]BIOFOOD–innovative, functional products of animal origin no. POIG.01.01.02-014-090/09 co-financed by the European Union from the European Regional Development Fund within the Innovative Economy Operational Programme 2007-2013.

Session 02 Poster 38

The lipid peroxidation in fish: from pond to plate

M. Hostovský and Z. Svobodová
University of Veterinary and Pharmaceutical Sciences Brno, Department of Biochemistry, Chemistry and Biophysics, Palackého tř. 1/3, 612 42 Brno, Czech Republic; hostovskym@vfu.cz

The integral parts of the aquatic environment are fish, which are endangered by many pollutants. The pollutants and their residues are widely found in surface and ground waters due to the intensity of agricultural and industrial production nowadays. One of the most frequently observed adverse effects of pollutants on fish is oxidative stress, which is the result of an imbalance of production and elimination of free radicals. There are numerous endogenous sources of oxyradical production but with regard to the environment the pollutants are currently in the focus of interest as they cause production of free radicals in fish organism. Fish are rich source of biologically valued food components. Fish meat is characterized by increased polyunsaturated fatty acids content, which very easily becomes a substrate for oxidation reactions. The oxidative changes of fish oils in the *in vivo* is compensated by enzymatic and non-enzymatic antioxidant systems but during the post-mortem technological processing of fish meat, especially thermal treatment, there is increased lipid peroxidation of fish oils. These oxidative changes are initiated by many physical, chemical, and biological factors. All components of the cell, including polyunsaturated fatty acids, are sensitive to oxidation and are continually attacked by free radicals. This secondary lipid peroxidation process leads to malondialdehyde production and its level is a known as a biomarker of oxidative sress. The most widely used method for determining malondialdehyde is the thiobarbituric acid test. Investigation of oxyradical damage in the fish lipids, acute and chronic effects of free radicals induced in fish *in vivo* or postmortem, and a potential bioaccumulation of lipid damage products must continue to the overall view on oxidative stress effects.

Session 02 Poster 39

Effects of zilpaterol hydrochloride and soybean oil on feedlot and carcass traits of lambs in summer

J.L. Dávila-Ramírez[1], U. Macías-Cruz[1], N.G. Torrentera[1], H. González-Ríos[2], R. Rojo-Rubio[3], S.A. Soto-Navarro[4] and L. Avendaño-Reyes[1]

[1]Universidad Autónoma de Baja California, Instituto de Ciencias Agrícolas, Ejido Nuevo León, Valle de Mexicali, 21705 Baja California, Mexico, [2]Centro de Investigación en Alimentación y Desarrollo, Alimentos, Carretera a Victoria Km. 0.6, Hermosillo, 83000 Sonora, Mexico, [3]Universidad Autónoma del Estado de México, Centro Universitario UAEM Temascaltepec, Km 67.5 Toluca-Tejupilco s/n, Temascaltepec, 51300 Estado de México, Mexico, [4]New Mexico State University, Animal and Range Sciences, 224 Knox Hall, Las Cruces, 88003 New Mexico, USA; lar62@hotmail.com

In order to evaluate the effect of zilpaterol hydrochloride (ZH) with or without soybean oil (SO) in finishing diets on feedlot performance and carcass traits under heat stress, forty Dorper × Pelibuey male lambs initially weighing 31.70±2.30 kg were used. A randomized complete block design was arranged as a 2 (0 and 10 mg ZH/d/head) × 2 (0 and 6% of SO/kg DM) factorial. Lambs were slaughtered after a 34-d feeding period. Climatic conditions were of moderate heat stress. Interaction ZH × SO did not affect ($P \geq 0.11$) any study variable. In the first 17 d, ZH increased ($P \leq 0.05$) BW, ADG and G:F without affecting feed intake ($P=0.40$); but between days 1 to 34 and days 18 to 34, feedlot traits were not affected ($P=0.18$) by ZH. Also, ZH decreased KPH fat ($P=0.01$) as well as ($P \leq 0.04$) dressing, LM area, LM pH at 24 h and leg perimeter. Meanwhile renal fat ($P=0.03$) decreased with ZH, others non-carcass components were not affected ($P \geq 0.06$) by ZH supplementation. Moreover, SO did not affect ($P \geq 0.20$) feedlot performance ($P=0.20$) and decreased ($P<0.01$) dressing without affecting ($P \geq 0.12$) the remaining carcass traits. In conclusion, under heat stress conditions, hair-breed male lambs did not improve feedlot performance nor carcass characteristics by the inclusion of ZH in combination with SO in finishing diets, or only SO. However, regardless the inclusion of SO, ZH increased growth traits only during the first 17 d of experiment.

Session 02 Poster 40

Effects of extruded linseed in ewe diets on the intramuscular fatty acid profile of suckling lamb

B. Gallardo[1], P. Gómez-Cortés[2], A.R. Mantecón[3], M. Juárez[2], M.A. De La Fuente[2] and T. Manso[1]

[1]Universidad de Valladolid, Area de Producción Animal, ETS Ingenierías Agrarias. Avd. Madrid s/n, 34004 Palencia, Spain, [2]Instituto de Investigación en Ciencias de la Alimentación (CSIC-UAM), C/ Nicolas Cabrera, 9. Universidad Autónoma de Madrid, 28049 Madrid, Spain, [3]Instituto de Ganadería de Montaña (CSIC-ULE), Finca Marzanas, 24346 Grulleros, León, Spain; beatriz.gallardo.garcia@uva.es

The aim of this study was to evaluate the effects of supplementing lactating ewe diets with extruded linseed on the fatty acid (FA) composition of intramuscular fat of suckling lambs. After lambing, twenty-four pregnant Churra ewes were fed daily 2.1 kg of a TMR containing lucerne and concentrate at a 40:60 ratio. Each ewe was assigned to one of two treatments: Control (with 70 g/day of FA from a calcium soap of palm oil, Magnapac®) and Lin (with 128 g/day of extruded linseed, Tradilin®). All lambs were reared exclusively on milk and were slaughtered when they reached 11 kg live weight. FA profiles of milk fat and intramuscular fat of suckling lamb were determined by Gas Chromatography. Data were evaluated by the GLM and MIXED procedure of SAS. The changes in FA composition of suckling lamb's meat were similar to those of milk from suckled dams. Lin treatment showed the highest ($P<0.001$ and $P<0.05$, for milk and meat respectively) proportions of polyunsaturated fatty acids (PUFAs). Extruded linseed supplementation caused an increase 3.5-fold and 4-fold in trans-11 C18:1 and an increase 2.4-fold and 3.5-fold in cis-9, trans-11 C18:2 in milk and intramuscular fats respectively compared to the control diet. The percentages of n-3 PUFA were higher in Lin suckling lamb meat, made up of mainly C18:3 n-3 ($P<0.001$), C20:5 ($P<0.05$), C22:5 ($P<0.05$) and C22:6 ($P<0.05$). Meat fat from Lin carcasses displayed a lower ($P<0.001$) n-6/n-3 ratio than control samples. This research concludes that dietary extruded linseed supplementation of lactating ewes enhances the nutritional quality of ewe milk fat and of suckling lamb´s meat.

Improvement of the selenium content in beef meat by feeding ingredients high in selenium

Y. Mehdi, O. Dotreppe, V. Robaye, L. Istasse, J.-L. Hornick and I. Dufrasne
University of Liege, Nutrition Unit, bd de Colonster 20, 4000 Liege, Belgium; ymehdi@doct.ulg.ac.be

In Belgium, beef meat is produced to a large extend from young growing fattening bulls of the double muscled Belgian Blue breed. They are offered high energy diets based either on whole plant maize silage or on sugar beet pulp. The present experiment was a comparison of a fattening diet in which three major ingredients (barley, spelt and linseed meal) were either high or low in selenium so that the selenium content in the diet was 106 µg/kg or 307 µg/kg. The other chemical components of the diet were unchanged, the average crude protein content being 16% and the ether extract content 3%. There were no differences in the animal performances and in slaughter characteristics (final live weight of 612 and 600 kg, total gain of 180 and 169 kg, duration of 123 and 121 days, cold carcass weight of 379 and 375 kg, killing out percentage of 66.4 and 66.6%, muscle proportion in the carcass of 73.4 and 73.8% and fat proportion in the carcass of 12.1 and 11.7%). There were no differences either in the chemical composition of meat (protein content of 89.8 and 91.0%, ether extract of 1.8 and 2.2%) and in the meat characteristics (cooking losses of 32.2 and 31.7%, tenderness WBSF of 32 and 34 N and color (L* of 41.0 and 42.3%)). By contrast the Se concentration in meat was about two times higher in the selenium group than in the control (325 and 509 µg Se/kg DM in the Longissimus thoracis muscle and 248 and 462 µg Se/kg DM in the Rectus abdominis muscle).

Effect of feeding system on fatty acid profile of lambs of three Veneto Region Alpine breeds

E. Pellattiero, F. Tagliapietra and A. Cecchinato
University of Padova, Department of Agronomy, Food, Natural resources, Animals and Environment, Viale dell'Università 16, 35020 Legnaro (Padova), Italy; erika.pellattiero@studenti.unipd.it

The aim of this trial was to investigate the effect of diet, breed, gender and the tissue on the fatty acid (FA) profiles of lambs of three native Italian sheep breeds. Thirty-six lambs, belonging to flocks subjected to an in situ conservation program (Legnaro-Villiago), were used. Six males and 6 female lambs for each breed: Alpagota, Brogna and Foza, were divided in three groups characterized by different feeding systems: pasture, penned in an open barn and fed with hay and concentrate and penned in the open barn and fed with hay and concentrate added with rpCLA supplement. At the age of 225 days the lambs were slaughtered, samples of muscles, fatty depots and liver were collected and analyzed for FA profile. FA were extracted and analyzed through GC. Data were analyzed by PROC MIXED of SAS considering the effect of the feeding system, breed, gender, age and tissue. The random effect of animal was used to test feeding system, breed, gender and age, whereas the effects of tissue and corresponding interactions were tested on the residual. Among the factors of variation of FA profiles feeding system was the most important one because, respect to the dry diets, pasture did not caused a variation of proportion among SFA, MUFA and PUFA, but changed the proportion within SFA (C18:0;C18:1vaccenic), and PUFA (C18:2c9 t11 CLA; >C18:3 α Linolenic; <C20:4n6 arachidonic), and decreased the omega6/omega3 ratio (3.28 vs. 6.17; P<0.0002). The supplementation of rpCLA on barn diet increased both CLA isomers (C18:2c9, t11, P=0.010, and C18:2 t10, c12, P=0.018). The tissue was the most important within-animal cause of variation and affected all the individual FA and their categories. Liver presented a FA profile very different from that of muscles and fatty depots and, among these last tissues, kidney fat was very peculiar.

Session 02 — Poster 43

Lidia bovine breed for production of high quality meat

C. Vieira, A.M. Fernández, R. Posado, D.J. Bartolomé and J.J. García
Instituto Tecnológico Agrario (ITACyL). Consejería de Agricultura y Ganadería de Castilla y León, Ctra. Burgos Km, 119, 47071 Valladolid, Spain; vieallce@itacyl.es

The Lidia bovine breed (otherwise known as the fighting bull) has similar zootechnical characteristics than other native beef breeds living in the Mediterranean forest ecosystem traditionally known as La Dehesa. Two years old females are subject to a *tienta* (trials, to test the behaviour characteristics of young bulls) and then are select to use as breeding female animals or are sold at low prices, because these animals are selected for genetic, morphology and behaviour but the carcass have low weight and yield which could improve by a short fattening period before slaughtering. Despite this potential, there are few studies related to this breed as beef producer. The objective was to evaluate the carcass and meat quality characteristics for females of Lidia breed. Eight bovine females two years old from Lidia breed were fattened for 100 days, controlling feed intake and growth rate. After slaughter, carcass characteristics and meat quality (moisture, fat, protein, haem pigments, shear force and sensory characteristics) were evaluated. Average weight gained during the fattening period was 52 kg and average cost of feed was 152.2 €, with a conversion rate of 10.5. Average carcass weight (136.5 kg), dressing percentage (50.6%), conformation (R) and fatness (2) scores rated on a 5-point scale, and also final pH (5.4-5.6) allow considering the animals suitable for meat production. The fat content (4.14% FM) was within the normal range but haem pigments content (7.57 mg/g muscle) was higher than beef meat average. After 10 days, shear force values (3.5 kg) corresponded with tender meat. The high scores given by the panelists also allow characterizing this meat as tender, juicy and intense aroma and flavor. Subjecting two years old females of Lidia breed to a short fattening period allow obtaining high quality meat. These results would provide to herds an economic added contribution to the production of these animals.

Session 03 — Theatre 1

Regulation of the hypothalamo-pituitary-ovarian axis in mare: what's new?

D. Guillaume, C. Decourt, J. Salazard-Ortiz, C. Briant and A. Caraty
Physiologie de la Reproduction et des Comportements, INRA, UMR85, CNRS, UMR6175, Université de Tours, Institut Français du Cheval et de l'Equitation, 37380 Nouzilly, France; dguillaume@tours.inra.fr

In equine species the success of insemination is maximal when insemination is performed around 24 h before ovulation. Therefore there is a need to predict and to induce ovulation accurately. In summer when the mares are cyclic, the follicular growth and ovulation, which can be directly monitored with ultrasound, are controlled by GnRH and gonadotropins secretions. The induction of ovulation with different biological extracts of LH is now routinely used but the risks of contamination linked to this method have motivates further researches for development of new strategies. The technical difficulties as well as the important cost in getting active recombinant LH have delayed its commercialization. GnRH or GnRH agonists, for pharmaco-kinetic reasons seems to be difficult to handle but very high doses are probably a good way to induce ovulation. Finally, another strategy to improve the yield of insemination in mare would be to delay ovulation till 24 h after insemination. GnRH antagonists are powerful to perform this task and are now ready for commercialization. Recently, the discovery of Kisspeptin molecules has opened a new avenue of research. In the brain, kisspeptine neurones appear to be the major regulator of GnRH secretion. Moreover they integrate information coming either from the body condition or the season. In ewes, this molecule has been shown to induce GnRH and LH secretion as well as ovulation in cyclic and anestrous acyclic ewes. In contrast, in mare, as for GnRH, a very high dose of peptide is required to stimulate LH secretion and to induce ovulation. Therefore more research is needed to determine the potential of this molecule to control ovulation in equine species.

The potential effects of social interactions on reproductive efficiency of horses

D. Burger[1] and C. Wedekind[2]
[1]Swiss Institute of Equine Medicine, ALP-Haras and University of Berne, Les Long-Prés 1, 1580 Avenches, Switzerland, [2]Department of Ecology and Evolution, University of Lausanne, Le Biophore, 1015 Lausanne, Switzerland; dominik.burger@vetsuisse.unibe.ch

The reproductive efficiency of domesticated horses is often lower than what could be expected from observations in feral herds. In the wild, stallions typically live with mares in harem bands, with other stallions in bachelor bands, or occasionally in mixed sex transitional bands. Hereby foaling rates of up to 90% are achieved, compared to approximately 75% in well-managed domesticated horse populations. An interdisciplinary view and analysis of the intra- and intersexual interactions of stallions and mares and their effects provide new potentially important informations which in the future might be used for optimization of breeding management. We provide here a summary of our present knowledge of female and male reproductive strategies in horses, leading to higher fertility and biodiversity.

Do stallions recognize the oestrous state by smelling the odor of mares?

C. Briant[1], A. Bouakkaz[2], Y. Gaudé[3], I. Couty[4], D. Guillaume[4], J.M. Yvon[3], Y. Maurin[5], B. Nielsen[5], O. Rampin[5] and M. Magistrini[4]
[1]IFCE, HN, 41043 Blois Cedex, France, [2]INSV, Université Ibn Khaldoun, Tiaret, Algeria, [3]INRA, UEPAO, 37380 Nouzilly, France, [4]INRA, PRC, 37380 Nouzilly, France, [5]INRA, NOPA, 78352 Jouy en Josas, France; christine.briant@ifce.fr

In some mammalian species, males recognize the estrous female on the basis of sex pheromones perceived through the reflex of flehmen. In horses the importance of olfactory signals is not clear. The aim of the present experiments was to evaluate if stallions can discriminate the estrous from the diestrous state by smelling mares or samples of urine or faeces of mares. In Exp1, 6 pony stallions were subjected to 4 different two-choice tests between an estrous and a diestrous mare: (1) in natural conditions; (2) stallions could not smell the mares; (3) stallions could not see the mares; (4) mares were sedated. In tests (1) and (2) stallions could differentiate between the two mares: more interactions with the estrous mare (sniffing/nuzzling, licking and biting), more flehmens, but could not in tests (3) and (4). In Exp2, 4 urine samples and in Exp3, 4 faeces samples (mare in estrus or diestrus, stallion, negative control) were presented to the stallions. The main observed behavior (sniffing, flehmens, early nasal discharge, shares of disinterest) showed that stallions do not discriminate the urines or faeces of estrus and diestrus but discriminate urine or faeces of mare and urine or faeces of stallion. In Exp3, analysis of serial blood samples showed that testosterone increases more after sniffing faeces of stallion and that prolactin increases more after sniffing faeces of mare. The time of semen collection tends to be reduced after sniffing faeces of mare. All these results suggest that stallions do not discriminate odors of estrus and diestrus but can distinguish odors of mare and stallion. Thus olfaction is not absolutely necessary for detection of the estrous mare but sight of the mare and interactions with her seem essential. However smell of the mare could stimulate sexual functions as ejaculation.

Session 03 — Theatre 4

Advances in the mechanism of sperm-oocyte interactions and cross-talk with the oviduct in the equine

G. Goudet[1], S. Mugnier[1], B. Ambruosi[1], C. Douet[1], P. Monget[1], F. Reigner[2] and S. Deleuze[3]
[1]INRA, UMR 85, Physiologie de la Reproduction et des Comportements, centre Val de Loire, 37380 Nouzilly, France, [2]INRA, UEPAO, centre Val de Loire, 37380 Nouzilly, France, [3]Université de Liège, Faculté de Médecine Vétérinaire, Liège, 4000 Liège, Belgium; ghylene.goudet@tours.inra.fr

Despite significant advances in our understanding of the interaction between spermatozoa and oocyte, this fundamental event remains largely enigmatic in the equine. To study the molecular mechanism underlying equine fertilization, we have developed a comparative strategy between the horse, with low *in vitro* fertilization (IVF) rates and no polyspermy, and the pig, with high IVF and polyspermy rates. Comparative studies between these two divergent models could identify conserved and species-specific molecular interactions that could highlight determining elements involved in fertilization. We have analysed the composition of the equine zona pellucida (ZP) and we have shown that it contains 4 glycoproteins, whereas the porcine ZP contains only 3. We have observed differences in the localization of the ZP glycoproteins and in the mesh-like structure of the ZP between equine and porcine species, which may correlate with the differences in the IVF rates. We have analysed the role of the secretions of the oviduct on sperm-oocyte interactions and we have shown that oviductal fluid induces a significant increase of the monospermic IVF rate in the equine and porcine species. In order to identify the proteins responsible for this positive influence, we have studied the effect of several candidates and we have shown that Deleted in Malignant Brain Tumor 1 is involved in the mechanism of fertilization in equine and porcine species. In conclusion, our comparative approach allowed us to identify key elements in the mechanism of equine fertilization.

Session 03 — Theatre 5

Oocytes and assisted reproductive technologies in the horse

S. Deleuze[1], C. Douet[2], S. Parrilla-Hernandez[1], F. Reigner[2] and G. Goudet[2]
[1]University of Liège Veterinary Faculty, Clinical Sciences-Small animal and equine reproduction, Bvd de Colonster, 20 B44, 4000 Lièeg, Belgium, [2]INRA-CNRS-Université de Tours-Haras Nationaux, UMR 85 Unité Physiologie de la Reproduction et des Comportements, Nouzilly, 37380 Nouzilly, France; s.deleuze@ulg.ac.be

Assisted reproductive technologies (ARTs) for the horse developed steadily over the last two decades and range from simple procedures such as artificial insemination to the complex production of clones. These technologies now enable salvation of otherwise lost genetics from either subfertile or even dead animals. Oocytes can be retrieved *in vivo* by follicular aspiration or *ex vivo* from abattoir ovaries. ARTs in the equine are only accessible to a minority of performing horses. Consequently, abattoir derived oocytes are almost exclusively of interest for research purposes. Yet, their recent shortage is now getting of concern as regards development of their use for clinical applications. Oocytes from small follicles can successfully undergo *in vitro* maturation (IVM). However, IVM success is commonly assessed by confirming meiosis resumption (nuclear maturation). To date, no tool is available to assess cytoplasmic maturation of the oocyte and predict its ultimate ability to undergo fertilization and develop into a viable embryo. Despite apparent satisfactory results IVM efficacy should be more accurately be evaluated to allow further improvement of IVM conditions. Use of oocytes for ARTs include Intra-follicular oocyte transfer and intra-oviductal Oocyte Transfer which involve transfer of oocytes, previously matured *in vitro* or not, in the preovulatory follicle or the oviduct of surrogate mares. They require availability of recipients, synchronisation between donor and recipient mares, surgery or specific equipment and skills. However, they potentially represent valuable alternatives to conventional *in vitro* fertilization which can be poorly efficient or to more technically demanding procedures such as Intra-Cytoplasmic Sperm Injection and cloning.

Genotyping and cryopreservation of equine embryos: new developments

F. Guignot
UMR INRA-CNRS-Université de Tours-IFCE, PRC, 37380 Nouzilly, France; guignot@tours.inra.fr

Embryo cryopreservation and transfer is a powerful tool for genetic selection and has extensive economical and clinical application in many species. Moreover, new knowledge about the mammalian genome and significant advances in embryo manipulation allow embryo selection on specific dispositions before transfer: termed preimplantation genetic diagnosis (PGD). Few biopsied embryonic cells are required for PGD. After whole genome amplification, several genes could be identified, related to disease, sex, coat colour, sporting capacity. Unfortunately, in equine species, the embryo cryopreservation remains problematic, due to the presence of an embryonic capsule and the large size of the embryo. Cryopreservation by freezing or vitrification of early equine embryos <300 µM in diameter has been reported with acceptable pregnancy rates (60-80%), but cryopreservation of expanded blastocysts >300 µM in diameter resulted in very low pregnancy rates after transfer (0-57%). The large amount of fluid within the blastocoele could be responsible for cryopreservation failures. In 2011, as previously performed in humans, induced blastocyst collapse allowed successful vitrification in equine embryos >300 µM in diameter. About 70% of the blastocoele fluid was aspirated by applying suction with a glass pipette attached to a Piezo drill. At the same time, trophoblast cells were aspirated for genotyping. Thereafter, the embryo collapsed and was vitrified. After embryo transfer to the recipient, acceptable results of pregnancy were obtained in the literature and in our lab (50-70%). Sex and 2 genetic diseases have been diagnosed in biopsied cells. Genotyping diagnosis was 82-100% efficient. Compared to control embryos, 85-100% of diagnosed genotyping were correct. This new approach is very exciting for equine embryo cryopreservation, especially for expanded blastocysts. Moreover it allows PGD on biopsied cells. Nevertheless, a skilled technician is required to perform the biopsy and embryo collapse. The procedure should be simplified before widespread clinical use.

Session 03 Theatre 7

Perspectives on stallion fertility and semen analysis

D.D. Varner
Texas A&M University, Department of Large Animal Clinical Sciences, College of Veterinary Medicine and Biomedical Sciences, College Station, TX 77843-4475, USA; dvarner@cvm.tamu.edu

Laboratory-based evaluations of stallion breeding potential have been conducted for many decades, but the results of such evaluations do not have an exact correlation with actual fertility. The reasons for this inequality are varied, but mare and management factors can certainly confound one's interpretation of the actual intrinsic fertility of a stallion. Features of stallion mating ability and method of breeding, e.g. with cool-transported or cryopreserved semen, can also negatively impact the fertility of stallions with good sperm output and initial semen quality. Despite these limitations, laboratory-based prognostication of stallion fertility remains a mainstay of veterinary practice, and is commonly used for pre-purchase examinations of potential or active breeding stallions, and to assess stallion reproductive potential prior to ensuing breeding seasons. Such evaluations are also conducted on stallions with low or declining fertility in an effort to determine the root cause. Semen quality and sperm output are important criteria to consider when predicting stallion fertility, estimating the number of mares that can be efficiently impregnated by a stallion during a given breeding season, or identifying underlying causes for subfertility. Standard tests of semen quality include evaluation of sperm motility and morphology. Ancillary semen tests that are becoming more commonplace in the United States include assessment of toxic components in seminal plasma that could impact longevity of sperm motility; sperm chromatin quality, plasma-membrane integrity, acrosomal integrity, and acrosomal responsiveness to stimulants of the acrosome reaction; and sperm quality following cooled or frozen storage. Numerous other assays have been developed for assessing various compartments or molecular features of sperm from a variety of species; however, the prognostic value of these assays remains largely unknown in relation to stallion fertility.

Session 03 Theatre 8

Stallion spermatozoa: putative targets for estrogens

C. Gautier[1], C. Delalande[1], I. Barrier Battut[2] and H. Bouraïma Lelong[1]
[1]Université de Caen Basse-Normandie, EA 2608 Oestrogènes Reproduction Cancer USC INRA 2006, Campus 1, Esplanade de la Paix, CS14032, 14032 Caen Cedex 5, France, [2]Institut Français du Cheval et de l'Equitation, Jumenterie du Pin, 62310 Exmes, France; helene.bouraima@unicaen.fr

Among the mammals, stallion appears as the male producing the largest amount of testicular estrogens. This synthesis occurs mainly in Leydig cells and seems to be submitting to a seasonal regulation. Indeed, Lemazurier *et al.* demonstrated the presence of higher amounts of estrogens in semen in April-May and June compared to December. To exert their effects estrogens use specific nuclear receptors (ESR1, ESR2), which could exert both genomic and non-genomic actions. Recently, we identified estrogen receptors in ejaculated stallion spermatozoa and in order to determine a putative seasonal response to estrogens a quantification of ESR1 was performed, on semen samples. Semen were obtained monthly between May 2012 and February 2013 from 5 stallions, aged from 10 to 23 years, housed at Jumenterie du Pin (IFCE). Seminal plasma was removed by centrifugation and spermatozoa were washed twice in Tyrode. ESR1 protein expression was studied by Western-blot, confocal analysis and flow cytometry with MC-20 antibody. Results were analyzed by ANOVA followed by Tukey-Kramer multiple comparisons test. Western-blot analysis showed a single 66 kDa band corresponding to the wild-type isoform, and then immunofluorescence analysis showed a flagellar staining. We described a high rate of spermatozoa from a sample positive for the detection of ESR1 between May and October then there is a progressive and regular decrease of signal from October to reach a nadir on January, followed by an increase of signal in February. So, we describe for the first time, a seasonal regulation of ESR1 associated to spermatozoa. Further studies will elucidate a putative relation between sperm ability to respond to estrogen and sperm quality.

Session 03 Theatre 9

Recent advances in processing of stallion sperm for cooled storage

H. Sieme[1], H. Oldenhof[1], G. Martinsson[2] and W.F. Wolkers[3]
[1]Clinic for Horses, University of Veterinary Medicine Hannover, Bünteweg 15, 30559 Hannover, Germany, [2]National Stud Celle, Spörckenstr 10, 29221 Celle, Germany, [3]Institute of Multiphase Processes, Callinstr 16, 30167 Hannover, Germany; harald.sieme@tiho-hannover.de

This presentation focuses on recent developments in processing of stallion sperm for cooled storage, with special emphasis on diluent composition, cooling regime, and centrifugation protocols. Sperm processing methods aim to overcome damages associated with ageing of sperm. Skim milk extenders are generally used to prolong sperm longevity, but their composition is not defined. Caseinates as present in such formulations seem responsible for protecting sperm during storage, and nowadays defined commercial extenders with caseinate are available (e.g. INRA96, EquiPro). Mitochondria mediated accumulation of high amounts of reactive oxygen species in sperm samples can be counteracted by use of zwitter-ionic buffers, and addition of antioxidants or nucleophilic thiols to extenders. Chilling injuries are described to result from lipid rearrangements and coinciding leakage. To avoid cold shock, sperm needs to be slowly cooled down to 5 °C. Containers like e.g. the Equitainer provide good cooling rates and minimal temperature fluctuations. Temperatures close to 0 °C are especially detrimental for sperm fertility. Recently; it was shown that stallion sperm tolerates a wide storage temperature range (4-15 °C) and various air exposures. Seminal plasma has detrimental effects on sperm quality, and can be removed using centrifugation. High speed cushioned centrifugation of semen has become increasingly popular, since sperm recovery is maximal without packing sperm in a pellet. Recently, 'Sperm Filter' has been developed, which uses a synthetic hydrophilic membrane to allow passage of seminal plasma but not sperm. In addition to removal of seminal plasma, density gradient centrifugation can be employed (e.g. EquiPure, Androcoll-E) to also select a sperm subpopulation.

Quality of stallion sperm processed with Androcoll-E and their relation with *in vivo* fertility

S. Gamboa[1], A. Quaresma[1], F. Castro[1], P. Bravo[1], R. Rebordão[1] and A. Rocha[2]
[1]Agricultural School, Animal Reproduction Laboratory, Bencanta, 3040-316 Coimbra, Portugal, [2]Instituto de Ciências Biomédicas Abel Salazar, Centro de Estudos em Ciência Animal, University of Porto, 4480 Vila do Conde, Portugal; scgamboa@esac.pt

Single layer centrifugation (SLC) with Androcoll-E™ enables a better sub-population of spermatozoa to be selected from stallion ejaculates. In stallions with fertility problems, SLC-treated sperm seems to result in normal pregnancy rates. Since no studies have been performed on the effect of Androcoll-E™ treatment on pregnancy rates obtained from fertile stallions, this study was designed to investigate whether SLC could be used to enhance fertility rates of normal stallion. Ejaculates obtained from a SIRE were used to study quality and longevity of the sperm recovered after centrifugation in Androcoll-E™ as well as pregnancy rates obtained after AI with and without SLC-centrifuged sperm. The sperm's characteristics not recovered by SLC-treatment were also investigated and results were compared by ANOVA. SLC-recovered sperm presented the highest percentage of live cells and membranes osmotically active (HOS test) while the highest percentage of sperm with high mitochondrial potential ($\Delta\Psi$mit; JC-1) and DNA integrity (SCD test) was retained in the upper layer. Sperm motility in SLC-selected samples cooled for 72 h didn't differ from non-SLC-selected sperm. A crosstabs procedure (chi-square test, χ^2) was used and per cycle fertility obtained with Androcoll-E™ treated semen (78%, 9 mares; 7 cycles DG+) didn't differ (χ^2(2)00; P=0.527; n=18) from per cycle fertility obtained with non SLC-treated sperm (89%; 9 mares; 8 cycles DG+). In conclusion, our results suggest that fertility is independent of treatment with Androcoll-E™. Answer to REviwers: We have no data with sub fertille stallions regarding *in vivo* and *in vitro* comparison.

Foetal programming and epigenetics: how important are they for the equine species?

P. Chavatte-Palmer, P. Peugnet and H. Jammes
INRA, UMR 1198 Biologie du Développement et Reproduction, Domaine de Vilvert, Bât 231, 78352 Jouy en Josas cedex, France; pascale.chavatte@jouy.inra.fr

The field of research on developmental programming has been increasing rapidly over the last 10 years, especially in order to understand the responsibility of programming and epigenetic adaptations in the currently observed epidemics of obesity, type 2 diabetes and cardiovascular diseases in the human population worldwide, leading to the birth of a scientific society centered around these aspects (Developmental Origins of Health and Disease, DOHaD, http://www.mrc.soton.ac.uk/dohad). These effects initially observed in humans and animal models are also important for production animals. In the horse, insulin resistance is recognized as an emerging disease linked to several major equine pathologies such as laminitis, hyperlipidemia, Cushing's disease, diabetes, endotoxemia and osteochondrosis. Our group is currently exploring the role of fetal developmental conditions on offspring growth, glucose homeostasis and osteo-articular health, with recent data demonstrating the role of fetal growth on insulin resistance in these species. It is now widely admitted that long term effects of adverse developmental conditions are mediated through epigenetic mechanisms that modulate gene expression without modifying the genome. The placenta and the gametes may be considered as key tissues for bearing the epigenetic memory of the prenatal developmental conditions and are therefore the target of our future investigations.

Prediction of foaling in mares based on changes in endocrine and cardiac parameters

C. Nagel[1], J.E. Aurich[2], R. Erber[1] and C. Aurich[1]
[1]Artificial Insemination and ET, Vetmeduni, 1210 Vienna, Austria, [2]Reproduction, Vetmeduni, 1210 Vienna, Austria; christina.nagel@vetmeduni.ac.at

Gestation length in the horse varies considerably. We tested the reliability of heart rate (HR), heart rate variability (HRV) and cortisol for predicting the onset of foaling in 24 mares. During the 15 days preceding parturition saliva for cortisol analysis was taken every 6 h and a daily ECG was made. Cortisol showed a diurnal rhythm which disappeared 3-5 days before foaling. Thereafter, cortisol increased until foaling ($P<0.001$). An increase of >2× the SD of concentrations on days -5, -4 and -3 occurred in 58-80% of mares at different times during the last 24 h before parturition ($P<0.001$). Until 24 h before foaling no changes in HR and HRV were found. When the cut-off point was calculated as >2×SD for HR this was exceeded in 10-46% of mares earlier than 24 h before foaling. During the last 2 h before foaling maternal HR did not increase. The number of mares with atrioventricular (AV) blocks and the number of AV blocks per mare increased before foaling ($P=0.001$). AV blocks indicate a strong parasympathetic influence. At 75 min before foaling and during the last 15 min before birth was completed 80% of the mares showed repeated AV blocks. In contrast to HR, HRV variables SDRR (standard deviation of beat-to-beat interval) and RMSSD (root mean square of successive RR differences) increased during the last 2 h before foaling. Increases in HRV indicate a decrease in sympathetic and/or increase in parasympathetic tone and indicate that the mare is non-stressed. Due to individual variations, only 57-62 (SDRR) and 56-67% of mares (RMSSD) reached the threshold value (>2×SD). In conclusion, cortisol increases before parturition and 80% of mares show cortisol levels >2 ng/ml during the last 24 h before parturition. Prediction of foaling cannot be based on HR and HRV due to frequent false positive events. Systems which detect AV blocks in mares may be able to predict parturition approximately one hour before birth of the foal.

Session 03 Theatre 13

Low levels of regulatory T-lymphocytes in blood of mares are associated with early pregnancy loss

C. Aurich[1,2], J. Weber[3], C. Nagel[1], M. Merkl[4], R. Jude[3], S. Wostmann[3], U. Baron[5], S. Olek[5] and T. Jansen[3]
[1]Lehndorff Institute, Neustadt State Stud, 16845 Neustadt (Dosse), Germany, [2]Artificial Insemination, Vetmeduni, 1210 Vienna, Austria, [3]Certagen, Certagen, 53359 Rheinbach, Germany, [4]Reproduction, Vetmeduni, 1210 Vienna, Austria, [5]Ivana Türbachova Lab for Epigenetics, Epiontis, 12489 Berlin, Germany; christine.aurich@vetmeduni.ac.at

A high rate of early pregnancy loss is an important reason for low fertility in the horse. Because in humans and mice, low numbers of naturally occurring regulatory T cells (Treg cells) have been linked to miscarriage, we have evaluated Treg cell levels in mares at the start of the breeding season. Results were related to the outcome of subsequent breeding. For identification and quantification of Treg cells, the Treg-specific demethylated region in the forkhead box transcription factor (FOXP3 TSDR) was adapted to the equine species and quantitatively determined by a q-PCR system. Pregnancy was followed until detection of early pregnancy loss (n=17), abortion without identification of an infectious or apparent cause (n=9) or birth of a viable foal (n=82). Treg cell levels between mares that conceived (82%; 1.50±0.04%) or did not get pregnant (18%; 1.45±0.10%) did not differ significantly. The Treg cell percentage at oestrus before breeding was significantly lower ($P<0.05$) in mares that underwent early pregnancy loss until day 40 of pregnancy (1.29±0.07%), than in mares that aborted (1.61±0.15%) or gave birth to a live foal (1.52±0.05%). It is suggested that low levels of Treg cells are related to early pregnancy loss in the equine species. Supported by the German Ministry of Economics and Technology, EP100479.

Oviductal secretions have a positive effect on fertilization in equine species, which involves DMBT1

C. Douet[1], B. Ambruosi[1], G. Pascal[1], P. Monget[1], C. Moros Nicolàs[2], U. Holmskov[3] and G. Goudet[1]
[1]INRA, UMR 85, Physiologie de la Reproduction et des Comportements, centre Val de Loire, 37380 Nouzilly, France, [2]University of Murcia, Cell Biology and Histology, Faculty of Medicine, Murcia, 30071 Murcia, Spain, [3]University of Southern Denmark, Institute for Molecular Medicine, Odense, 5000 Odense, Denmark; cecile.douet@tours.inra.fr

In the equine, several attempts to establish an efficient *in vitro* fertilization (IVF) technique were performed during the last decades. However, reported IVF rates remain quite low and no repeatable equine IVF technique is available yet. Oviductal secretions affect preparation of gametes for fertilization and fertilization itself. The aim of this study is to evaluate the effect of oviductal fluid and the possible involvement of Deleted in Malignant Brain Tumours 1 (DMBT1) on IVF in equine species. The presence of DMBT1 in the equine oviduct was shown by Western blot. We performed IVF after pre-incubation of oocytes with or without oviductal fluid supplemented or not with antibodies directed against DMBT1 or non-immune serum. We showed that oviductal fluid induces an increase of the IVF rate, and that this effect is cancelled by the addition of anti-DMBT1 antibodies whereas the non-immune serum has no effect. The presence of DMBT1 in cumulus-oocyte complexes was shown by Western blot analysis, and the localization of DMBT1 in the zona pellucida and cytoplasm of equine oocytes was observed using immunofluorescence analysis and confocal microscopy. Finally, a bioinformatics and phylogenetic analysis allowed us to identify the DMBT1 protein as well as a DMBT1-like protein in several mammals including the horse. Our results strongly suggest an important role of DMBT1 in the process of fertilization.

Genetic analysis of cryptorchidism in Swedish born Icelandic horses

S. Eriksson[1], K. Jäderkvist[1], A.-M. Dalin[2], J. Axelsson[1] and G. Lindgren[1]
[1]Swedish University of Agricultural Sciences, Dept. of Animal Breeding and Genetics, P.O. Box 7023, 75007 Uppsala, Sweden, [2]Swedish University of Agricultural Sciences, Dept. of Clinical Sciences, P.O. Box 7054, 75007 Uppsala, Sweden; susanne.eriksson@slu.se

Cryptorchidism, when one or both testes fail to descend normally into the scrotum, cause fertility problems, increased risk for tumors and costly castration surgery. Moderately high heritabilities have been estimated for cryptorchidism in dogs and pigs, but information on heritability for equine cryptorchidism has been lacking. The aim of this study was to estimate the prevalence and heritability of cryptorchidism in Swedish born Icelandic horses. A questionnaire was sent to 80 of the largest breeding farms. The breeders were asked for each stallion born 1990-2011 if the testes were down at the ages of 1, 6 and 12 months. The answering frequency was 57%, corresponding to 858 horses, from 230 sires and 471 dams. Many breeders did not know the status of the stallions younger than 12 months, some did not check before castration. The data indicated that the testes rather commonly descend later than six months after birth in Icelandic foals. At the age of 12 months, close to 9% of the 655 horses with information did not have both testes in the scrotum. Using logistic regression, probability of cryptorchidism was significantly influenced by breeding value for height at withers, farm and year of birth. Genetic parameters for cryptorchidism at the age of 6 (n=329, mean=0.25), 12 (n=655, mean=0.09) and 12 months or older (n=751, mean=0.06) were estimated using a linear animal model, including fixed effects of farm and birth year. The estimated heritabilities were 0.26 (S.E. 0.19), 0.14 (S.E. 0.12) and 0.08 (S.E. 0.10), respectively, on the visible scale and 0.45, 0.43 and 0.30 when transformed to the underlying continuous scale. The results support that equine cryptorchidism is heritable and could be selected against.

Lower reproductive success associated with locomotor stereotypies in mares

H. Benhajali[1,2], M. Richard-Yris[2], M. Ezzaouia[3], F. Charfi[1] and M. Hausberger[2]
[1]Université Tunis-ElManar, Unité de Biologie Animale et de Systématique Evolutive, Campus universitaire, Tunis, 2060, Tunisia, [2]Université Rennes 1, Ethologie Animale et Humaine, UMR CNRS 6552, Campus de Beaulieu, 263 av. Général Leclerc, 35042, Rennes cedex, France, [3]Haras national de Sidi Thabet, Sidi Thabet, 2020, Tunisia; martine.hausberger@univ-rennes1.fr

Stereotypies are often associated with suboptimal environments. However, their adaptive significance remains under debate. The aim of this study was to relate the occurrence of stereotypies in breeding mares to their reproductive fitness. The overall, first and second cycle conception rates, latency to first estrus, inter-cycle interval and the number of cycles per conception were compared between mares showing weaving (n=26) and control mares (n=31). Mares were mated by 9 stallions which were equally balanced between the two groups. Rectal palpation and ultrasound were used to monitor the follicular state of the mares and to confirm pregnancy. The relationship between the occurrence of stereotypies and reproductive efficiency was analyzed using a multivariate logistic regression. We used Mann-Whitney test to compare latency to first-cycle and inter-cycle intervals between stereotypic and control horses. χ^2 tests were used to compare independent variables' distribution between the two groups of mares. Stereotypic mares had a significantly lower overall conception rate (58% vs. 81%, P=0.034) and first-cycle conception rate (33% vs. 64%, P=0.005). Stereotypic mares showed a significantly higher number of cycles per conception (2.0±0.9 in stereotypic vs. 1.5±0.8 in control mares, P=0.037). There was no difference between stereotypic and control mares in latency to first estrus (45.7±25.9 in stereotypic vs. 45.7±23.2 in control mares, P=0.97), nor in inter-cycle intervals (26.1±7.7 in stereotypic mares vs. 23.9±6.6 in control mares, P=0.74). These findings suggest a lowered fitness in stereotypic mares.

Lusitano mare reproduction: contribution to the knowledge

F. Mata, J. Bourbon, A. Twigg-Flesner and L. Greening
Hartpury College, University of the West of England, Animal and Land Sciences, Hartpury House, Hartpury, GL19 3BE Gloucester, Gloucestershire, United Kingdom; fernando.da-mata@hartpury.ac.uk

A total of n=159 Lusitano mares and n=309 reproductive cycles were analysed. Data was collected retrospectively by kind permission from Uson Olaso, Lda, a veterinary clinic established in Alentejo, Portugal, and included: 'follicle size at ovulation', 'uterine oedema' (UO), and 'age of the mare' as covariates; and 'follicle side' (right/left ovary), 'season' (month), 'hormonal treatment with $PGF_{2\alpha}$ and/or hCG and/or Oxytocin' (yes/no), 'antibiotic treatment' (yes/no) and 'reproductive technique' (natural, fresh, frozen semen) as factors. The probability of successful pregnancy was fit with a logit model, with backwards stepwise selection of variables (P<0.05) after the Wald test; a t-test was used to analyse follicle size at ovulation; and a simple linear regression was used to relate 'age of the mare' and 'follicle size at ovulation'. IBM SPSS® Statistics 21 was the statistical package used in the analysis. There is an increase of 26.4 in the odds ratio of pregnancy success per increased unit of UO, and therefore a positive relation between pregnancy success and UO is shown (P<0.05). 'Follicle sizes at the time of ovulation' where found to be significantly different when hCG was used (P<0.05). No significant differences for $PGF_{2\alpha}$ (P=0.051), oxytocin (P>0.05) and antibiotics (P>0.05). Mean and 95% confidence interval with hCG treatment is 47.9 mm [45.9; 49.9] and without hCG treatment 45.2 mm [44.3; 46.0]. No significant difference (P>0.05) was found between left and right ovaries. The regression between age of the mare and follicle size at the time of ovulation was found to be significant (P<0.05). Follicle size at the time of ovulation correlates negatively with the mares' age (P<0.05) (Pearson's r=-0.168). No relation between mares' age and fertility was found (P>0.05). These results have implications for the way we monitor follicular growth and intervene in breeding, specific to the Lusitano breed, to accurately time artificial insemination.

The scientific development in horse reproduction in Russia

L.F. Lebedeva, V.A. Naumenkova and M.M. Atroshchenko
The All-Russian research institute for horse breeding, the laboratory of physiology of horse reproduction, Rybnoe, 391105 Ryazan region, Russian Federation; lebedeva-l18@yandex.ru

In the 1900s in Russia artificial insemination (AI) was worked out, thoroughly improved and widely introduced in farm animal breeding practice. In 1954 the first in the world foal after AI with frozen semen was born in the USSR. In the 1970-th the cryobank of stallion semen was organized in the Institute for Horse Breeding. The sperm freezing, storage and AI with frozen semen of outstanding stallions were included into selection programs as one of the main directions of horse reproduction. In 2009 3 foals were obtained by AI with cryopreserved sperm of Tersk breed stallion with name Samotsvet following 35 years of storage in liquid nitrogen. This is the longest period of horse semen storage in the world practice, pregnancy rate per cycle was 55% (5/9). Now cryobank contains about 3,000 sperm doses of 73 best stallions of different breeds. From 1974 the embryo transfer technology started to develop in the Institute for Horse Breeding and in 1982 the first foal after embryo transfer was born in the USSR. The recent 20 years were hard for the Russian agricultural science because of economic and politic instability, but researchers in the field of horse reproduction continue the work. Their scientific interests focus on embryo freezing, sperm ultrastructure and cryoresistance, horse embryo development, embryo stem cells and fundamental biological processes in horse reproductive system. In 2012 the first foals in Russia were born after embryo vitrification.

Horse embryo freezing in Russia: 2 aspects of technology

L.F. Lebedeva
The All-Russian research institute for horse breeding, the laboratory of physiology of horse reproduction, Rybnoye, 391105 Ryazan region, Russian Federation; lebedeva-l18@yandex.ru

The aim of the work was to study the effect of two methods of freezing (conventional slow cooling (Exp.1, n=5) and vitrification (Exp.2, n=9)) and three cryoprotective media (medium I (Exp.1, Exp.2), Vit-Kit-set and medium II (Exp.3, n=13)) on viability of 6.5-8 day horse embryos after thawing. Embryos (Exp.1 and Exp.2) were divided according to their diameter in 2 groups (Gr. 1<350 µm and Gr. 2 >350 µm). Only small embryos (<300 µm, n=13) were used in Exp.3. Embryo quality after thawing was evaluated by the percent of dead cells in the embryos stained with Evans Blue (0,05% in PBS Dulbecco) (Exp.1, Exp.2), or by embryo transfer to recipient mares (Exp.3). The Vit-Kit media (ethylene glycol, glycerol, galactose) were commercially available (EquiPro® Vit-Kit™, 'Minitube', Germany), but the medium I (ethylene glycol, DMSO, sucrose) and the medium II (ethylene glycol, glycerol, sucrose) were hand made. The results evidence that there is no significant difference in general damage effect on embryo cells by the use of both (Exp.1 and Exp.2) methods of cryopreservation (P>0.1) when using of medium I (37.4% (n=5) and 43.3% (n=6), respectively). But the zona pellucida was fractured in 3 of 5 embryos of conventional slow cooling method. The increasing of dead cells rate from small (<350 µm) to large (>350 µm) embryos was noticed in both freezing methods (Exp.1 (n=5): 31.7±15.2% and 46.0±9.0%; Exp.2 (n=9): 45.0±13.1% and 73.6±0.18%, respectively). The pregnancy rate after embryo transfer (Exp.3) for 'Vit-Kit group' and for 'medium II group' was 11.1% (1/9) and 50% (2/4) respectively. In 2011 the 1-st foal (Vit-Kit group) was born and died at the birth in the result of incident. The first two alive 'frozen' foals (media II group) were born in Russia in 2012. In conclusion, the media II can be used for horse embryo vitrification.

Session 04 Theatre 1

Breeding goals in the era of increasingly scarce and competing resources

E. Wall, R. Rees and D. Moran
SRUC, Carbon Management Centre, Edinburgh, EH25 9RG, United Kingdom; eileen.wall@sruc.ac.uk

Global changes have increased the challenges faced by livestock production. The demands placed on livestock production from the animal to the final consumer are vast and livestock keepers are increasingly expected to meet (and in some cases exceed) a multiple range of outputs (e.g. enhanced animal welfare, reduction in environmental impact, increased production efficiency, product quality and safety). Further, livestock is uses and competes for increasingly scarce resources including feed crops, land and water. These availability of these resources for livestock production could become a major limiting factor in ability of livestock production to respond to future demands. This talk will quantify the trade-offs and co-benefits of breeding goals across multiple criteria including profitability and production efficiency, environmental impact (e.g. GHG emissions, land use change, land degradation) and animal health and welfare. Further, we will explore the impact of resource limiting factors on breeding goals in cattle.

Session 04 Theatre 2

Improvement of a method to predict individual enteric methane emission of cows from milk MIR spectra

A. Vanlierde[1], F. Dehareng[1], E. Froidmont[1], N. Gengler[2], H. Soyeurt[2], P.B. Kandel[2], S. McParland[3], E. Lewis[3], M. Deighton[3] and P. Dardenne[1]
[1]Walloon Agricultural Research Centre, Gembloux, 5030, Belgium, [2]University of Liège, Gembloux Agro-Bio Tech, Gembloux, 5030, Belgium, [3]Animal and Grassland Research & Innovation Centre, Moorepark, Oak Park, Cork, Ireland; a.vanlierde@cra.wallonie.be

Besides being a greenhouse gas, enteric methane (CH_4) produced by ruminants during rumination is also associated with the loss of 6 to 12% of gross energy intake. Mitigation of those emissions could be based on combined actions on diet, herd management and animal genetics. In order to investigate easily the relationship between these parameters and the CH_4 emissions on a large scale, an equation to predict individual enteric CH_4 emissions from the whole individual milk mid-infrared (MIR) spectra was developed. To build this equation a total of 452 CH_4 reference data were obtained using the SF_6 method on Jersey, Holstein and Holstein-Jersey crossbred cows. In parallel a 40 ml sample of individual milk was collected at each milking (morning and evening) and was analyzed using MIR spectrometry. Then, these 2 spectra were averaged proportionally in function of the milk production to have one spectrum for one CH_4 measurement. Data were collected on 146 different cows (63, 36, 18, 29 animals in parity one to four+, respectively) receiving different diets. The calibration model was developed using Foss WINISI 4 software on spectral data after applying the first derivative and using PLS regression. The CH_4 emission prediction (g CH_4/day) showed a calibration coefficient of determination (R^2c) of 0.76, a cross-validation coefficient of determination (R^2cv) of 0.70 and the standard error of calibration was of 62 g/day. Results are very promising and showed the possibility to predict the eructed CH_4 from the milk MIR spectra. The relationship between measurements and predictions is linear and thereby allowing the distinction between low and high emitting cows.

Session 04 — Theatre 3

Future needs and challenges in dairy cattle breeding – based on a survey with Austrian farmers

F. Steininger[1], B. Fuerst-Waltl[2], C. Pfeiffer[2], C. Fuerst[1], H. Schwarzenbacher[1] and C. Egger-Danner[1]
[1]ZuchtData EDV-Dienstleistungen GmbH, Dresdner Str. 89/19, 1200 Vienna, Austria, [2]University of Natural Resources and Life Sciences, Dep. Sust. Agric. Syst., Div. of Livestock Sciences, Gregor-Mendel-Str. 33, 1180 Vienna, Austria; steininger@zuchtdata.at

Changing circumstances of production and the possibilities of genomic selection are the reason for the project 'OptiGene' to redesign breeding goals and breeding programs for the main dairy breeds in Austria. To observe the needs and challenges of Austrian cattle breeders, an internet survey with about 30 questions was conducted from March to August 2012. The main topics of the survey were the individual breeding goal of the respondent, usage of services offered by breeding organizations and also agricultural and socio-political topics. The results show that the individual breeding goals of Fleckvieh and Brown Swiss breeders have shifted mainly from dairy towards fitness traits during the last decade (dairy: -8, beef: -6, fitness: +10, conformation: +4 percentage points – on average). The ranking of the single traits in the individual breeding goal confirms this observation: The top 5 individual breeding goals are fertility, longevity, udder conformation, udder health and fat/protein-kg. Being asked for the most important traits for bull selection, the farmers answered: udder conformation, fertility, fitness index, longevity and somatic cell count. The farmers were also asked for the relevance of new traits in a future breeding goal. With little differences between breeds claw health, inter- and cross-sucking and metabolism ranked first followed by feed and energy efficiency. Asked for the biggest challenges of Austria cattle breeding in the next 10 years, breeders ranked increasing energy costs, low prices of agricultural products and increasing costs for crop first. The participatory elaboration of the breeding goal is the base to optimize the different steps in the breeding process to achieve the long-term genetic gain desired by the farmers.

Session 04 — Theatre 4

Breeding pigs for heat tolerance: challenges to face

J.-L. Gourdine[1], D. Renaudeau[1], J. Riquet[2], J.-P. Bidanel[3] and H. Gilbert[2]
[1]INRA, UR143 URZ, 97170, Petit Bourg, France, [2]INRA, UMR144 LGC, 31326, Castanet-Tolosan, France, [3]INRA, UMR1313 GABI, 78352, Jouy-en-Josas, France; jean-luc.gourdine@antilles.inra.fr

Heat stress (HS) impacts the efficiency of pig production systems, by decreasing animal performance and welfare. Above 24-25 °C, feed intake decreases to reduce body heat production, with subsequent negative impacts on pig performance and farmer profit. Finding solutions for improving heat tolerance (HT) in pigs is crucial, particularly with the expected effects of climate change. As observed in other livestock species, previous studies have shown genetic variability of HT in growing pigs and sows. Evaluating and potentially taking into account GxE interactions in HT is crucial, as they can reflect animal difference on the ability to cope with HS. A first strategy, to improve HT, already used by some international breeding companies, consists in selecting pigs for usual performance traits in hot environments. Another strategy would be to find indicators of HT and include them in the breeding goal or use as additional selection criteria so as to obtain robust pigs that maintain high performance levels in different environments. A major challenge for including HT traits in breeding schemes is to define relevant indicators of HT to be routinely recorded in most selection environments. Moreover, the correlation between HT predictors and performance traits must be assessed. Results generally indicate unfavourable correlations. Depending of the level of antagonism, different strategies could be used to implement HT breeding schemes. An estimation of the economic value of breeding for HT is needed to properly weigh them in selection index and to choose between selection strategies. Finally, genome-enabled technologies will facilitate the selection of HT pigs by providing significant associations with molecular markers. We will review these different aspects to propose strategies for breeding for heat tolerance in pigs.

Defining a breeding goal for sustainable farming of Atlantic salmon

H.M. Nielsen[1], I. Olesen[1,2], B. Gjerde[1,2], K.M. Grimsrud[3] and S. Navrud[4]
[1]Nofima, Osloveien 1, 1430 Ås, Norway, [2]Norwegian University of Life Sciences, Department of Animal and Aquacultural Sciences, 1432 Ås, Norway, [3]Statistics Norway, Pb 8131 Dep, 0033 Oslo, Norway, [4]Norwegian University of Life Sciences, UMB school of Economics and Business, 1432 Ås, Norway; hanne.nielsen@nofima.no

In order to breed for sustainable animal production, breeding goals must consider non-market values like social and ethical aspects in addition to economic values of traits. By including non-market values in the breeding goal, heritable traits related to e.g. animal welfare can be given appropriate social or strategic weight in order to avoid socially unacceptable deterioration due to intensive selection for production traits. This study provides an example of the consequences on selection response from including both economic and non-market values for traits related to welfare in a breeding goal for Atlantic salmon. Preferences of Norwegians households in terms of their willingness-to-pay (WTP) for breeding programs including welfare and social traits (frequency of deformities, frequency of injuries, resistance to salmon lice, and resistance to general disease) were estimated based on data from an internet survey of a representative sample of the Norwegian population. The survey asked respondents to choose among three different breeding programs for farmed Atlantic salmon. The average WTP expressed per kg of purchased salmon were 3.36, 5.21, 0.79 and 0.94€ for avoiding general disease, salmon lice, injuries and deformities, respectively. Based on these WTP estimates, we further estimated non-market values for the four traits. These non-market values were combined with economic values and genetic parameters for the four welfare traits in addition to growth to predict non-market selection response, market economic response, and total response (the sum of non-market response and market economic response). This study demonstrates how one may define a breeding goal for sustainable farming of Atlantic salmon.

Re-designing selection objectives to improve animal welfare

S.P. Turner[1], T.B. Rodenburg[2], S. Desire[1], E. Wall[1], L. Canario[3], R.B. D'Eath[1], C.M. Dwyer[1] and R. Roehe[1]
[1]SRUC, Animal and Veterinary Sciences, West Mains Road, Edinburgh, EH9 3JG, United Kingdom, [2]Wageningen University, Behavioural Ecology Group, Building 122, De Elst 1, 6708 WD, Wageningen, the Netherlands, [3]INRA, Department of Animal Genetics, 31326, Castanet-Tolosan, France; simon.turner@sruc.ac.uk

Selective breeding poses threats and opportunities to animal welfare. Modern breeding may accelerate the rate of desirable or undesirable change in correlated traits whilst re-focussed selection objectives to meet economic, food security and environmental concerns will demand heightened selection pressure on some existing traits and selection on novel traits. Animal welfare remains a consumer priority and improving welfare can contribute to improved economic and environmental sustainability, for example by improved neonatal survival. Predicted impacts on animal welfare should proactively inform future selection decisions, such as improving feed efficiency in ruminants. Many major welfare issues are long-standing and likely to persist as management solutions are difficult to implement. Breeding presents opportunities to break this deadlock and benefit welfare. Specific examples will illustrate that selection on traits ranging from neonatal survival to social behaviour is technically achievable. Behavioural traits are core to several persistent welfare issues and particular challenges exist in understanding how context-dependent the response to selection will be and how selection might affect animal experiences and other behaviours. Progress in understanding these issues will be illustrated using aggression and tail biting in pigs and feather pecking in hens as examples. Phenotyping costs remain a barrier to selection on complex welfare traits, even using genomic selection. Efficient but information-rich phenotyping may minimise these costs whilst evidence will be presented that kin, group or multi-level section on indirect genetic effects is showing promise for benefiting welfare traits without the need for additional phenotyping.

Session 04 Theatre 7

Consequences of selection for indirect genetic effects on growth for production traits in pigs

I. Camerlink[1,2], N. Duijvesteijn[2,3], J.E. Bolhuis[1], J.A.M. Van Arendonk[2] and P. Bijma[2]
[1]Wageningen University, Adaptation Physiology Group, P.O. Box 338, 6700 AH Wageningen, the Netherlands, [2]Wageningen University, Animal Breeding and Genomics Centre, P.O. Box 338, 6700 AH Wageningen, the Netherlands, [3]TOPIGS Research Center IPG B.V., P.O. Box 43, 6640 AA Beuningen, the Netherlands; irene.camerlink@wur.nl

A major challenge in livestock production is to improve animal welfare while sustaining productivity. Including indirect genetic effects (IGEs) in the breeding objective might contribute to a solution. IGEs, also known as associative or social effects, are heritable effects of individuals on phenotypes of others. IGEs can contribute to heritable variation and response to selection, but are largely ignored in current selection. Here we investigated the consequences of selection for IGE on growth in fattening pigs. In a one-generation selection experiment, a contrast of 3.6 g ADG in estimated IGE for growth during the finishing period was made. Offspring (n=480) were housed in a 2×2 setting with IGE (high vs. low) and housing condition (barren vs. straw) as factors on pen level. Pigs were housed with 6 per pen, giving an expected contrast of (6-1)×3.6=18 g ADG between IGE groups. Weight, ADG, FCR, and carcass quality were recorded. Surprisingly, high IGE pigs tended to have lower weight throughout the finishing period (P=0.08), and lower ADG from birth till slaughter (High 734 g; Low 749 g; P=0.07). ADG during the finishing period did not differ (P=0.28). High IGE pigs had lower carcass weight (High 92 kg; Low 95 kg; P=0.02) and less muscle depth (High 58 mm; Low 61 mm; P<0.01). There was no G×E-interaction between IGE group and housing condition. These results are opposite to our expectation. This might be due to a too small or over-estimated contrast, or more favourable animal management than in commercial farming. The estimated variance in IGEs suggests that they can contribute to sustainable genetic improvement. However, empirical validation is required in pigs, and the discrepancy observed here is further investigated.

Session 04 Theatre 8

Updating the breeding goals based on a bio-economic model in a French Holstein system

D. Pinard and D. Regaldo
Institut de l'Elevage, rue de Bercy, 75012 Paris, France; delphine.pinard@idele.fr

In order to improve the economic durability of herds, breeding goals have to be based on economic statement, taking into account future context elements such as the evolution of feed costs. To rank traits in the breeding objectives according to their economic incidence, a bio-economic model simulating costs and revenues associated with herd performances is very useful. And particularly, it allows taking into account the incidence of functional traits on the production costs. Breeding goals for the Holstein breed in France were defined in 2001 based on economic statement. It was updated in 2012 but only based on technical statement in order to take into account genomic selection implementation. To update the selection objectives of breeds a bio-economic model was developed to estimate the economic weights of 11 traits for the Holstein breed within a representative intensive system of Northern France described by the breeding networks of France. In order to validate our results we used the Ecoweight software. Using our model, the weight of functional longevity was increased a lot compared to 2001 and represented 40% of the breeding objective. Protein yield and fat yield accounted for respectively 21% and 19% of the breeding goal. Somatic cell score represented 6%, cow fertility and heifer fertility counted for 5% and 3% respectively. Each other trait counted for less than 3%. Using Ecoweight, the economic weight of longevity was smaller and counted for 30%. On the contrary, the economic value of cow fertility was a bit higher and accounted for 7% of the breeding goal. We can explain that by the influence of fertility and longevity on the lactation length, which induced a change in the average fat and protein contents in the French model. The Holstein model will be a base to develop other models for ruminants as part of the OSIRIS project (stands for Breeding goals for ruminants and total merit indexes) financed by the French Ministry of Agriculture (CASDAR fund) and FGE.

Session 04 Theatre 9

The cost of batch variability as a component of economic values for robustness traits

P. Amer[1] and S. Hermesch[2]

[1]AbacusBio Limited, P.O. Box 5585, Dunedin, 9058, New Zealand, [2]University of New England, Animal Genetics and Breeding Unit, Armidale, NSW 2351, Australia; pamer@abacusbio.co.nz

While selection for performance traits has been very successful in pigs, the resulting genotypes require improved management and also appear to have greater susceptibility to environmental fluctuations. A better understanding of the economic consequences of increased environmental sensitivity should incentivise development and use of new selection criteria and trait weightings in pig breeding programs that lead to improved long term outcomes of selection. The economic value of having more uniform performance across a batch of finishing pigs is not well understood. While continuous flow systems may mitigate the problem of batch variability, there are a number of bio security risks associated with mixing pigs. Adverse performance due to unfavourable social interactions is also likely. An alternative is to draft off heavier pigs as they reach target market weight. However, this leads to an inefficient use of the pig finishing facility, and there is wasted finishing capacity. The objective of this paper was to demonstrate the economic cost of batch variability in a pig finishing system whereby there is no mixing of batches, and pigs are drafted off for slaughter as they hit a target dressed carcase weight. A simple model of a pig finishing system is described and parameterised using information from the Australian pig industry. The model generates an optimum termination date at which point all remaining pigs in a finishing pen batch are slaughtered. Economic values for a 1 kg increase in the standard deviation of batch dressed carcase weight were found to be robust to the assumptions made about the price of feed and the base per kg carcase price. While these parameters have a large impact on the profitability per pig, a reduction in batch standard deviation does not create much opportunity to save feed costs. Reducing variability does create the opportunity to have more pigs sold at higher weights leading to greater average revenue per pig.

Session 04 Poster 10

A modeling tool to define breeding goals in cattle, sheep and goats

J. Guerrier[1], C. Experton[2], S. Patin[3] and F. Phocas[4]

[1]Institut de l'Elevage, Allée Pierre de Fermat, 63170 Aubière, France, [2]ITAB, rue de Bercy, 75012 Paris, France, [3]Races de France, rue de Bercy, 75012 Paris, France, [4]INRA, UMR1313 GABI, 78352 Jouy-en-Josas, France; jean.guerrier@idele.fr

Definition of breeding goals is the best way to select animal adapted to the actual and future contexts in terms of economic durability, sustainable development and animal quality product. The news challenges of animal production need to be considered in animal breeding goals, which are defined by the breeding organizations. The French project OSIRIS (Breeding goals for ruminants and total merit index) was launched in January 2012 for 3 years. One of the objectives of this project is to estimate the weights in breeding goals for new traits like quality product, longevity and resistance to diseases. To rank traits in the breeding objectives according to their economic incidence, a bio-economic model simulating costs and revenues associated with herd performances is very useful. Such models enable to create a total merit index (TMI) combining traits according to their economic weights. In France, the estimation of economic weights in cattle are more than 10 years old, and in small ruminants, breeding objectives are defined only on a technical basis. The first goal of OSIRIS is to harmonize the methods and tools, and to be able to update TMI regularly. We will develop a set of programming modules, suitable for beef or dairy, cattle, sheep or goats. Five breeds (one in each ruminant industry) and one organic system in Montbéliard breed will be used to build herd system simulation. During the last year of the project, the modeling tool will be applied progressively to other breeds. The second goal of the project is to estimate the economic value of new breeding goals such as the organoleptic quality of meat, the protein and fatty acids composition of milk, the length of productive life, the resistance to parasitism in small ruminants and to paratuberculosis in cattle. Financial support from the Ministry of Agriculture and FGE.

French 'Systali project' to new feed units for energy and protein in ruminants

D. Sauvant[1], P. Noziere[2] and J.L. Peyraud[3]
[1]INRA-AgroParisTech, 16 rue C. Bernard, 75005 Paris, France, [2]UMR-URH, INRA, Theix, France, [3]UMR-Pegase, INRA, St Gilles, France; sauvant@agroparistech.fr

An INRA steering group is working in France to update the feed energy and protein units by 2013. An ultimate target is to predict the absorbed flows of nutrients and the animal responses to diets, particularly extreme diets. Recent publications demonstrated the interest of meta-analysis of experimental databases to predict feeding practices influences on quantitative digestion of substrates and absorption of nutrients. In Systali, similar approaches were applied to obtain new models of responses and to update the previous feed units. For that, large data bases of digestion have been built: 'Bovidig' (cattle digestion; 800 Exp, 2106 Tr), 'Rumener' (calorimetric studies on ruminants; 186 Exp, 1100 Tr) and 'Ovidig' (sheep digestion; 116 Exp, 384 Tr). The major novelties included in these new units were: (1) transit outflow rates of forages, concentrates and liquids in function of dry matter intake, in % of live weight, (DMI%LW) and dietary proportion of concentrate (PCO); (2) digestive interactions, controlled by DMI%LW, PCO and rumen protein balance (RPB), were applied to OM digestibility, to CH_4 and to urine energy outflows; (3) Feed degradation predictions of protein and starch were based on in sacco measurements validated on *in vivo* duodenal flows; (4) fermentable organic matter (FOM) in the rumen was reconsidered to be closer to the true OM ruminal digestibility and to take into account digestive interactions; and (5) microbial protein flow at duodenum was redefined from FOM, PCO and RPB. The major responses of digestion were integrated into a simple mechanistic model of gut to check the consistency across all the equations. Moreover, these equations were also implemented under Excell to reconsider the values of the requirements and of the responses to dietary variations. Endly, a simulation tool is built to check the global consistency between supplies, requirements, responses and to enlarge the feeding context (grazing...).

Evaluation of fill unit systems used for dairy cattle

R.L.G. Zom, G. Van Duinkerken and A.M. Van Vuuren
Wageningen UR Livestock Research, P.O. Box 65, 8220 AB Lelystad, the Netherlands; gert.vanduinkerken@wur.nl

Predicting nutrient intake is essential to ensure that animals are fed in balance with requirements or genetic potential and requires accurate estimations of nutrient content and intake of the feed offered. During the last 60 years, several theories have been developed on feed intake regulation in ruminants. It is generally consented that the intake of low-quality feeds is restricted by the capacity of the gastro-intestinal tract to process these feeds, whereas for high-quality feeds intake is regulated through interactions between nutrients/metabolites and specific receptors, thereby triggering humoral or neural feedback mechanisms. Other factors are taste, environment and status of the animal (health, age, milk yield). Various attempts were made to include these theories into mechanistic models, but their prediction of voluntary feed intake for lactating dairy cattle is yet not accurate. Thus, models used to estimate feed intake are pragmatic, based on calibration datasets containing feed and animal characteristics, whereby the predicted feed intake results from a filling (satiety) index of the ration and a feed intake capacity of the animal. A selection of these models was evaluated, using an independent dataset. The behaviour of individual models was influenced by basic forage source (maize silage versus grass silage), forage:concentrate ratio, breed, genetic potential, parity and stage of lactation and gestation. It can be debated whether models including milk yield and bodyweight can be used to predict intake capacity because these animal characteristics may be highly correlated with nutrient intake. Amongst the tested models, the mean square prediction error (MSPE) varied between 4.0 and 14.9. Results showed that accurate predictions are possible without inclusion of performance data. Random error as proportion of MSPE for individual cows was large across all models, suggesting that models are better suited for groups than for individual cows.

The NorFor feed evaluation system

H. Volden[1,2]
[1]*TINE SA, Norwegian Dairies, Langbakken 20, 1430 Ås, Norway,* [2]*Department of Animal and Aquacultural Sciences, Norwegian University of Life Sciences, P.O. Box 500, 1432 Ås, Norway; harald.volden@tine.no*

Feed is one of the major expenses in modern cattle production. In addition to feed prices, its overall costs are affected by the efficiency of feed utilization and the output of animal products to be marketed. Hence, there is a clear need to evaluate feed quality in order to maximise profitability. In Denmark, Iceland, Norway and Sweden, the NorFor feed evaluation was introduced for use in practise in 2007. The NorFor system is a semi-mechanistic, static and science-based model, which predicts nutrient supply and requirements for maintenance, milk production, growth and pregnancy in cattle. The model can be divided into five parts: (1) an input section describing characteristics of the animal and feeds available; (2) module simulating processes in the digestive tract and the intermediary metabolism; (3) a module predicting feed intake; (4) a module predicting the physical structure of the diet; and (5) an output section describing nutrient supply, nutrient balances and production responses. In addition, NorFor has developed an IT and software system which is used can be used for ration formulation and economical optimization of diets. One of the major challenges in the NorFor system is the interactions and the non-linear relationships that are used to describe feed digestion and metabolism. This means that individual feeds do not have fixed feed values and that NorFor therefore must be considered as a ration evaluation system rather than a system used to evaluate individual feedstuffs.

Systool, a new calculation tool for the French Systali project

P. Chapoutot[1], *P. Nozière*[2] *and D. Sauvant*[1]
[1]*AgroParisTech, INRA, UMR 791, 16 rue Claude Bernard, 75231 Paris Cedex 05, France,* [2]*INRA, UMR 123 Herbivores, 63122 Theix, France; patrick.chapoutot@agroparistech.fr*

The new French feed unit system, developed by INRA in the Systali project, is based on digestive flows of nutrients and multiple animal responses. It integrates a mechanistic digestion model accounting for the effects of feeding level (FL), proportion of concentrate (PCO) and rumen protein balance (RPB), on digestive processes. In order to easily estimate the renewed nutritive values and nutrient flows and to allow their validation by comparing them to the animal responses described in published experimentations, a new calculation tool, has been developed. Systool, implemented in Excel, is linked to a new INRA feed table proposing feed values that are calculated with a reference value for FL and the hypothesis of PCO=0 and RPB=0. For each trial, after describing animals (body weight) and rations (feed and diet composition, *in vivo* measurements), the user can choose the feeds from the INRA table whose composition best matches the reported data. The nutritional values of these feeds are extracted from the table and Systool calculates the nutrient flows. For a given diet, the actual FL and PCO values are considered as input variables, giving initial estimates of digestive interactions. The RPB, calculated as an output variable, can alter these digestive interactions thanks to an iterative calculating process, which rapidly converges toward a final prediction. Systool allows comparison of the values of feeds initially proposed in the INRA table and finally obtained in the rations, as well as the estimated vs. the measured digestive efficiencies or nutrient flows of the diets. Moreover, the respective weight of the main factors involved in the digestive interactions can be evaluated. The results for all the treatments are gathered and can be stored in databases for comparison with animal responses and thus validation of the system.

Session 05a Poster 5

Evaluation on sheep of the INRA-Systali model of digestive interactions

P. Nozière[1], P. Chapoutot[2] and D. Sauvant[2]
[1]INRA, UMR 1213 Herbivores, Theix, 63122 St Genès Champanelle, France, [2]INRA-AgroParisTech, UMR 791 MoSAR, Paris, 75005, France; pierre.noziere@clermont.inra.fr

Renewing feeding systems to better predict animal responses to diets requires a more precise representation of digestive interactions. An aggregated model of digestion developed under the INRA 'Systali' project ensures consistency across empirical relationships obtained from large databases on digestion and metabolic balances. It allows predictions of the main digestive events involved in the renewed calculation of feed units. We aimed to evaluate the direct application on sheep of this model, mainly developed from data in cattle. A database gathering measurements of digestion in sheep (Ovidig, 116 Experiments, 384 Treatments), has been build, carefully coded according to the experimental factors, to allow dissociation between intra- and inter-experiment variations. Treatments were characterized according to INRA feed tables at the entry of the model, implemented under Excel ('Systool'). The predictions (P) of the model were compared to the observed (O) data. Special care was devoted to effects of feeding level (FL), percentage of concentrates (PCO), and rumen protein balance (RPB), which are the main factors of digestive interactions introduced in the model. The FL, PCO and RPB ranked from 0.9 to 5.1 (mean 2.2) %BW, 0 to 100 (mean 42) %, -63 to 174 (mean 18) g/kg DMI, respectively. The intra-experiment variations of OM digestibility (OMd, %) and OM truly digested in the rumen (OMtDR, g/kg DMI) are well predicted (slope between P and O not statistically different from 1). Differences P-O, on average (± SD) -3.9±6.3 points OMd and 77±79 g OMtDR/kg DMI, mainly negatively related to PCO, suggest that digestive interactions due to PCO are lower in sheep than in cattle. Predictions of duodenal N flows are on average satisfactory, with P vs. O=12.6±1.2 vs. 12.1±3.8 g microbial N/kg DMI and 22.1±2.6 vs. 20.9±3.8 g total non-NH_3 N/kg DMI.

Session 05b Theatre 1

Quest for novel feed resources

H.P.S. Makkar
FAO, Animal Production and Health Division, 00153 Rome, Italy; harinder.makkar@fao.org

By 2050 the world will need 70% more meat and milk, the drivers of which are the increasing population, income growth and urbanisation. This high and increasing consumption of animal products will impose a huge demand on livestock feed. Sustainability in feed production is increasingly becoming a challenge for agriculture due to growing concerns of land, soil and water scarcity, food-fuel-feed competition, on-going global warming and frequent and drastic climatic vagaries, along with increased competition for arable land and non-renewable resources such as fossil carbon-sources and minerals (phosphorus). To a large extent the increase in the demand of animal products will be met by the monogastric sector. In many currently used monogastric feeding systems, over 90% of the feed proteins compete with human food. Thus a huge increase in the feed protein requirements in the future could further adversely impact food security. This paper will attempt to identify some novel feed resources. The focus will be on co-products of the biofuel industry such as dry distillers grains and distillers grains with solubles from different starch-rich cereals, corn gluten meal, deoiled distillers grains, high protein distillers grains, vinasse, palm-based co-products, glycerol and fatty acid distillate; and the level at which these co-products could be safely used in diets of pigs, poultry, cattle and for some fish species. The potential of co-products from unconventional resources that are used for biofuel production, for example, seeds of toxic and non-toxic *Jatropha curcas*, *Ricinus*, *Pongamia*, Cramble, *Camelina* and Neem, sweet sorghum and micro-algae will also be presented. Detoxification of some seedmeals and cakes is necessary before they can be considered as feeds. Other feed resources that will form the subject of discussion are an array of insects and densely planted *Moringa oleifer* plant giving high biomass yield with protein quality and digestibility as good as of soybean protein. Future research areas for making the efficient use of these novel feed resources will also be identified.

Session 05b Theatre 2

Feasibility of novel protein sources for livestock

G. Van Duinkerken, P. Bikker, M.M. Van Krimpen, C.M.C. Van der Peet-Schwering and T. Veldkamp
Wageningen UR Livestock Research, P.O. Box 65, 8200 AB Lelystad, the Netherlands;
gert.vanduinkerken@wur.nl

The EU is dependent for more than 70% on imports of vegetable proteins as protein-rich feed ingredients. An alternative protein strategy, and an improved utilisation of biomass residues and left-overs is very important to further increase the sustainability of EU agro & food production. Cultivation of protein crops in Europe, conversion of low grade biomass, waste streams, and by-products to high quality proteins, and innovations in protein extraction and feed processing technologies are focal points in research targeted at a transition towards more sustainable protein chains in the EU. Within the category of proteins from oil seeds (defatted soybeans, rapeseed, sunflower seed) European soybean meal seems to be the most promising alternative for South American soybean meal. Nutritional value and protein digestibility of soybean meal is high. Soy varieties with an ultra-short growth season show potential to increase the protein yield as required for a sufficient economic feasibility. Within the category of grain legumes, peas seem the most favourable alternative for soybean meal, at least for the short-term. The protein yield of peas is reasonably high, but should be further improved. In the long-term, extraction of proteins from leaves (e.g. grass, sugar beet leaves) and novel aquatic proteins (e.g. microalgae, seaweed, duckweed) are promising, especially because they do not compete in land-use (or at least compete less) with traditional protein-rich crops. However, the low dry matter content of these novel protein sources is a disadvantage and there is an urge for research and innovations regarding protein separating techniques and nutritional value of these products. The use of insects (e.g. housefly larvae) to convert low-grade biomass into high-quality proteins appears quite promising. However, further progress should be made in the development of large-scale insect production techniques, economic feasibility and legislative and safety issues.

Session 05b Theatre 3

Seaweed and seaweed components as novel protein sources in animal diets

P. Bikker, M.M. Van Krimpen, A. Palstra, W. Brandenburg, A. López-Contreras and S. Van den Burg
Wageningen UR, Livestock Research, P.O. Box 65, 8200 AB Lelystad, the Netherlands; paul.bikker@wur.nl

The development of the world population stimulates the demand for animal proteins. This increases the competition of biomass for human and animal consumption or for biofuel, and urges an efficient use of available natural resources. Moreover, within the EU a more sustainable and less import based food production chain is required. In the past, coastal communities gathered seaweed onshore for use as feedstuff, mainly for ruminants. In intensive animal production in the EU, seaweed is not used to any significant extent. Nonetheless, seaweed (products) may be of interest, especially because seaweed cultivation does not compete in land-use with traditional arable crops. Therefore, we conducted a feasibility study into the production and use of macro algae in the North Sea area for inclusion in animal diets. Seaweed can be used in animal diets in complete form, as a residue of bioprocessing, or as a source of bioactive components and micronutrients. Mainly in young piglets, effects of seaweed components on immune competence and gut health are observed. Nutrient digestibility seems relatively low and may vary between animal species. More insight is required in the comparative feeding value of seaweed species, suitable for cultivation in the North Sea, in diets for ruminants, pigs, poultry, and fish. Attention should be given to the high ash content and the high seasonal and inter species variation in composition. Further research has to determine whether enzymatic or technological treatment of seaweed can improve nutrient digestibility and enhance the value of seaweed in animal diets. Our study suggests that combined use of seaweed fractions for food, non-food and feed applications through biorefinery allows the most promising opportunity for efficient use of resources and an economically viable business case. Therefore, future research should focus on the nutritional, functional and feed safety aspects of residue fractions in target species.

Effect of Yeast extract on growth performance and intestinal mucosa morphology of weanling piglets

X. Wu[1], C.Y. Xie[1], B. Li[2], B. Tan[2], J. Yao[3], Z.H. Li[3] and Y.L. Yin[1]
[1]Institute of Subtropical Agriculture, the Chinese Academy of Sciences, Key Laboratory of Agro-ecological Processes in Subtropical Region, Changsha, 410125, China, P.R., [2]Angel Yeast Co., Ltd, Yichang, 443003, China, P.R., [3]The Hubei Provincial Key Laboratory of Yeast Function, Yichang, 443003, China, P.R.; w.xin@foxmail.com

Spray dried plasma protein (SDPP) is an effective protein source for use in the postweaning diet for early-weaned pig, however, it may have potentially dangerous of infection source. This experiment was conducted to investigate the effects of Yeast extract (YE) replacing SDPP on growth performance and intestinal of early-weaned piglets. Ninety Duroc×Landrace×Yorkshire piglets from 10 pens (average pen weight 6.22±0.16 kg; weaned at 21±1 d) were grouped into 2 treatment groups with 5 replicates per group and 9 piglets per replicate, and fed one of the following diets for 14 days: a SDPP diet (SDPP, 30 g/kg), and a YE diet supplemented with 30 g/kg YE replacing SDPP. Feed intake (FI) was recorded, body weight was measured and average daily gain (ADG) was calculated. On day 15, five piglets were randomly selected from each replicate for blood samples, and then slaughtered for jejunum and ileum collections. The results showed that: (1) there were no differences in ADG and FI between the two groups (150.55 g/d vs. 147.89 g/d); (2) compared with the SDDP group, plasma phosphorous tended to be low in the YE group; plasma arginine were higher ($P<0.05$), while tryptophan and glutamate were lower ($P<0.05$), and both plasma glutamine and proline had low trend in the YE group ($P<0.05$); 3) there were no difference in both the villus height (VH) and crypt depth (CD) of jejunum, however, VH and CD of ileum were significantly lower in the YE group than that in the group SDPP ($P<0.05$). The results indicated that SDPP can be replaced by YE without any detrimental effect on growth performance in early-weaned pigs, although YE decreased the VH and CD in ileum. In conclusion, The use of YE in piglets feed may decrease the usage of SDPP.

Feeding value of condensed distillers solubles for growing-finishing pigs

S. Millet[1], J. De Boever[1], E. Teirlynck[1], M.C. Blok[2], L.O. Fiems[1] and S. De Campeneere[1]
[1]ILVO (Institute for Agricultural and Fisheries Research), Animal Sciences Unit, Scheldeweg 68, 9090 Melle, Belgium, [2]Product Board Animal Feed, Stadhoudersplantsoen 12, 2517 Den Haag, the Netherlands; sam.millet@ilvo.vlaanderen.be

The aim of this experiment was to determine the apparent ileal and fecal digestibility of condensed distillers solubles (CDS) nutrients and the variability between CDS from different origin. Five CDS from different origin were tested. Apparent fecal and ileal digestibility was estimated by comparing nutrient digestibility of the test diets including 30% (dry matter) CDS, with the nutrient digestibility of the basal diet. Each diet was tested with six pigs. The average apparent fecal digestibility coefficient (AFDC) of gross energy was 83.0±5.6% (mean±SD), with 76.6±2.3% for the least digestible and 88.8±1.6% for the best digestible CDS. The AFDC of crude protein varied between 71.9±4.3% and 83.8±1.0% (overall average 76.4±9.9%). The AFDC of ether extract varied between 75.2±5.9% and 87.9±7.8% (overall average 83.4±8.6%). The AFDC for gross energy was significantly ($P<0.05$) correlated with the AFDC of crude protein ($r=0.91$), organic matter ($r=0.98$) and NSP ($r=0.70$). The net energy content (MJ/kg dry matter) of CDS from different sources varied between 7.7±0.3 and 9.7±0.6 (overall average 8.9±0.9). The apparent ileal digestibility coefficients (AIDC) of crude protein varied between sources from 57.7±18.8% to 80.6±19.0%, with an overall average of 74.8±14.7%. The AIDC of lysine varied from 61.3±15.1% to 93.8±5.1%, with an overall average of 74.8±16.6%. The AIDC of the individual amino acids was well correlated with the AIDC of crude protein, with $r>0.85$, except for lysine ($r=0.71$). Aspartic acid showed the lowest AIDC (64.5±20.3), while arginine showed the highest AIDC value (85.3±10.5). It is clear that the digestibility coefficients and net energy content varied widely between CDS sources and source dependent matrix values will be necessary for accurate feed formulation.

The nutritional value of condensed distillers solubles for cattle

J.L. De Boever[1], S. Millet[1], M.C. Blok[2], L.O. Fiems[1] and S. De Campeneere[1]
[1]ILVO (Institute for Agriculture and Fisheries Research), Animal Sciences Unit, Scheldeweg 68, 9090 Melle, Belgium, [2]Product Board Animal Feed, Stadhoudersplantsoen 12, 2517 Den Haag, the Netherlands; johan.deboever@ilvo.vlaanderen.be

Condensed distillers solubles (CDS) is a by-product of bio-ethanol production from grains. Its quality may vary because of the use of different grains and grain fractions and of different production processes. In order to better valorize CDS in cattle nutrition, chemical composition and nutritive value of 5 batches originating from the 3 Belgian plants, 1 Dutch and 1 German plant were evaluated. CDS is a pasty product with a dry matter content varying from 248 to 325 g/kg and a low pH (3.5-4.7). Crude protein content was very high for the German CDS (495 g/kg DM) and varied from 238 to 315 g/kg DM for the other products. CDS is also rich in crude fat (61-95 g/kg DM) and sugars (97-154 g/kg DM). CDS contains little cell-walls (43-154 g NDF/kg DM). Crude ash content varied from 41 to 108 g/kg DM. According to the dairy cattle requirements CDS is high in P, K and Na but low in Ca. The organic matter digestibility of CDS fed to sheep in combination with maize silage (in a ratio of 40/60 on DM-basis) was high and varied moderately from 85.3 to 88.9%. From that a net energy value for dairy cattle (NEL) between 7.78 and 8.45 MJ/kg DM was derived. Rumen by-pass protein (%BP) and intestinal protein digestibility (%BPd) of CDS were estimated from *in vitro* digestibility with protease using regressions based on in situ nylon bag data from a set of 31 protein rich dry by-products. The %BP varied from 30.3 to 41.7% and %BPd from 81.5 to 99.1%. According to the Dutch protein system the content of protein digestible in the intestines (DVE) and the rumen degraded protein balance (OEB) was highest for the German CDS amounting to 235 and 204 g/kg DM, respectively. For the other products, DVE varied from 123 to 168 g/kg DM and OEB from 56 to 110 g/kg DM. It can be concluded that CDS is an energy and protein rich feed, but its nutritional value for cattle may vary considerably.

Enterolactone production and its correlation among body fluids in cows fed flax meal

H.V. Petit[1], A.L.B. Schogor[2,3], C. Benchaar[1], M.F. Palin[1] and G.T. Santos[3]
[1]Dairy and Swine R &D Centre, Agriculture and Agri-Food Canada, 2000 College St., Sherbrooke, QC J1M 0C8, Canada, [2]Aberystwyth University, Gogerddan, Aberystwyth SY23 3EE, United Kingdom, [3]Universidade Estadual de Maringa, Departamento de Zootecnia, Av. Colombo 5790, Maringa, PR, Brazil; helene.petit@agr.gc.ca

Concentrations of enterolactone (EL) in body fluids and correlations among them were assessed in cows fed increased levels of flax meal (FM). Eight cows were used in a replicated Latin square design with four 21-d periods and four diets: control diet with no FM or with 5, 10 and 15% FM. Samples of blood and ruminal fluid were collected on day 20, and milk and urine on day 21. Milk was taken from am and pm milkings, samples of urine and blood were collected, respectively, 2 and 6 h post-feeding, and ruminal contents were collected before (0 h), and 2, 4 and 6 h after feeding. Ruminal samples for the three post-feeding times were pooled within cow and period. Data on EL were analyzed as a replicated 4×4 Latin square using the MIXED procedure of SAS (2000; SAS Institute). The model contained time and time by treatment interaction for the repeated measurements. The Spearman's correlation test was used to determine strength of the relationships among samples with the CORR procedure of SAS. Concentrations of EL in urine, ruminal fluid (RF), milk and plasma increased linearly ($P<0.01$) with higher FM concentration. Correlation coefficients were statistically significant for all tested combinations except for correlation between EL concentration in urine and RF at 0 h. Correlation coefficients between EL concentration in RF 2 h post-feeding and EL in milk, EL in RF before and post-feeding, EL in plasma and urine, and EL in plasma and milk were, respectively, 0.76, 0.75, 0.64 and 0.61. These results show that EL concentrations are positively associated among body fluids. As EL has antioxidant properties, increased production of EL in the rumen resulting from flax supplementation may contribute to better health of animals and quality of milk.

Session 05b Poster 8

Intake, feed efficiency and milk composition of cows fed flax meal

H.V. Petit[1], A.L.B. Schogor[2,3], C. Benchaar[1], M.F. Palin[1], G.T. Santos[3] and R.M. Prado[3]
[1]Dairy and Swine R & D Centre, Agriculture and Agri-Food Canada, 2000 College St., Sherbrooke, QC J1M 0C8, Canada, [2]Aberystwyth University, IBERS, Gogerddan, Aberystwyth SY23 3EE, United Kingdom, [3]Universidade Estadual de Maringa, Departamento de Zootecnia, Av. Colombo 5790, Maringa, PR, Brazil; helene.petit@agr.gc.ca

Eight Holstein cows in mid-late lactation (686 (SE 35) kg of body weight) were used in a replicated 4×4 Latin square design with four 21-d periods and four diets to determine the effect of increasing levels of flax meal (FM) on dry matter (DM) intake (DMI), feed efficiency, and milk production and composition. The four isonitrogenous and isoenergetic diets were: control with no flax meal (FM), or with 5, 10 and 15% FM in the DM. Meals were offered twice daily in equal amounts for *ad libitum* intake and cows were milked twice daily. Feed intake and milk yield were measured daily throughout the experiment and data were averaged over the 7 d of the third week. Samples of diets were taken once weekly and pooled within period. On day 21, milk samples were taken from am and pm milkings and analyzed for fat, lactose, protein and urea by infrared spectrophotometry (procedure 972.16 AOAC). Data were analyzed as a replicated 4×4 Latin square using the MIXED procedure of SAS (2000; SAS Institute). Treatments were compared by contrasts in order to test the polynomial effects of FM. There was a linear effect of treatment (P=0.01) on DM intake as a result of higher intake with an increased level of FM in the diet. Concentration of FM in the diet had no effect on milk production and composition and yield of milk components, with the exception of lactose proportion in milk that showed linear (P=0.10), quadratic (P=0.03) and cubic (P=0.09) effects with an increasing level of FM in the diet. Feed efficiency, which is the quantity of milk produced (kg) per kilogram of DMI was similar among treatments. Flax meal may be used in practice to replace other protein supplements in the diet without detrimental effect on DMI and milk production and composition.

Session 05b Poster 9

Abomasal or ruminal infusion of citrus pulp and soybean oil on fatty acid and antioxidants in milk

G.T. Dos Santos[1], L.S. Lima[1], A.L.B. Schogor[1], J.C. Damasceno[1], F.E. De Marchi[1], N.W. Santos[1], F.S. Santos[1] and H.V. Petit[2]
[1]Universidade Estadual De Maringá, Zootecnia, Av. Colombo, 5.790 Campus Universitário, 87020-900 Maringa, PR, Brazil, [2]Agriculture and Agri-Food Canada, Dairy and Swine Research and Devolopment Centre, College rue, 2000, J1M 0C8, Sherbrooke, QC., Canada; gtsantos50@gmail.com

The aim was to investigate the effects of supplementing two products, soybean oil (SBO, 0.2 kg/d) or soybean oil + citrus pulp (SBO+CPP, 0.2+1.0 kg/d), at two different sites (rumen or abomasums) on performance and on the transfer of antioxidant properties to milk. Four ruminally fistulated lactating Holstein cows were assigned to a 4×4 Latin square design with a 2×2 factorial arrangement of treatments: (1) SBO administered in the rumen; (2) SBO infused in the abomasum; (3) SBO+CPP administered in the rumen; and (4) SBO+CPP infused in the abomasum. Intake of dry matter (DM) of the basal diet was decreased (P<0.05) due to administration of SBO+CPP in the rumen and infusion in the abomasum. Basal DM intake and total DM input were decreased (P<0.05) with abomasal infusion of SBO and SBO+CPP. Product and site of supplementation had no effect on milk production and composition. Concentrations of total polyphenols and flavonoids, reducing power and production of conjugated diene (CD) hydroperoxides in milk were not affected by products but infusion in the abomasum compared to administration in the rumen increased production of CD. Fatty acid (FA) prolife of milk was not affected by products. Cows infused in the abomasum compared to those administered in the rumen showed lower (P<0.05) proportions of short-chain and monounsaturated FA and higher (P<0.05) proportions of polyunsaturated, omega 3 and 6 FA, which resulted in enhanced (P<0.05) health-promoting index (HPI) of milk. There was no advantage to supplement SBO and CPP in the rumen or the abomasum on performance and milk antioxidant properties although rumen bypass of SBO increased the proportion of polyunsaturated FA in milk fat and enhanced milk HPI.

Session 05b Poster 10

Estimating degradability of purified protein sources using rumen *in vitro* gas production

G. Marín, G. Pichard and R. Larraín
Pontificia Universidad Católica de Chile, Facultad de Agronomía e Ingeniería Forestal, Departamento de Ciencias Animales, Vicuña Mackenna 4860, Macul, 6904411 Santiago, Chile; larrain@uc.cl

The *in vitro* gas production technique has been used to assess degradability of feeds based on the pattern of gas production when incubated with rumen fluid under anaerobic conditions. Making nitrogen (N) the first limiting nutrient to microorganisms growth could be a way to estimate availability of N from different sources. The goal of this study was to develop a method to estimate the availability to rumen microorganism of a purified protein source, based in an *in vitro* gas production technique. An *in vitro* ruminal fermentation system was designed to minimize all foreign sources of nitrogen. The incubation medium (42 ml) did not include ammonium bicarbonate, ammonium sulfate, casein hydrolysate and cystein-HCl. The substrate provided 200 mg of fermentable polysaccharides and sugars. Rumen inoculum and substrate provided 2.1 mg and 1.4 mg of N, respectively. Treatments were the addition of casein hydrolysate (assumed to be 100% digestible) equivalent to 0, 1, 2, 3, 4, 5, 6, 7 and 8% w/w CP in the substrate. Each treatment was replicated in 3 bottles, and the experiment was repeated 3 times. Kinetic parameters were determined using a logistic model. Potential gas production was selected as response variable and analyzed by ANOVA ($P<0.05$) using a complete block design. Potential gas productions (0 to 8% CP) were 27.5, 31.0, 36.3, 39.8, 41.7, 42.2, 43.1, 43.3 and 43.4±0.8 ml, respectively. The results indicated that it is possible to limit potential gas production by limiting N sources to rumen microorganisms and that adding 0 to 8% CP as casein hydrolysate produced a sigmoidal curve that adjusted to a logistic model with an R^2 of 0.99. We concluded that it is possible to estimate N availability of purified protein sources to rumen microorganisms by using an *in vitro* gas production technique with casein hydrolysate as standard.

Session 05b Poster 11

Mammary gene expression in Holstein cows upon flax meal supplementation

A.L.B. Schogor[1], M.-F. Palin[2], C. Benchaar[2], G.T. Santos[3] and H.V. Petit[2]
[1]Aberystwyth University, IBERS, Edward Llwyd Building, Penglais Campus, SY23 3DA, United Kingdom, [2]Dairy and Swine Research & Development Centre, 2000 College Street, J1M 0C8, Sherbrooke, QC, Canada, [3]Universidade Estadual de Maringa, Animal Science, Av Colombo 5790, 87020-900, Maringa, Parana, Brazil; analuizaschogor@hotmail.com

The objective of this study was to evaluate the effects of increased dietary levels of flax meal (FM) on the mRNA abundance of antioxidant enzymes and oxidative stress related genes in mammary tissue (MT) of Holstein cows. Eight cows were used in a replicated 4×4 Latin square design with four 21-d periods and four diets: control diet with no FM or with 5, 10 and 15% FM. Biopsies of mammary gland were taken on day 21 of each period and. Total RNA was extracted from MT and complementary DNA synthesis was performed. Relative mRNA abundance of genes was determined using real-time PCR amplifications. PCR amplification, detection and data analysis were performed using an ABI 7500 Fast Real-time PCR System and primer pairs were designed using the Primer Express software 3.0 (PE Applied BioSystems, USA). Specificity of amplified fragments was determined for all genes using the melting curve (dissociation curve) analysis. PCR amplifications were performed in triplicate and standard curves were established in duplicate for each gene. To obtain the relative mRNA abundance of studied genes, the standard curve method described by the Applied BioSystems User Bulletin #2 was used. There was a linear increase ($P=0.03$) in nuclear factor (erythroid-derived 2)-like 2 (NFE2L2) mRNA abundance in MT with increased FM level in the diet. A linear tendency ($P=0.09$) was observed for catalase (CAT), with increasing mRNA abundance observed with higher concentrations of FM. The mRNA abundance of CAT, glutathione peroxidase (GPx)1, GPx3, superoxide dismutase (SOD)1, SOD2, SOD3 and NFKB genes was not affected by treatment. These findings suggest that FM supplementation can improve the oxidative status of Holstein cows as suggested by increased NFE2L2/Nrf2 mRNA abundance in MT.

Session 05b — Poster 12

Effect of increasing amounts of corn/wheat DDGS in the diet of growing-finishing pigs

S. Millet, J. De Boever, E. Teirlynck, B. Ampe, L.O. Fiems and S. De Campeneere
ILVO (Institute for Agricultural and Fisheries Research), Scheldeweg 68, 9090 Melle, Belgium; sam.millet@ilvo.vlaanderen.be

With the increased production of bio-ethanol in Europe over the last years, a significant amount of wheat based DDGS is produced. In previous experiments, we determined the nutrient digestibility of European DDGS batches for pigs. However, apart from providing nutrients, growth depressing factors may affect performances if included in too high concentrations. The aim of this experiment was to determine the effect of an increasing amount of DDGS with known energy and digestible amino acid concentrations. This DDGS originated from the fermentation of corn (50%) and wheat (50%). A control diet and a diet including 30% DDGS were formulated with an equal net energy and digestible amino acid content. By mixing the two diets, 3 additional diets with intermediate DDGS level were created: 7.5%, 15% and 22.5%DDGS. A three phase feeding system (25-40, 40-70, 70-110 kg) was applied. For most parameters, it was not possible to find significant differences between groups. In the first feeding phase, performance results were best with 15% DDGS included. This effect disappeared over time. Over the whole fattening period from 25 to 110 kg, daily gain (g/day) was 728±21, 756±45, 763±55, 763±29 and 702±41 for respectively the 0, 7.5, 15, 22.5 and 30% DDGS group. Overall, feed conversion ratio (g/g) was 2.64±0.03, 2.64±0.13, 2.71±0.04, 2.63±0.02, 2.75±0.06. We suggest that an inclusion rate of DDGS up to 22.5% is possible without negatively affecting performances.

Session 05b — Poster 13

Potential role of rumen bacteria in the conversion of plant lignans into enterolactone

A.L.B. Schogor[1], H.V. Petit[2], G.T. Santos[3], N.D. Scollan[1], E.J. Kim[1] and S.A. Huws[1]
[1]Aberystwyth University, IBERS, Penglais Campus, SY23 3DA, United Kingdom, [2]Dairy and Swine R & D Centre, 2000 College Street, J1M 0C8, Sherbrooke, QC, Canada, [3]Universidade Estadual de Maringa, Av Colombo 5790, 87020-900, Maringa, Brazil; analuizaschogor@hotmail.com

Flax is the richest source of plant lignans, and secoisolariciresinol diglucoside (SDG) is the main flax lignan. In monogastrics SDG is converted into enterodiol and enterolactone (EL) by the colonic microbiota. However, the rumen microbiota responsible for this conversion is nonetheless unknown. This study aimed to identify the rumen bacteria potentially responsible for the conversion of flax lignans into the EL. Eight cows were used in a double 4×4 Latin square design and fed a control diet with no flax meal (FM) or a diet with 5, 10 and 15% FM. On day 21 of each period, ruminal contents were collected before morning meal, 2, 4 and 6 h post-feeding and strained. An aliquot was freeze-dried for EL analysis, whilst another fraction was also taken 2 h post-feeding for bacterial metagenomic analysis. Supplementation of FM linearly increased ($P<0.001$) EL concentration in ruminal fluid. UPGMA dendrograms obtained following Denaturing Gradient Gel Electrophoresis (DGGE) demonstrated that within four cows, bacterial communities clustered dependent on diet. For sequencing purposes, DGGE gels were run contrasting the CON treatment with 15FM for the four cows that demonstrated a clear effect of diet on ruminal microbiota. DNA bands that were absent under the CON diet but present on the 15 FM diet were exised and DNA extracted before cloning and sequencing for bacterial identification. Sequences revealed that bacteria associated with EL production belonged to uncultured bacteria classified as *Prevotella*, *Succinivibrionaceae*, *Alphaproteobacteria* and uncultured rumen bacterium *Succinivibrio*, *Lachnospiraceae*, *Bacteroidales*, *Anaerovorax* and *Prevotella*, and strain of *Fibrobacter succinogenes*. However, further studies are required to assess their role in EL production.

Session 05b Poster 14

Lipoperoxidation in body fluids of dairy cows fed flax meal

G.T. Dos Santos[1], A.L.B. Schogor[2], M.F. Palin[3], C. Benchaar[3] and H.V. Petit[3]
[1]Universidade Estadual de Maringá – funded by CNPq, Brazil, Zootecnia, Av. Colombo, 5.790,87020-900 Maringá, PR, Brazil, [2]Aberystwyth University, funded by CNPq, Brazil, Old College, King Street, SY23 3BF, Aberystwyth, United Kingdom, [3]AAFC, 2000, Rue du College, J1M 1Z3, Sherbrooke, QC., Canada; gtsantos50@gmail.com

The effects of increased dietary levels of flax meal (FM) on lipoperoxidation in plasma, milk and ruminal fluid (RF) were evaluated through the production of thiobarbituric acid reactive substances (TBARS) using the TBARS assay kit (OXI-TEK TBARS Assay Kit, Zepto Metrix Co., USA). Eight cows were used in a replicated 4×4 Latin square design with four 21-d periods and four diets: control diet with no FM (CON) or with 5 (5FM), 10 (10FM) and 15% FM (15FM). Milk samples were taken on day 21 from am and pm milkings and ruminal contents were collected on day 20 immediately before feeding (0 h), and 2, 4 and 6 h after feeding. Blood samples were collected before (0 h) and 3 h after feeding. Data were analyzed as a replicated 4×4 Latin square using the MIXED proc SAS (2000) and the model contained time and time by treatment interaction for the repeated measurements. TBARS production (in malondialdehyde equivalents) in milk was lower for cows fed 5FM and 10FM than for those fed CON and 15FM (quadratic ($P=0.009$) and cubic ($P=0.006$) effect of treatment). TBARS production in plasma was not affected by treatment ($P=0.43$) but was reduced ($P=0.04$) 3 h after feeding regardless of treatment, with mean values of 4.1 and 3.7 nmol/ml at 0 and 3 h, respectively. There was an interaction ($P=0.01$) between time and treatment for TBARS in RF; when cows were supplemented with FM, there was a linear ($P=0.01$) reduction in TBARS at 2 h after feeding. There were no treatment effects at 0, 4 and 6 h after feeding. Therefore, this could suggest that within the first few hours (i.e. <4) of feed consumption, antioxidants present in FM may have contributed to protect dietary lipids against oxidation in the rumen as suggested by the linear decrease in TBARS with the inclusion of FM in the diet.

Session 05b Poster 15

Feeding distiller's grains diets with an enzyme on muscle chemical composition of broiler chickens

N.B. Rano[1,2] and A.S. Chaudhry[2]
[1]Bayero University Kano, Animal Science, Kano, PMB 3011, Nigeria, [2]Newcastle University, Agriculture, Food & Rural Development, Agriculture Building, NE1 7RU, United Kingdom; nuhu.rano@ncl.ac.uk

This paper studied changes in the chemical composition of breast, thigh and leg muscles of broiler chickens as influenced by feeding wheat-based distiller's dried grains with solubles (wDDGS) supplemented with enzymes. A completely randomized design with 3×2 factorial arrangement (3 wDDGS levels; 0, 15, 30% and 2 enzyme levels; -,+ enzyme containing endo-1, 4-beta-xylanase of 9,200 U/g, alpha-amylase of 1,600 U/g and subtilisin as protease of 16,000 U/g). The enzyme was included at 0.25 kg/tonne of complete diet. Four replicate groups of 7 birds each per six treatments were raised in floor pens using the experimental diets. At 42 days of age, two birds per replicate were killed, defeathered, skinned, eviscerated and muscles were excised from breasts, legs and thighs. The muscle samples were minced and analysed for their chemical composition. The data were statistically compared for the effects of wDDGS, Enzyme and wDDGS × Enzyme interaction at $P<0.05$. The results showed that inclusion of 15% wDDGS supported high percentage of protein in breast muscles as influenced by enzyme inclusion ($P<0.05$). Similarly, enzyme inclusion also influenced higher ash content of breast and thigh muscles. The wDDGS × Enzyme interaction was significant ($P<0.05$) for the protein content of breast muscles which had greater protein for 30% wDDGS with enzyme than 30% wDDGS without enzyme. However, no significant effects were observed between the chemical constituents of leg muscles. The results revealed considerable influence of the experimental diets with enzyme addition on protein and ash contents of broiler muscles.

Session 06a Theatre 1

Meta-analysis of mineral requirements for farmed fish: comparison with data on terrestrial livestock

P. Antony Jesu Prabhu[1], J.W. Schrama[2], D. Sauvant[3], C. Mariojouls[3], I. Geurden[1] and S.J. Kaushik[1]
[1]INRA, UR 1067 Nutrition, Metabolism & Aquaculture, 64310 St Pée sur Nivelle, France, [2]Wageningen University, Wageningen Institute of Animal Sciences, 6700 AH Wageningen, the Netherlands, [3]AgroParisTech, 16, Rue Claude Bernard, 75231 Paris, France; pantony@st-pee.inra.fr

Fish are a diverse group of aquatic poikilotherms with few species being commercially cultured for food production. Minerals form a group of essential nutrients for fish as with other farmed animals and precise data on their requirements is essential for adequate and balanced dietary supply. Quantitative requirements for macro-minerals such as Ca, K, Mg, P and micro-minerals such as Cu, Fe, I, Mn, Se and Zn are established for one or more aquaculture species. We undertook a systematic analysis of quantitative data on mineral requirements of fish in order to analyse the effect of different response criteria, inter-species differences and other possible factors on the minimal dietary level estimates. Whole body or tissue mineral level of a well growing fish seemed the best criterion for evaluating requirements. For most minerals, the criterion weight gain gave lower estimates for the minimal dietary levels compared to other criteria. Environmental factors such as water mineral concentrations and salinity, species' physiological factors like gut anatomy and digestive physiology can significantly affect the minimal dietary levels for some minerals like Ca, Mg and P. Comparison of mineral requirements between fish and terrestrial livestock indicated similarities as well as differences depending on the mineral. Unlike in terrestrial livestock, mineral requirements of fish have been predominantly defined for the juvenile life stages; there is a need to assess specific requirements during critical windows of life cycle such as larvae and broodstock. More work is also required to identify integrated response criteria accounting for health, immune status and flesh quality of farmed fish.

Session 06a Theatre 2

Net energy evaluation of feeds: do fish differ?

J.W. Schrama[1], I. Geurden[2], L.T.N. Heinsbroek[1], J.A.J. Verreth[1] and S.J. Kaushik[2]
[1]Wageningen University, Aquaculture and Fisheries Group, P.O. Box 338, 6700 AH Wageningen, the Netherlands, [2]INRA, UR 1067, NuMeA, Nutrition, Metabolism and Aquaculture, 64310 St Pée sur Nivelle, France; johan.schrama@wur.nl

Energy utilization in fish is mostly evaluated on the feeds/ingredients' digestible energy (DE) basis, whereas in pigs, energy evaluation is often done on a net-energy (NE) basis. In NE systems, maintenance is assumed constant and the potential for energy retention is derived from the amount of digestible protein (DP), digestible fat (Dfat) and digestible carbohydrates (Dcbh) in feeds/ingredients. We assessed in 2 fish species, if the energy retention potential of diet could be estimated from the intake of digestible protein, fat and carbohydrate. A data set was used containing 8 studies with Nile tilapia (*Oreochromis niloticus*) and 9 with rainbow trout (*Oncorhynchus mykiss*), respectively with 23 and 45 different diets. In these studies, nutrient digestibility and complete energy balances was measured. Carbohydrate contents of feed and faeces were calculated as nitrogen- free extract (NFE). In tilapia, all digestible nutrients were linearly related to retained energy (RE), giving an equation of $NE=11.5(\pm0.82)*DP + 35.8(\pm1.18)*Dfat + 11.3(\pm0.63)*Dcbh$ ($R^2=0.99$). The efficiencies of energy utilization of digestible nutrients for energy retention were thus estimated to be 49%, 91% and 65% for respectively DP, Dfat and Dcbh, being slightly lower for DP and Dcbh compared to pigs. In contrast to tilapia, the quadratic polynomial of Dcbh was significant ($P<0.01$) in trout, resulting in the following equation for NE as $NE=13.7(\pm1.27)*DP + 33.2(\pm2.06)*Dfat + 34.23(\pm8.23)*Dcbh - 3.78(\pm1.38)*(Dcbh)^2$ ($R^2=0.92$). In tilapia, RE increased linearly with Dcbh, whereas in trout RE levelled off at a Dcbh intake of 3 to 4 $g/kg^{0.8}/d$. Both fish species did not differ in energy utilization of DP and Dfat. This study suggests that a NE evaluation for tilapia does not strongly differ from applied NE systems in pigs, but for trout, NE evaluation differs due to a deviating response to digested carbohydrates.

Carbohydrates in fish nutrition: issues and prospects

J. Biju Sam Kamalam, F. Médale, S.J. Kaushik and S. Panserat
INRA, UR 1067, Nutrition, Aquaculture & Genomics, 64310 St-Pée-sur-Nivelle, France; biju.kamalam@st-pee.inra.fr

European aquaculture is mainly focused on salmonids and marine fish belonging to high trophic levels, for which, the level of incorporation of dietary carbohydrates is generally limited to avoid growth retardation and nutritional pathologies. However, the necessary shift towards sustainable plant based feeds that are naturally rich in carbohydrates makes it imperative to understand the poor metabolic utilisation of dietary carbohydrates and devise facilitative strategies. Moreover, unlike other farmed animals, the extreme diversity in farming habitats, feeding habits, morpho-anatomical features and nutrient requirements of fish, comfounds the challenge. Carbohydrates are cheap energy sources, but non-essential and their utilisation varies between different species of fish. Compared to mammals, fish are considered to be glucose intolerant with persistent postprandial hyperglycemia, lower glucose turnover rates and poor disposal by the peripheral tissues. Nevertheless, they possess an active but less understood glucose homeostatic system and circulating blood glucose levels are sensitive to challenges. Hormones, nutrient sensors, key glycolytic enzymes and excess glucose storage pathways are responsive to carbohydrate rich diets, though not always exactly similar to in mammals. Besides traditional approaches focusing on maximum tolerable inclusion levels and ingredient processing techniques to improve digestion, prospects to address the poor carbohydrate utilisation includes, understanding the interactional influence of other dietary macro-nutrients, especially of lipids on glucose homeostasis; combining genetic selection and nutritional strategy to identify genotypes having better ability to utilise dietary carbohydrates; nutritional programming either with a high-carbohydrate stimulus during critical transition windows in early life stages or by modifying broodstock diets to produce a vertical carryover effect of better adaptation in offsprings and decoding the associated epigenetic changes.

Economically optimized selection index for the Finnish whitefish program

M. Janhunen[1], M. Kankainen[2], A. Kause[1], J. Koskela[2], H. Koskinen[2], M. Lidauer[1], J. Setälä[2] and H. Vehviläinen[1]
[1]MTT Agrifood Research Finland, Myllytie 1, 31600 Jokioinen, Finland, [2]Finnish Game and Fisheries Research Institute, Finland; matti.janhunen@mtt.fi

A multitrait selection index was developed for the Finnish whitefish (*Coregonus lavaretus*) breeding program, using information on the economic values and genetics for 5 production and 8 quality traits. The economic values (€/trait unit) were calculated for each trait using a bio-economic model that calculates profit across the whitefish food supply chain (fish farmers, processors, and retailers). Genetic parameters were estimated with data from three year classes and 11,206 pedigreed individuals reared in both freshwater nucleus (FW) and sea (SW) environment in a split-family design. Based on the economical values standardized by genetic SD of the traits, the four economically most important breeding objectives and their relative contributions in the index were: survival (2 different traits) 43%, body weight (BW, 2 traits) 34%, condition factor (g/cm^3) 14%, and fillet lipid% 9%. Feed efficiency, initially ranked the fourth most valuable trait, was excluded from the index because it is unrealistic to record individual feed intake. Survival was not genetically correlated between the two production environments (r_G=0.11±0.13; ±SE) and was therefore selected separately for FW and SW. Similarly, BW was defined as two separate traits for FW and SW (r_G across environments 0.54±0.11). In SW, gutted BW was used instead of round harvest BW, allowing more focused selection against visceral%, which in turn increases fillet yield. Gutted BW showed a strong and unfavourable positive genetic correlation both with condition factor (0.72±0.07) and lipid% (0.58±0.10), resulting in an economic trade-off between the traits. By economically optimal weighting of the traits the present selection index produces maximal profit across the whole supply chain by improving survival and growth potential of fish while restricting unfavourable changes in body shape and lipid deposition.

Session 06b Theatre 1

Measurement of body temperature in animal studies: how, when, where and why?

M.A. Mitchell
SRUC, AVS, the Roslin Institute Building, Easter Bush, Midlothian, EH25 9RG, United Kingdom; malcolm.mitchell@sruc.ac.uk

It is well established that the thermal micro-environment experienced by livestock during production and procedures such as transportation can constitute a major risk to their health and welfare. High thermal loads will cause hyperthermia and cold conditions hypothermia. Measurement of body temperature is regarded as a simple procedure for adequate assessment of an animal's thermoregulatory status and welfare. However, the selection of the most appropriate method and site for measurement of body temperature is a complex issue. Body temperature may be measured in the body core (DBT) or at some more easily accessible peripheral site. Surface or skin temperature may be measured and employed to predict core temperature by established relationships. Body temperature may be monitored continuously by devices attached to or implanted in the animal or may be measured at a single time point (e.g. rectal temperature). In a research environment methods may be more complex than in a commercial setting and may also be more invasive. These methods, however, must provide information that can be usefully extrapolated to commercial animal production settings. The basis for the identification of the most appropriate methods for assessment of body temperature must include consideration of the basic concepts of thermoregulatory physiology in relation to animal environments, thermal challenges and adaptations. The current studies have compared the application of several methodologies for the assessment of body temperature including infra-red non contact thermometry, radio-telemetry of deep body temperature, measurement of body temperature by implanted passive transponders and simple clinical thermometers as applied to pigs, cattle and sheep. The research has provided the basis for identification of 'best practice' in body temperature monitoring for animal studies and identifies the most suitable approaches for assessing animal welfare and thermal status in practical and commercial conditions and environments.

Session 06b Theatre 2

Validation of the RuminAct system for monitoring rumination in two breeds of beef cows

M. Jardstedt[1], P. Nørgaard[2], E. Nadeau[1] and A. Hessle[1]
[1]Swedish University of Agriculture, Animal Environment and Health, Box 234, 53223 Skara, Sweden, [2]University of Copenhagen, Department of Veterinary Clinical and Animal Sciences, Grønnegårdsvej 3, 1870 Frederiksberg C, Denmark; mikaela.jardstedt@slu.se

The RuminAct™ system uses Hi-Tag collars to register rumination by means of a sound sensor attached dorsally to the left side of the neck. The individual Hi-Tag collars record ruminating time in 2-hour intervals and the data are downloaded by infrared communication to a receiver. Rumination recordings by the Hi-Tag collars have previously been validated for dairy cows, but no corresponding study has been conducted to assess the reliability of the system for beef cows. The objective was to validate the measures of rumination time recorded by the Hi-Tag sensors (Hi-Tag) on two breeds of beef cows, fed different diets, against visual observations (VO). The study was conducted with Herford (n=24) and Charolais (n=24) cows fitted with Hi-Tag collars. Eight cows of each breed were allocated to one of three diets: grass-clover silage, reed canarygrass silage and whole crop oat silage, which were fed at *ad libitum* intake. The rumination time was visually observed during two specific 2-hour time intervals in the morning and in the evening within one day, which corresponded to the 2-hour time intervals for the individual Hi-Tags. The difference between the VO and Hi-Tag values (Dif_Rum) were statistically analyzed by use of the Proc Mixed procedure in SAS ver. 9.3. There was no significant effect of breed, diet or time of observation on Dif_Rum. The mean and std of the Dif_Rum values were 5.9 and 15 minutes, respectively. The Bland Altman plot showed that the Hi-Tag collars slightly underestimates the rumination time per 2-hour time interval. The Pearson correlation coefficient between the 96 Hi-Tag and the VO values was 0.81, $P<0.001$. The Hi-Tag system showed acceptable agreement with the VO values and can be considered as an alternative to visual observation.

Session 06b Theatre 3

Automatisms for phenotyping data gathering in animal farms

J.-F. Bompa, E. Ricard and L. Bodin
INRA, SAGA, Auzeville, 31326 Castanet Tolosan, France; jean-francois.bompa@toulouse.inra.fr

The INRA DGA has a great and long experience (since the 90s) in routinely phenotyping automatic measuring using electronic identification, in experimental farms. The goal was to develop the data collect reliability, to generate new data, to decrease the burden of this work in the field, to improve farm management and to introduce traceability for data. Each year, more than 10000 animals are electronically identified at birth, on different species on our different experimental farms. To manage this large data flow, a structured and centralized database is located on a specific server CTIG at Jouy en Josas (INRA). This only point of data storage is accessible by all the scientists and experimental farms making the relationship easier. To gather all this data, we use rugged handled computers and stationary antennas integrated as best as possible into the existing environments (weighing cages, feeding racks, milking parlour…). The work tools are neither bought from private industrial companies nor manufactured by ourselves, for very specific needs. At first the requests were mainly on reproduction, growing measuring, milking recording and feeding intake control for the geneticist needs. Now to answer the high throughput phenotyping and precision livestock farming we call on new sensors, instrumentations, and wireless communication technology (WiFi, BlueTooth). The fields of experimentations at DGA INRA are many and varied as: (1) temperature, cardiac rhythm, methane and carbone dioxide emanation, activity …; (2) social behaviour, heating, morphologic measurement… All our trials and experimentations show that animals adapt well to automatic functioning which using electronic identification. Furthermore the work of the personnel is less difficult and the animal more calm. Concerning the electronic identification systems, all the people involved in breeding from the farmer to the slaughterhouse, including veterinarians and technicians for the milk recording are very interest in it.

Session 06b Theatre 4

A model as a tool to describe the variability of lifetime body weight trajectories in dairy goats

O. Martin and L. Puillet
INRA, UMR 791 MoSAR AgroParisTech, 75005 Paris, France; olivier.martin@agroparistech.fr

Today, managing individual variability in dairy goat farming systems is a potential innovative way to face the challenges of efficiency and robustness in a changing and uncertain environment. Precision farming has primarily been developed through the automation of data acquisition but interpretative tools to capitalize on this raw material are lacking. The challenge is to design tools as translators of individual time series data on animal performance into phenotypic information providing quantification on variability and further useful benchmarks for decision support. In this study, we propose a model to quantify how dairy goats differ in their dynamics of body weight (BW) from birth to culling. The model is based on the decomposition of BW kinetics into elementary patterns of change associated with the sequence of physiological stages during growth and over repeated reproductive cycles throughout lifespan. The BW is thus formalized as the combination of the following processes: building a genetically driven and environmentally adapted mature weight, carrying and supporting litter growth, and storing and mobilizing body reserves. Model parameters provide a quantification of these elementary processes and the model is used to translate individual BW trajectories in a set of values, considered as a phenotypic vector. The model was implemented in the INRA RECORD/VLE software platform and fitted to individual time series data routinely recorded between 2005 and 2010 at the MoSAR experimental dairy goat station. We performed a statistical analysis of the individual phenotypic vectors to provide a synthetic view of the variability between goats, to quantify animal differences and to track types of BW trajectories. The present communication is intended to describe the model, to present the results of its application to depict the within-herd variability of individual BW trajectories and to put into perspective the use of such a model as a management tool.

The potential of milk MIR spectra to certify milk geographic origin

L.M. Dale[1], A. Laine[1], A. Goubau[1], H. Bel Mabrouk[1], H. Hammami[1,2] and N. Gengler[1]
[1]Gembloux Agro Bio Tech, Animal Science, 2 Passage des Déportés, 5030, Gembloux, Belgium, [2]National Fund for Scientific Research, 5 rue d'Egmont, 1000 Bruxelles, Belgium; lmdale@ulg.ac.be

Protecting and supporting local production systems, regional authorities, as well as producers, give a very important role to milk quality. Therefore, this study was aimed to investigate the potential of mid-infrared spectroscopy (MIR) for certifying the geographic origin of milk. Because milk MIR spectral databases and extra phenotypes (breed, testday, livestock herd and origin appellation of traditional products) were available in the Walloon Region of Belgium via European project OptiMIR (INTERREG IVB North West Europe Program), discrimination studies were conducted to distinguish the Ardennes region (which is linked to PDO 'Beurre d'Ardennes') from the rest of Wallonia. A total of 542,733 spectral records linked to their geographic origin coming from Wallonia milk recording were used (97,369 of MIR spectra -Ardennes region and 450,326 -rest of Wallonia). A mixed model (fixed effects: breed, year and month of record, random effects: herd × year) was applied to obtain predicted MIR spectral values for all testdays and prediction errors (residuals) representing the factors not present in the model. In order to test the MIR ability to milk authentication, chemometric tools, such as linear discriminant analysis were applied to residuals for three MIR spectral regions (e.g. 930-1,600 per cm, 1,710-1,810 per cm and 2,560-2,990 per cm). The classifications on corrected MIR spectral data were 95% and the cross-validation were 94% for Ardennes region. Results showed that after correction of MIR spectra, the discriminant function constructed on the residuals spectra allowed a good discrimination. Based on this results MIR spectroscopy techniques may provide useful fingerprints to detect geographic origin and could be potentially used in routine management decision and quality assurance tools.

Session 06b Theatre 6

Effect of water availability in grazed paddock on milking frequency and milk yield

I. Dufrasne, E. Knapp, V. Robaye, L. Istasse and J.-L. Hornick
Université de Liège, Nutrition Unit, bd de Colonster, 4000 Liege, Belgium; eknapp@ulg.ac.be

A herd of 48 Holstein cows was milked with an automatic milking system (AMS) located on a permanent pasture. The cows grazed on a rotational grazing. The cows were fetched twice per day in the waiting area in order to be milked. They can also go voluntary in the AMS. The effect of water availability was tested during 30 days at the end of the grazing season in August and September. The mean day in milk was 211. Water was available in a tin (1000 liters) next to the AMS. In the control paddocks (C), there was an extra individual automatic bowl but no one in the tested paddocks (N). The whole herd grazed successively in paddocks C and in paddocks N staying at least 3 days per paddock. The milk yield and the milking frequency were recorded in the AMS. Voluntary returns were calculated as follow: milking number + refused milking number + failed milking number – fetching number. The mean temperature during the trial was 17 °C and the mean distance between the AMS and paddocks was 150 m. The cows received daily 2.7 kg concentrate per cow. The milking frequency was higher in N paddocks than in C paddocks (2.3 vs. 2.0 milkings/cow; $P<0.001$). The voluntary returns were more than twice higher in N paddocks than in C paddocks (1.3 vs. 0.5 voluntary returns/cow; $P<0.001$). The refused milkings were higher in N paddocks (0.77 vs. 0.44 failed milkings/cow; $P<0.05$). The milk yield per milking was increased in N paddocks (8.9 vs. 7.6 kg; $P<0.001$) but the water availability had not influence on daily milk yield (18.3 kg; NS). The milking time per cow was shorter in N paddocks than in C paddocks (4 min 52 vs. 5 min 15; $P<0.001$). In conclusion, water can be used to stimulate the cows to visit the AMS but with no improvement in milk yield. However, in respect of animal welfare, grazing in paddocks without water is not possible when the climatic conditions are hot and dry.

The effect of hoof trimming on the locomotion score, neck activity and ruminating time of dairy cows

T. Van Hertem[1,2], E. Maltz[2], S. Viazzi[1], C.E.B. Romanini[1], C. Bahr[1], D. Berckmans[1], C. Lokhorst[3], A. Schlageter Tello[3], A. Antler[2] and I. Halachmi[2]

[1]Division M3-BIORES: Measure, Model & Manage Bioresponses, KU Leuven, Kasteelpark Arenberg 30, bus 2456, 3001 Leuven, Belgium, [2]Institute of Agricultural Engineering, Agricultural Research Organization (ARO), the Volcani Center, P.O. Box 6, Bet-Dagan 50250, Israel, [3]WageningenUR Livestock Research, P.O. Box 65, 8200 AB Lelystad, the Netherlands; tomv@volcani.agri.gov.il

Regular hoof trimming is often used to control lameness prevalence in the farm. The aim was to study the effect of hoof trimming on dairy cow gait, assessed by locomotion scores, and neck activity and ruminating time. A routine hoof trimming was performed in a commercial Israeli farm by 2 trained claw trimmers on all cows in the farm in a 10 day period at the end of November. All cows were trimmed in a tilt table. During 15 data collection sessions, cows were on-the-spot manually locomotion scored by a trained observer using the discrete 5-point numerical score [1=healthy; 5=severely lame]. Neck activity and ruminating time were measured online with a commercial sensor. A generalized linear mixed model was used to statistically test all main and interaction effects of hoof trimming, parity, lactation stage, and lesion presence on ruminating time, neck activity, milk yield and locomotion score (LS). Herd lameness prevalence ($LS \geq 3$) was higher ($P<0.001$) after hoof trimming (43%) than before hoof trimming (27%) Two months after the hoof trimming period, herd lameness prevalence was 33%, still bigger ($P<0.001$) than before the hoof trimming period. Hoof trimming affected activity level directly ($F(1,527)=4.821$; $P=0.029$), as activity level was higher before (mean ± standard error; 394±4 units) than after (387±4 units) trimming. The effects of hoof trimming on milk yield, ruminating time and locomotion score were indirect through interaction effects with other variables. The results show that hoof trimming affects dairy cow behaviour and performance. This study is part of the Marie Curie Initial Training Network BioBusiness (FP7-PEOPLE-ITN-2008), and contribution number 459-4398-951 funded by the Israeli Agricultural Research Organization (ARO).

Automatic measurement of the body condition of dairy cows with three-dimensional picture processing

U. Bergfeld[1], S. Pache[1], R. Hölscher[2], I. Steinhöfel[1] and R. Fischer[1]

[1]Saxon State Office for Environment, Agriculture and Geology, Animal Production, Am Park 3, 04886 Köllitsch, Germany, [2]Hölscher + Leuschner GmbH & Co. KG, Siemensstraße 15, 48488 Emsbüren, Germany; uwe.bergfeld@smul.sachsen.de

The body condition assessment is a key character to the herd management, feeding and health con-trol of dairy cows. In a research project a new sensor system (optiCOW) was developed to the auto-matic measurement of the body condition by dairy cows and tested in a field study with 1,500 cows. The sensor system has a high-resolution time of flight camera with a picture rate of 100 hertz. The camera automatically erects from every cow a 3D-video sequences with the entrance or abandonment of the milking stand. A special light system provides for high-contrast pictures. Especially developed algorithms process the 3D-picture dates and create 3D-models of the back area of the dairy cow. With the help of striking points from the 3D-models the body condition is estimated after the method by Edmondson *et al.* The continuous estimated values allow the representation of the individual fat mobilisation dynamics in the whole lactation. The sensor system is completed by an electronic animal identification, flow scales, to a delay gate and a process computer with Internet access and works fully automatically. The validation of the sensor system followed in the learning random check with 120 dairy cows of the teaching and research farm Köllitsch. For these cows there are continuous measuring rows to the body condition, back fat thickness, weight, size, metabolism and milk performance. The sensor system was also installed in two dairy farms. A good and plausible correspondence of the 3D-estimated values with the measured values of the learning random check could be achieved in the result of the method development. Subsequently the 3D-estimated values from the field study are examined concerning plausibility and exactness and are interpreted in connection with the performance characters.

Session 06b Poster 9

Labour requirement for feeding of dairy cows by automatic feeding systems

J. Mačuhová, V. Siefer and B. Haidn

Institute for Agricultural Engineering and Animal Husbandry, Prof.-Dürrwaechter-Platz 2, 85586 Grub (Poin), Germany; juliana.macuhova@lfl.bayern.de

The semi and fully automatic feeding systems (AFS) are used for feeding of forage to optimize feed management and improve the labour efficiency in dairy farming. The aim of this study was to create calculation models for estimation of labour requirement for feeding of dairy cows with different AFS as support tool for dairy farmers by the choice of feeding equipments. Therefore, labour studies (survey and labour input recording for individual task elements) were performed to determine the cause variables and labour requirement for single task elements on praxis farms. Calculation models were created for semi AFS with removing and transport of silage from silo to bunkers in a conventional way by different implements for tractor (front loader bucket, silage block cutter, silage cutting bucket and silage grip bucket) and for fully AFS with automatic silage pick up with rotary cutter from tower or deep silos. The calculation models enable to calculate the labour requirement for existing farms as well as for fictitious farms depending on herd size and desired farm design. The test calculations carried out for both systems (the semi and fully AFS) for the herd size between 40 to 300 heads show that the labour requirement can be reduced to third by fully automatic feeding in comparison with the most effective estimated technique (silage cutting bucket) by filling of bunkers in a conventional way every second day. Thereby, the labour requirement varies between 0.5 and 1.4 h and between 1.7 and 4.8 h per cow and year (decreasing with increasing herd size)) by semi AFS using silage cutter bucket and fully AFS, respectively. The capacity of pick up and transport equipments influences noticeably the labour requirement by semi AFS. However, also travel distances from silo and other feed stores to storage bunkers have also a considerable impact on overall efficiency.

Session 06b Poster 10

How age of automatic milking system and type of cow circulation impact production parameters?

D. Bizeray-Filoche[1] and B.J. Lensink[2]

[1]LaSalle Beauvais, CASE, Rue Pierre Waguet BP 30313, 60026 Beauvais, France, [2]Groupe ISA, 48 bd Vauban, 56046 Lille, France; dorothee.bizeray@lasalle-beauvais.fr

Automatic milking systems (AMS) are more and more popular in milking production, but few studies indicate how these systems evolve with time or which type of circulation (free or selective) implies the best production parameters. This study aimed to identify these factors influence milking parameters. Data of milking production were collected into 15 commercial french herds over one year. Data concerned individual milking performances (milking frequency, milk production/day and milk production/milking) and global AMS performances (nb of milkings/robot/day, total milk production/robot and nb of cows milked/day). Effect of age of service of the AMS (less or more than 2 years) and effect of type of cow traffic (free or selective circulation) were tested on milking production parameters. Cows in the most recent AMS system were milked significantly more frequently (2.3±0.01 vs. 2.2±0.02 milkings/day; $P<0.0001$) and individual milking production was higher (25.4±0.2 vs. 23.6±0.3 kg/day; $P<0.0001$). However, since milked cows were significantly less numerous in the youngest AMS systems (55±1 vs. 65±1; $P<0.0001$), total milk production was lower (1319±28 vs. 1488±20 kg/day/robot; $P<0.0001$). Effect of cow traffic was also evaluated. Cows with free access to the robot were milked more often (2.3±0.2 vs. 2.1±0.4 milking/day; $P<0.0001$) and were more numerous (64±1 vs. 57±1; $P<0.0001$) than cows with selective circulation. However they producted less milk/day (24.1±0.2 vs. 25.2±0.2; $P<0.01$) and per milking (10.3±0.06 vs. 11.8±0.1, $P<0.0001$). Then there was no significant difference between the free and selective traffic for total milking production/day/robot. Cow traffic and age of AMS influenced significantly milking production parameters. Free traffic implies more milkings per day with lower milk production. Cow behaviour may have more impact on performances when they are managed with free circulation than with selective circulation.

Adaptability of small ruminant farming facing global change. A north south analysis in Mediterranean

J. Lasseur[1], V. Allary[2], A. Aboul-Naga[3], P. Bonnet[2] and J.F. Tourrand[2]
[1]INRA, Sad-selmet, 2 place Viala, Montpellier, France, [2]CIRAD, Selmet, baillarguet, Montpellier, France, [3]APRI/ ARC, Dokki, Cairo, Egypt; lasseur@supagro.inra.fr

Global change (e.g. urbanization and demographic pressure in coastal zone, accentuation of extreme climatic events) induce new organizations of livestock farming systems. These reorganizations within space are part of family and landscape trajectories. They are responses to external constraints as well as family changes and socio political, economic and environmental opportunities (land use planning, change in public policies). We analyze adaptive capacities of small ruminants farming systems in Provence region(France) and Matruh region (Egypt). This research is based on two methodological frameworks: (1) analysis of socio ecological systems and their resilience (mainly in France); (2) analysis of families livelihoods and their vulnerabilities (mainly egypt). Maintaining of livestock farming activities is observed in case studies in Provence region. Adaptations mainly rely on increasing of economic dimension of farms and increasing of flock motilities. These allow farmers to take advantage of new resources (forages for the flock, market for the produces) relatively to contrasted and emerging abilities of coastal zones and hinterland areas of the region. These changes reinforce the development of farms but in a counterpart theyweaken interactions between livestock farming dynamics and local development surrounding farmland, compromising their co-evolution. In matruh region, adaptive capacities to last droughts have been operated through alternatives outside farming sector (mainly migration). But livestock farming remains a basis to secure family livelihoods and for functioning of Bedouin society on mid-terms. From a methological point this study highlights the necessity of cross scaled and long terms analysis to study adaptability. On these very contrasted situations, livestock farming remains at the core of adaptation even if the mechanisms involved differ.

Matching genetic resources and breeding objectives with the constraints in tropical farming systems

N. Mandonnet[1], M. Naves[1], S. Thevenon[2], G. Alexandre[1] and V.C. Yapi-Gnaoré[3]
[1]INRA UR0143, Domaine Duclos, 97180 Petit-Bourg, France, [2]CIRAD, UMR INTERTRYP, Montpellier, France, [3]DG-CIRDES, Bobo, Dioulasso, Burkina Faso; nahalie.mandonnet@antilles.inra.fr

A major challenge for the development of livestock in the tropics is the choice and the management of the various genetic resources available, to sustainably improve the productivity of farming systems. The intensification of international trade threatens the preservation of the local animal genetic resources (AnGR) in the tropics, increases the incidence of crosses or substitution with exogenous breeds. Therefore an important goal for tropical livestock production systems is to reconcile the preservation and improvement of the AnGR. Under limiting tropical conditions, the maximization of the productivity is an illusory challenge, and the multifunctionality of the animal and of the system allows passing through its failures. One difficulty is to objectivize the interest of domestic diversity, to give an economic value to adaptation traits for various productions systems, to multifunctionnality of animal resources. Beyond the objective interests of AnGR, sociological motivations of breeders should also be assessed as local breeds often suffer from pejorative connotation. Their image needs to be enhanced by the way, including tangible morphological changes in the aim of selection. Some projects already exist. As an example, in Guadeloupe island in the Carribeans, a participative action between the farmer cooperative, the extension services and INRA researchers is carried out in order to preserve and enhance the population of Creole goats. Farmers cooperative initiated the project to improve the economic viability of farms. Their approach will be presented. Furthermore, the expectations concerning genomic tools in such breeding conditions will be evoqued.

Survival and adaptability of indigenous Red Maasai sheep and their crosses in East Africa

J.M.K. Ojango[1], J. Audho[1], A.G. Marete[1], E. Zonabend[2], J. Philipsson[2], E. Strandberg[2] and A.M. Okeyo[1]
[1]International Livestock Research Institute (ILRI), Box 30709, 00100 Nairobi, Kenya, [2]Swedish University of Agricultural Sciences (SLU), Animal Breeding and Genetics, Box 7070, 750 07 Uppsala, Sweden; j.ojango@cgiar.org

Sheep breeds that are endemic to the arid lands of Eastern Africa, and kept in extensive pastoral systems include trans-boundary breeds such as the fat tailed Red Maasai sheep. Widespread indiscriminate crossbreeding with other exotic breeds, notably the Dorper has resulted in a rapid decline in numbers of Red Maasai sheep. Changing climatic conditions and severe droughts in 2008-2010 have decimated most populations of pure and higher grades of Dorper crosses in pastoral systems. However, the indigenous sheep breeds withstood the challenges much better. In 2003, the International Livestock Research Institute (ILRI) began a selection program for Red Maasai sheep and crosses with Dorper as part of a study on the genetic resistance to gastrointestinal nematodes. Following the droughts and loss of animals by pastoral livestock keepers in 2008-2010, the ILRI flock of 1,100 sheep has become a main source of breeding animals for communities living within the surrounding rangelands both in Kenya and neighboring countries. This paper presents genotype by environment interactions and genetic progress achieved in reproductive performance and survival of Red-Maasai Sheep and their crosses with Dorper from 2003-2012, and the implications for the conservation of indigenous animal genetic resources in the arid lands.

One hour of daily contact with sexually active males induce fertile ovulation in anestrous goats

M. Bedos[1], J.A. Flores[1], G. Duarte[1], J. Vielma[1], G. Fitz-Rodríguez[1], H. Hernández[1], I.G. Fernández[1], M. Keller[2], P. Chemineau[2] and J.A. Delgadillo[1]
[1]Universidad Autónoma Agraria Antonio Narro, Centro de Investigación en Reproducción Caprina, Periférico Raúl López Sánchez, Torreón, Coahuila, 27054, Mexico, [2]Physiologie de la Reproduction et des Comportements, UMR 7247 INRA-CNRS-Université de Tours, Nouzilly, 37380, France; joaldesa@yahoo.com

In subtropics, some breeds of goats display a reproductive seasonality provoking a seasonality of milk and meat production. To extend the availability of these products all the year round, some animals must be bred during the anoestrous period. The 'male effect' is a biostimulation technique that induces and synchronizes the sexual activity in does during the anoestrous period. The objective of the current study was to determine whether photo-stimulated males are able to induce the sexual activity of anestrous goats when duration of contact is reduced to less than 4 daily hours. Six bucks were subjected to long days (16 h of light per day) in an open pen from November 1st to January 15th to stimulate their sexual behavior during the sexual rest. Anovulatory goats were used. The control group remained isolated from males (n=20), and three other groups were exposed to photo-stimulated males for 4 (n=18), 2 (n=22), or 1 (n=21) h per day during 15 days (n=2 males per group). Proportions of females that ovulated and pregnancy rates were compared by $\chi2$ test. Most females exposed to the sexually active bucks ovulated (89%-100%), whereas only 5% did so in the control group ($P<0.001$). Proportions of females that ovulated did not differ among groups of females with decreasing time of contact with bucks ($P>0.05$). Pregnancy rates were not affected by this daily duration of contact with males (67%-91%; $P>0.05$). We conclude that one daily hour of contact with sexually active males is sufficient to stimulate ovulatory activity in seasonal anovulatory goats.

Session 07 Theatre 5

Genetic parameters for test day milk yields in crossbred Holstein × *Bos indicus* cattle in India

C. Patry[1], S. Gokhale[2] and V. Ducrocq[1]
[1]INRA, Domaine de Vilvert, 78350 Jouy-en-Josas, France, [2]BAIF, CRS, 412202 Uruli Kanchan, India; clotilde.patry@jouy.inra.fr

BAIF is a large Indian agricultural NGO which produces more than 7 million semen doses from buffaloes and bulls (Holstein, local bos indicus, their crossbreds) with a network of 88,272 villages in 12 Indian states. For the past 20 years, BAIF has been part of a national progeny testing program of Holstein and crossbred bulls where milk yield is collected in tiny herds (<4 cows). The current genetic evaluation is based on a contemporary comparison approach. The objective of the study was to develop a new model adapted to the huge variations in the environmental conditions during a lactation. About 120,000 daily milk yields on 6,675 crossbred cows and the roughly herd location (area covered by one AI technician) were available. A model based on test day records was implemented in two steps, using the WOMBAT software. First, the fixed environmental effects (AI area × year × month effect) were estimated using all records while accounting for a genetic + permanent environment cow effect. Then, the data set was reduced to 38,000 adjusted performances from cows with known sires and at least 5 recorded daughters. Breeding values were estimated for 193 bulls and 2,404 cows. The best model was based on a lactation curve partitioned into 10 lactation stages, described using splines for the fixed effect part and with reduced rank (rank=3) genetic covariance functions for the genetic and permanent environment part. A heritability of 0.12 was obtained for a whole lactation. This low value was expected, considering the data structure (missing information, tiny herds). Such a genetic evaluation would not only help ranking the sires of interest but also the dams to better benefit from the local genetic diversity in their crossbreeding scheme. If extended to a simplified milk recoding protocol, it would increase the genetic predictions reliability by including many more progeny per bull, without inflating the costs of the data collection.

Session 07 Theatre 6

Crossbreeding Creole cattle with european breeds in Guadeloupe: ambiguous results in the beef sector

M. Naves[1], G. Alexandre[1] and V. Angeon[2]
[1]INRA UR143, URZ, 97170 Petit Bourg, Guadeloupe, [2]UAG, CEREGMIA, Schoelcher, Martinique; michel.naves@antilles.inra.fr

Improvement programs of beef meat production in the French Overseas Departments have been based for years on the use of improved taurine breeds, purebred or crossbred with native breeds. In Guadeloupe, such a model has been set up and had very drastic effect on the genetic composition of the local herds. From about 95% in 1985, the local Creole cattle represent by now less than 40% of the total number of animals. This tendency is particularly clear in the young generations, and its acceleration is expected. This paper will discuss several aspects of the use of crossbreeding between local breeds and specialized beef breeds for meat production in the tropics, in a case study of the Creole cattle of Guadeloupe. The results are based (1) on experimental data on animal production in crossbred Limousin × Creole cattle in comparison with pure Creole cattle; (2) on surveys and interviews data on livestock systems and breeding strategies; and (3) on census of herd composition and animal production, during the last 3 decades. F1 Crossbreeding with Limousin cattle improve the animal production, in term of growth rate, weight at slaughter, and carcass characteristics; but this improvement was obtained with higher feed consumption and inputs. Crossbred animals are also more susceptible to tropical climate constraints (temperature, tick borne diseases). On the field, improvement of the animal management in Creole herds may achieve a better improvement of animal production. The promotion of crossbreeding led to a dramatic increment of uncontrolled crossbreeding, and a great heterogeneity of the herd. Therefore the production data available in the beef sector does not really reflect the investment made. In reaction to this situation, a breeding program of the Creole breed has been set up in 1995, aiming to improve and promote the local breed. It is expected that the tendency will be reversed in the future and the local breed will be preserved.

Session 07 Theatre 7

Multi-criteria evaluation of resources for livestock farming systems under Tropics

H. Archimede[1], C. Marie-Magdeleine[1], L. Rodriguez[2], D. Bastianelli[3], G. Alexandre[1], P. Lecomte[3], M. Boval[1], E. Gonzalez[3], G.W. Garcia[4] and D. Sauvant[5]
[1]INRA, UR143, Animal Science, Prise d'Eau Petit-Bourg, 97170 Petit-Bourg, Guadeloupe, [2]Finca Ecológica, Farming system, TOSOLY, AA48 Socorro, Santander, Colombia, [3]UMR INRA CIRAD, Animal Science, 2 place Viala, 34060 Montpellier, France, [4]University of the West Indies, Faculty of Science, Campus Saint Augustine, St Augustine, Trinidad and Tobago, [5]INRA, UMR791, Animal Science, 16 rue Claude Bernard, 75005 Paris, France; harry.archimede@antilles.inra.fr

Plant resources consumed as feed on farms have potentially multiple uses. However, the evaluation of these resources is primarily based on their potential to allow nutrients to maximize animal production. This mono-criteria evaluation is not satisfactory in the context of sustainable agriculture, animal welfare and bioenergy issues. To take into account these last evolutions in agriculture, the concept of multi-criteria evaluation of feed appeared. It is an ongoing concept which raises many questions: What should be the level of approach: territory, farm, animal? For a given level approach, which criteria have be taken into account? These criteria must be a minimum representative of different functions and/or services and easy to evaluate. Some concepts should also evolve. For example, concerning the classical feeding approach 'input/requirements' which tends to maximize animal production under economic constraint, it should move towards the concept of multiple responses. The choice of methods is also an open question. Among them, Life Cycle Analysis approaches as well as modeling seem adequate tools. It remains that the use of a standardized multi-criteria evaluation similar for all resources is questionable. Potential uses and functions of resources such as grasses, legumes, fodder trees, cereals, tubers, and byproducts are very different. The main objective of this paper is to realise a state of the art on these issues and identify opportunities to highlight. This analyse will be illustrated with field practices.

Session 07 Theatre 8

Nutritive value of four tropical forage legume hays fed to pigs in the Democratic Republic of Congo

B. Kambashi Mutiaka[1,2], C. Boudry[2], P. Picron[2], H. Kiatoko[1], A. Théwis[2] and J. Bindelle[2]
[1]University of Kinshasa, Animal sciences and nutrition unit, Mont Amba, Lemba, 243 Kinshasa, Zaire, [2]Gembloux Agro-Bio Tech/ University of Liege, Animal sciences and nutrition unit, Passage des déportés 2, 5030 Gembloux, Belgium; bkambashi@student.ulg.ac.be

The incorporation of fibrous feed ingredients such as legume forages in the diets of pigs can be economically interesting. They do not compete with human food. Nonetheless, information on their nutritive value is lacking. The digestibility of 4 tropical forage legumes (TFL) hays (*Vigna unguiculata*, *Psophocarpus scandens*, *Pueraria javanica* and *Stylosanthes guianensis*) was measured. Thirty-six Large White barrows were fed a corn-soybean meal based diet containing 0, 0.125 or 0.250 of one of the previous legumes hays. After 14 days adaptation to diet, the animals were placed in individual metabolic crates and urine and faeces were totally collected once a day for 5 days. The process was repeated a second time with barrows randomly allocated to another diet. Diets and faecal samples were analysed for DM (105 °C, for 24 h), energy (calorimeter), ash (AOAC 923.03) and NDF (Van Soest *et al.* using Ankom fiber Analyser) contents, and diet, feces and urine were analysed for crude protein (AOAC 981.10), in order to calculate total tract apparent digestibility and N retention. All 4 forage species decreased linearly for the total tract apparent digestibility (TTAD) from 0.76 to 0.65, 0.80 to 0.70, 0.54 to 0.44 and 0.58 to 0.34 except for stylosanthes (0.44), respectively for DM, N, NDF and N retention. Differences in digestibility ($P<0.05$) between species were observed only for 0.250 forage-based diets. TTAD (DM, N, NDF) was, on average, higher for Psophocarpus (0.65, 0.70 & 0.42), Stylosanthes (0.65, 0.71; 0.51 and 0.44) and Vigna (0.66, 0.72 and 0.44). Due to their negative effect on the overall digestibility of the diets, the contribution of TFL to the diet should not exceed 0.125 except for *Stylosanthes* whose N retention remained quite high (0.44) at the highest inclusion level (0.250).

Responses of growing ruminants to variable diets in harsh conditions: a meta-analysis

N. Nizar[1], D. Sauvant[2] and H. Archimede[1]
[1]INRA, UR143, Animal Science, Prise d'Eau Petit-Bourg, 97170 Petit-Bourg, Guadeloupe, [2]INRA, UMR791, Animal Science, 16 rue Claude Bernard, 75005 Paris, France; nizar.salah@antilles.inra.fr

The aim of this study, a meta-analysis, was to describe the responses (intake, total tract digestibility, growth (ADG), ...) of growing ruminants to varied nutritional factors. Published studies that reported growing sheep, goat and cattle were used for this analysis. To be included in this meta-analysis, the papers were selected on some criteria. At a minimum, trials should report data on animal performances and chemical composition of diets. In total, 590 publications representing 2,225 dietary treatments were pooled to be used in the present study. There were 325 publications on sheep with 1,287 treatments, 145 publications on goat with 544 treatments, 119 publications on cattle with 394 treatments. All the trials have been conducted in warm or tropical area. Genotype animals from warmer regions were differentiated from those of temperate regions. The meta-analysis has been performed following recommendations of Sauvant *et al.* (2008). Intra-publications regressions of diet characteristics (as explained variables) on animal responses have been performed. Both crude protein and percentage of concentrate have a quadratic and significant effect on dry matter intake (DMI) without any difference between species and genotypes. Maximum intakes were recorded with 10 to 15% crude protein (CP) in diet or 60% of concentrate. CP and NDF content of diets have a quadratic effect on organic matter digestibility. A quadratic increase of ADG is observed with increasing Metabolic Energy intake (MEI) and there was significant difference between species and genotype when MEI and ADG were expressed on LW-1. However, the difference was significant between small ruminant and cattle when MEI and ADG were expressed according to metabolic live weight (LW0.75).

Differences in feeding in stalls or at pasture may be linked to differences in feeding strategies

A. Agastin[1], D. Sauvant[2], M. Naves[1] and M. Boval[1]
[1]INRA, Unite de Recherches Zootechniques, Domaine Duclos, 97180 Petit-Bourg, Guadeloupe, [2]INRA, AgroParisTech UMR791, Cedex 05, 75231 Paris, France; aurelie.agastin@antilles.inra.fr

A meta-analysis including 112 publications (400 experiments) was conducted: (1) to provide an overview of the effects of 2 feeding environments (stalls vs. pasture) on the growth of cattle sheep and goat; (2) to determine if various feeding strategies could interfere with the effect of these feeding environments. Selected papers contained data on average daily gains (ADG g/kg LW) feeding strategies and diet characteristics: roughage source (green, not green or mixed), given *ad libitum* or not, addition of legume, addition of concentrate, concentrate main ingredient (barley, maize or other), level of complementation and concentrate given *ad libitum* or not. All these aspects were encoded. Relationships between the dependent variable and explanatory variables were studied with variance-covariance analyses, using the GLM procedure. First, we tested the feeding environment influences on ADG. Second, the effect of other feeding strategies (or diet characteristics), was additionally considered and added as co-variable. Feeding environments affected ADG, animals fed in stalls having a greater ADG than those fed at pasture (4.87 vs. 4.07 g/d, n=243, RMSE=1.3, $P<0.001$) and there was an effect of animal species on ADG ($P<0.001$). By considering if concentrate was added or not in the diet, the difference in ADG was attenuated (4.57 vs. 4.10 g/d, n=243, RMSE=1.3, P=0.051). By considering the percentage of concentrate in the diet (52.3±29.62%) when documented, feeding environment had no more effect on the ADG (P=0.991). The intra-experiment ADG response to %CO was ADG=0.021 %CO + 1.65 (n=24, R^2=96.5, RMSE=0.25). No more effect of feeding environment was observed when considering roughage source (P=0.500). Differences between feeding in stalls or at pasture appeared as mainly driven by complementation strategies. These results are encouraging for a better view of feeding at pasture.

Session 07 Theatre 11

Foraging behavior of Creole pigs kept outdoor under tropical conditions on sweet potatoes field

A. Burel, H. Archimède, M. Mahieu, A. Fanchone and J.-L. Gourdine
INRA, UR143, Duclos, 97170 Petit Bourg, Guadeloupe; harry.archimede@antilles.inra.fr

The trial was carried out to evaluate foraging behavior of Creole pigs kept outdoors under tropical conditions (Guadeloupe, Lat. 16.°N, Long. 61°W), on a sweet potato field. A total of 6 Creole pigs with initial weight of 31.0±2.4 kg and backfat thickness (P2 position) of 10.9±1.4 mm, were reared outdoors for 42 days (8 days of adaptation) with access to sweet potato leaves, stems and tubers) in a total area of 1,613 m^2, with an average free access of 11.20 m^2/animal/day. Based on growing pig nutrient requirements and on estimated potato leaf, stem and tuber biomass, it was estimated that the available feed was theoretically not limiting factor to reach moderate growth rate (150 g/d). Behavior of the pigs was studied during 12 h between 6 a.m. and 6 p.m. on two occasions at day 8 and day 22 after the adaptation stage. Pigs spent 44% of the time for resting and sleeping, 38% for eating and drinking and 18% for other activities. The 4.5 average hours of eating from 6 a.m. to 6 p.m. were divided in 45% of time for eating leaves and stems, 26% for potato tubers and 24% for other forages and 5% for drinking. During the 12 continuous hours of observations, the average distance traveled by pig was 380±40 m. There is a strong relationship between the hourly ambient temperature and feeding behavior (r=- 0.65) with higher feeding activities during the fresher periods of the day. The study is a preliminary step to evaluate the conditions to implement low input pig production systems in a context of mixed farming system. Further studies are needed to evaluate the growing performance and the economic gain, with the aim of developing a new niche market with local pigs reared outdoors.

Session 07 Theatre 12

Relative bioavailability of tropical volcanic soil-bound chlordecone in farm animals

C. Jondreville[1], S. Jurjanz[1], A. Fournier[1], S. Lerch[1], M. Lesueur-Jannoyer[2], H. Archimede[3], M. Mahieu[3], C. Feidt[1] and G. Rychen[1]
[1]Université de Lorraine, ENSAIA-UR AFPA, 2 Av. de la forêt de Haye, TSA 40602, 54518 Vandoeuvre cedex, France, [2]CIRAD, Le lamentin, Martinique, France, [3]INRA, Petit Bourg, Guadeloupe, France; guido.rychen@univ-lorraine.fr

The former use of chlordecone (CLD), a chlorinated polycyclic ketone pesticide, in French West Indies to fight against banana black weevil, has resulted in long-term pollution of soils. CLD may be transferred to animals through involuntary polluted soil ingestion. However, due to different properties of clays, tropical volcanic soils display variable capacities of pollutant retention: CLD is more persistent in andosol than in nitisol. The impact of soil type on CLD bioavailability has been assessed via relative bioavailability (RBA) studies in three farm animal species (laying hens, piglets and lambs). Thus, the response of CLD ingestion through andosol and nitisol was compared to the response obtained with CLD ingestion through oil, taken as a reference matrix. Our hypotheses were that: (1) CLD would be less available in soils than in oil; (2) CLD would be less available in andosol than in nitisol; and (3) RBA in soils may differ between animal species. The deposition of CLD in egg yolk (hens), in liver (piglets) and in serum (lambs) was measured in individually housed animals fed graded levels of CLD from polluted andosol, nitisol or spiked oil. Hens, piglets and lambs were exposed to CLD during 28, 14 and 15 days, respectively. For each animal species, the concentration of CLD in target tissue linearly increased with the amount of ingested CLD within each ingested matrix ($P<0.001$). However, the responses to andosol-diets, nitisol-diets and oil-diets could not be differentiated ($P>0.1$), indicating that CLD was equally bioavailable, irrespective of the matrix. These results demonstrate that: (1) soil does not modulate CLD availability; and (2) ingestion of polluted soils by farm animals contributes to farm animal contamination.

Session 07 — Theatre 13

Sustainable intensification of smallholder livestock production: fact and fiction

H. Udo
Wageningen University, Animal Production Systems group, P.O. Box 338, 6700 AH Wageningen, the Netherlands; henk.udo@wur.nl

Intensification of smallholder livestock production is widely advocated to contribute to improving livelihoods of rural households, to meet increasing demands for livestock products, and to reduce environmental impacts. Can smallholders live up to these expectations? This presentation discusses the impact of intensification using village poultry, integrated agriculture-aquaculture, small ruminant, and dairy case-studies. As livestock systems intensify, the relative importance of the various tangible and intangible functions of livestock changes. Smallholder dairying proved to be a good means to increase household incomes. A paradox is that poultry, pigs or small ruminants better fit the farming conditions of the poorest, but their contribution to household incomes remains relatively small. A major trade-off of livestock intensification is that only part of the smallholder farmers, particularly the better-off farmers, is able to take advantage of the increased demands for livestock products. A dual structure is developing with large-scale industrial pig and poultry systems producing for emerging urban markets and smallholder systems producing for local markets. Smallholder dairy cattle systems are competitive; they also supply the urban markets. In discussions about the future of smallholder livestock, it is often claimed that intensification is needed not only to meet increasing demands, but also to reduce environmental impacts per unit product. Our case-studies indicated that the relation between livestock and the environment is complex. The off-farm impacts of producing 'better diets' for intensifying livestock production need to be considered, together with impact allocation to the multiple functions of smallholder livestock. Can we expect smallholders to change their feeding practices in response to the global debate on greenhouse gas emissions? Innovations for sustainable smallholder livestock production will only be adopted if they fit farming household priorities and resources.

Session 07 — Theatre 14

Research issues for crop-livestock integration in mixed farming systems in the tropics: a review

F. Stark[1,2], H. Archimede[3] and C.H. Moulin[4,5]
[1]AgroParisTech, Centre de Montpellier, Montpellier, France, [2]CIRAD, UMR SELMET, Montpellier, France, [3]INRA, URZ, Guadeloupe, France, [4]INRA, UMR SELMET, Montpellier, France, [5]Montpellier Supagro, UMR SELMET, Montpellier, France; fabien.stark@supagro.inra.fr

Agriculture has to product more and better in a more constraint and changing world. In the tropics, agriculture is represented by a majority of familial small-scale mixed farming systems (MFS), which represent half of the world food production. These MFS could be characterised by agrobiodiversity and crop-livestock integration (CLI). Indeed, it is commonly accepted that agrobiodiversity – diversity of production activities on the same farm – is a pre-required for risk management and biodiversity. Moreover, CLI – agroecological complex integrated management between crop and livestock production – seems to be determinant too. By improving resiliency and efficiency of the whole farming system, CLI appears to be able to deal with the need for ecological intensification of agriculture, to produce more with less. Through a review of scientific literature, we identify research approaches concerning MFS and the key role of CLI for a sustainable development. Many studies deal with CLI and MFS, through strategic and methodological approaches, descriptive analysis or analytical case studies. However, few studies concerns systemic analysis of CLI and MFS. Indeed, studies concern the theoretically benefits of CLI integration, the systemic analysis of a specific CLI practice or the inventory of CLI in a geographic area. The fundamental link between farms structures, the potential of integration, the effective integrated functioning and the performances of the system, stay partially treated in the literature. All these elements bring us to revise research posture to understand the key role played by CLI to meet new agricultural aims. A framework to guide future researches is finally proposed, through a comparative analysis of CLI in three territories of the humid tropics (Guadeloupe, Cuba, Brazil), characterised by contrasted socioeconomic situations.

Session 07 Theatre 15

Supporting crop-livestock farmers in redesigning their production systems: the CLIFS approach

P.Y. Le Gal[1], N. Andrieu[1], N. Cialdella[1], P. Dugué[1], E. Penot[1], C.H. Moulin[2,3], C. Monteil[4], F. Douhard[5] and J. Ryschawy[6]
[1]CIRAD, UMR Innovation, 34398 Montpellier, France, [2]INRA, UMR Selmet, 2 Place Viala, 34060 Montpellier, France, [3]Montpellier SupAgro, UMR Selmet, 2 Place Viala, 34060 Montpellier, France, [4]Université de Toulouse, INPT-ENSAT, UMR Dynafor, 31326 Castanet Tolosan, France, [5]INRA, UMR Sad-Apt, 75231 Paris, France, [6]INRA, UMR Dynafor, 31326 Castanet Tolosan, France; pierre-yves.le_gal@cirad.fr

Modeling is often used by researchers to evaluate innovative production systems according to their own views and agendas. Based on a whole farm spreadsheet simulation tool called CLIFS (Crop Livestock Farm Simulator) our approach aims to support farmers in reconsidering their production strategies by evaluating ex-ante a range of alternatives in line with their objectives. Initial, Project, and Alternative Scenarios are successively designed and evaluated in a face-to-face interaction with the farmer. Each scenario corresponds to a system configuration, for which CLIFS calculates staple food, forage and organic manure balances, crop and livestock productions, and economic results according to the characteristics of crops (cropping pattern, technical sequence and yield per crop), herd (size, reproductive strategy and feeding practices per batch) and prices of inputs and outputs. The approach has been tested with small samples of farmers (2 to 15) in a range of French and Tropical contexts and of production systems. Simulated projects cover topics such as increasing milk production, improving feed autonomy, adopting conservation agriculture techniques, introducing new animal units, evaluating sensitivity of farm performances to climatic and economic shocks. Farmers appreciate the closeness of the simulated scenarios with their own context, the holistic approach of their farm and the reevaluation of their projects. Next step will consist in adapting this approach to advisers' working context and using it as a training tool with groups of farmers, professionals and students.

Session 07 Theatre 16

Use of agroecological concepts to design a crop-livestock system adapted to the French West Indies

A. Fanchone[1], F. Stark[1], G. Alexandre[1], J.L. Diman[2], J.L. Gourdine[1], H. Ozier-Lafontaine[2], J. Sierra[2], R. Tournebize[2] and H. Archimede[1]
[1]INRA UR 143 URZ, Petit-Bourg, 97170, Guadeloupe, [2]INRA UR 1321 ASTRO, Petit-Bourg, 97170, Guadeloupe; audrey.fanchone@antilles.inra.fr

Agroecology has been defined among others, as a set of practices that contribute to a more eco-friendly or sustainable agriculture). The general concepts of agroecology have been stated in several practices that contribute to increase sustainability of the farm. In the French West Indies (FWI), most of farms are integrated crop-livestock systems (ICLS). They are characterized by a high diversity. However, farmers integrate of few agroecological concepts in a conventional system. But, the specificity of agroecology came from the articulation of several agrecological concepts within the same farm. Some barriers appear to the adoption of such complex systems by farmers. A prototype of ICLS of 5 hectares has been designed using several agroecological concepts and set up in an experimental farm for demonstration and creation of technical, economical and environmental references. In this system, improvement of biodiversity has been achieved by using diversification of both animal (ruminant and monogastric) and vegetal production (export crops ie. sugar cane and banana, food crops ie. sweet potato and cassava, cover crops ie. legumes and vegetables). Moreover among the same crop several varieties have been used. Improvement of nutrient and energy cycling and concomitant decrease of the use of synthetic inputs has been achieved through maximization of integration between crop and animals by using: (1) dual purpose crops where production is sold and residues is used to feed animals; and (2) animal and green manure to fertilize crops. Improvement of soil quality was achieved through light tillage and use of restructuring crops as sugar cane and legume. This study aimed at illustrating how agroecological concepts contribute to design functional ICLS in the FWI.

Session 07 Theatre 17

Diagnosing constraints to market participation of small ruminant producers in Northern Ghana

K. Amankwah[1], L. Klerkx[1] and S.J. Oosting[2]

[1]Wageningen University, Knowledge, Technology and Innovation Group, P.O. Box 8130, 6700 EW Wageningen, the Netherlands, [2]Wageningen University, Animal Production Systems Group, P.O. Box 338, 6700 AH Wageningen, the Netherlands; kwadwo.amankwah@wur.nl

This study assesses why participation in markets for small ruminants is relatively low in northern Ghana. Specific objectives of this study were: (1) to describe the prevailing practices of small ruminant production and marketing in smallholder households; and (2) analyse the farm level and above higher (or institutional) level constraints that hinder smallholder small ruminant innovation in improving production and market participation. Qualitative case study design was employed. Upper West Region of Ghana was selected as the study area in a bigger programme to study livestock and food security. Methods employed included systematic sampling and interview of 53 compound household heads in five communities in Lawra and Nadowli Districts; interview of supply chain actors; and one-day multi-stakeholder workshops. The main constraints experienced by the smallholders are water shortages during the dry season, high ruminant mortality rates, and the theft of small ruminants. The constraints persist because of institutional and structural factors interacting at a range of levels and they block further developments by the majority of the smallholders. The study indicates that in the harsh conditions in which they live the smallholders seek resilience through diversifying their sources of livelihood, by low input investment in small ruminant production, and by keeping their animals as a capital stock and insurance. However, a few farmers ('positive deviants') have developed novel practices that enable them to overcome some of the constraints and to engage in market-oriented production of small ruminants. These novelties could provide the basis for diverse development pathways that open up a range of possibilities beyond purely market-led or purely technology-led change.

Session 07 Theatre 18

How do pastoralists arbitrate between accumulation and sales of animals: a case study in Senegal

C. Manoli[1], C. Corniaux[2], B. Dedieu[3], A. Ickowicz[2] and C.H. Moulin[2]

[1]ESA, URSE, 55, r. Rabelais, 49007 Angers, France, [2]SELMET, pl Viala, 34060 Montpellier, France, [3]INRA, Theix, 63122 St-Genes-Champanelle, France; c.manoli@groupe-esa.com

Pastoral herds are crucial for pastoral livelihood strategies in Sahel: sales of animals satisfy various familial cash needs, herds are accumulated as they are a rare possible saving account in pastoral areas. Therefore, decision making of sales in herd management is a complex process. Its understanding is a challenge related to both livestock contribution to growing markets and pastoral vulnerability. Our goal is to analyze how herders choose between sales and accumulation of cattle and sheep. We used a case study approach, within a stratified sample of ten pastoral settlements of Ferlo (Senegal), whose mixed herds of cattle, sheep, goats were monitored for one year. For analysis, we used numerical indicators, i.e herd numerical productivity, herd annual growth, saving rate, added value of each species, and also farmers practices recording and explanation. The system of activities, the settlement dynamics were also recorded as explanation variables. Results show that settlements were not always able to save animals, even in case of good climatic conditions. Both cattle and sheep contributed largely to herder's income, sheep generally contributing more than cattle. Sheep herds were all characterized by high off-take rates resulting in annual decrease of herd size, whereas some cattle herds were accumulated: herd growth was around 10% for 4 cases. Then, a detailed analysis shows different ways in arbitrating sales between species. Differences were related to the system of activities, familial trajectories and events occurred during the study year. These results weaken a classical representation of pastoral mixed herd management used in some pastoral studies where herd accumulation is considered as a dominant trait. It allows also a better understanding of decision making and existing rooms for manoeuvre of pastoralists between saving and selling sheep and cattle.

Strategic maize supplementation in grazing goats does not increases the birth weight of the kids

E.S. Mendieta[1], J.A. Bustamante[1], J.L. Covarruvias[1], L.M. Cedillo[1], J. Loya-Carrera[1], S. Ramírez[1], J.A. Flores[1], J. Vielma[1], G. Duarte[1], G. Fitz-Rodríguez[1], I.G. Fernández[1], M. Bedos[1], A. Terrazas[2], J.A. Delgadillo[1] and H. Hernandez[1]

[1]UAAAN, CIRCA, Periférico Raúl López Sánchez y Carretera a Santa Fe, 27054, Torreón, Coahuila, Mexico, [2]FESC-UNAM, Depto. de Ciencias Pecuarias, Km. 2.5 Cuautitlán-Teoloyucan, 54714, San Sebastián, Xhala, Mexico; estivemendieta@hotmail.com

The placental weight affects birth weight in ruminants. In sheep, the major placental weight occurs between 60 and 90 days of gestation. We assessed if inclusion of maize in the diet of semi-arid naturally grazing goats from 75 to 90 days of gestation and during late gestation could increase the kid's birth weight. Thirty-six multiparous local goats were allocated to one of the 3 groups: control group, does were only fed with the available natural vegetation in grazing areas (CG, n=12); mid and late gestation supplemented group, in addition to daily grazing each doe received 0.6 kg of maize/day from day 75 to 90 and again during last 15 days of gestation (MLG, n=12); late gestation supplemented group, each doe received the supplementation during the last month of gestation (LG, n=12). In all groups, 9 and 3 mothers gave birth to twin and single kids respectively. The kid's weight was recorded at birth. Placental weight and cotyledons number (in a 20×20 cm area) were recorded after whole placental expulsion. Data were analyzed using a one-way ANOVA. The birth weight of kids was not different (P>0.05) between groups (CG: 3.1±0.09 kg; MLG: 3.4±0.1 kg and LG: 3.1±0.08 kg). Also, the placental weight did not differ (P>0.05) between groups (CG: 0.6±0.04 kg; MLG: 0.7±0.07 kg and LG: 0.7±0.04 kg). Finally, the cotyledons number also was not different (P>0.05) between groups (5.3±0.3, 6.3±0.3 and 5.5±0.3 in CG, MLG and LG, respectively). We concluded that in subtropical does gave birth in summer, a short-term inclusion of maize in the grazing diet during mid and late gestation did not affect neither placental nor kids birth weight.

Impacts of a dual market on the relationship between dairy farmers and processors in Peruvian andes

E. Fuentes[1], G. Faure[1], E. Cortijo[2], E. De Nys[3], J. Bogue[4], C. Gómez[2], W. Mercado[2], C. Gamboa[2] and P.-Y. Le Gal[1]

[1]CIRAD, UMR Innovation, 34398 Montpellier, France, [2]Universidad Nacional Agraria léa Molina, Apartado 12-056, Lima, Peru, [3]World Bank, 1818 H Street, Nw., Washington DC, USA, [4]University college cork, Western Road, Cork City, Ireland; eduardo.fuentes_navarro@cirad.fr

Dairy supply chains in emerging countries are influenced by the prevalent participation of smallholder farmers, for whom milk production is considered a valuable source of income. This study investigates how a dual supply chain, operating within the same region through formal and informal markets, impacts on the interaction between smallholder farmers and processors at organizational level. Questionnaires and interviews were administered to gather information from farmers and processors in the Mantaro Valley between 2009 and 2012. The research reveals that two supply chains co-exist supplying consumers locally and also in the capital Lima. Formal markets mostly consist of supermarkets, which demand consistently high quality and safe products. Informal markets target consumers with lower purchasing power and focused on traditional cheeses. It is characterized by an absence of quality control throughout the chain. These two markets are supplied respectively by local industrial dairies and artisanal cheese-makers. Milk flows are observed between them, based on a mix of competition and cooperation depending on the time of the year and the level of consumer demand. Farmers frequently change processors, according to the milk price and services offered, as there is a lack of contracts between stakeholders along the chain. This supply chain organization is flexible enough to provide a broad range of dairy products and involve a diversity of stakeholders along the chain. However this favors opportunistic behavior, such as milk adulteration or non-payment of milk supplied. And it depends on the poor application of government quality regulations, which could disturb the informal sector.

Stability index of marandu grass pasture under different grazing intensities

S.S. Santana[1], L.F. Brito[1], P.M. França[1], M.E.R. Santos[2], A.C. Ruggieri[1] and R.A. Reis[1]
[1]Sao Paulo State University, Departament of Animal Science, Via de Acesso Prof. Paulo Donato Castellane s/n, 14884-900 Jaboticabal, SP, Brazil, [2]Federal University of Uberlandia, Departament of Animal Science, Av. João Naves de Ávila, 2121 Santa Mônica, 38.408-144 Uberlândia, MG, Brazil; sabrinazootec@yahoo.com.br

The effect of grazing intensity on the tillering of Brachiaria brizantha cv. Marandu under continuous stocking was evaluated at FCAV/UNESP, Brazil, from 12/ 2010 to 05/ 2011. Three grazing intensities were defined by sward heights of 15, 25 and 35 cm and they were maintained by Nellore young bull grazing under continuous stocking. The treatments were distributed in a completely randomized design with four repetitions (paddocks). Three areas (0.0625 m^2 each) were delimited in each experimental unit, representing the average initial condition of the pasture. At the beginning of the evaluation, all tillers contained in each frame were counted and marked with colored plastic coated wire. Each 30 days, all tillers were again counted and the new tillers were marked with different colored plastic coated wire. Tiller appearance rate (TA), tiller survival rate (TS) and stability index (SI) were calculated. The data obtained in early (ES) and late summer (LS) were compared by Tukey test (10%). The TA was higher in FV than in IV with average of 50.3% and 30.4%, respectively, which result is associated with the suitable climatic conditions, production and growth of Marandu grass. The TS was higher in LS (90.1%) and lower in the ES (80%), which was expected, because the tiller mortality rate was lower in LS (5.4%), which reflects a compensatory mechanism, with the objective of compensate a reduction in the TA. The SI was not influenced by grazing intensity and climate, with an average of 1.2. The pasture of Brachiaria brizantha cv. Marandu has a larger TA than TS during the same period, indicating stability of plant population, in all grazing intensities.

The characterization and sustainability of cattle production in the communal areas of Namibia

I.B. Groenewald[1], G.N. Hangara[2] and M.Y. Teweldemedhin[3]
[1]Centre for Sustainable Agriculture, University of the Free State, P O Box 339, 9300 Bloemfontein, South Africa, [2]Namibia Development Agency, 80 Dr F Indogo Street, Windhoek, Namibia, [3]Department of Agriculture, Polythecnic of Namibia, Private Bag 13388, Windhoek, Namibia; groenei@ufs.ac.za

The objective of this study is to characterize the cattle production systems followed in the research area and identify those factors which impair the sustainability thereof. Experimental methods employed included rapid rural appraisal techniques, questionnaires and structured interviews. A total of 570 farmers formed part of this study. These farmers are all registered as cattle producers with the Meat Board of Namibia. Trained enumerators were deployed to gather the research data under the supervision of the researcher (second author) over a period of two months. Data were analyzed using the SPSS. Results indicate that the average herd size per household is 66 animals. Steers are sold prior to maturity resulting in the breeding stock accounting for 74.3% of the total flock. The low calving percentage of 31.5% is brought about by an incorrect bull-to-cow ratio of 1:49. Where neighboring commercial farmers apply a 1:25 bull-to-cow ratio, their calving percentage is at least 75%. Poor weaning practices, lack of record keeping, droughts, diseases, straying and theft are compounding the sustainability of livestock production in that area. Contributing to the low viability of the production systems, is a total lack of proper support to the farmers. Respectively, a total of 73.5% and 63.5% of the respondents have not been visited by an extension nor a veterinary officer. The study conclude that proper and comprehensive training and access to credit will improve the livelihoods of these rural livestock producers.

Factors which hinder an effective supply chain for cattle producers in the communal areas of Namibia

I.B. Groenewald[1], G.N. Hangara[2] and M.Y. Teweldemedhin[3]
[1]Centre for Sustainable Agriculture, University of the Free State, P O Box 339, 9300 Bloemfontein, South Africa, [2]Namibia Development Agency, 80 Dr F Indogo Street, Windhoek, Namibia, [3]Department of Agriculture, Polythecnic of Namibia, Private Bag 13388, Windhoek, Namibia; groenei@ufs.ac.za

The objective of this study is to identify those factors which hinder an effective supply chain for cattle producers in the research area. The study further focussed on market information; the accessibility thereto and the management thereof. A total of 100 farmers, eight key informants of farmer's associations and co-operatives, an auctioneer and beef processor were interviewed in this study. Only farmers were selected who were selling livestock four or more times per year. Experimental methods used were comprehensive questionnaires and structured interviews. Trained enumerators were deployed to gather the research data which were analysed using the SPSS. Results indicate that 82% of the income derived from cattle sold were used for human food and basic needs, 16% were used for occasional needs i.e. funeral and only 2% was used for production inputs towards improving cattle production. This low investment in production cost might be a result of ignorance (63%) of the producers towards the quality criteria of the buyers. This resulted in 66% of the farmers not being happy with existing marketing systems. Producers claim ignorance of quality standards whilst auctioneers and processors claim the opposite, creating confusion. When asked which method of transferring marketing information is most appropriate, 93% of the respondents mentioned the radio, followed by farmer's associations and then the cellular phone. This study concludes the training of farmers in product quality, marketing plans and making optimally use of available information resources through farmer's associations and co-operatives. Access to more formal markets should be addressed institutionally through better means of transport, roads and auction facilities.

Session 07 Poster 24

Improving indonesia's beef sector, how is policy implemented on farmer family level

S. Gayatri[1], M. Vaarst[1], M. Eilenberg[1] and B. Guntoro[2]
[1]Aarhus University, Department of Animal Science, P.O. Box 50, 8830 Tjele, Denmark, [2]Gadjah Mada University, Department of Animal Science, Bulaksumur, Yogyakarta 55281, Indonesia; siwi.gayatri@agrsci.dk

Indonesian beef cattle farmers face difficulties in increasing the quantity and quality of their products. Indonesia was only able to supply around 50% of the nation's demand for meat. The government has launched a beef self-sufficiency program in 2005. The aim was to reduce imports of beef cattle to 10% of total demand by 2014. The aims of this study was to understand how is policy implemented and to know the role of different actors in Indonesia's beef cattle farming sector and to identify relevant theoretical political frameworks for understanding the dynamics of this production and how it can be more sustainable. The research used ethnographic methods and theories of political sciences and Actor Network Theory to understand the role of different actors and to analyze institutional activity and behaviour of the actors regarding the Beef Self Sufficiency Program in Indonesia. Quantitative and qualitative analysis was used to analyse data. The study indicate that the farmers must be reached out and should be made aware of the policies on beef cattle farming. All stakeholders must be encouraged to participate and must act together to be able to maximize resources and efforts in bringing about the necessary changes. Moreover the local government should improve the level of participation, not only of the community but other agencies as well as, in the development of programs on beef cattle development policy through people motivation and follow-ups. Local leaders should be mobilized to help in maintaining the sustainability of the beef development program. The government agents should be more careful in their contacts with rural communities and should avoid activities that can reduce the level of trust.

Session 07 Poster 25

Effects of experimental infection and diet supplementation on meat Creole goat performances

W. Ceï, M. Mahieu, H. Archimède, J.C. Bambou, A. Hiol and G. Alexandre
INRA, Animal Genetic UR 143, Duclos, 97170 Petit-Bourg, Guadeloupe; gisele.alexandre@antilles.inra.fr

In the Tropics the major constraint for goat production is gastrointestinal nematodes (GIN) infection. One promising alternative to chemotherapy is the improvement of host nutrition yet the underlying mechanisms remain unknown. The aim of the study was to assess the effects of GIN infection and supplementation on packed cell volume (PCV), average daily gain (ADG) and carcass quality of growing Creole kids. Sixty male goats were reared indoors following a 2×3 factorial design. The factors were the experimental infection levels IE (I: infected and NI: non-infected) and the diets D (G: grass only diet; B: grass plus dried banana and C: grass plus concentrate). Fecal egg counts did not vary among I groups (on average 2,200 ω/g). The PCV and ADG were improved (P<0.001) for NI animals vs. I ones. There was a D effect (P<0.001) and no I×D interaction was observed. There was no significant effect of EI upon the main carcass data, except liver and reticulorumen weights that increased slightly in I compared to NI goats (P<0.05). Same trend was observed for the breast proportion. The absolute values of abdominal fat (related to EBW), meat redness and water losses appear to be affected by IE levels (P>0.05). All carcass data increased significantly with the addition of supplement in the diet (P<0,001) except for carcass cut proportions. Obviously, the C groups performed better than the two others (whatever their EI levels). From one extreme group (GI) to another (CNI) there was an increase of 10.5% of carcass yield (P<0.01). Meat physical parameters are damaged when the I kids received the B diet with higher lightness and water loss than in the G and C groups. Given that the B diet contained less nitrogen (N) than the G one and that the GIN stress affect the animal N metabolism, it could be hypothesized that the IB kids may have suffered from a lack of N. Further studies are required to assess the N nutrition×parasitism interactions upon physiological features and carcass quality of goat.

Session 07 Poster 26

Contamination of grazing ducks by chlordecone in Martinique

C. Jondreville[1], A. Lavigne[2], F. Clostre[3], S. Jurjanz[1] and M. Lesueur-Jannoyer[3]
[1]Université de Lorraine, UR AFPA-INRA, TSA 40602, 54518 Vandoeuvre, France, [2]Fredon, Croix Rivail, 97224 Ducos, France, [3]Cirad, HortSys, PRAM, 97285 Le Lamentin, France; stefan.jurjanz@univ-lorraine.fr

Chlordecone (CLD) is an organochlorine pesticide used from 1971 to 1993 in French West Indies to fight against banana black weevil. The former application of this insecticide has resulted in long-term pollution of soils and in subsequent contamination of local waters and food resources. In this area, grazing poultry may be used to control weeds in orchards as an alternative to herbicides. However, CLD may be transferred into animal products through polluted soil ingestion. The question arises whether these grazing ducks may be consumed by the farmers, i.e. whether the concentration of CLD complies with the legal 20 µg/kg maximum residue limit (MRL). Muscovy ducks were raised on a guava orchard planted on a soil moderately contaminated (0.4 mg CLD/kg DM). Ducks were raised indoor up to 6 weeks of age and allowed to graze thereafter. Thirty-two females were sequentially slaughtered by groups of 3 or 4 ducks, either after a 4-, 16-, 19-, 22- or 26-week grazing period or after a 16-week grazing period followed by a 3-, 6- or 9-week decontamination period. During decontamination, ducks were raised indoor without any contact with the environment. After 4 weeks grazing, CLD concentrations in liver and in meat (leg with skin) were 258 and 60 µg/kg, respectively, already far over the MRL. These values increased with time to reach 1051 and 169 µg/kg, respectively, after 22-week grazing and remained steady thereafter. Ducks kept indoor gradually decontaminated with concentrations reduced down to 47 and 6 µg/kg, respectively, after 9 weeks. According to this decontamination curve, the time required for decontamination down to the MRL was estimated at 83 and 41 days in liver and in meat, respectively. Thus, the consumption of products from these ducks, especially liver, should be avoided, unless grazing and decontamination practices are improved.

Faecal near infrared spectroscopy to assess diet quality in tropical and temperate grassland

V. Decruyenaere[1], M. Boval[2], S. Giger-Reverdin[3], J.A. Fernández Pierna[1] and P. Dardenne[1]
[1]Walloon Agricultural Research Centre (CRAW), 9 rue de Liroux, 5030 Gembloux, Belgium, [2]INRA Centre West-Indies, Animal Production Unit (URZ), Domaine Duclos, 97170 Guadeloupe, France, [3]INRA-AgroParisTech, UMR 791 MoSAR, 75005 Paris, France; decruyenaere@cra.wallonie.be

Near infrared reflectance spectroscopy (NIRS) is a rapid technology currently used for predicting the forage chemical composition. It now appears possible to estimate *in vivo* organic matter digestibility (OMD) and dry matter voluntary intake (DMVI) using NIRS applied to faeces (F.NIRS). In this study, the F.NIRS potential is demonstrated by building a large faecal database in terms of forage (temperate, tropical) and animal (sheep, goat, cattle) for predicting OMD and DMVI at grazing. Both OMD and DMVI F.NIRS databases needs to generate diet-faecal pairs. So, OMD and DMVI reference values are obtained during digestibility trials. The final database contained 2214 faecal spectra (OMD range=0.303-0.849; DMVI range=24.0-163.0 g/kg metabolic weight (MW)) including 931 from tropical origin. Calibration equations have been built using two chemometric methods, a Modified Partial Least Squares (MPLS) procedure with cross validation and the Support Vector Machines (SVM) technique. MPLS model gives good precision for OMD (R^2>0.8 and standard error of cross-validation (SECV or prediction precision) close to 0.03 digestibility units, SECV/mean<4%). DMVI appears more difficult to predict as the MPLS model is poor (R^2<0.8, SECV>12 g/kg MW or more than 18% of the mean), probably due to a slightly non-linearity. For this reason, the SVM technique has been applied with as result an improvement of the precision of OMD and DMVI prediction models. With SVM technique, R^2 is higher than 0.85 for both parameters. The improvement is for DMVI, where the SECV became of 8.6 g/kg MW (13% of the mean). Based on these results, if appropriate chemometric equations are developed, it seems possible to use temperate and tropical merged F.NIRS databases for predicting *in vivo* diet characteristics.

Session 07 Poster 28

Ex ante evaluation of several scenarios of crop-livestock systems of Guadeloupe using IMPACT®

A. Fanchone[1], F. Stark[2], J.L. Kelemen[2], J.L. Diman[1,3], H. Archimede[1] and J.L. Gourdine[1]
[1]INRA UR 143 URZ, Petit-Bourg, 97170, Guadeloupe, [2]EPLEFPA, Baie-Mahault, 97122, Guadeloupe, [3]INRA UR1321 ATRO, Petit-Bourg, 97170, Guadeloupe; Audrey.Fanchone@antilles.inra.fr

To evaluate the impact of integration between crop and livestock on the sustainability of the system, five scenarios have been evaluated using the IMPACT®. In scenario 1 (S1), area of the system was 5 ha including 2.5 ha of sugar cane, 1.5 ha of banana, 0.5 ha of sweet potato and 0.5 ha of cassava. Crop residues produced allowed to fatten 5 cattle and 3 cohort of 10 pigs. In scenarios 2 (S2a and S2b), legumes were intercropped in crops of S1 to decrease need for weeding and fertilizing. In S2a, these legumes were kept in the field for green manure. In S2b, they were cut to feed animals inducing a diminution of the need for soybean meal as in S1. Scenarios 3 (S3a and S3b) were specialized in animal production. In S3a, the same animal number than in S1 was used, but they received only external feed (hay, and/or industrial diets). In S3b, the same land area than S1 was retained to feed 22 Creole cattle, whereas the 30 pigs were fed an industrial diet. IMPACT allowed to evaluate yearly, the economic and environmental performances of these scenarios. This ex ante evaluation showed that S1, S2a and S2b required on average 85% less chemical inputs (phytosanitary products and fertilizers) than S3a and S3b. However, S3a and S3b required on average 74% less work time than S1, S2a, and S2b, due to time needed to harvest and process crop residues to feed animals. To an economic point of view, S3a and S3b were economically more interesting than S1, S2a, and S2b, mainly due to the cost of work force. To an environmental point of view, nitrogen balance of S1, S2a and S2b was 6 time higher than that of S3a and S3b. IMPACT® did not take into account lixiviation of N in the soil. This study shows that innovation have to be produced in integrated systems to decrease the high need for labor linked to animal feeding.

Session 07 Poster 29

Milk composition and energy standardization of Arabian camel milk

R.S. Aljumaah[1], M. Ayadi[1], M.A. Alshaikh[1], R. Casals[2] and G. Caja[2]
[1]College of Food & Agriculture Sci., King Saud University, 11451 Riyadh, Saudi Arabia, [2]Group Ruminant Research (G2R), Universitat Autònoma de Barcelona, 08193 Bellaterra, Spain; alshaikh@ksu.edu.sa

A total of 108 dairy camels (*Camelus dromedarius* L.) of 4 Arabian breeds (Majahim, n=58; Maghatir, n=49; Shu'l, n=39; Sufer, n=34) were used throughout lactation (29 to 372 d in milk). Milk samples (n=720) were collected by hand at the morning milking. Milk components were analized using a Lacto Star (Funke-Gerber, Labortechnik, Berlin, Germany) calibrated for camel milk. Minerals were analyzed from white ashes (550 °C) by atomic absorption spectrometry (Analyst Spectrophotometer 300, Perkin-Elmer, Shelton, Connecticut, USA). A subset of 225 samples was freeze-dried and energy analyzed by adiabatic calorimetry (IKA calorimeter, Janke & Hunkel, Heitersheim, Germany). Simple and multiple correlations were calculated by the REG procedure of SAS v.9.1. Milk composition varied widely: fat (2.94±0.03%, 1.35 to 5.85), protein (3.45±0.01%, 2.45 to 4.40), lactose (4.98±0.02%, 3.56 to 5.99), ash (0.74±0.01%, 0.59 to 0.95), Ca (9.03±0.07 mg/l, 5.01 to 13.03), Na (3.57±0.04 mg/l, 1.08 to 8.01) and K (8.72±0.11 mg/l, 3.06 to 19.41). On average 79% milk samples showed inverted fat and protein ratio as a consequence of incomplete milkletdown. Correlations between components were low for overall data (r=0.21 to 0.84) but improved in the milk sample subset (fat vs. protein, r=0.39; lactose vs. Na, r=– 0.18; Na vs. K, r=0.61; fat vs. solids, r=0.82; protein vs. solids, r=0.84). Milk K:Na ratio (2.53±0.02) indicated a displaced equilibrium when compared to cow's milk. Milk energy was 626±6 kcal/kg (403 to 890). Equations of milk energy obtained by regression analysis from the measured (calorimeter; r^2=0.73) and the estimated (Perrin; r^2=0.89) data showed underestimation at the intercept. Proposed fat-corrected milk (FCM at 3% fat) equation for milk standardization (1 kg FCM3% = 642 kcal or 153 kJ) in dairy camels was: FCM3% = 0.197 × Fat (%) + 0.408.

Session 07 Poster 30

Growth and haematological values of indigenous Venda chickens fed varying dietary energy levels

O.J. Alabi[1], J.W. Ng'ambi[1], F.E. Mbajiorgu[2], D. Norris[1] and M. Mabelebele[1]
[1]Unversity of Limpopo, Deartment of Agricultural Economics and Animal Production, Private Bag X1106, 0727, Sovenga, South Africa, [2]Unversity of Limpopo, Medical Sciences Department, Private Bag X1106, 0727, Sovenga, South Africa; alabioj@gmail.com

The effect of feeding varying dietary energy levels on growth and haematological values of indigenous Venda chickens aged 8-13 weeks were evaluated. A completely randomized design was used. One hundred and sixty female indigenous Venda chickens (BW 362±10 g) were allocated to four dietary treatments. Each treatment was replicated four times and each replicate had ten chickens. Four maize-soya beans based diets were formulated. Each treatment had similar CP (180 g/kg DM) and lysine (9 g lysine/kg DM) but varying energy levels (11, 12, 13 and 14 MJ ME/kg DM). The birds were reared in a deep litter house; feed and water were provided *ad libitum*. Data on productivity and haematological values were collected. These data were analysed using one way analysis of variance. Duncan's test for multiple comparisons was used to test the significant difference between treatment means (P<0.05). Results showed that dietary energy level influenced (P<0.05) feed intake (99.82 g/bird/day), feed conversion ratio (3.85) and metabolisable energy values of chickens (10.48 ME MJ/kg DM). Birds on dietary energy level of 12 ME MJ/kg DM were superior to the other treatments. However, growth rate, live weight, N-retention and mortality were not influenced (P>0.05) by dietary treatments. Haematological results showed that birds on 12 MJ ME/kg DM had higher (P<0.05) values of white blood cell (26.17×10^3 µl), red blood cell (2.63×10^4 µl), mean corpuscular haemoglobin (46.80 pg) and mean corpuscular haemoglobin concentration (31.22 g/dl). However, pack cell volume (PVC) and mean corpuscular volume (MCV) were not influenced (P>0.05) by dietary treatments. The results of the present study suggest that dietary energy level of 12 MJ ME/kg DM at a CP level of 180 g was sufficient for indigenous chickens' production.

Role of mobility to face long drought (1995-2011) in the Coastal Zone of Western Desert (Egypt)

V. Alary[1], A. Aboul-Naga[2], P. Bonnet[1], M. Osman[2] and J.F. Tourrand[1]
[1]CIRAD, Campus international de Baillarguet, 34398 Montpellier Cedex 5, France, [2]Animal Production Research Institute, Nadi El Said, Giza, Egypt; adelmaboulnaga@gmail.com

Mobility is well known as a factor of flexibility of livestock systems and adaptation to harsh conditions. By enlarging the potential resources, the mobility can be considered as a way to reduce climatic risk. The objective of the study is to understand the role of mobility faced to the 15 drought years that have faced the Coastal Zone of Western Desert (CZWD) in Egypt (1995-2011). A farm survey has been conducted in 2011 among a sample of 120 farmers in the dry rainfed area from El-Alamein to Libyan border. The analysis is based on factorial and clustering methods based on synthetic indicators related to mobility (duration, distance), complementation (during and after the transhumance) and animal performance (reproduction performance and mortality rate) and profitability (net income by head of animal). A first typology shows that only the large breeders with more than 400 sheep and goats practice long transhumance during the last 15 drought years. The rate of profitability remains low due to high mortality. The other breeders adapt complementation procedure (mainly feed concentrates and grains) according to a strategy of maintaining their animal stock (maximizing reproduction rate) or a strategy of survival (by maximizing the profitability per animal) during the long drought. The lowest profitability was registered for breeders that maintain short mobility (less than 7 km). The natural feed intake cannot fulfill the animal need of energy to walk. So mobility is an adaptive mechanism to tough drought conditions to maintain the animal stock for the large breeders at the detrimental of profitability, but it has its limitations for small breeders with the objective of economic survival due the high rate of mortality. Only in normal climatic conditions, the mobility becomes a factor of profitability.

Empirical study of experimental designs for indirect genetic effects detection

P. Fullsack, C. Herbinger, B. Smith and R. Horricks
Dalhousie University, Department of Biology, 1355 Oxford Street, Halifax, NS B3H 4R2, Canada; philippe.michel.fullsack@gmail.com

Indirect genetic effects (IGEs) have been introduced by Griffing and Hamilton in the contexts of genetic evaluation and evolutionary games to account for heritable genetic variation related not to individuals but to groups of individuals. Individuals exposed to other individuals may have an influence on their social mates' traits, the best known and most studied case being the effect of mothers on their offsprings. Selection for traits under the influence of IGEs leads to ranking of individuals and coevolution of social and focal traits that cannot be accounted for in animal models based on direct effects only. Although a few recent experiments on trees, chicken, minks and pigs have successfully revealed the presence of IGEs, these effects remain difficult both to model and estimate. We carry out a series of simulations to explore the ability of various simple social genetic designs to retrieve IGEs of various magnitude. Our study highlights the role of design parameters such as the number of families per social group, number of families, number of social groups and block replicates, on statistical power. It also points out the difficulties that might be encountered under limited resources or more difficult conditions. taking the ongoing pilot experimental study of IGEs in aquaculture ('Social Arctic Charr Trials') as a basis for comparison. We conclude that estimation of IGEs requires special care and presents difficulties at the three levels of experimental setup, model selection and model estimation.

Session 08 Theatre 2

Direct and social genetic parameters for growth and fin erosion traits in Atlantic cod

H.M. Nielsen[1], B.B. Monsen[1,2], J. Ødegård[1], P. Bijma[3], B. Damsgård[4], H. Toften[4] and I. Olesen[1,2]
[1]Nofima, P.O. Box 210, 1430 Ås, Norway, [2]Norwegian University of Life Sciences, P.O. Box 5003, 1432 Ås, Norway, [3]Wageningen University, P.O. Box 338, 6700 AH, Wageningen, the Netherlands, [4]Nofima, P.O. Box 6122, 9291 Tromsø, Norway; hanne.nielsen@nofima.no

The aim of the study was to estimate direct and social genetic parameters for growth and welfare of Atlantic cod (Gadus morhua). In total 2,100 juveniles from 100 families (73 sires and 100 dams) at an average age of 200 days post hatching were used. Each family was separated in three groups containing 7 fish. The three groups within each family were distributed randomly in 100 experimental tanks, together with fish from two other families. The fish were fed restricted in order to facilitate social interactions. Recordings were performed at the start of the experiment, after two weeks and after six weeks. Individual body weight was measured at each recording. Fin erosions (0-100% in 5% intervals) of the first, second and third dorsal and the caudal fin were scored subjectively. Fin lengths of the fins were measured using digital image analysis. Variance components were estimated using a conventional animal model and a model containing a social effect in addition to a random effect of experimental tank and common rearing tank. Heritabilities for body weight ranged from 0.24 to 0.34, whereas heritabilities for fin erosion were 0.83±0.08, 0.17±0.04 and 0.01±0.04 for the first, second, and third dorsal fin and 0.06±0.07 for the caudal fin. Heritabilities for fin erosions varied from 0.05 to 0.80. Significant genetic social effects were found for fin erosion at the second dorsal fin and for length of the first, second and third dorsal fin. For e.g. length of the first fin direct effects contributed 6.1±2.1 to the total heritable variance, the direct-social covariance contributed 9.4 +3.2, and the social variance contributed 12.9±4.1. Results indicated that considering social breeding values when selecting cod may increase genetic gain in welfare traits.

Session 08 Theatre 3

Detecting genetic variance of social effects in aquaculture trials with Arctic charr

C.M. Herbinger, R. Horricks, P. Fullsack and B. Smith
Dalhousie University, Department of Biology, 1355 Oxford St, B3H 4R2 Halifax, Canada; christophe.herbinger@dal.ca

Access to genetically superior strains developed through efficient breeding programs is paramount to continued productivity gains in aquaculture. Interest has resurfaced recently for models that recognize that the value of the trait (phenotype) of an individual is also influenced by the placement of this individual in one (or a series of) groups of individuals. These models include additional 'heritable social effects' and lead to a different partition of the phenotypic variance. This is an exciting new approach which has been used with success with a few mammal and bird species, but has yet to be tested and exploited in aquaculture breeding programs. Properly accounting for social interactions could not only lead to more efficient breeding programs but to improved animal welfare as well. We ran a series of pilot experiments, 'Arctic Charr social trials', with juvenile Arctic charr (*Salvelinus alpinus*), a fish of aquaculture interest in Canada that seems to exhibit negative social interactions. In a first experimental set, fish were raised in small tanks containing 5 families (10 fish each), each family (24 in total) being identified with VIE tags. Analysis of fish length was performed with Wombat, either with a regular animal model, or with the extended model including social genetic effects. One data set did not show any detectable genetic variance for social effect, while another one did. A number of practical resource limitations were revealed, particularly related to the number of families and rearing units available, as well as to the difficulty to maintain the pedigree of small fish for an extended period with VIE implants. A second experimental trial is now on-going with larger fish being individually marked and followed using PIT tag, and with a different system of allocating families into tanks, based on results from an extended series of simulations.

Genetic selection can reduce aggression behaviour in group-housed mink

S.W. Alemu[1], L. Janss[1], P. Berg[1,2] and P. Bijma[3]
[1]Aarhus University, Department of Molecular Biology and Genetics, Blichers Allé 20, 8830, Tjele, Denmark, [2]NordGen, P.O. Box 115, 1431 Ås, Norway, [3]Wageningen University, Animal Breeding and Genetics, 6700 AH Wageningen, the Netherlands; setegnw.alemu@agrsci.dk

Group housing of mink is common. It is advantageous from a production perspective, but can lead to aggression between animals, raising a welfare issue. Bite marks on the animals are an indicator of this aggressive behaviour and selection to reduce bite marks would thus reduce aggression and improve animal welfare. A trait like bite marks also reflects the aggression of the group members, and for a successful breeding strategy it may be crucial to consider both direct and associative genetic effects on this trait. However, to date no study has investigated the genetic basis of bite marks in mink. We analysed bite marks on neck, body and tail, as well as their average, on 1,985 group-housed mink. An animal model with pedigree was used with both direct and associative genetic effects to estimate variance and covariance components. The total heritable variation expressed as proportion of phenotypic variance was more than two-fold greater than the heritability (h2) of a model with direct effects only. For instance, for average bite mark was 0.47 and was 0.18. This result is in line with survival days in laying chickens, and growth rate and feed intake in pigs. The genetic correlation between direct genetic and associative genetic ranges from 0.55 for neck bites marks to 0.99 for tail bite marks. A positive correlation suggests that an individual benefits from not biting others. Therefore, genetic selection can reduce bite marks and thus possibly aggressive behaviour in group housed minks. Including associative genetic effects would ensure a more efficient selection against frequency of bite marks.

Using pooled data to estimate variance components and breeding values for social interaction traits

K. Peeters, E.D. Ellen and P. Bijma
Wageningen University, Animal Breeding and Genomics Centre, De Elst 1, 6708 WD Wageningen, the Netherlands; piter.bijma@wur.nl

Through social interactions, individuals affect one another's phenotype. In that case, an individual's phenotype is affected by the direct (genetic) effect of the individual itself and the indirect (genetic) effects of the group mates. Using data on individual phenotypes, direct and indirect genetic (co)variances can be estimated. Together, they compose the total genetic variance that determines a population's potential to respond to selection. However, it can be difficult or expensive to obtain individual phenotypes, e.g. egg production and feed intake. Therefore, we investigated whether (direct, indirect and total) genetic variances and breeding values can be estimated from pooled data (pooled by group). In addition, we determined the optimal group composition, i.e. optimal number of families represented in a group to minimize the standard error on estimates. The study was performed in three steps. First, theoretical derivations were made. Second, a simulation study was conducted to look into the estimation of variance components and optimal group composition. Third, individual and pooled survival records on 12,944 purebred laying hens were analysed to look into the estimation of breeding values and response to selection. Through theoretical derivations and simulations, we showed that the total genetic variance can be estimated from pooled data, while the underlying components cannot. Moreover, we showed that the most accurate estimates are obtained when group members belong to the same family. Moreover, the correlation between estimated total breeding values obtained from individual and pooled data was surprisingly close to one. This indicates that, for survival in purebred laying hens, loss in response to selection will be small when using pooled instead of individual data.

Genome-wide estimates of coancestry and inbreeding depression in an endangered strain of Iberian pig

M. Saura[1], A. Fernández[1], L. Varona[2], A.I. Fernández[1], M.A. Toro[3], M.C. Rodríguez[1], C. Barragán[1] and B. Villanueva[1]
[1]INIA, Crta. La Coruña Km 7.5, 28040 Madrid, Spain, [2]Universidad de Zaragoza, Facultad de Veterinaria, 50013 Zaragoza, Spain, [3]Universidad Politécnica, ETS Ingenieros Agrónomos, 28040 Madrid, Spain; saura.maria@inia.es

In this study we have characterized the status of an ancient strain of Iberian pigs (the Guadyerbas strain) which is in serious danger of extinction, from a genomic perspective in a conservation context. The PorcineSNP60 BeadChip (Illumina) was used to obtain genome-wide estimates of coancestry (f) and inbreeding (F) that were then compared with those obtained from pedigree data. Rates of f and F were used to calculate the effective population size (N_e). We also evaluated the extent and decay of linkage disequilibrium (LD) and estimated inbreeding depression for two reproductive traits (number of piglets born alive and total number of piglets born). Molecular estimates of f and F were strongly correlated with genealogical estimates. Genealogical coancestry and inbreeding coefficients were very good predictors of molecular coefficients as expected, but also, and contrary to what it has been observed in previous studies, molecular coefficients were very accurate in predicting genealogical coefficients. The estimate of N_e was as low as 10 animals. The most rapid decline in LD (measured as r^2) occurred over the first 0.9Mb, with the mean r^2 decreasing by more than half over this period. The mean r^2 between non-syntenic markers was similar to that observed between syntenic markers at distances greater than 15Mb. Inbreeding depression was estimated using a linear mixed model, where molecular F was included as a covariate. Inbreeding depression resulted significant for a region of chromosome 13 where a QTL related to prolificacy has been previously described. Our results support that genomic information is more accurate and informative than genealogical information, providing insights of what is really occurring in the genome, which is of great value in management conservation programs.

Using dense SNP markers in runs of homozygosity as a tool to measure inbreeding

B. Hillestad[1], H. Grove[1,2], D.I. Våge[1,2], T. Meuwissen[1], J.A. Woolliams[1,3] and G. Klemetsdal[1]
[1]Norwegian University of Life Sciences, Department of Animal and Aquacultural Sciences, P.O. Box 5003, 1432 Ås, Norway, [2]Norwegian University of Life Sciences, CIGENE, Department of Animal and Aquacultural Sciences, P.O. Box 5003, 1432 Ås, Norway, [3]University of Edinburgh, The Roslin Institute and Royal (Dick) School of Veterinary Studies, Easter Bush Campus, Midlothian EH25 9RG, Scotland, United Kingdom; borghild.hillestad@umb.no

Runs of homozygosity (ROH) have been used to measure homozygosity and relatedness in humans since 1999. ROH are long, homozygote segments of an individual's genome that are traceable to the parents and might be identical by descent, with long segments indicating recent inbreeding, and shorter more historical inbreeding. With the advent of increasingly dense marker data, estimation of inbreeding coefficients (F) is in transition from probability calculations based on pedigree towards genomic patterns in the genome, like ROH. Here, ROH in 3289 Norwegian Red bulls were estimated for different SNP-densities ranging from 54 K to 777 K. A run had to exceed 500 Kb to be qualified as a ROH. The increased density leads to shorter average runs; 6.4 Mb in 54 K and 1.1 Mb in a 777 K density, respectively. However, total length of runs increased with higher density; from 113 Mb to 202 Mb. Based on average length of runs, the individual inbreeding coefficients were calculated (F_{ROHavg}). The correlation between F_{ROHavg} and F_{PED} was 0.35 (54 K) and 0.51 (777 K), respectively.

Session 08 Theatre 8

Study of environmental and genetic factors the variability of litter sizes of sheep

S. Fathallah, I. David and L. Bodin
INRA, SAGA, 24 Chemin de borde rouge Auzeville, 31326 Toulouse, France; samira.fathallah@toulouse.inra.fr

The litter size of sheep has increased in recent decades, due to improved breeding techniques and genetic progress, but this increase of the mean has been associated with an increased incidence of multiple births inducing higher lamb weakness and early mortality. The main objective for breeders is now to reduce the undesirable large litters (4 and more), which means canalize the litter size around the economic optimum. Litter size in sheep is a discrete ordered trait that can be analyzed by modeling the observed variable trough Poisson, binomial negative and multinomial models or modeling an underlying variable through threshold models. It can be also viewed as a continuous variable and treated by the classical linear mixed models. If several canalization models have been proposed for continuous traits by including genetic and environmental effects influencing the residual variance, nothing has been proposed for discrete traits and extensions of existing models should be done. Litter size can be also recodified in a set of exclusive binary variables attached to each litter class and associated to a vector of economic weights. According to this approach the breeder objectives is then to select for one of these class or to choose the weighing which maximizes the profit. The aim of this study is to discuss different possible models to analyze the litter size and its variability in sheep and to canalize it around the optimum value.

Session 08 Theatre 9

Responses of divergent selection for litter size residual variance in rabbit

M. Martínez-Álvaro[1], N. Ibañez-Escriche[2], M.J. Argente[3], M.L. Garcia[3] and A. Blasco[1]
[1]Universidad Politecnica de Valencia, Instituto de ciencia y tecnologia animal, C. de Vera, 46071 Valencia, Spain, [2]IRTA, Genètica i Millora Animal, C/Rovira Roure 191, 25198 Lleida, Spain, [3]Universidad Miguel Hernández de Elche, Departamento de Tecnología Agroalimentaria, Ctra de Benial km3.2, 03312 Orihuela, Spain; mamaral9@etsia.upv.es

Selection on LS has a low response due to its low heritability, and it can be increased reducing the environmental variance of LS. Besides, homogeneity in LS reduces cross-fostering, facilitating management. Several studies suggest that residual variance can be under genetical control. However, these studies are based on models highly parameterized that are not robust. Direct selection for residual variance has the advantage of being much less model dependent. Data from 1591 does of a divergent selection experiment for residual variance of LS in rabbits were used to estimate the response of 5 generations of selection. The selection criterion was residual variance of LS (Ve), calculated as the phenotypic LS variance within doe, using LS precorrected by year-season and lactation status. Residual variance was estimated using the minimum quadratic risk estimator: $1/(n+1) \sum^{i=n}_{i=1}(x_i - \bar{x})^2$, where x_i is the LS of parity i of a doe and n is the number of parities of the doe (n varying from 2 to 12). Each divergent line had approximately 125 females and 25 males per generation. The traits analyzed were Ve, residual variance without precorrecting LS, and LS. Residual variances were analyzed using a model having only the mean and the additive effect. The model for LS included the effects of generation, lactation status, additive effect and permanent effect. Bayesian methods were used for all the analysis. The results showed a response in Ve for both lines in all generations. The response in residual variance without precorrection was very similar, so correcting data had little effect. This confirms that residual variance is partly under genetic control. Selection for Ve showed a negative correlated response in LS.

Genetic parameters for male fertility, skatole and androstenone in Danish Landrace boars

A.B. Strathe[1,2], T. Mark[2], I.H. Velander[1] and H.N. Kadarmideen[2]
[1]Danish Agriculture & Food Council, Pig Research Centre, Axeltorv 3, 1609 Copenhagen V, Denmark, [2]University of Copenhagen, Department of Clinical Veterinary and Animal Sciences, Grønnegårdsvej 7, 1870 Frederiksberg C, Denmark; strathe@sund.ku.dk

The objective of this investigation was to study the genetic association between direct measures of male fertility and boar taint compounds in Danish Landrace. Concentrations of skatole and androstenone in the back fat were available for approximately 6,000 and 1000 Landrace boars, respectively. The litter size traits on female relatives of these boars, total number born (TNB), live piglets at day 5 (LP5) and survival until day 5 (SV5) were extracted from the Danish Landrace breeding program, yielding 35,715 records. Finally, the male fertility traits, semen volume, sperm concentration, sperm motility, and total number of sperms were available from 95,267 ejaculates. These ejaculates were collected between 2005 and 2012 and originated from 3,145 Landrace boars from 10 AI stations in Denmark. The traits were analyzed using single and multi-trait animal models including univariate random regression models. Skatole and androstenone concentrations were moderate to highly heritable (i.e. 0.33 and 0.59, respectively). The genetic correlation between the two compounds was moderate (0.40). Genetic variance of sperm production per ejaculate increased during the productive life of the boar, resulting in heritability estimates increasing from 0.18 to 0.31. The heritability (based on service-sire genetic component) of TNB, LP5 and SV5 was 0.02 and the correlation between these effects and the additive genetic effect on boar taint ranged from 0.05 to -0.40 (none of these correlations were significantly different from zero). Most importantly, the genetic correlations between boar taint and semen traits were low (i.e. 0.24 to -0.35) and increasingly favorable with age. In conclusion, the prevalence of boar taint may be reduced through genetic selection without negatively affecting important male fertility traits.

A novel method to visualize local variation of LD and persistence of phase for 3 Danish pig breeds

L. Wang[1], P. Sørensen[1], L. Janss[1], T. Ostersen[2] and D. Edwards[1]
[1]Aarhus University, Blichers Allé 20, P.O. Box 50, 8830 Tjele, Denmark, [2]Pig Research Centre, Danish Agriculture & Food Council, Axeltorv 3, 1609 Copenhagen V, Denmark; lei.wang@agrsci.dk

The extent of linkage disequilibrium (LD) is of critical importance for genomic selection and genome wide association studies in animal breeding. The objective of this study is to develop a novel method to investigate local LD pattern over chromosomes, and persistence of phase between pair breeds. We demonstrated the method on three Danish pig breeds (Duroc, Landrace and Yorkshire), which were genotyped by Illumina PorcineSNP60 Genotyping BeadChip. Further, we estimated local average LD level using a sliding window technique, and generated an LD map for each chromosome. Within each sliding window, we fitted pairwise LD (r^2) and distance (d) between marker pairs using a generalized linear model to estimate the average level of LD, then plotted the LD estimates against the physical position on the chromosome. We also calculated local persistence of phase between breeds by common marker sets in a similar way as LD. The regions with high average LD level also appeared as long LD blocks in LD heatmaps. Our LD map showed that LD varied considerably along chromosomes. When comparing local LD between breeds, we found that the LD patterns varied substantially between breeds; some regions with high LD level across all breeds may indicate selection processes. The persistence of phase for Landrace-Yorkshire at local level was much higher than that for Duroc-Landrace and Duroc-Yorkshire. Estimation and visualization of local LD and persistence of phase gives insight into how these quantities vary along chromosomes and across breeds. The results concerning persistence of phase imply that Landrace and Yorkshire are more closely related with each other than with Duroc.

Session 08 Theatre 12

Recent selection signatures in the Piétrain pig genome

P. Stratz[1], J. Bennewitz[1] and T.H.E. Meuwissen[2]
[1]Institute of Animal Husbandry and Animal Breeding/University of Hohenheim, Garbenstraße 17, 70599 Stuttgart, Germany, [2]Department of Animal and Aquacultural Sciences/ University of Life Science, Box 5003, 1432 Ås, Norway; patrick.stratz@uni-hohenheim.de

Under positive selection pressure allele frequency increases faster and the length of the haplotype blocks around the mutation is longer. In contrast, for a neutral mutation, it will take more generations to increase its frequency. During that time recombination breaks down the LD in the vicinity of the mutation. The objective of this study was to map putative regions of positive selection in German Piétrain pig breeding populations and to compare them with results of genome-wide association studies for growth, muscularity and meat quality. This sire line breed was selected for these traits in the recent decades. Around 1000 progeny tested boars were genotyped with the Illumina porcine 60K SNP-chip. After data filtering, markers were mapped to the *Sus scrofa* Build 10.2 and paternal haplotypes were reconstructed using the default parameters in fastPHASE. Core regions were identified using pairwise D' values that exceeded a predefined threshold. This results in 22,854 core haplotypes within 5,700 core regions. The genome-wide Extended Haplotype Homozygosity (EHH) test was applied for a particular core haplotype within a core region, by expanding it about a distance of 1 cM. Haplotypes out of 922 core regions which had a frequency above 0.7 were arranged in 6 frequency bins and their significance was tested based on the normalized Relative EHH (REHH)-values. Within each bin the P-value of the normalized REHH value was calculated for all haplotypes in the genome. 78 normalized REHH values were detected having a P<0.1 whereof 11 showing a P<0.02. On SSC8, SSC18, SSC6 and SSC1, the most signficant core regions (P<0.02) were found. In GWA studies significant SNPs were also found on SSC1 for carcass length content and on SSC6 for daily gain.

Session 08 Theatre 13

Whole genome scan to detect QTL for major milk proteins in three French dairy cattle breeds

M.P. Sanchez[1], A. Govignon-Gion[1], M. Ferrand[2], M. Gelé[2], D. Pourchet[3], M.N. Rossignol[4], S. Fritz[5], G. Miranda[1], P. Martin[1], M. Brochard[2] and D. Boichard[1]
[1]INRA, UMR GABI 1313, 78350 Jouy en Josas, France, [2]Idele, 149 rue de Bercy, 75012 Paris, France, [3]ECEL, Doubs-Territoire de Belfort, 25640 Roulans, France, [4]Labogena, CRJ, 78350 Jouy en Josas, France, [5]UNCEIA, 149 rue de Bercy, 75012 Paris, France; marie-pierre.sanchez@jouy.inra.fr

A whole genome scan was performed to detect QTL for milk protein composition in the three main French dairy cattle breeds i.e. Montbéliarde (MO), Normande (NO) and Holstein (HO). Protein composition was estimated from Mid-Infrared (MIR) spectrometry on almost 600,000 test-day milk samples from 116,495 cows in the first three lactations (PhénoFinlait programme). Among these cows, 8,080 (2,967 MO, 2,737 NO and 2,306 HO) were genotyped mainly with the Illumina 50k Beadchip. Individual test-day records were adjusted for environmental effects and then averaged per cow. After quality control, phasing, and missing genotypes imputation, QTL detection was carried out within breed by an approach combining linkage and linkage disequilibrium on clusters of 6 consecutive SNP. In each population, the most significant QTL regions were found on BTA6, 11 and 20 ($10^{-8}<P<10^{-11}$). The BTA6 QTL region, spanning from 80 to 95 cM, affected αs1, αs2, β and κ-caseins in milk. The BTA11 QTL (100 cM), had an effect on β-lactoglobulin in milk as well as on β and κ-caseins in protein. The QTL on BTA20 (55 cM) affected αs1-caseins in protein. The proportion of genetic variance explained by the most significant QTL was around 10-30% and reached 54% for BTA11 and β-casein in protein. In these regions, caseins (BTA6), β-lactoglobulin (BTA11) and GHR (BTA20) genes are good candidates. Moreover, other significant QTL effects ($P<10^{-5}$), partially overlapping across breeds, were highlighted on BTA1, 2, 14, 17 and 19. These first results pave the way to causal mutation identification. This project receives financial support from ANR, Apis-Gène, Ministry of Agriculture (CASDAR), Cniel, FranceAgriMer and FGE.

Profiling the architecture of genetic correlations between some fatty acids in ribeye of beef cattle

C. Diaz[1], D.J. Garrick[2], M. Saatchi[2] and J. Reecy[2]
[1]INIA, Departamento de Mejora Genética Animal, Ctra. de la Coruña km 7.5, 28040 Madrid, Spain, [2]Iowa State University, Department of Animal Science, Ames, IA 50011, USA; cdiaz@inia.es

The amount of intramuscular fat is a desirable trait in beef cattle because of its association with beef flavor. Breeders have been selecting for improved marbling for a long time. It could be reasonable to think that fatty acid composition (FA) may have been altered by corresponding changes in fat metabolism. The objective of our work is to gain insight into the genetic correlations between FAs by examining the correlation between the estimated genomic values for relevant 1Mb regions for different FAs. A total of 2,110 animals, born 2002-2008 were genotyped with the Illumina BovineSNP50 BeadChip and their FA (C14:0, C16:0, C18:0, C18: 1c9) profiles measured on a weight percentage basis. Bayes B with π=0.999 was used to estimate SNP effects for every FA. Genomic breeding values for every 1 Mb window were calculated for each FA. The correlation between the same 1Mb region for different FAs were calculated for those regions that accounted for the most variation. Correlations between FA adjusted for systematic effects ranged from -0.02 between C18:0 and C18:1c9 to 0.72 between C14:0 and C16:0. Different architectures were observed for the correlations between windows. FA correlations less than one were achieved by having in most regions a strong and positive correlation while in some regions there correlations were in the opposite direction. Mild negative correlations between FA were mostly obtained by having a balance toward negative correlations between windows. These results suggest that the overall correlation between pairs of FA could be changed by selection on particular 1 Mb windows.

Health and welfare of loose housed and free-range pigs

H. Mejer
University of Copenhagen, Thorvaldsensvej 57, 1870 Frederiksberg C, Denmark; hem@sund.ku.dk

Much of the world´s commercial pig production is carried out in intensive indoor systems favouring productivity at the cost of welfare. However, changes in consumer awareness and legislation have ensured management changes, providing pigs with better opportunities to express natural behaviour, potentially alleviating stress and improving wellbeing. Among the changes are loose/group housing of sows, bedding material or outdoor access to runs or pastures in free-range systems. Free-range systems provide pigs with the obvious welfare benefits of a more natural environment, but also impose less obvious constraints, e.g. increased risk of infectious diseases due to contact with wildlife. In addition, diseases may be more difficult to monitor and intestinal parasites are more common outdoors, which is attributed to the improved parasite survival on pastures. Heat and cold stress may also be more pronounced outdoors in some climates, but the often poor indoor air quality is avoided. Provision of food and water can be especially complex in areas with cold and wet winters, and a recurring issue in all regions is adequate water supply to sows in outdoor huts after farrowing. Indoors, loose housing may improve sow welfare compared to farrowing crates, but puts piglets in greater risk of being crushed, and pens have to be designed to counteract diverse problems. When kept in larger groups, sows can move about freely and engage in social behaviour, but aggression may happen and individual animals are more difficult to monitor. Especially in dynamic groups, bacterial, viral and parasite infections may spread more readily, due to the close contact between animals, than for single penned animals. Introduced as environmental enrichment, bedding material may help reduce some leg and foot problems, but appears to also increase parasite infection levels. Overall, there is still a need to better understand the complexity of how various management systems affect and interact with animal health, welfare and production results to ensure the sustainable development of pig production systems.

Similar farrowing progress among sows housed in crates and pens

J. Hales[1], V.A. Moustsen[2], A.M. Devreese[1] and C.F. Hansen[1]
[1]University of Copenhagen, Dept of Large Animal Sciences, Grønnegårdsvej 2, 1870 Frederiksberg C, Denmark, [2]Pig Research Centre, Axeltorv 3, 1609 Copenhagen V, Denmark; cfh@sund.ku.dk

The aim of this study was to investigate the effect of temporary crating on farrowing progress. We hypothesized that confinement would increase both farrowing duration and birth intervals. A total of 123 Landrace × Yorkshire sows from a commercial sow herd were split into two groups, either crated (C) or loose (L) prior to and during parturition. All farrowings were videorecorded and data was analysed in a mixed model using SAS, ver. 9.3. Mean (±SE) number of total born was 18.4±0.4 piglets per litter and mean parity was 3.4±0.2. Median farrowing duration from birth of first piglet (BFP) to birth of last piglet was 390 (P25=264; P75=646) minutes for C-sows and 417 (P25=234; P75=583) minutes for L-sows, whereas the median duration from BFP to birth of last liveborn piglet was 353 (P25=249; P75=528) minutes for C-sows and 390 (P25=225; P75=506) minutes for L-sows. Farrowing duration did not differ between C and L- sows. Birth duration was calculated as time from BFP to birth of the n'th piglet. The median value for birth duration was 188 (P25=94; P75=318) minutes for piglets born by C-sows and 168 (P25=86; P75=307) minutes for piglets born by L-sows. Birth duration did not differ between groups, but there was a tendency for birth duration to be shorter for stillborn piglets in L-litters compared to C-litters ($P<0.1$). Median birth interval was 653 (P25=286; P75=1480) minutes for piglets born by C-sows and 678 (P25=274; P75=1562) minutes for piglets born by L-sows. Birth intervals did not differ between groups, but were lower for piglets born by younger sows (parity one and two) compared to older sows (parity three or more) ($P<0.05$). In addition, birth intervals were longer for piglets born in litters with 7-16 total born piglets compared to litters with 21-28 total born piglets ($P<0.05$). In conclusion and contrary to our hypothesis farrowing progress was not affected by confinement of the sow.

Session 09 Theatre 3

Assessing fearfulness of gilts on farm: can QBA add information to standardised fear tests?

C. Pfeiffer, C. Leeb, A. Gutmann and C. Winckler
Division of Livestock Sciences, University of Natural Resources and Life Sciences Vienna, Department of Sustainable Agricultural Systems, Gregor-Mendel Straße 33, 1180 Vienna, Austria; christina.pfeiffer@boku.ac.at

Fear is an important emotional state which is related to autonomic and hormonal reactions that can subsequently influence the performance of different farm animals. In pigs, different approach tests have been developed for assessing fearfulness as a corresponding temperament trait. These tests consider quantitative parameters such as latencies to approach unfamiliar persons but so far neglect the qualitative component of behaviour. To examine whether qualitative aspects of behaviour during an approach test can complement quantitative measures, a voluntary human approach test was conducted with 44 gilts on two different organic pig breeding farms together with an additional qualitative behaviour assessment (QBA) during the test situation. Principal component analysis of qualitative attributes of behaviour revealed three dimensions: 'Emotional state', 'Proactiveness' and 'Selfconfidence'. Significant correlations between quantitative and qualitative components of behaviour were found. Gilts which entered the zone of 100 cm around the test person more slowly received more negative scores on the 'Emotional state' dimension ($P<0.05$), and gilts that approached faster received higher 'Proactiveness'-scores (first radius $P<0.05$; second radius $P<0.01$). No correlations between 'Selfconfidence' and quantitative parameters were found. Qualitative aspects of behaviour add meaningful information about an animal's emotional state that quantitative measures can not detect. QBA is a feasible and promising tool to gain more detailed information about gilts' emotional state in the context of fear tests.

Session 09 Theatre 4

Health and welfare challenges for laying hens kept in free-range systems

M. Gauly and G. Das
University of Göttingen, Department of Animal Sciences, Albrecht-Thaer-Weg 3, 37075 Göttingen, Germany; mgauly@gwdg.de

Impaired welfare, particularly due to behavioural restrictions, was the most potent argument against the use of the conventional cages for laying hens across the Europe. This paper focuses on frequent health and welfare problems of laying hens kept in non-cage housing systems, which have been expected to provide higher welfare standards in all respects. A special emphasis is given to re-emerging parasite infections in outdoor systems. Data obtained so far indicate that laying hens that are kept in free-range systems have a lower egg production level, but show more comfort behaviours. However, hens kept in these systems suffer indifferently from feather pecking and aggressive pecking leading to cannibalism, and have similar or even higher mortality rates when compared to those which were kept in conventional cage systems. Birds kept in large flocks may develop panic outbreaks, particularly when the birds are younger. Birds permanently kept outside are threatened by birds of prey, and may be chronically stressed. Almost all the hens kept in non-cage systems are infected with various endo- and ectoparasites. Among them, the gastrointestinal nematodes Ascaridia galli, Capillaria ssp., Heterakis gallinarum, certain cestodes and the red mite, Dermanyssus gallinae, appear to be the most prevalent parasites. Potential and already quantified effects of parasite infections on production, health and welfare of the hens are discussed, and possible solutions are documented. Such a solution may be seen in the use of genetically more disease resistant animals. In conclusion, laying hens in the non-cage systems are not facing only some of old health and welfare problems, but also being exposed to new challenges. There is a need to determine critical components of the non-cage systems which may lead to improvements in the health and welfare of the birds.

Session 09 Theatre 5

Suitability of egg-type cockerels for fattening purposes

F. Kaufmann and R. Andersson
University of Applied Sciences, Osnabrueck, Agricultural Sciences and Landscape Architecture, Animal Husbandry and Poultry Sciences, Am Krümpel 31, 49090 Osnabrück, Germany; f.kaufmann@hs-osnabrueck.de

Debates on culling day-old male egg-type chicks are growing. This study investigated suitability of male chicks for fattening purposes with regard to genetic background and rearing system. A total of 1,198 male layer-chicks of Lohmann Brown genotype were reared either in two modified mobile stable systems (LB1, LB2) or in a floor husbandry system (LB3). An additional 140 chicks of an experimental Lohmann Brown genotype (LBex) were kept in an equal floor husbandry system. The birds were reared for 80 days under organic conditions despite the outdoor access in the floor system groups. Body weight (BW) development, group feed consumption and mortality was recorded at regular intervals. At slaughter (d 80), randomly selected birds (n=30/group) were dissected to determine slaughter weight (SW) and carcass compositions. Average SW of the birds ranged from 952 to 1031 g/bird among groups. Final SWs of outside groups (LB1 and LB2) were higher ($P<0.05$) than that of the same genotype kept in the floor husbandry system (LB3), while both outside groups also differed ($P<0.05$). The LBex performed better in the floor husbandry system, reaching a higher slaughter weight (1,007 g) when compared with LB3 (952 g; $P<0.05$). Mortality rates were higher in the mobile systems (LB1: 12.5%, LB2, 5.6%) than those of LB3 (1.4%). Feed:gain was 2.5 in LB1 and LB2, whereas it was 2.7 in LB3. It can be concluded that rearing male-layers in mobile systems does not have adverse effects on growth performance of the animals which may be beneficial regarding marketing strategies. This is necessary as full-cost accounting analysis revealed five-time higher operating costs for the mobile stable system compared to conventional broiler production systems. However, carcasses of 80 day old egg type cockerels as a whole or processed may fulfill market requirements, offering an alternative to culling.

Assessing the effect of grazing on dairy cow welfare – using a multi-dimensional welfare index

T. Rousing, E. Burow, P.T. Thomsen and J.T. Sørensen
Aarhus University, Animal Science, Blichers Alle 20, P.O. Box 50, 8830 Tjele, Denmark; tine.rousing@agrsci.dk

Grazing is expected to benefit dairy cow welfare. Most studies have focused on single welfare indicators when assessing the welfare effect of grazing. Assessment of animal welfare at herd level is however not a trivial task – it is a multidimensional task. The present paper aims at: (1) discussing the theory that animal welfare assessment is a multidimensional task; and (2) discussing the effect of grazing on dairy cow welfare based on a multi-dimensional welfare index. A multi-dimensional animal welfare assessment based on 17 measures inspired by Welfare Quality® were carried out in 41 Danish dairy herds. The protocol mainly included animal behaviour and clinical health measures. Each of the herds were visited twice – once in the later winter and once in the summer, both during the year 2010. A herd Animal Welfare Index (AWI) aggregating the individual welfare effect of each of the protocol measures into one number were calculated for each of the herds for each of the two seasons. The principles of aggregation were: (1) additive aggregation of measures' prevalence; and (2) relative and linear scaling based on expert opinions. Summer and Winter AWI's where compared using non-paired students test. Comparison of the different components – one for each of the 17 animal welfare measures – led to the conclusion that animal welfare status of the individual herd would vary significantly depending on which specific welfare measures it is based on. Based on the suggested AWI positive effects of grazing on dairy cow welfare were found.

Variation factors of overall health score using Welfare Quality® protocol in French dairy herds

M. Coignard[1,2], R. Guatteo[1,2], I. Veissier[3], A. De Boyer Des Roches[3], L. Mounier[3], A. Lehebel[1,2] and N. Bareille[1,2]
[1]INRA, UMR1300 BioEpAR, CS 40706, 44307 Nantes, France, [2]LUNAM Université, Oniris, UMR BioEpAR, CS 40706, 44307 Nantes, France, [3]Clermont Université, VetAgro Sup, INRA, UMR1213 Herbivores, CS 10448, 63000 Clermont-Ferrand, France; maud.coignard@oniris-nantes.fr

Extensive information is available in the literature on the specific risk factors of the main health disorders afflicting dairy cattle herds. However, it remains difficult to manage a herd's overall health because measures controlling one risk factor can exacerbate the risk of another disease. To achieve good overall health, livestock systems and management practices need to simultaneously consider all of the main health disorders. We aimed to identify the systems and practices conducive to good herd health using the Welfare Quality® assessment protocol: Our objectives were: (1) to describe the distribution of herds' health scores in a French representative sample; and (2) to investigate systems and practices associated with variations of the overall health score. This protocol was carried out on 130 farms between December 2010 and March 2011. A multivariate analysis of variance was performed to investigate variation factors of the overall health score at the herd level. This score was classified as moderate for the majority of farms (95.4%) (mainly due to subclinical mastitis, dystocia and pain induced by disbudding) and varied little between farms. Some systems were associated with a higher overall health score: straw yards and milking parlors ($P<0.001$), highland versus lowland locations ($P=0.008$), Montbeliarde versus Holstein breeds ($P=0.07$). Some practices also were associated with a higher level of health: medium herd average parity ($P=0.01$), low proportion of dirty cows ($P<0.001$) and low proportion of cows with abnormal body condition ($P=0.07$). These results suggest that some systems contribute to better health and that improvement of health can be obtained in the short term by the modification of routine management practices.

Session 09 Theatre 8

Comparison of two types of salt licks located near or far the water: ingestion and cows behaviour

E. Knapp[1], A. Cheradome[2], T. Hetreau[2], L. Istasse[1], J.L. Hornick[1] and I. Dufrasne[1]
[1]Université de Liège, Nutrition Unit, bd de colonster, 20, 4000 liège, Belgium, [2]Centre d'Elevage, rte de l'Ecole d'Agriculture, 74330 Poisy, France; eknapp@ulg.ac.be

The aim of the study was to compare intakes and behaviour of cows offered two types of salt licks located at two different positions in the barn – far or near the water through. A dairy cow herd (n=80) was divided in two homogeneous groups. Two licks made of NaCl were available in the control group. In the second (enriched) group, the disposal was similar but with more brittle licks composed only with 19.5% of Na associated with major and trace minerals. All salt licks were weighed and changed 3 times per week. Video was used to observe 18 individuals in each group. The cows in the enriched group ingested significantly (P<0.001) more salt from the lick than the controls (60.1±16.4 vs. 40.2±12.4 g, cow/day) but less (P<0.001) Na (11.6±3.4 vs. 14.9±5.3 g, cow/day). In the enriched group, the cows spent significantly (P<0.05) less time than in the control group (2.4±1.1 vs. 3.5±1.6 visits, cow/day; 4.7±2.8 vs. 8.9±5.8 min, cow/day). The cows came more often to the licks between 24:00-01:00 and 15:00-16:00 with an average of 7.9 and 7.5 visits/h. There was also a peak of frequentation after milking. In the two groups, and regardless the position of the salt lick, the licking was usually followed by eating. When the cows visited the salt licks close to the water through, the drinking behaviour after licking was larger in the two groups. Activity behaviour after licking was observed for 20.6% and 10.9 of the cows in the enriched and control groups, respectively. There were only 0-6.5% of the cows visiting the automatic concentrates distributor after salt licking in the two groups. The classical salt lick, placed near the trough could be considered as an economical and efficient solution to provide sodium to the cows and stimulate water intake. However, there were many individual differences uncorrelated to milk production, days in milk or parity in the two groups.

Session 09 Theatre 9

The epidemiology and treatment of subclinical ketosis in early lactation dairy cattle

J.A.A. McArt[1], D.V. Nydam[2] and G.R. Oetzel[3]
[1]Colorado State University, Clinical Sciences, Fort Collins, CO 80523, USA, [2]Cornell University, Population Medicine and Diagnostic Sciences, Ithaca, NY 14853, USA, [3]University of Wisconsin, School of Veterinary Medicine, Madison, WI 53706, USA; jessica.mcart@colostate.edu

The objectives were to: (1) describe the epidemiology of subclinical ketosis (SCK) in early lactation cows; and (2) determine the effect of oral administration of propylene glycol (PG) in cows diagnosed with SCK on time to resolution of ketosis, development of displaced abomasum and removal from herd in the first 30 days in milk (DIM), conception to first service, and early lactation milk yield. Cows from 4 free-stall dairy herds were each tested 6 times for SCK from 3 to 16 DIM using a Precision Xtra meter. SCK was defined as a β-hydroxybutyrate (BHBA) concentration of 1.2-2.9 mmol/l. Mixed effects multivariable Poisson regression was used to assess risks of disease and reproductive outcomes. A semiparametric proportional hazards model was used to evaluate time to ketosis resolution, and repeated measures ANOVA was used to evaluate milk yield. A total of 741 of 1.717 (43.2%) eligible cows had a least one BHBA test of 1.2-2.9 mmol/l. Peak incidence and prevalence of SCK occurred at 5 DIM with a median time to resolution of 5 days. Cows developing SCK from 3 to 7 DIM were more likely to suffer from negative disease and production outcomes than cows that developed SCK from 8 to 16 DIM. Increasing BHBA concentration at first SCK positive test raised the risk of these outcomes. Treatment of SCK positive cows with PG reduced the risk of displaced abomasum development and removal from herd, improved conception to first service, and improved milk production over control cows. In addition, cows treated with PG resolved their SCK faster and were less likely to develop a more severe ketosis than control cows. These results show that time of onset and BHBA concentration of first SCK positive test are important indicators of individual cow performance, and that SCK positive cows benefit from treatment with PG.

Can sheep learn the virtual fencing system NoFence using operant conditioning?

E.I. Brunberg[1], K.E. Bøe[2], G.H.M. Jørgensen[3] and K.M. Sørheim[1]
[1]Bioforsk, Norwegian Institute for Agricultural and Environmental Research, Section of Organic Food and Farming, Gunnars veg 6, 6630 Tingvoll, Norway, [2]Norwegian University of Life Sciences, Animal and Aqua Cultural Sciences, P.O. Box 5003, 1432 Ås, Norway, [3]Bioforsk, Norwegian Institute for Agricultural and Environmental Research, Section of Arctic Agriculture, Parkveien, 8861 Tjøtta, Norway; emma.brunberg@bioforsk.no

The aim with the present project was to explore the possibility for sheep to learn the virtual fencing system NoFence with maintained welfare. A collar gives a sound warning followed by a mild electric shock if a sheep passes a GPS-based border. If the sheep turns back over the border on the sound, no shock is given. In experiment 1, 24 sheep were attracted with feed to cross a border three times. Only nine reached the learning criteria (max. two shocks out of three possible) and went on to the experiment 2. For three hours on two days, groups of three sheep were placed on a pasture (30×30 meters) on which one of four sides consisted of a NoFence border with a physical fence outside the virtual border. The sheep received fewer (P=0.02 and 0.02 respectively; W=28 and 28, Wilcoxon signed rank test) sound warnings (0.8±0.1, mean±S.E) and shocks (0.2±0.1) on day two than on day one (3.3±0.5 and 1.9±0.4 respectively). On day 4, the physical fence outside the border was removed and the number of sounds was not significantly different from day 3 (0.3±0.2) and no shocks were received. No sheep went outside the border during level two and three. Day 5, when the NoFence border was moved to the other side of the pasture, one group of animals ran out from the pasture. For the remaining two groups, a physical fence was present and the number of sound cues 1.3±0.6 and no shocks were received. In conclusion, individual sheep differ much in the ability to learn a virtual fencing system in a short time. The physical position of the border is important for the animals' understanding. This may cause problems and affect welfare if a border is moved.

Comparing traditional and modern methods for Arabian camel identification

G. Caja[1], E. Díaz-Medina[2], S. Cabrera[2], O. Amann[2], O.H. Salama[3], M.H. El-Shafie[3], H. El-Sayed[3], A.A.K. Salama[1,3], R.S. Aljumaah[4], M. Ayadi[4] and M.A. Alshaikh[4]
[1]Group of Ruminant Research (G2R), Universitat Autònoma de Barcelona, 08193 Bellaterra, Spain, [2]Oasis Park-Museo del Campo Majorero, 35627 La Lajita, Fuerteventura, Islas Canarias, Spain, [3]Animal Production Research Institue, Dokki, 12311 Giza, Egypt, [4]College Food & Agriculture Sci., King Saud University, 11451 Riyadh, Saudi Arabia; alshaikh@ksu.edu.sa

Traditional (red iron brands) and modern (plastic ear tags and radiofrequency boluses) identification (ID) systems were compared in 477 dromedaries of different breed, age and management conditions in Egypt (n=83), Spain (n=304) and Saudi Arabia (n=90). ID performances and camel wellbeing (apparent health and behaviour) were assessed during two years (d 0, 1 and 2; wk 1, 2, 3; mo 1 and 2; every 2 mo thereafter). A total of 16 camels died or were culled by causes not related with the experiment. Iron brands (3 digits) were done in the left flank of 45 yearlings. Plastic ear tags, rectangular flaps (15×50 mm, 3 g) or button (28.5 mm o.d., 3.5 g), were inserted in left ear of >1 yr camels. Five bolus types varying in specific gravity (s.g. 1.5 to 3.6), volume (5.2 to 22.8 ml) and weight (12.7 to 82.1 g), were also applied. Data were analyzed by the PROC MIXED of SAS v.9.1, using a Logit model. Iron brands showed healing problems, 38% brand digits being fully readable which misadvised on their use. Ear tag retention was lower in rectangular vs. button ear tags (66.0 vs. 81.1%; P<0.01). Boluses were safe at all ages, but 2 standard boluses (21×68 mm, 22.4 ml) were blocked at the diaphragmatic hiatus in 2 camel calves (70 kg BW) needing a probe to be unblocked. Bolus retention varied by s.g., the <2 being fully lost after 8 mo. Despite their volume and weight, s.g. >3 boluses were efficiently retained (99 to 100%) at all ages. In conclusion, >99% camel ID can be achieved by using high s.g. boluses. Boluses should be applied by trained operators and use of miniboluses is recommended in camel calves.

Horses living in an enriched environment have better welfare and stronger relationships with humans

M. Valenchon, F. Lévy, C. Neveux and L. Lansade
INRA of Nouzilly, UMR Physiology of Reproduction and Behaviours, Centre Val de Loire, INRA of Nouzilly, 37380 Nouzilly, France; lea.lansade@tours.inra.fr

Social isolation, confinement and unvaried food are common conditions that can impair horse welfare. The aim of this study was to test if an enrichment protocol could improve welfare, and have an impact on fearfulness and horse-human relationships. Ten-months-old Welsh ponies lived either in a standard (n=9) or in an enriched environment (n=10) for five weeks. In standard environment, horses lived in individual stables with wood shaving bedding, were fed concentrate pellets and left outside in individual paddocks thrice per week. In enriched environment, horses lived in individual stables with straw bedding during the day and in groups on a pasture during the night. Enrichment consisted of fractionating and delivering varied food all day long, offering social contacts, large stables and sensory stimulations (e.g. music, objects). The behaviour of the horses was recorded in the stable. We found many indications of welfare improvement in enriched conditions from the 1st to the 5th week. On the 5th week, horses kept in the enriched environment expressed less aberrant behaviour ($P<0.001$), alert postures ($P<0.001$), ears pointed backwards ($P<0.05$) and more lateral sleeping posture ($P<0.001$). At the end of the five weeks, tests of temperament showed that enriched yearlings were less fearful and closer to humans (e.g. glances at an unknown object: $P<0.01$; latency to return eating after a sudden event: $P<0.05$; sniffing and nibbling a passive human: $P<0.01$). In addition, horses kept in the enriched environment expressed less defensive behaviours toward humans during manipulations (e.g. escape, biting, head-butt: $P<0.001$). To conclude, such an enrichment program can be recommended to improve welfare and horse-human relationships.

Session 09 Poster 13

The effect of feeding practise on behaviour in group-housed pregnant sows

M.E. Caille
Chambre d'Agriculture de Bretagne, Av Borgnis Desbordes, 56009 Vannes, France; marie-estelle.caille@bretagne.chambagri.fr

Feeding practises in group-housed gestating sows have to assume adequate feed supply to each sow. The influence of meal frequency (one or two meals/day) and feeding designs (short stalls with head partitions (HP) or shoulder partitions (SP) at the trough) on the feeding behaviour was investigated. Animals were kept in groups of 6 sows. Two groups of sows were housed in HP and three groups in SP. They received alternately their dry meal once a day during two weeks and twice a day during two weeks. The order of the meal frequency was rotated between batches. The feeding duration spent by the group was measured. During the meal time, the presence of each sow at the trough was recorded by scan sampling at 30 s intervals, whereas their position changes at the trough (disturbance or voluntary changes) were recorded in continuous way during the meal. Data were analysed with ANOVA linear procedure of R, with the effects of the number of meals, the feeding design, the batches and all the interactions between the effects. The averaged duration of the meal for the sows fed once is 12.7 min vs. 5.8 min for the sows fed twice a day ($P<0.05$). The sows fed once a day spent 92.7% of their meal time at trough and the sows fed twice a day 87.6%. The frequency of changes at trough was lower for sows fed one meal per day (0.96/min vs. 1.61/min $P<0.05$). With head partitions at trough the differences between 1 and 2 meals were more pronounced ($P<0.05$) for time spent at trough (95.2 vs. 88% of meal time respectively) and changing places (0.89 vs. 1.86/min respectively). One large meal instead of two smaller meals per day reduced the meal disturbance for pregnant sows, especially with head partitions at trough.

Influence of free-range rearing on pig behaviour and carcass composition

V. Juskiene, R. Juska and R. Leikus
Lithuanian University of Health Sciences, Institute of Animal Science, R. Zebenkos 12, 82317, Lithuania; violeta@lgi.lt

The objective of this investigation was to compare behaviour and carcass composition of pigs raised in free-range and indoor systems. The study was carried out with 52 crossbred pigs of Lithuanian White (LW) × Swedish Yorkshire (SY) × English Large White (ELW). Pigs were allotted into two groups (n=26) analogous by origin, gender, weight and body condition score. The indoor group pigs were raised in the pens of 18.5 m^2 area. Free-range pigs were raised outdoors in enclosures of 850 m^2 area that were fitted with 7.5 m^2 shelters. The study indicated that the growth rate of pigs raised in free-range system was slightly higher during the whole fattening period, and their average daily gain was 12.7% (P=0.013) higher than those of raised indoors. The behavioural observations indicated that free-range system tended to increase pig activity by 1.5 times (P=0.085). Pigs raised indoors were more aggressive during the whole experiment. The free-range pigs were in better health. In this group only one pig was given medication, while three pigs were treated in the indoor group during the whole experimental period. Due to the fact that free-range pigs gained weight much faster, they had higher weights at slaughter in comparison with the pigs raised indoors. The carcass traits were not different between the groups, but in all cases the free-range pigs showed lower fat thickness. Free-range pigs had somewhat (P=0.06) higher warm carcass weight than indoor raised animals, but the dressing percentage did not differ between groups (P=0.50). It is concluded that the free-range system had a positive influence on pig growth rate, activity and health status whereas carcass traits remained mainly unaffected.

THI effect on the frequency of medical treatments of dairy cows in Central Europe

C. Sanker, C. Lambertz, S. Ammer and M. Gauly
Livestock Production Systems Group, Department of Animal Science, Albrecht-Thaer-Weg 3, 37075 Göttingen, Germany; sammer1@gwdg.de

This study investigated the effect of the temperature-humidity index (THI) on the incidence of medical treatments in lactating dairy cows in Central Europe. Records of all veterinary-treated cases per cow in two years (2003 and 2005) obtained from eight Holstein-Friesian dairy herds (55 to 170 cows per herd) kept in loose-housing systems were examined. Both farms were located in Lower Saxony, Germany, where temperate climatic conditions dominate. Management-dependent and preventive treatments such as vaccinations were excluded, and the remaining cases (n=5,547) were classified into four clusters: metabolism (37.4%), udder (32.9%), fertility (21.6%) and foot/leg (8.1%) without cow-individual distinction. Meteorological data were recorded hourly by the nearest weather station. THI was calculated and divided into four classes. The mixed model for data analysis included THI class, season and year as fixed effects and farm as random effect. In general, incidences were neither affected by the year (P>0.05) and season (P>0.05) nor by THI classes (P>0.05). With increasing THI, incidences of metabolic treatments increased and incidences of udder treatments decreased in tendency (P>0.05). Similarly, there was a non-significant trend (P>0.05) of season with greater incidences of metabolism and lower incidences of udder in summer compared to winter. The treatment-clusters fertility and foot/leg were neither influenced by THI class nor by season. In the present study, indications of moderate heat stress during summer months in Central Europe were found, whereas THI and season did not affect the different disease complexes significantly.

Economics of using genomic selection at the farm level

M. Calus[1], P. Bijma[2], J. Van Arendonk[2] and R. Veerkamp[1]
[1]Animal Breeding and Genomics Centre, Wageningen UR Livestock Research, P.O. Box 65, 8200 AB Lelystad, the Netherlands, [2]Animal Breeding and Genomics Centre, Wageningen University, De Elst 1, 6708 WD Wageningen, the Netherlands; mario.calus@wur.nl

Genomic selection is revolutionizing the design of breeding schemes, especially for dairy cattle. The fast uptake of genomic selection is a result of its potential to increase genetic gain considerably at an unprecedented rate by reducing the length of generation intervals up to three times. Predictions indicate that these decreased generation intervals increase genetic gain up to 100%, when implementing genomic selection in dairy cattle breeding schemes. Genomic selection in dairy cattle breeding schemes is currently applied in three of the four different selection pathways; i.e. selection of sires and dams of bulls, and selection of sires of dams. Our objective was to review the potential economic benefit of genomic selection at the level of a dairy farm, by optimization of replacement strategies with or without use of genomic information of the female animals in the dairy herd. In this respect, genomic tests are used as a management tool to support on-farm replacement decisions, rather than as a tool to identify the best animals for breeding purposes. The potential benefit of using genomic tests comes through more accurate selection of the best animals for replacement compared to selection based on their pedigree index. Considered replacement strategies include a range of different selection intensities by considering different numbers of heifers available for replacement as well as different replacement rates. Results include computation of break-even costs for genomic tests for heifers, i.e. the maximum allowed cost per test that justifies their use to support replacement decisions. All comparisons were conducted using selection index theory.

All cows are worth to be genotyped!

D. Boichard[1], R. Dassonneville[1,2], S. Mattalia[2], V. Ducrocq[1] and S. Fritz[3]
[1]INRA, UMR 1313, GABI, Domaine de Vilvert, 78350 Jouy en Josas, France, [2]Institut de l'Elevage, 149 rue de Bercy, 75595 Paris, France, [3]UNCEIA, 149 rue de Bercy, 75595 Paris, France; didier.boichard@jouy.inra.fr

In dairy cattle, genomic evaluation based on male reference populations has the same accuracy for males and females and a very similar accuracy for all traits. This provides a new opportunity to efficiently implement within herd selection. Much more than before, the farmer can customize his breeding goal due to the larger panel of available bulls and the accurate evaluation of females for all traits. The proportion of genotyped females is very sensitive to the evaluation cost. It is anticipated that, in European conditions, this technology can be generalized to a large proportion of the female population if the cost is below 40€. The genetic interest and the profitability of this choice are highly dependent on the selection intensity which can be applied within herd. With conventional reproduction, most females are needed for replacement and the interest is limited. With sexed semen, a technology expected to strongly develop, the upper half of the herd (based on the breeding objective of the farmer) can be oriented to produce female calves for replacement whereas the remaining part does not contribute to replacement and can be mated for other purposes, especially in crossbreeding. In practice, it is worth to start genotyping at least all young animals during their first year of life, allowing to gradually increase the proportion of genotyped animals in the herd. In addition to selection, genotyping females provides useful information for matings (today genetic defects, tomorrow optimal matings based on inbreeding minimization, QTL pyramiding, non additive effects…). Finally, it should be emphasized that these females will contribute the reference population of the future and there is a strong general interest to increase the proportion of genotyped cows and decrease the genotyping cost.

Which new traits are expected to be available in the near future?

C. Egger-Danner[1], J.B. Cole[2], J. Pryce[3], N. Gengler[4], B. Heringstad[5], A. Bradley[6] and K.F. Stock[7]
[1]ZuchtData, 1200 Vienna, Austria, [2]Animal Improvement Programs Laboratory, USDA, Beltsville, Maryland, USA, [3]Department of Primary Industries, Victoria, 3083, Australia, [4]University of Liège, Gembloux Agro-Bio Tech, Animal Science Unit, 5030 Gembloux, Belgium, [5]Norwegian University of Life Sciences, 1432 Ås, Norway, [6]Quality Milk Management Services Ltd, Somerset, BA5 1EY, United Kingdom, [7]VIT, 27283 Verden, Germany; egger-danner@zuchtdata.at

For several decades breeding goals in cattle were strongly linked to increases in milk production. Due to negative genetic correlations of milk yield with fitness traits there was an accompanying reduction in genetic merit for functional traits. Herd management is therefore challenged to compensate these effects in order to allow economically feasible milk production. Functional traits, such as direct information of cow health, have also increased in importance because of animal well-being and consumer demands for healthy and natural products. Sustainability and efficiency are also increasing in importance because of growing competition for high-quality, plant-based sources of energy and protein. For data recording efforts to succeed it is crucial that there is a balance of effort with benefits. To keep the labor associated with recording reasonable it is important that existing data sources be used. Examples include the use of milk composition data to provide additional information about the metabolic status or energy balance of the animals. Recent advances in the indirect use of mid infrared spectroscopy have to be mentioned. Other data sources already may exist in countries with compulsory recording of veterinary treatments and drug use. Additional sources of data outside of the farm include e. g. slaughter houses, and veterinary labs. On farm level huge amounts of data are increasingly available from automated and semi-automated milking- and management systems. In order to develop effective selection programs for new traits, the development of large databases is necessary.

How dairy farmers can benefit from new genomic tools

J.E. Pryce[1,2], B.J. Hayes[1,2,3] and M.E. Goddard[1,2,4]
[1]Dairy Futures Cooperative Research Centre, 5 Ring Road, Bundoora, VIC, 3083, Australia, [2]Biosciences Research Division, Department of Environment and Primary Industries, Agribio, 5 Ring Road, Bundoora, VIC, 3083, Australia, [3]La Trobe University, Agribio, 5 Ring Road, Bundoora, 3086, VIC, Australia, [4]University of Melbourne, Melbourne School of Land and Environment, Campus, Parkville, 3010, VIC, Australia; jennie.pryce@dpi.vic.gov.au

To date, genomic selection has been successfully applied to male pathways of selection in dairy breeding schemes. Farmers can already achieve higher annual rates of genetic gain through using genomically tested bulls in their herds. As genotyping costs continue to fall, it will likely become increasingly popular to capture extra value from genotyping females. Genotyping females can improve farmer profitability by assisting in: (1) the identification of elite females (potential bull mothers); (2) identifying the best heifers to become herd replacements; (3) providing better prediction of the true value of an animal's genetics, that may correlate to sale price; (4) achieving certainty of parentage of individual cows; (5) avoiding inbreeding through the use of genomic assisted mating plans, where relationships between animals are quantified at the genomic level; (6) avoiding genetic defects that could arise from mating cows to bulls that are known carriers of genetic diseases that are the result of a single lethal mutation; and (7) enabling selection for special interest genes, such as A2 or the red factor.

Session 10 Theatre 5

Is genomic selection really increasing the bull portfolio available to the breeders?

F. Miglior[1,2], J. Chesnais[3], M. Sargolzaei[4,5] and B. Van Doormaal[1]
[1]Canadian Dairy Network, Guelph, Ontario, N1K 1E5, Canada, [2]Canadian Dairy Network, 660 Speedvale Avenue West, Suite102, Guelph, Ontario N1K 1E5, Canada, [3]Semex Alliance, Guelph, Ontario, N1G 3Z2, Canada, [4]L'Alliance Boviteq, Saint-Hyacinthe, Quebec, J2T 5H1, Canada, [5]Centre for Genetic Improvement of Livestock, Department of Animal and Poultry Science, University of Guelph, Guelph, Ontario, N1G 2W1, Canada; miglior@cdn.ca

The fast adoption rate of genomics stemmed from its potential to significantly increase rates of genetic progress, with a belief that it would also bring a wider portfolio of bulls to dairy producers. The objective of this study was to verify if indeed a larger pool of bulls has been available for selection. Interbull pedigrees of bulls born between 2001 and 2010 were analyzed for countries that participated to the G-MACE test run of March 2013. Additionally, a comparison for North American (NA) bulls born in 2010 (pre-screened vs. entered AI) was performed. The number of sires of sons has increased more than 2 fold (2.2) with the advent of genomics globally and across all geographical areas. However, when only bulls that entered AI were analyzed, the increase was smaller (30% for NA bulls). In 2010, 23 bulls sired 50% of all sons, compared to 11 bulls on average between 2001 and 2007. However, when the analysis was restricted to sons that entered AI service, the increase was much smaller, since 8 bulls sired 50% of all sons in 2010 versus 7 on average from 2001 to 2007. Co-ancestry among bulls in the same birth year and across birth years was also investigated. In conclusion, AI organizations are pre-screening a very large number of young bulls through genotyping, so that the size of the genetic pool has more than doubled. However, genomic evaluations tend to favor a limited number of sire and maternal grand-sire families, so that bulls that enter AI service are drawn from a genetic pool that until today has only increased slightly compared to the years before genomics.

Session 10 Theatre 6

The optimal proportion of genomically selected young bulls in the herd AI mix

A.P.W. De Roos
CRV, P.O. Box 454, 6800 AL Arnhem, the Netherlands; sander.de.roos@crv4all.com

Dairy producers use artificial insemination to get their cows pregnant and improve the performance of their herd. Many dairy farmers use genomically selected young bulls as an alternative to daughter proven bulls, primarily because their breeding values or indexes are more attractive and they understand that this leads to greater genetic improvement. Other farmers, however, are more hesitative toward using bulls without daughter proofs. They often believe that genomic breeding values for many bulls are overestimated and using young bulls leads to more cows in their herd that are underperforming. This belief is supported by examples of bulls that dropped after daughter performance data became available, and examples of re-ranking due to changes in the evaluation system. To gain confidence it is firstly important that especially the best young bulls are generally not overestimated, i.e. regression coefficients of daughter proofs on genomic proofs should be very close to 1. Secondly, farmers should get a better understanding of quality and the variation among daughters of genomically selected young bulls, in their own herds, and from comparisons of daughter-based and genomic breeding values. In the coming years, the reliability of genomic breeding values will increase further and, because of increased rate of genetic improvement, the gap between the most elite young bulls and the best daughter proven bulls will be larger. As the confidence in genomics will probably grow simultaneously, genomically selected young bulls will replace daughter proven bulls over time. The pace at which this will occur, however, is uncertain and it may be quite slow.

Session 10 — Theatre 7

How genomics can be used to avoid inbreeding and increase diversity

T. Meuwissen
Norwegian University of Life Sciences, Box 5003, 1432 Ås, Norway; theo.meuwissen@umb.no

It is expected that genomic selection will reduce inbreeding substantially, because genomic selection mainly improves the prediction accuracy of within family genetic effects. This results in an increased weight of within family effects versus between family effects in the selection process, which reduces the co-selection of family members. The reduced co-selection of family members reduces future rates of inbreeding. There are two flaws to this argument. First, in dairy cattle, the implementation of genomic selection reduces the generation interval substantially, which in itself increases rates of inbreeding (reduces numbers of selected sires per generation and increases the turn-over of generations). Moreover, the progeny test results of bulls yielded accurate information on within (and between) family effects. Second, even when not co-selecting relatives genomic selection may be increasing the frequencies of specific regions in the genome, thereby increasing the homozygosity and thus inbreeding in these regions and in neighboring regions due to the hitchhiking effect. Thus, the co-selection of relatives may be less of a driver for inbreeding in genomic selection schemes than in traditional selection, and inbreeding may be increased by focusing selection on specific chromosomal regions across family lines. Thus, genomic selection calls for genomic control of inbreeding, where the increase in homozygosity in all regions across the genome is controlled. Methods for this will be described together with their results. This, change from family based to genomic control of inbreeding is perhaps a benefit in small populations where the numbers of families is small. The genomic control of inbreeding methods may also be used to revert the trend of 'holsteinization' in small breeds but this will increase the inbreeding.

Session 10 — Theatre 8

Prediction of expected variation in progeny groups and its application in mating programs

D. Segelke[1,2], F. Reinhardt[2] and G. Thaller[1]
[1]Institute of Animal Breeding and Husbandry Christian-Albrechts-University, Olshausenstraße 40, 24098 Kiel, Germany, [2]vit w.V., Heideweg 1, 27283 Verden, Germany; dierck.segelke@vit.de

Before genomic information was implemented into cattle breeding schemes the variation of sperm and ovum breeding values (transmitting ability) could only be predicted generally. With genomic information the estimation of expected variation in gamete breeding values is feasible. As a consequence the application of this information for specific mating is possible. For example breeding companies could select animals with a high variation of the gamete breeding values to increase the probability of extreme positive offspring. In contrast dairy farmers might be more interested in homogeneous offspring and therefore in bulls inheriting low variability. 59,664 animals were phased using Beagle. Afterwards meiosis events were simulated by either passing the maternal or the paternal inherited haplotype of the animal to its potential offspring. Additionally recombination hotspots were implemented to simulate crossing over. Simulated sperm haplotypes were multiplied with SNP effects resulting from routine genomic evaluation. 100,000 gametes of an animal were simulated and the mean and the variation of the gamete breeding values were calculated. Results showed that the variations of the gamete breeding values are different and normally distributed between animals. For the trait protein yield variation of gamete breeding values ranges from 0.21 to 0.61 genetic standard deviations between animals. The results demonstrate that the variation of gamete breeding values can be predicted for young candidates. A genomic mating program was developed in order to find optimum mating partners in respect to expected mean breeding value and variation of potential offspring. The probability of an offspring having a breeding value higher than a specific threshold can be derived. Since variation of the breeding values had a significant correlation to the genomic inbreeding coefficient the mating program can also be used to minimize genomic inbreeding.

Use of microsatellite markers in genetic management of Black Thibar sheep breed in small size flock

B. Brour[1], S. Bedhiaf Romdhani[2], M. Djemali[1] and A. Bedhiaf[1]
[1]Agronomic National Institut of Tunis, Avenue Charles Nicolle, 1082 Tunis Mahrajène, Tunisia, [2]National Institute of Agronomic Research of Tunisia, Rue Hédi Karray, 2049 Ariana, Tunisia; brour_basma@hotmail.fr

The wise management of animal genetic resources is becoming an even greater challenge for small size native breeds due to dramatic effects of inbreeding. This is the case of the Black Thibar which represents 1.8% of the Tunisian sheep population. The objectives of this study were to identify genetically distant future candidate rams to avoid inbreeding and to suggest a management breeding plan. The genomic DNA was extracted from 124 rams belonging to five flocks farms located in North of Tunisia. DNA samples were amplified by six microsatellites (BM1258, INRA0063, INRA0132, MAF0065, MCM0527, OarFCB0011). Within flocks, important levels of similarity was detected and the minimum coefficients of similarity found were 0,79; 0,70; 0,72; 0,63 and 0,65 respectively in the farm UCP Sidi Mbarek, UCP Montassar, UCP Ain Chalou, UCP Iadh and UCP Loubira. Based on these results, genetically distant candidate males were identified and proposed for a mating plan to reduce inbreeding in target flocks. For UCP Sidi Mbarek, 4 rams out of 9 individuals were selected because they were the more distant genetically. For UCP Montassar, out of 12 individuals, seven candidate rams were worth keeping. For UCP Ain Chalou, out of the 23 individuals genotyped, it was suggested to select only 13. Out of the 42 rams of UCP Iadh flock, 21 candidate rams were kept. Concerning UCP Loubira, only 18 were kept. It was found that exchanges between UCP Sidi Mbarek and UCP Montassar should be avoided because their individuals were closely related. Individuals of UCP Loubira should not be exchanged with other farms because they have a high similarity coefficient with other individuals. These results translated the intensity of inbreeding observed in the target flocks and demonstrated the usefulness of molecular techniques to optimize genetic management of flocks.

Session 11 Theatre 1

How sustainable animal breeding is helping to feed the world

D.E. Howard
European Forum of Farm Animal Breeders, rue de Treves 61, 1040 Brussels, Belgium; dawn.howard@effab.info

The Foresight report 'Global Food and Farming Futures' estimates that by 2050 our agricultural production systems will have to be capable of feeding a world population of over nine billion people. Policymakers face the challenge of balancing global food access whilst maintaining biodiversity and vulnerable ecosystems. Animal breeding and the selection of suitable breeds adapted for particular environments has a key role to play in helping to address the challenges of food security, sustainable production systems and ecosystem preservation. Modern animal production systems benefit from animal breeding improvements, where productivity gains are carefully balanced with health, welfare and genetic diversity. The genetics of animal breeds have evolved naturally over a long period of time and more recently have been managed within intelligent breeding programmes to deliver the range of breeds we see today. Maintaining sustainable production requires breeding programmes with a wide genetic base, avoiding inbreeding and allowing the continued introduction of improved genetics. Improvements in feed efficiency are delivering real environmental benefits, by reducing the land area and water needed to produce animal feed and thereby reducing GHG emissions. Developing economies face particular challenges, needing access to improved livestock genetics and productivity to secure their own food security within the limitations of local conditions. Better access to animal protein can help to lift local economies out of poverty; sustainable intensification becoming as important as for developed economies. Increasing urbanisation leads to a loss of both land and skills, requiring more efficient, easily managed breeds. The use of improved breeds provides major increases in productivity, improvements in disease resistance and quality of life. Sustainable animal breeding will continue to play an important role in securing future global food security.

Session 11 Theatre 2

Global layer breeding with special focus on sustainability

R. Preisinger
Lohmann Tierzucht GmbH, Am Seedeich 9-11, 27472 Cuxhaven, Germany; preisinger@ltz.de

Hybrids of multiple line crosses are used for table egg production worldwide. Commercial hybrids show outstanding fertility and livability as compared to local pure lines and their respective crosses. Commercial layers have to be bred to perform adequately in a variety of systems ranging from small scale free-range management to modern, fully air-conditioned and large intensive cage units – under different environmental conditions worldwide. Aside from better performance, commercial layer hybrids can either be feather-sexed or colour-sexed as day-olds. If commercials are used for breeding purposes, a significant drop in performance and the opportunities for feather or colour sexing would be gone. The gene pool of pure lines for poultry distributed globally are in the hands of private owners and customers can only buy sexed parent stocks. Local breeding programmes in Asia are based on European or North American genetic stocks which were purchased from commercial breeding companies years ago. In Africa, local chicken strains still contribute a significant share in the production of rural chicken where males are used for meat production and females for the production of eggs. As soon as performance data are recorded and balanced diets from local or imported raw materials are available, local strains would be replaced by imported parent stocks. The major difference between local strains and imported ones can be observed for feed efficiency. Breeders are faced with the need to forecast the demands of producers and consumers alike and to select stocks with special attributes at least five years ahead of market realisation. For the global business, diverse markets have to be served and each of these seeks different performance profiles of the commercial layers. This requires extensive gene pools comprising of elite lines which are combined to generate specific commercial products. The cost of maintaining and developing new lines, testing, selecting and reproducing primary stocks, imposes high fixed costs in the operation and very good skills in quantitative genetics.

Session 11 Theatre 3

The decreasing role of native genetic resources in modern Ukrainian pig production

O. Kravchenko[1], A. Getya[2] and O. Kodak[3]
[1]Poltava state agrarian academy, Production and Processing Technologies of Animal Products, str. 1/3 Skovorody, 36003 Poltava, Ukraine, [2]Ministry of agrarian policy and food of Ukraine, Department of animal production, str. Khreschatyk, 24, 01002 Kiev, Ukraine, [3]Kaposvar university, Animal science, str. Guba Sándor u. 40., 7400 Kaposvár, Hungary; oksanakravchenko@ukr.net

Conservation of animal genetic recourses is a global issue and the reported rate of breed extinctions is of great concerns. In the livestock sector of Ukraine genetic erosion concerns mostly local, native breeds which are replaced by a narrow range of high-yielding exotic breeds considered to have a competitive advantage in more intensive production systems. There is a big threat that Ukraine loses its national genetic recourses being replaced by breeds not well adapted to local climatic, environmental and technological conditions. Currently there are 6 native pig breeds in Ukraine: Mirgorodska (M), the Ukrainian white steppe (UWS), the Ukrainian spotted steppe (USS), Poltava meat (PM), Ukrainian meat (UM) and the Red white belt (RWBB). An analysis of occurrence and geographical distribution of Ukrainian pig breeds revealed for the years 2006-2012 a massive reduction in the number of breeding farms keeping those native breeds, namely: M: by 62.5%; UWS: by 72.7%; PM: by 68.4%; UM: at 43.5%; RWBB: by 41.2%. Most critical is the number of sows: Mirgorodska: 383 heads; Ukrainian white steppe: 334 heads; Poltava meat: 469 heads; and Ukrainian spotted steppe: 26 heads. Therefore the most urgent problem of Ukraine, is the development and implementation of strategies to maintain its autochthonous animal breeds, but a conservation program cannot be established without reliable characterization and monitoring of breeds. Breed certification, establishing definitions for what to be considered as pure bred, detection and timely elimination of crossbred animals, are currently considered to be the most urgent tasks.

Session 11 — Theatre 4

Crossbreeding dairy cattle: introduction and impact on tropical smallholder production systems

R. Roschinsky[1], J. Sölkner[1], R. Puskur[2] and M. Wurzinger[1]
[1]BOKU University of Natural Resources and Life Sciences, Department of Sustainable Agricultural Systems, Gregor Mendel Strasse 33, 1180 Vienna, Austria, [2]The World Fish Centre, P.O. Box 500, GPO 10670 Penang, Malaysia; romana.roschinsky@boku.ac.at

In many tropical countries exotic, high yielding dairy cattle breeds have been introduced by crossbreeding them with local, well adapted breeds. This is done to quickly increase dairy production to satisfy growing market demand for dairy products. Bodies responsible for crossbreeding introduction advocate the genetic potential of crossbreds. Apart from productivity increase, livelihood improvement of smallholders and increasing market participation are desired effects. Various challenges hinder a widespread adoption of crossbreeding and substantial changes of production systems have beneficial and detrimental effects. Success stories are rare. These facts question the sustainability of dairy cattle crossbreeding in the region. For this study empirical evidence has been collected from selected sites in Ethiopia, Uganda and India, where crossbreeding has been introduced. Interviews with 248 farmers using pre-tested questionnaires have been conducted in 2011 and 2012 and qualitative and quantitative data has been analysed using SAS software. We present three local contexts accommodating different production systems and intensity levels. We identify responsible drivers for the introduction of exotic genetics. These range from government extension and non-governmental organisations to progressive farmers. We discuss inherent challenges of introducing exotic dairy breeds and impacts of crossbreeding at farm level and farmers' perceptions on selected productive and reproductive traits of local and crossbred cows. We introduce which management and breeding practice changes take place. Our findings contribute to the current debate on sustainability of crossbreeding dairy cattle in the tropics by adding a farm level perspective to the discourse.

Session 11 — Theatre 5

Influence of Holstein Friesian and other varieties on Lithuanian native cattle

R. Sveistiene and E. Kauryniene
Institute of Animal Science Lithuanian University of Health Sciences, Animal Breeding and Genetics, R. Zebenkos 12, 82317, Lithuania; ruta@lgi.lt

Lithuanian Ash Grey (LAG) and Lithuanian White-Backed cattle (LWB), cattle were bred in Lithuania from the ancient times and are specific to the country. In 20th century these breeds were assimilated of Lithuanian Black-and-White and Lithuanian Red cattle's and where improved by foreign breeds. During last decades Lithuanian native cattle were intensively improved by using Duch, Danish, British, German Black and White cattle breeds and also American and Canadian Holstein-Fressian cattle breed. Restoration and separate registration of the LWB and LAG cattle have started in 1994. The conservation of native cattle is to coordinate conservation approaches with elimination of foreign genes from population what means low productivity. Some stakeholders are primarily interested in breed conservation, while other focuses to production increase. The aims of the study were to identify and to evaluate influence of Holstein Friesian and other varieties on productivity and population diversity of Lithuanian native LWB and LAG cows. The influence of the Holstein and other breeds on the whole cattle population was assessed using the contribution of the founders, the contribution of the ancestors. The genetic structure was studied from pedigree records. Data was performed using descriptive statistics in Statistica. In 2006, gene proportions in pedigree of cows from sire side reached 60% of different international breeds. 31% and 14% cows in LWB had Holstein and Germany's Black and White genes, respectively. LAG cows from sire side had 19% Holstein, 7% German, 7% Dutch Black and White and 5% British Friesian. LWB cows with unknown pedigree showed 1,533.3 kg and 2,054.8 kg lower milk yield than 75% and 87% Holstein cows, respectively. They also had lower milk fat and protein contents than cows with Holstein immigration.

Imported genetics as a basis of a sustainable development of beef cattle sector in Poland

P. Dakowski[1], B. Konopka[1], E. Martyniuk[2], J. Denkiewicz[1] and A. Strawa-Harasymowicz[1]
[1]Polish Association of Beef Cattle Breeders and Producers, ul. Rakowiecka 32, 02-532 Warszawa, Poland, [2]Warsaw Univeristy of Life Sciences, Department of Genetics and Animal Breeding, ul. Ciszewskiego 8, 02-786 Warszawa, Poland; pawel.dakowski@gmail.com

Traditionally, the cattle husbandry in Poland was based on utilization of dual purpose breeds that were selected mainly for milk performance, with beef being a secondary product from these primarily dairy herds. Initiation in the mid 1980s intensive use of HF led to a gradual transformation of the active dual-purpose population into a dairy type. In March 1994, a national beef cattle breeding programme was approved by the Minister of Agriculture and Food Economy, with the aim to initiate development of this sector in Poland, to provide high quality beef to consumers. At the time, there was no native beef breed in the country, and thus, the programme was based on extensive importation of high performance beef cattle breeds, mainly from France, the United States of America and Italy; later with imports from Canada, Germany, Sweden, and Hungary. Key imported breeds included: Aberdeen Angus, Charolaise, Hereford, Limousine, Piemontese and meat Simmental, other breeds, in smaller numbers, have followed over time. The aim, as defined in 1994, was to develop after the year 2000, a purebred population of 10,000 cows and backcrossed beef commercial population up to 1 million cows. At present, 14 beef cattle breeds are under recording scheme conducted by the Polish Association of Beef Cattle Breeders and Producers. In 2011, the total purebred population recorded in herd books included 16,216 cows while the backcrossed population 7,459 cows only. The Limousine breed proved to be the most popular among beef cattle breeders and producers, with the share of 69.75% and 89.19% of the purebred and crossbred population respectively. The paper provides analysis and lessons learnt from implementation of the programme and the SWOT analysis for the beef cattle industry in Poland developed on imported genetics.

Inbreeding and impact of foreign dairy cattle breeds in the German Angeln dairy cattle population

D. Hinrichs and G. Thaller
Institute of Animal Breeding and Husbandry, Christian-Albrechts-University, Olshausenstrasse 40, 24098 Kiel, Germany; dhinrichs@tierzucht.uni-kiel.de

The German Angeln dairy cattle breed is a small population located in Northern Germany. During the last decades this breed has been crossed with red dairy cattle breeds, e.g. Red Holstein. The aim of this study was the estimation of different inbreeding coefficients, i.e. classical- and ancestral-inbreeding, and to analyze the relationship between these different inbreeding measurements. The base year for the analysis of the inbreeding coefficients was 1950. Furthermore, the genetic contribution of other dairy cattle breeds to the Angeln dairy cattle breed was analyzed. Finally, the impact of the most important common ancestors was estimated. Therefore, two different reference populations were defined. Reference population 1 includes all animals born between 2000 and 2004 and reference population 2 summarizes animals born between 2005 and 2009. The complete pedigree file includes 75,264 animals and 42,469 of them were inbred. The mean classical inbreeding coefficient was 0.02 and the average ancestral inbreeding coefficient was 0.04. The overall correlation between classical and ancestral inbreeding was 0.14. During the last decades the genetic contribution of foreign dairy cattle breeds increased from 18.6% to 43.0%, and from 0.5% to 6.8% for Holstein and Swedish Red and White dairy cattle from 1980 to 2009, respectively. In reference population 1 the marginal genetic contribution of the 25 most important common ancestors could explain 58% of the genetic variation. The amount of genetic variation, explained by the marginal genetic contribution of the 25 most important ancestors increased in reference population 2 to 65%. All in all, the present study shows an increasing impact of foreign dairy cattle breeds. Therefore management of contributions from foreign breeds and management of inbreeding are important, and should be considered in the German Angeln dairy cattle breeding program.

Recovery of native genomes of conserved breeds with optimum contribution selection

R. Wellmann and J. Bennewitz
University of Hohenheim, Institute of Animal Husbandry and Animal Breeding, Garbenstraße 17, 70599 Stuttgart, Germany; r.wellmann@uni-hohenheim.de

In the course of the intensification of production, a small number of economically superior breeds have displaced many local varieties of livestock species and were often used to improve the economic value of the remaining ones. As migrant contributions increased, these crossing programs have eroded the gene pool the local varieties have inherited from their native ancestors, which makes the conservation worthiness of these breeds questionable. Thus, objectives of breeding programs for endangered breeds are not only to increase their economic values, but also to recover a large amount of the native genomes and to diminish migrant contributions. Breeding programs are required for these breeds that are able to account for these conflicting objectives. The conditional gene diversity of a breed is a suitable parameter for quantifying the diversity of genes inherited from native founders. The conditional gene diversity was defined as the conditional probability that two randomly chosen alleles are not identical by descent (IBD), given that both descend from native founders. It affects the number of genome equivalents originating from native founders. Native genome equivalents were defined as the minimum number of unrelated founders that would be needed to establish a new population whose gene diversity is as large as the conditional gene diversity of the breed under consideration. The aim of this study was to compare different approaches for optimum contribution selection with respect to their ability to reduce migrant contributions, to conserve gene diversity, and to increase native genome equivalents by using pedigree data of three German cattle breeds. A recommendable optimization approach is maximizing the probability that two alleles randomly chosen from the offspring are not IBD and that at least one of them descends from a native founder. A constraint may be added that ensures a sufficient level for the breeding values in the offspring population.

Toll-like receptor gene polymorphism in the indigenous Czech cattle breeds

K. Novak
Institute of Animal Science, Molecular Genetics, Pratelstvi 815, 104 00 Prague 22, Czech Republic; novak100@centrum.cz

The two historical Czech cattle breeds, Czech Red and Czech Pied, are conserved in frame of the national program of Conservation and Use of Farm Animal Genetic Resources. Both traditional breeds can potentially harbour rare variants of disease resistance genes thanks to the adaptation to the local conditions and the preserved intravarietal diversity. The value of additional resistance gene variants is given by the changing array of microbial pathogens, current limits on the antibiotic use, and decreasing genetic diversity in modern breeds. Screening has been initiated for the structural polymorphism of genes controlling innate immunity, namely, the genes coding for the Toll-like receptors that participate in the interactions with the bacterial pathogens. The survey is confined to 50 individuals of each breed. The capillary sequencing of the PCR-fragments of the TLR4 gene discovered at least four haplotypes, however, none of them was specific for the local breeds. The comparatively low diversity in the investigated locus is associated with the population bottleneck in both breeds and/or outcrossing with modern dairy breeds. On the other hand, the found substitution leading to the exchange Ile674Thr in the highly conserved TIR region of the receptor is known to be associated with increased infection resistance in the Canadian and Chinese Holstein populations. The validation of its effect for the historical breeds and the commercial herds of Czech Pied is in progress. In order to increase the screening efficiency in the remaining nine bovine TLR genes (TLR1-3, TLR5-10), the parallel sequencing of pooled PCR amplicons has been applied using the PacBio platform.

Session 12 Theatre 1

Prediction of polyunsaturated fatty acid content in bovine muscle

B.P. Mourot[1], D. Gruffat[1], D. Durand[1], G. Chesneau[2], G. Mairesse[2] and A. Lebert[3]
[1]INRA-VetAgro Sup, UMRH 1213, Theix, 63122 Saint-Genès Champanelle, France, [2]Valorex, La Messayais, 35210 Combourtillé, France, [3]Institut Pascal, UMR6602 UBP/CNRS/IFMA, 24 Avenue des Landais, BP 80026, 63171 Aubière, France; benoit-pierre.mourot@clermont.inra.fr

In the development of health oriented meat industry, characterization of fatty acid (FA) composition of muscles directly on bovine carcasses could have significant economic impact for all players in the sector. However, standard methods to evaluate this composition (lipids extraction from muscle and gas chromatography analyses) are not adapted to industrial rate. An innovate methodology, the near-infrared spectroscopy (NIRS), has been developed to provides fast, nondestructive, and cost-effective measurements. NIRS has potential for rapid routine measurements of FA composition. Previous studies reported a correct prediction of major FA (C16:0, C18:0, C18:1 cis 9, saturated FA, monounsaturated FA) in meat with NIRS whereas prediction of essential FA as polyunsaturated FA (PUFA) is low. Thus, to compensate NIRS's deficiencies, our study aimed to develop predictive equations for PUFA from major FA. For this purpose, a database integrating data from the literature or obtained in our laboratory on the impact of breeding factors on the FA content of beef was built (91 publications, 407 treatments). Multiple linear regression analysis were performed for each FA to obtain correlation coefficients. Currently, C18:2 n-6, C20:4 n-6 and total PUFA are correctly predicted with adjusted R-squared of 0.86, 0.86 and 0.88, respectively. Prediction models for other PUFA are still under investigation. Finally, these models will be validated with an external set of data. Concomitant use of prediction models and NIRS should facilitate determination of meat FA profile on carcasses in industrial slaughter chains and so optimized carcass sales through their nutritional quality.

Session 12 Theatre 2

Comparing environmental impacts of diverse beef production systems

I.J.M. De Boer
Wageningen University, Animal Production Systems Group, P.O. Box 338, 6700 AH Wageningen, the Netherlands; imke.deboer@wur.nl

Literature shows that beef production results in a relatively high environmental impact. My objective was to compare assessments of environmental impacts of diverse beef production systems in order to track down main differences. I reviewed eight studies that used life cycle assessment (LCA) to compare beef production systems. Systems were characterized mainly by origin of calves and type of feed used during calf fattening. Calves were bred by a dairy cow (dairy-based system) or a suckler cow (suckler-based system) and were fattened on mainly grass or hay (pasture based), on mainly concentrates (feedlot), or a combination of both. In suckler-based systems, maintaining the mother cows is the dominant contributor to all impacts, which is attributable to the low reproductive rate of cattle. This also explains why impacts were lower for dairy-based than for suckler-based systems. In a dairy-based system, the majority of the environmental impact for production and maintenance of the 'mother cow' is allocated to milk and not to beef. Using dual-purpose cows, therefore, might be a way to produce milk and beef in an environmental-friendly way. The environmental performance of suckler-based systems can be improved by reducing age at first calving and the calving interval, and by improving growth rate of the calf. The choice of grass-finished versus feedlot-finished beef is less clear. Studies showed that fossil energy use can be higher or lower for grass-finished than for feedlot-finished beef, depending on intensity of grassland management. Similarly, emissions of GHGs can be higher or lower for grass-finished beef than for feedlot-finished beef, depending on potential for carbon sequestration of grassland. Moreover, a comparison of grass-finished versus feedlot-finished beef is hindered because LCA results do not include environmental consequences of competition for land between humans and animals. This negatively affects performance of beef produced on grassland not suitable for crop production.

Sustainability index for beef production in Denmark and Sweden

A. Munk[1], E. Søndergaard[1], T. Kristensen[2], L. Mogensen[2], N.I. Nielsen[1], M. Trinderup[1], R.S. Pedersen[1] and H.B. Bligaard[1]
[1]AgroTech, Agro Food Park 15, 8200 Aarhus N, Denmark, [2]Aarhus University, Dep. of Agroecology, Blichers Alle 20, 8830Tjele, Denmark; evs@agrotech.dk

Production of beef in Denmark and Sweden derives from very varied production systems, from intensive bull production to extensive suckler cow systems based on grazing of seminatural areas. Beef production – in particular extensive suckler cow systems – is regarded as having a high negative impact on the climate compared to meat production from pigs and broilers but it may have some strengths in relation to e.g. biodiversity and animal welfare, which are not considered when only the climate impact is in focus. In order to make an overall assessment of the impact of different beef production systems, an indicator-based sustainability index was developed. The aim was to create an index based on data that already exist at the farm, or data that was relatively easy to collect. The index should be suitable for benchmarking both between farms and/or production systems, but also as documentation or certification to authorities, slaughterhouses or consumers. The index was developed based on scientific literature and practical experience from similar projects, and has been tested on 3 Danish farms. The index includes seven sub-indices including animal welfare, environmental impact, climate impact, use of resources, social responsibility, economy and biodiversity. An on-line platform was developed for handling data collection on farm, calculation and presentation of the index. The test showed that the index is able to assess the sustainability of various beef production systems. However, the index needs to be evaluated before it will be ready for general use on farms in Denmark and Sweden.

Environmental impacts of different beef production systems

M. Alig, F. Grandl and T. Nemecek
Agroscope, Reckenholzstrasse 191, 8046 Zurich, Switzerland; martina.alig@agroscope.admin.ch

The effects of meat production on the environment have become a topic of great concern. Especially beef production is under pressure. Therefore, the environmental impacts of beef production systems have to be analysed and quantified, in order to identify and mitigate their negative impacts on the environment. By using the LCA-method SALCA (Swiss Agricultural Life Cycle Assessment, developed by ART), we analysed three different beef production systems in three countries: conventional bull fattening based on feeds from arable land in Germany and Switzerland, organic bull fattening based on grassland and grassland based suckler cow systems (integrated and organic) in Switzerland as well as extensive pasture based beef production in Brazil. The following environmental impacts were analysed: non-renewable-energy demand, global-warming potential, ozone formation, demand for phosphorus and potassium resources, competition for land, competition for arable land, deforestation, blue water use, eutrophication, acidification, ecotoxicity and human toxicity. Additionally, potential impacts on biodiversity were assessed for the Swiss systems. The results per kg live weight showed that suckler cow systems had higher environmental impacts in most categories due to the fact that all environmental impacts were attributed to meat production, whereas in the bull fattening systems the fattening animals are treated as co-products of milk production. Still, the analysed suckler cow systems used less arable land (but more grassland) and had lower impacts regarding deforestation and ecotoxicity. A trade-off between productivity and biodiversity was found. In conclusion, none of the analysed systems presented a clear overall environmental advantage. Improvements regarding the environmental impacts of beef production systems require detailed analysis of the respective system as well as a clear definition of its objectives. One of the main factors influencing the environmental performance is the overall system efficiency.

Carbon footprint of typical beef production systems in Denmark and Sweden

L. Mogensen[1], T. Kristensen[1], N.I. Nielsen[2], M. Henriksson[3], C. Svensson[3], M. Vestergaard[1], P. Spleth[4] and A. Hessle[3]
[1]Aarhus University, Blichers Allé 20, 8830 Tjele, Denmark, [2]AgroTech, Skejby, 8200 Århus N, Denmark, [3]SLU, P.O. Box 86, 230 53 Alnarp, Sweden, [4]Knowledge Centre for Agriculture, Skejby, 8200 Århus N, Denmark; lisbeth.mogensen@agrsci.dk

The aim of this work was to assess the effects of typical beef production systems in Denmark (DK) and Sweden (S) on greenhouse gas (GHG) emissions. The existing beef production systems in DK and S were initially identified and defined, resulting in 5 typical Danish and 4 typical Swedish systems. For each system, typical feed rations and figures for daily gain, feed conversion and carcass yield, that represent average productivity of the given system, were collected. GHG emissions from the systems were calculated in a Life Cycle Analysis (LCA) perspective, which means that all emissions in the chain until the animals leave the farm were included. The total emission of GHG was calculated for each system over a year and expressed per kg carcass weight. Carbon footprints (CF) for feedstuffs were calculated based on typical national yields and cultivation systems in DK and S. Methane from enteric fermentation was estimated according to new equations developed from trials where typical Nordic diets have been fed. A CF was calculated for the input of 'a dairy calf' to the beef system. The manure collected was handled as an output in the way that each beef system got credit for this production corresponding to the fertilizer value of the manure. CF from production of 1 kg 'meat' showed a variation from 8.7 kg CO_2 in the most intensive system with fattening of bull calves coming from dairy herds up to 30.7 kg CO_2 in an extensive suckler cow system. Production of feed and enteric fermentation of feed were the major hotspots responsible for 28-58% and 28-56% of total emission, respectively. In the extensive suckler cow system, the highest single contribution was enteric methane, whereas in the intensive bull calf fattening system, it was GHG from feed production.

Variability among individual young beef bulls and heifers in methane emissions

G. Renand[1], E. Ricard[2], D. Maupetit[3] and J.-C. Thouly[3]
[1]INRA, GABI, 78352 Jouy en Josas, France, [2]INRA, SAGA, 31326 Castanet Tolosan, France, [3]INRA, Bourges-La Sapinière, 18390 Osmoy, France; gilles.renand@jouy.inra.fr

Enteric methane fermentation represents the principal contribution to climate change of beef production. Several strategies are currently explored for mitigating GHG emissions. The diet composition is obviously essential. Among animals fed the same diet, the estimation of the genetic variability requires a large number of animals being measured in adequate family structures. The measurement of beef cattle methane emissions is difficult. Direct measurement in respiratory chambers is possible only on a very limited number of animals and the SF6 tracer technique cannot be applied on hundreds of animals. Up to now, genetic parameters have been estimated only in dairy cattle for predicted methane emission or for methane emission rate during milking. In order to measure the methane emission rate on consistent number of beef cattle, three GreenFeed® stations were installed in an INRA experimental farm with Charolais cattle. Preliminary results were obtained with 18 young bulls aged 10-11 months fed with a medium energy pellet diet during 6 weeks. Gas emission was measured when animals visited the stations where pellet food was distributed in small fractional amounts so that the animals stay 5 minutes on average per visit. The measurements of the CH_4 concentrations while eating and of the air flow rate allowed calculating the mass of gas emitted per unit of time (g/day). The 7,860 methane flux measurements averaged 233 g/day. Significant differences were observed between visit hours (CV_h=10%). Among the 42 daily mean measurements, the date, animal and residual CV were respectively 9, 16 and 13% of the mean respectively. The repeatability of the animal effect averaged 0.72 within weekly periods and was 0.61 over the whole test period. Correlations between two consecutive week measurements averaged 0.89 and lowered to 0.55 between the first and the last two weeks.

Economic benefits of adopting Meat Standards Australia to the Beef Industry

D.W. Pethick[1], P. Mc Gilchrist[1], J.F. Hocquette[2] and J.M. Thompson[3]
[1]Murdoch University, School of Veterinary and Life Sciences, Murdoch, WA, 6150, Australia, [2]INRA-VetAgro Sup, UMRH 1213 Theix, 63122 Saint Genes Champanelle, France, [3]University of New England, School of Environmental and Rural Science, Armidale, NSW, 2351, Australia; d.pethick@murdoch.edu.au

The Meat Standards Australia (MSA) beef grading scheme uses commercial inputs from the production, processing and value adding sectors to predict eating quality of individual muscles which are graded into 4 categories (ungraded, 3, 4 and 5 star) specific to the cooking method. From small beginnings in 2000 with only 200,000 carcasses graded, the voluntary scheme has grown in Australia to over 2 million carcasses sourced from 22,794 registered producers in 2012. Overall a benefit cost ratio of about 4 to 1 out to 2020 has been attributed to the MSA research and adoption program. Further studies have estimated the financial benefits through the supply chain to the retailer, wholesaler and the producer which deliver on average $AUS 0.3 per kg carcase weight. These benefits were obtained despite the majority of retailers simply selling MSA 3 star beef and not dividing it further into quality categories. Taste panel studies consistently indicate that consumers are willing to pay substantial premiums for 4 and 5 star grades. Already supply chains are moving to market the 4 and 5 star graded cuts which have the potential to double the benefit. Increasingly there is pressure from industry to develop transparent feedback systems which link the payment for carcases to eating quality and saleable meat yield. Additional attributes of beef like healthiness, animal welfare and carbon footprint may also be important considerations for beef consumers, especially in the European market, but it is unclear if these attributes will be incorporated into value based trading or simply be demanded at no extra cost.

The EUROP carcase grading system does not predict the eating quality of beef

S.P.F. Bonny[1], I. Legrand[2], R.J. Polkinghorne[3], G.E. Gardner[1], D.W. Pethick[1] and J.F. Hocquette[4]
[1]Murdoch University, School of Veterinary and Life Sciences, 6150 Murdoch, Australia, [2]Institut de l'Elevage, Service Qualité des Viandes, MRAL, 87060 Limoges Cedex 2, France, [3]431 Timor Road, Murrurundi, NSW 2338, Australia, [4]INRA-VetAgro Sup, UMRH 1213 Theix, 63122 Saint Genes Champanelle, France; spfbonny@gmail.com

The European beef grading systems of EUROP and Carcase Fatness are used to determine carcase value throughout Europe. We investigated the relationship between EUROP class and eating quality. Six cuts (the Striploin, Outside, Rump, Tenderloin, Oyster blade and the Topside) from 18 French cattle were grilled, medium or rare. In total, 540 untrained French consumers rated the steaks for Tenderness, Flavour, Juiciness and Overall Liking, according to MSA protocols, and these scores were combined on a weighted basis (0.3, 0.3, 0.1, 0.3 respectively) to make a fifth term called Meat Quality score (MQ4). During processing the cattle were graded using the European 'EUROP' and 'Carcase Fatness' systems. The sensory scores were analysed using a mixed linear model with cut, age and EUROP score included as fixed effects, and animal ID as a random term. Due to the low numbers of animals, Carcase Fatness scores were selected from the limited range of 3-, 3= and 3+ and hence do not significantly influence eating quality. In all cases, the sensory scores differed between the EUROP classes. However, these differences followed no clear linear trend across EUROP classes within any of the cuts tested. Thus, while the EUROP system may adequately describe carcase muscling characteristics, it does not predict eating quality. The industry should consider using a system more related to eating quality to determine the monetary value of carcases, rewarding those producers supplying the meat most preferred by consumers.

Beef production in France: economic and environmental performances of suckler-cattle farms

P. Veysset
INRA-VetAgro Sup, UMR1213 Herbivores, 63122 Saint-Genès-Champanelle, France; veysset@clermont.inra.fr

French suckler herd provides two-thirds of the beef consumed in France, and is the main European supplier of weanling cattle to the Italian market. The Massif Central, a large disadvantaged grassland area of central France, owns 42% of French suckler cows. The farms number in the Massif Central, as everywhere in France, has decreased over the past decades (-25% from 2000 to 2010), while the utilised agricultural area remained stable and suckler cows number increased (+33% over 10 years). Labour productivity of suckler farmers increased by 50% over the last 20 years. These productivity gains have just allowed maintaining the annual farm income per worker (€ 22,000) at the cost of a high capitalization. Productivity gains on farms were redistributed as: (1) lower prices for beef (-25% from 1992 to 2011); and (2) purchase of equipment by the farmer. The global warming, and therefore the greenhouse gas emissions and non renewable energy (NRE) consumption, is a new challenge for the beef farmers. Techniques and practices aiming to reduce CH_4 emissions had a very low impact, even no, on the total GHG emissions at the farm scale. Systemic analysis of results from 59 farms showed that the big, diversified farms (mixed crop-livestock farming systems) have a more negative environmental impact than the moderate-sized, specialized (beef production) farms The less-GHG-emitting farms (8.00 kg eqCO_2/kg live weight) had an average size of 144 ha and 128 Livestock Units (LU), the most-GHG-emitting farms (12.05 kg eqCO_2/kg lw) kept 221 LU on 210 ha including 30 ha of cash crop. Animal productivity performances decrease with increasing herd size, and inputs use is below-optimal in the most strongly diversified farms. Through better animal productivity performances and lower use of inputs, the less-GHG-emitting farms also generate higher income per worker (+30%) while consuming less NRE. Our findings argue against the idea that size and diversification bring economic and environmental economies of scale and scope in suckler-beef production systems.

Limousin beef farms trajectories from 2000 to 2010: structural, technical and economic assessment

E. Sanne[1], S. Enée[2], R. Faron[3], D. Guichette-Debord[4], L. Aymard[5], S. Brisson[6], M. Besson[7] and S. Brouard[1]
[1]Institut de l'Elevage, 149 rue de Bercy, 75595 Paris, France, [2]Chambre d'Agriculture Charente, 16016 Angoulême, France, [3]Chambre d'Agriculture Corrèze, 19001 Tulle, France, [4]Chambre d'Agriculture Creuse, 23011 Guéret, France, [5]Chambre d'Agriculture Dordogne, 24060 Périgueux, France, [6]Chambre d'Agriculture Vienne, 86550 Mignaloux-Beauvoir, France, [7]Chambre d'Agriculture Haute-Vienne, 87017 Limoges, France; emma.sanne@idele.fr

Beef farms have been facing several troubles during the last decade: sanitary crisis, CAP reform, unforeseen climate events, economic condition fluctuations … Farms are continuously evolving due to the global context but also to farmers' choices. Thirty-seven beef farms located in the Limousin area (in the center of France) have been monitored from 2000 to 2010 according to the French Livestock Farm Network standards. Structural, technical and economic data were collected through the global system approach used by the stakeholders of this partnership e.g. farmers, Chambers of Agriculture and the French Livestock Institute. Structure of farms slightly changed, particularly in terms of size (hectares, number of cows, and number of calving per worker). However, the functioning of the production systems was not disrupted with regard to the type of animals sold and the stocking rate. From an animal point of view, culled cows' carcass weights increased by 10% but reproduction performances decreased by 3 points. As a consequence, there is no gain of animal productivity, assessed by the kilograms of live weight per livestock unit, at a farm level. As regard profitability, farms decreased their economic efficiency by 7 points in relation to costs development and farm management. Having said so, changes in structures and labor productivity rise minimized the loss. Finally, 13 specialized beef farms out of 33 improved their economic results from 20,470€ per farmer to 29,940€. This is thanks to gains in labor productivity combined with control of costs and investments.

Characterization of intensive beef production system of North East Italy

G. Cesaro, E. Sturaro, M. De Marchi, G. Bittante and L. Gallo
University of Padova (Italy), DAFNAE, Viale del l'Università 16, 35020 Legnaro, Italy; giacomo.cesaro@studenti.unipd.it

Young bulls and beef heifers are the main cattle categories of the Italian production of beef meat, accounting for over 70% of domestic cattle meat supply. The majority of this production is achieved in intensive farms located in the Po Valley, specialized in fattening of young animals imported from other European countries (mainly France). Animals are reared in pens and fed total mixed rations based on maize silage and concentrates. Although beef production in north Italy can be considered as distinguishing within European beef production scenario, benchmarks concerning performance and feeding traits are still scarcely known. On the basis of results from a survey on fattening units located in Veneto region (north east Italy), this study aimed to provide reference values on main productive characteristics of intensive young bulls production system of north Italy. Data were collected in 17 herds and included information on growth performance and diets characteristics such as ingredients and chemical composition. The reference unit of the survey was a farm batch, intended as a group of young bulls homogeneous for genotype, fattening period and diets. The final database included 237 batches with an average size of 67 young bulls. Productive data and diet composition were analyzed using ANOVA with farm and breed as fixed effects. Average weight at starting and finishing fattening period was 370 and 669 kg, respectively, average length of fattening period was 232 d and average daily gain was 1.31 kg/d. Diets were based on a common base of maize silage and soybean meal; CP, starch and NDF of the diets averaged 14.3%, 32.7% and 32.5%, respectively. Farm and genotype effects were significant sources of variation for productive and feeding traits. The future perspective is to approach the environmental sustainability of this production system by developing indicators able to consider the interaction between land use and carbon footprints of 'north Italy' beef meat.

Nitrogen excretion in fattening beef in conventional vs. extensive and sustainable farm systems

D. Biagini and C. Lazzaroni
Università degli Studi di Torino, Dipartimento di Scienze Agrarie, Forestali e Alimentari, via L. da Vinci 44, 10095 Grugliasco, Italy; carla.lazzaroni@unito.it

Extensive and sustainable productive systems, as the organic one, have the reputation to have several positive effects on environmental and socio-economic aspects, but also some negative ones could be pointed out, especially regarding rearing and nitrogen efficiency. N excretion has been studied in two groups of Piemontese beef cattle (10 animals each) fed according to conventional (2 kg/d of hay forage and 3-8 kg/d of concentrate) or organic (3.5-8 kg/d of hay forage, 60% of DM intake and 2-3 kg/d of concentrate) farming systems during the growing and fattening period (200-550 kg live weight). Monthly individual weights, average daily weight gain (ADG), daily feed consumption, and feed conversion rate (FCR) were recorded, and after slaughter (at about 16 and 20 months of age, according to feeding system) the nitrogen balance was calculated as 2.7% of weight gain (ERM/AB-DLO, 1999). The conventional rearing system showed better productive indices (ADG 0.96 vs. 0.85, $P<0.01$; FCR 6.41 vs. 9.18 kg DM/kg live weight, $P<0.01$; N-diet 131 vs. 140 g/d) and lower environmental impact considering individual nitrogen excretion (105 vs. 117 g/d) and efficiency (19.91 vs. 16.50%, $P<0.01$) than the extensive ones. Moreover, the number of animals allowed per surface unit in the organic farming could considerably reduce the soil nitrogen supply, causing a progressive reduction of soil fertility and organic matter content especially in the Mediterranean country for their soil and climatic conditions. In conclusion, livestock show several environmental functions, both positive and negative, changing in accordance to intensity, rearing systems and geographical areas. So the higher N excretion in extensive farm system should be evaluated considering all functions developed from livestock, especially in marginal areas subject to environmental risk and socio-economic decline.

Various dietary fat supplements modified simultaneously milk and beef lipids in lactating cows

J. Angulo[1], M. Olivera[1], G. Nuernberg[2], D. Dannenberger[2] and K. Nuernberg[2]
[1]University of Antioquia, Faculty of Agric. Science, Citadel Robledo, AA 1226 Medellin, Colombia, [2]FBN, Muscle Biology and Growth, Stahl-Allee 2, 18196 Dummerstorf, Germany; knuernbg@fbn-dummerstorf.de

This study aimed at a comprehensively characterising the effect of different lipid supplemented rations for lactating dairy cows on fatty acid composition and CLA profile of meat and milk from the same animal. The experiment investigated the effect of a ten-week feeding intervention of German Holstein cows (n=18, first lactation, 92 days in milk) with three different dietary fat supplements [saturated fat (SAT); linseed oil (LUNA) or sunflower oil (SUNA) plus DHA rich algae on fatty acid composition and CLA profiles of meat and milk. Meat quality and fatty acid data were analysed by the least-squares method using the general linear model procedures (GLM) of SAS with the fixed factor diet. Meat quality traits were not affected by the diets. Milk fat content was significantly lower in LINA and SUNA compared to SAT. Exogenous n-3 fatty acid supply (LINA=2.7% linseed oil + 0.4% DHA algae) caused significant higher concentration of n-3 PUFA in milk and intramuscular fat. Exogenous n-6 FA supply (2.7% sunflower + 0.4% DHA algae=SUNA) only increased n-6 fatty acids in muscle. Feeding plant PUFA to dairy cows decreased significant the content of saturated FA in milk. Contrary to this, there was only a tendency for a higher deposition of SFA in muscle lipids. SUNA feeding caused an accumulation of CLAtrans-10, cis-12 and CLAtrans-7, cis-9 in milk and muscle and a decrease of CLAtrans-12, trans-14, CLAtrans-11, trans-13 and CLAtrans-9, trans-11 proportion compared to SAT. Altogether, this study identified improvements of beef (higher n-3 FA, lower n-6/n-3 ratio) and milk (lower saturated FA) lipid profiles upon dietary fat supplementation without affecting meat quality traits in lactating cows. Effects were tissue-specific.

Session 12 Poster 14

N_2O emission by urine and feces of beef cattle in grasslands of Marandu grass during the winter

A. Cardoso, P. Pires, G. Madalena, M. Eliane, E. Januschkiewicz, R. Reis and A. Ruggieri
Unesp, Av. J.A.A. Martins, no. 801 apt 8, 14882-298, Brazil; abmael2@gmail.com

This study aimed to quantify N_2O emissions by cattle excreta in grasslands of Marandu grass pastures in central Brazil during dry season. The experiment was conducted at Campus of Jaboticabal of the Univ. Estadual Paulista. The soil is classified as Oxissoils and climate as tropical with dry winter and rainy summer. The evaluation of the flow of N_2O was performed using of static chambers where treatments were 1.5 l urine 1.5 kg fresh feces, urine l 0.75 + 0.75 kg of fresh feces and control not receive excreta. Samplings of N_2O were made in the morning followed by gas chromatographic determination. The evaluation period was of 110 days and the total emission of the period was obtained by weighted average. The uncertainty was calculated through standard error of the mean. Only treatment urine showed net emissions of N_2O. The total urine output represented by issuing N-N_2O adding 1% of the total. Half of N_2O emission factor default proposed by the IPCC. Already feces and urine and feces mixed emitted of 37.7 and 20.1 mg N-N_2O/m^2 respectively. This was due the predominance of anaerobic conditions. In conclusion, only the urine allowed net emission of N_2O during the evaluation period. Feces and urine + feces mixed occurred consumption of N_2O. The N_2O emission found in this study were less than the default emission factor of N_2O by direct excretion of cattle on pastures proposed by the IPCC for the tropics. The authors thank the FAPESP and CAPES for scholarships and financiering research.

Session 12 Poster 15

In vivo performances and slaughtering traits of Holstein young bulls fed with sunflower cake

F. Vincenti and M. Iacurto
Consiglio per la Ricerca e sperimentazione in Agricoltura, Centro di ricerca per la Produzione delle Carni e il Miglioramento genetico, Via Salaria 31, 00015 Rome, Italy; miriam.iacurto@entecra.it

The aim of this study was to evaluate the effect of the use of sunflower cake (by-product of biodiesel) as source of protein for cattle fattening. The study was carried out with 11 Holstein young bulls that were divided in two experimental groups: Group C (n=6), fed with a basis of maize silage and soybean as protein source; Group G (n=5) fed with the same diet of group C but 1/3 of the protein source was replaced with sunflower cake. The two diets had 0.96 UFV/kg DM in average and 14.50 PG/kg DM in average. Animals were fed for 140 days and were slaughtered at 540 kg of live weight. The average daily gains (ADG) did not differ significantly between the two groups even if the group G has a higher ADG than group C (1.36 kg/d vs. 1.18 kg/d). No significant differences were found about slaughter performances; carcass weights were 294 kg in average, dressing percentages were 54.5%, and SEUROP grip were O2. No significant differences were observed by Metii *et al.* on Marchigiana young bulls, while Mapiye *et al.* have shown that Nguni young bulls fed with sunflower cake had ADG and slaughter weights higher than control group but they had not significant differences on dressing percentages. Xazela *et al.* have shown that the effect of sunflower cake on goats is more related to the breeds than to the diets. In conclusion, it is possible to use sunflower cake as protein source without compromising *in vivo* and at slaughter performances. In addition it is possible to reduce the cost of diet, in fact the price of sunflower cake is 40% less than the price of concentrates with soybean source.

Session 12 Poster 16

Relationships of Infrared thermography data with environmental conditions in beef cattle

L.S. Martello, S.L. Silva, M.R. Mazon and P.R. Leme
FZEA/USP, Duque de Caxias Norte, 225, 13635-900 Pirassununga, SP, Brazil; martello@usp.br

Infrared thermography (IRT) has been used to measure the radiant energy from animal´s surface and it has been associated with important processes that involves body heat dissipation as feed efficiency and thermoregulation processes. The surface temperature is associated with changes of vascular circulation at different body areas, which results in temperature variation of these areas. Moreover, some factors related to environment, like air temperature, must be considered when using IRT as indicator of surface temperature because different body regions have distinct relationships with environmental and heat dissipation. Therefore, this study was conducted to evaluate the relationship of surface temperature measured by IRT (ST_{IRT}) at different body areas with air temperature (AT) and relative air humidity (RH) in feedlot fed *Bos indicus* cattle. ST_{IRT} was measured in 18 Nellore steers during ten consecutive days at 7, 12 and 16 h. IRT images were taken at multiple body locations which include frontal head, lateral head, eye, ribs, flank, rump and front feet. The AT and RH were also recorded simultaneously using data loggers. AT showed a quadratic association ($P<0.01$) with temperatures of frontal and lateral head, ribs, flank, rump and front feet areas, with coefficient of determination (R^2) ranging from 0.70 to 0.73. Conversely, AT was linearly associated with eye temperature ($P<0.01$; $R^2=0.38$). The quadratic association observed for AT with the most ST_{IRT} traits indicate a different daily pattern of relationship among these traits, where ST_{IRT} tended to be stabilized even with an increase of AT. The RH was also quadratically associated ($P<0.01$) with frontal head, lateral head, rump and front feet (R^2 ranging from 0.53 to 0.61) but was not associated with eye, ribs and flank. In conclusion, the temperature of different body locations, measured by IRT showed similar pattern among them and with AT, while for RH the association was not evident for all body regions.

A study on the determination of marketing margins in the Turkey beef market

Y. Aral[1], M.B. Cevrimli[1], C.Y. Kaya Kuyululu[1], M.S. Arikan[1], A.C. Akin[1], E. Aydin[2] and D. Ozen[3]
[1]Ankara University Faculty of Veterinary Medicine, Department of Animal Health Economics and Management, Diskapi, 06110 Ankara, Turkey, [2]Kafkas University Faculty of Veterinary Medicine, Department of Livestock Economics, Pasacayiri, 36100 Kars, Turkey, [3]Ankara University Faculty of Veterinary Medicine, Department of Biostatistics, Diskapi, 06110 Ankara, Turkey; yaral@veterinary.ankara.edu.tr

This study was conducted to determine intermediary margins (live-wholesale, wholesale-retail and live-retail) in the marketing system of cattle for beef production in Ankara, Turkey. The material of the study consisted in the records of sale transactions carried out between 2008 and 2011 in Ankara Commodity Exchange. Monthly average intermediary margins in marketing were calculated with current and fixed prices and in percentages, making use of the producer price index (PPI) and consumer price index (CPI) of the Turkish Statistical Institute. The study revealed that current annual average live weight prices of cattle (TL/kg) were 5.62, 6.28, 9.28 and 9.39. Wholesale carcass prices (TL/kg) were 9.79, 11.11, 15.62 and 14.61, and retail beef prices (TL/kg) were 15.60, 17.42, 23.61 and 22.26 in this period. Calculations revealed that average live-wholesale marketing margins of beef were 18.96, 19.94, 16.69 and 11.25% Wholesale-retail margins of beef were 38.94, 34.85, 32.41 and 35.75%, and live-retail margins of beef were 47.71, 45.34, 40.9% and 39.48%, respectively with fixed prices. Assessments based on monthly averages of live-retail margins showed that the share of producer in the retail beef price varied in a wide range, between 47.05 and 66.32%. On the other hand, a high level of correlation was found between retail beef prices and live cattle prices, wholesale beef prices, live-wholesale, wholesale-retail and live-retail marketing margins ($P<0.01$). In conclusion, it has great importance to immediately take measures that will eliminate the problems relating to production and organization and provide productive and rational marketing infrastructure.

Session 12 Poster 18

Factors affecting beef consumption and consumer preferences in Turkey: a case of Ankara province

Y. Aral[1], P. Demir[2], E. Aydın[2], A.C. Akin[1], E. Sakarya[1], E. Cavusoglu[1] and M. Polat[1]
[1]Ankara University Faculty of Veterinary Medicine, Department of Animal Health Economics and Management, Diskapi, 06110 Ankara, Turkey, [2]Kafkas University Faculty of Veterinary Medicine, Department of Livestock Economics, Pasacayiri, 36100 Kars, Turkey; acakin@ankara.edu.tr

The aim of this study was to identify the beef consumption profile of the consumers in Ankara and determine the factors affecting beef purchasing habits and consumption preferences. The material of this study consists of the data obtained from the surveys conducted face-to-face with 450 household heads. Stratified sampling method was used to determine the survey households, on the basis of the population density of different districts of Ankara in 2008. The data analysis consisted of weighted scoring for ranking of consumer preferences with the use of descriptive statistics, and the statistical comparisons between groups for the determined parameters with Mann-Whitney U and Kruskal-Wallis tests. One of the main findings of the research was the share of average monthly total food expenditures in the monthly total income, which was found to be 25.3% in the surveyed households (€318.90). In the monthly average meat consumption of households (3.48 kg), beef is the most preferred meat with a percentage of 80.43%. Furthermore, beef is preferred as the first, second and third choice by 46.7%, 16.0% and 15.3% of the households, respectively. The main reasons for the preference of beef as the first choice are taste, nutritional quality, fattiness, appearance, affordability and health benefits, in the given order. A significant difference between preference groups was found in terms of taste and price (at $P<0.001$). While no significant difference was found between the household income groups and the consumer preference orders of beef products, significant statistical differences was determined between groups in terms of the frequency of consumption of valuable beef products such as steak fillet, ribs loin, rib steak and chop steak (at $P<0.001$).

Session 12 Poster 19

CH_4 oxidation in grazed Marandu grass during the dry season in Central Brazil oxissoils

A.S. Cardoso, B. Quintana, G. Madalena, E.R. Januskiewicz, E.S. Morgado, L.F. Brito, T.T. Berchielli and A.C. Ruggieri
Unesp, Zootecnia, Av. José Adriano Arrobas Martins no. 801 apt 8, 14883-298, Brazil; abmael2@gmail.com

This study aimed to quantify the CH_4 emissions due to excretions of beef cattle and urea fertilizer on pasture of Marandu grass. The experiment was conducted at Campus of Jaboticabal of UNESP. The soil was Oxissoils and there was a local tropical climate with dry winter and rainy summer. The evaluation of the fluxes of CH_4 was performed using the static chambers where treatments were 1.5 l urine 1.5 kg fresh feces, urine 0.75 l + 0.75 kg of fresh feces, 80 g of N in the form of urea per m^2 and control. Samplings of gases were made in the morning and were followed by gas chromatographic determination. Evaluation period totaled 110 days and the total emission of the period was obtained by weighted average. Variability was estimated through standard errors of the means. During the whole trial period, we found negative CH_4 fluxes. Soils as good oxidizing conditions allowed biological oxidation of CH_4. With addition of urine net emissions of CH_4were found, the urine increased the biological oxygen demand hindering the action of methanotrophic. The CH_4 emissions were -724.1 (624.4), 660.8 (273.7), -112.3 (254.8) and 466.5 (253.3) mg C- CH_4/m^2 respectively by urea, urine, feces and urine + feces. The bovine urine and feces + urine addition allowed liquid flows of CH_4 while with the application of urea and feces oxidation of CH_4 was observed. For purposes of impact studies of CH_4 in pastures the oxidation of this gas should be included.

Session 13a Theatre 1

Compensatory feeding of gestating gilts: effects on mammary gland development and lactation

C. Farmer[1], M.-F. Palin[1] and Y. Martel-Kennes[2]
[1]Agriculture and Agri-Food Canada, Dairy and Swine R & D Centre, 2000 College St., Sherbrooke, QC J1M 0C8, Canada, [2]La Coop Fédérée, Animal Nutrition Division, St-Romuald, QC G6W 5M6, Canada; farmerc@agr.gc.ca

The impacts of compensatory feeding during gestation on mammary gene expression and development at the end of gestation or of lactation, and on lactation performance were determined. Gilts were fed a conventional (CTL; n=59) or an experimental (TRT; n=56) dietary regimen. The experimental regimen provided 70% (restriction) and 115% (compensatory) of the protein and DE contents provided by the CTL diet. The restriction diet was given during the first 10 weeks of gestation followed by the compensatory diet until farrowing. Some gilts (14 CTL, 14 TRT) were slaughtered on day 110 of gestation and the others were allowed to farrow. Of these, 28 (14 CTL, 14 TRT) were slaughtered on day 21 of lactation. Litters were standardized to 11 or 12 and piglets were weighed weekly until day 18. The MIXED procedure of SAS was used for statistical analyses with a univariate model. Body weight and backfat thickness of first-parity sows were lesser for TRT than CTL at the end of gestation ($P<0.01$) and body weight was also lesser in late lactation ($P<0.05$). There were no differences in piglet growth between CTL and TRT litters ($P>0.10$), yet mammary development and mammary gene expression were affected. There was less parenchymal tissue ($P<0.01$) at the end of gestation in TRT than CTL sows but its composition was not altered ($P>0.10$). Relative abundance of IGF-1 ($P<0.05$), ODC1 ($P<0.05$), STAT5B ($P<0.05$) and WAP ($P<0.01$) genes in parenchyma at the end of gestation were lower in TRT than CTL sows and the effect on WAP was still present at the end of lactation ($P<0.01$). Mammary composition at the end of lactation was unaffected by treatment ($P>0.10$). In conclusion, restriction and subsequent compensatory feeding in gestation had detrimental effects on sow body weight, backfat, mammary development and mammary gene expression but piglet growth rate was not affected.

Effects of high fiber intake in late gestating sows on colostrum production and piglet performance

F. Loisel[1,2,3], C. Farmer[4], P. Ramaekers[3] and H. Quesnel[1,2]
[1]INRA, UMR1348 PEGASE, Saint-Gilles, 35590, France, [2]Agrocampus Ouest, UMR1348 PEGASE, Rennes, 35000, France, [3]Nutreco R & D, 5832 AE Boxmeer, the Netherlands, [4]Agriculture and Agri-Food Canada, Dairy and Swine R & D Centre, Sherbrooke, QC, J1M 0C8, Canada; florence.loisel@rennes.inra.fr

Dietary fiber given during gestation may influence sow endocrinology and increase piglet weight gain during early lactation. The aim of the study was to determine whether dietary fiber given to late gestating sows induces endocrine changes that could modulate sow colostrum production and thus piglet performances. Twenty-nine Landrace × Large White nulliparous sows were fed diets containing 7.9% (HF, n=15) or 3.3% of crude fiber (CTL, n=14) from day 106 of gestation until the day of farrowing. Colostrum yield was estimated during 24 h, starting at the onset of parturition (T0), using piglets' weight gains. Blood samples were collected from sows from day 111 of gestation until 24 h after parturition (T24). Colostrum samples were taken at T0 and T24. Data were analyzed by ANOVA (MIXED procedure) or by a generalized linear model for mortality analysis (GENMOD procedure, SAS Inst.). The treatment did not influence (P>0.10) plasma concentrations of progesterone or prolactin in sows, colostrum yield (3.9±0.2 kg) or piglet weight gain during lactation. Colostrum intake of low birth weight (LBW) piglets (<900 g) was greater in HF than in CTL litters (216±24 g versus 137±22 g, P=0.02). During lactation, piglet mortality was lesser in HF than in CTL litters (6.3±2.7% versus 14.0±2.8%, P=0.01). Compared with CTL sows, colostrum from HF sows contained more fat at T24 (P=0.04) and less IgA at T0 and T24 (P=0.02). In conclusion, dietary fiber in late gestation affected sow colostrum composition but not colostrum yield, increased colostrum intake of LBW piglets and decreased pre-weaning mortality; these effects were not related to changes in peripartum concentrations of progesterone or prolactin.

Session 13a Theatre 3

Effects of dextrose and L-arginine in sow diet on litter heterogeneity at birth

H. Quesnel[1], N. Quiniou[2], H. Roy[3], A. Lottin[2], S. Boulot[2] and F. Gondret[1]
[1]INRA, UMR1348 PEGASE, Domaine de la Prise, 35590 Saint-Gilles, France, [2]IFIP-Institut du Porc, BP 35104, 35651 Le Rheu Cedex, France, [3]Chambre d'Agriculture de Bretagne, CS 74223, 35042 Rennes Cedex, France; helene.quesnel@rennes.inra.fr

Pre-weaning piglet mortality is partly attributed to within-litter variation of piglet birth weight (BW0). This study investigated the influence of different maternal feeding strategies on within-litter variation of BW0. Four batches of crossbred Landrace × Large White multiparous sows (parity ranging from 2 to 11) were used. Three treatments were compared: supplementations of dextrose from weaning to insemination (190 g/d) and of L-arginine (25.5 g/d) from day 77 of gestation until term (DEXA, n=26); supplementation of L-arginine from day 77 to term (ARGI, n=24); and no supplementation (CTL, n=23). Piglets were weighed at birth and at weaning. Data were analyzed by ANOVA using the MIXED procedure (SAS Inst.). Model 1 included treatment (ARGI, DEXA and CTL) as fixed effect and batch as random effect. In Model 2, the two groups of treated sows (ARGI+DEXA) were compared with CTL sows. The treatments did not influence the number or birth weight of total born (TB) or born alive (BA) piglets. The coefficient of variation of birth weight (CV_{BW0}) was lower in the ARGI and DEXA groups than in the CTL group (for TB: 21.7, 23.1 and 25.9%, P=0.06; for BA: 21.0, 22.2 and 25.6%, P=0.03, respectively). When litters were categorized according to litter size, the CV_{BW0} was lower in the ARGI than in the CTL group (17.6 versus 24.3%, P<0.05) and intermediate in the DEXA group (20.7%) in litters having less than 17 TB piglets. In larger litters, treatment had no significant effect on CV_{BW0}. Performances during lactation were similar in the 3 treatments. In conclusion, L-arginine supplementation during last third of gestation improved within-litter homogeneity of birth weight, with limited benefits in largest litters. There was no further advantage in simultaneous pre-insemination distribution of dextrose.

Phenotypic and genetic applications for total nutritional efficiency in pigs

L.M.G. Verschuren[1], H.A. Mulder[1], R. Bergsma[2] and E.F. Knol[2]
[1]Animal Breeding and Genomics Centre, Wageningen University, P.O. Box 338, 6700 AH Wageningen, the Netherlands, [2]TOPIGS Research Center IPG B.V., P.O. Box 43, 6640 AA Beuningen, the Netherlands; lisanne.verschuren@wur.nl

Selection changes animals and therefore the nutritional requirements change. At the same time, the main focus of selection is to increase nutritional efficiency in pig production. Our aim is to develop a phenotypic and genetic model to describe nutritional efficiency and to predict future trends. The model includes reproduction, stayability, survival, and birth weight traits, as well as adult sow weight, finishing carcass weight, feed intake, weight development and carcass composition. Total Feed Efficiency (TFE) is used for comparison. Phenotypic and genetic correlations were collected and, where necessary, estimated. Phenotypic results of Dutch farms showed that the average TFE improved by 0.045 kg feed per kg live weight per year in the last two years. This phenotypic progress is largely in line with the genetic trends, since based on genetic trends in sire and dam lines an improvement of 0.055 kg/kg per year is expected. Genetic progress in sire lines account for 56% of the improvement in TFE, mainly due to improved feed conversion ratio and average daily gain in grower-finisher pigs. The other 44% is explained by progress in dam lines, of which roughly two third is due to improvement in reproduction traits and the remainder in grower-finishers traits. The next steps are to develop a phenotypic regression model to evaluate management strategies to improve nutritional efficiency accounting for phenotypic correlations between traits and to develop a selection index for nutritional efficiency to evaluate different selection strategies to improve the trait accounting for genetic and phenotypic correlations between traits underlying nutritional efficiency. In conclusion, nutritional efficiency is already improving in conventional pig farms and this research will provide models to evaluate management and selection strategies to improve nutritional efficiency.

Multifactorial approach needed to optimize technical and economical results on sow farms via feeding

P.J.L. Beckers[1], B. Humphrey[2], A. Thompson[2] and A. Ascensao[1]
[1]Cargill Innovation Centre, Veilingweg 23, 5334 LV Velddriel, the Netherlands, [2]Cargill Innovation Centre, 10383, 165th Ave. NW, Elk River, USA; patricia_beckers@cargill.com

With increasing production the requirements of the sow change. If a farm average would increase in production level from 10 weaned piglets to 12, the SID lysine requirement would go up from 0.70 to 0.87% with the same 5.5 kg of feed intake and keeping all factors the same. How the requirements change will depend on many factors that are genetic and farm related. Not just the production differences lead to different nutrient levels in the feed; sow weight, body composition (protein, fat) and, one of the most important factors, actual feed intake are relevant too. Other factors are group housing and temperature in wintertime. Summertime emperature may have a negative impact on feed intake. Our research shows that significant differences in efficiency of nutrient utilization can be found between farms. All these factors make that there is not a static change in requirement when production level changes when using a specific genetic line. Average feed intake for instance can vary between breeds, as can their reaction to temperature. The SID lysine requirement of a sow with more than 11 piglets weaned would be 0.87% if she had a daily intake of 5.5 kg but 0.74% if she would have an intake of 6.5 kg and projecting 15 kg weight loss during the lactation period, -however-10 kg of weight loss with the same intakes would change the requirement to 0.92%SID lysine for 5.5 and 0.78% SID lysine for 6.5 kg feed intake. Since farming is an economic activity, there always has to be a balance between feed cost and maximizing production. So e.g. accept more weight loss if you can keep the production upright. The conclusion is that when feeding sows do not just consider their genotype, but more their phenotype. Given the complexity new mathematical computer models to design feeds that optimize farm production and economics are useful.

Interactions among sow prolificacy, sow lifetime productivity and efficiency of pork production

G.R. Foxcroft
University of Alberta, AFNS, 310A Agriculture-Forestry Centre, Edmonton, Alberta, T6G 2P5, Canada; george.foxcroft@ualberta.ca

Selection for increased lean growth has been linked to increased mature body weight. If sows need to achieve some proportion of their mature lean mass before becoming fully productive, this raises questions about appropriate gilt development strategies and thier relationship to overall efficiency of pork production. Selection for increased milk production and lactation efficiency has increased sow lifetime productivity but sows with increased milk production may be more or less inclined to use tissue mobilization for milk production when voluntary feed intake is limiting. Sub-populations of 'risk' sows support milk production at the expense of poorer embryonic development of their next litter and this antagonism is accentuated by the trend for catabolic sows to be increasingly insensitive to effects of catabolism on post-weaning fertility. Thus, selection for increased sow productivity in lactation may be linked to progeny with poorer post-natal growth potential. Finally, the reproductive trait that has responded most obviously to genetic selection is litter size. However, the component traits that determine total born, versus born alive and weaned, can only be subjected to indirect selection. Relationships between increased sow prolificacy and efficiency of pork production will vary among genetic programs, but also vary within populations. A repeatable low birth weight phenotype represents a GxE interaction in mature sows that is not consistent with efficient pork production and merits further attention. Overall, a more holistic goal of improving overall efficiency of pork production could drive refinements in genetic selection programs. Alternatively, refinements in gilt and sow nutrition, and in the post-natal management of the diverse progeny produced, can be targeted at reduced variability in pork production efficiency.

Effects of xylo-oligosaccharides and live yeast on lactation performance in lactating sows

X. Wu[1], C.Y. Xie[1], F.Y. Yan[1,2] and Y.L. Yin[1]
[1]Institute of Subtropical Agriculture, Chinese Academy of Sciences, Yuanda 2 Road, Changsha, 410125, China, P.R., [2]Hunan Jiuding-group, Changsha, 410120, China, P.R.; w.xin@foxmail.com

This study was conducted to investigate the effects and potential mechanisms of dietary xylo-oligosaccharides (XOS) and live yeast supplementation on the lactation performance of lactating sows. Thirty-six Landrace × Large White sows at d 110 of gestation were assigned randomly into 4 groups (a control group and experimental group 1, 2 and 3), 8 replicates in each group. The control group was fed with basic diet, and experimental groups were fed with basic diet supplemented with 10 g/kg XOS, 10 g/kg XOS+ 1 g/kg live yeast. Blood samples were obtained on d 7 after delivery to examine the plasma concentration of biochemical indices and free amino acids. The results showed that: compared with the control group, supplementation with XOS alone did not increase the number of piglets, however, supplementation with the combination of XOS and live yeast significantly increased lactation ($P<0.05$) and average body weight of weaned piglets by 8.70% ($P<0.05$), and there is an increased trend in serum glucose, but not significantly. These findings suggested that dietary supplementation with the combination of XOS and live yeast is able to increase the average weight of weaning piglets, enhanced lactation performance of sows, and the potential mechanism is that XOS increased live yeast in sows. In conclusion, the results of this trial can be taken as practical guidance to improve efficiency of sow performance and profitability of swine farms.

Evaluating environmental enrichment for pigs

S.A. Edwards[1] and N.R. Wainwright[2]
[1]Newcastle University, School of Agriculture, Food & Rural Development, Agriculture Building, Newcastle upon Tyne NE1 7RU, United Kingdom, [2]BPEX, Stoneleigh Park, Kenilworth CV8 2TL, United Kingdom; sandra.edwards@ncl.ac.uk

Current EU legislation requires all pigs to be given 'permanent access to a sufficient quantity of material to enable proper investigation and manipulation activities'. The interpretation of this requirement is still subject to debate, and a wide variety of different materials and objects are provided in commercial farms. To determine the adequacy of these enrichment strategies, different approaches have been suggested. Resource-based measures, such as are currently used in legislation and farm assurance, specify the nature of enrichment in terms of specific items, materials or presentation methods. The acceptability of these is generally determined by expert opinion, and models have been developed to formalise this process. In contrast, animal-based measures focus on the observed extent of use of the enrichment provided, or the occurrence of abnormal behaviours considered to reflect enrichment inadequacy. Using such an approach, short term experimental tests of different enrichment items relative to a standard benchmark can be devised, and examples are given. However, many confounding factors need to be considered and the diversity of items in current commercial use would be difficult to cover. The practical application of a simple observational test for on-farm monitoring of enrichment adequacy will be reviewed, with data from a large UK study relating the nature of the enrichment provided to an index of its utilisation, and their relationship with other welfare criteria.

Session 13b Theatre 2

Behavior and lesion scores of growing pigs raised on slatted floor with different objects

V. Courboulay
IFIP, BP 35104, 35651 Le Rheu cedex, France; valerie.courboulay@ifip.asso.fr

Three experiments (Exp) were conducted to evaluate different enrichment types for tail-docked group-housed pigs (10/pen), fed *ad libitum* on fully slatted floor from 25 to 115 kg. Two batches of animals were used in Exp 1 and 2 and a total of 6 pens were studied per treatment. One batch of pigs and 3 pens per treatment were compared in Exp 3. Direct behavioral observations were performed every 10 minutes during 2.5 hours every 2 or 3 weeks (3:00-5:30 pm), with additional observation performed in the morning in Exp1 (8:00-10:30 am). All pigs were scored for tail and body lesions on the week after behavioral observations (Exp 2 and 3) or 3 times during the fattening period (Exp1). In Exp1, permanent access to long straw in a rack (S) was compared to an object made of 6 plastic hoses reinforced by 6 metal chains fixed on the floor (P6). Pigs spent more time investigating straw than P6 (17 vs. 10% of observations, $P<0.01$), but no difference were obtained on negative social behavior, lesions or tail scores. In Exp2, P6 was simplified in P3 (3 hoses+chains) and compared to 3 different chain based objects: chain hung at the snout level (CS), 5 cm above the floor (CF), or an object made of 3 chains fixed to the floor (3C). The objects P3 and 3C were more often investigated than CF (15 and 13% vs. 9%, respectively, $P<0.01$), and CF more often than CS (9 vs. 6%, $P<0.05$). In Exp3, a chain was presented alone (CF) or ended with a 7 cm diameter plastic ball (CB), and compared to a wood log fixed to the floor (WF) or loose in the pen (WL). The objects CF and CB were used similarly (8%) but less often than WF (15%, $P<0.01$) and more than WL (3%, $P<0.01$). Pigs with CF tented to have more severe tail lesions than with CS. Type of object did not influence lesion scores in Exp 2 and 3, or tail lesions in Exp 3. In conclusion, it is suggested that the way to present objects may have a greater importance in inducing investigative behavior than the object itself. Investigation was more pronounced with straw but this system required manual removal of manure.

Session 13b Theatre 3

Reducing aggressive behaviour by an cognitive enrichment tool for young piglets

L.T. Sonoda, M. Fels, S. Rauterberg, M. Oczak, G. Ismayilova, S. Viazzi, M. Guarino, E. Vranken, D. Berckmans and J. Hartung
University of Veterinary Medicine Hannover, Foundation, Institute for Animal Hygiene, Animal Welfare and Farm Animal Behaviour, Bünteweg 17p, 30559, Germany; lilia.thays.sonoda@tiho-hannover.de

It is known that pigs raised in enriched environments express less abnormal and aggressive behaviour than pigs housed in barren pens. A new method of environmental enrichment based on cognitive challenges was studied at University of Veterinary Medicine in Hannover (Germany), where 78 suckling piglets in 8 entire litters, 25 days old, were trained in conventional farrowing pens to learn the link between a sound and a sweet feed reward given by an electronic feeder during a period of 8 days. After the training phase, the potential of the electronic feeder to interrupt aggressive interactions was tested in Resident-Intruder confrontations after weaning. When analysing the training data, ANOVA followed by post hoc test (SNK) was conducted. Data of Resident-Intruder test were compiled in contingency tables and chi square test was used. Piglets learned the link between sound and feed reward during 8 days of training and the number of piglets around the feeder awaiting chocolate candies after sound increased with consecutive training days ($P<0.05$). For Resident-Intruder test, 260 aggressive interactions were analysed. It was shown that on average 80% of aggressive interactions were broken by feeder activation ($P<0.05$). In 55% of stopped fights, the aggressor interrupted fighting. Regardless of whether the aggressor or the receiver responded to the feeder, an equally high number of fights were stopped (97% vs. 93%, respectively). We conclude that the electronic feeding system has the potential to be used as environmental enrichment for young piglets being able in principle to influence aggressive behaviour in later production stages.

Session 13b Theatre 4

Time evolution of suckling piglets' behaviour in farrowing crates

A. Bulens[1,2], V. Vanheukelom[2], S. Van Beirendonck[2], J. Van Thielen[2], N. Buys[1] and B. Driessen[2]
[1]KU Leuven, Department of Biosystems, Kasteelpark Arenberg 30, 3001 Leuven (Heverlee), Belgium, [2]KU Leuven|Thomas More Kempen, Kleinhoefstraat 4, 2440 Geel, Belgium; anneleen.bulens@khk.be

The European Directive 2001/93/EC states that every pig must have permanent access to a sufficient quantity of material to enable proper investigation and manipulation activities. This implies that also piglets in the farrowing crate should have access to suitable materials. Providing these materials can be considered as a form of environmental enrichment. Enrichment in the farrowing crate can be important in order to reduce frustration behaviour in piglets and to encourage play behaviour. The ethological question rises from what age on enrichment must be introduced to be used by the piglets. It is possible that during the first days, the piglets explore the environment only to a limited extent and they stay close to the sow. Later, the piglets focus their attention on the environment and pen mates too. To avoid harmful and frustration behaviour, enrichment could be important in this stage. In an attempt to determine the age on which enrichment can be introduced, 227 suckling piglets (116 males and 111 females from 22 sows) were observed in the farrowing crate until weaning. Behavioural observations were carried out using the scan sampling method. Data were analysed using the logistic mixed model. The results showed that social play behaviour increased with the age and the same was seen with exploring the crate. Piglets already showed explorative behaviour during the first week of life. Nosing and nibbling other piglets was also already seen during the first week of life. The frequency of nosing and nibbling the sow did not change with the age, except from a fluctuation in week 3. The results suggest that environmental enrichment can be introduced when piglets are one week old, in order to offer the piglets opportunities to perform explorative and play behaviour, without developing harmful behaviours.

Session 13b — Theatre 5

Effect of floor surface on behaviour and welfare of dairy cows

M. Speroni, L. De Matteis, C. Federici, A. Dal Prà and F. Abeni
Consiglio per la Ricerca e la Sperimentazione in Agricoltura, Centro di Ricerca per le Produzioni Foraggere e Lattiero-Casearie, via Porcellasco 7, 26100 cremona, Italy; marisanna.speroni@entecra.it

The work reports preliminary results of a project investigating the convenience of covering concrete slatted floor of dairy cows barns with slatted rubber mats. Both direct effects on claws and indirect effects on behaviour and welfare were estimated. In one of the two sides of a free stall barn, the existing slatted concrete floor (CONCR) of the feeding alley was covered with rubber mats (RUBB). Fifty-six cows (38 pluriparous and 18 primiparous) were blocked according to parity, days in milking and locomotion score; cows in blocks were randomly assigned to one of the sides of the barn. Claws were trimmed on three occasions during the experiment: C0, 32.5±4.9 days before the installation of rubber; C2, 142.5±10.5 days after C0; C3, 97±9 days after C2. Before each trimming session, the length of both lateral and medial claw or rear right and hind left feet were measured. After trimming, the length of all hooves was 7.5 cm. The part of claw length measured before a successive trimming that exceeded 7.5 cm was assumed to be the balance between hoof growth and hoof consumption. Number of cows standing, lying, feeding, drinking in feeding and resting areas of barn was scan sampled (1 h) in the interval between the two milkings (from 7 to 14 h) during 8 daily sessions. Body condition, milk yield, locomotion and body dirtiness were observed monthly. Variation of length of hooves was not demonstrated to be affected by treatment. Percentage of standing cows, not eating, not drinking, on the total of observed cows was higher ($P<0.05$) in RUBB than in CONCR, but milk yield and body condition were not affected by treatment. Dirtiness of anogenital area was lower in RUBB than in CONCR ($P<0.01$). Results of this experiment showed short-term effects of flooring slatted feeding alleys with rubber on behaviour of dairy cows.

Session 13b — Theatre 6

Effect of rubber covered slats on the welfare and behaviour of group housed sows

J.A. Calderón Díaz[1,2], A.G. Fahey[1] and L. Boyle[2]
[1]University College Dublin, School of Agric. & Food Sci., Dublin, 4, Ireland, [2]Teagasc Animal & Grassland Research & Innovation Centre, Pig Development Dept., Moorepark, Fermoy, Ireland; laura.boyle@teagasc.ie

Three experiments were conducted to evaluate the effect of rubber flooring on the welfare of group housed sows incorporating health and behaviour indicators. In a longitudinal study on a commercial farm, 164 replacement gilts were housed in groups of 8 in pens with free access feeding stalls (FS) during 2 parities. Lameness (0=normal to 5=severe), limb (0=normal to 6=severe) and claw (CL; 0=normal to 3=severe) lesions were recorded. Results showed that sows on rubber (RUB) slats mats were at lower risk of lameness (OR=0.6; CI=0.3-0.9; $P<0.01$), swellings (OR=0.4; CI=0.3-0.7; $P<0.01$), and wounds (OR=0.5; CI=0.2-0.7; $P<0.01$) on their limbs compared to sows on concrete (CON) slats. However, they were at higher risk of CL such as toe overgrowth (OR=3.17; CI=1.3-7.5; $P<0.01$) and white line disease (OR=6.9; CI=3.0-14.9; $P<0.01$). In the second experiment, 64 sows were housed in groups of 4 in pens with FS and a slatted group area (GA) from 28 d after service. In 2 pens the GA was covered with RUB and in the other 2 pens the GA was uncovered. Sows were video recorded for 24 h on 5 d during pregnancy. Videos were sampled every 10 min to record postural and spatial behaviour. Sows on RUB spent more time in the GA (76.3% vs. 53.3%±5.8; $P<0.01$) and stood less (19.1% vs. 35.6%±5.0; $P<0.05$) and lay more (80.1% vs. 62.5%±5.3; $P<0.05$) there compared with CON sows. This could reflect the preference for a comfortable surface for lying. We found few carry over benefits of housing on rubber during pregnancy on welfare in the farrowing crate. There were no treatment differences in the number of attempts, latency or time to lie-down ($P>0.05$); but RUB sows tended to make more postural changes (33.0 vs. 30.1±1.3; $P=0.10$) which could reflect their efforts to adapt to lying on a more uncomfortable surface. Cushioned flooring for pregnant sows improves their comfort during pregnancy and contributes to a reduction in lameness.

Innovative process for developing welfare-friendly alternatives to the farrowing crate in pigs

M.C. Meunier-Salaün[1], J.Y. Dourmad[1], L. Brossard[1], A. Bailliard[2], N. Lescop[2], A. Marien[2], O. Sergent[2], F. Serrurier[2], H. Lerustre[2] and J. Lensink[2]
[1]INRA-AgroCampus Ouest, UMR1348 PEGASE, 35590 Saint-Gilles, France, [2]ISA Group, CASE research group, 48 Bd Vauban, 59046 Lille, France; marie-christine.salaun@rennes.inra.fr

Designing farrowing and lactation environment which maximize sows and piglets welfare, while maintaining economic efficiency and sustainability, is a continuous challenge in pig production. Experimental investigations have been done on alternative housing systems combining environmental enrichment and freedom of sows, with performance (piglet survival) and economic evaluation. Another way is to use an innovation approach, based on the creative Concept-Knowledge theory. This approach involves a 'funnel step' to get the novel ideas, elaborate new concepts and identify lacking knowledge, and a 'tunnel step' to develop these concepts and elaborate prototypes. This work presents the first step assuring abundance of ideas around the concepts of 'housing, sow, farrowing/lactation', and promoting a diversity of views within focus groups (breeder, ethologist, ergonomist, doctor, vet, citizen, students, ...). Three methods were used according to the focus group: (1) 'the Advance Systematic Inventive Thinking' for naïve persons; (2) brainstorming for people with limited knowledge on the subject; (3)/ phone interview with experts from research, livestock, education and medical area. Concepts of farrowing/lactation housing emerged from the two first focus groups and their suitability for animal welfare and livestock issues was then evaluated by experts. A lot of ideas (74) and concepts (15), and housing prototypes have emerged, that meet the principles of maintaining the mother-young bond while protecting the young, and providing enriched living areas for sow and piglets. The following steps will be to assess the feasibility (building, practicality, cost) of the prototypes through advices of experts from livestock sector and research, and to test experimentally their impact on animal welfare and performance.

Effect of enriched housing on welfare, production and meat quality in lambs: the use of feeder ramps

L. Aguayo-Ulloa[1], G.C. Miranda-De La Lama[1], M. Pascual-Alonso[1], J.L. Olleta[1], M. Villarroel[2], M. Pizarro[1] and G.A. María[1]
[1]University of Zaragoza, Department of Animal Production & Food Science, miguel Servet 177, 50013, Spain, [2]Polytechnic University of Madrid, Av. Ramiro de Maeztu 7, 28040, Spain; mpascual@unizar.es

The aim of this study was to analyse the effect of enriched housing in finishing lambs on physiological welfare indicators, production and meat quality traits. Sixty Rasa Aragonesa lambs (male, 65 days old, 17.2±0.2 kg), were divided in 2 treatments, housed indoor for 5 weeks in 6 pens (2.9×3.3 m, density 0.45 m^2/lamb, 10 lambs/each, 3 replicates). Controls lambs were housed in pens that were similar to cooperative feed lots (CC) with straw or items. Enriched pens had straw as forage and bedding and a platform with ramps leading to the feed hopper. Concentrate consumption was recorded and lambs were weighed at the beginning and at the end of experiment to estimate average daily gain (ADG). Blood samples and IR temperatures were taken before slaughter and carcass and meat quality variables after slaughter. The neutrophil/lymphocyte (N/L) ratio and NEFA were significantly higher ($P<0.05$) in controls, indicating chronic stress. Enriched lambs had a higher ADG (+18.3%), heavier carcasses, higher fattening scores, lower values of meat pH at 24 h, higher values of meat colour (L*, b*, Hue) and lower texture than control lambs ($P<0.05$). Sensory meat quality was higher in enriched lambs, where lamb odour intensity and overall liking were also higher ($P<0.05$). A step-wise correlation analysis of sensory variables showed that tenderness was the most powerful parameter related with overall liking ($P<0.0001$). Results suggest that enriched lambs performed better and adapted better to the CC. Their enhanced performance was related to improved instrumental and sensory meat quality, confirming the importance of housing enrichment as a critical aspect to improve animal welfare. The study could be useful in developing animal welfare standards for sheep in the EU.

Effect of enriching housing on welfare, production and meat quality in lambs: the double bunk

L. Aguayo-Ulloa[1], G.C. Miranda-De La Lama[1], M. Pascual-Alonso[1], J.L. Olleta[1], M. Villarroel[2] and G.A. María[1]
[1]University of Zaragoza, Department of Animal Production & Food Science, miguel Servet 177, 50013, Spain, [2]Polytechnic University of Madrid, Av. Ramiro de Maeztu 7, 28040, Spain; mpascual@unizar.es

The aim of this study was to analyse the effect of enriched housing during the finishing phase of fattening on welfare, production and meat quality traits in lambs. Sixty Rasa Aragonesa male lambs (65 day old, 17.2±0.2 kg), were divided into two treatments, housed indoor for 5 weeks in 6 pens (2.9×3.3 m, density 0.45 m^2/lamb, 10 lambs/each, 3 replicates). The control (CG) was similar to cooperative feed lots (CC), without straw. The enriched system (EG) contained straw as forage and bedding and a double bunk with two ramps on each end. Concentrate consumption was recorded and lambs were weighed at the beginning and at the end of the experiment to estimate average daily gain (ADG). Blood samples and IR temperatures were taken before slaughter. Meat quality variables were measured after slaughter. EG had higher cortisol (+142%) and temperature (+0.57 °C) values (P<0.05), and significantly higher lactate (+74.4%), NEFA (+182%) and CK (40%). Results indicate that EG are more reactive due to less visual contact with handlers since they can use the double bunk as covered shelter. However, the higher level of N/L ratio (P<0.05) in CG suggest signs of chronic stress. There were no significant differences in productive performance. EG lambs had lower (P<0.05) cooking losses and colour values, but higher values of texture than CG lambs. Results did not confirm the hypothesis that enriched housing improves welfare, but EG lambs had better levels of immunity. Even though acute stress did not affect the productive performance of EG lambs, meat instrumental quality indicators were influenced. This study confirms the importance of analysing enriching elements carefully by species and production system to avoid economic losses at the commercial level. Data obtained can be useful for the development of animal welfare standards for sheep.

Effects of environmental enrichment on welfare and meat characteristics of Italian heavy pigs

A. Gastaldo[1], S. Barbieri[2], M. Borciani[1], A. Rossi[1], E. Bortolazzo[1], A. Bertolini[1], E. Gorlani[1], E. Canali[2] and P. Ferrari[1]
[1]Research Center For Animal Production (Crpa), Viale Timavo 43/2, 42121 Reggio Emilia, Italy, [2]Universita' Degli Studi di Milano, Department of Veterinary Science and Public Health, Via Celoria 10, 20133 Milano, Italy; p.ferrari@crpa.it

The aim of the study was to assess the effects of environmental enrichment on animal welfare and pork characteristics in the Italian heavy pigs. Two replicates of 324 pigs were housed in 9 fully concrete pens with slatted external alley of 36 animals each, from 80 to 175 kg. The experiment was conducted as a design with three treatments: Ts: straw offered in a rack; Tw: a hanged piece of wood; and Tc: no enrichment. Behavioural observations were performed two times using instantaneous and scan sampling method (10 min/72 h; 2 times of 72 h). The severity of body lesions was collected and divided into six regions. The novel object test (NOT) was performed and the interactions with the object were recorded at 2 four-minute intervals at the introduction and 30 minutes later. Lesion score and NOT were performed at the end of the fattening period. Meat quality and biochemical indicators (carcass lean meat %, water holding capacity, pH) were measured. Quality meat results were analyzed through 2-ways ANOVA with Duncan's separation of means test. Ts showed more exploration than Tc (P<0.01; Wilcoxon test) and less negative social and aggressive behaviour than Tw and Tc (P<0.01; Wilcoxon test). Ts presented the lowest % of pigs with lesions (ears and front), whereas Tw and Tc showed a higher percentage. NOT and pork quality were unaffected by treatments with conflicting results between replicates. These results indicate that pigs prefer substrates with manipulative characteristics, reducing redirect rooting behaviour and improving welfare by minimizing body lesions. Nonetheless straw on rack reported better results than a hanged piece of wood; it may be a cheap and practical solution in slatted floor housing systems.

Session 14 Theatre 1

The genetic basis of breed diversification: signatures of selection in pig breeds

S. Wilkinson[1,2], Z. Lu[2], H.J. Megens[3], A. Archibald[2], C. Haley[2,4], I. Jackson[2,4], M. Groenen[3], R. Crooijmans[3], R. Ogden[5] and P. Wiener[2]

[1]Scotland's Rural College, The Roslin Institute Building, Easter Bush EH25 9RG, United Kingdom, [2]University of Edinburgh, The Roslin Institute, Easter Bush EH25 9RG, United Kingdom, [3]Wageningen University, Wageningen, 6700 AH Wageningen, the Netherlands, [4]University of Edinburgh, Crewe Road, Edinburgh EH4 2XU, United Kingdom, [5]Wildgenes Lab, Royal Zoological Society of Scotland, Edinburgh EH12 6TS, United Kingdom; Samantha.Wilkinson@sruc.ac.uk

Following domestication, selection pressures for desirable traits resulted in phenotypically diverse pig breeds. Signatures of diversifying selection that account for phenotypic variation were studied in 13 European pig breeds genotyped at 60K SNP. Wright's F_{ST} was estimated in 13-SNP sliding windows and selected loci identified at the 99th percentile. Signals were found associated with production and morphological traits. Of all the signals, 13% were found across 12 breeds on SSC8, indicating historic selection on SSC8 during breed development. There were likely several targets of selection as numerous quantitative trait loci (QTL) and genes for carcass composition, coat colour and reproduction have been mapped to SSC8. Two commercial breeds, Duroc and Landrace, exhibited signals in genomic regions harbouring QTLs and genes associated with reproduction and fatty acid synthesis, respectively. At least 3 genomic regions were found associated with ear phenotypic variation in the pig breeds, one which was syntenic to a region associated with the same trait in dog breeds. SNPs in the region were near fixation for the derived flat-ear pig phenotype whilst near fixation for the alternate allele for the ancestral prick-ear pig phenotype. This suggests the presence of a locus important in the determination of ear structure. Signals in Gloucestershire Old Spots and Berkshire were associated with 2 coat colour genes, EDNRB and KITLG, respectively. The signatures of diversifying selection have revealed regions and genes central to the development and distinction of pig breeds at a phenotypic level.

Session 14 Theatre 2

Classical signatures of selective sweeps revealed by massive sequencing in cattle

S. Qanbari[1], H. Pausch[2], S. Jansen[2], S. Eck[3], A. Benet-Pagès[3], E. Graf[3], T. Wieland[3], T.M. Strom[3,4], T. Meitinger[3,4], R. Fries[2] and H. Simianer[1]

[1]Animal Breeding and Genetics Group, Department of Animal Sciences, Georg-August University, Goettingen, Germany, [2]Chair of Animal Breeding, Technische Universitaet Muenchen, Freising, Germany, [3]Institute of Human Genetics, Helmholtz Zentrum München, Neuherberg, Germany, [4]Institute of Human Genetics, Technische Universität München, Munich, Germany; sqanbar@gwdg.de

Domestication and subsequent breed formation have altered the patterning of variation within bovine genome. Unraveling gene variants underlying historical adaptation is of interest from the perspective of understanding how human interaction has influenced the cattle genome. In extension of some earlier low resolution studies, we employed a panel of nearly 15 million autosomal SNPs identified from re-sequencing of 43 German Fleckvieh animals to investigate past selection in cattle. Evidence for adaptation was investigated using several haplotype and allele frequency statistics in line with the features expected after a selective sweep. After a conservative screening of candidates, the genome wide scan revealed evidence of positive selection in 89 hitch-hiked regions. These regions contain genes with biological functions involved in immune system, blood clotting and particularly domestication phenotypes such as coloring pattern, sensory perceptions and neural system. We highlight several examples of adaptively evolving loci, including KIT, MITF, MC1R, NRG4, TMEM132D and GRIK3, among some others. To validate selection targets we conducted genome wide association analyses using appearance traits as additional evidence. This is the first comprehensive study for localizing signatures of past selection in cattle based on massive re-sequencing of entire genome. We provide signatures of widespread adaptation in cattle during speciation, domestication and breed formation exemplified with several striking selective sweeps of positive selection in appearance traits co-localized with major QTLs.

Using estimated allele frequency changes to map genomic regions under selection in farm animals

H. Simianer[1] and E.C.G. Pimentel[2]
[1]Georg-August-University, Department of Animal Sciences, Albrecht-Thaer-Weg 3, 37075 Goettingen, Germany, [2]University of Kassel, Nordbahnhofstr. 1a, 37213 Witzenhausen, Germany; hsimian@gwdg.de

This study is based on the assumption that alleles under positive selection increase in frequency over time. We used data of 2'294 HF bulls which were genotyped with the Illumina 54k SNP chip and for which accurate breeding values for 12 production, fitness, and fertility traits were available. SNP allele effects α were estimated for all 39,557 autosomal SNPs using random regression BLUP. For the positive allele at each SNP we estimated: (1) the allele frequency p_1 in the observed sample; and (2) the allele frequency p_0 in the founder population using the method suggested by Gengler *et al.* (2007) with a pedigree of 21,646 animals tracing back to 1906. The average number of discrete generation equivalents was 4.05. For each locus we calculated $\Delta p = p_1 - p_0$, which reflects the change in allele frequency over four generations. We found the average Δp being significantly >0 for milk yield, no significant deviation from 0 for fat and protein yield and for somatic cell score (SCS), and Δp significantly <0 for all other traits. Under positive selection, we expect that the allele frequency change Δp is higher for alleles with a large positive effect. To test this, we calculated the correlation ρ between Δp and α for all traits. We found $\rho>0$ for milk and protein yield, while correlations were negative ($-0.010>\rho>-0.047$) for the remaining production traits and SCS and even more negative ($-0.057>\rho>-0.137$) for cow fertility traits. Manhattan plots of the product $\alpha \times \Delta p$ reveal chromosomal positions in which selection has operated and allow annotation of underlying genes. The results suggest that selection in modern Holsteins was primarily targeted towards milk yield and has generated an undesirable correlated selection response in fitness (SCS) and fertility traits. The suggested method provides a novel approach to better understand natural and anthropogenic selection in farm animals.

Heritability of lamb survival on tick-exposed pastures

L. Grøva[1,2], I. Olesen[1,3] and J. Ødegård[3]
[1]Norwegian University of Life Sciences, Department of Aquaculture and Animal Science, P.O. Box 5003, 1432 Ås, Norway, [2]Bioforsk – Norwegian Institute for Agricultural and Environmental Research, Bioforsk Organic Food and Farming Division, Gunnars vei 6, 6630 Tingvoll, Norway, [3]Nofima, P.O. Box 5010, 1432 Ås, Norway; lise.grova@bioforsk.no

Sheep farming in Norway is based on grazing unfenced rangeland and mountain pastures in summer. A major welfare and economic issue in Norwegian sheep farming is the increase in lamb loss on such pastures. Tick-borne fever (TBF), caused by the bacteria Anaplasma phagocytophilum and transmitted by the tick Ixodes ricinus, is pointed out as one important challenge facing lambs during summer grazing. The objective of this study was to identify possible within breed genetic variation in lamb survival on tick-exposed pasture. Data on lambs within the normal distribution area of ticks, from flocks participating in ram circles with recordings in the Norwegian Sheep Recording System and registered with cases of TBF or using prophylactic treatment against ectoparasites at any time in 2000 to 2008 were included, making a total of 126,732 lambs with an average mortality of 3.8%. The data were analysed by a linear model, and the estimated heritability (on the observed scale) for the direct genetic effect on lamb survival was 0.220±0.005. The estimated maternal variance in proportion to phenotypic variance of lamb survival was 0.000±0.0004 indicating that maternal environment had very little effect on lamb survival. This indicates potential for genetic selection to improve survival in the studied population. The heritability cannot, however, be accurately attributed to resistance to A. phagocytophilum infection and TBF as the infection status of the lambs is unknown. Improved registrations of TBF and tick-exposure by farmers in ram circles in the Norwegian Sheep Recording System is therefore recommended to enable the use of such data for studying genetic variation in robustness in environments with and without ticks and a possible implementation in selection programs.

Effect of *H. contortus* infection on parasitological and local cellular responses of Creole kids

J.-C. Bambou
INRA, INRA Antilles-Guyane Domaine Duclos, 97170 Petit-Bourg, Guadeloupe; Jean-christophe.bambou@antilles.inra.fr

This study was carried out to evaluate the relationships between cellular changes in the abomasal mucosa and parasitological parameters, by comparing resistant and susceptible young Creole goats (kids) after experimental infection with *Haemonchus contortus* third stage larvae (L3). The kids were infected over 2 periods (challenge 1 and challenge 2) of 7 and 6 weeks respectively. Faecal egg count (FEC), blood eosinophilia and packed cell volume (PCV) were monitored weekly. At the end of both challenges a sub-group of kids were slaughtered for nematode burden measurements and analysis of inflammatory cells infiltration in the abomasal mucosa. A moderate anaemia was observed after the challenge 1 but not after the challenge 2, thus characterizing the infection as subclinical. Blood eosinophilia was higher in susceptible kids after both challenges. The FEC was significantly lower in resistant kids after both challenges. There was no difference in worm counts at necropsy between resistant and susceptible kids, but the number of immature worms and the means of female length were lower after challenge 2 whatever the genetic status. No differences were observed in the eosinophil and mononuclear cells infiltration between challenge 1 and 2 and resistant and susceptible kids. In contrast, globules leucocytes infiltrations were found higher after the challenge 1 in resistant kids.

Robustness in pigs

E.F. Knol
TOPIGS Research Center IPG, R&D, Schoenaker 6, 6641 SZ Beuningen, the Netherlands; egbert.knol@ipg.nl

Robustness, Easy2manage, Problem Free, User Friendly; all terms to indicate that pigs are adapted to contemporary pork production environments. Genetic adaptation of animals to a novel environment is straightforward and already described by Darwin. Problem for the pork chain is that the novel environment is a moving target. World trends (and differences) in climate (1), health challenges (2), feed quality (3) and cost of labor (4) have, so far, not lead to a uniform and constant production environment for which breeding companies can select the best animals. Temperature and disease tolerance, gut health, and easy to manage traits are therefore subjects of research, often as traits as such, but sometimes as proxy's for general robustness. TOPIGS has chosen to maintain line variation to accommodate lower challenged, higher potential environments on one side and more challenged environments on the other. Temperature tolerance measured in large crossbred populations of challenged sows yield upper critical temperatures in line with literature and yield estimates of exploitable genetic variation in temperature tolerance. An intensive piglet weighing/ cross-fostering/ survival program shows clear genetic variance in livability, extending before farrowing and after weaning, and in mothering ability. For disease the tolerance approach appears to be workable, that is to collect data from crossbred commercial sows in challenged environments, estimate crossbred EBV's and apply these in the selection index. Farrowing rate is an early indicator of environmental problems and a 'merry go round' concept, a sow cycle with a single insemination leading to a ready sow for the next cycle combines a number of robustness traits in one. The genomic toolbox and new approaches like social EBV's and epigenetic effects will help to further adapt pig populations to a uniform global or to regional specialized production environments.

Heat tolerance and reproductive performance in two sow lines

S. Bloemhof[1], E.F. Knol[1], I. Misztal[2] and E.H. Van der Waaij[3]
[1]TOPIGS research centre IPG, P.O. Box 43, 6640 AA Beuningen, the Netherlands, [2]university of georgia, animal and dairy science department, athens, 30605, USA, [3]Wageningen University, animal breeding and genomics centre, P.O. Box 338, 6700 AH Wageningen, the Netherlands; liesbeth.vanderwaaij@wur.nl

Pig breeding companies face the challenge to produce animals for a range of environments. One of the environmental challenges is elevated temperatures. Main research questions were: (1) does genetic variation for heat tolerance exist; (2) what is the correlation with reproductive performance; and (3) what could be underlying mechanisms for these correlations. Data was routinely collected for the TOPIGS breeding program in Spain and Portugal (high temperature zone), on Yorkshire (D-line) and Large White (ILW-line) pigs. Temperature recordings of local weather stations were used as approximation of the on-farm temperatures. Heat load was defined as the deviation upwards from the max temperature on the day of insemination from 19.20C. Farrowing rate (FR) was more affected by high temperatures in the D-line than in the ILW-line. The heritability for heat tolerance (HT) at 29.30C was higher in the D-line (0.04) than in the ILW line (0.02). The genetic correlation between FR and HT was 0.16 in the D-line and -0.36 in the ILW-line, and between litter size and HT -0.76 in the D-line and -0.10 in the ILW-line. Pearson rank correlations in the D-line in time reveal that heat stress during days 21 to 14 before insemination had the largest negative association with farrowing rate (-0.08), possibly related to oocyte quality. Heat stress between 7 days before to 12 days after insemination had the largest negative association with total number born (-0.05), possibly related to fertility and implantation. In conclusion, heat tolerance is heritable and of importance for reproduction in high temperature environments. Selection under temperate conditions results in higher reproductive performance and decreased heat tolerance. Taking day of insemination as point of reference is ok as the correlation with the most sensitive period is high (0.9).

Genetic parameters of thermoregulatory response in lactating sows

J.-L. Gourdine[1], D. Renaudeau[1], K. Benony[2], C. Anaïs[2] and N. Mandonnet[1]
[1]INRA, UR143 URZ, 97170, Petit Bourg, France, [2]INRA, UE1294 PTEA, 97170, Petit-Bourg, France; Jean-Luc.Gourdine@antilles.inra.fr

Selection for thermoregulatory response during heat stress (HS) might be a useful approach to reduce the magnitude of HS effects on lactating sow performances. The objective of this study was to estimate genetic parameters for thermoregulation traits and performance during lactation in Large White (LW) sows reared in a tropical environment (Guadeloupe, French West Indies). This area is characterized by having a tropical humid climate where the daily average ambient temperature (around 25.0 °C) is mostly above the lactating sows thermal comfort zone (about 22 °C). Sow rectal temperature (RT) and respiratory rate (RR) at 0700 and 1200, diurnal change of rectal temperature (dRT) were measured every 2-3 days during lactation. Sow average daily feed intake (ADFI), maternal body weight loss (BWL), and litter BW gain (LBWg) were collected. A total of 647 lactations from 224 LW sows were recorded between 2002 and 2012. Genetic parameters were estimated using univariate and bivariate animal models (ASREML program) with pedigree information available for 914 animals. Heritability estimates for mean RT at 0700 (RT07) and at 1200 (RT12) were moderate (0.31±0.09 and 0.39±0.10, respectively). The same trend was observed for mean RR at 0700 (RR07) and at 1200 (RR12) (0.30±0.10 and 0.23±0.07, respectively). Heritability estimates for mean dRT and for ADFI were low (0.17±0.04 and 0.10±0.06, respectively) but values for LBWg and BWL were moderate (0.28±0.05 and 0.33±0.05). Genetic correlations between mean RT and ADFI were negative (-0.05±0.32 to -0.12±0.31) but all low with large standard errors, suggesting that their antagonism could be manageable in a breeding objective. These results suggest that the individual lactating sow responses to HS is moderately heritable and genetic selection may enhance heat adaptation capacity in lactating sows.

Session 14 Theatre 9

Reaction norm for fat plus protein daily yield to evaluate genetic tolerance to heat stress in goats

A. Menéndez-Buxadera[1], H.M. Abo-Shady[1], A. Molina[1], M.J. Carabaño[2], M. Ramón[3] and J.M. Serradilla[1]
[1]Universidad de Córdoba, Producción Animal, Campus de Rabanales. Carretera. N IVa Km 396, 14014 Córdoba, Spain, [2]Instituto Nacional de Investigación y Tecnología Agraria y Alimentaria (INIA), Mejora Genética Animal, Carretera de La Coruña Km 7.5, 28040 Madrid, Spain, [3]Centro Regional de Selección y Reproducción Animal, Avenida del Vino 10, 13300, Spain; pa1semaj@uco.es

Genetic (co)variance components of daily fat plus protein yields (FPY) along the trajectory of a heat stress index (THI) were estimated by means of a Reaction Norm Model (RNM). FPY data were provided by the corresponding breeders associations and consisted of 126,825 and 154,240 test day (TD) from Florida and Malagueña goats, respectively. Climatic records register each TD were provided by the Spanish Meteorological Agency from weather stations located less than 20 km away from the farms. Daily maximum temperature and relative humidity were used to obtain a composite index of temperature and humidity (THI). The model included age at kidding, number of milking per day, a combination of flock-date of TD and a second order Legendre polynomial of THI on FPY as fixed effects, and a second order Legendre polynomial for the animal, a single individual permanent environmental effect and a residual as random effects. Results showed a comfort zone (THI from 6 to 21) where the heat stress effect was very small and a sensitive zone (THI from 21 to 31) having a significantly negative effect on FPY. Heritability estimated within the comfort zone ranged from 0.368 to 0.198 for Florida and from 0.294 to 0.159 for Malagueña. These ranges dropped from 0.185 to 0.112 and from 0.156 to 0.159, respectively, within the sensitive zone. Results also showed heterogeneity of genetic (co)variance and estimated breeding values across the trajectory of THI values in both breeds. Therefore, daily fat plus protein yield cannot be considered to be the same all over this trajectory. The RNM also allows for the identification of animals showing a robust response to high temperatures.

Session 14 Theatre 10

Genetic variation in macro- and micro-environmental sensitivity for milk yield in Swedish Holsteins

H.A. Mulder[1], L. Rönnegård[2], S. Wijga[1], W.F. Fikse[3], R.F. Veerkamp[4] and E. Strandberg[3]
[1]Animal Breeding and Genomics Centre, Wageningen University, P.O. Box 338, 6700 AH Wageningen, the Netherlands, [2]School of Technology and Business Studies, Högskolan Dalarna, 79188 Falun, Sweden, [3]Department of Animal Breeding and Genetics, Swedish University of Agricultural Sciences, Box 7023, 75007 Uppsala, Sweden, [4]Animal Breeding and Genomics Centre, Wageningen UR Livestock Research, P.O. Box 65, 8200 AB Lelystad, the Netherlands; han.mulder@wur.nl

Genetic variation in environmental sensitivity means that animals genetically differ in their response to environmental factors. Some factors (e.g. temperature) are known and called macro-environment, whereas other factors are unknown and called micro-environment. The objective of this study was to quantify the genetic variance in macro- and micro-environmental sensitivity and the genetic correlation between both types of environmental sensitivity for milk yield in Swedish Holstein cattle. We applied a double hierarchical generalized linear reaction norm model in ASREML on milk yield and its residual variance using average herd-year milk yield as environmental covariate for macro-environmental sensitivity. The heritabilities in macro- and micro-environmental sensitivity were as low as 0.007 and 0.012, respectively. The genetic variance in micro-environmental sensitivity was substantial, however: one genetic standard deviation would change micro-environmental sensitivity (=residual variance) by 21%. Genetic correlations between milk yield level and macro-environmental sensitivity and micro-environmental were 0.81 and 0.63, respectively, meaning that selection for higher milk yield increases both types of environmental sensitivity. The genetic correlation between macro- and micro-environmental sensitivity was 0.76, indicating that both types of environmental sensitivity are genetically similar. Due to the higher genetic variance, selection on micro-environmental sensitivity might be most useful to increase robustness against environmental disturbances.

Session 14 Theatre 11

How Corsican cattle breeders consider the adaptation of their breed: an exploratory approach

A. Lauvie[1], C. Rolland[1,2], C.H. Moulin[3] and F. Casabianca[1]
[1]INRA UR LRDE, Quartier Grossetti, 20250, France, [2]Montpellier SupAgro, Student, 2 place Pierre Viala, 34060 Montpellier cedex 1, France, [3]Montpellier SupAgro, UMR SELMET, 2 place Pierre Viala, 34060 Montpellier cedex 1, France; anne.lauvie@corte.inra.fr

To understand the genetic basis of adaptation, The GALIMED project proposes a multidisciplinary approach on 14 cattle breeds in the Mediterranean zone. This project combines population genetic, environmental, and livestock farming systems data to characterize the cattle and its management practices, quantify diversity, detect genetic variants and selection signals associated with adaptation to harsh conditions. This adaptation relies on both natural and artificial selection, linked with breeders' practices. We present here how the breeders considered adaptation of the Corsican cattle breed within their livestock farming systems. Our work is based on 20 semi structured interviews of breeders. They were chosen so as to cover the diversity of situation in the island (mountain and plain breeders, different geographical area, breeders in and out of the breed association). The interviews concerned the farming systems and the point of views of breeders about the breed and its adaptation. A thematic analysis of interviews has been conducted. We identify topics and traits of adaptation expressed by the breeders, and make counts to appraise their importance among their discourse. The topics for the characteristics of adaptation are, ranked by order of importance: autonomy for feeding, morphology and visible aspect, reproduction, adaptation to the environment, spatial and social behaviour, resistance to biotic and abiotic aggression. The criteria corresponding to each topic(abilities of the animals within the breed) are detailed and discussed. We finally discuss how such an approach could be improved to be applied on other breeds, and in particular for the other cattle breeds concerned by the GALIMED project, to implement a comparative approach.

Session 14 Poster 12

A preliminary investigation into GxE in first lactation South African Holstein cows

F.W.C. Neser[1], J.B. Van Wyk[1] and V. Ducrocq[2]
[1]University of the Free State, Animal Wildlife and Grassland Sciences, P.O. Box 339, 9300, Bloemfontein, South Africa, [2]INRA, UMR1313 Génétique Animale et Biologie Intégrative, 78352 Jouy-en-Josas, France; neserfw@ufs.ac.za

Genotype × environment interaction (G×E) in dairy cattle is a contentious issue that is usually ignored in genetic evaluations. It can, however, play a significant role in the accuracy of breeding values in different environments, if it exists, with a negative impact on genetic response when it is ignored. The purpose of the study was to investigate a possible genotype by environment interaction in first calf South African Holstein cows for both production and reproduction traits. Data from 100,975 cows on a total mix ration (TMR) and 22,083 pasture based cows were used. They were the progeny of 4,391 sires and 84,935 dams produced over a period of 11 generations. Traits analysed were milk production (corrected to a 305-day equivalent) and age at first calving (AFC). Both were recorded over a period of 30 years from 1980-2010. Production or AFC in each environment (TMR vs. pasture) was treated as a separate trait. Bivariate analyses, using an animal model in ASREML, were used to obtain genetic correlations between the traits measured in each environment. The genetic correlation for milk production measured in the two different environments was 0.90 (0.027) and that of age at first calving 0.28 (0.12). The heritability estimates for milk production were 0.23 (0.008) under the TMR system and 0.32 (0.015) for the pasture based system, while the estimates for AFC were 0.063 (0.005) and 0.055 (0.009), respectively. The rather large scale effect in the heritability (0.23 → 0.32) as well as the correlation of less than one for milk production between the two environments indicates that a GxE might exist. However, the low genetic correlation between the two environments for AFC is much more real and indicates that GxE should be taken into account when sire selection are done.

Genetic parameters of ability to tolerate once-daily milking in a Holstein × Normande population

H. Larroque[1], L. Heuveline[1], S. Barbay[2], Y. Gallard[2] and J. Guinard-Flament[3]
[1]INRA, UR631 SAGA, 31320 Castanet-Tolosan, France, [2]INRA, UE326 Domaine Expérimental du Pin, 61310 Exmes, France, [3]INRA, UMR1348 PEGASE, 35590 St-Gilles, France; helene.larroque@toulouse.inra.fr

To be well adapted to once daily-milking (ODM), a dairy cow has to exhibit low relative milk yield loss when switched to ODM and strong milk recovery when back to twice daily-milking (TDM). The aim of this study was to investigate genetic variability of relative milk yield loss and recovery and their genetic relationships with fat and protein contents (FC and PC, resp.) in order to evaluate predictive ability of milk composition. The study concerned 368 F2 crossed HolsteinXNormande dairy cows in 2nd lactation. The trial consisted of 3 successive periods: 1 wk with TDM (TDM1), 3 wks with ODM and 2 wks with TDM (TDM2). Genetic parameters were estimated by restricted maximum likelihood with an animal model. Milk yield averaged 28.3 kg/d during TDM1 (±5.4); it decreased by 8 kg/d (±2.9) (i.e. 28.2%) during ODM and increased by 4.0 kg/d (±2.5) (i.e. 20.5%) when switched back to TDM. FC and PC were 43.2 g/kg (±6.4) and 31.7 g/kg (±2.4) respectively during TDM1; they increased during ODM to 48.6 g/kg (±7.1) and 32.5 g/kg (±2.3) resp. Heritability was moderate for relative milk loss (0.26±0.08) and higher for relative milk recovery (0.43±0.06), with between them a correlation of -0.43 (±0.13). As expected, heritability of FC and PC were high on TDM1 (0.46±0.10) and on ODM (0.56±0.09 and 0.66±0.08, resp.). Relative milk loss was positively correlated to TDM1 FC and PC (0.28±0.15 and 0.50±0.13, resp.), whereas relative milk recovery was correlated to ODM FC and PC (0.24±0.18 and -0.59±0.08, resp.) and to TDM1 FC and PC but with very high SE (not shown). Despite the low staff and the use of a crossbred population this study has shown that relative milk yield loss and recovery in response to ODM are under genetic control. They are partially related together and to milk composition. This analyze will be carried on by a QTL detection.

Phenotyping goats on their feeding behaviour

C. Duvaux-Ponter[1,2], S. Giger-Reverdin[1,2], J. Tessier[1,2], E. Ricard[3], J. Ruesche[3], I. David[3] and L. Bodin[3]
[1]INRA, UMR791 MoSAR, 75005 Paris, France, [2]AgroParisTech, UMR791 MoSAR, 75005 Paris, France, [3]INRA, UR631 SAGA, 31326 Castanet Tolosan, France; christine.duvaux@agroparistech.fr

The evolution of farming systems due to societal demand for a more sustainable production will require adaptation of farming techniques as well as improvement of the adaptive capacity of production herbivores in the face of new challenges resulting from climate change. Animals can adapt to a new diet by modifying their feeding behaviour, and more precisely their intake rate. The aim of this work was to look for new, pertinent and repeatable criteria to evaluate this trait. Feeding behaviour was assessed at three different periods (1st gestation, 1st lactation, 2nd gestation + lactation). All the renewal goats born in January 2011 were tested. They were housed in individual crates with automatic measurement of the quantity of feed eaten every 2 min (3 days of measurement) and fed *ad libitum* a complete diet adapted to requirements (two feed allowances per day). Thirty-six goats completed the three periods of measurement. Statistical analyses were performed with the mixed procedure of SAS to test the individual fixed effects, and ASREML to estimate variance components. Four different phenotypes were analyzed: Q90 (quantity of diet consumed 90 min post evening feed allowance which corresponded to two thirds of the daily feed allowance), P90 (Q90/Total quantity of feed consumed after the evening feed allowance), $Area_{24}$ and $Area_{15}$ (average difference between the cumulative intake at each time point and the cumulative intake at time t; either over 24 h or over the evening feed allowance (15 h)). Intra-period individual repeatabilities were very high (~0.84) for the four criteria, while they were lower between periods (~0.30) but still significantly different from zero. These preliminary results on the variability of intake rate show that simple criteria could be found to phenotype goats on intake rate.

Approaches to reduce antibiotic resistance in the pork supply chain

L. Heres[1], S. Düsseldorf[2,3], D. Oorburg[1] and H.A.P. Urlings[1,4]
[1]VION Food Group, Noord Brabantlaan 303-307, 5657 GB Eindhoven, the Netherlands, [2]University Bonn, Katzenburgweg 7-9, Bonn, Germany, [3]Erzeuger Gemeinschaft Süd Ost Bayern, Gewerbering 13, Pocking, Germany, [4]Wageningen University, Zodiac, Wageningen, the Netherlands; lourens.heres@vionfood.com

Occasionally, use of antibiotics is necessary to treat diseased animals. Prudent use is however necessary, as antibiotic usage evokes the selection and propagation of antibiotic resistant bacteria. Antibiotic resistance in pigs is primarily a occupational risk for those who are in contact with pigs. Secondly, contamination of the environment and fresh food may expose the general population to resistant bacteria. Strict hygiene at slaughter can prevent the contamination of food with resistant bacteria. Recent experience in the Netherlands shows that obligatory reporting of antibiotic usage and accompanying benchmarking, can results in a marked reduction. Also a ban on use of critical antibiotics (fluoroquinolones and cephalosporins) appeared to be feasible. We hypothesize that good animal health and optimal biosecurity are crucial to further reduce the occurrence and propagation of antibiotic resistance. To support the farmers, veterinarians, and other advisors in these areas, we assumed that the collection of data on health parameters in the slaughterhouse has an added value. Pathological findings is 'classical' information about the health of the slaughtered animals. It was studied whether serological results from blood collected in the slaughterhouse, can be complementary information. We show that differences in *Salmonella*, *Mycobacterium avium* and *Toxoplasma* status can be used as a derivative of internal and external biosecurity. Serology on slaughterhouse blood for lung pathogens (e.g. PRRSV) provides additional information, which can support the animal health management. It is concluded that future challenges lay in exchange of easily accessible information collected in slaughterhouses, development of management alternatives based on this information, and development of additional serological methods.

Importance of herd management and building design on respiratory diseases in pig herds

C. Fablet, V. Dorenlor, F. Eono, E. Eveno, J.P. Jolly, F. Portier, F. Bidan, F. Madec and N. Rose
Anses, Swine epidemiology unit, B.P. 53, 22440 Ploufragan, France; christelle.fablet@anses.fr

A study was carried out in 143 french farrow-to-finish herds to identify noninfectious factors associated with pneumonia and pleuritis in slaughter pigs. Data related to herd characteristics, biosecurity, management and housing conditions were collected during a farm visit. Climatic conditions were measured in the post-weaning and finishing rooms where the slaughter pigs have been raised. A sample of 30 randomly selected pigs per herd was scored for pneumonia and pleuritis at slaughterhouse. Herds were grouped into three categories according to their pneumonia median score (class 1: $\leq$0.5; class 2: 0.5-3.75; class 3: $\geq$3.75). For pleuritis, a herd was deemed affected if at least one pig had extended pleuritis. Logistic regression models were used to identify factors associated with pneumonia and pleuritis. A short interval between successive batches, large finishing room size and high mean CO_2 concentration in the finishing room significantly increased the odds for a herd to be in class 2 for pneumonia. The same risk factors were found for class 3 and, in addition, a direct fresh air inlet from outside or from the corridor in the post-weaning room versus an appropriate ceiling above the pigs also increased the risk. The odds for a herd to have at least one pig with extended pleuritis was increased when the farrowing facilities were not disinsected, when tail docking was performed later than 1.5 days after birth and if the piglets were castrated when more than 14 days old. A short temperature range for the ventilation control rate in the farrowing room, a low mean temperature in the finishing room and large herd size were also associated with increased risk of pleuritis. Lung health may therefore be improved by implementing appropriate husbandry practices and correcting housing. In conclusion, several interconnected zootechnical pathways may be used to promote herd health management without using antibiotics.

Analysis of and approach to antibiotics use in period 2005-2012 in dairy sector in the Netherlands

A. Kuipers[1], H. Wemmenhove[2] and W.J. Koops[3]
[1]Expertise Centre for Farm Management and Knowledge Transfer WUR, P.O. Box 35, 6700 AA Wageningen, the Netherlands, [2]Livestock Research WUR, P.O. Box 65, 8200 AB Lelystad, the Netherlands, [3]Animal Production Systems WU, De Elst 1, 6708 WD Wageningen, the Netherlands; abele.kuipers@wur.nl

Careful use of antibiotics in animals is linked to human health, because of growing resistance against antibiotics in humans. Especially 3rd and 4th generation antibiotics (3/4GA) are essential for treatments in hospitals, but are also used in animals. National goal was to reduce antibiotic use by 50% in 2012 compared to 2009. A direct relation between veterinarian and farmer has been introduced, as well as a farm animal health plan. From 2012 on 3/4GA are only allowed in exceptional cases. To gain insight, antibiotic use was examined on 95 farms during period 2005-2012, linked to 30 Veterinary practices. Antibiotic use on farm level is expressed in daily dosages/cow/year (DD). Average DD over 2005-2010 was 6,07 with spread from 1-15, in 2011 5,80 and is presently analysed over 2012. Daily dosages were split up in contributions by dry cow therapy (43%), mastitis (25%), calves (3%), uterine and after birth treatments (3%) and other diseases (26%). 2/3 of antibiotics is going to the udder. Over 2005-2011 use for dry cow therapy and other diseases decreased a bit while use for mastitis was rather constant. Percentage use of 3/4GA decreased in same periods from 19 to 15%. Farm and herd factors were studied affecting antibiotics use, practising a step-wise regression procedure. Variation in total use was explained for 39% (R^2) and dry cow therapy use by 46% by factors like quota size, milk amount/ cow, health status, cell count and calving interval. A negative correlation of -0,55 between cell count and level of antibiotic use was found. Farmers expressing a good relation with veterinarian and the 'more entrepreneurial ' farmers tended to use somewhat more antibiotics than the other colleagues. A change in mindset is needed of farmers and veterinarians. Results over 2012 will show that this change is rapidly occurring.

Health monitoring concepts for long-term improvement of dairy health

K.F. Stock[1], D. Agena[1], R. Schafberg[2] and F. Reinhardt[1]
[1]Vereinigte Informationssysteme Tierhaltung w. V., Heideweg 1, 27283 Verden, Germany, [2]Martin-Luther-University Halle-Wittenberg, Institute of Agricultural and Nutritional Sciences, Theodor-Lieser-Strasse 11, 06120 Halle/Saal, Germany; friederike.katharina.stock@vit.de

Standardized recording of animal health data is prerequisite for reliable analyses across farms and measures to systematically lower disease incidences in the population. For cattle, the central key for health data recording has been set up as comprehensive reference allowing health monitoring (HM) with input from multiple sources of different expertise. Available infrastructure is not yet broadly used though in the German dairy sector, mainly because of lacking HM concepts that combine practical feasibility with short- and long-term benefits for management and breeding. Experiences from regional projects (GKuh: 2010-2012, 18,500 females/57 farms; THU: 2009-2012, 55,000 females/21 farms) with heterogeneous farm structures and implementation conditions were used to synergistically develop an integrated system for routine HM in German dairy farms. Recording by farmers via herd management software focused on disease diagnoses, with optional extension by health-related observations and measures. Interfaces made the system flexible with regard to additional input from e.g. claw trimmers and laboratories. All health data, relating to 24,500 (GKuh) and 241,000 (THU) diseases, were transferred to the central health data base with restrictive access rights to ensure data security. Management-oriented analyses included within-farm statistics and comparisons with regional averages, with individualized format of health reports according to the herd-needs (stand-alone HM protocol or supplement to already established within-farm analyses). Complete data from HM herds, including animal movements, performance and pedigree records, were used for genetic analyses. In the combined dataset, 2,219 sires were represented (6% with >50 daughters) implying improved conditions for genetic evaluations and selection for health traits in dairy cattle.

Animal based indicators for the implementation of a selective usage of anthelmintics in dairy cattle

N. Ravinet[1,2], A. Lehebel[1,3], C. Chartier[1,3], A. Chauvin[1,3] and N. Bareille[1,3]
[1]LUNAM Université, Oniris, UMRBioEpAR, CS 40706, 44307 Nantes, France, [2]IDELE, CS 40706, 44307 Nantes, France, [3]INRA, UMR1300BioEpAR, CS 40706, 44307 Nantes, France; alain.chauvin@oniris-nantes.fr

To optimize treatments for gastrointestinal nematode (GIN) in adult dairy cows by selective treatment, we need to identify cows with an improved milk production (MP) after treatment. The objective of this study was to evaluate the evolution of MP after treatment for GIN, and to identify herd and individual variation factors. A field trial involving 20 pastured dairy herds in Western France was conducted in autumn 2010 and autumn 2011. In each herd, lactating cows were randomly allocated to a treatment group (fenbendazole) (533 cows), or a control group (544 cows). Daily cow MP was recorded from 2 weeks before until 14 weeks after treatment. Individual serum anti-Ostertagia antibody levels (expressed as ODR), pepsinogen levels, egg output, and bulk tank milk ODR were measured at time of treatment. Moreover, in each herd, information regarding heifers' grazing history was collected to assess the time of effective contact in months (TEC) with GIN larvae before the first calving. MP data were analyzed using linear mixed models (herd and cow as random effects). The overall treatment effect on MP was significant but slight (maximum=+0.86 kg/d on week 6 after treatment). Cows from herds where the percentage of positive egg count was >22.6% responded better than those from herds where it was lower. Cows from low-TEC (<8 months) versus high-TEC herds also showed a better response, particularly when the bulk tank milk ODR was high; however, bulk tank milk ODR taken into account alone did not appear as a significant variation factor. Primiparous cows, cows with days in milk (DIM)<200 at time of treatment, and cows with low individual ODR (<0.38) responded better than multiparous cows, cows with DIM>200, and cows with higher ODR respectively. These results highlight promising key criteria for selective treatment for GIN in dairy cows.

Potential impact of the poultry red mite on health and welfare of companion and livestock animals

D.R. George, R. Finn, K. Graham and O.A.E. Sparagano
Northumbria University, Faculty of Health and Life Sciences, City Campus, NE1 8ST, Newcastle upon Tyne, United Kingdom; david.george@northumbria.ac.uk

The poultry red mite (PRM) *Dermanyssus gallinae* is best known as a pest of the laying-hen industry; adversely affecting production and hen health throughout the globe, both directly and through its role as a disease vector. Nevertheless, PRM is increasingly implemented in dermatological complaints in humans, suggesting its significance may extend beyond poultry. The main objective of the current work was to review the potential threat of PRM to companion animals and livestock. Results showed that as an avian mite, PRM is unsurprisingly an occasional pest of pet birds. PRM has been reported on other animals, however, including: cats, dogs, rodents, rabbits and horses. We conclude that although cases of PRM on mammals are rare, when coupled with the genetic plasticity of this species and evidence of permanent infestation on mammals, potential for host-switching exists. We report that host switching events have been recently documented in phytophagous pest insects (e.g. the diamondback moth) and Plasmodium falciparum (re: host-switching from birds to mammals), supporting the potential for such events per se. In poultry, infestations of PRM often proliferate regardless of standard management, promoted by the tendency of PRM to develop resistance to synthetic acaricides and difficulties in targeting these reclusive mites. A further aim of this study was to review work conducted in the NE of England on alternative management of PRM in poultry, evaluating which strategies could be of broader veterinary use. We report that over the last decade our research group has targeted the following control methods: vaccine development, plant- and marine-based acaricides, biopesticides and biological control. We conclude (based on potential drawbacks, existing practices and current commercial trends) that of these vaccination and novel acaricides hold most promise to target PRM in companion and livestock animals.

Session 15a — Poster 7

Coprophagous behavior of rabbit pups affects dynamic implantation of microbiota and health status

S. Combes[1], T. Gidenne[1], L. Cauquil[1], O. Bouchez[2] and L. Fortun-Lamothe[1]
[1]INRA, UMR 1289 Tandem, 31326 Castanet-Tolosan, France, [2]INRA, GeT-PlaGe, Genotoul, 31326 Castanet-Tolosan, France; laurence.lamothe@toulouse.inra.fr

The aim of this study was to investigate the role of coprophagous behavior in suckling rabbit on the implantation of caecal microbiota and to try improving the health status of rabbits modifying this behavior. Three groups were compared: in FM group (n=24 litters), pups had free access to maternal hard feces, in NF group (n=28), ingestion of hard feces was prevented, and in FF group (n=28), pups had access only to hard feces excreted by foreign females (n=5, 7 and 9 feces from 2 to 13 d, 14 to 17 d, 18 to 20 d, respectively). Pup mortality, excretion and ingestion of feces were measured daily. Bacterial composition was assessed by 454 pyrosequencing of the V3-V4 region of 16S RNA genes at 14, 35, 49 and 80 d of age. The total number of feces excreted by the does from 2 to 20 d after delivery ranged widely, but was similar among groups (16.1±12.6 feces doe-1). Ingestion of feces was 3 times greater in FF than in FM group (35.6 vs. 9.9, P<0.001). From 1 to 80 d of age the FF and NF groups exhibited respectively the lowest (9.3%) and highest (22.8%) pup mortality compared to FM group (15.5%, P=0.03). At age 14 d the caecal bacterial community was dominated by *Bacteroidetes phyla* (63.3%), *Bacteroidaceae* family (36.0%) and *Bacteriodes* genus (36.0%). With increasing age, Firmicutes phyla, Lachnospiraceae and Ruminococcaceae families became the dominant taxa (92.0%, 44.0% and 37.9%, respectively at 80 d of age). Impairment of feces ingestion delayed this ecological succession, with greater and lower relative abundance of *Bacteroidaceae* and *Ruminococcaceae* respectively in NF than the two other groups at age 35 d (P<0.1). In conclusion, the coprophagous behaviour of suckling rabbits is implicated in the maturation of caecal bacterial microbiota and stimulation of this natural behavior improve the health status of animals and could be used to limit the use of antibiotics.

Session 15b — Theatre 1

Prebiotics and probiotics for the control of dysbiosis, present status and future perspectives

R.V.A. Ducatelle, F. Van Immerseel and F. Haesebrouck
Ghent University, Pathology, Bacteriology and Avian Medicine, Salisburylaan 133, 9820 Merelbeke, Belgium; richard.ducatelle@UGent.be

Since the ban on antimicrobial growth promotors (AGP) in the EU, an increase in intestinal health problems has been noted in food producing animals, in particular in weaned piglets and in broiler chickens. These problems usually are referred to as dysbiosis or dysbacteriosis. Consequently, the feed industry is looking for alternatives to replace the AGP. Categories of products in this growing market of feed additives are: enzymes, herbs, herbal extracts and organic acids, and also the prebiotics and probiotics. Currently available probiotics usually contain one or more strains of *Lactobacillus* spp., *Bifidobacterium* spp., *Bacillus* spp., *Enterococcus* spp. or *Saccharomyces* spp. Fructo-oligosaccharides, xylan-oligosaccharides, mannan-oligosaccharides, beta-glucan oligosaccharides and galacto-oligosaccharides all are prebiotics with documented effects on growth of certain bacterial species. For several of these products a favorable effect has been shown on feed conversion (FC) in animals under stressed conditions. These strategies are based on the assumption that an optimization of the intestinal microbiota composition may improve FC and/or protect from intestinal health problems. A major constraint, however, is the lack of knowledge on the ideal intestinal microbiota composition and how beneficial microbes could favor intestinal health and FC. Recent technological developments, however, gradually allow a better insight in the underlying mechanisms. It appears that the cross-talk between certain components of the gut microbiota and the host is crucial to the development and the maintenance of oral tolerance, a gut-specific immune mechanism that is essential for the normal functioning of the gut. New pre- and probiotic strategies are currently being developed that specifically target the oral tolerance system.

Session 15b Theatre 2

Flavonoids and other plant substances enhance the immunity of the animal

W.-Y. Zhu
Nanjing Agricultural University, College of Animal Science and Technology, Weigang No.1, 210095 Nanjing, China, P.R.; zhuweiyun@njau.edu.cn

In view of side-effects of antimicrobial growth promoters, novel and sustainable dietary strategies are required to improve animal gut health and performance. Flavonoids and other plant substances are regarded alternatives to antibiotic feed additives. However, the biological activity of these compounds depends on their bioactive availability. Metabolism by the gut microbiota is an important factor for their bioavailability. Keeping in view all facts, *in vivo* studies were conducted on both pigs and broiler chickens to observe the multidimensional effects of flavonoids (genistein, daidzein and hesperidin) and other plant substances for growth performance, immunity and health of the animal, while *in vitro* studies were to investigate bioactive compounds-degrading bacteria and to examine the effects of dietary components on their bioavailability. In order to validate the antioxidant effects, *in vitro* studies were also conducted to further explore their synergistic activity and their interaction with gut microbiota. Results showed that these compounds could affect the immunity responses, oxidative status of the animal, with the effects age and dose dependent. Synergisttic effects were also observed between flavonoids and other plant substances. A number of flavonoid and other substances-metabolising bacteria were isolated from the intestine of the pig and chicken, leading to higher bioavailability of these compounds in the intestine. Dietary factors such as FOS could stimulate the bioavailability while others (such as organic acids) may reduce the availabilities of these compounds. Thus, there is a need to study the systemic bioavailability and metabolism of these compounds in animal gut, and microbial metaolites that are absorbed. Studies investigating the efficacy of flavonoids and other substances, and their metabolites, and their interactions (synergistic or antagonistic activity) within the gut may help understanding the role of plant bioactive substances.

Session 15b Theatre 3

Microbiota, metabolism and immunity: the potential for early-life intervention

M. Bailey, M.C. Lewis, Z. Christoforidou and C.R. Stokes
University of Bristol, School of Veterinary Science, Langford House, Langford, Bristo BS40 5DU, United Kingdom; mick.bailey@bristol.ac.uk

The potential for therapeutic manipulation of the intestinal microbiota is well accepted. There is now a considerable literature on the effects of probiotics and prebiotics in rodent models and humans, and an increasing amount of work carried out directly in pigs. Such work suffers from two major drawbacks: firstly, that it is often not clear what constitutes 'beneficial' and 'detrimental' effects; and secondly, that the established microbiota in adults is relatively stable, and interventions often need to be maintained for prolonged periods of time. However, we have proposed that birth and weaning are two hazard and critical control points in the development of young piglets at which manipulation of the environment, specifically the microbiota, may have long term effects on piglet performance. We have used three approaches to study intervention at these points. Firstly, we have established gnotobiotic piglets by caesarean section into sterile rearing environments, and examined the effects on the immune system of a defined, three-component microbiota given at birth. Secondly, we have used the human probiotic B. lactis fed in conventional environments from weaning. Thirdly, we have used our SPF isolator system to study the effects of prebiotics fed together with milk replacer from birth. Intervention had marked effects on intestinal physiology and on development of the immune and metabolic systems, providing support for the proposal that there is considerable potential for intervention at birth and at weaning aimed at improving pig health and performance throughout the growing and finishing periods.

Review of interaction of mycotoxins and endotoxins on inflammatory response

M.A. Rodriguez and J. Laurain
Olmix S.A., ZA du Haut du Bois, 56580, France; jlaurain@olmix.com

An effective immune system is determining for animal health and well being. Commercial animal production is based on balanced feed providing required nutrients and optimized immunosuppression, increased susceptibility to diseases and consequently decreased productivity of farm animals. Mycotoxins are one of the most immunosuppressive factors in animal diets. The purpose of this paper is to describe the influence of mycotoxins and endotoxins on inflammatory response. DON, a trichothecene mycotoxin common in cereal-based feeds, causes impaired growth in animals via a rapid induction of expression of pro-inflammatory cytokines, followed by up-regulation of several suppressors of cytokine signaling (SOCS), capable of impairing growth hormone (GH) signaling. Oral DON exposure perturbs GH axis by suppressing two clinically relevant growth-related proteins, IGFALS and IGF1. It also impairs the barrier function of the intestine by reducing the expression of claudin proteins implicated in the regulation of tight junction proteins and decreases trans-epithelial electrical resistance, thus resulting in an increased risk of trans-epithelial passage of both bacteria and endotoxins into the body. Endotoxins are derived from the cell membranes of Gram- bacteria. They are linked within the bacterial cell wall and are continuously liberated into the environment at cell death, during growth and division. Endotoxins act through activation of the immune system, with the release of a range of pro-inflammatory mediators, such as interleukin 6 and IL-1. This chain reaction leads to an increase of SOCS which have a negative action on GH-induced gene expression in liver, reducing the production of IGF1 and alleviate its many actions of growth hormone that have impact on productivity (growth, milk production, fertility…). Limiting the absorption of DON in the intestine by using interspaced clay, closes the door to endotoxins and pathogens and reduces its combined and dangerous effect on the immune response and IGF1.

Session 15b Poster 5

Exploration of the macrophage-virus interactions during a PRRSV infection by a modelling approach

N. Go[1,2], C. Bidot[2], C. Belloc[1] and S. Touzeau[3,4]
[1]Oniris, INRA, UMR 1351 ISA BioEpAR, 44307 Nantes, France, [2]INRA, UR341 MIA, INRA, F-78350 Jouy-en-Josas, 78352 Jouy-en-Josas Cedex, France, [3]Inria, Biocore, Inria, 06900 Sophia-Antipolis, France, [4]INRA, UMR1351 ISA, INRA, 06900 Sophia-Antipolis, France; natacha.go@jouy.inra.fr

Porcine Respiratory and Reproductive Syndrome Virus (PRRSV) infection is a major concern for swine industry. Our partial understanding of the interactions between the virus and the immune system is the major reason for the lack of efficient control measures. The PRRS virus replicates mainly in the pulmonary macrophages which: (1) are responsible for inflammation and viral destruction by phagocytosis; and (2) participate in the induction and orientation of the adaptive immune response. Consequently, macrophage infection hampers the whole immune response. The interactions between macrophages and virus during the first steps of infection have not been thoroughly investigated and their influence on the infection resolution is unknown. Here, we propose an original model simulating immune and infection dynamics to explore these complex mechanisms and test biological hypotheses. We highly detail the immune functions and infectious statuses of macrophages and take into account interactions between innate and adaptive responses and cytokines regulations. We use the model to study the relative influence of macrophage – virus interactions on the infection resolution comparatively to adaptive mechanisms.

Expression of selected immune-system genes in cow milk somatic cells of after Se supplementation

J. Jarczak, E. Kościuczuk, A. Jóźwik, D. Słoniewska, N. Strzałkowska, J. Krzyżewski, L. Zwierzchowski and E. Bagnicka
Institute of Genetics and Animal Breeding PAS in Jastrzębiec, Department of Animal Improvment, ul. Postępu 36A, 05-552 Magdalenka, Poland; j.jarczak@ighz.pl

We assumed that, the higher bioavailability of organic selenium from Se yeast should affect the expression of genes encoding antimicrobial peptides. The study was conducted on 16 Polish Holstein-Friesian (HF) dairy cows. Cows were equally divided into experimental and control groups. All cows were healthy and in the same stage of lactation of second parity. The experimental group was supplemented with organic selenium, while control group with the sodium selenite for 60 days. The milk samples were taken at the beginning, in 30 day and at the end of experiment. RNA was isolated from milk somatic cells. Several genes were selected to gene expression analysis using R-T qPCR method: bovine β-defensins LAP, TAP, BNBD4, BNBD7 and bovine cathelicidins CATH1, CATH2, CATH3, CATH4. Stage of lactation influenced the expression of two defensin genes: BNBD1 and LAP in both groups. Since the organism of the animal become more vulnerable to attack of pathogens at the end of lactation, the increased expression of both genes involved in antibacterial defense at that time may indicate activation of the immune system. Furthermore, the expressions of BNBD1 and LAP genes were much higher in experimental group. Therefore, the yeast supplementation, besides to be a source of selenium, may have exerted its effect because two sulfuric amino acids which are present in selenium yeast (methionine and cysteine) are also the element of defensin's structure. Thus, organic selenium may influence the activity of immune system stimulating expression of some genes. However, there were no differences between groups and during lactation in the expression of all cathetlicidin genes studied. Research was realized within the project 'BIOFOOD – innovative, functional products of animal origin' no. POIG.01.01.02-014-090/09 co-financed by the EU from the ERDF within the IEOP 2007-2013.

Hesperidin improves growth performance, immune response and antioxidant activity in pigs

S.K. Park[1], S.H. Lee[1], J.C. Park[1] and I.H. Kim[2]
[1]National Institute of Animal Science, Animal Nutrition & Physiology, Chuksan-gil 77, 441-706, Suwon, Korea, South, [2]Dankuk University, Animal Resource & Science, 126 Jukjeon-dong, 448-701, Cheonan, Korea, South; maiky@korea.kr

Hesperidin is a member of flavonoids and has beneficial effects on immune function and oxidative stress. Therefore, this study was conducted to investigate the effects of hesperidin supplementation on growth performance, immune function, and antioxidant activity of growing pigs. Total of 24 pigs weighed 20.4±0.6 kg were randomly allotted to 3 treatments with 8 replicates for 9 wk. Dietary treatments included: (1) basal diet (CON); (2) basal diet + 0.01% hesperidin (Hes-1); (3) basal diet + 0.02% hesperidin (Hes-2). Pigs fed Hes-1 and Hes-2 diets had higher ($P<0.05$) gain:feed (G:F) ratio compared to CON group. Blood creatinine concentration was lower ($P<0.05$) in Hes-1 treatment than that in CON group. *In vivo* antioxidant activity, represented by serum SOD activity, was increased ($P<0.05$) by Hes-2 compared to CON. To further confirm the immune function, pigs were i.p. challenged with lipopolysaccharide (LPS; 50ug/kg BW) and blood was analyzed. After 24 h LPS challenge, platelet concentration was lower ($P<0.05$) in hesperidin treatment group compared to CON group. However, immunoblobulin levels were not different among treatment groups. The results of current study indicate that administration of hesperidin has beneficial effects on growth performance and antioxidant activity in growing pigs. Furthermore, dietary supplementation of hesperidin reduced levels of serum creatinine as well as LPS induced platelet. However, further studies need to be performed to confirm the effect of hesperidin on stimulation of immune response in pigs.

Session 16 Theatre 1

Bayesian regression method for genomic analyses with incomplete genotype data

R.L. Fernando, D. Garrick and J.C.M. Dekkers
Iowa State University, Animal Science, Kildee Hall, Ames, IA 50011, USA; rohan@iastate.edu

High-density SNP genotypes are increasingly being incorporated into genetic evaluations. To obtain predictions that are optimal and not biased by selection, the conditional mean of the breeding value must be computed, given the genotypes, all other pedigree and all phenotypic data that were used for selection. When SNP effects have a normal distribution, single-step BLUP (SSBLUP) can be used to get this conditional mean. It requires computing the inverse of the matrix G of genomic relationships, which is dense and computationally infeasible as the number of animals genotyped increases. Computing G requires the frequencies of SNP alleles in the founders, which are not available in most situations, so frequencies from genotyped animals, which invariably have been subject to selection, are used. This results in biased evaluations, requiring ad-hoc corrections to G. Further, SSBLUP is limited to a model with normally distributed marker effects that is expected to perform poorly as marker densities increase relative to variable selection models such as BayesB and BayesC. Here, we present a single-step Bayesian regression (SSBR) method that combines all available data from genotyped and non-genotyped animals, as in SSBLUP, but accommodates a wider class of models. Our strategy is to use imputed genotype covariates for animals that are not genotyped, together with an appropriate residual to accommodate imputation errors. Under normality, SSBR is identical to SSBLUP but does not require computing G or its inverse and provides richer inferences. For example, prediction error variances are approximated in SSBLUP but obtained from the MCMC samples that are part of the analysis in SSBR. At present, Bayesian regression analyses seldom exceed ten thousand individuals. However, when SSBR is applied to all animals in a breeding program, there will be a 100 to 200-fold increase in the number of animals and associated computing challenges. Thus, parallel computing and alternative algorithms will be used to reduce computing time.

Session 16 Theatre 2

Genotype imputation accuracy in Holstein Friesian cattle in case of whole-genome sequence data

R. Van Binsbergen[1,2], M.C.A.M. Bink[1], M.P.L. Calus[2], F.A. Van Eeuwijk[1], B.J. Hayes[3], B. Hulsegge[2] and R.F. Veerkamp[2]
[1]Biometris, Wageningen UR, P.O. Box 100, 6700 AC Wageningen, the Netherlands, [2]Animal Breeding and Genomics Centre, Wageningen UR Livestock Research, P.O. Box 135, 6700 AC Wageningen, the Netherlands, [3]Department of Primary Industries, Biosciences Research Division, 1 Park Drive, Bundoora, 3038, Australia; rianne.vanbinsbergen@wur.nl

Despite falling costs of sequencing, sequencing a large number of individuals is still too expensive. A promising approach is to sequence the genomes of a core set of individuals and use these data to impute missing genotypes for individuals genotyped at lower density. The objective of this study was to investigate how imputation accuracy in Holstein Friesian cattle to whole-genome sequence was influenced by reference group size, and by number, location and minor allele frequency of the SNP. Whole-genome sequence data for BTA 1 and 29 of 114 Holstein Friesian bulls were provided by the 1000 bull genomes project. The Beagle software was used for imputation, accuracy was assessed via five-fold cross validation. For the validation individuals all SNP were set to missing, except for SNP that occur on the Illumina BovineSNP50 or BovineHD arrays. Imputation accuracy was calculated as the correlation between observed and imputed genotypes. For BTA 29 and for the largest reference group imputation accuracy from BovineSNP50 to whole-genome sequence was on average 0.37 and imputation accuracy from BovineHD was on average 0.80. For SNP with minor allele frequency above 0.25 the average imputation accuracy was 0.89. For SNP with a lower minor allele frequency this decreased to 0.13-0.38 (depending on reference group size). When distance to nearest genotyped SNP increased to 5,000 base-pairs the average accuracy dropped from almost 1 to 0.5, and dropped more rapidly at larger distances. We conclude that reference group size, and location and minor allele frequency of the SNP affect imputation accuracy, and that a 50k SNP chip is not sufficient to reach acceptable accuracy of imputation of sequence data.

Variation in genome sharing among non-inbred pigs

N.S. Forneris[1], J.P. Steibel[2], C.W. Ernst[2], R.O. Bates[2], J.L. Gualdrón Duarte[1] and R.J.C. Cantet[1]
[1]FAUBA, Av San Martín 4453, 1417 Buenos Aires, Argentina, [2]MSU, 1205 Anthony Hall, East Lansing, MI 48824, USA; rcantet@agro.uba.ar

Prediction of breeding values with a genomic relationship matrix depends on capturing the variability in genome sharing of relatives with the same pedigree relationship. The realized values of genome sharing deviate due to Mendelian sampling and linkage. We studied the empirical variation in genome sharing of relatives, from either identity-by-descent (IBD) or identity-by-state (IBS) based estimators. A total of 411 pigs from an outbred three-generation Duroc × Pietrain resource population were genotyped with the PorcineSNP60 Beadchip. For any pair of animals, the IBD-based estimator was computed as .5P1 + P2, where P1(P2) is the weighted average of the posterior probability of sharing 1(2) pair(pairs) of alleles IBD within the known pedigree, across all SNPs. The probabilities were inferred with the software PEDIBD. The IBS-based estimator used was VanRaden's genomic relationship coefficient. The estimated values were analyzed with SAS (CORR and UNIVARIATE procedures). Pearson correlation with the expected relationships was 0.87 and 0.79, for IBD and IBS-based estimates, respectively. For each pedigree relationship, the sampling standard deviation (SD) of genome sharing was compared against its theoretical SD, based on porcine genetic maps. For most relationships, both the IBD and IBS-based estimates were normally distributed; however, relationships with low degree of resemblance (e.g. half-cousins) were slightly skewed and had higher estimated coefficients of variation. The IBD-based estimated SD was always smaller than the IBS-based estimated SD, being on average, 18.52% and 70.68% higher respectively, when compared against the theoretical SD. For example, the IBD-based estimated mean and SD was 0.264±0.0353 compared with expectation 0.250±0.0352 for half-sibs. Our results suggest that the IBD-based method can detect more accurately the degree of genome sharing between relatives and could be used to compute realized relationships for genomic selection.

Impact of rare variants on the quality of genomic prediction in dairy cattle

T. Suchocki and J. Szyda
Wroclaw University of Environmental and Life Sciences, Department of Genetics, Kozuchowska 7,51-631 Wroclaw, Poland; tomasz.suchocki@up.wroc.pl

Predicting phenotypes from genotype data is important for plant and animal breeding, and evolutionary biology. Genomic-based phenotype prediction has been applied using data from single-nucleotide polymorphism (SNP) genotyping platforms. Usually a set of markers included in the final analysis is edited based on a minor allele frequency (MAF) and a call rate. Such filtering leads to the fact that additive effects of SNPs with rare genotypes are not considered in the analysis and impact of such markers into breeding value is not known. It is very interesting to check whether rare variants have impact on the quality of genomic prediction. Consequently, the major goal of this study is identification of SNPs with rare allelic variants i.e. for which minor allele frequency is lower than 1%, in the data set of bulls from the Polish Holstein-Friesian breed, and the comparison of the accuracy of breeding value prediction for data sets with and without rare alleles. The data set consisted of 2,854 proven bulls and 1,372 young bulls. Each bull was genotyped using 54K Illumina Bead Chip. Genotypic data was edited based on technical quality of the chip by removing single nucleotide polymorphisms with call rate lower than 95%. In our analysis protein yield, somatic cell score and the non-return rate of heifers were considered. The estimation of the additive effects of SNPs a SNP-BLUP model from the Polish genomic selection project was used. Using this model two evaluations were carried out: (1) with all available SNPs, including rare variants (52,122 SNP); (2) with common SNPs only, for which minor allele frequency exceeds 1% (46,267). Finally, statistical significance of the particular SNP estimates and the reliability of predicting breeding values were compared based on the two data sets.

Session 16 Theatre 5

Systematic differences in the response of genetic variation to pedigree and genome based selection

M. Heidaritabar[1], A. Vereijken[2], W.M. Muir[3], T. Meuwissen[4], H. Cheng[5], H. Megens[6], M. Groenen[7] and J. Bastiaansen[8]

[1]Wageningen University, Zodiac 122, 6700 AH, Wageningen, the Netherlands, [2]Hendrix Genetics, Research & Technology Centre, 5830 AC, Boxmeer, the Netherlands, [3]Purdue University, West Lafayette, IN 47907, USA, [4]Norwegian University of Life Sciences, Ås, 1432, Norway, [5]Avian Disease and Oncology Laboratory, East Lansing, MI 48823, USA, [6]Wageningen University, Zodiac 122, 6700 AH, the Netherlands, [7]Wageningen University, Zodiac 122, 6700 AH, the Netherlands, [8]Wageningen University, Zodiac 122, 6700 AH, the Netherlands; marzieh.heidaritabar@wur.nl

Selection with genetic marker data is an intense field of study in animal breeding. The term 'genomic selection' (GS) has been used to describe the simultaneous use in selection of up to hundreds or thousands of Single Nucleotide Polymorphisms (SNPs) across the whole genome. Using a 60K SNP chip across the whole genome of 3 lines of egg-laying chicken, we compared 2 selection methods (GS and pedigree BLUP selection), by evaluating allele frequency changes after 2 generations of selection. There were differences in effects of selection methods, both within and between lines. Within lines, changes in allele frequencies were on average 51% larger with GS than with BLUP selection. Between lines, the genomic regions with the largest changes in allele frequencies were different, probably reflecting differences in selection pressure and genetic architecture of the traits under selection. Empirical thresholds for allele frequency changes were determined from gene dropping in the real pedigrees, and differed considerably between GS (between 0.167 and 0.198) and BLUP selection (between 0.105 and 0.126). With GS, 70 selected regions were identified (empirical $P<0.05$) across the 3 lines, with 30 in line B1, 24 in line B2 and 16 in line W. With BLUP 35 selected regions were identified. The difference in the number of selected regions, and in the absolute allele frequency changes within these regions, indicate that GS applies much more directed selection than pedigree BLUP.

Session 16 Theatre 6

Quality of reconstructed haplotypes in cattle

M. Erbe[1], M. Frąszczak[2], H. Simianer[1] and J. Szyda[2]

[1]University of Goettingen, Department of Animal Sciences, Goettingen, Germany, [2]Wrocław University of Environmental and Life Science, Animal Genetics, Biostatistics Group, Wrocław, Poland; merbe@gwdg.de

Using data from genotyping platforms implies that there is no haplotype information available for the genotyped samples, which is required for the calculation of linkage disequilibrium or haplotype based association studies, etc. Several programs are available to reconstruct haplotypes and impute missing values based on linkage and/or linkage disequilibrium structure. While accuracy of imputation has been widely studied, this is not the case for the accuracy of phasing. The aim of this study was to assess the quality of phasing with different software tools (BEAGLE, findhap) in real cattle data. Various validation individuals were phased with two different sets of reference individuals, respectively. The two haplotypes obtained for the validation animals were then compared. We used Illumina 50K SNP genotypes of 5,501 Holstein-Friesian bulls. Different scenarios were modeled: bulls in the validation sets were chosen randomly, were the ones that were least related to the data set or had at least 5 genotyped sons in the data set. The number of individuals in the reference set varied from 50 to 2500. Criteria of comparison were number of positions ('jumps') where the phase changed between the haplotypes of the same individual obtained with the two reference sets, the percentage identically phased in both runs, and the probability of having an identical haplotype in both reconstructed phases given a specific haplotype length. BEAGLE generally performed better than findhap in terms of number of jumps (4.1 ± 4.4 vs. 16.6 ± 7.4 with 1000 random reference animals on BTA1), but worse in terms of percentage of positions equally phased (83.8 ± 18.0 vs. 92.8 ± 6.9 with 1000 random reference animals). To obtain a stable version of a reconstructed haplotype, a high relationship between reference and validation individuals is beneficial, but the number of genotypes used for reconstructing the haplotypes remains crucial.

Variance and covariance of actual relationships between relatives at one locus

L.A. Garcia-Cortes[1], A. Legarra[2], C. Chevalet[3] and M.A. Toro[4]
[1]INIA, Departamento de Mejora Genetica Animal, Carretera de la Coruna Km 7,5, 28040 Madrid, Spain, [2]INRA, UR 631 SAGA, 31326 Castanet-Tolosan, France, [3]INRA, UMR 444 LGC, INRA, France, [4]Universidad Politecnica de Madrid, Departamento de Produccion Animal, Senda del Rey s/n, 28040 Madrid, Spain; andres.legarra@toulouse.inra.fr

Genomic selection profits deviations of realized vs. average relationships as a consequence of Mendelian sampling and linkage. For instance, full-sibs may share 0, 1 or 2 alleles at one locus. These deviations are observed through molecular markers. However, the extent of those deviations is unknown for the general case, whereas understanding this variance is important to predict the benefits of genomic selection and properly model genomic relationship matrices. Further, realized relationships depend on previous realized relationships, and this generates covariances across realized relationships. The goal of this work is to develop this general formula for the one-locus situation. We provide simple expressions for the variances and covariances of all actual relationships in an arbitrary complex pedigree. The proposed method relies on the use of the nine identity coefficients and the generalized relationship coefficients; formulae have been checked by computer simulation.

The impact of selection on the genome

D.M. Howard[1], R. Pong-Wong[1], V.D. Kremer[2], P.W. Knap[2] and J.A. Woolliams[1]
[1]The Roslin Institute, Easter Bush, Midlothian, EH25 9RG, United Kingdom, [2]PIC, Ratsteich 31, 24837 Schleswig, Germany; david.howard@roslin.ed.ac.uk

In the course of selection, the rate of inbreeding (ΔF) observed at loci neighbouring QTL will be greater than expected for neutral, selection free loci, assumed for pedigree-related inbreeding. For populations with sustained directional selection, it is now feasible to quantify the fraction of genomes behaving in a way that is consistent with this 'standard' model. Therefore the study objective was to estimate this fraction in commercial pig breeding lines and examine the feasibility of identifying regions under selection within the timescale of 6 generations. SNP data obtained with the 54k Illumina Porcine Beadchip was used, with 1,500 diplotypes spanning 6 generations. The heterozygosity for individual i (H_i) was calculated for both individual loci and moving windows of 1 cM by simple counting of heterozygotes. For each locus/window the regression model $\log H_i = \alpha + \beta \log(1-F_i) + \varepsilon$ was fitted using a GLM, where F_i was the pedigree inbreeding coefficient for i. The null hypothesis H_0: $\beta=1$, was tested against an alternative H_1: $\beta>1$, with the one-sided alternative justified by the observation that for regions undergoing directional selection the ΔF experienced by loci will be increased cf. the standard model. Significance was judged by examining the distribution of β^<1 and, here, assumes symmetry. Initial comparisons were between SSC05 and SSC12: the medians per locus of β^ were 1.16 and 1.14 with means of 1.54 and 1.59 respectively. On both chromosomes 46% of loci had β^<1 and assuming symmetry in the error distribution for β^ suggests that >90% of loci appear consistent with the standard model. In this study thresholds of 1% chromosome-wide significance for excess loss of heterozygosity in 1 cM windows were taken as 4.36 for SSC05 and 4.77 for SSC12, with 4 regions and 2 regions exceeding the thresholds respectively. These regions may indicate proximity to QTL currently contributing significance variance.

Session 16 — Theatre 9

Across-breed genomic evaluation based on BovineHD genotypes, and phenotypes of bulls and cows

C. Schrooten[1], G.C.B. Schopen[1] and P. Beatson[2]
[1]CRV BV, P.O. Box 454, 6800 AL Arnhem, the Netherlands, [2]CRV Ambreed, NZ P.O. Box 176, Hamilton 3240, the Netherlands; chris.schrooten@crv4all.com

Most genomic evaluation systems have used a reference population consisting of bulls of one particular breed, genotyped with 50k SNP-chips. If the reference population is relatively small, and no opportunities exist to increase the number of bulls, other possibilities need to be explored. One option is to combine reference populations for individual breeds in a multi-breed genomic evaluation. Such an evaluation needs higher SNP-density than 50k, e.g. 777k. Another option is to add genotypes and phenotypes of cows to the reference population. The objective of this study was to estimate the effect on reliability of genomic breeding values, when single-breed reference populations are combined, and the reference population is augmented with high density genotypes and cow genotypes and phenotypes. High density genotypes (BovineHD, 777k) of 465 NZ Friesians (F), 227 Jerseys (J) and 57 crossbreds (F*J) were available. Genotypes of approximately 9,000 animals, obtained with 50k chips, were imputed to HD. After data edits, 9,486 animals were available for evaluation. Haplotype scores were obtained for 622k loci, 10% of these loci were eventually used in genomic evaluation for 26 traits. Depending on the trait, de-regressed proofs of 3,200-3,700 bulls and 1,300-2,600 cows were available. Phenotypes of the youngest bulls and their daughters were not used to estimate effects. The genomic prediction of these validation bulls was compared to their daughter-based phenotype, to derive reliabilities of genomic predictions. It was concluded that the across-breed evaluation, and including cow phenotypes and high density based haplotypes resulted in 2 to 3% higher reliability of genomic breeding values for Friesians, compared to the single-breed genomic evaluation based on 50k genotypes and phenotypes of bulls. For Jerseys, the increase in reliability was 2 to 6%. For all breeds, genomic evaluation using these data will be implemented.

Session 16 — Theatre 10

HD genotype imputation in 54k genotyped and ungenotyped Original Braunvieh and Brown Swiss cattle

B. Gredler[1], M. Sargolzaei[2], B. Bapst[1], A. Bieber[3], H. Simianer[4] and F. Seefried[1]
[1]Qualitas AG, Chamerstrasse 56, 6300 Zug, Switzerland, [2]L'Alliance Boviteq Inc., 19320 Grand Rang St. François, J2T-5H1 ST Hyacinthe, Canada, [3]FiBL Research Institute of Organic Agriculture, Ackerstrasse 21, 5070 Frick, Switzerland, [4]Georg August University Göttingen, Albrecht-Thaer-Weg 3, 37075 Göttingen, Germany; birgit.gredler@qualitasag.ch

In silico genotyping by imputation of unknown genotypes can be used to reduce the implementation costs of genomic selection. We evaluated the accuracy of genotype imputation from Illumina 54k to High Density (HD) in Original Braunvieh and Brown Swiss cattle in Switzerland. Genotype data consisted of 6,106 54k and 880 HD genotyped bulls and cows. Genotype data was checked for parentage conflicts and SNP were excluded if MAF was below 0.5% and SNP call rate was lower than 90%. The final data set included 39,004 SNP for the 54k and 627,306 SNP for the HD chip. HD genotypes of animals born between 2004 and 2008 (n=365) were set to unknown to mimic animals genotyped with the 54k chip. Population and pedigree (family) imputation methods were used as implemented in FImpute and Findhap V2. The accuracy of imputation was assessed by the squared correlation between true and imputed genotypes (R^2). Both programs resulted in high imputation accuracy. R^2 increased with increasing relationship between the HD genotyped reference population and 54k genotyped imputation candidates. Average R^2 for FImpute and Findhap were 0.98 and 0.97 when both parents of the 54k genotyped candidate were HD genotyped, respectively. R^2 was lower when no direct relatives were HD genotyped. FImpute and Findhap provide in silico genotypes for completely ungenotyped animals. Incorporating these genotypes in the reference population could be specifically beneficial for small breeds with low numbers of genotyped animals such as Original Braunvieh cattle in Switzerland. Therefore, next steps include the evaluation of accuracy of ungenotyped animals in Original Braunvieh and Brown Swiss cattle.

Session 16 Theatre 11

Validation accuracy of genomic breeding values with HD genotypes in Fleckvieh cattle

J. Ertl[1], C. Edel[1], R. Emmerling[1], H. Pausch[2], R. Fries[2] and K.-U. Götz[1]
[1]Bavarian State Research Centre for Agriculture, Institute of Animal Breeding, Prof.-Dürrwaechter-Platz 1, 85586 Poing, Germany, [2]Technische Universität München, Chair of Animal Breeding, Liesel-Beckmann-Straße 1, 85354 Freising, Germany; johann.ertl@lfl.bayern.de

Increasing the marker density is expected to improve accuracy and to reduce inflation of genomic breeding values. Medium-density (40,089; 50K) and high-density single nucleotide polymorphism genotypes (388,951; HD) of Fleckvieh bulls were used to predict genomic breeding values in milk, fat and protein yield, somatic cell count, milkability, muscling, udder, feet and legs score as well as stature with a linear model. Observed accuracy and inflation of estimated genomic breeding values (DGV) were evaluated with validation bulls by means of the correlation between DGV and daughter yield deviations and the weighted regression of daughter yield deviations on genomic breeding values. Theoretical accuracy was calculated from the model equations. The calibration and validation groups consisted of at least 5,324 and 1,321 bulls, respectively. A total of 1,485 bulls were actually HD genotyped, HD genotypes for the other bulls were imputed using the FImpute software. Validation accuracy with HD genotypes was compared to the distribution of observed accuracy in the validation group that resulted from repeated sampling of 50K markers out of the HD marker set. HD genotypes resulted in 0.011 larger validation accuracies, on average. This difference was statistically significant ($P<0.05$) for all analyzed traits. In contrast to this result, the theoretical accuracy of HD genotypes decreased by 0.022. Inflation was reduced with HD genotypes such that regression coefficients were by 0.036 closer to the expected values. The benefit in observed accuracy with HD genotypes was comparable to the theoretical expectation from the higher marker density.

Session 16 Theatre 12

Large scale genotype imputing for non-genotyped relatives in Holstein

H. Alkhoder[1], Z. Liu[1], F. Reinhardt[1], H. Swalve[2] and R. Reents[1]
[1]vit Germany, Heideweg1, 27283 Verden, Germany, [2]MLU Halle, Theodor-Lieser-Str. 11, 06120, Germany; hatem.alkhoder@vit.de

In genomic selection young candidates are usually genotyped on a large scale, but most genotyped old animals were bulls with daughters, e.g. in German Holstein, leaving many female ancestors non-genotyped, e.g. dams of genomic reference bulls. Using genotypes of those influential female ancestors could improve genomic prediction for the whole population. The aim of this study was to apply statistical methods to estimate genotypes of non-genotyped relatives using genotypes of all available animals. In first step, three methods were studied to estimate missing genotypes of animals on a SNP by SNP basis without exploring linkage disequilibrium information: using genotypes of sire, mate and direct progeny (M1), calculating genotype probabilities considering all genotyped ancestors (M2), and estimating genotypes with the gene content method (M3) by treating genotypes as phenotypic data in a conventional BLUP model. Genotype imputation was then performed in a second step for all SNPs using FImpute software after genotypes of some SNPs were estimated with the three methods. In order to validate the three methods, 2,500 genotyped animals selected from all genotyped (c.a. 68,000) were treated as if they were not genotyped. With M1, genotypes of about 10.6% of all SNPs were estimated with an accuracy of 99.9%. We could estimate, with M2, genotypes of c.a. 11% of SNPs with an accuracy of 99.5%, ignoring those SNPs with genotype probability below 0.99. A slightly lower accuracy, 98.4%, was found for the BLUP M3 method. However, M3 was able to estimate genotypes of twice more SNPs in comparison to M1 and M2. When the three methods were used jointly, genotypes of a total of 35% SNPs were estimated at an accuracy of 98.8%. For the genotype imputation in the second step, imputing accuracy depended on the number of estimated SNPs by the three methods. On average, alleles correctly imputed were 92.5%. Correlations of estimated direct genomic values ranged from 0.75 to 0.83 for all traits.

Session 16 — Theatre 13

Identification of six mutations responsible for prenatal mortality in dairy cattle

D. Boichard[1], A. Capitan[2], A. Djari[3], S. Rodriguez[1,3], A. Barbat[1], A. Baur[2], C. Grohs[1], M. Boussaha[1], D. Esquerré[4], C. Klopp[3], D. Rocha[1] and S. Fritz[2]
[1]INRA, UMR1313 GABI, 78350 Jouy en Josas, France, [2]UNCEIA, 149 rue de Bercy, 75595 Paris, France, [3]INRA, SIGENAE Bioinformatics Team, 31326 Castanet, France, [4]INRA, GeT Genomics Facility, 31326 Castanet, France; didier.boichard@jouy.inra.fr

Genomic regions harboring recessive deleterious mutations responsible for mortality during gestation were detected in three dairy cattle breeds by identifying frequent haplotypes (>1%) with a deficit of homozygotes. Material consisted in Illumina 50k Beadchip genotypes of 76,177 animals (47,878 Holstein, 16,833 Montbeliarde and 11,466 Normande). Forteen to 25 candidate regions (57 in total) were detected in each breed, including Brachyspina, CVM, HH1, and HH3 in Holstein breed. The two most frequent haplotypes (9 and 7%) were observed in Montbeliarde. A negative effect on calving rate, consistent in heifers and in lactating cows, was observed for 13 regions in matings between carrier bulls and daughters of carrier sires. Among these 13 regions, 6 new deleterious mutations were identified in GART, SHBG, NOA1, SLC37A2, UPF1, and SMC2 (responsible for HH3), after sequencing the genome of heterozygous bull carriers and control animals (36 animals in total). In addition, deleterious mutations for 3 other regions without any confirmed effect on fertility were also found. All of them were located in coding sequences, were non synonymous and had very damaging predicted effects. Previously reported causative mutations for Brachyspina and CVM were found with this approach. Finally we showed that some causative mutations were in the neighborhood of – but outside – these non homozygous regions, due to incomplete linkage disequilibrium. This study was funded by the ANR-10-GENM-0018 ANR-Apisgene Cartoseq project.

Session 16 — Poster 14

Bias in single-step genomic evaluations attributable to unknown genetic groups

S. Tsuruta, D.A.L. Lourenco and I. Misztal
University of Georgia, Animal and Dairy Science, 425 River Rd, Athens, GA 30602, USA; shogo@uga.edu

The objective of this study was to investigate bias in genomic evaluations using a single-step genomic BLUP (ssGBLUP) method due to unknown parent groups (UPG). Analyses involved final score in US Holsteins (Holstein USA, Inc., Brattleboro), 305-d milk yields in three parities in Israeli Holsteins, and several traits in pigs (PIC, Hendersonville). The US Holstein data consisted of 10,167,604 records for 6,586,605 cows and 9,602,031 animals in pedigree including 34,506 genotyped bulls with 42,503 SNP; the Israeli Holstein data consisted of 1,205,801 records for 713,686 cows and 829,437 animals in pedigree including 1305 genotyped bulls with 30,359 SNP; the pig data consisted of 2,923,141 records for 884,250 pigs and 906,660 animals in pedigree including 4853 genotyped animals with 63,219 SNP. Original unknown parent groups (UPG) were defined based on year of birth by sex, year of birth by sex by breed, and year of birth for US Holstein, Israeli Holstein, and pig data, respectively. Solutions to (G)EBV and UPG were calculated by BLUP and ssGBLUP. The UPG solutions differed mostly for the last UPG group for US Holsteins and for the last UPG group in the last parity for Israeli Holsteins. The differences with pig data sets were small. Regardless of methods, trends by UPG solutions showed high fluctuations. The UPG definitions were refined by merging groups with large changes. After refinements, differences in UPG solutions between BLUP and ssGBLUP were small. For US Holstein, Israeli Holstein, and pig data sets, correlations between GEBV from original and refined UPG models were 0.99, 0.95-0.97 and 0.97-0.99, respectively. Refinement of UPG improved convergence in GBLUP by 4, 55 and 35% for US Holstein, Israeli Holstein, and pig data sets, respectively. Refinement of UPG definitions is beneficial for both BLUP and ssGBLUP evaluations.

A simulation study of genomic evaluation combining pure and crossbred data in small populations

N. Ibáñez-Escriche, J.P. Sánchez and J.L. Noguera
IRTA, Animal Breeding and Genetics, Av. Alcalde Rovira Roure, 191, 25198 Lleida, Spain; noelia.ibanez@irta.es

The aim of this study is to evaluate under simulation the use of genomic information from purebred and crossbred lines using the Genomic Evaluation Single Step method in small populations, like Iberian pig. Three scenarios with two purebred lines (A and B) and a F1 cross were simulated: (1) and (2) two purebred lines have a common origin but they are 50 and 500 generations apart; and (3) two breeds without common origin. Trait phenotypic values controlled by 100 segregating QTL and with a heritability of 0.30 were simulated for the F1 individuals. Further, 3000 segregating markers on two chromosomes of 1 Morgan were chosen for analysis. The analyses were perform fitting a multivariate animal model in which records from A, B and F1 lines are considered as three correlated traits. The same amount of phenotypes (600 A, 600 B, 1200 F1), pedigree (800 A, 800 B, 1200 F1) and genotypes (600) data were used in all genomic evaluations. However, four different combinations of genotypes were exploited: without genotypes, purebred genotypes, only F1 genotypes and a mixing of purebred and F1 genotypes, respectively. The analyses were evaluated based on predictive correlations and mean square error (MSE) of purebred A individuals (fourth generation). Results showed an increase of predictive correlations (accuracy) and a reduction of the MSE when genotypes were included in the genomic evaluation. Moreover, it did not show a decrease of accuracy or increase of MSE when only F1 genotype information was incorporated in genomic evaluation. Additionally, an overestimation of the additive variance was found when purebred genotypes were used in the genomic evaluations. As conclusion, the results would strengthen of using genotype crossbred information to genomic evaluate small populations with the Single Step method. Nevertheless, further simulations with additive and dominance effects will be needed as well as studying alternatives to create the genomic matrix with proper frequencies.

Accuracy of genomic evaluation in pure and admixed populations

E.F. Mouresan, A. González-Rodríguez, C. Moreno, J. Altarriba and L. Varona
Universidad de Zaragoza, Área de Genética Cuantitativa y Mejora Animal, c/ Miguel Servet 177, 50013 Zaragoza, Spain; lvarona@unizar.es

The aim of this study is to compare the accuracy of the predicted breeding values obtained by genomic evaluation from various populations under different scenarios of marker densities, generations and type of divergence. An evolutionary process of 1000 generations of a population with effective size of 100 individuals was simulated. For each individual, 30,000 biallelic markers, evenly distributed throughout 20 chromosomes of 1 Morgan each, were simulated. One hundred of these markers were randomly selected as causative mutations for two traits with heritabilities 0.2 and 0.4. Later on, 4 different subpopulations were created, of 100 individuals each. Population A was under phenotypic selection for the trait 1, population B for the trait 2 and populations C and D reproduced randomly, all for 50 generations. At the end, the 4 populations were expanded to 2,000 individuals each. The 4 purebred populations (A, B, C and D) and 7 admixed populations (AB, AC, AD, BC, BD, CD and ABCD) with equal percentage of each purebred, were used as training sets for the genomic evaluation, which was conducted under a Bayesian Lasso model. The accuracy of the genomic evaluation was estimated by calculating the correlation between the predicted and the simulated breeding values in the validation sets (A, B, C and D). The results suggest that the accuracy is higher when training in a population that was not under selection (C and D), or was selected for another trait (B). When a 2-breed admixed population was used for training, the accuracy was around 0.35 for the populations implied and much lower for the ones not implied. When the training set was the 4-breed admixed population the accuracies for all validation sets were slightly better than those obtained from the 2-breed admixed training sets.

Selection footprints in the autochthonous spanish beef cattle populations

L. Varona[1], J. Altarriba[1], C. Avilés[2], J.A. Baró[3], J. Cañas[4], M.J. Carabaño[5], C. Díaz[5], A. González-Rodríguez[1], A. Molina[2], C. Moreno[1] and J. Piedrafita[4]
[1]Facultad de Veterinaria, Universidad de Zaragoza, 50013 Zaragoza, Spain, [2]Departamento de Genética, Universidad de Córdoba, 14071 Córdoba, Spain, [3]Escuela Superior de Ingenierías Agrárias, Universidad de Valladolid, 34004 Palencia, Spain, [4]Grup de Recerca de Remugants, Universitat Autonoma de Barcelona, 08193 Bellaterra, Spain, [5]Departamento de Mejora Genética Animal, Instituto Nacional de Investigación y Tecnología Agraria y Alimentaria, 28040 Madrid, Spain; lvarona@unizar.es

The aim of this study is to locate the genomic regions associated with selection and differentiation processes between five Spanish autochthonous beef cattle populations. For this objective, we selected 25 trios (sire-dam-individual) from each one (Asturiana de los Valles, Avileña Negra-Ibérica, Bruna dels Pirineus, Retinta and Pirenaica). The criterion of selection of the individuals from each population was to maximize the representativeness and to capture the available diversity within each population. All animals were genotyped for the 770K Bovine Illumina Beadchip. After editing, we computed the Fst statistic for 735,293 SNPs. From this information, we selected the genomic regions where at least 5 SNP markers provided an Fst estimate greater than 0.4. Those regions were located at BTA2, BTA5, BTA6, BTA11 and BTA18. A preliminary look at the genome browser (genome.ucsc.edu) suggests the following candidate genes: BTA2 (INPP1, HIBCH and C2H2orf88), BTA6 (LCORL), BTA11 (C1D, WDR92, PNO1, PPP3R1, CNRIP1, FBXO48, PLEK, APLF) and BTA18 (GPT2, DNAJA2 and ITFG1).

Development of a genomic evaluation for milk production for a local bovine breed

F.G. Colinet[1], J. Vandenplas[1,2], P. Faux[1], S. Vanderick[1], C. Bertozzi[3], X. Hubin[3] and N. Gengler[1]
[1]University of Liege, Gembloux Agro-Bio Tech, Gembloux, 5030, Belgium, [2]National Fund for Scientific Research, FNRS, Brussels, 1000, Belgium, [3]Walloon Breeding Association, Research and Development Department, Ciney, 5590, Belgium; frederic.colinet@ulg.ac.be

The dual purpose Belgian Blue breed (DP-BBB) is a vulnerable breed rooted in the tradition of the Walloon Region of Belgium. The aim of this study was to investigate the potential in the development of a single step genomic evaluation (ssGBLUP) for DP-BBB milk production using a Bayesian procedure to integrate the Walloon estimated breeding values (EBV) as a priori known external information. This procedure combined genomic, pedigree and EBV by considering a correct propagation of all the information and no multiple considerations of contributions due to relationships. The Bayesian approach allowed to combine EBV and associated reliabilities (REL) without computation of deregressed proofs (DRP). 99 different combinations of genomic (α) and additive (1-α) relationships into a merged covariance structure for ssGBLUP were tested. The optimal combination was chosen using EBV_{2008} (available for 2,457 animals) from genetic evaluation based on milk production recorded until 31/12/2008. The pedigree file included 4,777 DP-BBB animals. After usual editing, the SNP file contained 333 genotypes (from 155 cows and 178 bulls) of 34,531 useful SNP. In order to test the method, DRP_{2012} were computed from EBV from genetic evaluation based on milk production recorded until 31/12/2012 and associated REL for 2,462 animals. DRP_{2012} of 32 bulls (with EBV_{2008} equal to their pedigree index in 2008, with 10 EDC in 2012 and born after 2001) were regressed on genomically enhanced breeding values ($GEBV_{2008}$) obtained from each of the 99 ssGBLUP or on EBV_{2008}, with REL_{2012} as a weight. The α of 0.78 gave the best determination coefficient (0.494), while the determination coefficient for the regression DRP_{2012} on EBV_{2008} was 0.47. Thereby, these results showed the feasibility of a modified ssGLUP for a small breed.

SREBP gene expression in the mammary gland during the first stage of lactation in Sarda ewes

M.C. Mura, C. Daga, S. Bodano, G. Cosso, M.L. Diaz, E. Sanna, P.P. Bini, S. Luridiana and V. Carcangiu
Università di Sassari, Dipartimento di Medicina Veterinaria, Via Vienna 2, 07100 Sassari, Italy; endvet@uniss.it

SREBP-1 gene is considered a candidate gene and it plays a central role in the regulation of the synthesis of milk fat. Moreover this gene controls the expression of more than 30 genes involved in this process. The aim of this study was to examine the expression patterns of SREBP-1 gene in milk somatic cells and its association with milk fat synthesis during first stage of lactation in Sarda breed sheep. A sample of 20 Sarda ewes, aged between 4 and 5 years, in their third to fourth lactation were chosen. From each ewe 30, 60 and 90 days after lambing, milk yield was measured, and a 160 ml milk sample for the RNA extraction and to test somatic cells count and lactose, fat and protein contents was collected. From the obtained RNA, total cDNA was synthesized and the quantitative PCR was performed. The fat, proteins and lactose content showed many differences among the animals, but these variations were no correlated with the milk yield. The SREBP-1 gene expression resulted higher in the high milk fat producing (g/die) ewes. The expression level of this gene showed to be higher in the first compared to the last sample ($P<0.05$). The correlation analysis showed that the SREBP-1 expression level is related to the amount of milk fat ($P<0.001$), while the total RNA obtained from each sample was found to be related to the somatic cells number ($P<0.001$). Our data highlight that in sheep SREBP-1 gene is mainly expressed in the mammary gland during early lactation and it declines during the course of lactation. Moreover, the positive relationship between SREBP-1 gene expression and the milk fat yield suggest that SREBP-1 gene is required for the lipid synthesis in the sheep mammary gland.

Concordance analysis: from QTL to candidate causative mutations

I. Van Den Berg[1,2,3], S. Fritz[4], A. Djari[5], S.C. Rodriguez[3,5], D. Esquerré[6], C. Klopp[5], D. Rocha[3], M.S. Lund[1] and D. Boichard[3]
[1]Aarhus University, Department of Molecular Biology and Genetics, Blichers Allé 20, 8830 Tjele, Denmark, [2]AgroParisTech, 16 rue Claude Bernard, 75005 Paris, France, [3]INRA, UMR1313 GABI, Domaine de Vilvert, 78350 Jouy en Josas, France, [4]UNCEIA, 149 rue de Bercy, 75595 Paris, France, [5]INRA, SIGENAE Bioinformatics Team, Chemin de Bordé-Rouge, 31326 Castanet, France, [6]INRA, GeT Genomics Facility, Chemin de Bordé-Rouge, 31326 Castanet, France; irene.vanderberg@jouy.inra.fr

Although a large number of QTL have been detected for various traits in dairy cattle, the causative mutations underlying these QTL have only in a few cases been detected. Our goal was to narrow down from QTL region to a limited number of candidate mutations by a concordance analysis. The principle behind this analysis is that the causative mutation underlying a QTL should be homozygous when an animal is homozygous for the QTL and heterozygous when heterozygous for the QTL. QTL regions were selected using Bayes C for various quantitative traits, based on their posterior inclusion probabilities. For each selected region, QTL statuses were predicted using estimated marker effects. Polymorphisms in the region whose status was non-concordant with the QTL status were discarded. The success of the analysis depended on the accuracy of the QTL status prediction and the linkage disequilibrium in the QTL region. When several QTL are present in the same region, it becomes more difficult and sometimes impossible to predict the QTL status. Furthermore, for regions with strong linkage disequilibrium, most polymorphisms in the region will be in concordance and hardly any polymorphisms can be discarded. In such a situation, validation across breeds helps to reduce the amount of linkage disequilibrium. For other regions, however, it was possible to reduce the number of polymorphisms from more than 30,000 in the QTL region to less than 100, located in only one or two genes, showing the potential of the analysis in the search for causative mutations.

Session 16 Poster 21

Reconstruction of 777k SNP genotype of a founder using information from genotyped progeny

S.A. Boison[1], A.M. Perez O'brien[1], C. Nettelblad[2], Y. Utsunomiya[3], J.F. Garcia[3] and J. Sölkner[1]
[1]Univ. of Natural Resources and Life Sciences, Vienna, Division of Livestock Sciences, 1180, Austria, [2]Uppsala University, Dept. of Cell and Molecular Biology, 751 05, Sweden, [3]Sao Paulo State University, Dept. of Animal Production and Health, Aracatuba, 16050-680, Brazil; anita_op@students.boku.ac.at

Genotyping on a low density SNP panel and imputing to medium and high density has been deemed to be a cost effective way for genomic selection. Imputing genotypes have been shown to be very successful when haplotype libraries are built from individuals related to the ones been phased and imputed. Thus genotyping founders that highly contribute to the current population is useful. Our objective was to reconstruct the 777k HD genotype for a non-genotyped founder of the Nellore population. To reconstruct the genotypes, 10 (7 sons and 3 grandsons) individuals were genotyped for 777k HD and used for the analysis. Two approaches were adopted. In a 1-step approach, genotypes were imputed using pedigree and population based algorithms; AlphaImpute. For the 2-step approach, MERLIN was used to reconstruct the genotypes, the non-reconstructed part were imputed using BEAGLE, modified MaCH and AlphaImpute. To test the accuracy of called genotypes, 2 test bulls were selected for validation by mimicking the MAF distribution of the progeny used in reconstructing the founder. On average, imputation accuracy was 90.5±2.3% and 85.2±2.7% for the 2 test bulls using the 1-step approach. For the 2-step approach, the average percentage reconstruction from MERLIN was 88.0±1.9 with an accuracy of approximately 94.7% (83.2% of the total number of SNPs) and 88.0% (77.6%). Imputing the 12% non-reconstructed genotypes from MERLIN increased correctly called genotypes to 91.5% (BEAGLE), 93.8% (modified MaCH) and 91.3% (AlphaImpute) for the test bull that had the same MAF distribution like that of the sons of the founder. In conclusion, the 2-step approach increased correctly called genotypes and should be adopted in reconstructing founder genomes.

Session 16 Poster 22

On Mendelian variance and ancestral regression of breeding values with a genomic covariance matrix

R.J.C. Cantet and N.S. Forneris
Universidad de Buenos Aires, Facultad de Agronomía – CONICET, Producción Animal, Av. San Martín 4453, 1417 C.A.B.A. (Buenos Aires), Argentina; rcantet@agro.uba.ar

It has been hypothesized that the use of observed 'identity by descent' (IBD) relationships (GIBD) among related animals, rather than the usual expected relationships (RIBD), improves the accuracy of prediction of 'genomic' breeding values by reducing Mendelian residual variance. The GIBD is defined as the fraction of the genome shared IBD by any two individuals with one or more common ancestors, as done by S. W. Guo based on the idea of 'continuous IBD' by K. P. Donnelly. A framework for prediction of breeding value (BV) using a genomic covariance matrix is presented that generalizes the classic regression approach. In the latter, the BV of an individual is regressed to half the BVs of its parents plus a Mendelian residual. Using GIBD rather than RIBD results in regression coefficients for all genotyped ancestors up to the base generation, as all meiosis are informative to quantify segregation. The GIBD may be estimated using different algorithms based on Hidden Markov Models and the pedigree. A general expression is obtained for the Mendelian residual variance related to the Cholesky decomposition of the genomic covariance matrix when it is positive definite. GIBD between and individual and its grandparents are the informative key elements in reducing Mendelian variance when different from 0.25. When regression coefficients from all ancestors are accommodated in the rows of the square matrix P (order equal to the number of individuals in the pedigree file) so that (I – P) is lower triangular and non-singular, the genomic covariance matrix is then equal to $G = (I - P)^{-1} D (I - P')^{-1}$. Conditions under which D is diagonal are discussed. All developments are exemplified using small pedigrees.

Session 16 Poster 23

Strategy to simulate, analyse and predict longitudinal data with genomic random regression models

T. Yin, E.C.G. Pimentel, U. Von Borstel and S. König
Department of Animal Breeding, University of Kassel, Nordbahnhofstraße 1a, 37213 Witzenhausen, Germany; sven.koenig@uni-kassel.de

Availability of repeated measurements and genomic information allows for application of genomic random regression models (gRRM) and prediction of dairy cattle performances for environmental descriptors that are poorly represented in a data set. A longitudinal trait was simulated at five different levels for the environmental descriptor temperature × humidity-index (THI). These levels were THI 15, 30, 45, 60 and 75, with heritabilities of 0.30, 0.35, 0.40, 0.40 and 0.35, respectively. High and low linkage disequilibrium (LD) was combined with 5K and 15K SNP-chips to simulate different scenarios of genomic architecture. Two analysis-strategies were applied to calculate the accuracy of genomic predictions at an extremum of the environmental scale. (1) 100%, 80%, 50%, or 20% of phenotypes at THI 75 were deleted randomly, and the remaining dataset was used to predict the breeding value at THI 75 for non-phenotyped but genotyped cows. (2) Complete information (geno- and phenotypes) was available for 1600 cows, but phenotypes were missing for 400 genotyped cows at all THI levels. Without any phenotypic observations at THI 75, accuracy of prediction averaged over all scenarios was 0.22. If 20% of cows had phenotypic records at THI 75, accuracies were moderate with values ~0.60. Only a small proportion of phenotyped cows (i.e. 20%) in environments representing heat stress (=THI 75), is required to predict reliable genomic breeding values of cows without phenotypes at that environmental level. For the second strategy, even for low LD and a low density 5K SNP-chip, the average accuracy was 0.52 and higher than using pedigree relationships. Thus, replacing the pedigree-based genetic relationship matrix with the realized genomic relationship matrix in RRM can improve accuracy of genomic prediction, and can be used to predict genomic breeding values for traits measured late in life, e.g. longevity.

Session 16 Poster 24

Accuracy of genomic selection in a substructured population of Large White pigs

G. Ni[1], A. Haberland[1], C. Bergfelder[2], C. Grosse-Brinkhaus[2], M. Erbe[1], B. Lind[3], E. Tholen[2] and H. Simianer[1]
[1]Georg-August-University, Animal Breeding and Genetics Group, Albrecht-Thaer-Weg 3, 37083 Goettingen, Germany, [2]Friedrich-Wilhelms-University, Animal Breeding and Husbandry Group, Endenicher Allee 15, 53113 Bonn, Germany, [3]Foerderverein Biotechnologieforschung (FBF) e.V., Adenauerallee 174, 53113 Bonn, Germany; hsimian@gwdg.de

Other than in dairy cattle breeding, populations in pigs are of limited size and a considerable substructure exists even within one breed. In this study, we aimed at deriving genomic breeding values for the trait 'piglets born alive' based on ~1,300 Large White boars and sows from three different breeding programs, which were genotyped with the Illumina© PorcineSNP60v2 BeadChip. A principle component analysis revealed a non-overlapping separation between two of the three subpopulations. Estimated breeding values based on pedigree and progeny information (for boars) or repeated own performance (for sows) were deregressed and used as quasi-phenotypes. Genomic breeding values were estimated with a GBLUP procedure and empirical accuracies were evaluated in a five-fold random crossvalidation with 20 replicates. We compared accuracies obtained from training sets within or across subpopulations, the latter with or without including a fixed subpopulation effect in the model. We found that empirical accuracies evaluated across populations are apparently higher than expected (~0.80) given the limited training set size. Empirical accuracies within subpopulations were substantially lower (~0.60) and followed the different training set sizes within subpopulations. Including a fixed subpopulation effect in the model had no systematic impact on the observed accuracies within subpopulations. Overall the obtained accuracies of genomic breeding values are well above the level of accuracy of conventional breeding values of young boars and sows at the time of selection, so that genomic approaches appear promising in the complex population structures of pig breeding.

Session 16 Poster 25

The adjusted genomic relationships by allele frequencies within breeds and use in single-step GBLUP

M.L. Makgahlela[1,2], I. Strandén[1,2], U.S. Nielsen[3], M.J. Sillanpää[4] and E.A. Mäntysaari[1]
[1]MTT Agrifood Research Finland, Biotechnology and Food Research, Biometrical Genetics, 31600, Jokioinen, Finland, [2]University of Helsinki, Department of Agricultural Sciences, P.O. Box 27, 00014, Helsinki, Finland, [3]Danish Agricultural Advisory Service, Udkaersvej 15, 8200, Aarhus, Denmark, [4]University of Oulu, Departments of Mathematical Science and Biology and Biocenter, P.O. Box 3000, 90014, Oulu, Finland; mahlako.makgahlela@helsinki.fi

In many practical genomic evaluations, the construction of marker-based relationship matrix (G) is often carried out using simple allele frequencies (AF) within a single breed. The proposed methods when applied to multi-breed populations, do not account for differences in AF between breeds. In this study, we compare diagonal elements of G constructed using AF within breed (G_{WB}) and across breeds (G_{AB}) in an admixed population. Each matrix was calculated with estimated AF from the current genotyped animals (GAF) and also from the base (founding) population (BAF). Furthermore, we compared validation reliabilities of GEBV from 4 single-step GBLUP analyses, each with a different G. Individual breed proportions for 4,106 bulls (with genotypes for 38,194 markers) were derived from the pedigree. The GAF and BAF within breed were estimated by solving a multiple regression of gene content on breed proportions. The validation reliabilities of GEBV were obtained in a linear regression of deregressed breeding values on GEBV, weighted by the effective number of daughters. Diagonal elements from G_{AB} were higher than G_{WB}, both across breeds and within sub-populations. Standard deviations of diagonals from G_{WB} and G_{AB} across breeds reduced from 0.055 and 0.074 using GAF to 0.036 and 0.049, respectively, when using BAF. Similar patterns of diagonals were observed within sub-populations. The validation reliabilities of GEBV from all 4 matrices were however similar, indicating that while accounting for breed origin of alleles had an effect on G, the predictions remained the same.

Session 16 Poster 26

Comparison of additive and dominance models for genomic evaluation

M. Nishio and M. Satoh
Naro Institute of Livestock and Glassland Science, 2 Ikenodai, Tsukuba, Ibaraki 305-0901, Japan; mtnishio@affrc.go.jp

We used a genomic BLUP (GBLUP) model with a dominance relationship matrix (GBLUP-D) that was constructed from genome-wide dense biallelic SNP markers to estimate genetic variances and predict genetic merit. To assess the accuracies of the resulting estimates, we used simulated data and information regarding two actual traits (T4 and T5) in pigs. In the computer simulation, a 2,000-generation population that had a mutation rate of 0.0005 and effective population size of 100 was generated. After 2,000 generations, the population comprised 1,800 genotyped animals across six generations with 1,000 SNP markers randomly distributed on one chromosome of 1 Morgan. Potentially, 50 QTLs with both additive and dominance genetic effects (AGE and DGE) affect the phenotype were distributed randomly throughout the genome. AGE for each QTL was drawn from a gamma distribution with a shape parameter of 0.42. Degrees of dominance were drawn from a normal distribution with a mean of 0 and variance of τ^2. DGE for each QTL was determined as the product of the additive genetic value and the degree of dominance. A performance trait with broad-sense heritability of 0.3 was assumed. When the value of τ was 0.25, 0.5, or 1.0, GBLUP-D explained 90.6%, 61.1%, or 54.4% of dominance genetic variances and yielded 1.2%, 7.8%, or 24.7% higher accuracy of the predicted total genetic value, respectively, than did GBLUP. The dominance genetic variances in T4 and T5 accounted for 9.6% and 6.3%, respectively, of the phenotypic variances. Estimates of such small dominance genetic variances only minimally increased the accuracies of predicted total genetic values. Overall, estimates of AGE and its variance differed negligibly between GBLUP-D and GBLUP. We conclude that GBLUP-D is a feasible approach to improve performance in crossbred populations with dominance genetic variation and to identify mating systems with good combining ability.

Enlarging a training set for genomic selection by imputation of un-genotyped animals

M. Wensch-Dorendorf[1], E. Pimentel[2], S. König[2] and H.H. Swalve[1]
[1]University of Halle, Institute of Agricultural and Nutritional Sciences, Theodor-Lieser-Str. 11, 06120 Halle, Germany, [2]University of Kassel, Department of Animal Breeding, Nordbahnhofstr. 1a, 37213 Witzenhausen, Germany; monika.dorendorf@landw.uni-halle.de

Genomic evaluations routinely are conducted in all countries having large Holstein populations and essentially are based on genotyping of bulls while cows provide the phenotypes. As genotyping costs have decreased, recently genotyping of a larger number of cows has become feasible. This in turn enables schemes that include recording of new phenotypes in limited parts of the population. However, the number of individuals with phenotypes and genotypes will be high and costs are still a concern. In this study, a simple and fast imputation method based on Bayes' theorem is introduced and the impact of augmenting a cow training set with imputation of un-genotyped dams on the accuracy of genomic predictions for different populations, levels of h^2, and sizes of training sets is evaluated. The following scenario was assumed: cow, sire of cow and sire of dam are animals with known genotypes while the dam is not genotyped but the genotype could be imputed. Depending on the simulated genetic architecture such as low/high linkage disequilibrium and the absence/presence of selection (LowLD_NoSel, LowLD_Sel, HighLD_NoSel, HighLD_Sel) the correlation between true and imputed genotypes were 0.81, 0.86, 0.90 and 0.93, averaged across 10 replicates. Results with AlphaImpute are slightly better for the LowLD scenarios and slightly worse for the HighLD scenarios. In comparison with a real doubling of the number of genotypes (increase set to 100%), across all scenarios with varying h^2 (0.05-0.5) and varying number of genotyped cows (2,000 to 16,000), 63-93% of this increase can be achieved by doubling the number of genotyped cows via imputing un-genotyped dams.

Comparison of genomic selection approaches in Brown Swiss within Intergenomics

P. Croiseau[1], F. Guillaume[1,2], J. Promp[2] and S. Fritz[3]
[1]INRA, UMR 1313, GABI, Domaine de Vilvert, 78350 Jouy en Josas, France, [2]IDELE, 149 rue de Bercy, 75012 Paris, France, [3]UNCEIA, 149 rue de Bercy, 75012 Paris, France; pascal.croiseau@jouy.inra.fr

The European Brown Swiss federation, in collaboration with Interbull, funded and managed a project named Intergenomics. The goal of this project is to perform genomic evaluations of sires based on a joint analysis of all the genotypes collected around the world. To date, six countries are involved in Intergenomics and according to the country, between 3 and 15 traits are available. In this study, we propose to compare a panel of 5 genomic selection approaches to the pedigree-based BLUP (Best Linear Unbiased Predictor). Among these 5 methodologies, performances of the genomic BLUP (GBLUP) were compared to 2 bayesian approaches (Bayesian LASSO and Bayes Cπ), a variable selection approach (Elastic Net) and to the French genomic selection method (BLUP-QTL). Except the GBLUP, the other genomic selection approaches deal with the P>>n problem (number of Single Nucleotide Polymorphism or SNP (p) is much higher than the number of bulls (n)). We compare the correlations between observed and predicted deregressed proofs for the different traits, the different country scales and the different methods. Compared to the pedigree-based BLUP, genomic selection approaches allow a gain in correlation between 3.2 and 26.4%. Bayesian LASSO, Bayes Cπ and EN give the best results with a gain of correlation around 6% compared to a GBLUP and around 2% compared to the gMAS. The slope of regression was also investigated and the BLUP-QTL give the best results with a deviation to 1 of the slope of regression of 0.097 compared to 0.11 for the Bayesian LASSO, Bayes Cπ and EN, and 0.14 for GBLUP. The effect of the country scale was also investigated. In brown Swiss, the contribution of each country to the reference population is very different and we show that countries with a high contribution will convert CD of abroad bulls with a higher accuracy.

Session 16 Poster 29

Accuracy of genomic predictions in beef cattle with medium and high-density SNP panels

M. Gunia[1], R. Saintilan[2], M.-N. Fouilloux[3], E. Venot[1] and F. Phocas[1]
[1]INRA, UMR1313 GABI, Domaine de Vilvert, 78350 Jouy-en-Josas, France, [2]UNCEIA, 149 Rue de Bercy, 75595 Paris, France, [3]Institut de l'Elevage, 149 Rue de Bercy, 75595 Paris, France; melanie.gunia@jouy.inra.fr

Replacing pedigree-based BLUP evaluations by genomic evaluations (GBLUP) in beef cattle breeding schemes can result in greater accuracy and genetic gains. Because of a small number of seedstock animals in beef herds compared to dairy herds, the size of the reference population is a very limiting factor of the efficiency of genomic evaluation in beef cattle. Our first objective was to assess the gain in accuracy of genomic breeding values which occurs when doubling the reference population size of the French Charolais breeding scheme. A second objective was to compare the accuracy of predictions using 777k SNP (HD chip) versus 54k SNP (MD chip).The genotyping of 872 Charolais bulls with BLUP estimated breeding values (EBV) of high accuracy was performed using the HD Illumina beadchip; MD genotypes were extracted from the HD genotypes. Descendant Yield Deviations (DYD) of bulls (similar to the Daughter Yield Deviation of dairy bulls) were the phenotypes considered in a GBLUP analysis including the number of progeny records as weights. Those DYD were derived for the following 5 traits: birth condition, birth weight, weaning weight, muscle and skeletal development. The accuracy of the genomic prediction was the correlation between DYD and GEBV calculated for a validation population composed of the quarter of the youngest bulls, the three older quarters being the training population. When shifting from MD to HD genotypes, the average accuracy increased from 0.30 to 0.36 with some strong differences in gains across the 5 analyzed traits. Doubling the reference population size increased significantly the accuracy of GEBV for all traits, from 30% for MD genotypes to 38% higher accuracy for HD genotypes.

Session 16 Poster 30

A simulation study to evaluate different strategies of genomic selection in Italian heavy pigs

A.B. Samorè[1], L. Buttazzoni[2], M. Gallo[3], V. Russo[1] and L. Fontanesi[1]
[1]University of Bologna, Department of Agricultural and Food Sciences, Division of Aninal Production, Viale Fanin 46, 40127 Bologna, Italy, [2]Consiglio per la Ricerca e la Sperimentazione in Agricoltura, Via Salaria 31, 00016 Monterotondo Scalo (Roma), Italy, [3]Associazione Nazionale Allevatori Suini (ANAS), Via L. Spallanzani 4, 00161 Roma, Italy; luca.fontanesi@unibo.it

Pig breeding selection schemes have been designed to maximize their effectiveness dealing with specific biological features of this species (e.g. short generation interval, large litter size, etc.) and with the limited reproduction technologies available (e.g. reduced number of semen doses per boar). The introduction of genomic selection in pig breeding could provide additional advantages to the current schemes, further improving their efficiencies and overcoming some limits of the already established selection strategies. Simulations studies can give preliminary information on the impacts and effects of genomic selection on pig breeding. A few studies have already simulated different scenarios for the introduction of genomic selection in North European pig breeding systems. However, the Italian heavy pig breeding industry has specific peculiarities that should be considered in a simulation analysis. To this aim, we used the software QMSim to simulate a structurated pig population with a genome marked with 60K bi-allelic single nucleotide polymorphisms. The population was selected over 25 generations based on BLUP breeding values for a trait with heritability of 0.40. Different genotyping strategies and methods of predicting genomic breeding values were considered and compared. The introduction of genomic evaluations might increase the accuracy levels both in young piglets and in selected reproducers and the choice of male candidate would be more accurate. Nevertheless the evaluation of the return on investment in the Italian heavy pig breeding industry would determine the implementation of genomic selection in practice.

Impact of genetic markers information on breeding value accuracy and selection of replacement bulls

F.M. Rezende[1], J.B.S. Ferraz[2], F.V. Meirelles[2] and N. Ibáñez-Escriche[3]
[1]Federal University of Uberlandia, Genetic and Biochemistry Institut, Av. Getúlio Vargas, 230, 38700-128, Patos de Minas, MG, Brazil, [2]College of Animal Science and Food Engineering, Basic Science, Cx. Postal 23, 13650-900, Pirassununga, SP, Brazil, [3]IRTA-Cataluña, Genetic and Animal Breeding, Av. Alcalde Rovira Roure, 191, 25198 Lleida, Spain; frezende@ingeb.ufu.br

The proposal of this study was to evaluate the impact of a small molecular data set composed by 3,149 animals genotyped for 106 SNP on the breeding value accuracy and on the selection of replacement bulls of a well established Nellore beef cattle breeding program. Data of 83,404 animals measured for production traits, corresponding to 116,652 records on relationship matrix, were used. Breeding values were estimated by classical and marker assisted methods using MTDFREML software, under animal model. Markers effects were estimated by Bayesian ridge-regression methodology, using adjusted phenotypes as dependent variables. The impact on reliability was calculated as the mean percentage variation between the mean accuracy of breeding values estimated by classical genetic evaluation and the mean accuracy of breeding values estimated by marker assisted genetic evaluation, in two trait analysis. Selection conflicts were calculated as the divergences on the selection of replacement bulls based on classical and marker assisted breeding values, when the top 10% young bulls for each trait were selected. An increase of 6.6, 1.9 and 9.7% on the reliability of breeding values of genotyped young bulls for post weaning gain (PWG), scrotal circumference (SC) and muscle score (MS), respectively, due the inclusion of markers information on genetic evaluation, was observed. Divergences of 1.2, 0.53 and 1.1% on the selection of replacement bulls were estimated between classical and marker assisted selection, for PWG, SC and MS, respectively. Those outcomes demonstrated the potential of marker assisted selection on a beef cattle breeding program, even for a small set of SNP markers.

Session 16 Poster 32

Correlation between molecular breeding values estimated by different Bayesian methods

F.M. Rezende[1], J.B.S. Ferraz[2], F.V. Meirelles[2], J.P. Eler[2] and N. Ibáñez-Escriche[3]
[1]Federal University of Uberlandia, Genetic and Biochemistry Institut, Av. Getúlio Vargas, 230, 38700-128, Brazil, [2]College of Animal Science and Food Engineering, Basic Science, Cx. Postal 23, 13650-900, Pirassununga, SP, Brazil, [3]IRTA-Cataluña, Genetic and Animal Breeding, Av. Alcalde Rovira Roure, 191, 25198 Lleida, Spain; frezende@ingeb.ufu.br

Allelic substitution effects of SNP markers were estimated by six different methodologies: Bayesian multiple regression (BMR), Bayesian ridge regression (BRR), Bayes A (BA), Bayes B (BB), Bayes Cπ (BC) and Bayesian Lasso (LASSO), in order to compare differences among their estimates of molecular breeding values. Data on 3,149 animals belonging to a Nellore beef cattle selection program, measured for post-weaning gain (PWG), scrotal circumference (SC) and muscle score (MS), were used. All animals were genotyped for 300 SNP markers and only 106 SNP were considered after quality control that, among other criteria, considered MAF≥5%. Molecular breeding values were calculated as the sum of allele substitution effects of each marker for the three traits and six methods. Pearson's correlation coefficient and Spearman's rank correlation were used to measure the strength of the association of the molecular breeding values estimated by each method. When analyzing PWG, BMR's results correlated very well with BRR, BC and, perfectly, with LASSO, but poorly with BA and even worse with BB. BRR correlated almost perfectly with BC and LASSO, and BA, BB and BC correlated well between them. As related to SC, the results were similar to PWG, but BB's results were very poorly correlated with others. When MS analyses were performed, the results were higher correlated between all methods, except to BB. Spearman's rank correlation's results were almost identical to Pearson's correlations. Differences observed on the markers effects estimates were due the shrinkage process applied by each analyzed method. Under this study, estimates of molecular breeding value were almost equivalent when using BMR, BRR, BC and LASSO.

Session 16 Poster 33

Advantages of High-density genotyping for morphologic and production traits GWAS in Italian Holstein

P. Ajmone Marsan[1], M. Milanesi[1], L. Bomba[1], R. Negrini[1], L. Colli[1], E. Nicolazzi[2], S. Capomaccio[1], R. Mazza[3], N. Bacciu[3] and B. Stefanon[4]

[1]Istituto di Zootecnia, Università Cattolica del S. Cuore, Via Emilia Parmense, 29122 Piacenza, Italy, [2]Parco Tecnologico Padano, Via Cascina Codazza, 26900 Lodi, Italy, [3]Associazione Italiana Allevatori, Via Tomassetti, 00161 Roma, Italy, [4]Dipartimento di Scienze Agrarie e Ambientali -Università di Udine, via delle Scienze, 208, 33100 Udine, Italy; riccardo.negrini@unicatt.it

As a matter of facts, Genome Wide Association Studies (GWAS) are the state of the art of association studies for discovering variants associated with the traits of interest in livestock. This technique exploits the linkage disequilibrium (LD) and dense SNP panels to understand the allelic variation that underlies complex traits such as productive and linear traits. LD varies across cattle breeds depending on their history, management and effective population size: for example, it is estimated that Italian Holstein dairy breed has a low effective population size – less than 100 individuals – due to the wide use of artificial insemination (AI). In such contexts, LD is expected to be high, so that significant marker-trait associations can be detected at a distance of several hundred kb down and up stream of a causative mutation. However, LD varies also with respect to the genomic landscape and to the distribution of recombination hotspots along chromosomes. We genotyped 916 Italian Holstein bulls with the 800 K SNPchip and analyzed them at two resolutions (54 and 800 K) to investigate the effect of local LD on GWAS results. Single SNP mixed model, estimation of haplotype blocks and Bayesian approaches were evaluated for milk, fat, and protein yield as well as for height and udder conformation. Comparison results, including peculiarities and overlaps between the associations obtained from the two datasets with different statistical methods, are presented and discussed.

Session 17 Theatre 1

Innovations in dairy: regional feed centre, bedded pack barns and amazing grazing

P.J. Galama, H.J. Van Dooren, H. De Boer and W. Ouweltjes

Wageningen UR, Livestock Research, P.O. Box 65, 8200 AB Lelystad, the Netherlands; paul.galama@wur.nl

Innovations around regional feed centre, bedded pack barns and grazing systems have a great impact on the diversity and sustainability of dairy farms in the future. Participatory research with pioneers in the dairy will be illustrated. A regional feed centre in the north of the Netherlands receives feed crops grown from dairy and arable farmers, stores the feed, makes total mixed rations for 3,000 cows and delivers the rations to 26 dairy farms. The cost price for dairy farms and energy consumption on a regional level can be lower if the traffic movements are not too much increased. Critical success factors for a regional feed centre are the amount of cows per km2, the feed supply in that area, the roads along feed centre and good contracts between dairy farmers, feed centre and arable farmers. Since 2008 there has been done research on three experimental farms and three commercial farms with bedded pack barns. The animal welfare in bedded pack barns is increased compared to free stalls with cubicles. The commercial farms have a bedding of compost from a compost factory or a bedding of wooden chips with an aerating system, to stimulate the composting process. In total 30 of these are built in the Netherlands. The quality of the manure is different because more organic matter but slower release of nitrogen. The ammonia emission in the stable is higher but in the field lower. The total gasseous losses of nitrogen on farm level in these compost bedded pack barns are between 20 and 30%. But these beddings introduces new risks for the quality and safety of raw milk. Compost beddings are an important source of thermophilic aerobic sporeformers who might have impact on spoiling problems of commercial sterile dairy products. With a group of 30 dairy farmers new amazing grazing systems are identified, discussed and prioritized. New techniques to clean the pasture and new pasture based farm designs, with several milking robots and all weather systems, will be shown.

Environmental effects of dairy-farming: focusing the results of the DAIRYMAN-project

M. Elsaesser[1], J. Oenema[2] and T. Jilg[1]
[1]Landwirtschaftliches Zentrum Baden-Wuerttemberg (LAZBW), Grassland and Forage Production, Atzenberger Weg 99, 88326 Aulendorf, Germany, [2]University of Wageningen, Plant Research International, P.O. Box 616, 6700 AP Wageningen, the Netherlands; martin.elsaesser@lazbw.bwl.de

DAIRYMAN is an EU-Interreg IVb NWE project and focuses on the conditions and improvement of sustainable dairy-farming. In a network of 14 partners, 8 regions and 127 dairy-farms, a.o. the ecological effects of dairy-farming were investigated and evaluated with a specific sustainability index (DSI). For this index 18 parameters were selected and scored. Imbalances of nitrogen and phosphorus at farm-gate and different efficiencies by use of these nutrients could be observed depending on the project region. Feeding of concentrates is necessary in order to obtain the best milk performance and a high input of fertilizer seems to be necessary for high yielding grasslands with the risk of nitrogen leaching and phosphorus run-off. Farms running a high input strategy can reach the same efficient use of nutrients like those with a low input strategy and no purchased N- fertilizers. Greenhousegas-emissions were determined (IPCC) and the results show a wide range between regions and production systems (average of pilot farms: kg CO_2 equivalent/ha in Ge: 11,5; NL: 21,1; Ire: 11,2. Per kg CO_2 equiv./ton milk: Ge: 1,11; NL: 1,09; Ire: 1,27). Methane production for high performing cows had per kg ECM low GHG-emissions. Advantages for low input systems could be shown for energy balances and for the total amount of CO_2 emissions at farm level. Dairy farming influences moreover biodiversity in two ways. The lack of milk production in a region influences the structure of local landscape negatively and intensive use of grassland lowers the amount of plant diversity. Improving of ecological disadvantages is necessary and possible. Improvements could take place very much faster if farmers focus on the use efficiency of production factors and if they can learn from each other in special networks.

Analysis of dairy farmers' stategies and competences in three Central and Eastern European countries

M. Klopčič[1], A. Malak-Rawlikovska[2], A. Stalgiene[3], F. Verhees[4], C. De Lauwere[4] and A. Kuipers[4]
[1]Univ. of Ljubljana, Biotechnical Faculty, Dept. of Animal Science, Groblje 3, 1230 Domžale, Slovenia, [2]SGGW, Nowoursynowska 166, Warsaw, Poland, [3]LAEI, Kudirkos 18, Vilnius, Lithuania, [4]Wageningen UR, P.O. Box 35, Wageningen, the Netherlands; marija.klopcic@bf.uni-lj.si

As part of CEE project of WageningenUR combined with Leonardo da Vinci project by Warsaw University of Live Sciences, an analysis was performed of future development paths in Lithuania, Poland and Slovenia. A questionnaire was constructed with 49 main questions. 1039 questionnaires were received. With factor analysis no. of questions were reduced and cluster analysis was used to form farmer segments. On base of 10 strategies farmers indicated their 1st, 2nd and 3th most important strategies for development of farm. Farmers' segments identified were: Independent and Cooperating diversifiers, Specializing growers, New starters, Cooperating and Chain oriented and Wait&see farmers. Differences between countries will be discussed. Also, farmers indicated how important farming goals were on a 7 point likert scale. Factors identified were sustainable quality, succession, enjoy work, good management and financial management. Significant lower scores were identified with the Wait&see and Cooperating diversifiers segments of farmers. Likewise this was done for resources and opportunities and threats. Land and labour availability are the biggest problems for all farmers' segments. Particularly Wait&see farmers do not have enough labour available. Available resources vary mainly per country and not by segment. Opportunities and threats gave a diverse picture. Farm performance was based on future expectations. Regression analysis of farming goals, resources, opportunities and threats, and strategies on future expectations took place. Efficiency, land and capital availability, attitude towards EU-membership and interest in new techniques explained part of the variation in performance. Two classes of competences were identified, which were consequently linked to all factors studied.

Session 17 Theatre 4

Future dairy sector and herd management from American perspective

S.A. Larson
Hoards Dairyman magazine, Fort Atkinson, WI, USA; steveandleotalarson@gmail.com

Increasingly, the United States' dairy industry is being shaped by participation in world markets, environmental and regulatory factors, changing consumer preferences and shifts in federal dairy policy. This presentation will be an overview of the dairy industry in the U. S. in light of those developments. It will cover trends in recent years as well implications for the future. There will be discussion of structural changes in all segments of the dairy industry with emphasis on the production sector. The presentation will relate trends in feeding, housing, and caring for herds of various sizes including adaptation of technology. How the business model for many U.S. dairy operations is changing will be addressed. There will be discussion of the role that specialization of labor and management plays in dairy business strategy. The presentation will review farmer-funded dairy promotion efforts, the status of family owned and operated farms, interest in on-farm milk processing, the role of immigrant dairy farm employees and provide an update on federal dairy research priorities.

Session 17 Theatre 5

Dairy sector in a non-quota environment

H. Versteijlen
European Commission, Directorate-General for Agriculture and Rural Development, Director Economics of agricultural market and single CMO, Brussels, Belgium; abele.kuipers@wur.nl

In 2015 the 'temporary' quota system, installed in 1983 will seize to exist. Already now quota have much less relevance due to the soft landing policy where additional quota were made available. In 2011/2012, 21 Member States did not use their quota to the full extent while only 6 Member States exceeded their quota and have to pay super levy. In 2012 some additional instruments were created to allow farmers to better negotiate milk prices and to reinforce transparency in the market. Milk producers and milk processors will have to improve their ability to adapt supply to demand in a non-quota environment in order to avoid too much volatility. Prospects are very good since, due to subsequent reform and developments at the world market, EU and world market prices have converged and the EU is exporting important quantities without export refunds. The challenge is to sustain the obtained competitiveness at reasonable prices for milk producers. The future challenges and perspectives will be discussed.

Strengths and weaknesses of the French dairy sector and of its main competitors in Europe

C. Perrot, J.M. Chaumet and G. You
Institut de l'élevage, Economie, 149 rue de Bercy, 75 595 Paris Cedex 12, France; christophe.perrot@idele.fr

The 2006-2011 period marked by high volatility of milk prices and costs for dairy farms proved to be an informative test for assessing the competitiveness and resilience of national dairy sectors in Europe. Micro-economic analysis conducted from the FADN European data base (and its national premises) on specialized dairy farms including cross-country comparison of key time series (milk prices, charges by categories, labor productivity) helped to explain the dramatic changes in the remuneration of labor in most countries during this period. Interpretation has been achieved through field surveys and a literature watch in Germany and the Netherlands, in Denmark, Ireland and in the UK. Market signals (prices and demand volume) did not come at the same time or with the same intensity to the dairy farms in different countries. The delayed effects generated by smoothing rules established in France when the markets were regulated have created major consequences on the competitiveness of the French dairy sector (upstream in 2007, then downstream in 2009). In 2010 and 2011, milk deliveries per farm and farm incomes increased strongly in lowland areas thanks to the phasing out of milk quota system while mountain dairy farming (25% of dairy farms, 16% of dairy production) has been specifically supported in the Health Check of the CAP (specifically designed in France). In the Netherlands, 2009 will appear as a single misstep in a march to profitability unmatched in Europe for milk production, except by the strong profitabilty of the Irish grass-based system at the end of the period. In Germany, the situation is more complex, because more diverse, but the crisis, short and sharp with a dip in milk price, strengthens territorial contrasts. Meanwhile the Danish model (large dairy units with capital-labor substitution pushed to the top, full mixed ration without pasture, hired labor) has plummeted as a result of the financial (2008) and dairy (2009) crises and has shown little sign of a significant recovery since then.

The development of milk prices paid to producers in the past and the future

W. Koops
Dutch Dairy Board, Policy Affairs, P.O. Box 755, 2700 AT Zoetermeer, the Netherlands; w.koops@pz.agro.nl

To give dairy farmers more insight into the international market for farm milk and more transparancy in prices paid to producers the LTO International Milk Price Comparison was started in 1999. Since a comparison of prices paid by 17 large European dairies is published every month at www.milkprices.nl. After each calendar year has ended a report is presented with the calculated milk prices paid for that year inclusing additional analyses and remarks. This milk price comparison resulted in a data base with unique time series of prices paid to producers. Based on these data conclusions can be drawn about the the past (from EU support driven to market oriented) and for the future (more volatility, more influence world market). Also the question if the phenomenon of a pork cycle is now also coming into the dairy market is analysed. Driven by the increased milk price volatility and the abolition of the quota in 2015 dairy farmers are becoming more and more interested in future contracts as a risk management tool as is shown in a recent study supported by the Dutch Dairy Board. Also there have been interesting developments regarding the systems for paying for milk, like fixed milk price contracts, milk price guarantee certificates and two price (A and B) systems.

Session 17 — Poster 8

First experience with AMS on Slovak large scale farms

S. Mihina[1], J. Broucek[2] and A. Haulikova[1]

[1]Slovak University of Agriculture, Faculty of Engineering, Department of Production Engineering, Tr. A. Hlinku 2, 94976 Nitra, Slovak Republic, [2]Animal Production Research Centre Nitra, Institute of Breeding Systems and Animal Welfare, Hlohovecká 2, 951 41 Lužianky, Slovak Republic; stefan.mihina@uniag.sk

Economic, social and managerial changes on dairy farms are expected in connection with the EU milk market rules modifications after 2015. This certainly will not avoid technological changes in housing and milking targeted to increase economic efficiency and welfare of animals. On Slovak dairy farms intensive modernisation is undertaken more than two decades already. Way of housing is more or less similar to the Western European one. However, because of majority of large scale farms, automatic milking stations with typical Western European design suitable for family farms went into production only recently. Advantages of AMS as labour saving, stability and quality of milking process respecting cows 'wish', wider scope for data collection and others have to balance possible disadvantages as are high initial costs, reduced contact between farmer and herd and others. As a matter of fact both advantages and disadvantages are influenced by user's skills and his or her experiences and practises used before on farm without AMS, in addition with very high number of cows in one place. In recent years this is the case in Slovakia. In this presentation data are presented, that are obtained from farms both of 'family' size and large scale ones, examining the economic, managerial and social aspects.

Session 17 — Poster 9

ProAgria CowCompass: a novel operation mode in advisory services for dairy farms in Finland

T. Huhtamäki, S. Nokka and H. Wahlroos

Association of ProAgria Centres, Urheilutie 6, P.O. Box 251, 01301 Vantaa, Finland; tuija.huhtamaki@proagria.fi

CowCompass is a novel operation mode which has been established 2011. The aim of CowCompass is to enhance advising of dairy farms in feeding management, to create more value to customer and to consider entirety of functions at farm. The farms have been segmented according to their future aspects of production. A farm visit consists of herd observations, analysis of production results and of feeding management. Feeding plans are e.g. based on optimisation of difference of milk revenues and feed cost taking into account the specific requirements of animal groups the aims of the farmers. Thus economy, milk production and animal welfare are considered simultaneously. Repetitive customer surveys are being made to monitor customer satisfaction and future development. Approximately 70% of the Finnish dairy farms are using the CowCompass management tool for feeding management. Merely a half of the farms claim to have utilised the economic aspects of the new feeding plans. What they value most is analysis of feeding management and herd observations made by the advisors. They do not consider co-work between advisors on separate fields important but appreciate an integral view on the farm. Farms with TMR/PMR have not experienced benefits as much as the farms where feeds are given separately. However, generally around 80% of the farms are satisfied with the new operation mode in feeding management and the new reports from CowCompass tools. The customer surveys give us a strong recommendation of segmentation of advisors more efficiently which requires improvements also on the operational level. A specific effort must be made to create more added value to farms with TMR/PMR via enhanced advising. CowCompass tools will be developed also for plant and economy sectors, which allows more efficient co-operation between different sectors of farm advisory work.

Session 17 Poster 10

Impact of extension team trained for improve milk quality on small farms in Southeastern of Brazil

L.C. Roma Jr, A.C.S. Gonçalves, M.S.V. Salles and F.A. Salles
APTA, Av Bandeirantes 2419, 14030-670, Brazil; lcroma@apta.sp.gov.br

Milk production has a significant contribution in small farm´s activity. However, the farmer´s aim, recently, is not to focus only on yield but also on quality. Regulations in the dairy sector are very demanding, especially for small producers, in order to ensure fulfillments on the requirements for milk quality control. During 12 months, 240 small farms from Southeast of Brazil were evaluated on milk quality, such as: fat, protein, solids non fat (SNF), somatic cell count (SCC) and total bacterial count (TBC). Among this total, 60 farms were supported by extension assistance by a technician trained to focus on milk quality and the other 180 farms were just evaluated without any extension training for milk quality. The variables were analyzed by Proc Mixed ($P<0.001$) and rejection rate by Chi-square test ($P<0.001$). Over the four first months, there was no difference in milk quality among farms trained or not. On the other hand, after this period of adaptation the difference between trained or not was significant for all variables, especially for SCC and TBC. The difference in average during the year was 102,000 cels/ml for SCC and 443,000 cfu/ml for TBC. In concordance with the Regulation of Milk Quality, there was difference between both treatments (trained or not) in relation to milk samples rejected for at least one parameter. The rejection rate for non trained group was 59% and 49% for trained group. A significant effect was found between variables on the rejection study, and the most important was SCC (32%), followed by TBC (25%), SNF (13%), fat (11%) and finally, protein 5% of total number of milk samples. As conclusion, extension is important to improve milk quality, provided that technicians are trained to focus on milk quality specifically. Furthermore, it was seen that the main problem for small farms were SCC and TBC, whereas the rejection was at least 25% of the milk samples, according to regulation for milk quality standard in Brazil. (Financial support by FAPESP 10/20893-1)

Session 18 Theatre 1

The 'new equine economy' in the 21st century

R. Evans[1] and C. Vial[2]
[1]Hogskulen for landbruk og bygdeutvikling, 213 Postvegen, 4353 Klepp Stasjon, Norway, [2]Institut Français du Cheval et de l'Équitation, UMR 1110 MOISA, 34000 Montpellier, France; vialc@supagro.inra.fr

Across Europe, most estimates of equine economic activity demonstrate that traditional activities such as racing and betting generate the largest amount of income. There is, however, a growing realization that leisure use of horses in riding, whether for stress reduction, tourism or therapy is a field of economic activity which is growing along with the approximate seven percent annual growth of horses and riders. This significant growth involves evolutions of the equine sector and new questions about the role of equines in economic dynamism, culture, social links and rural development -- questions which reflect major changes in Society. Little is known about these new kinds of activities which involve new consumer demand for equestrian services, different needs for horse qualities/characteristics, questions about horse welfare linked to its evolving status , to changes in human-horse relations, land management and physical planning, and to the Common Agricultural Policy. Given the importance of, and challenges faced by the horse industry, the number of studies devoted to the sector has begun to multiply internationally. What is this New Equine Economy? What questions does it raise? What new possibilities does it present to those who wish to make a living from their relations with horses? This presentation addresses these social and economic issues and marks a beginning to the task of defining the New Equine Economy in the 21st century.

The French horse industry in 2030: scenarios to inform decision-making

C. Jez[1], B. Coudurier[1], M. Cressent[2], F. Méa[2] and P. Perrier-Cornet[1]
[1]INRA, 147 Rue de l'Université, 75338 Paris, France, [2]IFCE, 83-85, Bd Vincent Auriol, 75013 Paris, France; christine.jez@paris.inra.fr

In France, the horse population has been expanding since 1995 to reach 950,000 in 2010, representing about 15% of the total European horse population today. This growth is the result of the development of pony-riding for children and the increasing interest of French people in recreational riding and betting. These changes offer new advantages to different horse sectors, especially in the context of declining State support, increasing international competition in the horse market, societal changes regarding animal welfare, decreasing horse-meat consumption and the harmonization of regulations at European level. To understand new possible directions for research and public policy, and also to help equine industry stakeholders to prepare for coming changes, the French National Institute for Agricultural Research (INRA), together with the French Institute for Horse and Horse Riding (IFCE), have conducted a collective scenario-building exercise for the French equine industry to 2030. The study is based on 'morphological analysis', a method which explores past and current trends and potential shifts in order to consider possible future developments. It has led to four strongly contrasting long-term scenarios ('Everybody on a Saddle', 'Horses for the Elite', 'The Citizen Horse' and 'The Companion Horse'), characterized by differing horse usage, different horse populations, and differing employment. However, they all raise shared concerns in terms of: (1) the relationship between humans and horses; (2) economic efficiency; (3) environmental issues; (4) the preservation of breeds and pressure on land use; and (5) health, welfare and caring for animals up to and beyond death. These questions call for new research development in the fields of animal behaviour, economic and social sciences, breeding and genetic improvement. They also emphasize the need to enhance knowledge and innovation transfer.

How horse business professionals can adapt to the new consumer demand?

S. Pussinen and T. Thuneberg
HAMK University of Applied Sciences, Degree Programme in Agricultural and Rural Industries, Mustialantie 105, 31310 Mustiala, Finland; sirpa.pussinen@hamk.fi

The Equine industry has grown in Finland during the last decades, but studies concerning market conditions in equine business are very few. The equine industry in Finland employs 15,000 people. The number of stables that provide services as a business is approximately 3,000. HAMK, the University of Applied Sciences, carried out a small-scale survey in 2009 to investigate the demand and profitability of the equine services in the near future. A variant of the survey was implemented during spring 2013 so that the results can be compared to an earlier study, and possible changes can be discovered. The survey involved questions about demand and profitability. The entrepreneurs evaluated the demand situation from their own enterprises' perspective at the moment, and within the next five years. Entrepreneurs, who offer horse tourism, riding and livery services, foresee the demand to grow. Generally, the respondents believe that demand would increase rather than decrease. The focus in breeding and trotter training is to maintain activities at least at the present level. The profitability, which was subjective estimate of the entrepreneur, correlated to the demand. The entrepreneurs, who make a living in the horse sector, tend to consider their income to be, at least, moderate. Horse businesses are often known more as a way of life than a mere economic enterprise. New customer groups may press for specialized services and challenge the entrepreneur's know-how and willingness to be of service. In finding new customers, image and marketing communications must be emphasized. Even though the entrepreneurs are quite optimistic about the future there is a need for them to respond to the changes of the markets. The question is how to retain or create more demand when competition between multiple pastime activities and people's actual leisure time becomes harder. Like any branch of business, the equine industry needs more marketing and brand-building to find new customers.

Session 18 Theatre 4

Equine business: the spectacular growth of a new equine economy in France

S. Chevalier and G. Grefe
Univesité d'Angers, Maine et Loire, 7 allée Fançois Mitterand, 49004 Angers, France; sylvine.chevalier@univ-angers.fr

Since the end of the 20th century, the equine leisure sector has experienced dramatic growth in France. The number of riders registered (FFE) has multiplied by almost five between 1984 and 2012. This increase is boosting the equine economy in breeding and in riding schools, but also in the more general equine business. The popularity of riding supports the development of a large number of products and services in a mature market. The equine market is characterized by a rich diversity responding to the plurality of riding practices at different levels, from beginners to professionals, and through varied budgets thanks to the democratization of an ancient upper-class sport. This diversity creates a spectacular growth of products, which is strengthened by the feminization of this ex-male dominated activity. The fact that most of the riders in France are women (82% in 2012) and more especially girls (70% are 18 years old or less) implies also the rapid growth of fashion onto the equine market, upsetting the old utilitarian orientation. Our objective will be to understand better this spectacular growth of the equine business, by: Analyzing the spread of the equine business in France, studying through statistics (Equiresources, Annuaire du Cheval) their number, activity and localization in France. The results will be presented with a map. Analyzing the adaptation of equine business entrepreneurs – do they follow the evolution of the market or do they create it? Through interviews with the directors of 12 equine business enterprises in France we will study their perception of the evolution of the market, their strategy in marketing and in hiring (do they need changing competences?). Analyzing the functioning of the only bachelor focused on Equine business in France, in Saumur at the University of Angers. Through statistics from the University, we will profile the new generation of sellers, where they come from and their entry into the market.

Session 18 Theatre 5

Assessing economic impact of equine activities in Norway and Sweden using input-output modelling

G. Lindberg[1], A. Spissoy[2] and Y. Surry[3]
[1]Nordic Centre for Spatial Development, Box 1658, 11186 Stockholm, Sweden, [2]Norwegian Agricultural Economics Research Institute, Box 7317, 5020 Bergen, Sweden, [3]Swedish University of Agricultural Sciences, Economics, Box 7013, 75007 Uppsala, Sweden; arild.spissoy@nilf.no

The purpose of the paper is to analyse economic impacts of horse-related activities in Norway and Sweden. Care is taken to examine differences between effects at national and regional levels. Understanding the overall economic impacts of potential growth in the equine industry is important to the industry as well as to policymakers. An IO (input-output) model is our tool for analysing such linkages and impacts. Unfortunately, horse related information is currently either missing or distributed across different sectors of the economy in national IO accounts. We examine earlier attempts to separate a horse sector in IO models, and propose a simple method for disaggregating such accounts based on different sources of data including surveys, interviews and disaggregated sector data. We have developed IO coefficients for horse-related activities in both countries. Moreover, such coefficients are used to derive multipliers showing the impacts on the economy of expanding different horse activities. We also examine the regional structure of the multipliers. Taking the example of Sweden, the results indicate that the highest multipliers are found for riding schools (3.19) and breeders (2.90). The reason for such high multipliers can be that the enterprises with such activities use most of the revenue for purchasing inputs and spend it in the supply chains and hence do not make much profit. Activities needing to allocate some revenues to returns on capital and wages show somewhat lower multipliers; e.g. boarding enterprises (2.86) and professional trainers (2.61). However, if we look at the employment impacts of 'one-more-person working' the results are reversed, indicating that the more business-oriented activities have a stronger potential to create employment if they were to expand.

Effects of the 2008 recession on aspects on the UK horse industry

C. Brigden, S. Metcalfe, S. Mulford, L. Whitfield and S. Penrice
Myerscough College, St Michael's Rd, Preston PR3 0RY, United Kingdom; cbrigden@myerscough.ac.uk

Britain's economy has suffered since the 2008-2009 recession. Nervous consumers with less disposable income threaten leisure industries like the horse industry. The objectives were to explore the recession's effect on marketing strategies, participation in British Showjumping (BS) & horse price. 133 equine businesses completed an online survey, recording business characteristics, marketing strategies and recession related outcomes. BS participation was assessed by memberships, registrations, number of events and entries in 2006-2010. Competitor (n=144) and venue (n=32) surveys explored perceptions. Horse price was sampled in adverts of typical types in 2007, 2009 & 2011. Five elite equine sales firms were interviewed. Surveys investigated vendor/purchaser experiences (n=74). 45% of businesses reported recession effects. Services were less affected than retail or manual aspects. 58% of large (>21 staff) businesses saw recession effects, compared to 40% of small (<5 staff). 51% used a new marketing method, mostly increasing customer base (81%). 98% reported new customers. BS memberships fluctuated, but number of events reduced and number of entries/member fell by an average 3.64/year. Fuel price and the recession were significantly associated ($P<0.001$) in influencing decisions to enter competitions. Training was also reduced by riders who entered fewer events ($P<0.05$). Venues concurred that fuel prices had the most influence on entries. 66% perceived a reduction in entries, but few altered strategies. Median horse prices reduced between 2007, 2009 and 2011 (£7,500, £6,500 & £6,250). Prices varied with location from £10,500 to £750. 73% of respondents saw reduced horse prices during this time and related it to lower disposable incomes. Elite sales saw a reduction in average prices in 2009, but increases in 2011. In conclusion, recession effects differ according to the nature of business. The leisure aspect appears to be more vulnerable and participants' spending seems to be more strategic.

Promoting slaughtering of horses and consumption of horse meat

M.T. Saastamoinen
MTT Agrifood Research Finland, Animal Production, Equines, Opistontie 10 a 1, 32100 Ypäjä, Finland; markku.saastamoinen@mtt.fi

The number of horses in Europe is approximately 6,000,000. It can be calculated, based on data from Sweden and Finland that about 5% of hobby and athletic horses die or put down every year mainly due to illness or old age. This means that about 300,000 horses in Europe die yearly (+ those raised for meat production). Of these, 20-30% are slaughtered in their home countries, (e.g. about 4,000-5,000 in Sweden, 15,000 in Germany and 2,000 in Finland). Many horses are transported to be slaughtered in other countries and many are buried, legally or illegally, causing a risk to the environment if not done properly. Horse meat has many advances: its nutritional properties are good, and the ecological footprint is low. One benefit of increased local horse meat production is that it will promote the establishment of small scale slaughterhouses in rural areas. Further, specialized quality horse meat production based on native local breeds may be one way to support these breeds and, thus the diversity of horse populations in cases where the breeds are endangered. There are many examples of this in European countries. Both long transportation and the burying of bodies is ethically and ecologically unsustainable. Thus, promoting horse owners to sell their horses to slaughterhouses and consumers to use more horsemeat may be good for horse welfare at the end of a horse's life. In Finland, two projects promoting use of horse meat and encouraging horse owners to sell their horses to slaughterhouses have been carried out between 1998 and 2009. The main problems which arose were the safety of the raw meat (identification, medication), low slaughter price, long transportation distances, high slaughtering costs and the unwillingness of owners to sell horses for food production because of uncertainty of the handling of the horse during slaughtering. Special attention should be paid to the transportation length and circumstances (temperature, humidity, space, risk of injuries, etc.) in the vehicles used for slaughter transportations.

The 2012 Kentucky equine survey

R.J. Coleman[1], M.G. Rossano[1], C.J. Stowe[1], S. Johnson[1], A. Davis[1], J. Allen[1], A.E. Jarrett[1], G. Grulke[2], L. Brown[3] and S. Clark[3]
[1]University of Kentucky, College of Agriculture, Lexington, KY 40546-0276, USA, [2]Kentucky Horse Council, 1500 Bull Lea Rd. Suite 214 C, Lexington, KY 40511, USA, [3]USDA-NASS, P.O. Box 1120, Louisville, KY 20102-1120, USA; jill.stowe@uky.edu

Little is formally known about the U.S. equine industry. Even Kentucky's renowned equine industry faced many challenges in navigating the new economy following the 2008 Great Recession. The 2012 Kentucky Equine Survey was conducted to provide data for decision-making; the objectives were to obtain an inventory of all equine and estimate the industry's economic impact. The National Agricultural Statistics Service (NASS) conducted the inventory study. Extensive list-building efforts resulted in a database of over 31,000 equine operations. Questionnaires were mailed to 15,000 operations, stratified by operation size and geographic location. Non-responders were contacted by telephone or field enumerators. The response rate was 71.4%. The economic impact analysis included collecting expenditure data from equine event attendees (n=1,500). An additional survey, designed to capture the non-monetary benefits of the industry's presence, was mailed to randomly selected Kentucky residents (n=8,200). Data were used to test the hypothesis that residents are willing to pay to preserve the horse industry. NASS utilized standard statistical sampling methods for inventory estimates. Economic impact results were estimated using IMPLAN and multivariate regression analysis. Kentucky has 242,400 horses on 35,000 equine operations. The total value of equine and equine-related assets was $23.4 billion. Total equine-related sales and income were about $1.1 billion, and total equine-related expenditures were about $1.2 billion. The equine economic cluster has a significant impact on the state's economy, and most Kentucky residents were supportive of the industry and willing to incur an annual tax to preserve it. Results will be used to identify areas for growth within the industry as well as rural development.

A study on equestrian tourists motivation and involvement

J. Wu
The University of Hull, Business School, Cottingham Road, Hull, United Kingdom, HU6 7RX, United Kingdom; jie.wu@2007.hull.ac.uk

With an increasing diversity in leisure interests in 21st century, individual tourism demand is changing from General Interest Tourism or Mixed Interest Tourism to Special Interest Tourism , which can be defined as the provision of customized leisure and recreational experiences driven by specific expressed interest of individuals and groups (Derrett, 2001). Within the SIT literature, there is no unified model looking into tourists' motivation and retention, and in the case of the equine industry, there is a total lack of research on equestrian tourists' motivations. This study will use a modified version of the theory of planned behaviour integrating motivation and involvement to predict tourists' behaviour intentions. Involvement is conceptualised as a multidimensional construct consisting of the variables of attraction, centrality, identity affirmation, identity expression and social bonding. Data was collected by online survey from Nov., 2012 to Feb., 2013. The data result shows that main push factors of equestrian tourists are relaxation, social bonding, escape, learning and self-development, while a friendly atmosphere is the most important factor to pull equestrian tourists to participate in riding activities. Further, equestrian tourists have high leisure involvement and 'attraction' is the strongest dimension among the involvement construct. This study which focuses on analysing equestrian tourists' motivation is likely to enhance the understanding of equestrian tourism as a product with the potential to develop into a viable sector in future.

Session 18 Theatre 10

The impact of horses on farm sustainability in different French grassland regions

G. Bigot[1], S. Mugnier[2], G. Brétière[1], N. Turpin[1], C. Gaillard[2] and S. Ingrand[3]
[1]IRSTEA, UMR Métafort, BP 50085, 63172 Aubière, France, [2]AgroSup Dijon, BP 87999, 21079 Dijon, France, [3]Inra, Theix, 63122 Saint-Genès-Champanelle, France; genevieve.bigot@irstea.fr

In France, horse rearing is often associated with other farming activities such as beef or dairy productions which prevail in grassland areas. To analyze the impact of horses in the sustainability of farming systems, we surveyed a hundred farms in four major areas of horse breeding, which were chosen according to their production systems, representative of regional agricultural systems. Regions differ by their agro-climatic context associated with a type of horse production: saddle horses in oceanic and continental plains, and draught horses in two mountainous regions in Central and Eastern France. Farmers have been questioned about the role of equine production in economic, environmental and social functioning of their farming system. The results from this sample show that the number of horse varies from less than 10% to 100% of the total livestock. In mountainous areas, draught horses grazed in grasslands at the same time or after cattle. These hardy horses need a little labor but the low income of this production limits its development. In plains, farmers raise saddle horses either alone or with dairy or beef cattle. In farms specialized in horses, breeders develop services linked to horses such as taking horses in livery. In mixed herds, farmers spend comparatively more time on horses than on cattle, whether beef or dairy production. The impact of saddle horses on gross production depends on their number and on the age they are sold. In conclusion, horse rearing can present a low profitability, especially for draught horses, or an unpredictable one for saddle horses but the environmental and social impacts are always important. Whatever the type of animal productions, horse grazing improves the maintenance of grassland areas and farmers raise horses because they like to take care of these animals which attract volunteer labor.

Session 18 Poster 11

The REFErences network, an actor in the economic knowledge of the French horse industry

X. Dornier[1], P. Heydemann[1] and B. Morhain[2]
[1]French Institute of Horse and Horse Riding, Economic and Social Observatory, Route de Troche, BP3, 19231 Arnac Pompadour Cedex, France, [2]French Livestock Institute, 9, Rue de la Vologne, 54520 Laxou, France; xavier.dornier@ifce.fr

Since 2006, the French Institute of Horse and Horse Riding, the French Livestock institute, the Horse Councils and Chambers of Agriculture have partnered in forming the Economic Network of the Equine Industry (REFErences) to develop tools for the techno-economic diagnosis and support of industry actors. This network combines complementary approaches in macroeconomics and microeconomics. The macroeconomic approach focuses upon activities and market indicators. The economic model is based upon a chronological series of studies consolidating data from nearly 50 partners (governing body of sport and racing, agriculture statistics, auction companies…). This database is supplemented with surveys on businesses and horse buyers. Recent trends show a declining production of horses in France while buyers turn increasingly to foreign horses. In the meantime, betting and horse riding seems to stagnate after many years of growth, while employment is still progressing. The microeconomic approach aims at gaining a better understanding of the way that equine businesses are managed, both technically and economically. In order to collect data, 250 farms (breeding farms, equestrian establishments) are audited several times a year. Breeding enterprises turn out to be less economically efficient than equestrian centers (riding schools, livery yards, equestrian tourism centers) though businesses with complementary activities have better economic performance. Room for improvement has been identified to reduce costs, especially of feeding (eg. using local resources) and health management (eg. reasoned deworming). Future work will contribute to improve the control of production costs and the match between supply and demand, based on innovative solutions with increased transfer of results to professionals.

Jobs in the french equine sector

E.F. Farman, C.C. Cordilhac and F. Clément
Equi-ressources, le Tournebride, 61310 Le Pin au Haras, France; francoise.clement@haras-nationaux.fr

In France, the equine industry generates about 77,000 jobs. The definition of 'equine jobs' is built on closeness with horses: Direct jobs (45,430) represent the jobs 'in contact' with the animal (breeder, trainer ...). Indirect jobs (29,288) represents the jobs of those who play a role in the economic sector but have no direct link with the horse (administrative staff of an institution dedicated to the horse, etc.). There are a very large number of jobs in the equine industry. Most require basic skills but some positions require high skills (engineers, project or study). Hence, the scope of opportunities is very large, covering a lot of fields such as agriculture, sports, government, commerce, leisure and even art. Within the context of the economic crisis and the increasing rate of unemployment in France, the employment rate in the equine sector is dynamic. The profile of the population in the industry is rather young and feminized. Half of employees are in precarious situations and turnover is relatively high. Jobs in the equine industry are structured as following: there are about 30 jobs in the equine industry, 5 of them accounted 70% of this sector overall. As always, in a dynamic industry, tensions can occur: some are of 'quantitative nature' (lack of applications to satisfy an offer (riding instructorss, training riders, farriers)), some are of 'qualitative nature' (despite many candidates, employers do not find the skills they are looking for (professional equestrian riders, grooms, farriers for some specialties, etc.). The equine industry has created a specialized equine institution: Equi-ressources. Founded in 2007 in partnership with many institutions of the industry, Equi-resources aims at making the supply and demand meet in terms of employment. Equi-resources is also a 'lab' which aims at observing and analyzing employment in the equine market. Finally, Equi-resources stands as the institution which can provide a young public with information, advice and orientation, which is to choose training or starting a career in the equine industry.

Equine entrepreneur's well-being

T. Mustonen and T. Thuneberg
HAMK University of Applied Sciences, Degree Programme in Agricultural and Rural Industries, Mustialantie 105, 31310 Mustiala, Finland; terhi.thuneberg@hamk.fi

Horse businesses such as breeding, riding activities and trotter training, are very labour-intensive and physical enterprises. The duties of an entrepreneur are various, and include a many different kinds of know-how: in addition to horse care, an entrepreneur should master e.g. economic management and business administration. The challenge of the work and the workload with long working hours can cause problems with well-being. Farm workers' long-lasting stress can lead to burnout as well as other physical or mental disorders and illnesses. Overall, mental health problems have increased rapidly over the last years and are one of the major reasons for the premature retirement of the farm workers. HAMK University of Applied Sciences implemented a survey of equine entrepreneurs in 2011. Its aim was to find out how entrepreneurs themselves determine their well-being and work strain. The inquiry was targeted at 196 entrepreneurs. The response rate was 33%. The majority (78%) were full-time entrepreneurs offering, in most cases, horse riding services. The results indicated that Finnish equine entrepreneurs feel well, in general, but they face many challenges in their work that could have an impact on their well-being and coping. This type of entrepreneurship gives a certain liberty in organizing the job, but the flip side of the freedom is horses, which need 24 hours responsibility. Technical developments, such as feeders, are quite rarely used in stables. Respondents evaluated their work in five-stepped scale (from light=1 to extremely strenuous=5); half of them defined their work as at least quite strenuous mentally (mean 3.4), and two thirds physically (mean 3.7). The capacity for work reduced during their careers and 17% were unsatisfied with their health. Nevertheless, 66% regarded the quality of life good. Partners, family and friends are important support for the entrepreneur. Network of other entrepreneurs is also a significant factor in maintaining their well-being and the management of enterprises.

Trends in the sport horse industry's economical & socio-economical contribution to the Irish ecomony

A. Corbally[1], K.M. Brady[1], A.G. Fahey[2] and D. Harty[1]
[1]Horse Sport Ireland, Beech House, Millennium Park, Osberstown, Naas, Co. Kildare, Ireland, [2]UCD, School of Agriculture & Food Science, Belfield, Dublin 4, Ireland; acorbally@horsesportireland.ie

Over the past 20 years a number of studies have estimated the extent to which the sport horse industry contributes to the Irish economy. These studies have been based on surveys of active participants in the sport horse industry including those in the breeding, competition and leisure sectors combined with information supplied by organisations operating within the industry and data provided by the Irish Central Statistics Office. In the 1996 it was estimated that the sport horse industry contributed £80,000 to the economy in Ireland, in 2007 this had increased to €400m and most recently in 2012 in a detailed sector analysis it was reported that the sport horse industry contributed in excess of €708m to the Irish economy. This increasing trend has been due to changes within the sport horse industry, such as population size, markets and the development of elements within the competition and leisure sectors. The importance of the sport horse sector in terms of its socio-economic value has been evident and increasing from 1996 to 2012. Sport horses are located in every county and region of Ireland and are a major benefit to the communities in the less populated and rural areas of Ireland. For example, the sport horse element in 140 country shows contributed over €35 million to local rural economies in 2012. In 2012 it was estimated that 47,096 people were involved in the sport horse sector and the industry contributed to the incomes of 29,295 households. This level is significant given the size of the Irish agricultural sector with 139,860 farm holdings recorded in the 2010 Census of Agriculture.

Economic impact and social utility of equestrian events, examples from France

C. Vial[1], E. Barget[2] and J.J. Gouguet[2]
[1]Institut Français du Cheval et de l'Equitation, Institut National de la Recherche Agronomique, 2 Place Pierre Viala, 34060 Montpellier, France, [2]Centre de Droit et d'Economie du Sport, 13 Rue de Genève, 87100 Limoges, France; vialc@supagro.inra.fr

In France, equestrian sports and leisure have been growing since the 1990s. Consequently the number of equestrian events has multiplied. For example, 120,000 equestrian competitions were organized in 2012 and their number has increased by 111% in 10 years. As a result we wonder in what ways these events could participate in territorial development. To answer this question, we undertook a research program to analyze the economic, social and environmental impacts of equestrian events. Our first results are presented here. Using Economic Base Theory, we studied the economic impact of two international equestrian events which took place in 'Le Pin', a national stud in the ''Basse-Normandie' region. The first one, 'the Equirando', is one of the largest European gatherings of horsetrekking riders. The second one, the 'Grand Complet', is an international competition of Eventing. The total economic impact of these events for the county was evaluated at €190,000 for the Equirando and €240,000 for the Grand Complet. We also studied social impacts during 7 other competitions and shows which took place in two other national studs in the 'Lorraine' and 'Bourgogne' regions. Our final aim is to study social utility thanks to a cost-benefit analysis using Economic Welfare Theory. The work done in 2012 was just a first step to this end. It enables us to evaluate what kind of attributes spectators and local inhabitants associate with equestrian events. Our findings highlight the social, patrimonial and territorial values of equestrian events for local dynamism and people's welfare. Our aim to develop the methodology is to take into account not only the short term economic impacts of equestrian events but also their social and environmental externalities and long term impact on territorial development.

Price determinants for horse boarding in Norway and Sweden

Y. Surry[1], A. Milford[2] and H. Andersson[3]
[1]Swedish University of Agricultural Sciences, Economics, Box 7013, 75007 Uppsala, Sweden, [2]Norwegian Agricultural Economics Research Institute, Box 7317, 5020 Bergen, Norway, [3]Swedish University of Agricultural Sciences, Economics, Box 7013, 75007 Uppsala, Sweden; yves.surry@slu.se

The objective of the paper is to determine the factors influencing the prices charged by boarding horse farms in Norway and Sweden. Farmers in both countries are becoming more pluriactive as they seek to develop alternative business activities to supplement farm incomes following stagnating incomes in agriculture. In this context, developing horse-related activities such as horse boarding is suitable for farm owners who have grazing opportunities and a farm building structure in place. Boarding horse services implies that different sets of qualities and services are offered by farmers at a price that usually corresponds to the quality and service level provided. A hedonic price model linking the logarithm of the price of horse boarding to a set of explanatory variables representing structural, locational, environmental and social characteristics of the farms are estimated econometrically by ordinary least squares. The model uses data collected in a sample survey of boarding horse farms which was conducted in both countries. The survey consists in 53 questions to yield information about the economic performance both in general and in relation to boarding, the pricing system of boarding, managerial and technical performance, customer details, as well as vital background information of both the business structure and the farm owner. We obtained 330 responses with information on the price of horse boarding which were used in the econometric estimates. Preliminary econometric results show that the selected model provides good explanatory power. Several variables representing the characteristics of the farm and its stables are statistically significant. Further, locational variables such the distance from the closest city, also significantly influences the price of horse boarding.

Session 18 Poster 17

Current acceptability of horse meat through consumer surveys

R. Álvarez[1], L. Perona[1], M. Valera[1], A.R. Mantecón[2] and M.J. Alcalde[1]
[1]University of Seville, Dept. Agricultural and Forestry Science, Ctra. de Utrera km. 1, 41013 Sevilla, Spain, [2]Insituto de Ganadería de Montaña (CSIC-ULE)., Finca Marzanas, 24346 Grulleros (León), Spain; ralvarez1@us.es

Horse meat consumption in Spain is marginal in comparison with other types of meat. However, the high price of feedstuff and maintenance of the horses during the current economic crisis have led to a rise in horse-meat production in Spain. Three hundred sixty consumers (balanced in sex and age) from Southern Spain were asked about their consumer habits regarding horsemeat consumption. Frequency distribution and contingency table with chi-square tests were carried out in order to evaluate their answers. 93.3% of respondents admitted that they do not consume horse meat at all. However, 88.1% have tasted it at least once, mostly on special occasions (41.5%). The main reasons to not taste it were aversion, as horses are considered pets and never having consumed it previously (56.9%). It is considered as an expensive meat (77.6%) and hard to find (95.0%), but healthy (92.0%). It is considered good for animal welfare (72.3%) and environment (82.5%) as it is considered as a friendly production. In addition, it was considered as a nutritive meat by both consumers who have tasted it (100%) and those who have never done so (88.1%). Moreover, it was considered as a high quality meat (82.5%) with adequate fattening (70.6%). Regarding the sensory characteristics, consumers think that horse meat has a mild flavor (65.9%) and good appearance (84.8%). 88.4% of consumers who had tasted the meat thought it is tender and juicy, whereas 74.5% of those who have not tasted it thought it is tender and 72.6% considered it juicy. These answers differ significantly ($P<0.01$) between both types of consumers. Thus, it could be said that horse meat is better valued by consumers who have tasted it before. In addition, 84.6% of respondents who had tasted it pointed out that horsemeat has similar sensory characteristics to beef.

The acceptance of complementary therapies in equine communities and what therapists need to know

J. McKeown
Aberystwyth University, School of Management and Business, S9 Cledwyn Building, Penglais Campus, Aberystwyth, SY23 3DD Ceredigion, United Kingdom; jum1@aber.ac.uk

One new and growing aspect of the equine economy in the 21st century is the provision of complimentary therapies (CAM) to horses, such as herbology, chiropractic, acupuncture and so on. This study looks at the acceptance of therapies in Wales amongst horse owners. It focuses on the provision and horse owners' adoption of equine CAM, decision and choice processes over which therapies to purchase, when and from who and therefore the impact this has on equine therapists in the region. Introspection was used to develop a survey which was sent to horse owners across Wales. A sub-set of the respondents were then interviewed in depth. Further interviews were had with Vets and an owner of an Equine Spa. Interviewees were chosen specifically to understand decisions taken and attitudes towards CAM in a cross section of different equine communities. Alongside this, desk research was done going back 10 years in the popular equine press, in order to understand the level of advertising and articles written about CAM, in magazines that the respondents read. Initial findings show that there are many more women than men in the equine communities and these are more likely to accept CAM, especially those aged above 30. Owners practising Natural Horsemanship were also more open to CAM adoption. Of the 'Big 5' therapies (acupuncture, chiropractic, herbal medicine, homeopathy, osteopathy), homeopathy was the least accepted and chiropractic/osteopathy the most accepted. Herbs and massage were not always considered to be CAM by the respondents. The more 'scientific' the modalities, the more likely they were to be accepted. Articles were more likely to lead to adoption of CAM than adverts but recommendation by trusted peers within the community was the most powerful adoption tool of all. What constitutes CAM in the eyes of the horse owners is subjective and appears to be based on belief structures.

Initial approach to define the potential market of recent biotechnologies: the case of cloning

A.P. Reis[1], E. Palmer[2] and M. Nakhla[3]
[1]AgroParistech, SESG, 16 r Claude Bernard, 75231 Paris, France, [2]Cryozootech, 16 r André Thome, 78120 Sonchamp, France, [3]Mines Paristech, CSG, 60, bd St Michel., 75272 Paris, France; alline.depaulareis@agroparistech.fr

Biotechnologies evolved from a broad market (artificial insemination; AI) to models increasingly oriented to niche, elitist markets (ET, ICSI, cloning). Cloning is the most recent biotechnology, intended to reproduce the best performers. However the extent of its potential market is not well established. The question concerning the potential market for horse cloning is highly complex. The present work is a preliminary study with the aim of supplying players in the horse industry with a judgment tool to help in identifying and geographically defining the potential market for this technology by defining the target population. The research question was restricted to the FEI, WBFSH and FIP data related to jumping, dressage, eventing, endurance and polo. This first research is a numerical evaluation of the target population around the world based on individual performances. Only the 0,3% superior class of the rankings was considered as potential candidate because of its real potential to contribute to genetic improvement based on performance. The target population identified comprises 146 to 240 individuals: 100 jumping, 24 dressage, 11 events, 8 endurance and 24 to 100 polo horses. Different poles of development, with breeding/exploitation specific to each sport, were identified. The target population is dynamic because an individual remains on top for 2 to 5 years. This preliminary study identified the potential market for cloning based on performance traits. However, this market is singular and socio-economic factors including culture, technological misunderstandings (GMO), value of cloning and clones and financial potential can influence the real potential market for the technology. This study must be continued to establish a tangible socio-economic apparatus helping to understand the potential market of cloning.

A comparative study into the impact of social media in the equine and agriculture industries

C.F. Martlew
Myerscough College, St. Michaels Rd, Bilsborrow, Preston, Lancashire, PR3 0RY, United Kingdom; c.f.martlew@gmail.com

The popularity and significance of social media (internet-based applications centred on communication and file-sharing) has increased exponentially in the last decade, changing the way businesses and professionals communicate and market themselves. There is little academic research into how this has impacted equine and agriculture industries, both of which are heavily based on tradition. The aim for this study was to compare how equine and agriculture have embraced social media both in industry and commercial capacities. The objectives were to survey a variety of participants to determine their uses and perceptions of social media sites; compare the responses of agriculture and the equine sectors and determine how social media is impacting these industries. The mixed methodology included online questionnaires (407 equine and 110 agriculture responses); e-mail interviews (3 equine and 2 agriculture professionals who work with online communications); and two focus groups with agriculture and equine degree students. Survey results show trends in use of social media for businesses/professions differed in the media-sites used (agriculture preferred Twitter by 69.77% and equine Facebook by 91.26%); the frequency they were visited (68% of agriculture respondents visited sites <5 times/day and 86.67% equine visited >10 times/day) and the attitudes toward it (more agriculture than equine respondents felt their business/profession had benefited (78.79% to 67.50%)).Qualitative responses in the surveys and interviews in both industries showed a positive attitude toward using social media sites for fast information sharing, improving communications with customers and broadening professional networks. Other trends from the surveys, interviews and, in particular, focus groups revealed limitations in social media as having a lack of reliability and risks of reputation-damaging comments. Overall, a variety of uses and attitudes were shown with gaps in knowledge and awareness that both industries could potentially improve on.

Session 19 Theatre 1

dam nutrition, affects the lifetime milk production of ewe offspring

A.M. Paten[1], N. Lopez-Villalobos[1], P.R. Kenyon[1], D.S. Van Der Linden[2], A.M. Adiletta[1], S.W. Peterson[1], C.M.C. Jenkinson[1], S.J. Pain[1] and H.T. Blair[1]
[1]Massey University, Institute of Veterinary, Animal and Biomedical Sciences, International Sheep Research Centre and Gravida: National Centre for Growth and Development, PB 11222, Palmerston North, 4442, New Zealand, [2]AgResearch, Grasslands, Animal Nutrition and Health Group, PB 11008, Palmerston North 4472, New Zealand; a.m.paten@massey.ac.nz

A previous study showed that ewes born to dams fed maintenance (M) levels during pregnancy (days 21 to 140 of gestation) produced greater milk, lactose and crude protein yields in their first lactation when compared with ewes born to dams fed *ad libitum* (Ad). In the same study; ewes born to heavier (Hv; live weight (LW): 60.8±0.18 kg) dams produced greater milk and lactose yields when compared to ewes born to lighter (Lt; LW: 42.5±0.17 kg) dams. While first lactation performance was altered, it is unknown if dam size or gestational nutrition has lasting effects on progeny milk production. The objective of this study was to analyse the lifetime lactation performance of the previously mentioned ewes. Using daily milk yield and composition data, lactation yields were calculated for a 42-d period for each year for milk (MY), fat (FY), crude protein (CPY), true protein (TPY), casein (CY) and lactose (LY) using a Legendre orthogonal polynomial model. Statistical analysis comparing the lactation yields, carried out for five lactations (the typical productive lifetime) showed that ewes born to Hv-dams produced greater MY (P=0.02), LY (P=0.01), CPY (P=0.02), TPY (P=0.02) and CY (P=0.03) than offspring born to Lt-dams. After adjustment for ewe LW, the differences in MY (P=0.05) and LY (P=0.02) remained. Dam nutrition during pregnancy did not affect lifetime milk yields or milk composition of the offspring. These results suggest that maternal nutrition during pregnancy only affects first-lactation performance of offspring while dam size has lifelong effects on milk production of ewe progeny.

Session 19 Theatre 2

Effects of different energy and protein supply during pre puberty on goat's growth and lactation

L. Drouet[1], F. Dessauge[2] and S. Duboc[1]
[1]Sanders, R&D, Centre Affaires Odyssée, 35172 BRUZ, France, [2]INRA, UMR1348 PEGASE, Domaine de la Prise, 35590 St. Gilles, France; laurent.drouet@sanders.fr

To reach live weight superior to 35 kg at first breeding, goat's farmers increase energy and protein supply before puberty. Heifer's studies demonstrated a negative effect of high growth before puberty on mammary gland developement and lactating performance. We study incidence of high supply on goat's growth and lactation. 4 groups of 10 alpine kid goats between weaning (10 weeks old) and breeding (30th weeks old) from INRA herd (Le Rheu) were used. 4 concentrates were distributed (values as feed): AL-L: 1.56 NEL Mcal-17% protein-*ad libitum* AL-H: 1.56 NEL Mcal-19% protein-*ad libitum* R-L: 1.43 NEL Mcal-16% protein-100 g/d by month old R-H: 1.44 NEL Mcal-17% protein-100 g/d by month old. Every 2 weeks, goat kids were weight and consumption during 24 h was recorded. Between 2nd and 6th month of lactation, milk production was evaluated 2 times a week and protein and fat content once a month. Growth data were analysed using ANOVA procedure (SPSS). Lactation data were analysed using a linear mixte model repeated by period. Kid goat fed *ad libitum* had higher consumption (925 vs. 536 g/d) and grew significantly faster (125.5 vs. 185 g ADG/d). NELingested/100 g ADG was better for limited feed intake groups (-0.173 NEL Mcal/100 g). There was no significant live weight difference after kidding between groups. AL-L milk production was significantly higher than AL-H and R-L (respectively +0.23 and +0.19 kg/d). Unlike heifers, high ADG before puberty didn't seem to limit development of milk secretion tissue in udder even udder looked fatty. Greater growth performance reached with *Ad libitum* groups didn't seem to limit later milk production. Histological and molecular analysis of mammary gland will be realized in next experiences to get precise knowledge on udder goat development, with potential difference with heifers.

Session 19 Theatre 3

Effect of extended photoperiod on ovulatory activity and milk yield in dairy goats

K.J. Logan[1], F.R. Dunshea[1], V.M. Russo[1], A.W.N. Cameron[2], A.J. Tilbrook[3] and B.J. Leury[1]
[1]The University of Melbourne, Agriculture and Food Systems, Royal Parade, Parkville 3010, Australia, [2]Meredith Dairy, 106 Camerons Road, Meredith 3333, Australia, [3]SARDI, Livestock and Farming Systems, Roseworthy 5371, Australia; brianjl@unimelb.edu.au

Short day length is associated with reduced milk yield from dairy animals. Dairy cows, sheep and goats have been successfully housed under lights throughout winter to extend the day length and increase milk yield. However, very little data exists on the effect of extended day length on the subsequent reproductive status of dairy goats. The objective of this study was to examine the effect of long day photoperiod (LDPP) and exposure to bucks on milk production and plasma progesterone and prolactin in dairy goats. The study was conducted in 122 non-pregnant lactating dairy goats over an 18 week period from April to August (autumn and winter in the Southern hemisphere). The goats were kept in open sided sheds in which the control treatment received natural lighting while the LDPP treatment received 16 h of light, part of which was artificial. In June, July and August synchronised does were randomly assigned each month to the presence or absence of a buck. Plasma prolactin was increased (0.095 vs. 1.33 ng/ml, $P<0.001$) while progesterone was reduced (0.73 vs. 0.46 pmol, $P<0.001$) in LDPP goats. The latter response was particularly pronounced in the latter stages of the study in August (0.58 vs. 0.004 pmol, $P<0.001$) indicating a lack of functional corpora lutea. While there was no overall effect of buck exposure on progesterone there was an interaction such that progesterone was increased ($P<0.05$) by exposure to bucks in LDPP goats in late lactation. Milk yield was increased in LDPP goats, particularly in the latter stages of the study (1.55 vs. 1.82 l/d, $P<0.05$). Also, persistency of lactation was greater ($P<0.05$) in LDPP goats. These data indicate that LDPP can increase milk yield and persistence while decreasing ovulatory activity in dairy goats.

Responses to a heat stress episode in lactating Saanen and Alpine goats

L.S. Jaber[1], C. Duvaux-Ponter[2,3], S. Hamadeh[1] and S. Giger-Reverdin[2,3]
[1]American University of Beirut, Riad El Solh 1107-2020, 110236-Beirut, Lebanon, [2]AgroParisTech, 16 rue Claude Bernard, 75005 Paris, France, [3]INRA, 16 Rue Claude Bernard, 75005 Paris, France; lj01@aub.edu.lb

This study aims to assess the effect of a sudden heat episode on lactating Saanen and Alpine goats bred under temperate climate (Paris area, France). Eight Saanen and eight Alpine goats were included in the experiment. They were offered *ad libitum* water and a total mixed ration twice daily; feed and water intake were recorded daily. The study extended over two 5-day periods: the first period served as control with a maximum thermal heat index (THI) of 64.5 and the second 5 days period represented the heat episode with a minimum THI of 68.0. Selected physiological parameters were analyzed including rectal temperature, plasma glucose, NEFA and urea, and blood pCO_2, Na^+ and HCO_3^-. Milk production and composition was also individually assessed. Data were analyzed using the mixed procedure of SAS® for repeated measurements. Blood analysis showed that the does resorted to hyperventilation to dissipate the extra heat load. In addition, heat stressed animals drank more water probably to compensate for water lost for cooling especially since rectal temperature increased. Finally, milk production was maintained, although milk fat and protein content dropped during the heat episode. The data were also subjected to PCA analysis which revealed a significant effect of the period and of the breed with no interaction between these two factors. In conclusion, the experiment showed that although the animals were born and raised under a temperate climate, they could handle a short heat wave with minimal physiological disturbances.

Once-daily milking ability of the Lacaune ewes: synthesis of the results of a 4 years French study

F. Barillet[1], P. Hassoun[2], C. Allain[1], E. Gonzalez-Garcia[2], J.P. Guitard[3], P. Autran[4], M.R. Aurel[4], O. Duvallon[4], D. Portes[4], E. Vanbergue[5], F. Dessauge[5] and P.G. Marnet[5]
[1]INRA, UR631, 31320, France, [2]INRA, UMR868, 340600, France, [3]Lycée agricole de St Affrique, La Cazotte, 12400, France, [4]INRA, UE321 La Fage, 12250, France, [5]INRA, UMR1080, 35590, France; francis.barillet@toulouse.inra.fr

In France, dairy sheep breeders aim at reducing the milking labor workload. The research program ROQUEFORT'IN included 10 experiments using Lacaune ewes bred in two experimental flocks (La Fage and La Cazotte) from 2009 to 2012, for a total of 574 lactations. Half of the ewes was milked twice a day (TDM) and the other half milked once a day (ODM – morning) from 50 days in milk to the end of the lactation. The ODM ewes were, either fed *ad libitum* individually or in batches, either fed according to the actual milk yield level of the batch or at higher levels up to the feeding of the TDM ewes. On average milk yield of the ODM ewes decreases significantly by 18% (from 10% to 25% depending on the experiment) with no significant differences between primiparous and multiparous ewes. Without feeding restriction, ODM ewes do not adjust their feed intake to their reduced milk yield. Compared to TDM ewes, milk protein content of ODM ewes tends to increase slightly, mainly due to soluble proteins increase. Milk fat content decreases significantly with at libitum feeding ODM ewes, while milk fat content is not significantly different from TDM ewes when feeding of ODM ewes is adjusted to their milk yield decrease. Thus overfeeding must be avoided and even more so feeding adjustement to the actual milk yield of ODM ewes does not reduce milk yield more. Milk somatic cells count of ODM ewes is not significantly different to TDM ewes (udder health) and milk flow is significantly higher for ODM ewes (milking duration preserved).These results show a good ODM ability of the Lacaune ewes. This research was supported by the ROQUEFORT'IN contract funded by FUI, Midi-Pyrénées region, Aveyron & Tarn departements & Rodez town.

Exploring udder health by infrared thermography at milking in dairy ewes

A. Castro-Costa[1], G. Caja[1], A.A.K. Salama[1], M. Rovai[1] and J. Aguiló[2]
[1]Group of Ruminant Research (G2R), Universitat Autònoma de Barcelona, 08193 Bellaterra, Spain,
[2]Group of Biomedical Applications, Universitat Autònoma de Barcelona, 08193 Bellaterra, Spain;
gerardo.caja@uab.es

A total of 83 lactating ewes (Manchega, n=48; Lacaune, n=35) were used for assessing on using infrared thermography (IRT) for detecting intramammary infections (IMI). In Exp. 1, ewes were milked twice-daily and IRT pictures taken (IRI 4010 camera; Irysis, Northampton, UK) before and after milking, at 46 and 56 DIM. Udder skin temperature (UST) was measured from pictures and IMI detected by milk bacterial culture, resulting in 85.5% healthy udder halves and 14.5% IMI. No UST differences were detected by udder health (healthy vs. IMI; P=0.484) nor side (left vs. right; P=0.879), but UST varied by effect of breed (P=0.003), milking (P=0.014) and schedule (P<0.001). UST increased linearly with ambient temperature (r=0.88). In Exp. 2, 9 Lacaune ewes milked once-daily in late lactation (155 DIM), were used for evaluating the acute response to an *Escherichia coli* O55:B5 endotoxin challenge (0.083 µg/kg BW). Treatments were: (1) control (C00, both halves untreated); (2) treated-control (T10 and C01, one udder half treated and the other untreated); and (3) treated (T11, both halves treated). Temperature, milk yield and milk composition were monitored for 3 d. Local and systemic signs of IMI, as well as milk changes (flakes, CMT, SCC and composition) were observed from h 6 (P<0.05 to 0.001). For all treatments, UST increased after challenge, peaking at h 6 in T11 (P<0.001) and decreasing thereafter without treatment effects. No differences were detected in fat and protein milk contents, but lactose content and SCC in milk were different between treated vs. untreated udder halves (P<0.05) throughout the challenge. UST and ambient temperature correlated (r=0.60). In conclusion, despite the accuracy of the camera (±0.15 °C) and the SEM obtained for UST (±0.05 to ±0.24 °C), we were unable to discriminate between healthy and IMI udder halves in dairy ewes.

Changes in the milk and cheese fatty acid profile of ewes fed extruded linseed

E. Vargas-Bello-Pérez, R. Vera, C. Aguilar and R. Lira
Pontificia Universidad Católica de Chile, Ciencias Animales, Av. Vicuña Mackenna 4860, Casilla 306, Santiago, P.O. Box 6904411, Chile; evargasb@uc.cl

Sheep milk and cheese are considered gourmet products in Chile and there is an increasing demand from a growing and diversified market, that includes an incipient sector of functional and nutraceutical foods. This study was carried out to enhance 18:3n-3 and rumenic acid (RA) levels in ewes' milk and cheese fat under field conditions (commercial farm) by dietary means (extruded linseed) in a short period of time. During 26 days, a group of lactating ewes (n=9) (Latxa × Milchaf × Corriedale) in mid-lactation were managed under grazing conditions and supplemented during each manual milking with 50% corn + 50% oats (1000 g/ewe/d) during the first 6 d (control; TC), from day 7 to 20 the supplement was based on 25% corn + 25% oats + 50% extruded linseed (1000 g/ewe/day) (extruded linseed; TEL) and finally, from day 21 to 26 the ewes were fed TC. The data were subjected to analysis of variance, with treatment as an independent variate and the milk and cheese FA analyses as dependent variates. Saturated fatty acid (FA) content in milk was reduced (3.6%) and monounsaturated, polyunsaturated and n-3 FA (7%, 10% and 25%) were increased when ewes were supplemented with extruded linseed (TEL). A modest correlation of rumenic acid with vaccenic acid (R^2=0.68) was found in milk fat from ewes fed TEL. In TEL cheeses, contents of n-3 and polyunsaturated FA were increased (21% and 7%). Color and texture of cheeses made with TEL did not substantially differ from those made with non-supplemented ewe's milk. In conclusion, supplementation of extruded linseed in ewes under grazing conditions could be an alternative lipid source that can result in dairy products from ewes with nutritional added value.

Milk production and body composition of East Friesian × Romney and Border Leicester × Merino ewes

T.E. Hunter[1], F.R. Dunshea[1], L.J. Cummins[2], A.R. Egan[1] and B.J. Leury[1]
[1]The University of Melbourne, Agriculture and Food Systems, Royal Parade, Parkville 3010, Australia, [2]Ivanhoe, Bulart Bridge Road, Cavendish 3314, Australia; brianjl@unimelb.edu.au

While there is a move toward incorporating new breeds into meat lamb breeding program in Australia, the productivity of the East Fresian (EF) is largely unknown under Australian conditions. This study was conducted to measure the potential milk production and body composition in EF × Romney (EF×R) and dual purpose Border Leicester × Merino (BL×M) ewes. Eight second parity single-bearing EF×R ewes and eight second parity single-bearing BL×M ewes were selected from a flock that had been mated to EF rams. After lambing, ewes were offered concentrate pellets *ad libitum* and 250 g/d of oaten chaff. Potential milk yield was determined using the iv oxytocin method 2X/wk for the first 9 wk of lactation. Body composition was determined by dual energy X-ray absorptiometry at 1, 3, 5 and 9 wk. Potential milk production was higher in EF×R than BL×M ewes (2.57 vs. 1.92 kg/d, P=0.03) and declined (P=0.005) as lactation advanced. Potential milk protein, fat and lactose yield were all higher (0.02<P<0.06) in EFxR ewes. Energy intake and energy balance increased (P<0.001) over the first 4 wk of the study before reaching a plateau. There was no significant difference in energy intake (33.3 vs. 29.2 MJ ME/d, P=0.14) or energy balance (5.99 vs. 8.24 MJ ME/d, P=0.43) between breeds. Changes in tissue energy between DXA scans were highly correlated (R^2=0.41, P<0.001) to estimated energy balance, although the intercept of the regression line for EF×R ewes was higher (P=0.05) than for BL×M ewes suggesting that actual milk yield was overestimated by the oxytocin method in the EF×R ewes. There was no effect of breed on birthweight (P=0.85) or daily gain (P=0.97) of lambs. These data indicate that EF×R ewes have a higher potential milk yield than BL×M ewes although this may not be expressed in single-bearing ewes with moderate suckling intensity.

Lactation length and lactation milk yield in Alpine goat in Slovenia

D. Kompan
University of Ljubljana, Biotechnical Faculty, Jamnikarjeva 101, 1000 Ljubljana, Slovenia; drago.kompan@bf.uni-lj.si

Slovenia has no longer tradition with goat production since implemented again after earlier nineties The aim of this thesis was to analyse the effects year and month of kidding, breeder, parity, litter size, age at kidding, lactation length ant interaction between year and breeder on milk, protein and fat yield, protein, fat and lactose content and lactation length of Alpine goat breed. The analysis included 132,691 data from 47 breeders in the period from 1994 to 2011. The average lactation lasts 254.26 days and average milk yield was 527.07 kg. The longest lactation was in 5th parity (259.67 days), also the highest milk yield (575.06 kg). Goats with litter size 3 kids had on average 12.06 days longer lactation and milked 174.36 kg more milk than goats with one kid. Lactation lengths shorten from 1994 to 2011, while milk yield was fluctuating. The longest lactation was in 1999 (271.43 days) and the shortest in 2008 (235.34 days). Goats kidding in January milked 232.65 kg more milk and had for 80.46 days longer lactation than goats kidded in summer season.

Session 19 Poster 10

Effect of non-genetic factors on goat milk yield and milk composition

M. Marković and B. Marković
Universiry of Montenegro, Biotechnical Faculty, Department for Livestock Science, Mihaila Lalica 1, 81000 Podgorica, Montenegro; bmarkovic@t-com.me

The objective of this study was to estimate an effect of some non-genetic factors on milk trait parameters (lactation length, milk yield, daily milk yield, content of fat, protein and non-fat solid) of Balkan goat breed reared under extensive management system. Total number of 419 doers distributed in four flocks, every of them from different region, was included into the investigation. In order to estimate the effect of non-genetic parameters on variation of these traits, the mixed model methodology was applied. Based on 529 completed lactations, lasted 204 days in average, the milk yield was 140.54 litres in average. Variation of average milk yield was affected significantly (P<0.01) by fixed factors such as flock and subsequent lactation, but not by strains of goats. Average daily milk yield was 0.638 litres, determined on 3,221 individual measurements (test days) during regular milk recording procedure. Based on the analysis of the same number (3,221) of milk samples the average content of milk fat percentage was 3.38%; protein 3.30% and solids non-fat 8,38%. These traits were significantly (P<0,01) affected by fixed effect of flock and lactation (daily milk yield and solids non-fat only), as well as the random regression effect of animal, test day and interaction of flock × test day.

Session 19 Poster 11

Olive oil by-products improves fatty acid profile of milk and cheese in ewe's diets

E. Vargas-Bello-Pérez, R. Vera, C. Aguilar and R. Lira
Pontificia Universidad Católica de Chile, Ciencias Animales, Av. Vicuña Mackenna 4860, Casilla 306, Santiago, P.O. Box 6904411, Chile; evargasb@uc.cl

Olive oil extraction yields by-products rich in oil that can be used in animal diets as a source of monounsaturated fat. The objective of this study was to evaluate the effect of dietary supplementation of olive oil by-products (lampant olive oil and olive cake) on milk and cheese fatty acid (FA) profiles in ewes. In experiment 1, dietary treatments were supplemented with 0, 36 and 88 g of lampante olive oil/kg of DM. In experiment 2, diets were supplemented with 0, 281 or 751 g/d of dry olive cake. Each experiment was conducted separately using lactating ewes (n=9) in a replicated (n=3) 3×3 Latin square design. In experiment 1, dry matter intake, milk yield and milk composition (fat and protein) were not affected by dietary treatments. In experiment 2, except for total solids, dry matter intake, milk yield and milk composition were not affected by dietary treatments. In experiments 1 and 2, oleic acid and monounsaturated FA gradually increased (P<0.05) as the saturated FA and atherogenicity index decreased (P<0.05) in milk and cheese as the content of lampante olive oil and olive cake were increased in dietary rations. Overall, from the human health standpoint, FA profile of milk and cheese from sheep can be naturally improved by supplementation of olive oil by-products in lactating ewe diets.

The role of stockperson beliefs and behaviour in the welfare of extensively managed sheep

C.M. Dwyer, M. Jack and K.M. McIlvaney
SRUC, Animal and Veterinary Sciences Research Group, King's Buildings, West Mains Road, Edinburgh EH9 3JG, United Kingdom; cathy.dwyer@sac.ac.uk

In intensively managed livestock, stockperson attitudes and behaviour are known to influence animal welfare and productivity. In extensively managed species, where animals may have infrequent contact with humans, much less is known about how stockperson behaviour can influence welfare. We hypothesised that shepherds can affect the lives of their animals directly, through their interactions with sheep, and indirectly through their management decisions, e.g. decisions on provision of supplementary feed. We carried out two studies to investigate these two elements of stockpersonship. In the first study a detailed questionnaire was answered by 37 hill sheep farmers about management and their beliefs about sheep stress during movements (scored on a numerical rating scale from 1=no stress to 5=extreme stress). In the second study the nature and quality of interactions with animals were recorded for 18 shepherds whilst moving sheep. Survey data were analysed by Mann Whitney and Spearman's Rank correlation tests, observation data were analysed by cluster analysis. Most (59%) shepherds considered gathering sheep to cause no or low stress to sheep (median=2), and rated their own stress as greater than that of the sheep ($P<0.05$). Stress scores given to sheep and to shepherds were, however, highly correlated ($r^2=0.55$, $P<0.001$). Shepherds scored stress at shearing/ handling more highly than gathering (median=3), and equivalent to their own stress for this procedure. Cluster analysis of stockperson behaviour revealed three main styles of interacting with sheep: (1) use of gentle physical interactions only; (2) soft vocalisations and hand gestures without physical contact; and (3) frequent moderate to hard contacts and loud voices. Styles 2 and 3 were most frequently used (38 and 50%, respectively). These data suggest there is significant variation in both shepherds beliefs about sheep stress and their handling of sheep, which may impact on sheep welfare.

Identifying stress in lambs: an investigation of behavioural measures related to saliva cortisol

C.L. Dodd[1,2], J.E. Hocking Edwards[2,3], S.J. Hazel[1] and W.S. Pitchford[1]
[1]The University of Adelaide, School of Animal and Veterinary Sciences, Roseworthy 5371, Australia, [2]Cooperative Research Centre for Sheep Industry Innovation, University of New England, Armidale 2351, Australia, [3]South Australian Research Development Institute, Struan Research Centre, Naracoorte 5271, Australia; cathy.dodd@adelaide.edu.au

Measurement of stress in sheep is important for their management to improve welfare and productivity. There is evidence that the behaviour of sheep in response to stress has an economic impact on sheep productivity in terms of growth and maternal ability. Additionally, there is increasing social pressure to reduce stress and improve welfare, thus necessitating the ability to practically measure stress in sheep. This study aimed to identify behavioural measures in sheep that are indicative of stress. Behavioural measures were conducted on 251 lambs within six weeks of weaning, at mid-growth and within three weeks of slaughter at 40-50 weeks, to assess repeatability and practical applicability. Behavioural measures included agitation (the amount of movement of a lamb whilst visually isolated from flock mates), scores of movement in the weigh crate and exit speed from the weigh crate. Saliva samples were collected at the mid-growth and pre-slaughter testing times for analysis of cortisol levels. The behaviours measured are moderately repeatable. Movement level in the agitation test is associated with saliva cortisol, with increased movement correlated with lower post-handling cortisol levels ($P=0.03$). Weigh crate exit speed is related to pre-handling cortisol levels ($P=0.002$) and rate of cortisol increase over time ($P=0.004$). This supports results from previous, much larger trials which indicate that agitation and exit speed are likely to measure different components of behavioural reactivity. These results provide evidence that the behavioural measurements investigated may be practical indicators of stress in lambs, allowing the targeted management of flocks to improve welfare.

Lamb mortality in France: frequency of exposure to risk factors

J.M. Gautier[1,2] and F. Corbiere[1,3]
[1]UMT Small Ruminants Heard Health, 23 Capelles path, BP 87614, 31076 Toulouse Cedex, France, [2]Institut de l'Elevage, BP 42118, 31321 Castanet Tolosan Cedex, France, [3]National Veterinary School of Toulouse, 23 chemin des Capelles, BP 87614, 31076 Toulouse Cedex, France; jean-marc.gautier@idele.fr

It is the aim to obtain French reference data for lamb mortality rate and causes and frequency of well-known risk factors related to ewes, lambs and environment. 54 suckler sheep farms, belonging to three major management systems (two lambing periods, accelerated system, pastoral system) where followed up during one year between July, 2011 and July 2012. Farmers were asked to record all births (including late abortion and stillbirths) with date, litter size, gender and dam id. Records regarding lamb mortality were collected until weaning, based on a standardized protocol. Also the selenium status of dams was evaluated in each farm and temperatures were recorded every hour in animal houses. Finally, for each lambing period, farmers were asked about their management practices through a questionnaire. The median lamb mortality rate until weaning was 13.7% (lower quartile 8.6%; upper quartile 18.3%). 51.9% of deaths were related to abortions or stillborn lambs and to lambs dying within their first 2 days of life, while deaths after 10 days of life accounted for only 8.4% of the overall recorded mortality. The main cause of early death observed by farmers was related to the 'lamb vigour'. In more than 10 days of lamb life, infectious causes such as respiratory disorders, diarrhoeas, enterotoxaemia were most frequently reported. As suspected, farmers did not apply all good farming practices to reduce mortality risks. Differences in risk factors frequency could be evidenced between farming systems. This study may help developing a diagnosis and identifying the main points for which education of farmers would be beneficial such as the control of colostrum intake, ambience and hygiene in barns, feeding of ewes in late pregnancy.

The effect of ewe prolificacy level on number of lambs born, lamb birth weight and lamb mortality

P. Creighton[1], F. Kelly[1] and N. McHugh[2]
[1]Teagasc, Animal & Grassland Research and Innovation Centre, Teagasc, Athenry, Co. Galway, Ireland, [2]Teagasc, Moorepark, Fermoy, Co Cork, Ireland; philip.creighton@teagasc.ie

The main challenge for pasture based systems of sheep production is to increase the output of lamb from grassland. One of the most important factors influencing this is ewe prolificacy. The aim of this study was to investigate what effect prolificacy level can have on the number of lambs born, lamb birth weight and lamb mortality. Two groups of animals (180 primiparous two-tooth ewes in each group) differing in prolificacy were assembled: a medium prolificacy group (Suffolk × ewes aiming to wean 1.5 lambs/ewe, MP) and a high prolificacy group (Belclare × ewes aiming to wean 1.8 lambs/ewe, HP). The effect of ewe prolificacy level on the number of lambs born, lamb birth weight and lamb mortality was measured. Data were analysed by analysis of variance using Proc Glm in SAS and odds ratios were also calculated using Proc Genmod. Odds ratios were derived by acquiring the exponent of the partial regression coefficients. Prolificacy level had a significant effect ($P<0.05$) on the number of lambs born with a greater number of lambs born per ewe in the HP group. When odds ratios were calculated the HP group were 1.83 times more likely to have a greater number of lambs born compared to the MP group. Lamb birth weight was significantly effected by ewe prolificacy level with lambs in HP weighing 0.14 kg less at birth compared to the MP group (4.15 vs. 4.29 kg, $P<0.05$). Mortality at birth was significantly effected by litter size ($P<0.01$) with the HP group 1.1 times more likely to have a higher mortality level compared to MP. When total lamb mortality from birth to weaning was examined there was a 1.4 times greater probability of a lamb not surviving to weaning in group HP compared to MP. Ewe prolificacy level can effect on lamb birth weight and mortality levels however increasing prolificacy does significantly increase output.

Behavioural studies on Icelandic leadersheep

Ó.R. Dýrmundsson[1], E. Eythórsdóttir[2] and J.V. Jónmundsson[1]
[1]Farmers Associaton of Iceland, Bændahöllin, Hagatorg, 107, Iceland, [2]Agricultural University of Iceland, Keldnaholt, Reykjavík, 112, Iceland; ord@bondi.is

Icelandic leadersheep, a unique strain within the North European short-tailed Iceland breed, are known for their strongly inherited urge to walk or run in front of their flock. Leadersheep, normally kept in small numbers in each flock, are now found in just over 400 out of 2,200 sheep flocks in Iceland, numbering only 1,500 head out of the national sheep population of 475,000. Emphasis has been placed on conservation measures since the late 1950s, mainly through AI and individual recording. Although the difference between leadersheep selected for their outstanding behavioural abilities and intelligence, and other sheep of the same breed mainly selected for meat production characteristics, has been well known for centuries, experimental evidence acquired under controlled conditions has been lacking until recently. This paper presents the main results of a series of standardized trials in five sheep flocks so as to determine the willingness of ewes, rams and ewe lambs of the leadersheep strain to walk or run in front of groups of 'ordinary' sheep driven a certain distance from and to sheep houses known to them. The same two observers recorded behavioural events in all five trials and one of them was filmed. The trials were conducted in late October and early November of the same year, just before the onset of winterhousing. The results of the repeated tests were unequivocal in demonstrating the clearly pronounced and intrinsic leading instinct. Thus there was a perfect harmony in the behaviour pattern of all the leadersheep tested. It can be concluded that this first experimental evidence has substantiated previous knowledge of the unique characteristics attributed to Icelandic leadersheep.

Session 20a Theatre 6

Genetic effects on lamb survival traits

D. Francois[1], C. Raharivahoaka[1], F. Tortereau[1], J.P. Poivey[2] and L. Tiphine[3]
[1]INRA, GA, UR 631 SAGA CS 52627, 31326 Castanet Tolosan, France, [2]INRA, UMR 112 SELMET, 2 place Viala, 34398 Montpellier, France, [3]Institut de l'Elevage, MNE, 149 rue de Bercy, 75595 Paris cedex 12, France; dominique.francois@toulouse.inra.fr

Profitability of sheep industry is conditioned by the number of lambs produced by flocks. Both fertility and ewe prolificacy traits determines the number of born lambs. Then the survival rate influences the productivity of the flock. Survival traits are determined by different effects. Study of genetic effects on lamb survival has been conducted at two INRA experimental flocks composed of prolific Romanov ewes and of Romane ewes (former INRA-401 line=50% Romanov, 50% Berrichon du Cher). Survival was registered with codification of the mortality causes when possible. Survival data of 4,215 Romanov lambs and 22,428 Romane lambs has been studied at birth and from birth to 60 days of age on linear models with direct effects or with direct and maternal effects. Survival rate was 83.3% at birth for Romanov lambs and 69.5% at 60 days while in Romane the rate was 94.5% at birth and 87.4% at 60 days. Genetic parameters estimation was 0.047 direct heritability for survival at birth and 0.048 for maternal heritability in Romanov, as direct was higher (0.087) than maternal (0.006) in Romane. Romane maternal repeatability (c^2) was 0.101. From birth to the age of 60 days, direct heritability was 0.059 and maternal heritability was 0.074 in Romanov, as direct was higher (0.067) than maternal (0.020) in Romane. Romane maternal repeatability (c^2) was 0.060. In Romane, phenotypic correlation between survival at birth and birthweight (BW) was 0.182. Genetic correlation between survival direct effects and BW direct effects was -0.072 while that between survival direct effects and BW maternal effects was 0.020. At 60 days, phenotypic correlation between survival at 60 days and 30 days liveweight (LW) was 0.277. Genetic correlation between 60 d_survival direct effects and LW direct effects was -0.124 while that between 60 d_survival direct effects and LW maternal effects was 0.230.

Session 20a Theatre 7

Assessment protocol for measuring and monitoring sheep welfare in long distance transportation

E.N. Sossidou[1], S. Messori[2], B. Mounaix[3], W. Ouweltjes[4], C. Pedernera, [9995], V. Vousdouka[1] and H. Spoolder[4]
[1]Hellenic Agricultural Organization-Demeter, Veterinary Research Institute, Nagref Campus, 57001-Thermi, Thessaloniki, Greece, [2]Istituto G. Caporale, Torre di Cerrano, 64025-Pineto, Italy, [3]Institut de l'Elevage, 149, rue de Bercy, 75012 Paris, France, [4]Wageningen UR Livestock Research, Edelhertweg 15, 8200 AB Lelystad, the Netherlands, [5]IRTA, Institut de Recerca i Tecnologia Agroalimentàries, Passeig de Gràcia, 08007 Barcelona, Spain; sossidou.arig@nagref.gr

To date no protocol existed to objectively assess the welfare status of the sheep transported in long distance routes throughout Europe. This paper presents the protocol developed for the assessment of sheep welfare during long journeys in the framework of the project 'Development of an EU wide animal transport certification system and renovation of control posts in the European Union- SANCO/2011/GR/CRPA/SI2.610274'. Available literature together with the EC Regulations and the Welfare Quality ® protocols were used as base references. The protocol is divided into three parts: (1) animal based measures and management based measures recorded during unloading; (2) resource and transport parameters concerning the journey itself; and (3) a checklist for drivers on the assessment of fitness to travel at departure. This covers most of the adverse effects identified for the transport hazards and addresses 12 welfare criteria grouped into four main principles (good feeding, good housing, good health and appropriate behaviour). The protocol is currently being tested in practice both in Greece and in Italy to assess the welfare of sheep transported through the main routes of Europe. Preliminary results indicated that the main parameters affecting sheep welfare are overcrowding, handler's behaviour and deficiencies of truck equipment (such as ramps dimensions). It is also noted that temperature monitors and digital tachographs are often failing or absent.

Session 20a Theatre 8

Factors associated with Awassi lambs mortality in two production systems in Jordan

A. Abdelqader
The University of Jordan, Department of Animal Production, Faculty of Agriculture, Amman 11942, Jordan; a.abdelqader@ju.edu.jo

Less or no information is available on lamb mortality of Awassi sheep, a local breed in Jordan that is known as highly adaptive to the harsh desert conditions and arid environments. This study aimed at investigating factors affecting pre-weaning mortality in Awassi lambs reared in extensive and semi-extensive production systems. Data were recorded on 12,080 lambs descending from 20,133 ewes, born between October 2011 and January 2013. The study included 120 sheep flocks from the different agro-ecological zones in Jordan. Records of each flock were controlled by a commercial sheep flock monitoring system. The overall mean of lamb mortality from birth to weaning (at 60 days of age) was 19.3%. Most of the deaths (16.2%) occurred within the first week of life, while a much smaller percentage of lambs (3.1%) died from second week of life to weaning. The major causes of mortality were hypothermia and cold exposure (28.8%), diarrhoea and other digestive disorders (22.1%), starvation (9.1%) and respiratory problems (8.8%). The lamb losses were significantly ($P<0.05$) higher in females (11.6%) than in males (7.7%). Compared to lambs born in semi-extensive systems, lambs born in extensive system were 3.2, 2.7 and 2.8 times more likely to die due to diarrhoea, hypothermia and starvation, respectively. When compared single lambs, twins were 2.3 and 2.1 times more likely to die due to hypothermia and starvation, respectively. Lambs born from triples litters were under considerable high ($P<0.05$) risk to die due to all previous causes of death, compared to lambs born from single litter. The results suggest that large litter sizes are associated with low birth weight and increased mortality. Therefore it is recommended to avoid enhancement of large litter size, particularly in extensive production systems. These factors can be considered by sheep farmers to increase lamb survival rate. This study was funded by the Scientific Research Support Fund. Project No. Agr/1/03/2011.

Session 20a Poster 9

Effects of partial suckling technique on growth and behaviour of Sarda replacement ewe lambs

S.P.G. Rassu[1], C. Carzedda[1], G. Battacone[1], A. Mazza[1], I. Dimitrov[2] and G. Pulina[1]
[1]University of Sassari, Dipartimento di Agraria, viale Italia 39, 07100 Sassari, Italy, [2]Agricultural Institute, 6000 Stara Zagora, Bulgaria; battacon@uniss.it

The effect of partial suckling management on lamb growth and behaviour was studied. From day 5 postpartum, 22 Sarda female lambs were divided into two groups: 11 lambs (TS group) followed the dams outdoors and 11 lambs (PS group) were housed indoors separated from their dams, during the grazing time. Lambs from both groups stayed indoors with their dams the rest of the time. PS lambs were supplemented with a concentrate to compensate for the lower milk availability. An Arena test (180 seconds/lamb) was performed at weaning to evaluate the effects of breeding method on lamb behaviour, by measuring their level of confidence in stockperson. Behaviour data were analysed by ANOVA. Lamb body weight (BW) and average daily gain (ADG) were measured and analysed by GLM, considering group, sampling and their interaction as fixed effects. Lamb groups did not differ for BW at the start of the trial and at weaning, therefore ADG was similar (TS 241 g/d and PS 231 g/d; P=0.288). However, PS lambs suffered in the first week of treatment, when they had lower growth than TS lambs (196 vs. 397 g/d; P<0.001), due to low adaptation to lower milk availability and concentrate intake. Milk removed from the mothers was higher in ewes of PS lambs than in those of TS lambs (641 vs. 103 g/head/d; P<0.001). Of the 6 lambs per group that sniffed the stockperson, latency time towards him was significantly lower in PS lambs (58 vs. 113 s; P=0.048). In conclusion, partial suckling does not affect the growth of lambs, but seems to improve the way the lambs approach stockperson and favors commercial milk production from ewes during suckling period. Research funded by MiPAAF (OIGA-Belat project)

Session 20a Poster 10

Identity profiles related to behavioural, morphologic, physiological and cognitive features in goats

M. Pascual-Alonso[1], G.A. Maria[1], W.S. Sepúlveda[1], L. Aguayo-Ulloa[1], M. Villarroel[2] and G.C. Miranda-De La Lama[1]
[1]University of Zaragoza, Animal Production and Food Science, Miguel Servet, 177, 50013 Zaragoza, Spain, [2] E.T.S.I.A. Polytechnic University of Madrid, Department of Animal Science, Av Ramiro de Maeztu, 7, 28040 Madrid, Spain; mpascual@unizar.es

The aim of this study was to analyse morphological, physiological and cognitive features in goats in order to understand their social strategies. Social interactions of 33 goats were recorded over a period of 16 days for 96 h. Blood samples and morphological measures were taken from each goat and they were put through a T-maze test. Seven variables of social interactions were analysed using the factor test, providing the 'avoider', the 'non-agonistic' and the 'agonistic' factors. Subsequently, a hierarchical cluster analysis was performed in order to identify groups of similar animals which could help to explain the possible association between social strategies and index of success, social and individual behaviour and the characteristics studied. The results suggest the existence of four identity profiles which were termed 'aggressive', 'affiliative', 'passive' and 'avoider'. The aggressive profile included dominant animals with high values for the agonistic factor. The affiliative profile consisted of half dominant animals with high values for the non-agonistic factor. The avoider profile included low dominance animals, with high values for the avoider factor. Finally, the passive profile included low dominance animals which had negative values for all three social strategies. In conclusion the identity profiles comprise behavioural, physiological, morphological and cognitive features, associated with social strategies to create, adjust and use a series of behavioural solutions to adapt to the productive environment. Studies of identity profiles in farm animals could have implications for selecting for traits that confer adaptive advantages under specific production conditions.

Session 20a Poster 11

Effect of stocking density on blood parameters and meat quality of commercial lambs during transport

B. Teke[1], B. Ekiz[2], F. Akdag[1], M. Ugurlu[1] and G. Ciftci[3]

[1]Ondokuz Mayis University Faculty of Veterinary Medicine, Department of Animal Breeding and Husbandry, 55200 Atakum, Samsun, Turkey, [2]Istanbul University Faculty of Veterinary Medicine, Department of Animal Breeding and Husbandry, 34320 Avcılar, Istanbul, Turkey, [3]Ondokuz Mayis University Faculty of Veterinary Medicine, Department of Biochemistry, 55200 Atakum, Samsun, Turkey; bulentteke@gmail.com

One of the most important factors influencing welfare during transport is stocking density. Few studies are available about the effects of stocking density during short distance lamb transportation. This study was aimed to determine the effects of stocking density on some blood parameters and meat quality characteristics during short distance transportation. Fifty five Karayaka lambs were divided into two groups, one of which had high stocking density (HD) and the other had normal stocking density (ND). The half of the same vehicle was loaded at HD (0.20 m^2/lamb) and the remaining was loaded at ND (0.27 m^2/lamb). The distance was 130 km and it took 2 hours 15 min. The blood samples were taken from both groups before and after transportation. The lambs were slaughtered 1 hour after they were unloaded. In order to determine the effects of stocking density on blood parameters and meat quality characteristics, paired t-test was performed. Glucose, LDH, CK and ALT values of lambs of HD group were higher than lambs of ND group and difference between groups was significant ($P<0.05$; $P<0.01$). pH_{ult} values of the lambs in HD and ND groups were 5.55 and 5.56, respectively and difference between groups was not significant ($P>0.05$). In the study, the effects of stocking density on meat quality characteristics (L*, a*, b*, C*, H*, WHC, Cooking loss and WBSF) was not significant ($P>0.05$). In conclusion, the meat quality characteristics of the lambs transported with stocking densities 0.20 and 0.27 m^2/lamb along 130 km were not affected negatively, while blood parameters of the lambs transported with high stocking density were affected negatively.

Session 20a Poster 12

Effects of different lairage durations on some blood parameters and meat quality of lambs

B. Teke[1], B. Ekiz[2], F. Akdag[1], M. Ugurlu[1] and G. Ciftci[3]

[1]Ondokuz Mayis University Faculty of Veterinary Medicine, Department of Animal Breeding and Husbandry, 55200 Atakum, Samsun, Turkey, [2]Istanbul University Faculty of Veterinary Medicine, Department of Animal Breeding and Husbandry, 34320 Avcılar, Istanbul, Turkey, [3]Ondokuz Mayis University Faculty of Veterinary Medicine, Department of Biochemistry, 55200 Atakum, Samsun, Turkey; bulentteke@gmail.com

Lairage is a common process applied to animals before slaughter to recover from transportation stress. Few studies are available about optimum lairage duration of lambs after short distance transportation with rough road conditions. This study was aimed to determine the effects of different lairage durations (0, 2 and 4 h) after short distance transportation on some blood parameters and meat quality characteristics. In this study 25 Karayaka lambs were transported for 30 km at rough road condition. The journey took approximately 30 min. The lambs were divided into three groups (L0, L2 and L4), according to their lairage durations. Blood samples were taken after lairage period and then they were slaughtered. In order to determine the effects of lairage time on blood parameters and meat quality characteristics, one way ANOVA was performed. In the study, the effect of lairage duration on Glucose, LDH, CK and ALT values and on pH_0, pH_{45} and pH_{ult} were not significant ($P>0.05$). The difference between the values a^*_{1d} (15.74; 14.47; 13.90) ($P<0.01$), c^*_{1d} (16.83; 15.68; 15.08) ($P<0.01$), c^*_{3d} (17.81; 16.26; 14.78) ($P<0.001$) were significant. As the lairage duration increased, some meat quality characteristics changed negatively while blood parameters didn't change after short distance transportation. In conclusion, waiting before slaughter is not recommended after short distance transportation in order to prevent the adverse effects of transportation on meat quality.

Breeding value estimation for fertility in Swiss sheep breeds

A. Burren[1], C. Hagger[1], C. Aeschlimann[2], G. Schmutz[2] and H. Jörg[1]
[1]Bern University of Applied Sciences, School of Agricultural, Forest and Food Sciences, Länggasse 85, 3052 Zollikofen, Switzerland, [2]Swiss Sheep Breeding Association, Industriestrasse 9, 3362 Niederönz, Switzerland; alexander.burren@bfh.ch

The Swiss Sheep Breeding Association made breeding values for weight gain between birth and 40 days since 2009 for the Swiss White Alpine sheep, the Brown Headed Meat sheep, the Swiss Black-Brown Mountain sheep and the Valais Blacknose sheep. It is a goal to extend breeding value estimation also for fertility traits in these breeds. We evaluated 19 traits for fertility in these four breeds. The analyses were performed on herd book data from the last ten years and variance components were estimated. The following four traits were selected for breeding value estimation: age at first lambing, lambing interval, litter size 1 and litter size 2. The litter size 1 and litter size 2 are the number of lambs at 40 days for the first and second lambing, respectively. The heritability estimates for age at first lambing ranged from 0.17 to 0.28 in these four breeds. The ones for lambing interval, litter size 1 and litter size 2 ranged from 0.05 to 0.10. The values correspond to the known values and are comparable between the breeds. Among the selected traits there aren't any traits for weights, since they are involved in the breeding values for weight gain. The litter size at 40 days compared to litter size at birth was able to take in account loses of lambs during the first days of life. Litter sizes of lambing 3 and further lambings had a high genetic correlation to litter size 2 and were therefore not used directly for breeding value estimation. The four breeding values for fertility in combination with the breeding values for weight gain serve as a basis for a breeding index with breed specific weightings.

French sheep-for-meat production: state of the art and perspectives for sustainable farming systems

I. Sneessens[1], G. Brunschwig[2] and M. Benoit[1]
[1]INRA, EGEE, Route de Theix, 63122 Saint-Genès-Champanelle, France, [2]VetAgro Sup, Sybel, Avenue de l'Europe, 63370 Lempdes, France; ines.sneessens@clermont.inra.fr

Over the last thirty years, French sheep-for-meat production has decreased by 50%. Nowadays, with a self-sufficiency rate of 46%, French production is much smaller than domestic consumer demand. Despite recent improvements in profitability, long run sustainability remains uncertain in view of the expected climatic, economic and environmental challenges. We will focus on production systems from plainlands areas, where the competition with other agricultural sectors is toughest and exacerbates the need for innovative strategies, apt to safeguard this production. The study of the evolution of French sheep-for-meat production systems in France allows one to uncover the main strategies that may enhance long term economic and environmental performances: high numerical productivity and low dependence on external inputs. These two features are shown to be influenced by the livestock management system, itself largely constrained by the socio-economic and political context. For instance, low self-sufficiency systems with autumn lambing are promoted by cooperatives in order to enhance French sheep meat competitiveness. Improving the sustainability of these systems can thus be interesting too. It is also shown that joint production of crop and livestock should improve self-sufficiency of sheep-for-meat production systems. However, the effects of crop-livestock integration on the performance of the entire farming system are not so well-known. We propose a framework to better characterize and understand these effects. Our objective is to build a simulator that enables us to analyze the correlation between production strate-gies, economic and environmental performances and their relative sensitivity to various economic situations. The tested scenarios will be based on the above mentioned farm features that a priori seem to improve sustainability of sheep-for-meat production systems in plainland areas.

Session 20b Theatre 2

The role of sheep and goat breeds' value chains in the rural development in the Mediterranean Region

I. Tzouramani[1], E. Sossidou[2], F. Casabianca[3], G. Hadjipavlou[4], A. Araba[5] and C. Ligda[2]
[1]Agricultural Economics and Policy Research Institute, Ilisia, 115 28, Athens, Greece, [2]Veterinary Research Institute, P.O. Box 62272, 57 001 Thessaloniki, Greece, [3]INRA, SAD LRDE, Quartier Grossetti, 20 250 Corte, France, [4]Agricultural Research Institute, P.O. Box 22016, 1516 Lefkosia, Cyprus, [5]IAV II Hassan, BP 6202, 101 01 Rabat, Morocco; chligda@otenet.gr

Previous studies have shown that the incorporation of innovative methods in the traditional sheep and goat production systems and the commercialization of their products can increase the profitability of the sector. Therefore, in the frame of the DoMEsTIc project (EU FP7 ARIMNet) the economical aspects of these systems are addressed. More specifically, in four participating regions of the partner countries (Greece, France, Cyprus and Morocco), information related with the distribution of the products through the value chain and the role of the different stakeholders were collected by semi-structured personal interviews with representatives from local actors. The analysis of the value chains aims to identify the different actors, assess the governance mechanisms in the value chain, analyze the opportunities for upgrading within the chain by different chain actors and distribute the gains along the chain. This approach provides a framework to analyze the nature and determinants of smallholder competitiveness in market chain and to determine core points for designing and implementing appropriate development practices and policies. In this context, the institutional environment is also studied, with reference to the legal and regulatory environment, the policies and their impact on the value chain. All types of the supply chain were included in the survey form local/ informal and semi-regulated to formal supply chains. This work provides applicable ways to improve and propose alternative policies and optimal interventions in the value chain of local sheep and goat breeds.

Session 21a Theatre 1

Feedipedia: an open access international encyclopedia on feed resources for farm animals

V. Heuzé[1], G. Tran[1], D. Bastianelli[2], H. Archimède[3] and D. Sauvant[1,4,5]
[1]AFZ (Association Française de Zootechnie), 16, rue Claude Bernard, 75231 Paris Cedex 05, France, [2]CIRAD, SELMET, Campus de Baillarguet, 34398 Montpellier Cedex 5, France, [3]INRA, UR143 URZ, Prise d'Eau, 97170 Petit-Bourg, France, [4]AgroParisTech, 16, rue Claude Bernard, 75231 Paris Cedex 05, France, [5]INRA, UMR MoSAR, 16, rue Claude Bernard, 75231 Paris Cedex 05, France; valerie.heuze@zootechnie.fr

The demand for livestock products has been growing steadily in emerging and developing countries and with it the need for information about animal feeds. However, users of these countries often have to resort to feed data that are either obsolete or from temperate countries. The Feedipedia program led by INRA, CIRAD, FAO and AFZ aims to create an updated and comprehensive encyclopedia comprised of more than 600 datasheets on fodders and raw materials. The datasheets provide information such as physical descriptions, feed availability, potential constraints of use and environmental impact, as well as feeding recommendations and nutritional values for the main species of farm animals. The first goal of the project is to better identify and characterize the local feed resources in order to improve the technical and economic performance of farms. Nutrition modelling, collaborations between research teams and identification of gaps in knowledge are part of the scientific objectives. The datasheets are created by a group of 25 scientists and engineers, who rely on a massive collection of scientific literature and experimental data to write qualitative and quantitative reviews (via methods such as meta-analysis) and build representative and consistent tables of composition and nutritional values. The Feedipedia website opened in October 2012 with 200 updated datasheets and more than 600 tables of composition and nutritive value. Reception of the website has been very positive and Feedipedia is becoming a go-to reference for technical and scientific information about feed resources.

Pre-processing of animal feed data: an essential step

F. Maroto-Molina[1], A. Gómez-Cabrera[1], J.E. Guerrero-Ginel[1], A. Garrido-Varo[1], D. Sauvant[2], G. Tran[2], V. Heuzé[2] and D.C. Pérez-Marín[1]
[1]Universidad de Córdoba, Ctra. Nacional IV km 396, 14014 Córdoba, Spain, [2]AgroParisTech, 16 rue Claude Bernard, 75005 Paris, France; gilles.tran@zootechnie.fr

Data about the nutritional aspects of feed are systematically obtained in laboratories. Some initiatives are collecting them into large databases, but this information is highly variable, so data pre-processing is essential to ensure reliable results. The objective of this paper is to evaluate the performance of various methods regarding 2 aspects of pre-processing: outliers and missing data. A database containing nutritional data on roughly 18,000 alfalfa samples was used. Several methods, both univariate and multivariate, were examined for detecting outliers: Z-score, Chauvenet's, regression residuals, principal components, adjusted Wilks' and Local Outlier Factor. Detected outliers were traced and characterized. Various methods for handling missing data were tested as well, both deletion methods (listwise and pairwise) and imputation methods, univariate (mean substitution) and multivariate (regression imputation, Expectation-Maximization and Data Augmentation). Such evaluation was based in the comparison of the outputs obtained from a reference complete dataset with the outputs obtained from 4 simulated incomplete datasets: 2 types of missing data (at random and not at random) × 2 loss intensities (1/3 and 2/3). There were important differences in the number and type of outliers detected by different methods. Such tests did not allow discrimination between outliers and errors, but a heuristic approach based on several methods enabled certain recurring error patterns to be identified. This information could be used to design ad hoc routines for error detection. Relating missing data, imputation methods were found to perform better than deletion methods, both in terms of maximizing information use and minimizing bias. The importance of the differences among the evaluated methods depended on the type of missing data.

Development of infrared reflectance spectroscopy databases for efficient livestock managements

V. Decruyenaere[1], M. Boval[2], P. Dardenne[1], N. Gengler[3] and P. Lecomte[4]
[1]Walloon Agricultural Research Centre, CRAW, 5030 Gembloux, Belgium, [2]INRA French West Indies, Tropical Animal Production Unit (URZ), Domaine Duclos, 97170 Guadeloupe, France, [3]ULg, Gembloux Agro Bio Tech, 5030 Gembloux, Belgium, [4]CIRAD, UMR SELMET, 34398 Montpellier, France; decruyenaere@cra.wallonie.be

There is a strong demand towards livestock for high quality products and services while limiting the impact on the environment. That goes for different management systems as outdoor for which consumer expectation is high. Decision Support Tools (DST) are therefore needed to achieve various food production and other agricultural resources, as manure. This synthesis aim is to highlight the potential of Near and Mid-Infrared Reflectance Spectroscopy (NIRS and MIRS) as tool for enhancing the value of livestock systems. NIRS and MIRS are non-destructive technologies that estimate simultaneously several parameters as chemical composition, nutritive value of various products, feeding characteristics (digestibility, intake), animal physiological status (pregnancy), detection of metabolic disorders (acidosis, mastitis) or enteric greenhouse gas emission (GES). A lot of products can be analysed by IRS: forages and feeds, effluents, faeces, milk, meat. All spectral information can be used in qualitative and quantitative ways to develop DST able to improve livestock system sustainability, herd management in regard to diet and animal welfare, to select efficient animals and this, for various animal species. Many developments are underway to merge databases of several research centres. So, NIRS is used to predict forage chemical composition; to estimate, from faeces analysis, parameters reflecting diet feeding value, in tropical and temperate areas. MIRS analysis of milk is another example of pooling databases for dairy system management. Prediction of new parameters as GES production from milk analysis and performances of livestock grazing on natural areas are studied. These examples illustrate the potential of NIRS and MIRS for the development of effective DST.

Session 21a Theatre 4

The Swiss feed data base: a GIS-based analysis platform

A. Bracher[1], P. Schlegel[1], M. Böhlen[2], F. Cafagna[2] and A. Taliun[2]
[1]Research Station Agroscope Liebefeld-Posieux, Postfach 64, 1725 Posieux, Switzerland, [2]University of Zürich, Department of Informatics, Binzmühlenstr. a4, 8050 Zürich, Switzerland; patrick.schlegel@agroscope.admin.ch

The Swiss Feed Data Warehouse (www.feed-alp.admin.ch/feedbase) is a public service for farmers, consultants, feed industry and governmental and research institutions that provides detailed and up-to-date information about multi-species feeding values for both, raw materials and roughage. Three data sources contribute to the data pool: research activities, surveys and feed industry. At the technical level, the key challenges to tackle are the processing of irregularly collected, incomplete, multi-granular and unclassified data. Over 600 feed types are defined which are searchable on an aggregated level or on an individual sample basis. The nutrient measurements are enriched with geographical, temporal, biological and technical information. We propose a solution that offers a fast, effective and intuitive approach to query and analyze large amounts of high dimensional data. An interactive web-application enables dynamic query construction with multiple charts to visualize the spatio-temporal variability of feed data. As a novel approach, geo-referenced sample origins and corresponding nutrient contents are transformed using two-dimensional Kernel density estimation and regression. The density value of given coordinates is then converted into a color value which is displayed as a map overlay representing either local sample density or alternatively local nutrient density. This technique allows the detection of patterns in feed quality. The data set based on the yearly hay survey gives evidence for spatial quality patterns. Particularly carbohydrate and mineral content of hay correlate with altitude, region and local animal density. Historical data contained in the scatter plot visualizes possible time trends. Our goal is to provide all the facts needed to fine-tune animal feeding at the farm level which contributes to high resource use efficiency.

Session 21a Theatre 5

The French feed database: an efficient private-public partnership

G. Tran
AFZ (Association Française de Zootechnie), 16 rue Claude Bernard, 75231 Paris Cedex 05, France; gilles.tran@zootechnie.fr

The French Feed Database is an information and expertise centre on the quality of feed materials, managed by the French Association for Animal Production (AFZ). Created in 1989, it is supported by a network of 16 member organisations including feed manufacturers, producers of raw materials, R&D institutes and trade organisations. Its core mission is to disseminate information about the chemical composition and nutritive value of feedstuffs. The database stores 1.9 million composition and nutritional data for 2,500 feed materials and 800 parameters on more than 440,000 samples. One major feature of the French Feed Database is that most of the data are provided by its members, who choose the nature, quantity and periodicity of the data they want to share. While the mostly private origin of the data imposes certain constraints on their nature (mostly chemical data), acquisition (delay) and dissemination (confidentiality), this gives the French Feed Database a unique overview of the nature and quality of feed ingredients actually used in feed manufacturing. Data from public sources (INRA, CIRAD and international literature) are present in smaller numbers in the database but are extremely valuable as they are usually derived from bioavailability assays. All the data are registered with the relevant information such as geographical origins, providers, methods of analysis, literature references, etc. The main users of the database are its members, who have a privileged access to several tools (desktop database and on-line database) and services (requests for raw or processed data, literature reviews, consultancy in feed data management). For more than two decades, the French Feed Database has been serving the needs of the feed sector, by providing technical data to feed professionals, by assisting private organisations to develop feed data management systems and by helping public institutions to disseminate feed information (INRA-AFZ tables, Feedipedia...) and to take informed policy decisions about animal feeds.

Authentication of ruminant meat, milk and cheese produced in grassland based production systems

S. Prache and B. Martin
Institut National de la Recherche Agronomique, Physiologie Animale et Systèmes d'Elevage, INRA Theix, 63122 Saint-Genes-Champanelle, France; bruno.martin@clermont.inra.fr

A number of factors have contributed to research interest in authentication of ruminant products from grassland based production systems. They include the increasing consumer demand for assurance about mode of production because of several food scares, the evidence that pasture-feeding can impart beneficial effects on meat and milk from a nutritional perspective and the interest in production practices which are environmentally sustainable. To meet these interests, producers and retailers develop specifications via quality certifications. There is therefore a need for analytical tools that may guarantee that the specification commitments have been fully met or to help with constructing them. We review the different approaches that have been investigated, some leading examples concerning the discrimination of contrasting feeding situations, together with the persistence of some diet markers in the event of changes in animals' diet. The nature of the diet strongly influences the composition of the animal tissues and products, which is due to specific compounds that are directly transferred from the feed to the end product or that are transformed or produced by rumen micro-organisms or the animal's metabolism under the effect of specific diets. Some of these compounds can therefore be used as diet markers, such as carotenoids, terpens, phenolic compounds, fatty acids, and ratios of stable isotope. Moreover, differences in meat and milk composition induce differences in their optical properties, and therefore in their spectral features, which can also be used for diet authentication. These techniques have already allowed discriminating among products obtained in contrasting feeding conditions (diets based on pasture vs. preserved forages or concentrates). Intermediate situations, such as modifications of the animal's diets or diets composed of different forages, may be less easily recognized and may require a combination of tracing methods.

Session 21b Theatre 2

Effect of fattening practices on meat quality from Maine-Anjou Protected Denomination culled cows

S. Couvreur[1], G. Le Bec[1], D. Micol[2] and B. Picard[2]
[1]LUNAM Université Groupe ESA, UR Systèmes d'Elevage, 55 rue Rabelais, 49007 Angers, France, [2]INRA VetAgroSup, UMR1213 Herbivores, Theix, 63122 Saint-Genès-Champanelle, France; s.couvreur@groupe-esa.com

Our objective was to assess if the diversity of fattening practices allowed by specifications affect meat quality from Maine Anjou (MA) PDO culled cows. Longissimus thoracis (LT) and Rectus abdominis (RA) muscles were sampled in an industrial slaughterhouse at 24 h post-mortem on 97 MAPDO culled cows. Physico-chemical traits were measured out on both muscles and sensory quality traits on RA muscle. Fattening practices were individually recorded. A clustering analysis and a general linear model were led to create and assess the effects of fattening groups on meat quality. Five fattening groups were defined: Long fattening period (142 d) based on hay/haylage + 5.8 kg/d concentrate (LgF, n=17); Short fattening period (80 d) based on hay + 7.3 kg/d concentrate (HF, n=26); average fattening period (107 d) based on hay with 9.7 kg/d concentrate (CcF, n=18); Short fattening period (80 d) based on haylage and 9.1 kg/d concentrate (HlF, n=15); Short fattening period (86 d) based on pasture + 7.3 kg/d concentrate (PtF, n=21). LT: Compared to other groups, PtF tended to have higher a* and b* color index and IIA fiber proportions (62.1 vs. 51.7 to 56.3%) at the expense of IIX fibers (7.0 vs. 10.9 to 14.6%); and CcF showed higher fat/muscle ratio (38.0 vs. 29.1 to 30.9%) without any differences in IMF content. RA: fattening practices did not modify IMF content, color, fiber type and size, and enzymes. Compared to other groups, CcF tended to have lower total and insoluble collagen contents and improved tenderness (4.9 vs. 4.5 to 4.6/10). Juiciness and bovine flavors were higher for LgF group compared to others (3.6 and 4.7 vs. 3.1 to 3.3 and 4.3 to 4.4, respectively). The diversity of fattening practices can modify muscle physico-chemical traits and meat sensory quality in interaction with muscle type. It opened new possibilities to modify diet specifications.

Session 21b — Theatre 3

The effect of lysine restriction in grower phase on carcass and meat quality of heavy pigs

J. Suárez-Belloch, J.A. Guada and M.A. Latorre
IUCA, Universidad de Zaragoza, 50013 Zaragoza, Spain; jsuarezbelloch@gmail.com

A total of 200 Duroc × (Landrace × Large White) pigs, 50% barrows and 50% gilts, of 26.3±0.50 kg and 73±3 d of age, were used to study the influence of dietary lysine content during the growing period on carcass characteristics and meat quality. Animals were allotted by sex and weight to 40 pens, with 5 pigs per pen, and these allocated to 4 dietary treatments. For the growing period (45 days), experimental diets contained 3.260 Kcal ME/kg and 1.1, 0.91; 0.78 and 0.52% Lys (24.0, 19.3, 16.2 and 14.9% CP, respectively). For the finishing period (until 123.0±2.35 kg BW), all pigs were fed a common diet. At the slaughterhouse, a total of 160 carcasses (4 per pen) were used to study the carcass characteristics and 80 of them (2 per pen) were used to study the quality of the longissimus dorsi muscle (n=20 and 10, respectively). Carcass weight was not affected by sex or dietary treatment (P>0.10). Carcasses from gilts had lower fat depth between the 3rd and 4th ribs (P<0.01) but were longer (P<0.001) than those from barrows although the ham length and circumference were similar for both sexes (P>0.10). Also, gilts had higher loin yield than barrows (P<0.01) but no differences were found in shoulder or ham percentage (P>0.10). The dietary Lys restriction at early age increased linearly the fat thickness (P<0.05) without affecting carcass or ham size (P>0.10). Also, the shoulder proportion decreased linearly as Lys was reduced in the grower diet but no effect was observed on loin or ham yield (P>0.10). Sex did not affect meat composition (P>0.10) but Lys restriction decreased linearly the protein content (P<0.05) and tended to increase linearly the intramuscular fat content (P=0.07) of meat. The shear force value was not affected by sex or dietary treatment (P>0.10). In conclusion, a decrease of dietary Lys content from 1.1 to 0.52% during the growing period improved some carcass and meat characteristics which are desirable in pigs intended for dry-cured ham production.

Session 21b — Theatre 4

Improving winter milk fatty acid profile by linseed supplementation to conventional and organic cows

S. Stergiadis[1], C. Leifert[1], M.D. Eyre[1], H. Steinshamn[2] and G. Butler[1]
[1]Newcastle University, Nafferton Ecological Farming Group, School of Agriculture Food and Rural Development, Nafferton Farm, Stocksfield, NE43 7XD, Northumberland, United Kingdom, [2]Norwegian Institute for Agricultural and Environmental Research, Bioforsk, Gunnars veg 6, 6630 Ting, Norway; sokratis.stergiadis@ncl.ac.uk

Many studies show considerable changes in milk fatty acid (FA) profile between summer and winter. This study investigated the impact of linseed supplementation of winter diets on milk FA profiles in both organic and conventional herds. Two herds (conventional, organic) were divided into two groups of 20 animals, receiving two different diets (control, linseed-2 kg/cow per day) over a 6-weeks period, with milk sampled on three occasions. Analysis of variance was performed by linear mixed effects models in R, using 'management', 'diet' and 'sampling date' as fixed factors and individual cow as random factor. Milk FA profiling was carried out by gas chromatography. Cows in the (1) organic herd and (2) linseed group produced milk with higher (P<0.001) concentrations of nutritionally beneficial individual FA (vaccenic; 47.0% and +85.1%, α-linolenic; +72.1% and +67.4%, and rumenic; +39.7% and 55.9%) and FA groups (monounsaturated FA; +15.9% and +27.7%, polyunsaturated FA; +41.5% and +41.1%, and omega-3 FA +53.1% and +85.4%) and lower concentrations of saturated FA (-8.6% and -12.3%) when compared with conventional system and control diets respectively. Beneficial eicosapentaenoic was higher under organic than conventional management (+24.9%) but decreased when cows ate linseed rather than control diets (-36.1%). Although both herds responded to supplementation, those fed organic diets (with grass clover silage and slightly higher forage content) showed a greater response (P<0.05). The consequences of linseed on milk fat quality appear dependant on the basal diets and in this study the beneficial impact of the organic feeding and linseed were complimentary.

Modifying milk composition trough dairy cow's diet to improve human nutrition and health

M.S.V. Salles[1], A. Saran Netto[2], L.C. Roma Junior[1], M.A. Zanetti[2], T.S. Samóra[1] and F.A. Salles[1]
[1]APTA, Department of Agriculture and Supply of São Paulo State, Av Bandeirantes, 2419 Ribeirão Preto, 14030-670, Brazil, [2]FZEA, The University of São Paulo, Av. Duque de Caxias Norte, 225, Campus da USP, Pirassununga, SP, CEP 13635-900, Brazil; marciasalles@apta.sp.gov.br

Health nutrition is a challenge for modern man and a preoccupation of most of the world's population, therefore the importance of animal science projects in conjunction with nutrition and human health to obtain a better milk nutrient composition. The aimed was to study the effect of sunflower oil as a source of fat to improve the fatty acid profile of milk, combined with the effects of the antioxidants vitamin E and selenium added to the diet of lactating cows on milk composition, productivity and health. Thirty-two lactating Jersey cows from APTA/Brazil, during 75 days in a randomized block design (early and mid lactation) were allocated in four treatments: C (control); C + A (2,5 mg of organic selenium + 2,000 IU of vitamin E/day); O (4% of sunflower oil in DM diet); O + A (4% of sunflower oil in DM diet + 2,5 mg of organic selenium + 2,000 IU of vitamin E/day. Cows were fed with a total mixed ration of: 0.50 of concentrate, 0.42 of corn silage and 0.08 of coast-cross hay (DM). The daily milk yield and milk composition parameters were taken once a week. Data were analyzed using SAS PROC GLM. Milk production and milk fat, protein, lactose, and non-fat solids did not differ among treatments. Cows supplemented with C + A had lower DMI at 30 days (11.55, 9.92, 10.82 and 10.01 kg/day for C, C + A, O, O + A, P=0.08). The remaining data of selenium, vitamin E, fatty acid composition of milk and blood as well as health parameters of the cows are still being analyzed. The milk produced from this project was offered to elderly and evaluated by medical researchers from USP/Brazil. Preliminary data suggest that the supplementation of these nutrients in the diets of lactating cows provided the same milk production with a lower food intake. Financial support by FAPESP.

Canola oil, organic selenium and vitamin E in steers rations and Se levels in animals and humans

M. Zanetti, L. Correa, A. Saran Netto, S. Garcia and J. Cunha
University of Sao Paulo, FZEA, Animal Science, Rua Duque de Caxias Norte, 225, 13 633-000 Pirassununga, SP, Brazil; mzanetti@usp.br

The objective of this research was to study the effect of canola oil, organic selenium and vitamin E inclusion in feedlot steers ration, upon Se and vitamin E serum and meat animal's levels and the effect of meat Se on human serum Se levels in the people that ate the fortified meat. Forty eight Nellore steers were allocated in four treatments (twelve animals per treatments), in individual pens: C (control); C + Antioxidants (2, 5 mg of organic selenium/kg of DM + 1000 IU of vitamin E/day); Oil (3% of canola oil in DM diet); Oil + Antioxidants (3% of canola in DM diet + 2.5 mg of organic selenium in DM diet + 1000 IU of vitamin E/ day). The experimental period lasted 82 days. In the weeks 0, 4, 8 and 12 the steer blood was sampled for vitamin E and selenium analysis. At the end of the trial the animals were slaughtered for the assessment of vitamin E and selenium in the meat. Part of the meat was offered to humans during 90 days to study the effect of selenium and vitamin E in the meat on human blood serum. Statistical analysis was for a completely randomized design using the mixed procedure (SAS). The selenium supplementation in steers ration increased ($P<0.01$) the selenium serum levels and the selenium meat levels ($P<0.001$). The vitamin E was not yet analyzed. In the supplemented steer meat, the selenium increased from 39.3 μg/kg in the control group to 667 μg/kg in the group that received selenium, vitamin E and canola oil in the ration. At the end of the trial, the selenium blood levels in the steers were 42 μg/ml for the control and the oil groups, 105 μg/ml for the antioxidant group and 103 μg/ml for the antioxidant plus the oil group. The selenium in the meat increased the human serum selenium ($P<0.05$). It was concluded that the selenium supplementation during 82 was enough to increase significantly the selenium in the steers serum and meat. People who ate meat with selenium during 45 days had higher serum selenium levels.

Session 21b Poster 7

Effect of lysine restriction in grower phase on growth performance and carcass fatness of heavy pigs

J. Suárez-Belloch, J.A. Guada and M.A. Latorre
IUCA, Universidad de Zaragoza, 50013 Zaragoza, Spain; jsuarezbelloch@gmail.com

A total of 200 Duroc × (Landrace × Large White) pigs, 50% barrows and 50% gilts, 26.3±0.55 kg BW and 73±3 d of age, were used to study the influence of dietary lysine (Lys) content during growing phase on growth performance and carcass fatness. There were 8 treatments with 2 sexes and 4 levels of total Lys which were replicated 5 times in a randomized block design with 5 pigs per pen as the experimental unit. For the growing period (45 days), diets were based on corn, barley, wheat and soybean meal and contained 3,260 kcal ME/kg and 1.1, 0.91, 0.78 and 0.52% Lys (24.0, 19.3, 16.2 and 14.9% CP, respectively). For the finishing period (until 123.0±2.35 kg BW), all pigs were fed a common diet. During the growing phase, Lys restriction reduced average daily gain (ADG) ($P<0.0001$) and average daily feed intake (ADFI) ($P<0.01$) and increased feed conversion ratio (FCR) ($P<0.0001$), being all effects quadratic. No significant interaction sex × Lys content was detected but barrows grew faster than gilts ($P<0.05$). During the finishing phase, the previous Lys restriction carried out linearly a compensatory growth ($P<0.0001$) associated to a higher ADFI ($P<0.10$) and a lower FCR ($P<0.01$) particularly for the two lower Lys levels. Barrows grew faster than gilts and ate more feed ($P<0.0001$) independently of dietary treatment. For the overall trial, ADG showed a quadratic reduction ($P<0.05$) as Lys content decreased in the grower diet while FCR increased ($P<0.001$) for pigs fed 0.78 and 0.52% Lys. Also, barrows had higher ADG and ADFI than gilts ($P<0.0001$) irrespective Lys level. Although no effect of sex or Lys was detected on carcass yield, fat thickness at gluteus medius muscle was wider in carcasses from barrows ($P<0.001$) and increased as Lys content was reduced ($P<0.01$). It can be concluded that 0.91% Lys during the growing period optimized the growth performance variables and increased carcass fatness which is desirable in heavy pigs intended for dry-cured products.

Session 22a Theatre 1

Assessing the multiple services provided by livestock: a French case-study

J. Ryschawy[1], C. Disenhaus[2], S. Bertrand[3], G. Allaire[4], C. Aubert[5], O. Aznar[6], C. Guinot[7], E. Josien[6], J. Lasseur[8], S. Plantureux[9], E. Tchakérian[10] and M. Tichit[1]
[1]INRA, UMR SAD-APT, 75321 Paris, France, [2]Agrocampus-Ouest, UMR PEGASE, 35042 Rennes, France, [3]CNIEL, 75314 Paris, France, [4]INRA, US ODR, 31324 Toulouse, France, [5]ITAVI, 22440 Ploufragan, France, [6]VetAgro Sup, 63370 Lempdes, France, [7]Interbev, 75587 Paris, France, [8]INRA, UR ECODEV, 84914 Avignon, France, [9]INRA, UMR LAE, 54500 Nancy, France, [10]Idele, 34397 Montpellier, France; julie.ryschawy@agroparistech.fr

Even if its contribution to food security is recognized worldwide, livestock production is often criticized for its negative impacts on environment and nature. The role of livestock in soil fertility or as a creator of employment is poorly dealt with. There is a pressing need to fill that knowledge gap with regards to the multiple services provided by livestock to human society. The objective of this study was to develop a framework for the assessment and the recognition of the social, economical and ecological services provided by livestock. Combining expert knowledge and literature review, we first defined the set of services provided by livestock. We then selected indicators to quantify each service. We finally assessed services and their relationships (synergies or tradeoffs) on a gradient of French regions. Four main classes of services were revealed: provisioning (e.g. food quantity and quality), environmental quality (e.g. biodiversity, diversified landscape), territorial vitality (e.g. employment, rural dynamism) and cultural heritage (e.g. gastronomy, landscape quality). The analysis of the spatial distribution of services showed a nonrandom distribution; some services were regularly matched or not. The main ecological, geographical and socio-economic determinants were identified. Further work will include semi-directive surveys to connect the services provided by livestock to the expectations of end-users. Our results should contribute to inform policy makers and society about these services and the relationships existing among them.

Farmers and citizens perceptions of ecosystem services and sustainability of mountain farming

A. Bernués[1], R. Ripoll-Bosch[2], T. Rodríguez-Ortega[2] and I. Casasús[2]

[1]Department of Animal and Aquacultural Sciences, Norwegian University of Life Sciences (UMB), P.O. Box 5003, 1432 Ås, Norway, [2]Centro de Investigación y Tecnología Agroalimentaria de Aragón (CITA), Avda. Montañana 930, 50059 Zaragoza, Spain; trodriguezo@cita-aragon.es

Public goods derived from agro-ecosystems need to be valued for better design of agri-environmental policies and payments for ecosystem services (ES). In this context, 5 focus groups (2 with farmers and 3 with citizens; n=33) were organized in north-eastern Spain to gain information on the spontaneous knowledge and perceptions of farmers and citizens about ES provided by mountain agriculture. Discussions were guided according to 5 general questions, lasted around 1.5 hours, were video recorded and transcripts were written for text analysis. Items appearing in texts were organized according to the type of ES they referred to (provisioning, regulation, habitat, cultural) or other sustainability issues (farm economics, family/ social issues, socio-economic context, policy/regulatory context). The ES considered more important by participants were (number of times mentioned in descending order): aesthetic (landscape/ vegetation); gene pool protection (biodiversity maintenance); disturbance prevention (forest fires); lifecycle maintenance (nutrient cycling, photosynthesis); raw materials (firewood, forage); water purification/ waste management; spiritual experience; recreation/ tourism; soil fertility/ erosion prevention; and culture/ art. Other important sustainability issues were (in descending order): ethics of food production, farming abandonment and rural development, product quality, wildlife conflicts and agricultural policy. Differences between farmers and citizens were observed: farmers gave more importance issues relating their own farming activity or local circumstances (provisioning and regulating ES; policy and legal context), whereas citizens showed in general more global concerns (cultural ES; socio-economic context, in particular ethical concerns on food production).

A framework for the assessment of the global biodiversity performances of livestock production

F. Teillard[1,2] and P. Gerber[2]

[1]INRA-AgroParisTech, UMR 1048 SADAPT, 16 rue Claude Bernard, 75005 Paris, France, [2]FAO, Animal Production and Health Division, Viale delle Terme di Caracalla, 00153 Rome, Italy; felix.teillard@fao.org

Livestock production is facing a challenge: satisfy an increasing demand for animal products while improving its environmental sustainability. Widely recognized quantitative assessments of its environmental performances are pivotal to reveal improvement options. Most existing assessments focused on greenhouse gases emissions; yet, the environmental impacts of livestock production are not restricted to this component. As a major user of land resources, livestock have a strong impact on biodiversity which has not been quantified so far. This study aimed at providing a framework for quantifying the impacts of livestock production on biodiversity at global scale. An extensive review of the methods, indicators and data addressing the relationships between livestock production and biodiversity was conducted. We propose a framework linking: (1) indicators of pressures on biodiversity related to livestock activities on farm, upstream and downstream; to (2) indicators of the state of biodiversity. We identified key drivers of impacts and showed that they differed among livestock sectors, production systems and climates; therefore, the set of relevant pressure indicators also varied among these categories. Importantly, pressure indicators also considered that livestock can have a positive influence on biodiversity. State indicators should ideally be applicable at global or continental scale. Several indicators fulfilling this requirements were found, including species richness, abundances, population trends and metrics based on remotely sensed vegetation data. Several focus case studies around the globe could be used to perform a statistical analysis of the relationships between pressure and state indicators. We discuss options for the improvement of biodiversity performances and their potential conflicts with mitigation options already proposed in the context of greenhouse gases emissions.

Assessment of gradual adaptations of a low input mixed farming system for improved sustainability

M. Marie[1,2], J.-L. Fiorelli[1] and J.-M. Trommenschlager[1]
[1]INRA, ASTER-Mirecourt, 662 Avenue Louis Buffet, 88500 Mirecourt, France, [2]Université de Lorraine, ENSAIA, 2 Avenue de la Forêt de Haye, TSA 40602, 54518 Vandoeuvre lès Nancy cedex, France; michel.marie@mirecourt.inra.fr

A dairy mixed farming system, established since 2004 as organic and low input, has been progressively mended by introduction of practices designed for improving its autonomy and sustainability. These changes concern the diversity of crops and intercrops (mixture of wheat cultivars, introduction of more rustic cereals such as triticale and rye, cereals-proteaginous legumes associations, and multipurpose intercrops), cultivation technics (reduced ploughing, direct sowing, adapted weeding machines), and grassland management and herd feeding (reduced alfalfa grazing, but rather use for hay production, dairy cows mixed feeding at early lactation, improved calving grouping). The system has been assessed for 2004-2006 (establishment of the system), 2007-2009 (introduction of first changes) and 2010-2012 (all changes introduced) periods by a quantitative multicriteria method covering environmental (abiotic: P losses and availability, NO_3 and NH_3 emissions, greenhouse gases production, organic matter, energy, and biotic: ecologic regulation areas, ground beetles abundance, cultural management, vegetal diversity), social (external: products quality, landscape management, employment, services, and internal: work hardness, hygiene and security, decisional autonomy) and economic (viability, financial independency, transferability, economical efficiency) components of sustainability. The results were compared to a qualitative assessment based on expert knowledge, showing that the changes had a positive effect on crops, livestock and farmers decisional autonomy, as well as on agronomic quality, livestock management and production, farm economy and supply chain development, at the expense of work load. As a consequence, the system appears thus to be more suited to adapt to climate or market fluctuations.

Free-range pigs foraging on Jerusalem artichokes

A.G. Kongsted, K. Horsted and J.E. Hermansen
Aarhus University, Agroecology, Blichers Allé 20, 8830 Tjele, Denmark; anneg.kongsted@agrsci.dk

Free-range pig production in Northern Europe is characterized by high inputs of concentrate on grassland. This increases risk of nutrient leaching, increases feed costs and puts a pressure on land resources. Pigs' unique ability to find a part of their food directly in the field where they are kept should be taken into consideration. One below ground field forage characterized by very high yields and with great potential as fodder source is Jerusalem Artichokes (JA) tubers. The nutritional contribution from free-range foraging, growth, feed conversion and behaviour were investigated in 36 growing pigs foraging on JA and fed concentrates restrictedly (30% of energy recommendations) or *ad libitum*. Behavioural observations were carried out weekly over the entire experimental period of 40 days. The average daily consumption of concentrate was 51 MJ and 11 MJ ME/pig for pigs fed *ad libitum* and restrictedly, respectively. Compared to the *ad libitum* fed pigs, the pigs fed restrictedly had a significant lower daily gain (560 vs. 1,224 g/pig), improved feed conversion ratio (17.6 vs. 42.8 MJ ME concentrate/kg live weight gain) and spent more time foraging JA tubers (7.9 vs. 1.1%). Body conditions were comparable between the two treatments. It is estimated that pigs fed restrictedly found approximately 60% of their energy requirement from foraging in the range and consumed 1.3 kg to 1.6 kg DM of JA/pig per day. The results indicate good possibilities for substituting a large proportion of concentrates with home-grown JA tubers biological harvested by foraging pigs. Future studies are needed to reveal the most appropriate concentrate feeding regime when combined with foraging JA in relation to consequences for growth and feed conversion, but also for meat quality, animal health and nutrient balances.

Conditioned aversion to vines for grazing sheep in vineyards

C.L. Manuelian[1], E. Albanell[1], M. Rovai[1], A.A.K. Salama[1,2] and G. Caja[1]
[1]Group of Ruminant Research (G2R), Universitat Autònoma de Barcelona, 08193 Bellaterra, Spain, [2]Animal Production Research Institute, Dokki, 12311 Giza, Egypt; gerardo.caja@uab.es

Sheep grazing is a sustainable and environmentally friendly alternative to the traditional vineyard weed control. However, vineyard grazing has its drawbacks; sheep are attracted by grape leaves and sprouts. Sheep grazing usually damages vines, compromising grape quantity and quality. With this in mind, 2 mid-term experiments were conducted with 12 Manchega and 12 Lacaune ewes, consisting on: exp. (1) Aversion induction to grape leaves (novel food) and persistence evaluation under simulated grazing conditions; and exp. (2) Persistence validation in a commercial vineyard. In exp. 1, ewes were allocated into 4 groups (6 ewes/group and breed) in which grape leaves intake was measured after dosing LiCl (AV, 225 mg/kg BW) or water (C, control). Aversion was created after offering individually 100 g of grape leaves (var. Tempranillo) in the barn for 30 min, and orally giving a single LiCl dose post-consumption; validation was done for 3 consecutive days. Aversion persistence on a simulated vineyard (2 kg leaves in wood frames on an ryegrass prairie) was tested during 30 min in 11 sessions (d 5 to 375). Control ewes avidly ate the grape leaves in the barn for 4 d (90.0±4.9 g/d and ewe) and in the simulated vineyard (1.6±0.1 kg/group). On the contrary, AV ewes fully rejected the leaves and sprouts in both cases. In Exp. 2, AV ewes were moved to a commercial vineyard (var. Tempranillo) in the Penedes county (Spain) and allowed to rotationally graze, according to grass availability, for 3 h/d during 10 d (d 401 to 410). AV ewes reduced 70% grass cover between vine lines but, they started to bite leaves and sprouts when grass was scarce (d 403 to 410); no significant damage in the vines was appreciated. In conclusion, aversion to vines persisted in the AV ewes for 1 yr but, in practice, the use of a reinforcing LiCl dose after this time is recommended for assuring an effective aversion behavior in grazing sheep.

Exploring mitigation potential of GHG emissions from livestock farming systems at the global level

A. Mottet, C. Opio and P. Gerber
FAO, AFGAL, Viale delle Terme di Caracalla, 00153 Rome, Italy; anne.mottet@fao.org

Livestock farming systems (LFS) need to mitigate their environmental impact while answering a growing demand. The sector is an important consumer of natural resources, water, land and nutrients and a larger emitter of greenhouse gas (GHG). Its growth will need to reconcile with the increasing scarcity of natural resources and the need to address climate change. But significant reductions can be achieved. This analysis aims at exploring mitigation scenarios of GHG emissions from different LFS at the global level. This analysis is based on a life cycle assessment approach using the Global Livestock Environmental Assessment Model developed by FAO. This model quantifies GHG emissions arising from production of meat and milk from cattle, sheep, goats and buffalo; meat from pigs; and meat and eggs from chickens. The model calculates total emissions and production for a given LFS within a defined area, from which the emissions per unit of product can be calculated for combinations of different commodities/LFS/locations at different spatial scales. Total emissions from the livestock sector in 2005 were estimated at 7.1 GT of CO_2e, in line with previous estimates. Results also confirm the relative contributions of different species and LFS, cattle production systems being the main contributors, and the main sources: feed production, ruminants' digestion and manure management and storage. It is estimated that the sector's emissions can be reduced by approximately 30% thanks to a much wider appliance of existing technologies and practices that generate efficiency gains. This include feed production, animal husbandry and manure management technologies and good practices used by the most efficient 10 to 25% of production units, but also technologies which are now only marginally used such as biogas or energy saving devices. This study explores 5 different mitigation scenarios, assessing the impact of various packages of options for different species, LFS and regions of the world.

Session 22a Poster 8

Farmers objectives and environmental concerns in Aurland coastal mountains, Norway

B.L. Hylland[1], L.O. Eik[1], M. Clemetsen[2], F. Alfnes[3] and A. Bernués[1]
[1]Norwegian University of Life Sciences (UMB), Dept. of Animal and Aquacultural Sciences, P.O. Box 5003, 1432 Ås, Norway, [2]Dept. of Landscape Architecture and Spatial Planning, P.O. Box 5003, 1432 Ås, Norway, [3]UMB School of Economics and Business, P.O. Box 5003, 1432 Ås, Norway; alberto.bernues@umb.no

Small-scale sheep and goat farming in Norway is an important activity in remote areas located in north-alpine and coastal (fjord) mountains. This activity represents a minor and decreasing contribution to the local economy, which is largely dependent on tourism. However, sheep and goat farming is essential to maintain cultural landscape and provides other public goods (cultural heritage, quality food products). Twenty-seven farms (48% of total) in Aurland were surveyed with a face to face questionnaire collecting data on farming recent changes, prospects for changes in the future, farmers objectives and opinions, and farmers perceptions on relationships between farming and the environment. Five objectives got an average figure of 4 or more in a 5-points Likert scale of importance (in decreasing order): give farm in good conditions for children, increase quality life of family, have a good relationship with neighbours, improve product quality and improve/renew facilities. Other important objectives were: minimise production costs, be more environmentally friendly and reduce workload. Six opinions about farming got an average figure of 4 or more in the degree of agreement: connect animal farming to landscape and nature, proud of being a farmer, society needs to know more about farming, subsidies are essential for animal farming, updated information to take good decisions and use more grazing. Open questions on positive impacts of farming on the environment resulted in the following aspects: traditional/beautiful landscape, open landscape/vegetation, rural development and economic activity, local food/quality, use of local resources, cultural heritage and social issues, maintenance of soils and biodiversity.

Session 22a Poster 9

Cattle transhumance to summer farms: milk yield, pasture management and biodiversity conservation

E. Sturaro, G. Bittante, L. Marini and M. Ramanzin
University of Padova, DAFNAE, viale dell'università 16, 35020 Legnaro (PD), Italy; enrico.sturaro@unipd.it

This study aimed to analyze the interactions between cattle husbandry, pastures management and arthropods biodiversity in the summer farms of Trento province (Italian Alps). Twenty-one summer farms used by dairy cattle were sampled during the summer 2012. Data collected were: utilized pasture surface (UPS, georeferenced on GIS), herd size and composition (breed and livestock categories), supplementary feeding (quantity and quality), milk yield and quality. In addition, in each pasture butterflies and grasshoppers were sampled and counted, and weed encroachment was estimated at the end of the summer season. Average UPS was 82.8 ha (SD=70.5), average elevation 1,687 m asl (SD=307) and average slope 16.7° (SD=4.6). The average herd size was 65 Livestock Units (LU) (SD=34), with a prevalence of Brown Swiss and Simmental breeds. All the farms with cows on milk used supplementary feeding, with an average of 4.0 kg of concentrates/cow per day (SD=1.2). The stocking rates varied widely from 0.4 to 2.1 LU/ha (average: 1.1 LU/ha). The amount of supplementary feeding was positively correlated with milk yield; for both these variables no correlations were found with stocking rate, weed encroachment and insect biodiversity indexes. Pastures used only by heifers showed higher levels of weed encroachment than those used by dairy cows. Insect biodiversity indexes were weakly influenced by differences in average stocking rate and weed encroachment between pastures, but within pastures increased with increasing distance from the farms buildings. In conclusion, livestock productivity seems more supported by supplementary feeding than by pastures management and therefore cannot be directly related to pastures conservation, as indexed by weed encroachment, and to insect biodiversity.

Session 22a Poster 10

Prospects, objectives and opinions of livestock farmers in the area of a Pyrenean ski resort

I. Casasús, J.A. Rodríguez-Sánchez and A. Sanz
Centro de Investigación y Tecnología Agroalimentaria del Gobierno de Aragón, Avda. Montañana 930, 50059 Zaragoza, Spain; icasasus@aragon.es

Tourism plays a major role in socio-economic development of some European mountain areas, where it may provide some synergies but also compete with farming activities for the use of resources such as land or labour. In order to determine the effect of a ski resort on livestock farming systems in its surroundings, a structured interview was conducted with all farmers whose cattle or horse herds grazed during the summer on pastures within a ski station in the Pyrenees (Northern Spain). Information was gathered on their production systems and prospective changes under the current socio-economic circumstances and different potential scenarios, namely total decoupling of EU subsidies, increased income from agri-environmental schemes, reduced income from off-farm work, and increased price of purchased feedstuffs. They were also asked about the relative importance for them of different technical, economic and social objectives, and also their opinions about the future, tourism and environmental issues. The continuity of the farms was ensured on a medium term, but might be low in the long run due to lack of succession. Farms were stable in terms of size and management, and few changes were envisaged in the future in the current conditions. However, if the socio-economic situation changed or different agricultural policies were implemented, farm structure and technical management may be considerably modified. Concerning the relative importance of various objectives, they considered that economic aspects and those related to their family's quality of life were crucial, while technical objectives were less important. They considered that the ski resort had been beneficial for the valley population, and also indirectly for them and their farming activity, mostly because it provided alternatives to diversify their income. In turn, they were aware that the ski resort profited from the ecosystem services provided by livestock grazing in its area.

Session 22b Theatre 1

From local to global: mental models of local people about livestock sector

J.F. Tourrand[1], C. Barnaud[2], P. Valarie[1], T. Bonaudo[3], A. Ickowicz[1], C.H. Moulin[4], L. Dobremez[5] and B. Dedieu[2]
[1]CIRAD, 2, p. Pierre Viala, 34070 Montpellier, France, [2]INRA, SAD, Theix, 63122 Saint Genes Champanelle, France, [3]Agroparistech, 16 rue Claude Bernard, 75005 Paris, France, [4]SupagroM, 2 place Pierre Viala, 34070 Montpellier, France, [5]Irstea, Territoires, domaine universitaire, 38000 Saint Martin d'Heres, France; tourrand@aol.com

MOUVE project and the LIFLOD network are developing a research focused on the better understanding of the mental models about livestock at local scale, with the goal to report the results at global scale. The authors present the results based on a set of 13 sites selected according to: (1) the interest of the local team; (2) significance of the local context, and its representation of the diversity at global scale. The sites are located in South America, Europe, North and West Africa, South-East Asia and China. Data collection used interviews with local people involve in livestock sector: farmers, traders, unions, agro-industries, development agencies, local governance or regional policymakers, NGO representatives. The collected information concerns: functions of livestock, points of view on past, current situation and future of livestock sector in the area, main factors of change, themes of debate at local scale, livestock farming systems in the future and position on environmental issues. Results relates to: (1) the diversity of the functions of livestock; (2) the diversity of the mental models between the sites linked to the local contexts and the contrasts between the sites; (3) the similarities between the sites, more particularly between the groups of local people (i.e agro-industries) but not within farmers; (4) common representations: environmental sector has a critical position about the livestock impacts, livestock subsector usually wants to improve the farming productivity, and local governance tries to find a consensual positions. The results give a great overview about the diversity of the mental models, especially the difference into and between the sites and the groups of local people.

Session 22b — Theatre 2

Tradition and innovation in Caribbean LFS: old rhum in new barrels

G. Alexandre[1], J. Factum-Sainton[2] and V. Angeon[3]
[1]INRA, URZ Antilles-Guyane, Petit-Bourg, Guadeloupe, [2]UAG-CRREF, Morne Ferret, Abymes, Guadeloupe, [3]UAG-CEREGMIA, Campus Schoelcher, Schoelcher, Martinique; gisele.alexandre@antilles.inra.fr

There is a growing concern about the lack of adoption of certain technologies at the farm level. It can be hypothesized that the context, as a whole, may influence the successful use of innovative technology. The transfer of technology policy is criticized, particularly in tropical regions, for its inadequacy to the socio-cultural context of the LFS. Meanwhile, livestock keepers have steadily accumulated indigenous experiences that have built sustainable LFS. Some of them have helped designing innovations, showing synergistic interactions between traditional and modern knowledge:- using the traditional male effect practice for small ruminant (SR), is studied as an efficient reproduction management alternative increasing the herd productivity without any hormonal treatment;- feeding animals with by-products in crop-livestock system is empowered through the agro-industrial sector that deliver pellets combining these diverse sources owing to research on animal nutrition and technological process;- determining, through an ethnoveterinary enquiry, the bio-resources employed to cure the animals, has lead to biochemical and parasitological studies in order to enhance the use of neutrociticals for soft health control in SR; managing the land and pasture resources through traditional tethering practices has become a very relevant and valid experimental tool to assess individual intake at pasture. Other traditional practices, relevant to the local farmers in officious context, are rejected in an official context. While exogenous innovations are supported by the formal sector but do not last. Interdisciplinary research and participatory action approaches are needed to introduce innovations embedded in the biotechnical and human context. But we need to analyze this appropriation-rejection process in terms of human values (identifying the cognitive and historical sources).

Session 22b — Theatre 3

Evaluating innovative scenarios in partnership to enhance mixed crop-livestock farms sustainability

J. Ryschawy[1], A. Joannon[2], J.P. Choisis[1], A. Gibon[1] and P.Y. Le Gal[3]
[1] INRA, UMR Dynafor, 31324 Castanet-Tolosan, France, [2]INRA, UR SAD Paysage, 35042 Rennes, France, [3]CIRAD, UMR Innovation, 34398 Montpellier, France; julie.ryschawy@agroparistech.fr

Mixed crop-livestock farms are regaining interest worldwide as a way to reduce environmental problems while allowing a productive and economically viable agriculture. However, these farms have been regressing in Europe. This study aims at evaluating scenarios including technical innovations that could enhance sustainability of mixed crop-livestock farms. Scenarios were defined through a participatory process with farmers and other stakeholders of the Coteaux de Gascogne, a French unfavoured area where crop-livestock farms still exist. Technical innovations were selected in line with two farmer adaptative strategies that allowed mixed crop-cattle farms to last over the long term: 'maximizing farm autonomy' and 'diversifying productions'. We adapted the whole farm simulation tool CLIFS (Crop Livestock Farm Simulator) to evaluate each scenario along with two farmers. CLIFS calculates feed and manure balances, based on characteristics of animals and crops systems within the farm, and economic results. Innovative scenario 'maximizing farm autonomy' was based on sowing forage legumes between two cash crops to achieve autonomy for herd feeding while maintaining soil fertility. Innovative scenario 'diversifying productions' was based on fattening and selling heifers in a short chain. Implementing these innovations allowed an increase of the overall gross margin per hectare of cultivated area by respectively 8.9% and 17.1%. These two scenarios did not offset the drastic shocks resulting from two contrasted political and economic futures: (1) heightening of the current globalisation trends; and (2) political and market incentives for a relocation of production and consumption. Political support would be needed to maintain mixed crop-livestock farms. This approach allows to strongly involve local actors through collective brainstorming on future adaptative strategies.

Session 22b Theatre 4

Farm processing and short chains: solutions to territorialise livestock farming in mountain areas?

S. Cournut[1], M. Millet[2] and A. Dufour[3]
[1]VetAgro Sup, UMR Métafort, 89 av de l'Europe, 63370 Lempdes, France, [2]INRA, LRDE, Quartier Grossetti BP 8, 20250 Corte, France, [3]Isara Lyon, Aster, 23 rue Jean Baldassini, 69364 Lyon, France; sylvie.cournut@vetagro-sup.fr

In France, in mountain regions where the soil and climate conditions limit the productivity of livestock farms, farm processing and short chain distribution are solutions proposed by the stakeholders to maintain livestock farming in their territory. This is the case in Livradois-Forez, a rural territory in the east of the Massif Central, where some people see this evolution of livestock farming as being an alternative to enlarging the farm or as a way of making establishment outside the family framework easier in a context where land is fragmented and heterogeneous. It is also a possible option when collecting milk or meat is no longer possible because it is too expensive (mountainous area). To understand how this dynamic is expressed at farm level (functioning and link with the territory), we interviewed 16 cattle farmers in the Livradois-Forez region, who process and market their animal products in short distribution chains. We show that this activity does not fulfil the same aims nor does it have the same impacts on how the farm operates; it depends on whether it is part of the establishment project, or if it follows conversion to organic farming or accompanies the installation of a spouse. For all the farmers interviewed, more importance is given to the use of natural resources, and the livestock farmers' discussion networks are widened thanks to links with consumers and with other people necessary to their activity (tourism, local authorities, grocery stores). Although marketing in short distribution chains only very rarely concerns the whole of the farm's livestock production, it has a strong influence on practices and modifies the link with the territory of the farm.

Session 22b Theatre 5

Combined used of three whole farm simulation tools for the design of innovative production strategie

A.W. Sempore, N. Andrieu and P.Y. Le Gal
CIRDES, N°559, rue 5-31 angle avenue du Gouverneur Louveau; 01 BP 454 Bobo-Dioulasso, 454 Bobo-Dioulasso, Burkina Faso; pierre-yves.le_gal@cirad.fr

Sub-Saharan Africa is facing a sharp increase in the number of cattle. This is associated with the increase in the cultivated area that has led to a higher pressure on natural rangelands, a reduction in the availability of fodder for animals, and consequently a decrease in animal production. In this context farmers have to find strategies allowing improving livestock production systems through better planning of farming activities. The objective of this paper is to assess the complementarity between three applications at the farm scale to help farmers developing innovative production strategies. This study was conducted in the western zone of Burkina Faso with 9 farmers planning to improve their production strategy. After a first training phase of farmers on modeling, we used sequentially three models: a linear programming model (Optimcikɛda), a spreadsheet model simulating the technico-economical performances of the farm (Cikɛda), and a rule-based model simulating the impact of different strategies on long-term performances of the farm (Simflex). The linear programming model was first used with the farmer in order to assess the distance between his project and the optimal allocation of resources allowing maximizing farm income, calculated by the model. Cikɛda model was then used with the farmer to simulate the impact on the performance of the farm of strategies that rose from the previous discussion process. Simflex was used to assess the sensitivity of the strategy selected by the farmer to the variability of his climatic and economic environment. Optimcikɛda model allowed farmers to design their cropping pattern while Cikɛda helped refining the plan under various technical and economic assessments. Simflex allowed farmers to improve their decision thresholds in relation to climate risk and the variability of input prices. The articulation of three whole-farm modeling tools is a novel method for the conception of innovative strategies.

Change of cattle breed: dairy specialized farmers' motivations for Montbeliarde or Simmental breeds

C. Gaillard[1], A. Gérard[1], S. Mugnier[1], M. Courdier[1], S. Moureaux[2] and E. Verrier[2]
[1]UMR Métafort, BP 87999, 21079 Dijon Cedex, France, [2]UMR 1313 GABI, Allée de Vilvert, 78352 Jouy-en-Josas, France; c.gaillard@agrosupdijon.fr

Montbeliarde and Simmental breeds are two dairy cattle breeds from Eastern France. Since the last decades, their cattle number went on increasing, out of their historical nucleus, especially towards Western France. Thus, in this region, some farmers choose to change their cattle breed, from the Holstein breed to less specialized breeds, namely Montbeliarde or Simmental breeds. This study focuses on the trajectories of such farms to understand the motivations and conditions of herd evolution. For that, we surveyed 40 dairy farmers who have introduced more than 25% of Montbeliarde or Simmental cows in their herds (n=20 with Montbeliarde, n=20 with Simmental). The interview related the origin of breed change, the interests of farmers and described farm management. By using multivariate analyses, a farm classification was realized according to motivations, farm trajectories and substitution level in herd. Four groups were identified: (1) more sustainable systems defined by low inputs and less maize in cows diet, with complete substitution of herds breed over a long period; (2) advantage of better functional traits of new breeds to face limits of Holstein cows, especially mastitis or low fertility, and varied levels of herd substitution; (3) dual purpose qualities well-adapted to milk and beef production with partial or total substitution of breed on a short time; (4) opportunity of change on a short time and only partial substitution of herd. Finally, two major situations were pointed out to explain Montbeliarde or Simmental breeds integration: (1) because of better functional traits, it was seen as an opportunity of improving herd management, facing to technical difficulties; and (2) because of dual purpose abilities and hardy animals, breed change occurred at the same time as farming system evolution as it allowed adaptations to more sustainable systems.

Mediterranean biodiversity as a tool for the sustainable development of the small ruminant sector

C. Ligda[1], E. Sossidou[1], A. Araba[2], A. Lauvie[3] and G. Hadjipavlou[4]
[1]Veterinary Research Institute, P.O. Box 62272, 57 001 Thessaloniki, Greece, [2]IAV Hassan II, BP 6202, 101 01 Rabat, Morocco, [3]LRDE, INRA SAD, Quartier Grossetti, 20 250 Corte, France, [4]Agricultural Research Institute, P.O. Box 22016, 1516 Lefkosia, Cyprus; chligda@otenet.gr

The DoMEsTIc project, funded under EU FP7 ARIMNet, aims to enhance our knowledge on pastoral and rangeland sheep and goats production systems. Case studies from Greece, Cyprus, France and Morocco are analysed through field surveys, focusing on livestock farming systems, the genetic management of the breeds, and economical aspects. Surveys are conducted, with personal interviews of farmers, on the basis of a detailed questionnaire. In Greece, the field work is carried out in Epirus with a significant sheep and goat sector that faces various challenges towards a sustainable future. In Cyprus, the survey includes the two main local breeds, the fat-tailed sheep and the Machaeras goat, and the locally adapted Chios sheep and Damascus goats. Sheep and goats are an important part of livestock production, but at present, the sector does not fulfill its full potential. Sheep production in Corsica is mainly organized around milk production, with the dominant Corsican breed being characterized by the production of the PDO Brocciu cheese, while a PGI is pursued for the lamb meat. Extensive dairy sheep and goat breeding, is a distinctive characteristic of regional identity, and despite a marked decline, remains as the only 'productive' activity maintaining the mountainous areas in the island. In Morocco, the surveys are conducted in the province of Boulemane, characterized by the diversity of its ecosystems. In the region, three production systems are identified: sylvopastoral, agro-pastoral and oasianery. Comparative analysis will lead to assessment of associations between the structure of the farming systems, farmer practices and the characteristics of the sheep and goat breeds with the sector's resilience, competitiveness and overall sustainability.

Conservation of Walloon poultry diversity sustained by short chains

M. Moerman[1], S. Dufourny[1], N. Moula[2] and J. Wavreille[1]
[1]Walloon Agricultural Research Center, 8 rue de Liroux, 5030 Gembloux, Belgium, [2]Faculty of Veterniary Medicine, Univ. of Liege, 20 blvd de Colonster, 4000 Liège, Belgium; s.dufourny@cra.wallonie.be

The project of conservation of Walloon poultry breeds is started by the Walloon Agricultural Research Center in 2011 with financial support of SPW-DGARNE. This work is justified by the Belgian wealth in poultry breeds and the critical status of most of them. Interventions are therefore requested to maintain these breeds connected to cultural value, possessing a great genetic diversity and being a source of diversification for agricultural farms through direct sale of local products. First the situation of poultry breeds livestock in Wallonia was conducted through surveys and interviews. A SWOT analysis has characterized the sector, the risk status of each breed was estimated i.a. based on the population sizes and the number of breeders (FAO's method). The result shows that 27% of the breeds are under critical status, 65% in danger and 9% under vulnerable status. Sheets resuming the history and characteristics of breeds were drafted. These documents have provided necessary informations to choose the breeds to prioritize for conservation namely: phylogenic links, geographic distribution, authenticity and potential of valuation. Six breeds were selected: the Bassette, the Brabançonne, the Fauve de Hesbaye, the Herve, the Naine Belge and the Naine du Tournaisis. In situ conservation as a network of breeders with rotary coupling scheme has been planned. Once the choice of breeds and device settled, the genetic characterization has started. Blood samples of 10 to 20 unrelated individuals per variety of each breed are analyzed with 29 microsatellite markers. Phenotypic measures will complete these data. At the same time, the networks of breeders will be settled. They will be officially recognized by Wallonia and will produce individuals compliant with poultry standards maintaining a low inbreeding level. The individuals will be drained to educational/organic farms and to the general public.

Influences of local policies and opportunities on farmers' strategies and grassland management

G. Martel[1], F. Herzog[2] and O. Huguenin-Elie[2]
[1]INRA, SAD-Paysage, 65, rue de Saint Brieuc, 35000 Rennes, France, [2]Agroscope, Reckenholzstrasse 191, 8046 Zürich, Switzerland; gilles.martel@rennes.inra.fr

In mountainous landscapes most of the agricultural surfaces are pastures, which point out the importance of heterogeneous management to maintain agricultural landscape complexity. Little research has focused on possible drivers of management heterogeneity of grasslands in mountainous areas others than natural factors. The production strategy at farm level as well as local policies and opportunities might nevertheless deeply affect plot management and its heterogeneity. In order to investigate the role of these factors we studied three areas within similar natural contexts (folded Jura), but differing in terms of political and opportunity contexts (Doubs department in France, and Cantons of Vaud and Neuchâtel in Switzerland). The areas differ on cheeses AOC productions, on farm land classification and on ecological policies. We surveyed 33 farmers to collect their strategies and practices on herd and plots. A multifactorial analysis with hierarchical clustering identified 16 types of plot management and 9 types of production strategy at farm level. Three types of strategies were found in all areas, whereas two strategies are specific to the canton of Vaud, one to France and one to the canton of Neuchâtel. The specificity of strategies is mainly explained by the AOC policies (e.g. Gruyère policies on distribution of milk production over the year forbid the 'French' strategy), but also by occasional opportunities for marketed products and by local dynamics of agriculture. We also found that most of the strategies induce one or two over/under represented plot management types (e.g. 'no use in fall' plot management is overrepresented in the 'French' strategy) and so affect grassland management at the plot scale. Thus, this study points out a clear link from policies and opportunities to plot management in mountainous landscapes, which highlights several levers that may increase the diversity of plot management.

Session 22b Poster 10

Between local and global: changing in interactions concerning dairy factories: LFS-territory

M. Napoléone[1], C. Corniaux[2], F. Alavoine-Mornas[3], V. Barritaux[4], J.P. Boutonnet[1], S. Carvalho[5], S. Cournut[6], A. Havet[7], M. Houdart[4], A. Ickowicz[2], S. Madelrieux[3], H. Morales[8], R. Poccard[2] and J.F. Tourrand[2]

[1]INRA, SAD Selmet, Supagro Montpellier, 34000, France, Metropolitan, [2]Cirad, ES Selmet, Supagro, Montpellier, 34000, France, [3]IRSTEA, St Martin d'hières, 38402, France, [4]IRSTEA, UMR 1273, Aubières, 63172, France, Metropolitan, [5]EMBRAPA, Guama, Belem, Brazil, [6]VetAgro Sup, UMR 1273, Clermont-Ferrand, 63000, France, Metropolitan, [7]INRA, SAD, Thiverval Grignon, 78850, France, Metropolitan, [8]IPABA, Montevideo, 3802, Uruguay; martine.napoleone@supagro.inra.fr

The changes affecting farming, operating between local and global, are important concerning milk processing. How these changes express themselves locally? What consequences concerning strategies of processing firms and concerning livestock farming systems? Developing diachronic studies we analyze interactions between marketing strategies of milk processing factories, livestock farming systems and consequences of these changes at local scales. Comparing contrasted situations (France, South America, Vietnam and Senegal), we found mostly two contrasted dynamics: (1) a process of globalisation (concentration of operators, increasing importance of long food chains, location of farms in high productivity areas, intensification of farming practices); (2) a process of strengthening relationship between product and local specificity (emphasizing local image and the link between product and the 'terroir', handicraft processes of transformation), increasing of local marketing channels, farming practices valorizing a diversity of local resources. Intermediating between production and distribution, dairy factories are at the core of these two dynamics. We will present the methodologies employed to characterize the studied situations. From examples we will illustrate how these two dynamics are emerging and operate. We will discuss then how in a same place complementarity, tensions or exclusions may appear between these two dynamics.

Session 22b Poster 11

Determinants of crop rotation choices by pig farmers in Brittany

G. Martel[1], E. Tersiguel[2], J.-L. Giteau[2] and Y. Ramonet[2]

[1]INRA, SAD-Paysage, 65, rue de Saint Brieuc, 35000 Rennes, France, [2]Chambre d'agriculture de Bretagne, 4, avenue du Chalutier Sans Pitié, 22195 Plérin, France; gilles.martel@rennes.inra.fr

Pig farms in Brittany have on average 65 ha of cultivated area (UAA). Although these farms can represent the main land use of an area, there is very little data on how pig farmers use their UAA and on the links between crop rotations and production system. The aim of this study is to review the diversity of crop rotations of pig farms and to identify factors influencing their choice. A survey was carried out in 28 swine farms in Brittany. The survey sample is heterogeneous in size (number of animals and UAA), in effluent management and in existence of a feed production plan on farm (FPP). Content analysis of surveys shows that crop rotation are related to several structural criteria (farm feed supply, effluent management, plot pattern), but also to economical, agronomical and workforce characteristics. A principal component analysis with hierarchical clustering brings out three types of crop rotation: (1) one mainly composed of wheat and maize (80% of the UAA), found in specialized farms with FPP; (2) one more diverse with forage crops and mainly found in mixed herds of cattle and pigs; (3) one with a greater presence of oilseed rape and a variety of grains but less maize, and corresponding to specialized pig farms without FPP. Our study therefore highlights a limited range of crop rotation which are mainly explained by the farm structure (presence of cattle, FPP and to a lesser extent the plot pattern) and regulatory constraints on the management of livestock manure. We did not found a relation with the farm size (UAA and number of animals) which can be explain by a generalized over capacity production in the sample (most of the farms should have at least twice their UAA to be able to feed their herd). Our results suggest the interest of further discussion of the evolution of land use in areas with a high density of pig production, especially concerning a self-sufficiency orientation.

Is it possible to alleviate the negative effects of heat stress on reproduction of dairy cows?

M.A. Driancourt
MSD Animal Health Innovation, BP 67131, 49071 Beaucouze, France; marc-antoine.driancourt@merck.com

Heat stress of dairy cows is a severe problem in several parts of the world (South of the US, Middle East countries) that may increase with the expansion of dairy farming in developing countries as well as with global warming. Heat stress reduces food intake (hence worsening energy balance) and strongly interferes with fertility of dairy cows, not only during the periods of heat stress, when high heat and humidity generate a THI (Thermo humidity index) exceeding 76, but also during the weeks following the return of THI to acceptable levels (below 70), when carry over effect of heat stress occur. Periods of the reproductive cycle when sensitivity to heat stress is maximal have been identified and include the follicular phase (when high THI causes impaired follicular dominance, impaired estradiol production, interferes with nuclear and cytoplasmic maturation of the oocyte and generates shallow estrus signs and a blunted LH surge), the early luteal phase (when heat displays direct detrimental effects on early embryonic development, increases apoptosis in the ICM of the embryo and reduces progesterone production par the CL) and around luteolysis (when heat may increase prostaglandin production). Strategies to allow successful reproduction during the heat stress periods include a combination of cooling strategies, optimized feeding practices and hormonal treatments whereby AI may be timed (following GnRH administration) without the need for estrus detection. Induction of the formation of accessory corpora lutea during the luteal phase (on days 5 or 12 post AI) has been shown to be useful in some cases. Finally, embryo transfer, when using embryos produced during the cool season, may also have potential, as such embryos have gone past the stages most sensitive to heat. Strategies to minimize the carry over effects of heat stress have also been proposed. They include the removal (by superovulation or follicular aspiration) of the 'heat stress affected follicles' before breeding is attempted.

Effects of high temperature on the reproductive physiology the sow

S. Boulot and N. Quiniou
IFIP, 35, La Motte au Vicomte, 35651 Le Rheu, France; sylviane.boulot@ifip.asso.fr

High temperatures have various detrimental effects on sow reproduction, including delayed puberty or post-weaning estrus, anestrus, lower farrowing rate, abortions, small litters... Impaired reproduction is reported during hot seasons, both in temperate or tropical regions. Though photoperiod, housing, feeding, genotype, parity or management may also be involved, heat stress is recognized as a main component of seasonal infertility. When mechanisms involved in thermoregulation fail to maintain body temperature, hyperthermia directly alters sow ovarian function, hypothalamic-pituitary-ovarian axis or acts indirect via activation of the adreno-corticotrope axis. Post-weaning fertility disorders are mainly attributed to endocrine and metabolic adaptations in relation with reduced feed intake and negaltive energy balance. Alterations include changes in gonadorophin synthesis or release (LH, GnRH), poor follicular growth, reduced ovarian steroidogenesis (oestrogens, progesterone), linked to alterations in the concentrations of insulin-IGF1, in expression of ovarian heat shock proteins and modified oxidative processes. Specific negative effects of heat stress on the oviduct, uterine environment and placental or foetal development (thermal imprinting) have been poorly investigated in pigs. The lactating sow is at highest risk in warm farrowing rooms (>25 °C), owing tof low upper critic temperature (22 °C). Conception losses are higher when sows and gilts are heat-stressed around insemination or within 2-3 weeks (implantation). After 30 d, pregnant sows become less sensitivetill close to farrowing, high risks then resume with sudden death and disturbed birth process, and more stillborn piglets or savaging. Better knowledge of mechanisms associated to heat-stress reproductive disorders could support more efficient use of alleviating solutions (environmental management, feeding strategies, hormonal treatments, genetic selection).

Adverse effects of heat stress on reproduction in lactating dairy cows and strategies for mitigation

J. Block and P.J. Hansen
University of Florida, Department of Animal Sciences, P.O. Box 110910, Gainesville, FL 32611, USA; blockj@ufl.edu

Heat stress causes significant reductions in fertility in lactating dairy cows. Several aspects of reproductive function are adversely impacted in lactating dairy cows exposed to heat stress. Among these, effects of elevated temperature on the oocyte and early embryo are quite deleterious and have been well characterized. The magnitude of exposure to elevated temperature on oocyte competence and early embryo development is dependent upon genotype, stage of development, redox status and presence of cytoprotective molecules in the reproductive tract. Strategies that exploit these determinants of oocyte and embryonic responses to elevated temperature can be used to mitigate the effects of heat stress on fertility of lactating dairy cows during summer. One effective strategy to improve pregnancy rates during heat stress is to utilize embryo transfer to bypass the effects of elevated temperature on the oocyte and early embryo. The success of embryo transfer during the summer can be improved by culturing embryos in the presence of insulin-like growth factor-1. Other cytoprotective molecules that can enhance fertility during heat stress are bovine somatotropin and various antioxidants although an effective method for delivery of these molecules has not been identified. Genes in cattle exist for regulation of body temperature and also cellular resistance to elevated temperature. Incorporation of these genes into dairy breeds through selection, cross-breeding or on an individual-gene basis is another potential strategy for improving fertility in lactating cows during summer. In summary, fertility in lactating dairy cows is compromised by heat stress in part by the deleterious effects on the oocyte and early embryo but there is potential for enhancing fertility during the summer through either physiological or genetic manipulation of the cow to increase oocyte and embryonic resistance to elevated temperature.

Session 23a Theatre 4

Assessing climatic effects on the reproductive performance of sows in a temperate climate

C. Lambertz, K. Wegner and M. Gauly
Livestock Production Systems Group, Department of Animal Science, Albrecht-Thaer-Weg 3, 37075 Göttingen, Germany; clamber2@gwdg.de

The present study aimed to investigate climatic effects during different time periods on the reproductive performance of sows. Therefore, the temperature and humidity in 6 commercial multiplier farms located in northern Germany were recorded hourly from July 2011 until August 2012 and the temperature-humidity index (THI) was calculated. In each farrowing, servicing, and waiting compartment 2 data loggers were installed. The reproductive performance of 8,250 farrowings of 4,743 sows including the litter size at birth and weaning, number of stillborn piglets, pre-weaning mortality, and the weaning-to-service interval were assigned to the average as well as maximum THI during different time periods, namely 1, 3 and 6 days before and after artificial insemination (AI) and farrowing, respectively. Average and maximum THI values were divided into the 4 classes <67, 67-69, 69-71, >71 and <68.5, 68.5-70.5, 70.5-73.5, >73.5, respectively. Throughout the year THI values in the 3 compartments ranged between 70 and 87 with a mean of 75 ± 4 (SD). With increasing THI values before AI the litter size at birth increased in tendency from 14.6 to 15.0 ($P>0.05$), whereas an effect following AI was not observed. Similarly, the climatic conditions peri- and post-partum did not affect the litter size at birth and weaning. In contrast, the number of stillborn piglets was highest in the class with the lowest average THI before farrowing. From the lowest to the highest average THI class one day after farrowing the pre-weaning mortality decreased from 15.1 to 13.4% ($P<0.05$). In contrast, the thermal conditions in the peri-partum period did not affect the mortality rate. With increasing maximum THI before as well as after AI the weaning-to-service interval increased by about 3 days from the lowest to the highest class ($P<0.05$). In conclusion, climatic effects on the reproductive performance of sows kept in indoor housing systems under temperate climates are low.

Effect of heat stress during intrauterine development on litter size in sows

C. Sevillano Del Aguila[1,2], S. Bloemhof[2], E.H. Van der Waaij[1] and E.F. Knol[2]
[1]Wageningen University, Animal Breeding and Genomics Centre, P.O. Box 338, 6700 AH Wageningen, the Netherlands, [2]TOPIGS Research Center IPG, P.O. Box 43, 6640 AA Beuningen, the Netherlands; sevillano.claudia@gmail.com

Heat stress affects sows reproductive performance. Differences in heat stress tolerance have been previously observed between sow lines. Developmental programming might be an important source of phenotypic variation to heat tolerance later in life. During developmental programming, early in life structure and function of organs can be influenced by environmental insults such as heat stress. The effect of heat stress during intrauterine development on sows reproductive performance later in life, measured as litter size, was investigated. The data set consisted of 21,403 records of litter size performance from Yorkshire sows on 16 farms in Spain and Portugal. Correlations were estimated between the maximum temperature for each day of the sow's own intrauterine development, and her subsequent reproductive performances to determine critical periods where heat stress during intrauterine development has larger effects on the sow's future litter size performance. A critical period was identified around day 80-100 of sows intra-uterine development. The interval 80-100 days was then divided in three heat classes: 'No', 'Moderate' and 'Severe' (<20 °C; 20-29 °C and ≥30 °C, respectively). 'Moderate' heat stress during late intra-uterine development improved sows heat stress tolerance later in life. 'Severe' heat stress during late intra-uterine development compromised sows performance in terms of litter size. These sows produced smaller litters, irrespective of the level of heat load during performance.

Expressions of HSPs mRNA in different tissues of sow during late gestation in continuous hot weather

Z.Y. Fan[1], C.M. Long[1], Y.H. Chen[1], J.H. He[1] and X. Wu[2]
[1]College of Animal Science and Technology, Hunan Agricultural University, Changsha, 410129, China, P.R., [2]Institute of Subtropical Agriculture, Chinese Academy of Sciences, Key Laboratory of Agro-ecological Processes in Subtropical Region, Changsha, 410125, China, P.R.; w.xin@foxmail.com

Heat shock proteins (HSPs) are a class of functionally related proteins whose expression is upregulated when cells are exposed to elevated temperatures or other stress factors. To study expression of HSPs mRNA in different tissues of sows during late pregnancy, when maintained continuously under hot temperatures, twelve Landrace × Large White sows were selected, with six repeats in each group. All the sows were slaughtered on d 90 or 110 of gestation respectively, and the brain, heart, liver, kidneys, spleen, lung, and ovary samples were collected aseptically over liquid nitrogen and frozen at -80 °C. Then total RNA isolation, reverse transcription, followed by quantitative real-time PCR (qRT-PCR) were performed to determine the amountsof HSP70 and HSP90 mRNA. The results showed that HSP70 mRNA expression amounts in different tissues of sows at d 90 of gestation could be ranked as follows: lung > spleen > liver > brain > kidneys > adrenals = ovary > heart. For HSP90, the following ranking could be made: lung > kidneys > ovary > brain > spleen > liver > heart; on d 110, HSP70: lung > spleen > kidneys > brain > liver > adrenals=ovary > heart, and HSP90 with: lung > ovary > spleen > kidneys > brain > liver > heart. Compared with the d 90 values, the relative expression of HSP70 and HSP90 mRNA were lower in lung (P<0.01), and HSP90 mRNA was lower in those of d 110 (P<0.01). These findings documenting expression of HSP70 and HSP90 mRNA demonstrated differennces in expression of both HSPs between tissues, with highest expression of HSP70 and HSP90 mRNA in lung.

A metabolomic approach for evaluating heat stress in growing pigs

D. Renaudeau[1], M. Tremblay-Franco[2], P.H.R.F. Campos[3], J. Noblet[3], C. Canlet[2] and J. Riquet[4]
[1]INRA, UR143 URZ, 97170 Petit Bourg, France, [2]INRA, UMR 1331 Toxalim, 31027 Toulouse, France, [3]INRA, UMR 1348 PEGASE, 35590 St Gilles, France, [4]INRA, UMR 144 LGC, 31326 Toulouse, France; david.renaudeau@antilles.inra.fr

Heat stress (HS) negatively impacts pig performance. Thus, there is a need to have a better understanding of the mechanisms underlying thermal adaptation and how these responses jeopardize animal performance. Metabolomic is an explorative methodology which aims at identifying discriminatory metabolites or biomarkers for different physiological situations. A total of 16 LW pigs from two lines divergently selected for residual feed intake were used in this study. Pigs were exposed for 7 days (d) at 24 °C and thereafter to 30 °C for 14 d. Plasma samples were obtained during the first week (wk1, d-5 and d-1; 24 °C), the 2nd wk (wk2, d1 and d2, 30 °C), and the 3rd wk (wk3, d7 and d13; 30 °C). A total of 96 samples were analyzed by ^1H NMR spectroscopy and NMR spectra were reduced into 728 buckets. These data were analyzed using the A-SCA method (ANOVA-simultaneous component analysis). The time (wk, n=3), line, and time × line interaction sub-models explained 27.1% ($P<0.001$), 3.3% ($P=0.098$) and 2.0% ($P=0.823$) of the total variation in the data, respectively. The first principal component (PC1) explained 97% of the variation of the factor 'time'. From the 121 chemical shifts with a significant contribution to the PC1, 22 metabolites were identified. Concentrations of betaine, choline, glycerol, and isoleucine were lower at 30 °C than at 24 °C ($P<0.05$). In contrast, glutamine, glycine, valine, alanine and histidine concentrations were greater at 30 °C. Lipids and glutamate plasma concentrations were significantly higher only in wk2. Creatine and creatinine concentrations increased only in wk3. In conclusion, thermal HS altered the plasma metabolomic profile in pig. These changes can be suspected to result from both direct and indirect effects (i.e. via a reduced feed intake) of high temperature on pig metabolism.

Acute phase proteins as biomarkers of disease and stress in pigs

M. Piñeiro, J. Morales and C. Piñeiro
PigCHAMP Pro Europa S.L., Sta Catalina 10, Segovia, Spain; carlos.pineiro@pigchamp-pro.com

The acute phase proteins (APP) are blood proteins which modify their concentration because of tissue damage, infection, stress or neoplastic growth. Changes in APP levels are part of the physiological changes taking place during the acute phase response, the rapid, generalized, reaction of the organism, directed to fight against any attack to its integrity. APP, as general and unspecific markers of inflammation, can be used to detect pathological states and are considered valuable for evaluating the health status of pig herds. The concentration of APP increases in presence of bacterial, viral or parasitic infections, lesions or injuries such as tail or ear biting. The magnitude of the increase is dependent on the severity of the underlying conditions. In the pig, major APP include CRP, SAA, haptoglobin and pig-MAP. As a biomarker pig-MAP has the advantage of relatively low variability in its normal state compared to other APP such as haptoglobin, which facilitates the establishment of a threshold to differentiate between normal and pathological states. Studies performed in the EU project APP in pigs showed that pig-MAP had the highest sensitivity for the global assessment of disease among the individual APP. The detection ability can be increased by the use of an APP index. APP have been found to be useful biomarkers for the detection of both clinical and subclinical disease, and are also increased in the presence of stressors that affect well-being and reduce growth rate, such as low space allowance, mixing with other animals, extreme temperatures or inadequate handling of feed. The increase of APP was associated with a decrease in the productive performance. The concentration of APP can also augment significantly after transportation and reflect the quality of transport conditions. In conclusion, APP such as pig-MAP, can provide a general view of the health status and welfare of pigs, and might be useful parameters in monitoring programmes aimed to improve the quality and sustainability of pig production.

Haptoglobin in milk: immunologic biomarker for monitoring health status by on-farm analysis

U. Bergfeld[1], T. Möllmer[1], S. Pache[1], K. Zoldan[2] and R. Fischer[1]
[1]Saxon State Office for Environment, Agriculture and Geology, Am Park 3, 04886 Köllitsch, Germany, [2]Fraunhofer Institute for Cell Therapy and Immunology, Perlickstr.1, 04103 Leipzig, Germany; uwe.bergfeld@smul.sachsen.de

The objective of the investigation was to monitor the health status of dairy cows. A new method to analyze the immunological biomarker Haptoglobin (Hp) directly in milk on farm was developed. Haptoglobin is associated to the group of acute-phase proteins and can show the onset of an inflammatory response by increase of concentration. Various studies show that the Hp concentration in plasma and in milk increases during a mastitis infection. The aim of this work was to test the new on-farm-analysis regarding: (1) sensitivity and specificity of prediction of an inflammatory response of systemic or local infection; and (2) control of individual response curves depending on the types of disease. Raw milk samples were collected from 100 Holstein Friesian cows in the 1st to 6th lactation from day 2 post partum (pp)and up to day 44 pp in a 2 or 3-day rhythm. All animals were included in a program of usual veterinary observation. The measurement of Hp concentration was performed with the on-farm device eProCheck® based on the principle of a sandwich ELISA. With the test it is possible to detect the Hp-concentration in the raw milk directly on a dairy farm. 1,300 milk samples were analyzed to estimate sensitivity and specificity to predict inflammatory diseases. Further information regarding various disease complexes (uterus, udder, limbs) was collected. We also included data about veterinary diagnoses and treatments, as well as data from the monthly milk recording and the daily milk yield. The individual variation of Hp values where investigated with analysis of variance. A threshold value of 1.2 µg/ml Hp in milk to classify cows as either 'ill' or 'healthy' was used. The estimated sensitivity was between 0.67 and 0.69 and the specificity between 0.84 and 0.89.

Physiological biomarkers for prevention of production diseases in dairy cows

K.L. Ingvartsen[1], K.M. Moyes[2], T. Larsen[1] and L. Munksgaard[1]
[1]Aarhus University, Dept. of Animal Science, Foulum, 8830 Tjele, Denmark, [2]University of Maryland, Dept. of Animal and Avian Sciences, College Park, MD 20742, USA; kli@agrsci.dk

Traditionally skilful dairy farmers have checked dairy cows for diseases, both infectious and non-infectious, at each milking. This is still the case in many herds but this type of surveillance has been challenged due to a structural development that has resulted in a rapid increase in the average herd size, now >150 cows in Denmark, but also to the fact that milking is taken over by automatic milking systems. In these larger units the farmer or farm staff have to overlook an increasing number of animals concerning production diseases and other factors important for production and welfare and consequently for risk management and optimization. A prerequisite for preventing production diseases is identification of the disease and its potential risk factors. The majority of the production diseases occur around calving or during early lactation and often the problems are related to inappropriate feeding and/or management during the previous lactation, dry period or late pregnancy causing reduced ability of the cow to adapt to the challenges of early lactation. This causes physiological imbalance or subclinical states and a major challenge is to combat these states associated with increased risk of disease and suboptimal performance and reproduction. By their nature these subclinical states are difficult for the farmer or farm staff to identify. However, early detection of physiological imbalance or subclinical states is crucial for proactive risk management in order to prevent risk of disease development and loss of efficiency. It is argued that proactive risk management needs to be carried out at individual dairy cow level and at real-time to allow proactive management. This calls for easily accessible data or samples, e.g. milk that can be collected and analysed automatically in-line and used real-time. Aspects of ideal physiological biomarkers, including physical and behavioural biomarkers, are discussed.

Session 23b Poster 4

Social support attenuates behavioural and neuroendocrine responses to isolation stress in piglets

E. Kanitz, B. Puppe, T. Hameister, A. Tuchscherer and M. Tuchscherer
Leibniz Institute for Farm Animal Biology, Wilhelm-Stahl-Allee 2, 18196 Dummerstorf, Germany; ekanitz@fbn-dummerstorf.de

There is increasing evidence that social relationships can alleviate the impact of stressful life experience. This study investigated whether social support in postnatal pigs by familiar or unfamiliar conspecifics could buffer behavioural and neuroendocrine stress responses of a 4 h social isolation. Piglets were classified into four treatment groups (control without isolation, isolation alone, isolation with familiar or unfamiliar piglets) and examined at 7, 21 or 35 days of age. The behavioural responses were analysed in repeated open-field/novel-object (OF/NO) tests, and the mRNA expression of glucocorticoid receptor (GR), mineralocorticoid receptor (MR), 11β-hydroxysteroid dehydrogenase 1 and 2 (11β-HSD1 and 11β-HSD2) was quantified by real-time RT-PCR in stress-related brain regions. Data were evaluated by ANOVA using the MIXED procedure of SAS. Piglets isolated alone were more excited in the OF/NO test and showed higher ACTH and cortisol concentrations compared to controls. Furthermore, these piglets also displayed a lower MR/GR ratio and a higher 11β-HSD2 mRNA expression in different brain regions. Social support during isolation reduced the behavioural activity in the OF/NO test and diminished the release of stress hormones compared to isolation of piglets alone. Moreover, the imbalance of MR/GR mRNA expression was reversed, and the 11β-HSD2 mRNA expression was not affected in socially supported piglets. With respect to the level of familiarity, the behavioural and neuroendocrine responses of piglets isolated with a familiar conspecific were similar to that of the controls. In conclusion, social support can buffer the negative consequences of psychosocial stress in pigs indicated by diminished arousal and neuroendocrine stress response. These results should be considered in livestock practices so as to improve welfare and emotional experience.

Session 23b Poster 5

Comparison of maternal behaviour and biomarkers for stress in beef and dairy cows

K. Hille[1], S. Theis[1], M. Piechotta[2], U. König V. Borstel[1] and M. Gauly[1]
[1]G-A.U. Göttingen, Albrecht-T. Weg 3, 37075 Göttingen, Germany, [2]TiHo Hannover, Bünteweg 2, 30559 Hannover, Germany; koenigvb@gwdg.de

A total of 40 cow-calf dyads were tested in the present study. They either belonged to a breed focused on beef (B) production (Simmental; n=20) or on dairy (D) production (German Black Pied Cattle; n=20). Animals were housed under similar conditions at two research farms. Observations included different aspects of maternal behaviour such as the number (MN) and intensity (MI) of interactions between cow and calf, the defensiveness (DS), overall behaviour (BS) and agitation (AS) of the cow during separation and handling of her calf (measured on different scales with higher values indicating stronger reactions) as well as cows' heart rate (HR) and cortisol levels during these tests. Each test was conducted twice: once on the 2nd and once on the 3rd day of the calf's life. Contrary to our hypothesis, B cows were not more maternal than D (e.g. MI, BS, mixed model: P>0.1; MN: B=8.1±1.4 vs. D=13.9±1.4 interactions; P=0.005). However, B were more agitated during separation (AS: B=4.6 vs. D=1.7 scores; P=0.001), more aggressive towards the handler (DS: B=1.3±0.1 vs. D=0.83±0.1 scores; P=0.002), and they had higher HR than D (B=92.0±2.2 vs. D=79.6±2.2 bpm; P=0.0002). In contrast, D showed higher levels of stress than B during handling of their calves (B=1.2 vs. 1.7 ng cortisol/ml saliva; P=0.04). The majority of parameters decreased or tended to decrease from the 2nd to 3rd day of calf's life (e.g. HR: 2nd = 87.7±1.8 vs. 3rd = 83.9±1.8 bpm; P=0.05). Parity influenced cows' heart rate (-2.0±0.7 bpm per additional parity; P=0.005) but none of the other parameters (P>0.1). Neither heart rate nor cortisol levels correlated significantly with any of the behavioural parameters (P>0.1). In conclusion, beef cows do not seem to be per se more maternal than dairy cows, although they appear to be more aggressive when humans attempt to handle their calves. Neither heart rate nor cortisol seems to be a good biomarker for maternal behaviour.

Session 23b Poster 6

Cortisol and testosterone levels in neuroreflection types of pigs according to habituation

J. Petrák, O. Debrecéni and O. Bučko
Slovak University of Agriculture in Nitra, Department of special animal production, Tr. A. Hlinku 2, 949 76 Nitra, Slovak Republic; juraj.petrak@uniag.sk

The dynamics of a reaction to load can be assessed on the grounds of excitability of the nervous system. Neuroreflection types in terms of arousal of the nervous system are divided into highly excitable EHb +, medium EHb° and low excitable EHb-types. As a marker of a stress load we chose cortisol. Increased levels of cortisol impair cognitive processes of organisms. On the other hand, one of the hormones that has the neuroprotective effect is testosterone which controls the activity of Brain-derived neurotrophic factor – BDNF. The aim of this work was to monitor the concentrations of selected steroid hormones in relation to excitability type of pigs. The animals were divided into particular excitability types (EHb+, EHb°, EHb-) on the grounds of quantity of motion in habituation chamber. We tested 45 Large White pigs (barrows and gilts). Test of habituation was performed at a body weight of 30-35 kg. Saliva was collected using a gauze swab. EHb+ type included 6 pigs, EHb° type 31 pigs, and EHb- type 8 pigs. Blood was collected immediately after slaughter at a body weight of 105 kg. The concentration of cortisol in serum and testosterone in saliva was determined by ELISA. The highest concentration of cortisol was measured in EHB- type, lower concentration in EHb+ type and the lowest concentration in EHb° type. The highest concentration of testosterone was measured in EHb-, lower in EHb° and the lowest concentration in EHb+ type. The differences in concentrations of testosterone and cortisol among particular excitability types were not statistically different. However, the results suggest that the EHb- type involves hypothalamic-pituitary-adrenocortical axis (HPA axis) adaptation, which permanently increases the basal level of cortisol. On the other hand, the highest level of testosterone in the EHb- pigs probably ensures its neuroprotective function.

Session 23b Poster 7

Concentrations of IGF-I and cortisol in serum for assessing the sensitivity of pigs to load

J. Petrák, O. Bučko, O. Debrecéni and M. Margetín
Slovak University of Agriculture in Nitra, Department of special animal production, Tr. A. Hlinku 2, 949 76 Nitra, Slovak Republic; juraj.petrak@uniag.sk

One of the growth factors involved in the repair of the body during and after load is insulin-like growth factor-I (IGF-I), which protects cells from apoptosis and promotes their growth and proliferation. As a marker of a stress load we chose cortisol. Increased levels of cortisol impair cognitive processes of organisms. Changes in concentrations of IGF-I in the circulation during acute stress load in individual excitability types of pigs (EHb+ highly, EHb° medium and EHb- low excitable types) are not known. We know that EHb+ and EHb° types are more resistant to stress than EHb- type pigs. The aim of our study was to measure the levels of IGF-I and cortisol immediately after the application of a mental form of load by test of habituation. We tested 45 Large White pigs (barrows and gilts). Test of habituation was performed at a body weight of 30-35 kg. EHb+ type included 6 pigs, EHb° type 31 pigs, and EHb- type 8 pigs. Blood was collected immediately after slaughter at a body weight of 105 kg. The concentration of IGF-I and cortisol in serum was determined by ELISA. The highest concentration of cortisol was measured in EHb- type, lower in EHb+ and the lowest in EHb° type. The highest concentration of IGF-I was measured in EHb+, lower in EHb° and lowest in EHb- type. Significant differences were observed for IGF-I levels ($P \leq 0.05$) between excitability types EHb+ and EHb° and also between EHb° and EHb-. The results may suggest that the EHb- type makes adaptation in the hypothalamic-pituitary-adrenocortical axis which leads to a sustained increase in basal cortisol level. We also speculate that the higher concentration of IGF-I in serum of EHb+ pigs may allow its better access to cells and tissues and thus making the pig better overcome a stress load.

Herding practices and livestock products in France for 6,000 years: contribution of archeozoology

M.-P. Horard-Herbin
Université François Rabelais de Tours, UMR 7324 du CNRS Citeres, BP 60449, 37204 Tours cedex 03, France; marie-pierre.horard-herbin@univ-tours.fr

Squeletal remains found in archaeological excavations are a primary source of information for studying past herding practices and livestock products. This conference will deal with the wild origins and the diffusion of major domestic animals. It will concentrate on when and how they were transported far from their areas of origin toward the West – to France, while accounting for recent studies on animal genetics. Another significant phenomenon lies in visible modifications, like size decrease or growth, and morphological transformations, which affect all the domestic species through history. In particular, modifications of cattle from the Iron Age to the Roman period are studied. Then, some examples of adaptation of livestock farming systems to the economic, social and environmental conditions, as well as cultural demand of meat production from the Neolithic period to the Modern period will be introduced. With the domestication of pasture-fattened animals, man gradually assumed control of the animal world for the purpose of livestock production – a major revolution with many consequences. Meat production, in particular, progressively concentrated on a few domesticated species – cattle, pigs, sheep, goats and poultry, and these became the almost exclusive suppliers of the meat consumed in the Ancient World; a situation which indisputably lasted for several thousand years. The conference will also seek to document preparation and processing techniques from the live animal to the meat food product. It will also focus on the social organisation of butchery activities as well as their evolving economic importance through time and space. Finally, various contexts of meat consumption will be examined, such as food offerings in burial rites during inhumation and cremations, and changes in the consumption of two species, dogs and horses, in their social and symbolic meaning. The purpose of this conference is to show how the study of animal bone assemblages from prehistorical and historical sites provides substantive contribution to our understanding of past diet and food production systems.

Session 24 Theatre 2

How selective sweeps in domestic animals provide new insight into biological mechanisms

L. Andersson
Science for Life Laboratory, Department of Medical Biochemistry and Microbiology, Uppsala University, Box 582, 751 23 Uppsala, Sweden; leif.andersson@imbim.uu.se

Domestic animals provide unique opportunities for exploring genotype-phenotype relationships. Firstly, selective breeding during thousands of years has enriched for mutations that have adapted domestic animals to a new environment, i.e. farming under various environmental conditions. Secondly, the population structure is often favorable for genetic studies, large families and more or less closely related subpopulations (breeds). Thirdly, strong positive selection leaves genomic footprints that facilitate positional cloning. The combined use of whole genome resequencing, linkage mapping and linkage disequilibrium (LD) mapping within and between breeds provides a powerful approach for positional identification of both monogenic and multifactorial trait loci. The successful use of this approach for identifying genes underlying phenotypic traits will be illustrated on the basis of our research program in chickens, pigs, dogs, horses and rabbits. Several emerging features as regards the phenotypic evolution of domestic animals will be illustrated including: (1) the importance of tissue-specific regulatory mutations; (2) the importance of structural changes (duplications, deletions, inversions); and (3) evolution of alleles at loci under strong directional selection.

Session 24 Theatre 3

Estimations of nitrogen and phosphorus excretion and emissions by livestock in EU-27

O. Oenema[1], L. Sebek[2], H. Kros[1], J.P. Lesschen[1], M. Van Krimpen[2], P. Bikker[2] and G. Velthof[1]
[1]Wageningen University, Alterra, P.O. Box 47, 6700 AA Wageningen, the Netherlands, [2]Wageningen University, Livestock Research, P.O. Box 65, 8200 AB Lelystad, the Netherlands; oene.oenema@wur.nl

Livestock production utilizes approximately 60 to 70% of the agricultural land in the European Union (EU-27), and has a relatively large share in emissions to air and waters. Especially, excrements are a main source of ammonia, methane and nitrous oxide to air, and for nitrogen (N) and phosphorus (P) leached to groundwater and surface waters. However, the amounts of nitrogen and phosphorus in animal excrements are not accurately known, mainly because of methodological artefacts and limited data availability. Here, we report on a review of the diversity of methods and data collection – processing – reporting procedures used in Member States of the EU-27 for the estimation of N and P excretion. We examined reports submitted by Member States to EU and UN bodies in the framework of policy reporting, and we screened the literature. Most methods are based on a mass balance, i.e. Excretion = Intake – Retention. The difficulty though is the lack of accurate data for estimating the N and P intake in the animal feed and the N and P retention in milk, meat and egg accurately, as function of animal category, animal productivity, and management. Another difficulty relates to up-scaling; for which farming systems and areas are the estimated N and P excretion representative? Based on the review of the Member States reports and the scientific literature, we propose a common methodology and common data collection - processing - reporting procedure for estimating N and P excretion across EU-27. This common methodology is a compromise between the wish of having spatially detailed, accurate and scientifically sound estimates on the one-hand, and low-cost data collection - processing – reporting procedures on the other hand. In the presentation, we will present data, information and results of current methodologies for estimating N and P excretions and emissions in Member States of EU-27 and we will outline the approach towards a common methodology and common data collection - processing - reporting procedure for estimating N and P excretion to be used across EU-27.

Session 24 Theatre 4

Prospects from agroecology and industrial ecology for animal production in the 21st century

B. Dumont[1], L. Fortun-Lamothe[2], M. Jouven[3], M. Thomas[4] and M. Tichit[5]
[1]INRA, UMR1213 Herbivores, 63122 Saint-Genès-Champanelle, France, [2]INRA, UMR1289 Tandem, 31326 Castanet-Tolosan, France, [3]SupAgro, UMR868 Selmet, 34060 Montpellier, France, [4]UL-INRA, USC0340 AFPA, 54505 Vandoeuvre-les-Nancy, France, [5]INRA, UMR1048 SADAPT, 75231 Paris, France; bertrand.dumont@clermont.inra.fr

Agroecology and industrial ecology can be viewed as complementary means for enhancing food security while reducing the environmental footprint of animal farming systems. Agroecology stimulates natural processes. Industrial ecology closes system loops, thereby reducing demand for raw materials, lowering pollution and saving on waste treatment. Animal farming systems have so far been ignored in most agroecological thinking. Here, we propose five principles for the design of ecology-based animal production systems: (1) adopting management practices improving animal health; (2) decreasing the inputs needed for production; (3) decreasing pollution by optimizing the metabolic functioning of farming systems; (4) enhancing diversity within animal production systems to strengthen their resilience; (5) preserving biological diversity in agroecosystems by adapting management practices. We then review case studies from different production systems (ruminants, pigs, aquaculture) and analyze the combination of these principles. Alternatives to chemical drugs have only recently been investigated and the results are seldom transferable to farming practices. Integration of cropping with livestock systems decreases some of the inputs needed for production and limits nutrient fluxes to the atmosphere and hydrosphere. The persistence of a number of ecological functions and ecosystem services depends on preserving biological diversity in agroecosystems. Finally, we highlight that the development of such alternatives implies changes in the positions adopted by technicians, extension services, researchers and policy makers. It calls for animal production systems being not only considered holistically but also in the diversity of their local and regional conditions.

Session 24 Theatre 5

International issues in dairy cattle genetics

V. Ducrocq
INRA, UMR 1313, 78352 Jouy-en-Josas, France; vincent.ducrocq@jouy.inra.fr

International trade of dairy cattle germplasm has been growing steadily for several decades. Semen is marketed on the basis of predictions of genetic merit which must be both accurate and unbiased. National evaluations of a large number of relevant traits have been developed. Since the nineties, Interbull has contributed to the international recognition of national predictions. Performing meta-analyses of national results, Interbull routinely provides international bull rankings on the national scale of each participating country. The fast development of genomic prediction and genomic selection is transforming this context substantially. A pessimistic perception of its impact would argue that the genomic revolution: (1) favors the largest (groups of) countries able to jointly assemble large reference populations; (2) favors the already world dominant breed – the Holstein; (3) essentially promotes a unique national scale (the American (G) TPI); (4) leads to the illusion that exhaustive data recording or improvement of genetic evaluations are no longer crucial; (5) considerably enlarges the gap between two groups of countries: those that have genomic and genetic evaluations versus those that do not; (6) seriously undermines both the future quality of regular genetic evaluations and the role of Interbull; (7) opens a new era for new players (e.g. pharmaceutical companies) and for patented genetic tests. On the contrary, genomic innovation can be viewed as a unique opportunity to make genetic and genomic evaluations more robust, to share genotypes and new phenotypes internationally in a win-win context, to generate a more balanced genetic gain and to maintain genetic diversity within and across breeds. It represents the necessary tool to provide accurate predictions adapted to local conditions, properly accounting for genotype × environment interactions, especially in developing countries. In other words, genomic selection may – or may not – become the decisive path towards more sustainable dairy cattle breeding.

Session 25 Theatre 1

Improving and adding value to the pig genome sequence

A.L. Archibald[1], L. Eory[1], T. Hubbard[2], S.M.J. Searle[2], P. Flicek[3], L.B. Schook[4] and M.A.M. Groenen[5]
[1]The Roslin Institute and R(D)SVS, Easter Bush, Midlothian, EH25 9RG, United Kingdom, [2]Wellcome Trust Sanger Institute, Cambridge, CB10 1SA, United Kingdom, [3]EMBL-EBI, Cambridge, CB10 1SD, United Kingdom, [4]University of Illinois, Urbana, IL 61801, USA, [5]Wageningen University, 6708 WD Wageningen, the Netherlands; alan.archibald@roslin.ed.ac.uk

The Swine Genome Sequencing Consortium (SGSC) have recently published a draft reference pig genome sequence. However, establishing its sequence is only the first step in characterizing a genome. Identifying the functional elements within the genome sequence is essential for understanding the phenotypic consequences encoded in the genome. Annotation of the pig genome is currently limited to gene models deduced from alignments with expressed sequences (cDNA, ESTs, RNAseq) and some sequence variation (SNPs). The value of the genome sequence could be enhanced through further research and analysis. The BAC clone based framework of the reference genome provides a template for locus-specific improvement. The sequencing of multiple individual pigs reveals the extent of genetic variation in pigs at nucleotide resolution. Next-generation sequencing technologies have transformed the ease with which functional DNA elements can be identified on a genome-wide scale at dramatically reduced cost. By focusing on a subset of assays – RNAseq, Transcription Start Sites (CAGE), histone marks and methylation states – and by coordinating efforts to minimise redundant activity it should be possible to make significant progress towards enhanced annotation analogous to the outputs from the ENCODE project. I will describe how this could be achieved in part through community based efforts and the work of the Ensembl team to make the enhanced annotation accessible. Please refer to the pig genome paper for a full list of those who have contributed to the pig genome project. This research was funded by many sources, as acknowledged in the paper, including USDA, BBSRC and the EC-funded FP7 Project Quantomics-222664.

Pleiotropic effects of a QTL region for androstenone level on pig chromosome 6

A.M. Hidalgo[1,2], J.W.M. Bastiaansen[1], B. Harlizius[3], E. Knol[3] and M.A.M. Groenen[1]
[1]Wageningen University, Animal Breeding and Genomics Centre, P.O. Box 338, 6700 AH Wageningen, the Netherlands, [2]Swedish University of Agricultural Sciences, Department of Animal Breeding and Genetics, P.O. Box 7023, 750 07 Uppsala, Sweden, [3]TOPIGS Research Center IPG, P.O. Box 43, 6640 AA Beuningen, the Netherlands; andre.hidalgo@wur.nl

Androstenone is one of the main compounds causing boar taint in boars. A 1.94 Mbp region on pig chromosome 6 (SSC6) was found to affect androstenone level. Within this region, two major haplotypes (high- and low-androstenone) can be distinguished. Analysis of sequence data showed that the low-androstenone haplotype originated from Asian breeds and was found at different frequencies in European commercial breeds. Because androstenone levels were not taken into account in typical breeding programs, we hypothesize that this haplotype accumulated indirectly by selection for another correlated trait. In this study, three pig lines were used for the estimation of pleiotropic effect of the 1.94 Mbp region on SSC6: sow line 1 (~1,450 animals); sow line 2 (~1,300 animals); and a boar line (~900 animals). Phenotypes were available for eight traits: birth weight, backfat thickness, growth rate, total number born, litter birth weight, teat number, sperm motility, and number of spermatozoa per ejaculation. Association between phenotypes and haplotypes were tested using ASReml v3.0. For both sow lines, a favorable effect for teat number (+0.11) was detected of the low-androstenone haplotype (P=0.058; P=0.013 respectively). In sow line 2, a favorable effect of the low-androstenone haplotype on number of spermatozoa per ejaculation was detected (P=0.023). No effects were identified on production and female reproduction traits. These results show that pleiotropic effects for androstenone level on SSC6 are favorable and have low magnitude. The absence of unfavorable pleiotropic effects suggests that selection for low-androstenone levels at this location is possible without negative effects on other traits.

Session 25 Theatre 3

Using NGS data to characterize genetics of meet-type and egg-type chicken lines

P.F. Roux[1,2], F. Lecerf[1,2,3], A. Djari[4], D. Esquerre[5], S. Marthey[6], T. Zerjal[7], E. Le Bihan-Duval[8], C. Klopp[4], M. Moroldo[6], J. Estelle[7], C. Désert[1,2,3], B. Bed'hom[7], M. Tixier-Boichard[7], O. Demeure[1,2] and S. Lagarrigue[1,2,3]
[1]Agrocampus Ouest, UMR1348 Pegase, 35000 Rennes, France, [2]INRA, UMR1348 Pegase, 35590 Saint-Gilles, France, [3]Université Européenne de Bretagne, UEB, 35000 Rennes, France, [4]INRA, SIGENAE, 31320, France, [5]INRA, GENOTOUL, 31320, France, [6]INRA, UMR1313 GABI, CRB GADIE, 78350 Jouy-en-Josas, France, [7]INRA, UMR1313 GABI, 78350 Jouy-en-Josas, France, [8]INRA, UR83 Recherches Avicoles, 37380 Nouzilly, France; pierre-francois.roux@rennes.inra.fr

New sequencing technologies called 'NGS' allow the analysis of full genomes. In this new context, our study aims at characterizing at the genome scale polymorphisms within INRA broilers and layers experimental lines divergent for abdominal fatness (G/M) and feed efficiency (R+/R-), respectively. Genomes of about 15 animals per line were sequenced either in pools (R+/R-) or individually (G/M), using a HiSeq 2000 and with a depth of 19X. On those 4 lines, we identified 11 million SNPs. In addition, 386,177 SNPs discriminating R+ from R- and 510,577 SNPs discriminating G from M were identified. About 1.5% of these SNP are localized in genes, possibly with a functional modification, which is currently analyzed. A fine study of genes functions leads to new hypotheses on how the divergence between lines occurred. Finally, these approaches allowed us to identify one mutation that might be functional in a candidate gene for a QTL affecting abdominal fatness.

Association study of candidate genes selected in QTL regions for immune responses in chickens

M. Siwek[1], A. Slawinska[1], M. Rydzanicz[2], J. Wesoly[2], M. Fraszczak[3], T. Suchocki[3] and J. Szyda[3]
[1]University of Technology and Life Sciences, Animal Biotechnology, Mazowiecka 28, 84-085, Poland, [2]Adam Mickiewicz University, Institute of Molecular Biology and Biotechnology, Department of Human Molecular Genetics, Umultowska 89, 61-614 Poznan, Poland, [3]Institute of Animal Genetics, Wroclaw University of Life Sciences, Biostatistic Lab, Kozuchowska 7, 51-631 Wroclaw, Poland; siwek@utp.edu.pl

Chicken experimental cross (WLZk) obtained by mating White Leghorn (WL) males with Green-legged Partidgelike (Zk) females has been subjected to a QTL study. QTL regions for a non-pathogenic antigen keyhole limpet hemocyanin (KLH), and two environmental antigens: lipopolisaccharide (LPS) and lipoteichoic acid (LTA) were detected on GGA9, GGA14 and GGA18. In silico analysis of positional and functional candidate genes was performed in the selected QTL regions. The function of candidate genes was defined based on the data from: NBCI, KEGG and Gene Ontology. The goal of this study was to verify in silico candidate gene selection with SNP genotyping and association study. Therefore a set of 384 SNPs located in 18 genes was selected based on the Biomart data base. All together 480 individuals from WLZk cross were genotyped using Golden Gate Illumina protocol, custom SNP assay and Bead Chip array. SNPs were genotyped using Genome Studio software. Out of 384 SNPs used: 17% didn't give a positive signal, 31% were homozygous and 52% SNPs were informative. This data set was further used for the association study. Three methods of SNP additive effects estimation were compared: (1) a series of single SNP mixed models including a random polygenic effect and a fixed SNP effect; (2) a multi-SNP mixed model with random SNP effects; and (3) a model free approach based on the CAR score statistics. The research was supported by the National Science Centre in Krakow (Poland), grant no. NN311558640 and NN311609639. The genotyping experiment was performed in IMBB, in Genome Analysis Laboratory equipped within 'NanoFun' project POIG.02.02.00-00-025/09.

A comparison of methodologies to locate autosomal recessive genetic diseases using SNP chip genotype

G.E. Pollott
Royal Veterinary College, Royal College Street, NW1 0TU London, United Kingdom; gpollott@rvc.ac.uk

A range of methods to locate the site of mutations causing autosomal recessive genetic diseases using SNP chip data have been suggested. This paper compared five methods with the recently proposed autozygosity-by-difference (ABD) approach. These were chi-squared (CS), ASSHOM, ASSIST, HOMOYGOSITY MAPPER and the -homozyg option in PLINK. Data on the well-documented Lavender Foal Syndrome (LFS; a single base deletion at position 138,235,715 on ECA1, found using 6 cases and 30 controls) were analysed by all 6 methods. The original published position of this condition was located with a 3×2 SNP genotype-by-diseases status CS approach using Fisher's Exact Test (CSF) to a 10.5Mb region of ECA1 containing 14 significant SNP. This was refined to a 1.6Mb length by haplotype analysis. These results were replicated in this study, using CSF in PLINK. The ABD method suggested that the mutation was found in a 1.56Mb region from positions 136,812,666 to 138,375,254 of ECA1 with the highest scores; the same region as suggested in the original report after using haplotyping to narrow down the region suggested by CSF methods. Both ASSHOM and ASSIST suggested a different chromosome may contain the mutation although Chromosome 1 did receive high scores. ASSHOM indicated a 0.8Mb region on ECA6 and ASSIST suggested a 0.1Mb region on ECA2. The homozyg option in PLINK found 13 regions involving all 6 cases, of which 7 were monomorphic. Of these, the same 1.56Mb region on ECA1 was identified as the longest consensus run. HOMOZYGOSITY MAPPER also located the same segment of ECA1. Technically 4 of the 6 methods compared here all found much the same segment of ECA1 containing the LFS mutation; the exceptions being ASSHOM and ASSIST. Both these methods, and the CSF method, rely on cases being monomorphic and controls being polymorphic at the mutation site, which was not the case in this dataset. The situations where specific methods may be preferable are outlined, using illustrations from other datasets, and the utility of the ABD method highlighted.

Session 25 Theatre 6

FST as an indicator of selective sweeps using admixed animals as a control

A. Frkonja[1], G. Meszaros[1], I. Curik[2], J. Solkner[1], U. Schnyder[3] and B. Gredler[3]
[1]University of Natural Resources and Life Sciences Vienna, Department of Sustainable Agricultural Systems, Gregor Mendel strasse 33, 1080 Vienna, Austria, Austria, [2]University of Zagreb, Faculty of agriculture, Department of Livestock Sciences, Svetosimunska 25, 10000 Zagreb, Croatia, [3]Qualitas AG, Chamerstrasse 56, 6300 Zug, Switzerland; frkonjica@gmail.com

Indication of a selective sweep involves local reduction in variation within a selected gene and in adjacent SNPs. The degree of genetic differentiation between subpopulations, fixation index (F_{ST}) can be used for identification of selective sweeps in one of the populations. Aim of this study was to derive selective sweeps using fixation index (F_{ST}) between Simmental (SIM) and Red Holstein Friesian (RHF) cattle, which are the ancestral populations of the composite breed Swiss Fleckvieh (SF). Illumina BovineSNP50 Beadchip of 493 animals (87 pure SIM, 101 pure RHF, 305 composites) with 37836 SNPs after quality control were available. The composites were divided in groups according to RHF admixture levels of 0.01-0.50, 0.50-60, 0.60-0.70, 0.70-0.80, 0.80-0.99, according to whole genome admixture. F_{ST} of pure breeds was calculated for sliding windows of size 2-20 SNPs. Chromosomes where F_{ST} window peaks were found were: 3, 5, 6, 10, 15 and 18. The top signal was found on chromosome 6, in a region of 2.6 Mb. The reported region includes 41 SNP and 22 of them were monomorphic in SIM. The KIT gene responsible for coat patterns was situated in this window. The finding was surprising as there is no obvious difference in coat patterns of SIM and RHF. The signal is very close to the gene coding for the insulin-like growth factor binding protein 7. Levels of allele frequencies in this particular region on chromosome 6 followed closely the genome-wide admixture levels of crossbred animals.

Session 25 Theatre 7

Introgression of European *Bos taurus* genome in Ugandan taurine and zebuine cattle breeds

R. Negrini[1], M. Milanesi[1], E. Eufemi[1], A. Stella[2], S. Joost[3], S. Stucki[3], P. Taberlet[4], F. Pompanon[4], F. Kabi[5], V. Muwanika[5], C. Masembe[5] and L. Colli[1]
[1]Università Cattolica del Sacro Cuore, Ist. Zootecnica, Via E. Parmense 84, 29122 Piacenza, Italy, [2]Parco Tecnologico Padano, Via Einstein Loc. Cascina Codazza, 26900 Lodi, Italy, [3]EPFL, route cantonale, 1015 Lausanne, Switzerland, [4]UJF-CNRS, France, rue des martyrs 25, 38042 Grenoble, France, [5]Makerere University, P.O. Box 7062, Kampala, Uganda; riccardo.negrini@unicatt.it

One of the objectives of the NextGen EU project is to investigate vector-borne-disease resistance in Ugandan cattle. To date 915 animals have been collected from 52 grid cells covering the whole country. In total 9 populations belonging to Ankole (crossbred between *Bos indicus* and *Bos taurus*), Zebu and Ankole-Zebu crosses were sampled together with GPS data and health status information. These data, together with disease vector distribution, disease prevalence, 50K and 800K SNP genotypes, and whole genome sequences, will be used to identify relevant resistance genes by an integrated GWAS and selection signatures approach. Here we describe the genetic structure of the samples collected. All samples had some level of admixture, that cannot be estimated from animal phenotype. By running a Bayesian clustering approach with the ADMIXTURE software on Ugandan cattle analysed with 50K SNPs together with reference African and European *B. taurus* and *B. indicus* breeds, all ancestral genomic components of the animals sampled could be traced back to their origin. About 20% of the genome of Zebus is African taurine, confirming previous data on the origin of African Zebu populations. The European taurine contribution is a minor component, rare in Zebu and evenly spread in Ankole and Ankole-Zebu crosses. The genomic components have clear geographical structures: indicine genome is prevalent in north-eastern Uganda, while taurine predominates in the south-western area. Holstein Fresian introgression is present mostly in south-western Uganda.

Genome wide association study of insect bite hypersensitivity in two populations of Icelandic horses

M. Shrestha[1], L.S. Andersson[1], F. Fikse[1], T. Bergström[1], A. Schurink[2], B.J. Ducro[2], S. Eriksson[1] and G. Lindgren[1]
[1]Swedish University of Agricultural Sciences, Animal Breeding and Genetics, Gerda Nilssons väg 2, Ultuna, Box 7023, 75007 Uppsala, Sweden, [2]Wageningen University, Animal Breeding and Genomics Centre, Zodiac building number 122 De Elst 1, 6708 WD, Wageningen, the Netherlands; merina.shrestha@slu.se

Equine Insect Bite Hypersensitivity (IBH) is a chronic, pruritic, recurrent seasonal dermatitis. It is caused by an allergic reaction to protein in the saliva of biting midges of *Culicoides* spp and sometimes *Simulium* spp. It leads to discomfort and disfigurement that impairs the quality of life of the horse and economical loss for the horse owner. Severely affected horses are sometimes euthanized. The worldwide prevalence ranges from 3% to 60%. There are no effective preventive measures and cures for IBH. Environmental and genetic factors are responsible for its development. Heritability for IBH in Swedish born Icelandic horses is estimated to be around 0.10 (observable scale) and 0.33 (sd=0.19, liability scale); in Dutch Shetland breeding mares around 0.24 (sd=0.06) and 0.16 (sd=0.06, liability scale) for Friesian broodmares. A genome wide association study was performed using Illumina 50K SNP chip data of 209 Icelandic horses. GenABEL package in R was used in analyses fitting a single marker effect at a time. A significant association was observed on chromosome 23. Odds ratio for IBH development of the unfavorable allele was 23.4 and the unfavorable allele had a frequency of 0.1 in cases and 0.004 in controls. This tentative association will be confirmed by re-genotyping the horses included in the study and also genotyping additional horses. Borderline associations were observed on chromosomes 3, 10, 17, 18 and 32. For further confirmation analysis using multi marker association model, based on a Bayesian variable selection method, will be done. A combined GWAS will also be performed on data of 146 and 209 Icelandic horses from the Netherlands and Sweden respectively, using Bayesian methodology.

Session 25 Theatre 9

Influence of lactation during pregnancy on epigenetic regulation of genes in dairy cattle

A. Bach[1,2] and A. Aris[1]
[1]IRTA, Department of Ruminant Production, Caldes de Montbui, 08140, Spain, [2]ICREA, Barcelona, 08010, Spain; alex.bach@irta.cat

The objective of this study was to explore potential differences in the methylation degree of the entire genome of calves born either to primiparous (no co-existence of lactation and pregnancy; PPC) or to multiparous (co-existence of lactation and pregnancy) cows (MPC). A blood sample was obtained from 10 female Holstein calves (27±5 d old and 42.9±6.87 kg of BW) daughters of the same sire born to either a PPC (n=5) or to a MPC (n=5). Genomic DNA was extracted from isolated peripheral blood mononuclear cells. The methylation status of all chromosomes was assessed using the Agilent Genome CpG Island Chip Array prior immunoprecipitation of methylated DNA. The methylation degree of each region was determined using Batman calls, and differences between groups using MethLab. There were no differences between calves born to PPC and MPC on the overall methylation degree (45.8%). There were significant differences and totally opposite degree of methylation between the 2 groups for 4 distinct CpG sites. Furthermore, there were 70 regions of the genome that had different methylation status in all 5 calves within parity group of the dam. Out of these 70 regions, there were 30 known genes. Many of these genes are involved in cellular assembly and organization (such as ARL88, BNIP2, RAB8A, TCC15, TBL2, KTN1, RHOC, SNX8, PRKCA, MAPK1, and GPR17). Interestingly, all calves born to PPC had the ZPBP gene (involved in fertilization) unmethylated, whereas those born to MPC had it methylated. Contrary, all PPC calves had the MACROD1 gene (involved in estrogen signaling) methylated and MPC unmethylated. Furthermore, all 5 calves born to PPC had the ATP8A2 gene (an ATPase) unmethylated and the ALDH4A1 gene (interconnects the urea and tricarboxylic acid cycles) methylated, whereas all 5 calves born to MPC had them methylated and unmethylated, respectively. To our knowledge, this is the first evidence of epigenetic differences in genomes of calves as induced by the co-existence of lactation and pregnancy.

Session 25 Theatre 10

A novel method allows accurate identification of key ancestors within populations

M. Neuditschko[1], R. Von Niederhäusern[1], H. Signer-Hasler[2], C. Flury[2], M. Frischknecht[1,3], T. Leeb[3], E. Jonas[4,5], M.S. Khatkar[5], H.W. Raadsma[5] and S. Rieder[1]
[1]Agroscope Liebefeld-Posieux (ALP-Haras), Swiss National Stud Farm, Les Longs-Prés, 1580 Avenches, Switzerland, [2]School of Agricultural Forest and Food Sciences, Länggasse 85, 3052 Zollikofen, Switzerland, [3]University of Bern, Institute of Genetics Vetsuisse Faculty, Bremgartenstrasse 109a, 3001 Bern, Switzerland, [4]SLU, Department of Animal Breeding and Genetics, P.O. Box 7070, 750 07 Uppsala, Sweden, [5]University of Sydney, Reprogen, Animal Bioscience, Faculty of Veterinary Science, 410 Werombi Road, 2570 Camden, Australia; markus.neuditschko@agroscope.admin.ch

Identification of key ancestors is important for population genetics studies and conservation. Based on Principal Components Analysis (PCA), we present a novel approach using genome wide SNP data which identifies key ancestors within populations without using any prior ancestry information. We demonstrate with a sheep (1,430 individuals and 44,693 SNPs) and a horse (1,077 individuals and 38,124 SNPs) dataset that our approach effectively allocates key ancestors within a population, whilst defining fine-scale population structures. Our novel method successfully identified four influential foundation (F1) sires in an Awassi × Merino resource population as the most informative individuals, despite the close relationship between animals and multiple generations of back- and intercrossing. Our method also performed exceptionally well in the horse dataset, and stallions descending from main lineages have been assigned with highest informative scores. Our approach allows both the selection of informative individuals and the characterization of fine-scale population structures. This method can also be applied to analyse fine-scale population structures in indigenous and wild populations, where ancestry information is not readily available. Furthermore, our approach will be useful for assembling most informative resource populations for full genome sequencing initiatives, to facilitate accurate genotype imputation across populations.

Session 25 Poster 11

Identification of candidate polymorphism in a QTL region by combining eQTL mapping with NGS data

P.F. Roux[1,2], Y. Blum[1,2,3], C. Désert[1,2,3], A. Djari[4], D. Esquerre[5], E. Le Bihan-Duval[6], B. Bed'hom[7], P. Le Roy[1,2], F. Lecerf[1,2], M. Moroldo[8], S. Marthey[8], C. Klopp[4], S. Lagarrigue[1,2,3] and O. Demeure[1,2]
[1]Agrocampus Ouest, UMR1348 Pegase, 35000 Rennes, France, [2]INRA, UMR1348 Pegase, 35590 Saint-Gilles, France, [3]Université Européenne de Bretagne, UEB, 35000 Rennes, France, [4]INRA, SIGENAE, 31320 Castanet Tolosan, France, [5]INRA, GENOTOUL, 31320 Castanet Tolosan, France, [6]INRA, UR83 Recherches Avicoles, 37380 Nouzilly, France, [7]INRA, UMR1313 GABI, 78350 Jouy-en-Josas, France, [8]INRA, UMR1313 GABI, CRB GADIE, 78350 Jouy-en-Josas, France; pierre-francois.roux@rennes.inra.fr

In this study, we propose a genetical genomic approach aiming at characterizing QTL affecting adiposity in a chicken model, combining hepatic gene expression and whole genome re-sequencing data. After performing a classical linkage analysis on our meat-type chicken design with abdominal fat weight as the target phenotype, we identified 3 QTL regions. Carrying out hepatic eQTL mapping on those QTL regions, we identified no local acting eQTL, suggesting that causal mutations underlying QTL were located on coding region. Using NGS data, we then selected SNPs having an impact on coding region. Secondly, we selected SNPs impacting highly conserved regions among vertebrates, which is conditional to a functional role. These successive filters allowed us to reduce the number of candidate genes in the 3 QTL regions. One of them was coding for acetyl-CoA carboxylase (ACACA), well known for its implication in fatty acids biosynthesis. Moreover, the two SNPs associated to this gene were missense and one of them was located in the protein carboxyl catalytic domain. These two observations make these two SNPs in ACACA strong candidate mutations responsible for adiposity in our scheme. This original approach, combining NGS and QTL and eQTL mapping, could be applied to identify candidate mutations for any complex trait in any species. S. Lagarrigue and O. Demeure contributed equally to this work.

Session 25 Poster 12

Genetic variation in choice consistency for cows accessing automatic milking units

P. Løvendahl[1], L.P. Sørensen[1], M. Bjerring[2] and J. Lassen[1]
[1]Aarhus University, Dept. Molecular Biology and Genetics, Research Centre Foulum, 8830 Tjele, Denmark, [2]Aarhus University, Dept. Animal Science, Research Centre Foulum, 8830 Tjele, Denmark; peter.lovendahl@agrsci.dk

Dairy cows milked in automatic milking systems (AMS) with more than one milking box may as individuals have preference for specific milking boxes if allowed free choice. Estimates of quantitative genetic variation in behavioural traits of farmed animals have previously been reported with estimates of heritability ranging widely. The aims of this study were to obtain estimates of genetic and phenotypic parameters for choice consistency in dairy cows milked in AMS herds. Data was obtained from five commercial Danish herds (I – V) having two AMS milking boxes (A, B). Only data from 'Undisturbed' milkings were then used for the statistical analysis in order to fulfill a criterion of 'free choice situation' (713,772 milkings, 1,231 cows). The lactation was divided into 20 segments covering 15 days each, from 5 to 305 days in milk. Choice Consistency scores (CCS) were obtained as percentages of milkings without change of box for each segment. Data were analyzed for one part of lactation at a time using a linear mixed model, for first parity cows alone, and for all parities jointly. Choice consistency was found to be only weakly heritable (h^2 0.02 to 0.14) in first as well as and in later parities, and having intermediate repeatability (t=0.27 to 0.56). Especially the heritability was low at early and late lactation states. These results indicate that consistency, which is itself an indication of repeated similar choices, is also repeatable as a trait observed over longer time periods. However, the genetic background seems to play a smaller role compared to that of the permanent animal effects, indicating that consistency could also be a learnt behavior. It is concluded that consistency in choices are quantifiable but is only under weak genetic control.

Session 25 Poster 13

Comparison of the immune responses of crossbred line of mice selected for two different immunities

T. Miyazaki, D. Ito, Y. Miyauchi, T. Shimazu and K. Suzuki
Tohoku University, Graduate School of Agricultural Science, 1-1 Tsutsumidori-Amamiyamachi, Aoba-ku, Sendai, 981-8555, Japan; k1suzuki@bios.tohoku.ac.jp

In this study, we investigated whether selective breeding for disease resistance could be successful in mice. The mice were selected for peripheral blood immunity, and phagocyte activity (PA) and antibody production (ABP) were used as determinants for peripheral blood immunity. The mice were selected for high PA, high ABP, and both high PA and high ABP, and the selected lines were named N line, A line, and NA line, respectively. In addition, a control line (C line) without selection was included in the study. To examine how immunity is inherited during crossbreeding, we compared the immune response to vaccinations among the crossbred lines, N, A, NA, and the C line. A crossbred line was created by reciprocal crossing of the A and N lines. In total, 72 mice of the 6 male and 6 femalefrom 6 lines were used. Sheep red blood cell (SRBC) was injected into the selected and crossbred lines as an antigen. The body weight, phagocyte activity, SRBC-specific antibody levels, total white blood cell number, peripheral-blood lymphocyte/granulocyte ratio, and T cell subset ratio were determined before and after SRBC injection. The phagocyte activity, plasma IgG1 level, total white blood cell number, number of T cells, B cells, myeloid cell, CD4+ T cells, and CD8+ T cells of the crossbred lines were intermediate between those of the N and A lines. On the other hand, the peripheral blood IgA level of the crossbred line was identical to that of the A line, which shows that the A line is completely dominant over the N line. Moreover, the IgG2a levels and the B cell number were significantly different among the crossbred lines. These results show that most of the immunity-related traits in the crossbred lines created by crossing the lines selected for high PA and high ABP were intermediate to those of the parent lines. However, heterosis was observed in a few traits.

Genetic relationships of lactation persistency with test-day milk yields and somatic cell scores

T. Yamazaki[1], K. Hagiya[1], H. Takeda[2], O. Sasaki[2], S. Yamaguchi[3], M. Sogabe[3], Y. Saito[3], Y. Nakagawa[3], K. Togashi[4], K. Suzuki[5] and Y. Nagamine[6]
[1]NARO Hokkaido Agricultural Research Centre, Sapporo, 062-8555, Japan, [2]NARO Institute of Livestock and Grassland Science, Tsukuba, 305-0901, Japan, [3]Hokkaido Dairy Milk Recording and Testing Association, Sapporo, 060-0004, Japan, [4]Livestock Improvement Association of Japan, Tokyo, 135-0041, Japan, [5]GSAS, Tohoku University, Sendai, 981-8555, Japan, [6]Nihon University, Fujisawa, 252-0880, Japan; yamazakt@affrc.go.jp

The genetic correlations between milk production traits (daily milk yield and lactation persistency as the difference between milk yields at days 240 and 60) and daily somatic cell score (SCS) within and across first and second lactations in Holstein cows were estimated by using a two-trait, two-lactation random regression test day (TD) animal model. The data set consisted of 200,095 TD milk and SCS records from 21,238 cows in their first lactations and 143,051 records from 15,281 cows in their second. Genetic correlations between milk production traits and daily SCS were estimated from additive genetic variance component estimates of random regression coefficients. Correlated responses of daily milk yield and SCS to selection on EBV of persistency in first and second lactation were predicted. Heritabilities of persistency in first and second lactation were 0.13 and 0.21, respectively. Genetic correlations between daily milk yield in first lactation and daily SCS in both lactations were positive and peaked in early lactation stage. The average genetic correlation between daily SCS and persistency were negative: -0.23 (0.08 to -0.34) and -0.22 (0.23 to -0.53) in the first and second lactations, respectively. Average correlated responses of daily milk yield to the selection on persistency were positive in both lactations. The average responses of daily SCS were negative, especially in second lactation. These results suggested that selection for lactation persistency could help to increase lactation yield in the first and second lactations without increasing SCS.

The polymorphism of DGAT1 gene in Polish maternal PL, PLW and native Puławska breeds

M. Szyndler-Nędza[1], K. Piórkowska[1], K. Ropka-Molik[1] and T. Blicharski[2]
[1]National Research Institute of Animal Production, ul. Sarego 2, 31-047 Krakow, Poland, [2]Institute of Genetics and Animal Breeding of the Polish Academy of Sciences, Jastrzębiec, ul. Postępu 1, 05-552 Magdalenka, Poland; magdalena.szyndler@izoo.krakow.pl

The diacylglycerol O-acyltransferase 1 (DGAT1) gene, which has been identified in mammalian cells, plays a major role in lipid metabolism. The DGAT1 gene is expressed mainly in the small intestine. It is involved in intestinal synthesis of triglycerides and their transport to the lymphatic system. In pig, on chromosome 4, to which the DGAT1 locus was assigned, almost 200 QTLs were identified: QTLs associated with backfat thickness, weight gain, carcass fatness, fatty acid composition, etc. The aim of the study was to determine the DGAT1 gene polymorphism in pigs used as maternal component: Polish Landrace (PL) and Polish Large White (PLW), and also in the Puławska, which is included in conservative breeds. A total of 50 PL, 47 PLW and 51 Puławska sows were investigated. Single nucleotide polymorphism in the gene was identified by PCR-RFLP. The PCR reaction was performed using primers that amplify specific gene fragment and restriction enzymes to detect point mutations (acc. to Nonneman and Rohrer). The 257-bp PCR product was digested with AvaII enzyme. This enzyme detects a single A or G substitution (rs45434075 dbSNP) at position 5,504 (ensemble ENSSSCG00000005918) of intron 2. Three polymorphic forms of the DGAT1 gene were found in both maternal breeds. The PL breed was characterized by a similar number of homozygotes with DGAT1AA (frequency 26%) and DGAT1GG genotypes (22%). Most animals in this breed were heterozygous (52%). In the PLW breed, most sows were of the DGAT1AA (46.8%) and DGAT1AG genotypes (44.7%), and the DGAT1GG was least frequent (8.5%). Practically one polymorphic form of this gene was found in the Puławska breed. Out of the 51 Puławska analysed animals, 49 were of DGAT1AA genotype (96.1%) and only 2 had the DGAT1AG genotype.

Lactoferrin content in milk of dairy cows in relation to gene polymorphisms and udder health status

G. Sender, A. Pawlik, A. Korwin-Kossakowska and J. Oprzadek
Institute of Genetics and Animal Breeding Polish Academy of Sciences, Jastrzębiec, 05-552 Magdalenka, Poland; g.sender@ighz.pl

Lactoferrin (LF), protein present in mammalian milk, is known for its antibacterial properties. Due to its role in the innate immunity, LF gene is among potential candidate genes for mastitis resistance. The aim of the study was to associate changes in LF content in milk with LF gene polymorphism and udder health status. Milk samples were collected from 95 Polish Holstein-Friesian cows to determine LF content (ELISA method) and somatic cell count (SCC). Udder health status was assessed based on the SCC in milk. The threshold value of 200,000 SCC/ml of milk was chosen above which cows were categorized as sub-clinical mastitis and below which they were considered to be healthy. All cows were genotyped with PCR-RFLP method for polymorphisms in LF gene promoter and 5'UTR region at position -926 bp (transition G/A) and at position +33 bp (transition C/G), respectively. The analysis of variance of LF content in milk fitted LF genotype, stage of lactation, udder health status and interaction between stage of lactation and udder health status. The health status affected at P<0.01 LF content in milk. The lowest value of LF content was found in the group of healthy cows in full lactation. There were significant differences in the content of LF in milk between cows of different LF genotypes. Both polymorphisms influenced LF content in milk at P<0.05 for LF -926 bp and at P<0.01 for LF +32 bp. GG genotype at position +32 bp increased the content of LF in milk, and this polymorphism appeared to be the main factor of all fitted in the model.

Identification of reproductive trait loci on chromosomes 7 and 9 of Large White pigs through WGAS

R. Otsu[1], K. Watanabe[1], T. Shimazu[1], T. Matumoto[2], E. Kobayashi[3], S. Mikawa[2] and K. Suzuki[1]
[1]Tohoku university, agriculture, 1-1 Amamiya-machi, Tsutsumidori, Aoba-ku, Sendai, Miyagi, 981-8555, Japan, [2]National Institute of Agrobiological Sciences, Tsukuba, Ibaraki, 305-8602, Japan, [3]National Institute of Livestock and Grassland Science, Tsukuba, Ibaraki, 305-0901, Japan; k1suzuki@bios.ac.jp

Reproductive traits of swine are difficult to improve because records are available only for female pigs and because of low heritability. Reproductive performance traits include total number born (TNB), number born alive (NBA), and piglet weight at weaning (PWW). Genomic improvement in pig reproductive traits requires detailed whole-genome association studies (WGAS) for identifying chromosomal regions and genetic markers that control the variation in these traits. In the present study, we used PorcineSNP60 BeadChip to identify the genes related reproductive traits. Using 13,912 records of litters maintained by the Pacific Ocean Breeding Company, we calculated the TNB, NBA, and PWW for each pig and selected 434 females. DNA was extracted from the peripheral blood of these selected female pigs. DNA samples of 700-1000 ng with a ratio of A260/280 higher than 1.80 and a concentration >20 ng/l were used for genotyping. PLINK was used for the quality control of the identified SNPs. The exclusion criteria for SNPs were as follows: minor allele frequency, <0.01; call rate, <0.95; and Hardy–Weinberg equilibrium, <0.001. Of the 61,565 SNPs analyzed, 37,943 SNPs were selected for association analyses. In these analyses, the first step corrects phenotype by accounting for the fixed effects such as the effects of farm, parity, year, and month and calculates the breeding value; the second step tests the SNP effect by using simple linear regression; and the third step corrects the P-values via the GC method by using the PLINK program. Average litter size was 12.6 piglets and average parity was 2.8. We detected significant SNP regions for TBN and NBA on SSC7 and for PWW on SSC9.

Genomic prediction of milk production traits in German Holstein cows incorporating dominance

C. Heuer and G. Thaller
Christian-Albrechts-University Kiel, Institute of Animal Breeding and Husbandry, Olshausenstraße 40, 24098 Kiel, Germany; gthaller@tierzucht.uni-kiel.de

Genomic selection has mainly been implemented to improve the selection accuracy of young AI-bulls by estimating genome-assisted breeding values. These phenotypes do not contain information other than additive genetic effects. It is almost certain that purely additive models do not fully reflect the nature of complex traits. When investigating phenotypes measured directly on the individual level using genotype information, the incorporation of non-additive genetic effects in prediction models seems promising. A total of 794 German Holstein cows genotyped with a standard 50k SNP-chip were available for a genomic prediction approach as follows: A mixed model using genomic additive and dominance relationship matrices was applied to predict additive and dominance effects of validation groups by 5-fold cross validation. Furthermore, the impact of penalized estimators of both relationship-matrices was analysed using the variable selection approach Carscore. It can be shown that both, the incorporation of dominance effects and penalized estimators of genomic relatedness, are able to improve prediction accuracies of direct phenotypes in dairy cows. Especially the traits fat and protein percentage benefit from variable selection. For fat percentage the correlation between genomic predicted and actually observed phenotypes increased from 0.42 to 047, if penalized relationship matrices were used.

Developing the index of productive value for beef breed bulls in Poland

Z. Choroszy and B. Choroszy
National Institute of Animal Production, Department of Animal Genetics and Breeding, Sarego Str. 2, 31-047 Kraków, Poland; zenon.choroszy@izoo.krakow.pl

The Polish population of beef breed and beef-type cattle stands at around 70,000 animals, of which 25,000 are performance tested. Due to population size, herd structure (25 cows per herd on average) and the use of mostly natural mating, around 500 breeding bulls are needed every year. Individual performance testing can only be used to obtain such a large number of tested bulls from the national population. This test is based on the Total Index (WOZ), which consist of two subindices: the Meatiness Subindex (WM) and the Development Subindex (WR). The WM and WR subindices were constructed based on estimated genetic parameters of traits measured on live animals: body weight (kg) standardized to 210 (M210) and 420 (M420) days of age, measurement of chest circumference (OKLP) and withers height (WKL) in cm, and ultrasound measurement of Eye Muscle (Longissimus dorsi) Depth (USG). Meatiness Subindex (WM)=-32.821 + 0.176 × WKL + 0.170 × M420 + 3.056 × USG. The Meatiness Subindex WM was estimated as a multiple regression equation that determines meat percentage in primal cuts depending on the traits tested. Development Subindex (WR) = 24.99 × M210 + 0.51 × M420 – 1.73 × WKL + 4.89 × OKLP. The Total Index of Productive Value (WOZ) was developed after determining proper weightings for both subindexes. WOZ = 0.6 × WM + 0.4 × WR. The Total Index, introduced into breeding practice, forms the basis for ranking beef breed bulls used for reproduction. To date, a total of 1,624 beef breed bulls raised in Poland have been tested using this method, of which 75% are Limousin bulls in accordance with the breed structure of the population. Biological material is collected from the tested bulls for future use in genomic evaluation.

Session 25 Poster 20

Genetic correlations between type traits of young Polish Holstein-Friesian bulls and their daughters

W. Jagusiak[1], A. Otwinowska-Mindur[1], E. Ptak[1] and A. Zarnecki[2]
[1]Agricultural University, al. Mickiewicza 24/28, 30-059 Krakow, Poland, [2]National Research Institute of Animal Production, ul. Krakowska 1, 32-083 Balice, Poland; rzptak@cyf-kr.edu.pl

The objective of this study was to estimate the genetic correlations of conformation traits of Polish Holstein-Friesian bulls and similarly defined type traits of their daughters. Young bulls were evaluated as required for registration in the herd book and for entering progeny testing. Data were 7 linearly scored (1-9 scale) and 3 descriptive (scored from 50 to 100) conformation traits of 933 young bulls born between 2005 and 2008, and the same traits evaluated in their 65,479 daughters. A two-trait animal model was used to estimate genetic correlations between the type traits of bulls and their daughters. (Co)variance components were estimated by a Bayesian method via Gibbs sampling. Two linear models were used: the linear model for bulls included fixed linear regressions on age at evaluation, fixed effects of herd and classifier, and random additive genetic effect; the linear model for cows contained fixed effects of herd-year-season-classifier, lactation stage, fixed linear regression on age at calving, and random additive genetic effect. Estimates of bulls' heritabilities for all analyzed traits ranged from 0.07 for feet and legs to 0.25 for body depth. Heritabilities of cows were lowest for rear legs rear view (0.05) and foot angle (0.06), and highest for size (0.43). The genetic correlations between similarly described traits of bulls and their daughters were moderate to high (0.42-0.91). The lowest genetic correlation (0.42) was for chest width, and the highest (0.91) for rump angle. The magnitude of genetic correlations between pairs of type traits of sires and daughters were sufficiently high to suggest that bulls' own conformation evaluations could contribute to breeding value estimation of bulls based on relatives.

Session 25 Poster 21

Estimation of dominance variance with sire-dam subclass effects in a crossbred population of pigs

M. Dufrasne[1,2], V. Jaspart[3], J. Wavreille[4] and N. Gengler[2]
[1]FRIA, 5 Rue d'Egmont, 1000 Brussels, Belgium, [2]Animal Science Unit, Gembloux Agro-Bio Tech, University of Liege, 2 Passage des Déportés, 5030 Gembloux, Belgium, [3]Walloon Pig Breeding Association, 4 Rue des Champs-Elysées, 5590 Ciney, Belgium, [4]Walloon Agricultural Research Centre, 9 Rue de Liroux, 5030 Gembloux, Belgium; marie.dufrasne@ulg.ac.be

The most important nonadditive effect is probably dominance. Prediction of dominance effects should allow a more precise estimation of the total genetic merit, particularly in populations that use specialized sire and dam lines, and with large number of full-sibs, like pigs. Computation of the inverted dominance relationship matrix, D^{-1}, is difficult with large datasets. But, D^{-1} can be replaced by the inverted sire-dam subclass relationship matrix F^{-1}, which represents the average dominance effect of full-sibs. The aim of this study was to estimate dominance variance for longitudinal measurements of body weight (BW) in a crossbred population of pigs, assuming unrelated sire-dam subclass effects. The edited dataset consisted of 20,120 BW measurements recorded between 50 and 210 d of age on 2,341 crossbred pigs from 89 Piétrain sires and 169 Landrace dams. A random regression model was used to estimate variance components. Fixed effects were sex and date of recording. Random effects were additive genetic, permanent environment, sire-dam subclass and residual. Random effects, except residual, were modeled with linear splines. Only full-sib contributions were considered by using uncorrelated sire-dam classes. Estimated heritability of BW increased with age from 0.40 to 0.60. Inversely, estimated dominance decreased with age, from 0.28 to 0.01. Ratio of dominance relative to additive variance was high at early age (58.3% at 50 d) and decreased with age (2.6% at 200 d). Those results showed that dominance effects might be important for early growth traits in pigs. However, this need to be confirmed and dominance relationships will be included in the next steps.

Session 25 Poster 22

Paternal genomic imprinting and maternal animal models

A. González-Rodríguez[1], E.F. Mouresan[1], J. Altarriba[1], C. Díaz[2], C. Meneses[2], C. Moreno[1] and L. Varona[1]
[1]Universidad de Zaragoza, Genética Cuantitativa y Mejora Animal, Miguel Servet, 177, 50013, Spain, [2]Instituto Nacional de Investigación y Tecnología Agraria y Alimentaria, Mejora Genética Animal, 28020 La Coruña, Spain; gr.aldemar@gmail.com

Maternal animal models are used for genetic evaluation of beef cattle. However, it is usual to report a strong negative genetic correlation between direct and maternal effects. In a previous study, we showed that the potential presence of non-considered paternally imprinted genetic effects may generate an inflation of the additive and maternal genetic variance components and produce a negative genetic correlation between them. In order to clarify this statement we analyse three datasets of weaning weight in Spanish Beef Cattle populations, Pirenaica (PI), Rubia Gallega (RG) and Avileña-Negra Ibérica (ANI). We used 17,106 weaning weight for PI, 41,248 for ANI and 55,631 for RG. The pedigrees consisted of 34,418, 60,136 and 80,166 individuals for PI, ANI and RG, respectively. Up to eight statistical models were assumed, that includes exclusively a direct additive genetic effect (D), direct and paternal effects, with (DP) and without (DPnull) correlation between them, direct and maternal (DM and DMnull), paternal and maternal (PM and PMnull) and direct, paternal and maternal effects (DPMnull). Overall an important presence of paternal effects was observed in models that include it (DP, DPnull, PM, PMnull and DPM). These estimates were always greater than maternal effects and even than direct genetic effects. Further, models that do not consider paternal effects (DM) reports a relevant overestimation of direct and maternal variances, jointly with a strong negative covariance between them. These phenomena can be explained as a consequence of the redistribution of the paternal variance over the effects included in the model of analysis. The results may indicate the presence of sire genomic imprinting effects in beef cattle weaning weight and may explain the presence of a negative genetic correlation between direct and maternal effects.

Session 25 Poster 23

Effect of calving ease and calf mortality on functional longevity in Polish Holstein-Friesian cows

M. Morek-Kopeć[1] and A. Zarnecki[2]
[1]University of Agriculture in Krakow, Al. Mickiewicza 24/28, 30-059 Krakow, Poland, [2]National Research Institute of Animal Production, ul. Krakowska 1, 32-083 Balice, Poland; rzmorek@cyf-kr.edu.pl

Survival analysis was applied to evaluate the effect of first and later calvings on cow functional longevity. Calving ease (CE) and calf mortality (CM) scores of 744,852 first and 1,418,646 later single calvings registered in 2006-2012 were used. CE was scored as without assistance (23.8% in first parity, 40.0% in later parities), with assistance (70.5%, 57.6%), with veterinary assistance (5.5%, 2.3%), difficult calving (0.17%, 0.08%) and caesarean section (0.05%; 0.02%). The last two classes were pooled. CM scores were: live (91.8%, 95.5%) and stillborn or died within 24 h (8.2%, 4.5%). Calving data were merged with longevity records of 1,734,002 cows (50.2% with censored data). Functional longevity was defined as length of productive life corrected for production. Classes of CE or CM scores × parity (1, $\geq$2) × sex of calf were time-dependent fixed effects in the Weibull proportional hazards model together with year-season, parity-stage of lactation, annual change in herd size, fat yield and protein yield, random herd-year-season and time-independent fixed effect of age at first calving. Likelihood ratio tests showed highly significant effects of CE and CM on functional longevity. Difficult calvings were associated with increased risk of culling dams (RRC). In first-parity cows, difficult birth of bull or heifer increased RRC respectively 2.2 or 1.3 times as compared with calvings without assistance. In later parities, RRC related to difficult calving category was 2.0 times (for male calves) and 1.3 times (for female calves) higher than RRC associated with calvings without assistance. Calf mortality showed a negative impact on longevity in both heifers and cows. First-parity stillbirth increased RRC depending on sex of calf by 18% in females and by 15% in males. A smaller increase of RRC (respectively 7% and 9%) was observed in later parities.

Estimation of causal structure of growth and reproductive traits in Syrian hamsters

T. Okamura, M. Nishio, E. Kobayashi and M. Satoh
NARO Institiute of Livestock and Grassland Science, 2 Ikenodai, Tsukuba, Ibaraki, 305-0901, Japan; okamut@affrc.go.jp

Phenotypic and genetic relationships among reproductive traits have traditionally been studied by using standard multiple trait models. Causal structure may exist between phenotypes of such economic traits in livestock, but there have been few studies of causal structure in this context, even though understanding causal structure is very important for animal breeding. Valente *et al.* have developed techniques for estimating causal structure by using a structural equation model and inductive causation (IC) algorithm. Here, we estimated the causal structure between growth and reproductive traits in Syrian hamsters randomly selected over four generations. Five traits of 757 females were used: body weight at 8 weeks (BW8), total number born alive (NB), litter weight at birth (WB), total number at weaning (NW), and litter weight at weaning (WW). To estimate causal structure, graphical modeling was performed with structural equation models and an IC algorithm. For NB, WB, NW, and WW, generation was considered as a fixed effect and age of dam at birth as a covariate. For BW8, generation was considered as a fixed effect and litter number as a random effect. These effects were calculated preliminarily by using a restricted maximum likelihood with an animal model, and the adjusted phenotypic values were used for graphical modeling. After fitting of the multitrait animal model, the posterior distribution of the residual covariance matrix was obtained from 150,000 iterations with a Gibbs sampler. Causal structure was estimated by using this matrix and the IC algorithm. The causal structure consisted of four relationships: BW8→NB (the magnitude of the change in NB due to a 1 g increase in BW8: 0.065 heads/g), NB→NW (0.29 heads/head), NB→WB (1.9 g/head), WW→NW (0.026 heads/g), where the arrow represents a direct effect. The causal relationship between NW and WW was in conflict with conventional beliefs about animal breeding that NW affects WW.

Identification of QTL for prognostic ketosis biomarkers in primiparous dairy cows

J. Tetens[1], C. Heuer[1], M.S. Klein[2], G. Thaller[1], W. Gronwald[2], W. Junge[1], P.J. Oefner[2] and N. Buttchereit[1]
[1]Christian-Albrechts-University Kiel, Inst. of Animal Breeding & Husbandry, Olshausenstr. 40, 24098 Kiel, Germany, [2]University of Regensburg, Institute of Functional Genomics, Josef-Engert-Str. 9, 93053 Regensburg, Germany; jtetens@tierzucht.uni-kiel.de

Ketosis is one of the most common metabolic disorders in dairy cattle. The pronounced energy deficit in early lactation leads to excessive lipomobilization and subsequent accumulation of a critically high amount of ketone bodies. The concentration of these metabolites in milk can be applied to the diagnosis of existing clinical and subclinical ketosis. A recent study, however, using NMR metabolomic data revealed that the glycerophosphocholine (GPC) to phosphocholine (PC) ratio in milk can be used as a prognostic biomarker for the risk of ketosis. In this study, healthy animals had significantly higher levels of milk GPC and lower levels of PC in early lactation than animals suffering from ketosis. It was hypothesized that high GPC/PC ratios reflect higher rates of blood phosphatidylcholine (PtC) breakdown. These animals might thus be able to utilize more blood PtC as a fatty acid source for milk lipid synthesis reducing the need for lipomobilization. Within the current study, a mixed-model GWAS for GPC, PC and GPC/PC was performed in 237 primiparous HF cows using the Illumina 54k Chip. A genome-wide significant QTL for PC was identified on BTA19. A second QTL on BTA25 was significant for GPC as well as GPC/PC. Assuming increased PtC breakdown to cause the elevated GPC/PC ratios, it is plausible that GPC and GPC/PC are influenced by the same single QTL. The second QTL solely influencing PC is, however, unexpected. Analyses regarding possible interactions between the QTL will be conducted to further unravel the genetic architecture and the respective QTL regions will be inspected for potential candidate genes involved in phospholipid metabolism.

Zebu admixture proportions in two taurine cattle populations of Burkina Faso

I. Álvarez[1], A. Traoré[2], I. Fernández[1], M. Cuervo[3], H.H. Tamboura[2] and F. Goyache[1]
[1]SERIDA-Deva, Camino de Risoseco 1225, 33394 Gijón, Spain, [2]INERA, BP 8645 Ouagadougou 04, Burkina Faso, [3]SERPA S.A., Luis Moya Blanco, 261, 33203 Gijón, Spain; fgoyache@serida.org

Introgression of Sahelian zebu genes into taurine cattle populations of West Africa jeopardise trypanotolerance. Here we estimate zebu admixture coefficients (ZAC) in either the Lobi (LBF; 34 samples; provinces of Noumbiel, Dano and Poni, South eastern Burkina Faso) and the N'Dama (NBF; 33 samples; province of Mangodara, South-western Burkina Faso) cattle breeds of Burkina Faso using 38 microsatellites and the program LEADMIX. Parental (P) populations used were as follows: (1) P1; East African zebu (EAZ; 32 individuals including Bororo, Danakil, Raya Azebo and Boran breeds) or West African zebu (WAZ; 92 samples belonging to the Peul, Azawak and Goudali breeds); and (2) P2; N'Dama from Congo (NCo; 42 samples), South African Bonsmara (BON; 34 samples) or European cattle (EC; 83 individuals of 21 different breeds). The NBF samples were also used as P2 for LBF. When the same pair P1-P2 was used, ZACs were consistently higher for NBF than for LBF. ZACs were also higher when P1 was WAZ. The estimated ZACs increased using sequentially NCo, BON and EC as P2 except for the LBF breed and the pairs EAZ-NCo (0.223) and EAZ-BON (0.170). When NBF was used as P2 for the LBF breed, the ZACs estimated were negligible (0.0005) regardless of P1 was WAZ or EAZ. Here we show that introgression of Sahelian zebu cattle genes into southern Burkina Faso follows a complex pattern that may not follow a simple latitude-linked pattern. The estimate of ZACs largely depens on the populations assumed to be parental: two geographically close populations (LBF and NBF) may share most neutral genetic background while two populations of the same breed (NBF and NCo) may not; WAZ, geographically closer than EAZ to the derived populations, gives higher ZACs; the EC used as P2 gave the higher ZACs (higher than 0.750); the composite BON (Afrikaner, Hereford and Shorthorn) gave intermediate results.

Investigation of genes related to lipid metabolism as candidate for sexual precocity in Nellore

M.M. Dias[1], H.N. Oliveira[1], F.R.P. Souza[2], L.G. Albuquerque[1], L. Takada[1] and I.D.P.S. Diaz[1]
[1]State University of Sao Paulo, Animal Science, Via de Acesso Prof. Paulo Donato Castellane s/n, 14884-900, Brazil, [2]Federal University of Pelotas, Rua Gomes Carneiro, 96001-970, Brazil; marina.mortati@gmail.com

The aim of this study was to evaluate possible associations between known polymorphisms in genes related to adipose tissue and sexual precocity in Nellore cattle. 1,085 precocious and non-precocious heifers belonging to Delta G Connection (Conexão Delta G) breeding program were analyzed. A subset of SNPs from the panel of High Density Bovine SNP BeadChip of 777,000 SNPs was evaluated. This subset of SNPs is located within a region of candidate genes with a distance up to 5 Kb, since it is considered that in this distance there is linkage disequilibrium (LD). Only 445 precocious and non-precocious heifers were genotyped, for the remained 640, the average number of copies of each allele from the genotyped population, was used. The statistical evaluation was made by linear models. To analyze the reconstruction of haplotypes and LD presence, the fastPHASE and GenomeStudio softwares were used, using r^2 procedure. In total, 54 candidate genes and 443 SNPs were analyzed. Among these SNPs, 370 formed 83 haplotypes while the remained SNPs were studied separately. The statistical analyses revealed that only two sets of haplotypes, formed by two and four SNPs located on FABP4 and PPP3CA, and one isolated SNP on PPP3CA gene, had significant effect ($P<0.05$) for the sexual precocity trait. These results indicate that FABP4 and PPP3CA genes have an influence on sexual precocity of the animals and should be considered in selection breeding programs in Nellore cattle to evaluate this trait.

Session 25 Poster 28

Detection of QTL influencing egg production in layers receiving various diets

H. Romé[1], A. Varenne[2], F. Hérault[1], H. Chapuis[3], C. Alleno[4], A. Vignal[5], T. Burlot[2] and P. Le Roy[1]
[1]INRA, UMR1348 PEGASE, 35590 Saint-Gilles, France, [2]Novogen, Novogen, 22800 Le Foeil, France, [3]SYSAAF, URA, 37380 Nouzilly, France, [4]Zootests, Zoopole, 22440 Ploufragan, France, [5]INRA, UMR0444 LGC, 31326 Castanet-Tolosan, France; helene.rome@rennes.inra.fr

Egg production in layers may well be affected by the diet composition. Behind this phenotypic observation, the relative genetic values of candidates to selection may vary. Furthermore, the genetic architecture of traits, i.e. the quantitative trait loci (QTL) influencing traits, could be different according to various environments. A population of 440 sires was genotyped using a high-density SNP Affimetrix chip (600K) and used to search for QTL affecting laying intensity and egg quality traits. A total of 31,539 crossbred daughters, issued from the 440 roosters, were phenotyped in 3 hatches in collective cages of 12 hens. An half of them were feed with a high energy diet (2,881 Kcal) and the other with a low energy diet (2,455 Kcal). Egg number per cage was registered every day from 18 to 75 weeks of age and eggs were collected 2 times, i.e. at the age of 50 and 70 weeks, for quality measurements. Laying intensity of each cage was calculated for 3 periods, i.e. 18-30, 31-50, 51-75 weeks of age, as the ratio between egg number and hens*day in the cage during the period. Egg quality measurements concerned the yolk ratio, the Haugh unit, the egg weight, the egg shell color (L*a*b Minolta coordinates), the fracture force, the shape index, diameter and deformation. All these data were adjusted for the hatch effect using the SAS-GLM procedure. Two 'phenotypes', one per diet, were calculated for each sire as the mean of his daughters' phenotypes. The PLINK software was used to filter the genotyping data and to carry out GWAS analyses. The QTL stability according to diet is variable. These results highlight the interest of taking into account genetic by environment interactions in genetic evaluation of layers.

Session 25 Poster 29

Improving genetic evaluation of litter size and piglet mortality by combining marker information

X. Guo[1], O.F. Christensen[1], T. Ostersen[2], D.A. Sørensen[1], Y. Wang[3], M.S. Lund[1] and G. Su[1]
[1]Aarhus University, Dept. of Molecular Biology and Genetics, Blichers Allé 20, 8830 Tjele, Denmark, [2]The Danish Agricultural and Food Council, Pig Research Centre, Axeltorv 3, 1609 Copenhagen, Denmark, [3]China Agricultural University, College of Animal Science and Technology, Yuanmingyuan West Road No. 2, 100193 Beijing, China, P.R.; xiangyu.guo@agrsci.dk

A single-step model allows genetic evaluation using information of genotyped and non-genotyped animals simultaneously. This study compared traditional BLUP and single-step model for genetic evaluation of litter size and piglet mortality for non-genotyped animals in the Danish Landrace population. The data used in current study consisted of 701,588 records of 281,642 Danish Landrace sows born between 1998 and 2011. The individuals were divided into a training population and a validation population by the birth date April 1st, 2007. The traits in the analysis were total number of piglets born, litter size at five days after born and piglet mortality rate. Among 300,186 animals in the pedigree data, 704 boars and 1,079 sows were genotyped using Illumina PorcineSNP60 BeadChip. Two prediction models were used. One was a BLUP model with pedigree-based relationship matrix, and the other was a single-step model with a combined relationship matrix constructed from marker and pedigree information. The reliabilities of EBV (r^2_{EBV}) were measured as squared correlation between EBV and corrected phenotypic value, divided by the heritability of the trait. It showed that r^2_{EBV} from the single step model were higher than those from the conventional BLUP not only for genotyped animals but also for non-genotyped animals in the test data. There was a trend that r^2_{EBV} decreased with increasing distance from training population. The trend was more profound for EBV from the conventional BLUP model than the single step model. The results indicate that even only small number of genotyped animals, r^2_{EBV} for all candidates increases considerably when using a single-step model.

QTL detection for growth and carcass quality traits thanks to a high density SNP chip in pig

F. Hérault[1], M. Damon[1], J. Pires[2], J. Glénisson[2], C. Chantry-Darmon[3], P. Cherel[2] and P. Le Roy[1]
[1]INRA, UMR1348 PEGASE, 35590 Saint-Gilles, France, [2]Hendrix-Genetics, RTC, 45808 Saint Jean de Brayes, France, [3]LABOGENA, 78352 Jouy-en-Josas, France; frederic.herault@rennes.inra.fr

Genetic improvement of carcass quality in pig implies traits measurements on related animals. In such situation, marker-assisted selection could lead to greater genetic gain than phenotypic selection. A family structured population of 454 F2 pigs was produced as an inter-cross between 2 commercial sire lines in order to detect quantitative trait loci (QTL) for growth and carcass quality traits. Animals from the 3 generations of the experimental design were genotyped using the Porcine 60k SNP Illumina Beadchip. Linkage Disequilibrium and Linkage Analyses were performed according to a maximum-likelihood interval mapping methodology using the QTLMap software. A total of 77 QTL were detected at the 5% significance chromosome-wide level. Four of these QTL exceeded the genome-wide 5% significance threshold. Thirty nine QTL influence growth or body composition traits and 34 QTL influence meat quality traits. QTL affecting the average daily gain were detected on SSC4, 6, 14 and 15. QTL affecting carcass composition traits were detected on all chromosomes, except SSC10, 17 and 18. Finally, QTL were detected for early and ultimate pH, colour measurements (L*a*b Minolta coordinates), shear force measured on raw and cooked meat, intramuscular fat content and glycolytic potential, on all chromosomes, except SSC12 and 13. Several QTL were co-localized suggesting pleiotropic effects for some chromosomal regions. Thus, significant QTL were detected in the present study for a large scale of production traits. Additionally, a transcriptome analysis of LM and SM samples, obtained shortly after slaughter, was realized to detect expression QTL (data not shown). These data may be useful to identify causal polymorphisms of QTL and to exploit them in efficient marker-assisted selection programs.

Candidate genes for fatty acid composition assessed with FT-NIR spectroscopy in heavy pigs

D. Guiatti, B. Gaspardo, M. Fanzago, S. Sgorlon, C. Fabro and B. Stefanon
Università di Udine, Dipartimento di Scienze Agrarie ed Ambientali, via delle Scienze 206, 33100 Udine, Italy; bruno.stefanon@uniud.it

The variability in fatty acid compositions of adipose tissue and muscle has several effects on meat quality. The fatty acid composition influences the firmness of adipose tissue and the oxidative stability of muscle, affecting flavour and colour and represents one the main targets of the Italian pig breeding industry. This paper evaluates putative associations of candidate genes for the fatty acid composition of adipose tissue and muscle in a population of 800 pigs, crosses of Italian Duroc × Large White (400 pigs) and Commercial hybrid × Large White (400 pigs). Pigs were reared in commercial farms, with similar feeding and environmental conditions. Fat and muscle samples were collected from tights at slaughterhouse and analyzed with Fourier transform near infrared (FT-NIR) spectroscopy. A total of 159 fat samples and 157 muscle samples were used as calibration set to define a principal components regression model, applying partial least square regression algorithm with full cross validation as internal validation. External validation was performed in 42 and 36 unknown samples respectively. Carcass weight, tight weight, lean percentage calculated with FOM, mm of lean and mm of backfat were also recorded and a muscle sample was collected for DNA analyses for each animal. The polymorphism of 103 SNPs in the promoter regions of 52 candidate genes already known for association with fat traits were selected by in silico analysis. On the basis of allele frequencies, 67 SNPs resulted segregating in both hybrids, while 6 SNPs segregated only in Duroc × Large White. Association analysis was carried out with Pearson correlation test and Bonferroni correction and evidenced significant correlations ($P<0.01$) between fat traits and analyzed polymorphisms.

New SNP in calpastatin gene associated with meat tenderness and frequency in different cattle breeds

L.P. Iguácel[1], J.H. Calvo[1,2], J.K. Kirinus[3], M. Serrano[4], G. Ripoll[1], I. Casasús[1], M. Joy[1], L. Pérez-Velasco[1], P. Sarto[1], P. Albertí[1] and M. Blanco[1]
[1]CITA, Avda. Montañana 930, 50059 Zaragoza, Spain, [2]ARAID, C/ María de Luna, 11, planta 1ª, 500018 Zaragoza, Spain, [3]Universidade Federal de Santa Maria, Av. Roraima, 1000, 97105-900, Santa Maria, Brazil, [4]INIA, Crta. de la Coruña, km 7,5, 28040 Madrid, Spain; icasasus@aragon.es

Calpastatin (CAST) inhibits µ- and m-calpain activity and, therefore, regulates post-mortem proteolysis, being some SNPs in CAST associated with meat tenderness. In this work, a new SNPs located at exon 7 (position BTA29: 98535683 on UMD 3.0) was associated with meat tenderness (P=0.001) in Parda de Montaña cattle breed. The frequencies of this SNP were 45.8% AA, 45.1% AG and 9.0% GG. This mutation changes the amino acid sequence at position Thr182Ala and could affect the electrostatic charges localized in the interacting regions between the calpastatin L-domain and calpain. Moreover, heterozygous genotypes did not show differences with intermediate tenderness, indicating an autosomal recessive inheritance effect of the Thr182Ala mutation for this trait. The effect of the genotype of the Thr182Ala mutation on tenderness was higher (0.84 SD) than the effect of other SNPs found in the CAST gene. Furthermore, samples of different cattle breeds (Parda de Montaña, Pirenaica, Bruna dels Pirineus and Holstein-Friesian) were collected to study the frequency of this new variant, finding similar genotype frequencies in these breeds. Furthermore, the alignment of the sequences of CAST deposited in the GenBank database revealed that this new variant was present in domestic cattle (*Bos taurus*; Hereford breed) and in Yak (*Bos grunniens*). Funcional studies are necessary to test the effect of the CAST Thr182Ala genotypes on calpastatin activity to confirm the effects of this new polymorphism found in the current work.

An example of nutrigenomics and nutrigenetics in ovine: Stearoyl-CoA Desaturase (SCD)

L. González-Calvo[1], E. Dervishi[1], M. Serrano[2], G. Ripoll[1], F. Molino[1], M. Joy[1] and J.H. Calvo[1,3]
[1]CITA Aragón, Avda. Montaña 930, 50059 Zaragoza, Spain, [2]INIA, Ctra. La Coruña, Madrid, Spain, [3]ARAID, M. Luna, Zaragoza, Spain; lgonzalezc@aragon.es

Two experiments were conducted to study the effect of feeding system and a polymorphism located at SCD promoter on the SCD gene expression. In the first experiment 44 Rasa Aragonesa male lambs were grouped according to the feeding system in grazing alfalfa (ALF, n=22) and indoor concentrate feeding (IND, n=22). In the second experiment 58 Rasa Aragonesa male lambs were allocated in 2 groups: 8 lambs with their dams feed with grazing alfalfa (ALF) and the rest (n=50) feeding commercial concentrate supplemented with 500 mg of dl-α-tocopheryl acetate/kg for different days before slaughter (IND). Lambs were slaughtered at 22-24 kg live-weight, and a piece of Semitendinous (ST) and L. thoracis (LT) muscles from experiment 1 and 2, respectively, were collected to gene expression and SNP genotyping. SCD gene expression levels were determined by real time-PCR, and normalized using 3 housekeeping genes in each tissue: GAPDH, ACTB, and B2M in ST and RPL19, B2M and YWHAZ in LT. In the first experience IND group showed higher levels (7.7-fold) of SCD expression comparing with the ALF lambs (P=0.03), finding significant differences among the genotypes of the polymorphism located at the SCD promoter. Thus, AA animals (n=1) express 72 fold more than CC animals (n=10) (P=0.02) and CA (n=11) animals express 5 fold more than the CC ones (P=0.04). However, in the second experiment only significant differences were found in relative SCD gene expression in LT between the AA (n=7) and CA (n=25) genotypes in animals feeding concentrate (P=0.03). Results indicate that the feeding system acts as modulator of the effect of the polymorphism located at the SCD promoter over the gene expression in both muscles. In alfalfa lambs the genotype had not effect on gene expression while in lambs feeding concentrate the genotype seems to be implied in the regulation of the gene expression.

Session 25 Poster 34

Next generation sequencing and *de novo* assembly of a Nelore (*Bos indicus*) bull genome

L.C. Cintra[1], A. Zerlotini[1], F.P. Lobo[1], F.R. Da Silva[1], P.F. Giachetto[1], P.K. Falcao[1], L.O.C. Silva[2], A.A. Egito[2], F. Siqueira[2], N.M.A. Silva[3], S.R. Paiva[3], M.E.B. Yamagishi[1] and A.R. Caetano[3]
[1]Embrapa Informática Agropecuária, CP6041, 13083-886, Campinas, SP, Brazil, [2]Embrapa Gado de Corte, Av Rádio Maia, 79106-550 Campo Grande, MS, Brazil, [3]Embrapa Recursos Genéticos e Biotecnologia, CP 02372, 70770-917 Brasília, DF, Brazil; alexandre.caetano@embrapa.br

Bos indicus cattle breeds present several natural adaptations to biotic and abiotic stresses found in the tropics and have been extensively used for dairy and beef production in these regions of the world. A *B. indicus* genome assembly represents an essential tool which will be vital to help identify and understand the underlying genetic variations that distinguish taurine and indicine cattle, which have diverged >250,000 years ago, as well as facilitate the work of breeder associations striding towards incorporating genomic tools into ongoing genetic evaluations and breeding programs to improve productivity and beef and milk quality traits. DNA obtained from semen from a Nelore bull born in 1987, with an estimated cumulative inbreeding coefficient of 29.4%, and that can be traced to animals imported from India, was used to produce 100 bp paired-end sequences from short (300 and 700 bp) and long insert (3, 5 and 10 kbp) libraries, with an Illumina HiSeq platform. A total of 1,201 Gbp were sequenced, corresponding to 45× raw coverage of the genome. The SOAP de novo assembler was used to build contigs and scaffolding. Several parameters sets were evaluated to obtain the best assembly based on the number of scaffolds, number of bases in scaffolds, N50, and total gap length. The best assembly obtained so far contains 2.7 Gbp, 15,103 scaffolds with N50 of 649 Kbp and 756 Mbp of gaps. Current results are being used to target additional sequencing of specific libraries to improve scaffold assembly. In addition, additional data generation using different sequencing technologies is underway to improve sequence assembly quality before comparisons with the reference *B. taurus* sequence are performed.

Session 25 Poster 35

Associations of SNPs in leptin and Pit-1 genes with long-life milk production traits in cattle

N. Moravčíková, A. Trakovická and R. Kasarda
Slovak University of Agriculture in Nitra, Department of Animal Genetics and Breeding Biology, Tr. A. Hlinku, 94976 Nitra, Slovak Republic; nina.moravcikova1@gmail.com

The leptin (LEP) and pituitary specific transcription factor (Pit-1) genes were studied as genetic markers of long-life milk production traits in Slovak Simmental and Pinzgau cattle in this study. Use of genetic markers can aid on the identification of animals with better breeding value in dairy cattle. The total numbers of blood samples were taken from 288 samples of Slovak Simmental and 85 Pinzgau cows. Genomic DNA was used in order to estimate LEP/Sau3AI (BTA 4, inron 2) and Pit-1/HinfI (BTA 1, exon 5) genotypes by means of multiplex PCR-RFLP method with using Sau3AI restriction enzyme (LEP) and HinfI restriction enzyme (Pit-1). In Slovak Simmental and Pinzgau cows were the allele frequencies 0.839/0.694 and 0.161/0.306 for A and B LEP variants, and 0.226/0.353 and 0.774/0.647 for A and B Pit-1 variants, respectively. Our data shows relatively median level of polymorphic information content of loci across breeds based on the heterozygosity (0.38). The standard Nei's genetic distance Ds and the genetic distance Da between populations based on loci frequencies were 0.0231 and 0.0129, respectively. The statistical analyses show in both populations significant effect of LEP/Sau3AI genotype on long-life milk production traits ($P \leq 0.05$), with A as a desirable allele. In Slovak Simmental cows were the milk, protein and fat yield significantly higher in the AA compared with BB genotype cows with differences 5,505.5, 180.1 and 217.9 kg, respectively. The milk, protein and fat yield in Pinzgau cows were similarly significantly higher in cows with AA genotype compared to AB genotype (differences 5,488.3, 190.3 and 210.8 kg). The statistical analysis of Pit-1/HinfI genotypes effect on production traits were in both populations non significant, but results indicates potential positive effect of heterozygous cows on milk production.

Accuracy of genomic prediction in Chinese triple-yellow chicken

T. Liu[1,2], H. Qu[1], C. Luo[1], D. Shu[1], M.S. Lund[2] and G. Su[2]
[1]Institute of Animal Science, Guangdong Academy of Agricultural Sciences, Dafeng 1st Street 1, Wushan, Tianhe District, Guangzhou 510640, China, P.R., [2]Department of Molecular Biology and Genetics, Aarhus University, Blichers Allé 20, P.O. Box 50, 8830 Tjele, Denmark; tianfei.liu@agrsci.dk

This study investigated the efficiency of genomic prediction in Chinese triple-yellow chicken. The data comprised 511 birds from 8 half-sib families. The birds were genotyped using the Illumina Chicken 60K SNP Beadchip. Two growth traits and three carcass traits were analyzed, i.e. body weight at 6th week, body weight at 12th week, eviscerating percentage, breast muscle percentage, and leg muscle percentage. Genomic prediction was assessed using a four-fold cross validation procedure for two validation scenarios. In the first scenario, each test data set comprised two half-sib families (family sample). In the second scenario, the whole data were randomly divided into four subsets (random sample). Genomic breeding value (GEBV) was predicted using a genomic BLUP model, a Bayesian LASSO model, and a Bayesian mixture model with four distributions. Pedigree index for the test birds was also estimated using a conventional BLUP model. Accuracy of GEBV was evaluated based on correlation between GEBV and corrected phenotypic value (Cor(GEBV, Yc)). Using the three models, Cor(GEBV, Yc) ranged from 0.45 to 0.47 for the two growth traits, and from 0.18 to 0.26 for the three carcass traits in the scenario of family sample, and the correlations were between 0.47-0.52 and between 0.33-0.45 in the scenario of random sample. The differences between prediction accuracies of the three models were very small. The correlations between conventional pedigree index and corrected phenotypic value in general were close to zero in the scenario of family sample, and ranged from 0.11 to 0.35 in the scenario of random sample. The results indicate that genomic selection can greatly increase accuracy of selection, compare with conventional selection.

Genetic determinism of sexual development and boar taint in Pietrain and Pietrain × Large White pigs

C. Larzul[1], A. Prunier[2], N. Muller[3], S. Jaguelin[2], R. Comté[2], C. Hassenfratz[4] and M.J. Mercat[4]
[1]INRA, UMR1313 GABI, Domaine de Vilvert, 78350 Jouy-en-Josas, France, [2]INRA/Agrocampus, UMR1348 PEGASE, Domaine de la Prise, 35590 Saint-Gilles, France, [3]INRA, UETP, Domaine de la Motte, 35653 Le Rheu, France, [4]IFIP, Domaine de la Motte, 35653 Le Rheu, France; catherine.larzul@jouy.inra.fr

The European pig industry is engaged in a voluntary abandonment of surgical castration of male piglets by 2018. A condition of this abandonment is to solve different problems related to breeding boars, especially meat quality. One of the best ways is that of genetics, including genomic selection. A resource population has been implemented using French Pietrain pigs raised in a control animal testing station. About 1500 Pietrain type or cross-type Pietrain × Large White boars were raised and slaughtered to determine the characteristics of sexual development, body composition and above all meat quality traits. Plasma testosterone concentration differed from one genetic type to another, but the differences were generally small. This concentration was not affected by the batch or by the weight of the animals or by their age at measurement. Androstenone content in backfat was not affected by the batch but the average almost doubled from one genetic type to another, the highest levels being observed both in one Pietrain type and in one crossbred type. Androstenone content increased with slaughter weight and plasma testosterone concentration. The effect of slaughter age depended on the genetic type. Skatole content in backfat was little affected by the concentration of testosterone, but varied from one batch to another, from one genetic type to another and decreased with age at slaughter. Genetic parameters were estimated in the different populations. They were estimated within the range of values usually observed for these traits.

Genotypes imputation as a supporting method in pedigree control in Polish Holstein-Friesian cattle

K. Zukowski[1], A. Gurgul[2], D. Rubis[3] and A. Radko[3]
[1]The National Research Institute of Animal Production, Department of Animal Genetics and Breeding, 1, Krakowska Street, 32-083 Balice, Poland, [2]The National Research Institute of Animal Production, The Laboratory of Genomics, 1, Krakowska Street, 32-083 Balice, Poland, [3]The National Research Institute of Animal Production, Departament of Animal Cytogenetics and Molecular Genetics, 1, Krakowska Street, 32-083 Balice, Poland; dominika.rubis@izoo.krakow.pl

The imputation methods are widely used to predict genotypes of SNPs, but can be also applied to multiallelic markers like microsatellites (MS). Because the MS markers were used for cattle parentage testing through years, in this study we attempt to utilize information from MS and SNPs along with imputation methods, in order to predict the MS genotypes from selected SNPs panels. For this purpose, 11 MS and three different panels of SNPs selected from Illumina Bovine50SNP BeadChip assay of 321 Holstein-Friesian individuals were used. 10 and 20 SNPs spanning MS, as well as additional panel based on SNPs proposed by ICAR/ISAG for parentage testing were chosen for imputation procedure. The Phase and Beagle software were used to impute genotypes for the three models of missing observations (subsets with 100% (no.ms) and 50% MS (ms) missing and combined 50% MS and 50% SNPs missing (snp.ms)). In general, performance of imputation methods was better for Phase than Beagle for 20 SNP window. The mean proportion of genotypes imputed correctly ranged from 18.58% to 74.54%, wherein the mean proportion of haplotype phase imputed correctly ranged from 36.09% and 86.76%. The best results in both cases were observed for combination of: Phase, 20 SNP window and 'ms' subset and were in opposite to Beagle, in which the worst performance was found for parent window and 'no.ms' subset. The highest number (>90%) of genotypes imputed correctly was obtained for 6 out of 11 chromosomes included in analyses for subset consisting of 20 SNPs.

Genomic-wide scan of ovulation rates in beef heifers and cows

A. Vinet[1], J.L. Touzé[2], F. Guillaume[1], J. Sapa[1] and F. Phocas[1]
[1]INRA, UMR1313 GABI, Domaine de Vilvert, 78352 Jouy-en-Josas, France, [2]INRA-CNRS, UMR0085 PRC, Centre de Recherche INRA de Tours-Nouzilly, 37380 Nouzilly, France; aurelie.vinet@jouy.inra.fr

Because ovulation rate is a sex-limited trait that is difficult to collect in commercial farms this trait is an ideal candidate for gene-assisted selection. On a biological perspective, one may wonder whether the same biological mechanisms underline ovulation rate in heifers and cows. The objective of the study was to perform a genome-wide scan for quantitative trait loci (QTL) detection for heifer and cow ovulation rates in an INRA twinning herd of Maine-Anjou cattle. Both LDLA and Bayesian methods were used for the QTL detection. Our findings showed that, despite a genetic correlation close to 1, ovulation rates in heifers and cows can not be considered as the same phenotype. Indeed analyses revealed putative QTL of ovulation rate in heifers and cows in common chromosomic regions but the major part of QTLs, and particularly the most significant QTLs were detected in different chromosomic regions for heifers and cows.

Genetic study of mortality rate in Danish dairy cows: a multivariate competing risk analysis

R.P. Maia[1], P. Madsen[1], J. Pedersen[2] and R. Labouriau[1]
[1]Aarhus University, Molecular Biology and Genetics, P.O. Box 50, 8830 Tejle, Denmark, [2]Knowledge Centre for Agriculture, Agro Food Park 15, 8200 Aarhus, Denmark; rafaelp.maia@agrsci.dk

Dairy cow mortality has been increasing during the last two decades in Denmark. This study aims to verify whether genetic mechanisms are contributing to this increase. To do so, the records of 880,480 Holstein, 142,306 Jersey and 85,206 Red Danish dairy cows calving from 1990 to 2006 were retrieved form the Danish Cattle register. Two causes of culling were considered: death and slaughtering. Bivariate competing risk genetic models with a sire model structure were used to model the death and the slaughtering rates simultaneously. The models included two random components: a sire random component with pedigree representing the sire genetic effects and a herd-year-season component. Moreover, the coefficient of heterosis and the sire breed proportions were included in the models as covariates in order to account for potential non-additive genetic effects due to the massive introduction of genetic material from other populations. The correlations between the sire components for death rate and slaughter rate were negative and small for the 3 populations, suggesting the existence of specific genetic mechanisms for each culling reason and common concurrent genetic mechanisms. In the Holstein population the effects of the changes in the level of heterozygosity, breed composition and the increasing genetic trend act in the same direction increasing the death rate in the recent years. In the Jersey population, the level of heresozygosity and breed proportion effects were small (non-significant), and only the increasing genetic trend can be pointed as a genetic cause to the observed increase in the mortality rate. On the other hand, in the Red Danish population neither the time-development pattern of the genetic trend nor the changes in the level of heterozygosity and breed composition could be causing the observed increase in the mortality; thus, there must be non-genetic factors causing this negative development.

A single nucleotide polymorphism in the 5´UTR of ovine FASN gene is associated with milk fat yield

A. Sanz[1], C. Serrano[1], J.H. Calvo[2], P. Zaragoza[1], J. Altarriba[3] and C. Rodellar[1]
[1]Universidad de Zaragoza, Laboratorio de Genética Bioquímica (LAGENBIO). Facultad de Veterinaria., Miguel Servet 177, 50013 Zaragoza, Spain, [2]Centro de Investigación y Tecnología Agroalimentaria (CITA), Unidad de Tecnología en Producción Animal, Avda. Montañana 930, 50039 Zaragoza, Spain, [3]Universidad de Zaragoza, Grupo de Mejora Genética, Facultad de Veterinaria, Miguel Servet 177, 50013 Zaragoza, Spain; rodellar@unizar.es

Fatty acid synthase (FASN) is a complex homodimeric enzyme that regulates de novo biosynthesis of long-chain fatty acids. Ovine FASN gene maps on OAR11, where a QTL affecting the fatty acid composition of milk sheep has been identified. Besides, some SNPs in the bovine FASN gene have been also associated with milk-fat content and with fatty acid composition of milk and beef. Hence, it is a candidate gene for fat content in milk animals. In this study we screened the 5´ untranslated region of the ovine FASN gene, and we identified a G>A substitution in the untranslated exon 1 (g.982G>A) according to the bovine FASN genomic sequence (AF285607). The SNP was tested in Assaf individuals grouped for milk fat production, including 50 animals with the highest milk-fat content and 50 animals with the lowest milk-fat content. Allele frequencies differed significantly (P=0.0121) between fat content groups. Since the mammalian FASN gene is regulated at both the transcriptional and post-transcriptional levels; this polymorphism could modify the putative Sp1, Sp2, Sp3 and Sp4 transcription factors binding site in the untranslated exon 1 and also could alters FASN mRNA stability. The prediction of RNA secondary structure within the 5´UTR predicts that allele G may produce a more stable folding of the 5´UTR because it has a lower free energy (-43.0 kcal/mole) with respect to alelo A (-36.9 kcal/mole). Those evidences suggest the implication of this polymorphism in the variation in milk-fat content in sheep. We propose that the ovine FASN gene is a candidate gene for a milk-fat content QTL.

Session 25 Poster 42

Immune response of the chicken lymphocytes activated with KLH, LPS and LTA antigens

M. Siwek[1], A. Slawinska[1], K. Sikorski[2] and H. Bluijssen[2]
[1]University of Technology and Life Sciences, Animal Biotechnology Department, Mazowiecka 28, 85-084 Bydgoszcz, Poland, [2]Adam Mickiewicz University, Department of Human Molecular Genetics, Institute of Molecular Biology and Biotechnology, Umultowska 89, 61-614 Poznań, Poland; slawinska@utp.edu.pl

The aim of this study was to unravel details of various types of immune responses in chicken by activating B lymphocytes with a non-pathogenic antigen keyhole limpet hemocyanin (KLH), and two environmental antigens: lipopolisaccharide (LPS) and lipoteichoic acid (LTA). These antigens trigger various signaling pathways, leading to up- or downregulation of the immune-related genes. Their effects were analyzed in a chicken B cell lymphoma cell line (DT40), upon *in vitro* stimulation. This avian cell line is widely used to study gene and protein function and is considered an alternative to animal research. DT40 cell line (DSMZ, Germany) was cultured in RPMI medium supplemented with foetal bovine serum, glucose, sodium pyruvate and β-mercaptoethanol. Stimulation of the B lymphocytes *in vitro* with the given antigen (LPS, LTA and KLH) was conducted in order to induce a respective immune response. Stimulation was performed within a range of doses (1, 5 and 10 µg/ml) and specific time points (3, 6, 9 hrs). Gene expression was assessed with qRT-PCR, performed with EvaGreen chemistry and ubiquitin as a reference gene. Target genes included a panel of cytokines, such as interleukins (i.e. IL-4, IL-6, IL-18) and interferons (i.e. IFN-α, IFN-β and IFN-ɣ). Results were analyzed using ddCt method. The experimental LPS-, LTA and KLH-stimulated groups were compared to the control group with the t-test. Preliminary results indicate that LPS induced expression of IL-18 and IFN-ɣ. These genes may be important regulators of the T_H-2 type immune response in chickens, generated by contact of the B lymphocytes with LPS-containing Gram-negative bacteria. Gene expression study will help to elucidate signaling pathways *in vitro*.

Session 25 Poster 43

Comparison of mapping accuracy between methods predicting QTL allele identity using haplotypes

L.J. Jacquin[1], J.M. Elsen[1] and H. Gilbert[2]
[1]INRA, UR631 SAGA, 31326 Castanet Tolosan, France, [2]INRA, UMR444 LGC, 31326 Castanet Tolosan, France; julien.jacquin@toulouse.inra.fr

Detecting haplotypes in association with trait variability usually exploits local similarities between chromosomes: locally similar chromosomes have better chance to harbor identical QTL alleles than chromosomes showing local differences. Predictors of allelic identity based on 6 marker haplotypes were compared for their ability to correctly predict QTL allelic identity and for their mapping accuracy. The tested methods were: (1) identity by state (IBS) at each tested marker; (2) IBS between the haplotype pairs surrounding the tested position (IBS_hap); (3) a similarity score between haplotype pairs (SCORE); (4) an IBD probability between haplotype pairs derived from a coalescence process; (5) an identity predictor derived from local clustering between haplotypes pairs (BEAGLE); and (6) a predictor trained on the dataset (TP). These predictions were collected in (2n×2n, n being the number of individuals) matrices M^P_{id} and compared to the actual QTL identity between the chromosomes assembled in a 2n×2n matrix M_{QTL}, by calculating the normalized distances between matrices M^P_{id} and M_{QTL} for each tested position. Secondly, mean square errors of the QTL position estimate, obtained from mixed models including a random haplotypic effect with covariance described by the MPid matrices, were computed. Algebraic developments were applied to the normalized distance between matrices to infer the predictors' ability to predict the allelic identity at the QTL depending on the local linkage disequilibrium (LD). Numerical comparisons were performed on simulated datasets based on a set of 235 porcine chromosomes and a 26 generations pedigree. The algebraic developments and simulations showed that IBS_hap was expected to be best when the LD is high. When LD diminishes, the best predictor tended to be SCORE or BEAGLE. Distributions of the normalized distances depending on LD were examined to get further insights on these comparisons.

Session 25 Poster 44

Polymorphism of SNPs dedicated for parentage testing in two Polish cattle populations

D. Rubis[1], A. Gurgul[2], A. Radko[1] and K. Zukowski[3]

[1]The National Research Institute of Animal Production, Department of Animal Cytogenetics and Molecular Genetics, 1, Krakowska Street, 32-083 Balice, Poland, [2]The National Research Institute of Animal Production, The Laboratory of Genomics, 1, Krakowska Street, 32-083 Balice, Poland, [3]The National Research Institute of Animal Production, Department of Animal Genetics and Breeding, 1, Krakowska Street, 32-083 Balice, Poland; dominika.rubis@izoo.krakow.pl

Recent years have seen increasing interest in the use of SNPs (Single Nucleotide Polymorphisms) not only in genomic studies but also in parentage testing. The continuous development of molecular biology techniques enables increasingly rapid and accurate determination of a large number of SNPs. Among the techniques used for high-throughput SNP analysis, particular attention is given to the microarray method. The present work is an attempt at preliminary determination of the parameters of an ISAG SNPs panel for parentage verification of cattle in two genetically distinct Polish cattle breeds. The study was performed with samples obtained from 192 Polish Red cows (RP) and 71 Polish Holstein-Friesian cows (HO). Genotypes were determined using BovineSNP50 BeadChip assay (Illumina). Minor allele frequency (MAF) for the SNPs in Holstein cattle ranged from 0.1 (except for two monomorphic markers) to 0.5, with mean and median values of 0.380 (±0.093) and 0.3873, respectively. For Polish Red cattle, these parameters were similar: mean MAF of 0.3834 (±0.097) and median of 0.4089. The observed heterozygosity for the analysed markers ranged from 0.2113 (HO) and 0.0781 (RP) to 0.6479 (HO) and 0.5625 (RP), with a mean value of ~0.46. For the analysed panel, very low PI values were observed for both populations: $1.84E^{-47}$ for RP and $1.75E^{-47}$ for HO. CPE1 and CPE2 values were 0.99999835 and 0.99999999 for RP cattle, and 0.99999842 and 0.999999999, respectively, for HO cattle. It is concluded that the tested SNP panel, except for 2 monomorphic loci in HO cattle and 1 in RP cattle, could be successfully used for parentage tests in these breeds.

Session 25 Poster 45

Old and recent inbreeding impact in litter composition in Gazella cuvieri

M.B. Ibáñez[1], I. Cervantes[2], F. Goyache[3], E. Moreno[1] and J.P. Gutiérrez[2]

[1]CSIC, Ctra/ Sacramento s/n, 04120 Almería, Spain, [2]Universidad Complutense de Madrid, Avda/ Puerta de Hierro s/n, 28040 Madrid, Spain, [3]SERIDA, Cam Rioseco 1225, 33394 Gijón, Spain; gutgar@vet.ucm.es

Conservation of endangered species through captive breeding programmes focuses on the preservation of genetic variability to avoid any negative impact of inbreeding. Increased inbreeding level exposes harmful recessive deleterious genes to selection and has been associated with reduced fecundity, offspring viability, and individual survivorship; the so called inbreeding depression. However, the genetic consequences of inbreeding by purging deleterious alleles during the bottlenecks benefit the population. The objective of this study was to ascertain the effect of both the old and the recent inbreeding in litter composition in a captive population of *Gazella cuvieri*. Genetic parameters for litter composition (combination of the sex of newborn and the sex of littermate) using 700 calf records were estimated. The total pedigree contained 740 individuals. The trait was assigned to the individual, to the mother, to the father or to both parents. Models included age of the mother (linear and quadratic), year of calving and maternal experience as systematic effects. Maternal and permanent environmental were included as random effects, when necessary. After choosing the best model fitting the data, the old and recent inbreeding were also included as covariates in order to separate their effects, considering the old inbreeding at different pedigree depths (3, 4 and 5 generations). Heritability ranged between 0.05 and 0.29. The recent inbreeding had always a negative effect on the trait whereas old inbreeding in most of the cases gave a positive regression coefficient. Provided that deleterious recessive genes were previously present, these results could suggest the presence of purge in this captive population. Further research is needed in order to elucidate purging occurrence in this population.

Sire × contemporary group to model genotype by environment interaction in genetic evaluations

J.P. Eler[1], M.L. Santana Jr[2], A.B. Bignardi[2] and J.B.S. Ferraz[1]
[1]College of Animal Science and Food Engineering, University of São Paulo, Veterinary Medicine, Av Duque de Caxias Norte, 225, 13630-900 Pirassununga, SP, Brazil, [2]University of Mato Grosso, Campus Rondonópolis, MT-270, km 06, 78735-901 Rondonópolis, MT, Brazil; joapeler@usp.br

The objectives of the present study were: (1) to evaluate the importance of genotype × production environment interaction for the genetic evaluation in a population of composite beef cattle; and (2) to investigate the importance of sire × contemporary group interaction (S×CG) to model genotype by environment interaction (G×E) in routine genetic evaluations. Analyses were performed with different definitions of production environments. Thus, WW records of animals in a favorable environment were assigned to either trait 1, in an intermediate environment to trait 2 or in an unfavorable environment to trait 3. The (co)variance components were estimated using Bayesian approach in single-, bi- or three- trait animal models according to the definition of number of production environments. In general, the estimates of genetic parameters for WW were similar between environments. The mean genetic correlation between direct effects was 0.63 in favorable and unfavorable environments for WW, a fact that may lead to changes in the ranking of sires across environments. When S×CG was included in two- or three-trait analyses, all direct genetic correlations were close to unity, suggesting that there was no evidence of a genotype × production environment interaction. Furthermore, the model including S×CG contributed to prevent overestimation of the accuracy of breeding values of sires, provided a lower error of prediction for both direct and maternal breeding values, lower squared bias, residual variance and deviance information criterion than the model omitting S×CG. Thus, the model that included S×CG can therefore be considered the best model based on these criteria and so, this S×CG interaction should be included in the genetic evaluation of this composite population.

Estimation of genetic parameters of racing traits of Arabian horses in Algeria

S. Tennah, N. Kafidi, N. Antoine-Moussiaux, C. Michaux, P. Leroy and F. Farnir
Faculty of Veterinary Medicine, University of Liege, Department of Animal Production, Unit of Genetics, Biostatistics and Rural Economics, Boulevard le Colonster, B43 Sart Tilman 4000 Liege, Belgium; safia.tennah@doct.ulg.ac.be

The results of the flat races organized in Algeria from 1995 to 2007 by the Algerian Horse Racing Society, were used to estimate genetic parameters of racing performances of Arabian horses. The data consist of 36,492 race records, obtained from 913 horses. The pedigree file of the horses includes 1,812 animals from 166 stallions and 392 mares. The analysis was performed on two traits: the logarithm of average annual virtual earnings per start (LAEV/S) and a normalized ranking (PERF). To identify the fixed effects to be included in the genetic model, a preliminary analysis was conducted using the General Linear Models (GLM) procedure from SAS software. The effects of age (3 to 8 years and older), sex (male or female), year (1995 to 2007) and the interaction between year of the race and age and between sex and age were included in the model for both traits. In addition, two random effects, a direct genetic effect of the animal and a permanent environmental effect were included in the mixed model. The variance components and genetic parameters were estimated by the restricted maximum likelihood (REML), procedure using the MTDFREML program. The analysis, using a repeatable animal model, led to the following estimation of genetic parameters: for LAEV/S, heritability was 0.23 (±0.04), while estimate of repeatability was 0.34 (±0.04). The heritability for the normalized ranking was higher, 0.37 (±0.05), indicating that this trait might be more appropriate for breeding programs of Arabian horses in Algeria. The repeatability estimate for the normalized ranking was 0.59 (±0.04) and the genetic correlation between this trait and LAEV/S was 0.79.

Economical weighting of breeding objectives and definition of total merit indexes in BMC sheep breed

A. Cheype[1], J. Guerrier[2], F. Tortereau[3], D. François[3], J.P. Poivey[4], K. Chile[5] and J. Raoul[6]
[1]Institut de l'Elevage, Boulevard des Arcades, 87060 Limoges, France, [2]Institut de l'Elevage, 9 allée Pierre de Fermat, 63170 Aubière, France, [3]INRA UR 631 SAGA, CS 52627, 31326 Castanet-Tolosan Cedex, France, [4]UMR SELMET, 2 place Viala, 34060 Montpellier, France, [5]ROM Selection, Paysat-Bas, 43300 Mazeyrat d'Allier, France, [6]Institut de l'Elevage, BP 42118, 31321 Castanet-Tolosan Cedex, France; agathe.cheype@idele.fr

Breeding goals of the French Blanche du Massif Central (BMC) sheep breed scheme have been updated by economical weighting. New weights were estimated by the expected change in profit resulting from a change of one physical unit in that trait. Inputs and outputs of a flock were modeled. The breeding objective is now composed as follows: fertility (21%), prolificacy (21%), viability and suckling ability (29%), fat depth (16%), dressing percentage (7%), conformation (4%) and growth (2%). Consequently the weights of traits in total merit indexes have been updated. Those merit indexes are used at different levels of the breed selection scheme. New traits ranking have been set based on economical weights used in breeding objectives and on genetics correlations between traits and theirs estimated breeding values. Weights of the total merit index for the on farm evaluation, which consists in prolificacy and sucking ability traits, have been updated. Rams' meat capacities are selected through performances testing in performance test stations (PTS) and through progeny testing. Maternal traits estimated on ancestry and meat traits estimated on individual performances have been combined in a new total merit index published in PTS. Thanks to a ten-year database on PTS rams, expected response to selection is composed of 37% conformation, 17% fat depth, 17% weight, 12% growth, 9% prolificacy, 8% suckling ability. The total merit index used in progeny testing has been updated on the same way. Introduction of these new indexes in the selection program of the French Blanche du Massif Central sheep breed is in progress.

Inclusion of correlated random effects in proportional hazards frailty models with The Survival Kit

G. Mészáros[1], J. Sölkner[1] and V. Ducrocq[2]
[1]University of Natural Resources and Life Sciences, Vienna, Division of Livestock Sciences, Gregor-Mendel-Str. 33, 1180 Vienna, Austria, [2]INRA, UMR 1313 Génétique Animale et Biologie Intégrative, Domaine de Vilvert, 78352, Jouy-en-Josas, France; gabor.meszaros@boku.ac.at

Frailty models are an extension of the standard survival analysis models which account for unobserved random heterogeneity by including random effects. When two random effects are considered in frailty models, these can be independent from each other or related to some degree. In this case, the variances of the two random effects need to be estimated along with their correlation coefficient. Typical examples in dairy cattle are the modeling of the sire's influence on culling in early and late life or early and late lactation of their daughters, leading to estimations of 'time dependent sire' effects. We demonstrate the use of a Weibull frailty model with the Survival Kit in such cases using simulated data sets. Set1 and Set2 considered 50 and 100 levels of the random effects, respectively, with 100 records associated with each level, resulting in sample sizes of 5,000 and 10,000 individuals. The true values of the correlation coefficient were set to -0.2, -0.6 or 0.6. Two models assuming no correlation or estimating the correlation coefficient were run 200 times each. Resulting variances were somewhat underestimated (respectively overestimated) when the true correlation was negative (respectively, positive) and the correlated nature of the random effects was ignored. With the complete model, the correlation estimate was nearly unbiased. The differences between models were small, but they may be much larger when the two random effects are not as perfectly cross-classified.

Session 25 — Poster 50

Strategies to determine the necessary number of phenotyped candidates in genomic selection

T.O. Okeno[1], M. Henryon[2,3] and A.C. Sørensen[1]
[1]Aarhus University, Molecular Biology and Genetics, P.O. Box 50, 8830 Tjele, Denmark, [2]University of Western Australia, School of Animal Biology, 35 Stirling Highway, Crawley WA 6009, Australia, [3]Danish Agriculture and Food Council, Pig Research Centre, Axeltorv 3, 1609 Copenhagen V, Denmark; tobias.okenootieno@agrsci.dk

We reason that marginal returns in genetic gain diminish as we increase the proportions of selection candidates that are phenotyped for traits under selection in genomic breeding programs. This is an important area because: (1) phenotypes increase the accuracy of breeding values predicted for phenotyped selection candidate, and (2) phenotyped candidates are added to reference populations, thereby increasing the accuracy of all selection candidates. We will test our premise using stochastic simulation of schemes resembling those used in pig breeding. The genetic architecture of the founder population will mimic the linkage disequilibrium observed in the Danish pigs. Two single-trait selection schemes, one for a highly-heritable trait (h^2=0.4) and one for a lowly-heritable trait (h^2=0.1), will be considered. In each scheme, different proportions of selection candidates ranging from 0-100% will be phenotyped. The genetic gains and rates of inbreeding realised in the two selection schemes will be estimated. This study will provide guidelines for the proportions of the selection candidates that should be phenotyped with genomic selection when high and low heritability traits are considered.

Session 25 — Poster 51

Joint estimation of recombination fraction and linkage disequilibrium for phasing long haplotypes

L. Gomez-Raya[1], A. Hulse[2], D. Thain[2] and W.M. Rauw[1]
[1]Instituto Nacional de Investigación y Tecnología Agraria y Alimentaria (INIA), Departamento de Mejora Genética Animal, Ctra A Coruña Km. 7.5, 28040 Madrid, Spain, [2]University of Nevada, Reno, 1664 North Virginia Street, 89557 Reno, NV, USA; rauw.wendy@inia.es

An E.M. algorithm for the joint estimation of recombination fraction and linkage disequilibrium in half-sib families was developed. Monte Carlo computer simulations showed that the new method is accurate if true recombination fraction is 0. For example, the estimates of recombination fraction and linkage disequilibrium were 0.00 (SD 0.00) and 0.19 (SD 0.03) for simulated recombination fraction and linkage disequilibrium of 0.00 and 0.20. The method was used for phasing long haplotypes in 36 half-sibs using the 50k Illumina BeadChip by the following strategy: (1) to establish linkage phase in the parent for each of two consecutive SNPs; (2) to determine fragments inherited in progeny; (3) to reconstruct phases of parent for each fragment; and (4) to reconstruct phases of the progeny. Homozygous SNPs in progeny allowed determination of the fragment inheritance from the sire and, consequently, recovering of haplotype information from fragments inherited from dams. Genotyping errors and likely assembly errors were detected. A total of 613 recombination events were detected after linkage analyses between fragments. SNPs for which the sire and calf were heterozygotes became informative (over 90%) after the phasing of haplotypes. Comparison of maternal or paternal gametes for regions of identity between individuals is proposed as a measure of relatedness. The mean of whole dam contribution homology for 630 comparisons with regions of at least 20 SNP in common was 0.11 with a maximum of 0.29 and minimum of 0.05. Efficiency of genomic selection might improve by the use of long haplotypes generated by the phasing approach because inheritance of long haplotypes better reflects the nature of the transmission of genetic information from parents to offspring than the usual assumption of many unlinked loci.

Non-additive genetic effects increase additive genetic variation and long-term response to selection

H. Esfandyari[1], M. Henryon[2,3], P. Berg[4], J.R. Thomasen[1,5], P. Bijma[6] and A.C. Sørensen[1]
[1]Center for Quantitative Genetics and Genomics, Department of Molecular Biology and Genetics, Aarhus University, P.O. Box 50, 8830 Tjele, Denmark, [2]School of Animal Biology, University of Western Australia, 35 Stirling Highway, Crawley WA 6009, Australia, [3]Danish Agriculture and Food Council, Pig Research Centre, Axeltorv 3, 1609 Copenhagen, Denmark, [4]Nordic Genetic Resource Cente, Raveien 9, P.O. Box 115 Ås, Norway, [5]VikingGenetics, Ebeltoftvej 16, 8860 Assentoft, Denmark, [6]Animal Breeding and Genomics Centre, Wageningen University, P.O. Box 338, Wageningen, the Netherlands; hadi.esfandyari@agrsci.dk

We hypothesized that genetic-variance components and genetic gain are affected by the presence of non-additive genetic effects in finite-locus models when a trait is subjected to truncation selection. More specifically, we expected maintain more additive genetic variance (V_a) and realize more long-term genetic gain when non-additive genetic effects are present, compared to a completely additive trait. A genetic model including additive, dominance, and additive-by-additive epistatic effects was simulated. The simulated genome for individuals consisted of 25 chromosomes, each with a length of 1M. One hundred biallelic QTL, four on each chromosome, were considered. In each generation, 100 sires and 100 dams were mated, producing five progeny per mating. The population was selected for a single trait (h^2=0.1) for 100 discrete generations with selection on phenotype or BLUP-EBV. Our results showed that V_a increased over time, for few generations, if non-additive genetic effects influenced the trait. However, in the additive genetic model, V_a decreased over time. In all genetic models, BLUP-EBV realized more genetic gain than phenotypic selection. It also resulted in a greater fixation of favorable alleles, leading in a more rapid loss of additive variance. We conclude that non-additive genetic effects can have a considerable impact on both change of V_a over time and response to selection and ignoring them might bias predicted rates of change.

Inbreeding and homozygosity in the Thoroughbred horse

L.A. Khrabrova
All-Russian Research Institute for Horse Breeding, Laboratory of Genetics, Divovo 35-29, Rybnoe, Ryazan district, 391105 P/O Institute for Horse Breeding, Russian Federation; l.khrabrova@yandex.ru

Thoroughbred horse breed originated from 3 famous sires and 70 foundation mares recorded in General Stud Book (1791). A closed system of breeding for more than 40 generations contributed to the genetic consolidation of the breed and inevitably resulted in increase of inbreeding. The study was carried out to assess the level of homozygosity of TB horses on STR-markers depending on the degree of inbreeding in pedigrees. Microsallite genotypes were determined at each of 13 panel loci (AHT4, AHT5, ASB2, HMS1, HMS2, HMS3, HMS6, HMS7, HTG4, HTG6, HTG7, HTG10, VHL20) for a sample of 1945 horses registered in the Russian Stud Book (182 stallions, 566 dams and 1197 foals). Inbreeding level was estimated by Wright`s inbreeding coefficient (Fx) in 5 generations in pedigrees. Over the past 40 years Fx in Russian population of TB increased by 0.1% (from 0.3% to 0.4%). The most of tested horses (58.8%) resulted from outbreeding method, 28.8% of horses had Fx<1.0%, 8.3% had Fx=1.1-2.0% and only 0.8% had Fx >4.1%. Proportion of inbred horses in groups of stallions, mares and foals was similar. Degree of homozygosity horses tested for microsatellite loci had normal distribution (P<0.01) with 0-79.6% interval and 30.8% modal class. Degree of homozygosity didn't change significantly in Fx<3.0%. Only in Fx >3.0 there was insignificant increase of this parameter from 31.2-32.6% to 35.6%. So it is suggested that the use of moderate inbreeding doesn't lead to increase of homozygosity in the TB horse. STR and other DNA-markers may be used for assessment of homozygosity of horses as outbred pedigrees don't guarantee high heterozygosity of animals.

Polymorphism evaluation of microsatellite markers in native Russian horse breeds

L.A. Khrabrova and M.A. Zaitceva
All-Russian Research Institute for Horse Breeding, Laboratory of Genetics, Divovo 16A, Rybnoe, Ryazan district, 391105 P/O Institute for Horse Breeding, Russian Federation; l.khrabrova@yandex.ru

The State Registry of selection achievements of Russia includes 44 horse breeds; half of them are unique native populations. It was carried out the population genetic analyses on 17 microsatellite loci for 316 horses of 10 native breeds, including Altai, Bashkir, Buryat, Khakasskaya, Mezenskaya, Pechorskaya, Tuvinskaya, Vyatka, Yakut and Zabaykalskaya. Genetic diversity within the populations was evaluated by the total number of allele variants (Na), effective number of allele (Ae), number of allele per loci (NV), observed (Ho) and expected heterozygosity (He), Fis and Fst coefficients calculated using GENEPOP 1.3. The compared mean values of microsatellite variation were insignificantly higher for the inspected native breeds than for the most cultural breeds. Average values of populations indices were: Ae: 4.00±0.16; Ho: 0.725±0.21; and NV: 5.87±0.36. Population diversity varied in inspected breeds by Ae from 3.61 to 4.47, by Ho from 0.605 to 0.776. In horses of Altai, Bashkir, Khakasskaya, Mezenskaya, Pechorskaya and Yakut breeds Ho exceeded He, resulting in negative Fis values. The highest values of Na (141) including private alleles ASAB17D, CA425E, HMS2D and HMS2G were found in Bashkir horse. The studied populations differed in their genetic structure and degree of differentiation (Nei`s genetic distances were in interval 0.15- 0.66).The analysis showed that the native breeds of horses from Eastern Europe and Asia form overall cluster to which the branch of Tuvinskaya breed isolated throughout a long time can be carried. Our previous study showed that stud horse breeds and native breeds form two quite different subclusters that evidence two different ways of microevolution of the two horse breed types. All the native populations of horses revealed rather high resources of genetic variability permitting successful implement of programs of their conservation.

Genetic components of piglet production and sow production in the Chinese-European line Tai Zumu

M. Banville[1,2], J. Riquet[2], L. Canario[2] and M. Sourdioux[1]
[1]GENE+, Rue du Moulin, 62134 Erin, France, [2]INRA, LGC, Chemin de Borde-Rouge, 31326 Castanet-Tolosan, France; maxime.banville@toulouse.inra.fr

The Tai Zumu composite line was created in 2001 by the French pig organization GENE+ to produce and select 50% Meishan grand-maternal dams. We studied the genetic components of piglet and sow production in that population bred in 3 nucleus herds. From 2008 to 2012, 25,763 piglets were weighed at birth (T0) and 17,604 piglets were weighed at 3 weeks of age (T3). Sow production traits (age AG, backfat thickness BF) were recorded at 100 kg from 19,763 sows. Numbers of piglets born alive (NBA) and weaned by a sow (NW) were recorded. Growth traits included the mean piglet weight (MW), the direct (d) and maternal (m) genetic effects on piglet weight and the standard deviation of weights within the litter (SD). Variance components were estimated using the restricted maximum likelihood methodology applied to animal models. The heritability for MW was higher at T0 than at T3 (0.51 and 0.26). Heritability estimates were lower for NBA and NW (0.20/0.06) and piglet weight SD_{T0} and SD_{T3} (0.19/0.15), and were highest for sow production traits (0.67 for BF, 0.38 for AG). Heritability for d and m on piglet weight were similar at T0 (hd^2=0.05; hm^2=0.16) and T3 (hd^2=0.08; hm^2=0.14) and a genetic antagonism between the two effects was obtained at T3 (-0.38±0.17). A strong unfavorable genetic correlation was found between NBA and MW_{T3} (-0.73±0.13); whereas genetic correlations of SDT3 with NBA and MW_{T3} were lower (0.10±0.20; 0.28±0.20). Genetic correlations of AG and BF were favorable with MW_{T3} (-0.39±0.09;-0.32±0.09) but unfavorable with SD_{T3} (-0.35±0.12;-0.39±0.11). The results suggest that selection for NBA might increase NW (rg=0.39±0.19) and that the breeding objective should also include homogeneity of piglet weights and piglet growth during lactation with use of both direct and maternal effects. This strategy will be assessed by analysis of response to selection with comparison of alternative breeding programs.

Session 25 — Poster 56

Genomic analysis of water holding capacity of meat in a porcine resource population

H. Heidt[1], M.U. Cinar[1], S. Sahadevan[1], C. Looft[1], D. Tesfaye[1], E. Tholen[1], A. Becker[2], A. Zimmer[2], K. Schellander[1] and C. Große-Brinkhaus[1]

[1]University of Bonn, Institut of Animal Science, Endenicher Allee 15, 53115 Bonn, Germany, [2]University of Bonn, Institute of Molecular Psychiatry, Sigmund-Freud-Str. 25, 53127 Bonn, Germany; cgro@itw.uni-bonn.de

In this study lean meat water-holding capacity (WHC) of a Duroc × Pietrain (DuPi) resource populations with corresponding genotypes and transcriptomes were investigated using the approaches of genetical genomics. WHC was characterized by drip loss measured in M. long. dorsi. Genotypes of 169 F2 DuPi animals were identified by the 60k Illumina SNP chips. Whole genome tanscriptomes of muscle were available for 132 F2 animals using the 24k Affymetrix expression microarray. Performing genome wide associations studies (GWAS) of transcriptional profiles which are correlated with phenotypes allows elucidating cis- and trans-regulation. Expression levels of 1,228 genes were significantly correlated with drip loss and were further analyzed for enrichment of functional annotation groups as defined by gene ontology and KEGG pathways. A hyper geometric gene set enrichment test was performed and revealed glycolysis/glyconeogenesis, pentose phosphate pathway and pyruvat metabolism as most promising pathways. For 267 selected transcripts eQTL analysis was performed and revealed 78,576 significant ($P<0.01$) associations in total. Because of positional accordance of the gene underlying transcript and the eQTL location it was possible to identify 59 eQTL that can be assumed to be cis-regulated. Comparing the results of gene set enrichment and the eQTL detection tests molecular networks and potential candidate genes which seem to play key roles in the expression of water holding capacity were detected. One gene was the glucose-6-phosphate transporter SLC37A4 which was assumed to be cis-regulated and involved in the glucose metabolic process. This approach supports to identify trait-associated SNPs and to understand the biology of complex traits.

Session 25 — Poster 57

Estimation of genetic parameters and breeding values for the major Swiss dairy goat breeds

B. Bapst[1], J. Moll[1], C. Baes[1] and U. Herren[2]

[1]Qualitas AG, Chamerstrasse 56, 6300 Zug, Switzerland, [2]Schweizerischer Ziegenzuchtverband (SZZV), Belpstrasse 16, Postfach, 3000 Bern 14, Switzerland; beat.bapst@qualitasag.ch

Three years ago the Swiss Goat Breeders Association developed a genetic evaluation process for the three major Swiss dairy goat breeds (Brown Alpine, Saanen and Toggenburg). Since then, routine evaluation of milk yield and fat and protein percentage has been implemented. Here we present the statistical model used as well as estimated genetic parameters and breeding values for these populations. A BLUP (best linear unbiased prediction) animal model, including a fixed lactation number effect, a fixed kidding × season effect, a random herd × time period effect, a random permanent environment effect and a random animal effect, was implemented. Multiple lactations of individual animals were considered repeated observations in the model. Phenotypic inputs included 100-day yield as an auxiliary trait and 220-day yield as the main trait. The number of lactation records used for analysis were as follows: 40,799 records distributed over 1,086 herds (Brown Alpine), 43,409 records distributed over 1,324 herds (Saanen) and 15,594 records distributed over 646 herds (Toggenburg). Genetic parameters were estimated, whereby heritabilities for milk yield were breed-dependant and ranged between 0.13 and 0.18. Fat percentage ranged between 0.38 and 0.48; protein percentage ranged between 0.42 and 0.56. As expected, negative correlations between yield and percentage traits and positive correlations between fat and protein content were observed. These results are similar to those of other published investigations. The estimated breeding values (EBV) are indexed with a mean of the base population of 100 and a standard deviation of 10. The base population contains all goats 4 to 6 years old with at least one observation for the auxiliary trait; this definition holds for all breeds. EBVs are estimated annually before the mating season begins. A positive genetic trend during the last 10 years is observed.

Gene expression phenotypes for cattle and sheep management

B.P. Dalrymple[1], N.J. Hudson[1], A. Reverter[1], N. Dejager[1], K. Kongsuwan[1], B. Guo[1], P. Greenwood[2] and R. Barnard[3]
[1]CSIRO, 306 Carmody Road St, St Lucia, QLD 4067, Australia, [2]NSW DPI, JSF Barker Building, University of New England, Armidale, NSW 2351, Australia, [3]University of Queensland, St Lucia, QLD 4067, Australia; brian.dalrymple@csiro.au

The objective of this work was to identify gene expression signatures for intramuscular fat (IMF) deposition, average daily gain (ADG), nutritional status and trenbalone acetate and estradiol (HGP) use. Gene sets were identified from gene expression (micro array) data from the longissimus muscle (LM) of animals in three experiments. Briefly (1) Piedmontese and Wagyu × Hereford cattle during a time course of LM development, (2) Brahman steers grown at two experimental sites, New South Wales and Western Australia, with and without HGP treatment; and (3) White Suffolk × Merino wethers with and without HGP treatment. Expression of the gene encoding oxytocin (OXT) in the LM and circulating levels of OXT was diagnostic of HGP-treated cattle, but much less diagnostic of HGP-treated sheep. However, expression of a number of genes in the LM, including GREB1 and WISP2 was diagnostic of HGP-treatment in both species. The expression of genes encoding triacyl-glyceride (TAG) synthesis and storage proteins in muscle was correlated with IMF%. At the level of groups of animals the expression of these genes predicts the effect of different experimental sites and HGP-treatments on IMF% in the LM using a much smaller number of animals than by using IMF% measurements alone. The same set of genes was also applicable to IMF% in sheep LM muscle. The expression of a set of genes encoding proteins involved in the cell cycle was correlated with ADG in the Brahman cattle. Very dynamic expression of PDK4 and a number of other genes encoding proteins involved in the Randle cycle was observed in both species. In other mammals the expression of these genes responds to the balance of the use of glucose and fatty acids by muscle and thus their expression in cattle and sheep LM may indicate the nutritional status of the animals.

Detection of population genetic structure when groups of relatives exist

S.T. Rodríguez-Ramilo[1,2], M.A. Toro[3] and J. Fernández[1]
[1]INIA, Mejora Genética Animal, Ctra. La Coruña Km. 7,5, 28040 Madrid, Spain, [2]CONAFE, Ctra. de Andalucía, Km. 23,6, 28340 Madrid, Spain, [3]ETSI Agrónomos. UPM, Producción Animal, Ciudad Universitaria, 28040 Madrid, Spain; jmj@inia.es

Bayesian clustering algorithms have emerged as a potential tool to infer population genetic structure from molecular marker information. Basically, these methodologies minimise both Hardy-Weinberg and linkage disequilibrium within subpopulations. As a consequence, most of these approaches assume that sampled individuals are not related. However, close relatives could be sampled from the same subpopulation if large families are present. This is particularly probably in species with a high fecundity or when the population size is small. Consequently, assumptions of Hardy-Weinberg and linkage equilibrium could not be satisfied in Bayesian clustering algorithms, which could reduce the accuracy of these methodologies to infer population genetic structure. In this study, two methodologies for the inference of population genetic structure have been evaluated using simulated data with different degrees of related individuals. The first method (STRUCTURE) implements a Bayesian approach to minimise Hardy-Weinberg and linkage equilibrium within subpopulations. The second methodology (CLUSTER_DIST) maximises the genetic distance between subpopulations and does not make Hardy-Weinberg and linkage equilibrium assumptions. The results indicate that the second approach is less influenced by the presence of related individuals, and is more appropriated when close relatives are supposed to be present within the individuals subject to classification.

Genetic variability of the equine casein genes

J. Schwarz, G. Thaller and J. Tetens
Institute of Animal Breeding and Husbandry, Christian-Albrechts University of Kiel, Hermann-Rodewald Str. 6, 24118 Kiel, Germany; jtetens@tierzucht.uni-kiel.de

The genetic diversity of the casein proteins (αs1-, β-, αs2-, κ-casein) of dairy species like cattle and goat has extensively been studied. The genetic variability of the milk protein genes results in different protein variants, which may influence the processing properties and the nutritive value of milk. Although mare milk has become more and more important for human nutrition because of positive health effects, studies about equine milk protein variation are still limited. The aim of the current study was to identify new equine casein variants and to establish a nomenclature for the equine casein gene variants. For this purpose, the open reading frames of all four equine casein loci (CSN1S1, CSN2, CSN1S2, CSN3) were resequenced in 192 horses from eight different breeds, which are used for mare milk production in Germany. All known equine casein variants were also found in our study. Additionally, at least ten putatively new variants were identified based on DNA mutations with a predicted impact on amino acid sequence, which are subject to confirmation on the protein level. A coherent nomenclature was established for the variants and the evolution of equine casein genes was discussed. The knowledge about the variability of equine casein genes developed within the current study might provide an important resource for further research regarding putatively beneficial effects of mare milk consumption on human health and the nutritive value of mare milk.

Genetic parameters of immune response in pigs and covariation with growth and carcass traits

J.P. Bidanel[1], D. Desson[1], Y. Billon[2], I. Oswald[3], J. Estelle[1] and C. Rogel-Gaillard[1]
[1]INRA, UMR 1313 GABI, 78350 Jouy-en-Josas, France, [2]INRA, UE967 GENESI, 1770 Surgères, France, [3]INRA, UMR 1331 ToxAlim, 31027 Toulouse, France; jean-pierre.bidanel@jouy.inra.fr

The heritability of 29 innate and adaptive immune response (IR) traits and their genetic correlations with average daily gain between 10 and 22 weeks of age (ADG) and backfat thickness at 22 weeks of age (BFT) were estimated on 911 Large White pigs using the restricted maximum likelihood method applied to a multivariate animal model. IR traits were measured at 9 weeks of age after vaccination against Mycoplasma hyopneumoniae and/or ovalbumin. Heritability values ranged from 0 to 0.68 depending on the IR trait. Blood count traits and immunoglobulin concentrations had similar average heritability (0.31), higher than that estimated for the levels of cytokines after stimulation (0.17 on average). The relative proportions of the different leukocyte populations and production of the cytokines IL-1B, IL-2, IL-4, IL-8, IFNG, TNF had low and non- significant genetic correlations with ADG and BFT. Conversely, significant genetic correlations with ADG and BFT were estimated for red blood cell counts, platelets (PLT), immunoglobulin A and M concentrations, IL-6 and IL-12 interleukin levels after stimulation. Genetic correlations with ADG were positive, except for PLT and IL-6 and IL-12 levels. Genetic correlations with BFT were negative, except for PLT and red blood cell counts. These results are very encouraging, but need be confirmed on a larger scale.

Estimation of longevity breeding values using sire-maternal grand sire or animal model

J. Jenko[1] and V. Ducrocq[2]

[1]Agricultural institute of Slovenia, Animal science department, Hacquetova ulica 17, 1000 Ljubljana, Slovenia, [2]INRA, UMR 1313 Génétique Animale et Biologie Intégrative, Allée de Vilvert, 78350 Jouy-en-Josas, France; janez.jenko@kis.si

Sire-maternal grandsire (MGS) and animal models were tested for the genetic evaluation of longevity in Slovenian Brown cattle population characterized by small herds. Three genetic models were compared: a sire-MGS model, an animal model, and an approximate animal model based on the sire-MGS model results. In addition, modeling the contemporary group effect was defined as either a herd or a herd-year (HY) effect. Estimated heritabilities were 0.096 for model with herd and 0.098 for model with HY random effect in case of sire-MGS model. For the animal model the obtained heritabilites were higher and were 0.134 for model with herd and 0.210 for model with HY random effect. Correlations between the breeding values estimated with animal model and approximate animal model were large: 0.94 for cows and sires when a herd effect was fitted and 0.88 for cows and 0.92 for sires when HY effect was fitted. The average correlation between estimated breeding values and survival at different stages was 0.43 for both sire-MGS and animal model. To avoid confounding and reduce computational requirements, it is suggested that the approximate animal model is an interesting compromise for practical applications of genetic evaluation of longevity in cattle populations.

Estimation under random designs: the case of pedigrees

P. Fullsack, B. Smith and C. Herbinger

Dalhousie University, Department of Mathematics and Statistics, 6316 Coburg Road, Halifax, NS B3H 4R2, Canada; philippe.michel.fullsack@gmail.com

Estimation of statistical models is often conducted under the hypothesis of perfectly known and deterministic designs. This needs however not be the case. Partially or imperfectly observed designs, or randomness built-in models, e.g. as latent variables in hierarchical models, lead to cases where designs must themselves be a target for the estimation. We outline three perspectives on the particular example of probabilistic pedigrees. We show, through a synthesis of past work and our own findings how estimation of selection index or genetic diversity under probabilistic pedigrees can benefit from: (1) joint estimation procedures in which genetic and phenotypic data help estimating posterior pedigree and breeding value parameters; (2) large deviation theory analysis, which we use to quantify fluctuations of estimates; and (3) non-parametric bayesian mixture models which can add robustness to the estimators We also discuss the role of probabilistic pedigrees as a tool to quantify genetic diversity and inbreeding accumulation.

Session 25 Poster 64

Pedigree based monitoring of the effective population size through the PopREP web service

E. Groeneveld[1], C. Kehr-Apelt[1], R. Fischer[2] and M. Klunker[3]
[1]Institute of Farm Animal Genetics (FLI), Animal Breeding and Genetic Resources, Höltystr. 10, 3535 Neustadt, Germany, [2]Saxon State Office for Environment, Agriculture and Geology (LfULG), Am Park 3, 04886 Köllitsch, Germany, [3]HTW University of Applied Sciences, Pillnitzer Platz 2, 01326 Dresden, Germany; eildert.groeneveld@fli.bund.de

Management of breeding populations and conservation actions usually require knowledge about the actual effective population sizes N_e, A number of different Methods for the computation of N_e exist with very different properties and different suitability for the monitoring objective of obtaining the most recent changes in population size. In all, six methods have been implemented in the new PopREP monitoring module. The first is based on census data, while the other five foot on the rate of inbreeding based of different settings for computing the ΔF. These are current generation versus direct parents, the rate of inbreeding based on current and previous generation cohorts, regression of coancestry on time, the log regression on individual inbreeding coeffients and those based on the equivalent complete generation procedure. The methods differ in the time span or number of generations to be included and are, thus, not equally well suited for the objective of estimating most recent changes in population size. Depending on the pedigrees some methods may not yield results in the parameter space, while others do. Here, a decision tree has been developed to chose the estimate most suitable for monitoring purposes. The results are presented on one page with five graphs and a table with the six N_e values along with the proposed best estimate of the current effective population size. Input to PopREP is an ASCII text file with columns animal, sire and dam ID, birthdate and sex. The file gets uploaded to the poprep web site and processed right away. Results in pdf and spreadsheet format are returned through email. The service is free and operational around the clock. The new extension is available under http://poprep.tzv.fal.de.

Session 25 Poster 65

Fine mapping of QTL of carcass and meat quality traits in a chicken slow-growing line

S. Allais[1,2,3], C. Hennequet-Antier[4], C. Berri[4], M. Chabault[4], F. D'Abbadie[5], O. Demeure[2,3] and E. Le Bihan-Duval[4]
[1]Université Européenne de Bretagne, UEB, 35000 Rennes, France, [2]INRA, UMR1348 Pegase, 35590 Saint Gilles, France, [3]Agrocampus Ouest, UMR1348 Pegase, 35042 Rennes, France, [4]INRA, UR83 Recherches Avicoles, 37380 Nouzilly, France, [5]SASSO, Route de Solférino, 40630 Sabres, France; sophie.allais@agrocampus-ouest.fr

Chicken slow-growing lines were developed in the sixties for the high-quality Label Rouge production. Whereas this production represents a large part of the whole carcass market (56% in 2011), the market share of Label Rouge chicken for cuts and processed products is still limited (11% in 2011). To adapt this production to the cuts market, a better knowledge of the genetic determinism of growth, body composition and breast meat quality traits was needed. Therefore a QTL detection was performed on birds originated from a slow-growing line selected by the SASSO breeding company since 1994. More precisely, 764 chicken and their parents (10 sires and 87 dams) were genotyped on the Illumina chicken SNP 60K Beadchip. Measures of body weight, breast meat yield, abdominal fat percentage, leg percentage, pH, meat color, drip loss, shear force and intramuscular fat content were registered. Different methodologies of fine mapping were tested: Linkage Disequilibrium and Linkage Analysis (LDLA) or pure LD analysis. For the LDLA analysis, haplotypes were inferred for all the population. Results suggest that several markers could be used in selection to adapt the slow-growing line to the cuts and processed market. The authors O. Demeure and E. Le Bihan-Duval contributed equally to this work.

Genetic parameters of milk coagulation traits in the first three lactations using random regression

D. Pretto[1], M. Vallas[1,2], H. Viinalass[1,2], E. Pärna[1,2] and T. Kaart[1,2]
[1]Institute of Veterinary Medicine and Animal Sciences, Estonian University of Life Sciences, Kreutzwaldi 1, Tartu, 51014, Estonia, [2]Bio-Competence Centre of Healthy Dairy Products, Kreutzwaldi 1, Tartu, 51014, Estonia; denis.pretto@emu.ee

The objective of this study was to estimate heritabilities and genetic correlations for milk coagulation property (MCP) traits (milk coagulation time, RCT, curd firmness, E_{30}) among first three lactations in Estonian Holstein cows using a multiple-lactation random regression animal model. A total of 40,154 test-day records from 5,216 Estonian Holstein cows in 98 herds across the country were collected from April 2005 to May 2010. Measurements of RCT and E_{30} were determined with the Optigraph (Ysebaert, Frepillon, France). The cows had in average more than 3 records per lactation and 71.5% of the animals had records from at least two different lactations. The model for parities 1 to 3 treated as separate traits accounted for fixed effects of herd, year-season of calving and test-day, sample age as days from sample collection to MCP analysis, and random effects of animal and permanent environment modelled by 3rd order Legendre polynomials. A pedigree of four generations in total with 20,791 animals was used. The MCP traits had a moderate to high heritability over the lactations with average daily heritabilities of 0.40, 0.39 and 0.39 for RCT, and of 0.48, 0.41 and 0.52 for E_{30}, for parities 1 to 3 respectively. Genetic variance of MCP traits was lower in the range of 60 to 120 days in milk suggesting that the contribution for total variance in this stage of lactation is more affected by environmental effects. For RCT the average daily genetic correlations between parities 1 and 2, 1 and 3, and 2 and 3 were 0.91, 0.81 and 0.98, and for E_{30} 0.95, 0.92 and 0.96, respectively. The genetic correlations among parities were strong suggesting that if a genetic evaluation for MCP is developed, regular collection of 1st lactation records should be sufficient for that.

Is there interest in implementing genomic evaluations in a pig male line nucleus? A simulation study

T. Tribout, C. Larzul and F. Phocas
INRA, UMR1313 GABI, 78350 Jouy-en-Josas, France; ttribout@jouy.inra.fr

Replacing pedigree-based BLUP (pBLUP) evaluations by genomic evaluations (GE) in pig breeding schemes can result in greater accuracy and genetic gains, especially for traits with limited phenotypes, but would generate additional costs. Our goal was to determine whether any additional expenditure would be more profitably devoted to implementing GE or increasing phenotyping capacity while retaining pBLUP evaluations. A stochastic simulation was used. The study population contained 1,050 breeding females and 50 boars. It was selected for 10 years for a breeding goal including two uncorrelated traits whose heritabilities were 0.4. The reference breeding scheme was based on phenotyping 13,770 candidates for trait 1 and 270 relatives from 10% of the litters for trait 2 per year, and selection was based on pBLUP estimated breeding values (EBV). Increased expenditure was allocated to either increasing the phenotyping capacity for trait 2 while maintaining pBLUP evaluations, or implementing genomic selection. The genomic scheme was based on two training populations: one for trait 2 made up of relatives whose number increased from 1000 to 3,430 over time, and the second for trait 1 made up of candidates. Several genomic scenarios were tested, where the size of the training population for trait 1 and the proportion of genotyped candidates pre-selected based on their parental EBV, varied. Both approaches resulted in higher genetic trends and lower inbreeding rates compared to the reference scheme. However, even a very marked increase in phenotyping capacity for trait 2 could not match the improvements achieved by genomic selection when the number of genotyped candidates was large. Genotyping a limited number of pre-selected candidates significantly reduced the extra costs while preserving most of the benefits relative to genetic trends and inbreeding. Implementing GE was the most efficient approach when major expenditure was possible, while increasing the phenotypes was preferable under limited resources.

Which quality indicators could be the success key for the rare breeds development?

L. Markey[1] and C. Couzy[2]
[1]Institut de l'Elevage, Castanet Tolosan, 31321, France, [2]Institut de l'Elevage, Agrapole, 23 rue J. Baldassini, 69364 Lyon Cedex 07, France; lucie.markey@idele.fr

Rare breeds (less than 5,000 cows, 8,000 ewes or goats, 1000 sows) are very small part of French livestock even they are 30% of French breeds number. Most of these breeds reach a critical step in the seventies before to be saved by conservation programs, which support them for new development. The increase was as well for the animals number than for the breeders number including professional breeders. It is necessary to find and strengthen outlets for economical valorization to continue the development of these less productive animals without any genetic selection. VARAPE project (rare breeds valorization with short supply chains) is coordinated by Institut de l'Elevage with 7 technical partners and 13 associated breeds and is a study about success factors and limits of a collective project to develop short supply chains. The project leading is based on the 13 breeds survey (production and marketing inventory, local committee and exchange meetings…) as well as the analysis of 16 breeds approaches with a significant products marketing experience. The project shows that the studied breeds are waiting for better marketing structure, often farms number increasing and also their products protection. We observe that protected designation of origin (PDO) approaches, based on territory / product / breed, are regularly presented as the main economical success key for these breeds. Indeed we consider in France that the PDO brings directly a reputation to the product and can protect the breed. But the analysis results of VARAPE project show that PDO indicator can only succeed in particular conditions: for example the farmers group has to be enough big and dynamic to assume the approaches heaviness, the inspection price need quite big production volume to be not too much expensive… With these observations the question is to know if it would be better that the breed managers use other quality indicators as traditional speciality guaranteed (TSG) to promote limited numbers breeds products?

Genotype imputation in Nelore cattle

R. Carvalheiro[1], J. Sölkner[2], H. Neves[1], Y. Utsunomiya[1], A. Pérez O'Brien[2], S. Boison[2], M. Silva[3], C. Tassel[4], T. Sonstegard[4], J. McEwan[5], F. Schenkel[6] and J. Garcia[1]
[1]Universidade Estadual Paulista, São Paulo, 16050-680, Brazil, [2]University of Natural Resources and Life Sciences, Vienna, BOKU, Austria, [3]EMBRAPA-CNPGL, Juiz de Fora, MG, Brazil, [4]USDA-ARS, Beltsville, MD, USA, [5]AgResearch, Invermay Agricultural Centre, 50034, New Zealand, [6]University of Guelph, Guelph, ON, Canada; anita_op@students.boku.ac.at

Genotype imputation efficiency in Nelore cattle was assessed using Illumina Bovine HD (777K) as the reference chip. From a total of 995 bulls HD genotyped, 202 younger had their genotypes masked except from those SNPs present in the tested lower density (LD) chip. Different commercial and customized LD chips were tested, with densities varying from 6K to 50K. Customized LD chips were designed taking into account minor allele frequency, linkage disequilibrium and distance among SNPs. Imputation was performed considering or ignoring pedigree information and using a milder (call rate>0.98) or a more stringent (call rate>0.98; maf>0.02) quality control (QC) on the genotypes of the reference population. The criteria adopted to evaluate imputation efficiency were the percentage of correct imputed genotypes (PERC) and the correlation between imputed and observed genotypes (CORR), assigned as 0, 1 or 2. Imputation was carried out using FImpute software. Commercial Illumina 6K and 50K chips provided an average (minimum) PERC and CORR equal to, respectively, 90.56 (76.85) and 0.926 (0.800), 97.14 (87.24) and 0.978 (0.904). PERC and CORR average (minimum) values for the customized 15K chip were equal to 97.85 (89.34) and 0.984 (0.921). Omitting pedigree information had no effect on 15K or higher density chips, but reduced imputation efficiency from 6K to HD. Considering monomorphic SNPs (milder QC) didn't improve imputation efficiency. In conclusion, the adoption of LD chips (e.g. 15K) and imputation methods can improve cost effectiveness in Nelore cattle genotyping, compared to the strategy of using just HD chips.

Session 25 Poster 70

Comparison on GWAS using different models and different markers data sets

X.P. Wu[1,2], G. Su[1], G. Sahana[1], B. Guldbrandtsen[1], D. Sun[2], Q. Zhang[2] and M.S. Lund[1]
[1]Department of Molecular Biology and Genetics, Aarhus University, Blichers Allé 20, P.O. Box 50, 8830 Tjele, Denmark, [2]College of Animal Science & Technology, China Agricultural University, 2 Yuanmingyuan West Road, Haidian District, 100193 Beijing, China, P.R.; xiaoping.wu@agrsci.dk

Mastitis is the most common disease in dairy cattle causing substantial economic losses. The aim of this study was to compare the GWAS for refining the six previously detected QTL affecting mastitis, using two different models and three maker densities based on the 54K and imputed 770K (HD) marker data as well as genome sequence data. The dataset consisted of 4,496 Danish Holsteins. After quality control, there were a total of 790, 8,260 and 77,853 SNPs available within the six regions, respectively. The mastitis response variable was the de-regressed proof of the Nordic mastitis index. Two models were used in the analysis: one was a linear mixed model (LMM) with random polygenic effects and fixed effect of a single SNP, which performs GWAS once per SNP (SNP-wise GWAS). The other was a Bayesian variable selection model (BVS) including polygenic effects and SNP effects as random variables, where the model fitted all SNP simultaneously. The current study compared the pattern from association signals using the two models (LMM and BVS) and the three marker densities. The QTL identified in Danish Holsteins will be compared with Nordic Red and Danish Jersey populations.

Session 25 Poster 71

Population level genome-wide association study for calving traits in Holstein cattle

X. Mao[1], G. Sahana[1], D.J. De Koning[2] and B. Guldbrandtsen[1]
[1]Aarhus University, Department of Molecular Biology and Genetics, Blichers Allé 20, P.O. Box 50, 8830 Tjele, Denmark, [2]Swedish University of Agricultural Sciences, Department of Animal Breeding and Genetics, Inst för HGEN, Box 7023, Gerda Nilssons väg 2, 75007 Uppsala, Sweden; xiaowei.mao@agrsci.dk

The decrease in calving performance in dairy cattle causes considerable economic loss as well as reduction in animal welfare. Recently, genome-wide association analysis has been applied to identify quantitative trait loci (QTL) contributing to calving traits, utilizing 50k Single Nucleotide Polymorphism (SNP) panel. To further narrow down the QTL interval with less cost, 50k SNP genotyped cattle was imputed to high density SNP genotyped cattle by IMPUTE2 software. The imputed data consists of 664,204 SNPs which are all on 29 bovine autosomes and 14 calving traits from progeny-tested Holstein bulls. A mix model was fit to detect associations between these SNPs and calving traits. In our study, rs136283363 in chromosome 18 located in exon 1 of LOC786539 gene was discovered to be the most significant SNP ($P=3.57\times10^{-46}$) associated with calving performance. Besides, the result agrees with previous studies using lower density SNPs. SNP rs109478645 in chromosome 18, which was reported to be the most significant in other studies, was also retrieved in our analysis ($P=1.48\times10^{-23}$). In general, the confidence interval of the QTL was smaller compared to that in previous studies. To validate the results the effects in Holstein cows were examined using raw phenotypes. Our study demonstrates the advantage of utilizing high density SNPs through imputation and provides more evidence on the path to identify the candidate genes underlying calving performance.

Session 25 Poster 72

Combination of two polymorphisms in leptin gene influences milk performance traits in Holstein cows

J. Tomka, D. Vašíček, M. Bauer, K. Vašíčková, M. Oravcová, J. Huba and D. Peškovičová
Animal Production Research Centre Nitra, Hlohovecká 2, 95141 Lužianky, Slovak Republic; tomka@cvzv.sk

Several single nucleotide polymorphisms in leptin gene were identified and associated with milk performance traits. In our study we have analyzed two published SNP located in exon2 (C1180T, C/T transition resulting in change from arginine to cysteine) and exon 3 (C3100T, C/T substitution resulting in change from alanine to valine). The performance data (fat yield and fat content, protein yield and protein content, milk yield during the 305 days lactation) from 175 Holstein primiparous dairy cows from one farm were gathered during one season. Cows originated from 54 sires. The linear models with individual polymorphism and sire effect were applied. Also the effect of genotypes combination (C1180T × C3100T) was estimated. The frequencies of C1180T allele were almost similar (0.54 vs. 0.46). The frequencies of C3100T alleles were different (0.67 vs. 0.33). Linear models with individual polymorphisms and sire effect fitted explained up to 48% of overall variability of studied performance traits. However there was no significant effect of single polymorphisms on those traits. The effect of genotypes combination (C1180T × C3100T) was significant in models for fat yield, protein content and protein yield and milk yield. The models with genotypes combination (C1180T × C3100T) and sire effect explained up to 53% of overall variability. These findings indicate the importance of two studied SNPs combination on milk performance traits.

Session 25 Poster 73

Microsatellite DNA polymorphism in some naked neck village chicken genotypes

L. Mercan and A. Okumus
Ondokuz Mayis University Agriculture Faculty, Agricultural Biotechnology, Ondokuz Mayis University Agriculture Faculty Agricultural Biotechnology Dept., 55139 Atakum/Samsun, Turkey; lmercan@omu.edu.tr

The genetic polymorphism of naked neck village chickens in the Blacksea region of Turkey was analyzed using 20 SSR markers with wild type genotypes. Genotypes selected from 445 village chickens were compared with each other to find the genetic variability. The number of polymorphic alleles observed in the naked neck village chickens ranged from 4 to 8; LEI 0094, LEI 0166, MCW 0067, MCW 0078, MCW 0098, MCW 0103, MCW 0123, MCW 0206 and MCW 0222 loci had 4, MCW 0014, MCW 0034 and MCW 0037 loci had 5, ADL 0268, MCW 0111, MCW 0183 and MCW 0248 loci had 6, ADL 0112, ADL 0278 and MCW 0080 loci had 7, and MCW 0069 locus had 8 alleles. They were observed not to have unique alleles that were different from normal genotypes. Also it was determined that number of naked neck chickens was very limited in the study area. The existence of naked neck chicken genotypes was a sign of genetic interactions but the results showed that all the genotypes had the same genetic structure with their wild type relatives in terms of studied microsatellite loci. It was evaluated that the naked neck chickens only differs in terms of naked neck gene (Na) according to normal feathered chickens.

The correlation matrix of additive marker effects in full sibs

S. Bonk, F. Teuscher and N. Reinsch
Leibniz Institute for Farm Animal Biology, Institute of Genetics and Biometry, Wilhelm-Stahl-Allee 2, 18196 Dummerstorf, Germany; bonk@fbn-dummerstorf.de

Within families the additive effects of densely spaced markers are correlated, due to close linkage. These correlations were theoretically investigated for full-sib families, where parents are arbitrarily homozygous or heterozygous at marker loci, assuming parental diplotypes known. It is shown in detail how the correlation matrix can directly be computed from the genetic map. Exact formulas were derived as a generalization of results from controlled crosses. Pair-wise correlations are functions of the distance between markers and the combination of parental linkage phases. Resulting correlation matrices are positive-semidefinite, as double homozygous markers do not contribute to genetic variability in offspring. Results can be applied in the field of mate allocation in e.g. dairy cattle, where this correlation matrix is needed for calculating the expected genetic variability of different sib ships. Details (computing time, treatment of X-chromosome) will be illustrated by examples. Further applications are in gene mapping via a heterozygous parent approach.

Session 25 Poster 75

A method of separating genetic variance of litter weight into direct and maternal genetic variances

M. Satoh
National Institute of Livestock and Grassland Science, Ikenodai 2, Tsukuba-shi, 305-0901, Japan; hereford@affrc.go.jp

Litter traits such as litter weight (LW) are generally regarded as performance traits of the dam. However, if genetic performance expressed as LW could be divided into the growth ability of the offspring (direct effect) and the nursing ability of the dam (maternal effect), it would be possible for the genetic characteristics of LW to be improved more effectively. In the present study, we propose a method of dividing variance of the genetic characteristics of LW into direct and maternal genetic variances, using the body weights of the individual progeny. We treat the phenotype (P) for individual body weight as the sum of fixed effects (F), direct (D) and maternal (M) genetic, common environment (C), and residual error (E) random effects. Thus, P=F+D+M+C+E. Assuming no covariances between C or E and other random effects, the variances on both sides are $\sigma_P^2=\sigma_D^2+\sigma_M^2+2\sigma_{DM}+\sigma_C^2+\sigma_E^2$. The phenotype of LW is represented as the sum of the phenotypes of each individual body weight in a litter. Thus, the predicted phenotypic variance of LW is $[(n+1)/2]\sigma_D^2+n^2\sigma_M^2+2n^2\sigma_{DM}+n\sigma_C^2+n\sigma_E^2$, where n is the number of progeny in a litter. Each variance and covariance components can be estimated using a versatile REML program. Number n can be calculated using estimates of each variance-covariance components. Estimates of direct and maternal genetic variances for LW are, thus, $[(n+1)/2]\hat{\sigma}_D^2$ and $n^2\hat{\sigma}_M^2$, respectively. In one example, a total of 30,990 records of individual body weights at weaning from 4,792 first-parity litters of Syrian hamsters were used to estimate σ_D^2, σ_M^2 and σ_{DM} for LW at weaning. Heritability estimated for LW at weaning by conventional methods was 0.39 (SE=0.03). However, based on the present study, the direct and maternal heritabilities and correlation between D and M for LW at weaning were estimated to be 0.10 (0.02), 0.21 (0.02) and 0.66 (0.09), respectively. In conclusion, if each of the individual body weights in a litter is measured, it is possible to estimate direct and maternal heritabilities for LW.

Genetic parameters of faecal worm egg count and objective wool traits in the Tygerhoek Merino flock

P.A.M. Matebesi-Ranthimo[1], S.W.P. Cloete[2,3], J.B. Van Wyk[4] and J.J. Olivier[2]
[1]National University of Lesotho, P.O. Roma 180, Roma, Lesotho, [2]Institute for Animal Production: Elsenburg, Private Bag X1, Elsenburg 7609, South Africa, [3]Stellenbosch University, Private Bag X1, Matieland 7602, South Africa, [4]University of the Free State, P.O. Box 339, Bloemfontein 9300, South Africa; vanwykjb@ufs.ac.za

The cost of internal parasite control in grazing sheep calls for an alternative approach to ovine helminth infestation. Faecal worm egg count (FEC) may be a suitable criterion to select for resistance to nematode infestation. Genetic parameters for FEC and objectively measured wool traits were derived, using data from Merino sheep from a selection experiment in the South African winter-rain cropping-pasture area. The number of records ranged from 3,842 for staple strength (SS), 5,473 for FEC and 6,717 for most other traits. The pedigree file included 7,100 animals, the progeny of 554 sires and 2,483 dams born from 1989 to 2010. Individual rectal faecal samples were taken at 13-16 months of age after drenching was withheld >10 weeks. Eggs were counted using the McMaster technique, at a sensitivity of 100 eggs per gram of wet faeces. Only additive animal affected the data for FEC. Heritability estimates were 0.10 (0.02) for untransformed FEC, 0.15 (0.02) for cube root transformed FEC and 0.16 (0.02) for Log (FEC + 100). Genetic correlations of Log (FEC + 100) with SS, staple length, coefficient of variation of fibre diameter and standard deviation of fibre diameter were all favourable. The genetic correlations of Log (FEC + 100) with wool weight traits were unfavourable in absolute terms, but not significant. Selection for FEC is unlikely to result in marked unfavourable correlated responses to wool traits in South African Merinos, except for wool weight. It is thus important to proceed with genetic research on resistance and/or resilience to gastro-intestinal nematodes in other parts of South Africa. The usage of selection for reduced FEC for the breeding of more resistant animals should also be assessed.

Evaluation of genetic diversity in the Slovak Simmental breed using SNP analyses of genetic markers

A. Trakovická, N. Moravčíková and A. Navrátilová
Slovak University of Agriculture in Nitra, Department of Animal Genetics and Breeding Biology, Tr. A. Hlinku, 94976 Nitra, Slovak Republic; anna.trakovicka@uniag.sk

The aim of this study was detection of polymorphisms in leptin (Sau3AI), leptin receptor (T945M), growth hormone (AluI) and pituitary specific transcription factor (HinfI) genes in population of Slovak Simmental cows. Bovine genes as molecular markers are chosen for study on the basis of known relationships between physiological processes and production traits. The total numbers of blood samples were taken from 353 samples of Slovak Simmental cows. Genomic DNA was isolated by phenol-chloroform extraction method and analyzed by PCR-RFLP method. After digestion with restriction enzymes were detected in population alleles with frequency: LEP/Sau3AI A 0.83 and B 0.17 (±0.0141); LEPR/BseGI C 0.958 and T 0.042 (±0.0076); GH/AluI L 0.695 and V 0.305 (±0.0173) and Pit-1/HinfI A 0.249 and B 0.751 (±0.0163). Based on the observed vs. expected genotypes frequencies was population in Hardy-Weinberg equilibrium (P>0.05). In population were predominant homozygous genotypes for three markers: LEP/Sau3AI AA genotype (0.686), LEPR/T945M CC genotype (0.915), Pit-1/HinfI BB genotype (0.751) and for marker GH/AluI was dominant heterozygous LV genotype with frequency 0.467. The low or median polymorphic information content of loci (average 0.241±0,115) were also transferred to the low expected heterozygosity 0.282, 0.081, 0.424 and 0.374 for locus of leptin, leptin receptor, growth hormone and pituitary specific transcription factor, respectively. Until now has been confirmed effect of these polymorphisms on cattle production performance in many studies. Combination of molecular and statistical analyses of genes polymorphisms effect can be strong tools in future breeding dairy cattle programs.

Session 25 — Poster 78

Estimation of genetic parameters for stillbirth of Japanese Black cattle in Japan

S. Maeda[1], T. Yonekawa[2], Y. Furukawa[2] and K. Kuchida[1]
[1]Obihiro University A&VM, Inada-cho Obihiro-shi Hokkaido, 0808555, Japan, [2]Tokachi ikeda-cho Agricultural Cooperatives, aza toshibetu Ikeda-cho Nakagawa Hokkaido, 0830034, Japan; s22330@st.obihiro.ac.jp

Economic losses caused by stillbirth are serious problem for breeding farmer of Japanese Black cattle because calves price is expensive (average is about 3,400 EUR/head). Objective of this study is to estimate genetic parameter for stillbirth of Japanese Black cattle. Calving records collected from Japanese breeding farm between January 2009 and June 2012. The number of calving record was 4,212. Pedigree records were traced back 3 generations and total number of sire was 70. The mathematical model for genetic analysis included year of birth, twin birth, season, herd size and birth year of dam as fixed effects. Farm and additive genetic effect were also included as random effects. Heritability was estimated by the THRGIBBS1F90 program with sire threshold model. Average stillbirth rate was 3.6% so stillbirth does not occur with high frequency for Japanese Black cattle. Heritability of stillbirth was estimated 0.22(±0.15). The value is slightly elevated above the past study using Holstein. The little high heritability might be cause by sires have high stillbirth rate. Median of sire's stillbirth rate was 3% but 4 sires had over 10% stillbirth rate. In addition, variance of farm effect (0.029) was small relative to sire additive genetic variance (0.064). Low stillbirth rate eliminated the distinction between farms and made variance of farm effect small. Relatively-high heritability of stillbirth might be cause increasing stillbirth if stillbirth has a genetic correlation with other favorable traits and if a sire has high genetic capacity of stillbirth. Therefore, collecting stillbirth records and genetic evaluation are needed to reduce stillbirth for Japanese Black cattle.

Session 25 — Poster 79

Comparison of the survival analysis with random regression model for genetic evaluation of herd life

O. Sasaki[1], M. Aihara[2], A. Nishiura[1], H. Takeda[1] and M. Satoh[1]
[1]NARO Institute of Livestock and Grassland Science, Tsukuba, Ibaraki, 305-0901, Japan, [2]Improvement Association of Japan, INC., Koto-ku, Tokyo, 135-0041, Japan; sasa1@affrc.go.jp

Survival analysis (SUV) with a piecewise hazard function, which is divided by parity and lactation stage, would be suitable for estimation of the genetic ability of herd life. In SUV, genetic ability of herd life cannot be easily evaluated by direct use of the genetic information of other traits. Random regression animal model can adapt to the estimation of genetic ability of herd life as well as SUV. Furthermore, the multiple-trait random regression model (MTRRM) can use information of genetically related traits, such as milk yield, somatic cell score (SCS), and conformation traits. This study aimed to compare SUV and MTRRM for estimation of genetic parameters and genetic abilities of herd life. The study included 290 Holstein herds of Japan with at least 5 third-parity cows on each test day from 2006 to 2010. The data set contained 2,085,253 test-day data up to the first 5 parities from 2001 to 2010. Data of 79,698 cows were used for the sire-MGS model of SUV, including piecewise Weibull hazard function that is divided by parity. Data of 79,909 cows were used for MTRRM with repeatability effect, and they consisted of survival rate, milk yield, and SCS records. In SUV, the heritability estimate on a logarithmic scale and effective heritability of herd life were 0.066 and 0.100, respectively. The heritability estimates of the survival rate (standard deviation), milk yield, and SCS in MTRRM were 0.03-0.12 (0.002-0.009), 0.12-0.38 (0.005-0.012) and 0.10-0.14 (0.004-0.011), respectively. These values increased with the day after calving. Genetic correlations of survival rate with milk yield and SCS ranged from 0.08 to 0.19 and from -0.25 to -0.43, respectively. The absolute value of correlation coefficient between the breeding value of MTRRM and the transmitted genetic ability of SUV was 0.67. The MTRRM is as promising as the SUV in predicting the genetic ability of herd life.

β-defensin genes expression in udder secretory tissue infected with coagulase-positive Staphyloccoci

E. Bagnicka, E. Kościuczuk, J. Jarczak, P. Lisowski, A. Jóźwik, N. Strzałkowska, D. Słoniewska, J. Krzyżewski and L. Zwierzchowski

Institute of Genetics and Animal Breeding PAS in Jastrzębiec, ul. Postępu 36A, 05-552 Magdalenka, Poland; e.bagnicka@ighz.pl

The main mastitis pathogens are *Staphylococcus aureus* and *Streptococcus agalactiae*. Bacteria that overcome external barriers of the udder, encounter immune cells – macrophages and neutrophils armed with antimicrobial arsenal such as β-defensins. The aim of this study was to measure the transcripts levels of β-defensin genes in cow mammary gland secretory tissue infected with coagulase-positive Staphyloccoci (CPS). The study was conducted on 40 Polish Holstein-Friesian (HF) dairy cows of Black and White type. Cows were culled at the third stage of lactation because of reproduction problems. The secretory tissue samples from each quarter of udder were collected. Milk samples were taken from each quarter two days before slaughter and examined for the presence of bacteria. The samples were divided into four groups according to parity and health status of the udder. Two groups consisted of samples collected from cows (one per cow) with infections caused by CPS in $1^{st}/2^{nd}$ lactation (n=13) or in $3^{rd}/4^{th}$ lactation (n=14). The control groups (n=9 in each group) consisted of samples (one per cow) collected from pathogen-free udders but only when all quarters were not infected. Expression analysis of bovine β-defensin1, 4, 5, 10 and LAP genes (BNBD1, 4, 5, 10 and LAP) was done using R-T qPCR method. GAPDH and ACTB genes were used as references. The expression of all studied genes was shown to be much higher in tissues infected with CPS than in tissues from healthy udders regardless parity with exception of BNBD10 gene, expression of which was the same in infected and non infected tissues of cows in $3^{rd}/4^{th}$ lactation.

Genetic parameters for calf survivability for beef cattle in the Czech Republic

L. Vostry[1], B. Hofmanova[1], Z. Vesela[2], I. Majzlík[1] and M. Milerski[2]

[1]Czech University of Life Sciences Prague, Kamycka 129, 165 21 Prague, Czech Republic, [2]Institute of Animal Science, Pratelstvi 815, 10401 Prague – Uhrineves, Czech Republic; vostry@af.czu.cz

The aim of this study was to estimate the genetic relationship among calf survivability (CS), calving ease (CE) and birth weight (BW). Eight pure breeds of beef cattle were included in the analysis: Beef simental (11%), Gasconne (1%), Hereford (10%), Aberdeen angus (26%), Charolais (36%), Limousine (9%), Blonde d'Aquitane (4%), Pimontese (3%). A total of 81,092 records were obtained from the Czech Beef Breeders Association field database. The three traits were analysed as traits of the calf fitting a multivariate linear mixed model. All tested models included fixed effects of year of birth, age of dam, sex and breed of calf, whereas random effects included direct and maternal genetic effects, maternal permanent environment effect, contemporary group (herd × year × season) and residual error. Calving ease and birth weight were modelled as linear trait. The linear logit model was used for a analysis of calf survivability. Estimates of heritability (SE in brackets) for the direct genetic effects (CSd, CEd and Bwd) were 0.04 (0.03), 0.12 (0.01) and 0.13 (0.01) respectively, whereas the estimates for maternal genetic effects (CSm, CEm and BWm) were 0.04 (0.04), 0.01 (0.001) and 0.05 (0.02). Genetic correlations between direct, maternal genetic effects involving CS, CE and BW was low and negative, expect for the pair BW-CE (positive), ranging form -0.06 to 0.20. The genetic correlations for CSd-CEm, and for BWd-BWm were moderate and negative, whereas for CEd-CEm was low and positive. Results suggest that low additive and maternal genetic variances of survival and low genetic correlations (direct and maternal) between CS, CE and BW limit the possibility to be exploited in a specific breeding programme.

Session 25 Poster 82

Genetic structure in four selected pig populations of Czech Republic using microsatellite markers

I. Vrtkova, L. Stehlik, L. Putnova, L. Kratochvilova and L. Falkova
Mendel University in Brno, Laboratory of Agrogenomics, Zemedelska 1, 61300 Brno, Czech Republic; irenav@mendelu.cz

The aim of the study was to investigate the genetic diversity of four Czech pig breeds (large white – father line – LWF, large white – mother line – LWM, czech meat – CM, prestice black pied, genetic resource – PC), explain their genetic relationship and assess their integrity and degree of admixture. 357 individuals from these breeds were genotyped for 10 microsatellite markers (S0068, S0107, SW24, S0355, S0386, SW353, SW936, S0070, SW72 and TNFB). In general, high genetic diversity – observed heterozygosity ranging from 0.628±0.016 to 0.712±0.015, and moderate breed differentiation (F_{ST}=0.095) were observed, F_{IS} index of each population was as following LWF=0.019, LWM=0.051, CM=- 0.007, PC=0.025. The negative intrapopulation index F_{IS} value (- 0.007) was observed in population CM, which suggesting an excess of heterozygotes due to non-random mating. Furthermore, the analysis of population structure indicates there is very little admixture among breeds, with each one being identified with a single ancestral population. Results of this study confirm that all breeds of pigs represent a very interesting reservoir of allelic diversity. However, this requires a synergic management strategy at the farm level to maintain diversity and avoid inbreeding mainly in a small population of pigs. On the basis of these results, we propose that these microsatellite markers may be used with reliability for studying the genetic diversity.

Session 25 Poster 83

Sire effects on longevity depending on POTS in Holstein population in Japan

Y. Terawaki[1], A. Nomura[1], H. Nagata[1], S. Yamaguchi[2], Y. Gotoh[3] and V. Ducrocq[4]
[1]Rakuno Gakuen University, 582 Midori_machi Bunkyo_dai, 069-8501 Ebetsu, Japan, [2]Hokkaido Dairy Cattle Milk Recording and Testing Association, 4-1-1 Chuo_ku, 060-0004 Sapporo, Japan, [3]Holstein Cattle Association of Japan, Hokkaido Branch, 15_5 Kita_ku, 001-8555 Sapporo, Japan, [4] INRA, UMR 1313, Génétique Animale et Biologie Intégrative, 78352 Jouy en Josas, France; terawaki@rakuno.ac.jp

Sire effects on functional longevity were compared depending on the proportion of cows having a type score (POTS) in the herd. The data was provided by the Hokkaido Dairy Milk Recording. The study period was from January 1, 1991 to December 31, 2007. Records of cows still alive at the end of the study period were considered as censored. Cows having a first calving before January 1, 1991 or after September 30, 2007 were excluded from the analysis. The final data set had 1,176,931 cow records. Longevity was defined as the number of days from first calving to culling or censoring date (LPL). The records were divided into 9 subsets depending on POTS (A:0%, B:0-10%, C:10-20%, D:20-30%, E:30-40%, F:40-50%, G:50-60%, H:60-80%, I:>80%) in the herd. Sire effects on LPL were estimated using the Survival Kit (V5.0) software. Sire variance was smallest (0.017) in subset B and largest (0.034) in subset I. A total of 505 sires appeared in these two subsets. Rank correlation between sire effects estimated in each subset was 0.364. Regression coefficient of sire effects in subset B on sire effects in subset I was 0.245. Three groups of 271, 63 and 15 sires were created by requiring 10, 100 or 500 daughters' uncensored records, respectively, in each subset. For these groups, regression coefficients of sire effects estimated in subset B on sire effects estimated in subset I were 0.311, 0.476 and 0.615, respectively. The corresponding rank correlations were 0.482, 0.754 and 0.764. These results show that the genetic component of LPL strongly differs between herds with low or high proportion of cows with a type score in Japan.

Genetic analysis of milk fatty acids composition of Italian Brown Swiss cows

A. Cecchinato[1], F. Tagliapietra[1], S. Schiavon[1], M. Mele[2], J. Casellas[3] and G. Bittante[1]
[1]University of Padova, Department of Agronomy, Food, Natural resources, Animals and Environment (DAFNAE), Viale dell'Università 16, 35020 Legnaro, Padova, Italy, [2]Università di Pisa, Department of Agricolture, Food and Environment, Via del Borghetto, 80, 56124 Pisa, Italy, [3]Universitat Autònoma de Barcelona, Departament de Ciència Animal i dels Aliments, 08193 Bellaterra, Spain; alessio.cecchinato@unipd.it

The aim of this study was to infer variance components and heritabilities for 47 milk fatty acids (FA) (including conjugated linoleic acid) and for unsaturation ratios in the Italian Brown Swiss population. A total of 1,271 cows were sampled once in 85 herds. Milk samples were collected during the evening milking concurrently with the monthly test-day milk recording. Each sample was analyzed for single FA percentages (computed as FA weight as a proportion of total fat weight) by gas chromatography. A Bayesian univariate animal model was implemented via Gibbs sampling. The model accounted for the effect of days in milk, parity, herd and the animal additive genetic effect. Flat prior distributions were assigned to systematic effect and to genetic effect. Heritability (h^2) estimates (SD) for unsaturated FA ranged from 0.03 (0.02) for C18:2 trans-15, cis-11 to 0.44 (0.10) for C14:1 cis-9. For the saturated FA the corresponding estimates varied from 0.05 (0.03) for C:22 to 0.43 (0.09) for C:16. Polyunsaturated FA ($h^{2=}0.28$; SD=0.08) had numerically higher estimates than did monounsaturated ($h^{2=}0.17$; SD=0.07) and saturated FA ($h^{2=}0.22$; SD=0.08). Heritability for index of desaturation (C14:1 cis-9/C14:0 + C14:1 cis-9) and C18:2 cis-9 trans-11 (CLA) were 0.51 (0.11) and 0.18 (0.07), respectively. Results highlight the existence of important and exploitable genetic variations in milk fatty acids composition, which could be used to improve the nutritional properties of milk fat by selective breeding.

Genetic parameters for body conformation scores and heifer pregnancy in Nelore cattle

J.P. Eler[1], M.L. Santana Jr[2], D.C. Cucco[3], A.B. Bignardi[2] and J.B.S. Ferraz[1]
[1]College of Animal Science and Food Engineering, University of São Paulo, Veterinary Medicine, Av Duque de Caxias Norte, 225, 13630-900 Pirassununga, SP, Brazil, [2]University of Mato Grosso, Campus Rondonópolis, MT-270, km 06, 78735-901 Rondonópolis, MT, Brazil, [3]Santa Catarina State University, Chapecó, 89815-630 Chapecó, SC, Brazil; joapeler@usp.br

Genetic parameters and associations between finishing precocity score (PRE), bone score (BONE), and probability of pregnancy at 14 months (HP) were estimated using records of Nelore animals born between 1984 and 2008 on 12 farms from three Brazilian states. The Bayesian linear-threshold analysis via the Gibbs sampler was used to estimate the (co)variance components applying a multi-trait animal model. Posterior mean estimates of heritability for PRE, BONE, and HP were 0.26±0.01, 0.26±0.01 and 0.50±0.02, respectively. Therefore, the genetic improvement of these traits is possible from selection programms. The genetic correlations among traits studied were: 0.85±0.01 between PRE and BONE; 0.25±0.03 between PRE and HP; and 0.03±0.05 between BONE and HP. High genetic correlation was observed between visual scores. Therefore, the simultaneous inclusion of PRE and BONE in a selection index does not seem to be necessary for the present population. Furthermore, the genetic correlations indicated that selection for traits related to body conformation has not a big effect on HP in Nelore cattle.

Casein genes polymorphism in some Egyptian and Italian goat breeds

S. Agha[1], M. D'Andrea[2], F. Pilla[2], A.R. Elbeltagy[3], A.Z.A. Salam[4] and E.S. Galal[1]
[1]Ain Shams University, Agriculture, Animal Production, Haddayek Shoubra, 68, Cairo, Egypt, [2]Molise University, Department of Animal, Agricultural and Environmental Sciences, Campobasso, Molise University, Campobosso, Italy, [3]Animal Production Research Institute, Ministry of Agriculture, Department of Animal Biotechnology, Dokki, Cairo, Egypt, [4]Ain Shams University, Agriculture, Department of Genetics, Haddayek Shoubra, 68, Cairo, Egypt; ahmed_elbeltagi@yahoo.com

Two Egyptian goat breeds, Zaraibi and Barki and two Italian breeds, Nicastrese and Aspromontana were used to investigate the genetic structure of the casein genes in these breeds, evaluate the genetic variability at the alpha S1 casein and study the relationships among and within the studied Egyptian and Italian goat breeds. Genotyping of CSN1S1 showed high frequency of strong alleles, A and B, in the Egyptian breeds. CSN1S1-01, null allele, has been detected only in Barki with low frequency (0.045). Genetic differentiation (FST) showed similarity between Egyptian breeds. CSN1S2-0 allele was only observed with low frequency (0.04) in Nicastrese. There was no polymorphism found at CSN2 locus in Barki and Zaraibi. The Egyptian breeds carried only the CSN2-A allele. In kappa casein, genotype CSN3B/B appeared in all the tested Egyptian animals showing no polymorphism. Lower genetic distance estimates were observed among the Egyptian and the Italian breeds. Results indicate that Egyptian goat breeds carry favorable alleles, associated with high and normal casein in high frequency. Those results make Egyptian goats potential dairy animals for milk and subsequently for cheese production. Selection for alleles associated with high and normal casein is recommended when developing breeding programs for Egyptian goats.

Session 26a Theatre 1

Cattle debate session

B. Whitelaw
Roslin Institute, University of Edinburgh, Easter Bush Campus, Midlothian EH25 9RG, United Kingdom; bruce.whitelaw@roslin.ed.ac.uk

The cattle sector regularly deals with a continuous bombardment of new techniques such as robotic milking, precision agriculture, on-line measurement, automatic feeding. More generally, changes in farming systems for dairy e.g. intensification, family versus large farms, confinement versus grazing, medication, etc are subjects of continuous societal interest. The consuming public is generally accepting new farming techniques. However, the rapid development of new breeding tools such as genomic selection, gene modification and editing requires an informed scientific debate on their efficacy in solving societal problems, their suitability for doing so, the possible trade-offs that need to be considered and the potential competing technologies that could provide similar benefits. In this session experts from various fields of interest will make a short introductory statement and the session leader will then facilitate a debate on the use of breeding techniques in modern agriculture. (1) Dr. Ferry Leenstra, expert in Poultry, Livestock Research Wageningen UR will give her view on the dairy sector as an outsider. (2) Dr. Roswitha Baumung / Irene Hoffmann, Animal Production and Health Division, FAO presents some views. (3) Dr. Mike Coffey, expert in cattle breeding, SRUC, UK, will react as a representative of the cattle sector. (4) Professor Bruce Whitelaw will outline some emerging technologies in animal breeding.

Grazing and dairy payments: developments in the Netherlands

A. Elgersma
independent scientist, P.O. Box 323, 6700 AH Wageningen, the Netherlands; anjo.elgersma@hotmail.com

Changing societal drivers and consumer demands require systems that provide desired human foods produced through sustainable production systems. This paper aims to show effects of grazing system on milk fatty acid (FA) composition in Dutch farming practice and to analyse current developments in grassland utilization and dairy payments. Trends in grazing systems were analyzed using statistical data. Farm milk samples were collected and analysed. Questionnaires provided associated information on soil type, feeding system and hours cows spent at pasture. Milk composition has changed with higher milk solid contents and lower unsaturated fatty acid concentrations. Research data illustrate effects on FA composition in farm milk from different grazing systems. Despite beneficial effects of fresh herbage in the dairy production chain, the trend in the Netherlands is that cows are more indoors and consume less fresh grass. However large regional differences occur, related to soil type, diet, and grazing system and this is reflected in regional differences in milk FA composition as will be shown. The declining number of cows at pasture is visible in the landscape. Action groups have triggered the public debate on indoor versus outdoor cows, mainly from the viewpoint of animal welfare. This has provoked much discussion and raised concern in the public opinion and in politics. Thus sentiments and marketing rather than scientific evidence dictated the political agenda. In 2011 the major Dutch dairy company changed its policy in favour of promoting grazing, mainly to preserve the natural image and for providing dairy farmers a societal license to produce. Farmers who graze their cows for at least 6 h/day during at least 120 days receive a premium price of 50 ct/100 kg. Milk and dairy from grazed cows is sold with a special ('Weidemelk') logo.

An empirical study of strategies for organic dairy farms in Austria

S. Kirchweger and M. Eder
Institute of Agricultural and Forestry Economics, BOKU, Feistmantelstraße 4, 1180 Vienna, Austria; stefan.kirchweger@boku.ac.at

The success in dairy farming is mainly influenced by the milk price as well as the produced milk yield per cow. Whereas farmers can barely control the milk price, they influence the level of the milk yield through operative and strategic management. Especially under the increasing volatile markets strategic management becomes more and more important. There are two main strategies in dairy which are the most promising: high output and low input strategy. Therefore we look at these strategies in Austrian organic dairy farms and compare those over the past few years of volatile markets. In order to assess these strategies in organic dairy farming we use a dataset from 2005 to 2010 of 70 voluntarily bookkeeping farms which show farm income per labour above the average. We apply cluster analysis to identify homogenous farm groups which differ in their strategies regarding intensity. Whereas this is done with data from 2010, we then compare the development of these groups over the time period 2005 to 2010. The cluster analysis identifies two out of six clusters which show the most different values in the used variables representing a high output strategy (n=14) and a low input strategy (n=8). First results show that these two clusters differ in milk yield per cow, total farm output and total farm input but are similar in utilised agricultural area and in farm income but. This leads to different developments during the time of volatile prices. In years of high prices, the total farm output rises on high output farms much more than on low input farms but, which is the same for total farm inputs. Therefore, the volatility of farm income for low input farms is much lower over the years observed. One might argue that farm income of low input farms is stabilized by public payments, but we find that these are similar for both groups. All in all our results show that the low input strategy is competitive with the high output strategy, especially we farms have to cope with volatile markets.

Supplementary crude protein and phosphorus levels: effect on spring milk production in dairy cows

M. Reid[1,2], M. O'donovan[1], C. Elliott[2], J. Bailey[3], C. Watson[3], J. Murphy[1], F. Coughlan[1] and E. Lewis[1]
[1]Teagasc, Animal & Grassland Research & Innovation Centre, Moorepark, Co Cork, Ireland, [2]Queens University Belfast, Institute for Global Food Security, Stranmilis Road, Belfast BT9 5AY, United Kingdom, [3]Agri-Food & Bioscience Institute, Newforge Lane, Belfast BT9 5PX, United Kingdom; michael.reid@teagasc.ie

Milk protein composition is affected by dietary concentrations of crude protein (CP) and minerals, notably phosphorus (P). Excess dietary CP can increase milk urea nitrogen (MUN) concentration, which is associated with reduced milk processability. A P-deficient diet can reduce milk protein concentration. The aim of the study was to investigate differing levels of dietary CP and P on milk production and animal P status in early lactation. Forty-eight spring-calving dairy cows were allocated to four treatments: high CP high P (HPrHP), medium CP high P (MPrHP), low CP high P (LPrHP), low CP low P (LPrLP), for 8 weeks. All treatment groups were offered 13 kg DM grass and 4 kg DM concentrate feed per cow daily. The HPrHP concentrate contained 240 g/kg CP, 57 g/kg P, the MPrHP 160 g/kg CP, 57 g/kg P, the LPrHP 70 g/kg CP, 57 g/kg P and the LPrLP 70 g/kg CP, 0.096 g/kg P. Data were analysed using the PROC MIXED and PROC GLM statements of SAS. Milk yield, milk fat and milk protein concentration did not differ significantly between treatments. In the first 7 weeks of the study, the HPrHP MUN concentration was significantly higher than that of the LPrHP and LPrLP treatments. In weeks 1, 2 and 5 the LPrHP and LPrLP MUN concentrations were lower than the MPrHP. There were no significant MUN concentration differences between LPrHP and LPrLP. Faecal P concentration and blood P concentration were lower in the LPrLP treatment than in the HPrHP, MPrHP and LPrHP treatments. The study suggests that a grazed grass diet with a low CP low P concentrate can reduce P excretion, without affecting milk yield or composition and that offering a low CP concentrate alongside grazed grass can reduce MUN concentration.

Genetic evaluation of in-line recorded milkability from milking parlor and automatic milking systems

C. Carlström[1], G. Pettersson[2], K. Johansson[3], E. Strandberg[1], H. Stålhammar[4] and J. Philipsson[1]
[1]Swedish University of Agricultural Sciences, Department of Animal Breeding and Genetics, P.O. Box 7023, 750 07 Uppsala, Sweden, [2]Swedish University of Agricultural Sciences, Department of Animal Nutrition and Management, Kungsängen Research Center, 753 23 Uppsala, Sweden, [3]Växa, Sverige, P.O. Box 7023, 750 07 Uppsala, Sweden, [4]Viking Genetics, P.O. Box 64, 532 21 Skara, Sweden; caroline.carlstrom@slu.se

Selection against slow milking cows is of great interest in order to save time and to make efficient use of expensive equipment in dairy herds with modern milking systems. In Sweden 28% of the cows are milked in automatic milking systems (AMS) and another 48% in conventional parlors (CMP). The aims of this study were: (1) to estimate heritabilities and genetic correlations for milkability traits based on observations from commercial herds with AMS or CMP; (2) to estimate genetic correlations for milkability across milking systems; and (3) to study the opportunities for an efficient genetic evaluation model using data from both milking systems. Data came from 19 herds with AMS and 74 herds with CMP. In total, information from 13,176 Swedish Holstein cows with 1,335,589 observations and 9,303 Swedish Red cows with 1,358,849 observations were included. Average flow rate was used as a common milkability trait for both systems, whereas milking time and box time was used from CMP and AMS, respectively. Univariate and bivariate repeatability models were used to estimate heritabilities within system, and genetic correlations across traits, lactations and systems. Estimated heritabilities were in the range of 0.25-0.51 and similar for both systems. Even though the traits were differently defined in the two systems, the corresponding traits were genetically closely related (0.93-1.00). The results clearly indicated that it is possible to achieve accurate genetic evaluations of milkability traits, not only for sires of cows but also for individual cows across milking systems.

Integrating pasture into a saturated robotic milking system: 4 years of experiments in Derval

T. Huneau[1], V. Brocard[2] and M. Fougere[1]
[1]Chambre d'Agriculture de Loire-Atlantique, Ferme expérimentale de Derval, La Touche, 44590 Derval, France, [2]Institut de l'Elevage, BP 85225, 35652 Le Rheu Cedex, France; thomas.huneau@loire-atlantique.chambagri.fr

Automatic milking system (AMS) have an exponential increase with about 2,800 running stalls in France. AMS technology is touching new customers wishing to keep grazing in their system. Though studies made in France show an important decrease in grazing after the purchase of an AMS because farmers lack technical advice to integrate AMS and grazing. The experimental farm of Derval (western France) bought a Delaval AMS in 2008 to study various strategies to optimize grass management. The grazable area reaches 0.40 ha per cow for a herd of 73 Holstein cows. Tracks and paddocks have been designed to match optimal grass management and farmers targets (simplified rotational grazing, 3 paddocks); various organisations during grazing season have been tested and assessed through their impacts on milking frequency, milked produced by the AMS, production per cow and margin over feed. During transition period, with buffer feed inside, a specific organisation is implemented due to the saturation of the AMS: cows milked after 12 pm are sorted out at 8 am and pushed outside, the other cows go out after milking one by one. Cows have free access to building between from 12am to 6 pm. Cows still grazing are fetched at 6 pm to get maize silage inside. During the 100% grass period, the whole herd is fetched at 6 pm and cows are allowed outside after 9 pm, after each milking. No come back is allowed before 8 am to make sure all the cows inside have time to be milked. In 2012, the milking frequency dropped from 2.10 in winter period to 1.86 in 100% grass period and dairy production from 29.9 kg to 27.6 kg/cow/d in full grazing period.; the concentrate intake decreased by 1.7 kg/d which led to an increased dairy margin. The total intake of grazed grass reached 1,200 kg DM/cow/yr. More experiments will be led in the coming years to study activity and traffic of the animals (Autograssmilk FP7 European project).

The temperament estimation in the dairy cattle

P. Juhás and P. Strapák
Slovak University of Agriculture, Department of Animal Husbandry, Tr. A. Hlinku 2, 949 76 Nitra, Slovak Republic; peter.juhas@uniag.sk

Aim of the study was to identify reliable behavior characteristics in the dairy cattle suitable for temperament estimation. Temperament is measured as the behavior response to unknown cue – duration of approaching the unknown cue, number of stops, distance of first stop and duration of first stop. As unknown cue was used the red LED lamp for bicycles, blinking in high frequency placed at the 120 cm above ground at the end of 11 meters long passage. The 30 dairy cows were tested. Recorded duration of passing was corrected by distance and duration of first stop. Corrected duration of passing was analyzed by Hierarchical cluster analyze. Differences among clusters were tested by Kruskall-Wallis test. Reaction to unknown cue can invoke three types of response – spontaneous passing without stops, spontaneous passing with stops, forced passing of cow non-able approach the unknown cue and pass the passage. Corrected duration of passing ranged from 11 seconds and 8 minutes and 2 seconds. Hierarchical cluster analyze divided group to 3 clusters. The cluster 1 contains 8 cows with corrected duration of passing from 1 minute and 28 seconds to 4 minutes and 10 seconds. All cows with forced passing were in this cluster, recorded distance of first stop was from 6 to 11 meters, duration of first stop was from 2 seconds to 1 minute 40 seconds, number of stops ranged from 2 to 9. The cluster 2 contains 9 cows with corrected duration of passing from 11 seconds to 23 seconds. All cows pass spontaneously, two cows make the stop during passing, maximal distance of first stop was 4 meters and duration 5 seconds. The cluster 3 contains 11 cows with corrected duration of passing from 29 seconds to 53 seconds. All cows pass spontaneously but only one without stops. The distance of first stop was from 4 to 8 meters, duration from 2 to 12 seconds. Differences among clusters in recoded behavior traits are considerable ($P<0.001$). The research was supported by VEGA No.: 1/2717/12 and ECOVA and ECOVA Plus projects.

Descriptive analysis of milk official recordings and phenotypic trends by breed in Italy

M. Cappelloni[1,2], N. Guzzo[2], C. Sartori[2] and R. Mantovani[2]
[1]Italian Animal Breeders Association (AIA), Via G. Tomassetti, 9, 00161 Rome, Italy, [2]University of Padua, Agronomy Food Natural Resources Animals and Environment, Viale dell'Universita', 16, 35020 Legnaro (PD), Italy; roberto.mantovani@unipd.it

The aim of this study is to review the official milk recordings gathered in Italy for cattle by the Italian Animal Breeders Association. Based on milk records registered in year 2011 at farm level (18,882 records; 15,369 farms), the Italian geographic distribution of farms, cow and milk yield was obtained. Using all data from 2004 to 2011 (135,834 records; 18,642 farms), a cluster analysis was conducted in order to obtain the degree of proximity between 8 purebreds (Friesian, Brown, Simmental, Jersey, Alpine Grey, Rendena, Oropa Red Pied, and Aosta Red Pied) and crossbred animals, for the main traits recorder (i.e. yearly average milk yield, fat and protein percentage, age at first calving, days open, and no. of inseminations/conception). Using a subset of records (10 samples; average 7,648 records; 1,228 farms) obtained by stratifying data on farm within breeds, the (co)variance structure between yearly repeats and phenotypic trends were analyzed under a mixed model accounting for breed (B, 9 levels), farm within B, year (Y, 8 levels) and interaction B*Y. Geographical distribution of 2011 records indicated a strong concentration of dairy activities in the north of Italy (79% herds and 87% of cows and milk yield). For milk yield the cluster analysis indicated that the cosmopolitan breeds and crossbred animals resulted markedly different from autochthonous breeds. Phenotypic trends showed a significant gap in milk yield between Friesian and all the other breeds, while Jersey, Brown and Aosta Red Pied had marked differences in milk quality as compare to the others. The Friesian breed showed the worst fertility records. Focusing on production and reproduction data in crossbred animals, a great similarity to Friesian performances appeared, suggesting a most likely proximity between crossbred and Friesian than with other breeds.

Genetic gain in the breeding program of Pinzgau cattle under restriction of increase of inbreeding

R. Kasarda, E. Hazuchová, I. Pavlík and O. Kadlečík
Slovak University of Agriculture in Nitra, Animal Genetics and Breeding Biology, Tr. A. hlinku 2, 94976 Nitra, Slovak Republic; radovan.kasarda@uniag.sk

The aim of the work was to simulate scenarios with the main focus on restriction of inbreeding and development of genetic gain. Restriction was made under maximum avoidance of inbreeding (MAI) mating strategy, and compare it with a random mating alternative. The parameters of the simulation were based on the structure the Slovak Pinzgau active population of 2,868 animals (930 purebred cows). Simulated was selection under a total merit index (TMI) covering the milk, survival and the live weight breeding value estimation results. The heritability of TMI (h^2=0.09) was estimated using a REML single trait animal model. The changes in genetic gain patterns were then studied over 20 consecutive generations. A truncation selection principle with avoidance of mating relatives was used. to limit inbreeding by reducing increases in average relationships. In separate simulation runs, the number of sires of sires was set at either 2, 3, 4 or 5, mated with 40 dams of sires in all cases. The sex ratio of the offspring was assumed to be 50/50 male/female. Ten consecutive generations were simulated for both random and maximum avoidance of inbreeding mating. Significant positive differences in genetic gain was observed in the MAI mating system with 3 (diff.=0.87**), 4 (0.30**) or 5 sires (0.14**) in comparison to random mating design. When using MAI, significantly lower inbreeding was observed with 3 (diff.=0.54**), 4 (0.31**), 5 (0.25**) sires. Simulation results showed that the use of a maximum avoidance of inbreeding mating strategy would lead to significantly decreased rates of inbreeding while maintaining suitable levels of genetic gain in the Slovak Pinzgau population.

Relationships between Igenity scores and measured growth and carcass traits in Finnish beef bulls

M. Pesonen[1], A. Huuskonen[1] and M. Hyrkäs[2]
[1]MTT Agrifood Research Finland, Animal Production Research, Tutkimusasemantie 15, 92400 Ruukki, Finland, [2]MTT Agrifood Research Finland, Animal Production Research, Halolantie 31A, 71750 Maaninka, Finland; maiju.pesonen@mtt.fi

Identification of SNP markers tied to economically important traits with beef production are included in commercial DNA tests. Relatively little work has been done to validate the marker panels currently sold and marketed by commercial genotyping companies. The objective of this study was to assess the relationships of growth and carcass traits (EUROP-classification) to Igenity panel scores. Data and hair samples were collected from Atria beef breed bull test station from 193 bulls of five different beef breeds (59 angus, 50 charolais, 41 hereford, 34 simmental and 9 limousin). Across all breeds the average daily gain between 0-365 days was 1.4 kg/d. The average slaughter weight across all breeds was 387 kg, the carcass conformation score (EUROP) 9.0 (R+) and EUROP-fat class 2.9. Data were used to assess the relationships of measured growth and carcass traits and Igenity panel scores for average daily gain (ADG), fat thickness, red meat yield and rib eye area (REA). The statistical analyses were performed using the SAS CORR procedure. Because many of the variables were not normally distributed Spearman's correlation coefficients was used for all variables. Surprisingly, results revealed a low, but statistically significant (P<0.05), negative correlation (-0.18) between measured daily gain and Igenity ADG score. No significant correlations between the Igenity panel scores for red meat yield and fat thickness with observed EUROP conformation and EUROP fat score were perceived. However, a significant (P<0.001) correlation (0.35) between the Igenity panel score for REA with EUROP conformation score was observed. The limited material in the present study did not show a clear evidence of functionality of the commercial genetic test with measured traits in Finnish production system.

Some factors affecting milk production during post-partum in cattle breeding in northern Algeria

M. Laouadi[1], S. Tennah[2] and M. Madani[3]
[1]University Amar Telidji of Laghouat, Algeria, Agronomy, BP 37G Road of Ghardaïa, Algeria, 03000, Algeria, [2]Superior National Veterinary School of Algiers, BP 161 Hassen Badi El-Harrach, Algiers, Algeria, [3]University Ferhat Abbes, faculty of sciences, Department of Agronomy, Setif, Algeria; laouadi_vet1000@yahoo.fr

This work aims to study the effect of some factors (race, origin of breed, age at first calving, parity and season of calving) on the variation of milk production (quantity and fat content). A total of 27 dairy cows were followed over a period of one year. The level of milk production is modest (16.53±3.88 liters at peak lactation), demonstrating the limits imposed by the environment on the expression of genetic potential. The highest performance of milk production (quantity) is observed in the cows calving in winter and spring compared to those calving in summer (P=0.007) and during the first three months of lactation. The decrease in milk production during the summer season has resulted in a parallel increase in fat content values (concentration phenomenon). Calving season exerts its effect mainly through diet that remains a limiting factor of the expression of genetic potential of dairy cows raised in Southern Mediterranean conditions that limit therefore the development of dairy farming in Algeria.

Session 26b Poster 11

Analysis of monthly recorded body measurements of Wagyu cattle by random regression models

T. Munim[1], T. Oikawa[2], T. Ibi[1] and T. Kunieda[1]
[1]Okayama University, Graduate School of Environmental and Life Science, 1-1-1 Tsushima-Naka, Okayama, Japan, 700-8530, Japan, [2]University of the Ryukyus, Faculty of Agriculture, 1 Senbaru, Nishihara-cho, Okinawa, Japan, 903-0213, Japan; tkroikawa@gmail.com

Japanese Black (Black Wagyu) cattle has breed characteristic of lower growth rate and narrow body structure but high meat quality, especially marbling. Objective of this study was to evaluate the possibility of random regression model (RRM) to a dataset of limited number of animals having intensively measured records. Number of records was 9,045 for 887 calves. Records included wither height (WH), hip height (HH), chest depth (CD), body length (BL), hip width (HW), and body weight (BW) measured from 1978 to 2008 at an experimental station of Okayama University, Japan. Variance components of body measurements were estimated by VCE602 with multiple trait model (MTM) and RRM. Low to moderate heritabilities were estimated for HH, CD BL, HW and BW through 0 to 12 mo by MTM analysis; showing no explicit trend along with ages. Estimated maternal heritabilities were low to moderate, except BW was low to high, showing clear downward trend; the highest estimate at 1 mo and the lowest estimate at 12 mo. By RRM analysis with 3-order Legendre polynomials estimated heritabilities of direct genetic effects were (0.64 to 0.71), (0.62 to 0.76), (0.51 to 067), (0.45 to 0.63), (0.54 to 0.74) and (0.49 to 0.76) for WH, HH, CD, BL, HW and BW, respectively, showing clear upward trend, whereas estimated heritabilities of maternal genetic effects were (0.02 to 0.12), (0.05 to 0.21), (0.02 to 0.08), (0.10 to 0.22), (0.11 to 0.25) and (0.12 to 0.22) for those, respectively and variance ratio of permanent environmental effects (c^2) were (0.04 to 0.07), (0.02 to 0.07), (0.01 to 0.06), (0.01 to 0.04), (0.03 to 0.08) and (0.00 to 0.11) for those, respectively, showing clear downward trend. The results of RRM generally agreed with MTM results. Therefore, it suggested that the accuracy of selection would be higher with RRM due to consistency of the estimates.

Session 26b Poster 12

Herd-test-day variability of methane emissions predicted from milk MIR spectra in Holstein cows

M.-L. Vanrobays[1], P.B. Kandel[1], H. Soyeurt[1], A. Vanlierde[2], F. Dehareng[2], E. Froidmont[2], P. Dardenne[2] and N. Gengler[1]
[1]University of Liege, Gembloux Agro-Bio Tech, 5030 Gembloux, Belgium, [2]Walloon Agricultural Research Centre, 5030 Gembloux, Belgium; mlvanrobays@ulg.ac.be

The aim of this study was to estimate the herd-test-day (HTD) effect on milk yield, fat and protein content, and methane (CH_4) emissions of Walloon Holstein first-parity cows. A total of 412,520 test-day records and milk mid-infrared (MIR) spectra of 69,223 cows in 1,104 herds were included in the data set. The prediction equation developed by Vanlierde *et al.* (R^2 of cross-validation=0.70) was applied on the recorded spectral data to predict CH_4 emissions (g/d). Daily CH_4 emissions expressed in g/kg of milk were computed by dividing CH_4 emissions (g/d) by daily milk yield of cows. Several bivariate (a CH_4 trait with a production trait) random regression test-day models including HTD and classes of days in milk and age at calving as fixed effects and permanent environment and genetic as random effects were used. HTD solutions of studied traits obtained from these models were studied and presented large deviations (CV=17.54, 8.93, 4.68, 15.51 and 23.18% for milk yield, fat and protein content, MIR CH_4 (g/d), and MIR CH_4 (g/kg of milk), respectively) indicating differences among herds, especially for milk yield and CH_4 traits. HTD means per month of milk yield and fat and protein contents presented similar patterns within year. The maximum of monthly HTD means corresponded to the spring (pastern release) for milk yield and to the winter for fat and protein contents. The minimum corresponded to the month of November for milk yield and to the summer for the other traits. For MIR CH_4 (g/d), monthly HTD means showed similar patterns as fat and protein content within year. MIR CH_4 (g/kg of milk) presented maximum values of monthly HTD means in November and minimum values in May. Finally, results of this study showed that HTD effects on milk production traits and on MIR CH_4 emissions varied through herds and seasons.

Behaviour patterns of buffaloes on pasture in autumn season

B. Barna and G. Holló
Kaposvár University, Guba S. street 40, 7400 Kaposvár, Hungary; brighitte@vipmail.hu

Our study was carried out in a Buffalo Reservation in the area of Balaton Uplands National Park. We monitored total of 39 buffaloes (24 cows, 1 bull and 14 calves) in one herd. The age of animals were younger than 10 month (calves) to 25 years (cows and bull). There were uniparous and multiparous cows in the herd. The monitoring started at 9 am and ended 15 pm for a week in autumn. We recorded the behaviours of animals in every in quarters. The behaviors divided into poses and activities. We didn't disturb the natural behavior of the animals. The most salient observation was that the animals moved at the same time on the pasture. We can explain this because when animals graze these animals are surely on the move. We observed a kind of hierarchy between the calves. Certain calves lie close to each other, while others lie little further somewhat separated from this group. The buffaloes were grazing scattered on the pasture. The calves often went on a journey of discovery. The animals grazed all day during the experimental periods. The animals used various objects for the scratching. The partner grooming had bigger role among calves than among adult animals.

Comparative feeding strategies for dairy bulls in a 19 month production system

B. Murphy[1,2], P. French[3], A.K. Kelly[1] and R. Prendiville[2]
[1]University College Dublin, University College Dublin, School of Agriculture and Food Science, Belfield, Dublin 4, Ireland, [2]Teagasc, Livestock systems, Teagasc, Animal and Grassland Research and Innovation Centre, Grange, Dunsany, Co. Meath, Ireland, [3]Teagasc, Livestock systems, Teagasc, Animal and Grassland Research and Innovation Centre, Moorepark, Fermoy, Co. Cork, Ireland; brian.murphy@teagasc.ie

Animals were assigned to a 3 finishing strategy (bulls supplemented with 5 kg of concentrates dry matter (DM) at pasture for 80 days and finished indoors on concentrates *ad libitum* (HC), pasture only for 80 days and finished indoors on concentrates *ad libitum* (MC) and pasture only for 80 days and finished on 5 kg DM of concentrates at pasture (LC)) × 2 concentrate feeding allowance (2 kg/head/day or 1 kg/head/day at pasture during their first grazing season) factorial arrangement of treatments. Bulls were finished over a 94 day period. Data were available from 80 spring born Holstein-Friesian bulls; 29 HC, 24 MC and 27 LC, respectively. Estimated individual concentrate DM intake for the finishing period was estimated to be 1.65, 1.27 and 0.47 tonne for the HC, MC and LC groups, respectively. Large differences in live weight at slaughter ($P<0.01$) and carcass weight ($P<0.05$) were observed. Live weight at slaughter was similar for HC and MC (600 kg and 578 kg, respectively) and lowest with LC (522 kg). Carcass weight followed a similar trend; 314 kg, 297 kg and 270 kg for the HC, MC and LC, respectively. Conformation score was greater ($P<0.01$) for HC (5.51) compared with LC (4.42) with MC intermediate (5.01). Fat class was greater ($P<0.001$) with the HC and MC groups compared with LC (7.09, 6.80 and 4.09, respectively). Kill out proportion was lower for MC (512 g/kg) compared with HC and LC (522 and 518 g/kg, respectively). Average daily gain during the finishing period was similar for HC and MC groups (1.78 and 1.72 kg/d) but lower for LC (1.42 kg/d). The influence of varying levels of concentrate supplementation during the first season at pasture proved non-significant.

Session 26b Poster 15

Effect of growing cattle grazing management on further feedlot performance and carcass traits

A. Simeone[1], V. Beretta[1], J. Elizalde[2], C.J. Caorsi[1] and J. Franco[1]
[1]University of the Republic, Ruta 3 km 363, 60000 Paysandu, Uruguay, [2]Consultant, Rosario, 2000, Argentina; beretta@fagro.edu.uy

An experiment was conducted to evaluate the effect of grazing management (GM) during beef cattle growing phase on further finishing feedlot performance and carcass traits. Forty-eight Hereford steers (278±34 kg; 11 months) were randomly allocated to a native (NP) or improved pasture (IP, *Festuca arundiancea*, *Lotus corniculatus*, *Trifolium repens*) grazed from spring to fall (219 days) followed by a winter finishing feedlot period (84 days). Cattle within GM were then randomly allocated to 6 pens (n=6/treatment; 4 steers/pen) and fed *ad libitum* a total mixed ration (85% concentrate) and slaughtered at a fix date. Liveweight (LW) was recorded every 28 days (grazing) or 14 days (feedlot), and LW gain (LWG) estimated by regression. Longissimus dorsi area (LDA) and subcutaneous back fat (SBF) were measured at the beginning and end of the grazing season. Feed/ gain ratio (FG) was estimated for the feedlot phase and carcass traits were recorded at slaughter. Statistical model included GM effect and initial records as covariates. The NP showed lower mean available forage biomass and quality (2225 kg/ha, CP 8.5%, TDN 56.3%) compared to the IP (3660 kg/ha, CP 10.4% TDN 61.4%). IP steers showed higher LWG during the grazing season (0.62 vs. 0.28 kg/d, P<0.01), were heavier by the end of this period (420 vs. 346 kg, P<0.01), had higher LDA (63 vs. 53 cm^2, P<0.01) and SBF (5.8 vs. 4.1 mm, P<0.01) compared to NP. However, during the feedlot phase, steers that had grazed NP showed higher LWG (1.53 vs. 1.19 kg/d, P<0.01) and better FG (6.6 vs. 10.1, P<0.01). NP steers had lower carcass weight (263 vs. 294 kg, P<0.01), but no differences were observed in SBF, LDA, pH, or fat and muscle colour parameters (P>0.05) compared to IP steers. Results suggests that restricted LWG of growing cattle during the grazing season may be compensated during the finishing feedlot phase, reducing differences in carcass and meat traits.

Session 26b Poster 16

Effect of feed delivery system on feeding behaviour of lot-fed calves

V. Beretta[1], A. Simeone[1] and J. Elizalde[2]
[1]University of the Republic, Ruta 3 km 363, 60000, Uruguay, [2]Consultant, Rosario, 2000, Argentina; beretta@fagro.edu.uy

An experiment was conducted to evaluate the effect of feed delivery system (FDS) on feeding behaviour of lot-fed weaned calves. Forty eight Hereford calves (148±26 kg) were randomly allocated to 8 pens outdoors to receive 1 of 2 treatments (T) (n=4 pens/T), consisting in an total mixed ration (6.3% rice hulls, 60% sorghum grain, 15% wheat bran, 14% sunflower meal, 4.7% premix) offered *ad libitum*, daily in 3 meals (DF) or using a self-feeder placed in each pen (SF). After diet introduction calves were fed for 8 weeks. Dry matter intake (DMI) was measuredon 3 consecutive days (D) per week (W). On day 42, all calves were observed (08:00 to 19:00 h), recording every 15 minutes eating, drinking, ruminating and idling activity. Intake rate (g/min) was estimated as DMI/ time spent eating. DMI was analysed through a generalized model for repeated measurements: $Y_{ijkl}=m+T_i+e_{ij}+W_k+(TW)_{ik}+D(W)_l+(TxD(W))_{il}+e_{ijkl}$, and behaviour data was submitted to LOGIT transformation assuming binomial distribution and analysed as the probability to find an animal on a specific activity. DF calves showed higher DMI (3.68 vs. 3.19 kg/100 kg liveweight; P<0.01). DMI was stable between-week and between-day, independent of treatment (P>0.05). Eating activity was higher in DF (0.25 vs. 0.19, P<0.01) while in compensation idling was reduced (0.67 vs. 0.72; P<0.01). No differences were observed in ruminating activity and intake rate (P>0.05). Although feed in DF was delivered 3 times a day, steers showed higher intake activity compared to SF only between 0800 and 1100 h (0.47 vs. 0.24; P<0.01) reducing activity between 1100 and 1300 has (0.20 vs. 0.28; P<0.05) and between 1300 and 1700 h (0.20 vs. 0.30; P<0.05). No differences were observed by the end of the day (0.12 vs. 0.18; P>0.05). Results suggest that the SF system for lot-feeding beef calves with all-concentrate diets would be a safe system, given the observed controlled intake and stable DMI between-day and within-day.

Effects of production system on early maturing dairy cross steers

R. Prendiville[1], B. Swan[2] and P. French[3]
[1]Teagasc, Animal and Grassland Research and Innovation Centre, Grange, Dunsany, Co. Meath, Ireland, [2]Teagasc, Crops, Environment and Land Use Programme, Johnstown Castle, Wexford, Co. Wexford, Ireland, [3]Teagasc, Animal and Grassland Research and Innovation Centre, Moorepark, Fermoy, Co. Cork, Ireland; robert.prendiville@teagasc.ie

The study comprised of 32 February born early maturing dairy cross steers. Animals were assigned to a 2 breed group (Aberdeen Angus (AA) and Hereford (HE)) × 2 production system (slaughtered at 21 (21MO) and 23 months (23MO) of age) factorial arrangement of treatments. Steers in the 21MO were slaughtered off pasture at the end of the second grazing season while those in the 23MO were finished indoors during the second winter. Animals in the 21MO were supplemented with 2.5 kg of concentrates dry matter (DM) per day for the final 60 days. Those in the 23MO were offered grass silage *ad libitum* with 5 kg concentrates DM/day for 88 days pre-slaughter. Live weights at slaughter were 533 and 572 kg for the 21MO and 23MO, respectively (P=0.0584). No difference in carcass weight was observed between the production systems, 278 and 289 kg for the 21MO and 23MO, respectively. Kill out proportion was greater (P<0.001) for steers in the 21MO than 23MO, 521 and 505 g/kg, respectively. No difference in conformation score and fat class was observed; 5.44 and 7.38 for the 21MO and 5.93 and 8.07 for the 23MO, respectively. Average daily gain during the finishing period was greater (P<0.001) for 23MO (0.98 kg/day) than 21MO (0.57 kg/day). Both production systems produced acceptable carcasses. No difference in animal performance was observed between the breed groups. Further research with a larger dataset is warranted to discern potential differences between the production systems.

Marker assisted selection of carcass traits in Nellore cattle

J.B.S. Ferraz[1], R.C.G. Silva[2], F.M. Rezende[3] and J.P. Eler[1]
[1]Center of Animal Genetic Improvement, Biotechnology and Transgenesis, FZEA/University of Sao Paulo, Veterinary Medicine, Rua Duque de Caxias Norte, 225, Campus da USP/FZEA/ZMV/NAP-GMABT, 13635900 Pirassununga SP, Brazil, [2]Merial Saude Animal, Animal Health, Av. Carlos Grimaldi, 1701 Edifício Galeria Corporate, 4o. andar, 13091-908 Campinas, S.P., Brazil, [3]INGEB/Federal University of Uberlandia, INGEB, Campus Avançado da UFU em Patos de Minas, 38705-108 Uberlandia MG, Brazil; jbferraz@usp.br

Brazilian beef industry competes for the leadership of beef exports and it's necessary to improve productivity of the beef herd. This study aimed to evaluate the effect of inclusion of the molecular breeding values, estimated from a very low density of genetic markers (Nellore Profile IGENITY® V3) panel on marker assisted selection of Nellore cattle. Data of 9,749 animals measured with ultrasound for rib eye area, fat thickness and rump fat thickness, with a relationship matrix of 39,687 animals were analyzed. Two models were used: single trait model was performed only for each observed phenotypes and two-trait models were performed for phenotypes, considered as one trait, and molecular breeding values for the same trait, considered as a correlated attribute. Inclusion of molecular information of 3,033 animals increased accuracies of predicted breeding values of genotyped animals, mainly, for replacement young bulls, where the difference reached up to 66.6% of original accuracy. Divergences of selection for 20% best animals, classified by 1-trait breeding values when compared to the selected by the 2-trait breeding values demonstrated highest divergence for sires and replacement young bulls. The genetic progress rate on the 2-trait model increased 2.4% for rib-eye area, 0.9% for fat thickness and 1.9% for rump fat thickness. The results suggest that the inclusion of molecular breeding values, even when estimated from very low density genetic markers panels contribute to the increment of both accuracy of predictions and genetic change rate on animal breeding programs for Nellore cattle.

Influence of calving difficulty on rumination and lying time in Holstein dairy cows

M. Fustini, L. Mammi, A. Palmonari, N. Negri and A. Formigoni
DSMVET, via Tolara di Sopra 50, 40064, Ozzano Emiila, Italy; mattia.fustini3@unibo.it

Several researches suggest that a difficult calving could compromise productivity and welfare in dairy cows. Aim of this study was to evaluate the effects of dystocia on rumination and lying behavior in dairy cows. A total of 25 Holstein dairy cows, 9 first calving heifers and 16 multiparous, was monitored from the 7th month of pregnancy to 80 days of lactation. Calving behavioral data were video recorded continuously, lying and rumination time were obtained via accelerometers placed on the cow leg and neck. The ratio among cow body weight (adjusted for a 3.5 Body Condition Score – BCS) and calf body weight was defined as criteria to classify calving difficulty: <14 difficult (D), >17 easy (E), within medium (M). Behavioral data were matched with milk production and quality, body weight and BCS. Data were then processed for statistical analysis using one-way ANOVA model with the software Statistica (v.6.0, StatSoft Italia S.r.l., Padova, Italy). Time from appearance of the amniotic sac or feet to birth longer than 60 minutes was associated with heavier calves (45 kg average weight), compared to the others (39.3 kg). No first calving heifers have been classified as group E. Group D shown an higher drop in rumination time after calving ($P<0.05$), both for heifers and multiparous cows. First calving heifers score D had a lower resting time in the first week after calving compared to the group score M ($P<0.05$). Milk production was not different among groups, while milk fat tend to be higher in the first week after calving in D groups, both for primiparous and multiparous cows. Group D in primiparous cows had the most serious drop in lying time ($P<0,05$) compared to the group M. Moreover in the first week after cows where moved to the lactation group (on average 10 days after calving), primiparous have an higher reduction in lying time compared to multiparous cows. In conclusion rumination time, along with lying time, could be usefully applied as a welfare index.

Performance and slaughter value of suckling male buffalo calves

B. Barna and G. Holló
Kaposvár University, Guba S. street 40, 7400 Kaposvár, Hungary; brighitte@vipmail.hu

The aim of this study was to evaluate the slaughter value of suckling buffalo male calves. Animals (n=18) were born in June (n=6), in August (n=6) and in September (n=6) 2011 and were slaughtered on two occasions (in February and in March 2012) at the same live weight average of 195.24 kg. The animals were kept in a cow-calf herd on the pasture, during winter the suckling calves were given as supplementary feeding only roughage. The average live weight production of calves was 903 g/day. The EUROP meat and fat score were, 4.83 (O0) and 4.59 (2+) resp. The dressing percentage was about 51%. In slaughter traits there were no significant differences among birth groups. The percentage in the right half carcasses was significant higher (65.53%) in August born buffalo calves opposite to those born in June (61.21%) and in September (62.49%).The fat and bone content showed an opposite trend. The highest fat percentage had buffalo calves born in June. The 12th rib sample composition measured by CT showed similar tendency, the highest muscle (65.75%) and the lowest fat content (10.65%) were detected in August born buffalo calves. The intramuscular fat level of longissimus muscle varied between 0.61-0.65%, showed similar tendency as fat content in the carcass. Significant differences were detected in fatty acid composition for cis-9 trans-11 CLA and alfa-linolenic acid percentage among groups. The fatty acid profile of meat from suckling male buffalo calves is prosperous concerning CLA content and n-3 fatty acids level. The P/S and the n-6/n-3 ratio were on average 0.31 and 2.96.

Building of biomimetic structures in order to reproduce the outer membrane of bull spermatozoa

J. Le Guillou[1], M.-H. Ropers[2], D. Bencharif[1], L. Amirat-Briand[1], S. Desherces[3], E. Schmitt[3], M. Anton[2] and D. Tainturier[1]
[1]Oniris, 44, Site de la Chanterie, route de Gachet, 44300 Nantes, France, [2]INRA, 44, Rue de la Geraudiere, 44316 Nantes, France, [3]IMV Technologies, 44, ZI n°1 Est, 61300 L'Aigle, France; chalawak@yahoo.fr

The knowledge about the mechanisms of action of protective extenders used to preserve bull semen for frozen or chilled is essentially empirical. The purpose of this study was to create biomimetics structures which can reproduce these mechanisms. To achieve this goal, a structure which reproduces the outer membrane of spermatozoa has been built. The biomimetics structure chosen for experiments is a lipid monolayers at the air-water interface formed on a Langmuir balance. Composition of subphase is controlled during experiments and can be changed to test interactions between biomolecules and biomimetic membrane. The purpose of these tests is to highlight biomolecules implicated in the protective effect on the semen. First, a lipids mix deposit was done, then the barriers positions was modulated to get the desired molecular compression with a controlled pressure. Then, molecules with protective effect were introduced in the subphase. The monolayer changes were monitored. Each experiment was replicated twice. Miscibility studies at 34 °C and 8 °C shows the formation of homogeneous domains of sphingomyelin and cholesterol, located in fluids domains composed of phosphatidylcholine. Complex biomolecules extracted from egg yolk like Low Density Lipoprotein was incorporated into the monolayer, contrary to other purified molecules like egg phospholipids. Contact tests with the monolayer were conducted with bull seminal plasma. Seminal plasma was injected either alone or associated with protective molecules which seems to inhibit the effect of seminal plasma on the monolayer. This model could be an opportunity for further studies where monolayer composition can vary as lipids composition depending the species studied. Other protective molecules, as well as the composition of the subphase can be also tested.

Genetic parameters for carcass traits at different slaughter age in Japanese Black cattle

K. Inoue[1], T. Osawa[1], K. Ishii[1], T. Katsuta[2] and M. Aoshima[3]
[1]National Livestock Breeding Center, Fukushima, 961-8511, Japan, [2]Wagyu Registry Association, Kyoto, 604-0845, Japan, [3]Japan Meat Grading Association, Tokyo, 101-0063, Japan; k1inoue@nlbc.go.jp

The purpose of this study is to investigate the changes of genetic parameters for carcass traits at different slaughter ages in Japanese Black cattle. Genetic parameters were estimated for carcass weight (CW), rib eye area (REA), rib thickness, subcutaneous fat thickness, yield estimated percentage (YEP) and meat quality traits (beef marbling standard number (BMS), beef color standard number (BCS), brightness, firmness and texture of beef) using a total of 522,037 records of Japanese Black cattle. Sire model (on GIBBS3F90 program) was used to estimate the genetic parameters. The model included fixed effects of sex, fattening farm × slaughter year, slaughter month (slaughter house × year × month for meat quality traits) and sire effects with random regression on slaughter age (26-32 month) using second-order Legendre polynomials and residual effects assuming heterogeneous variances at each slaughter age (26-32). In addition, 541 pedigree information was used. The averages of heritability estimates within the traits were 0.28 (BCS) to 0.52 (YEP) and the variation of those were between 0.01 (REA) and 0.10 (BMS). The trends of heritability estimates with aging showed various patterns: an increase with aging in BMS, a quadratic curve in CW and so on. The change of heritability estimates in CW was caused by bigger residual variances at early and late ages of slaughter in spite of unchanged genetic variances at all ages. By contrast, genetic variances increased with aging as residual variances did in BMS. Genetic correlations between slaughter age at 29 month (average slaughter age) and the other months within the traits were above 0.90 in all the traits.

Session 26b Poster 23

Myosin heavy chain expression in bovine muscles was detected at the single muscle fiber level

M. Oe[1], K. Chikuni[1], I. Nakajima[1], K. Ojima[1], M. Shibata[2] and S. Muroya[1]
[1]NARO Institute of Livestock and Grassland Science, Tsukuba, Ibaraki, 305-0901, Japan, [2]NARO Western Region Agricultural Research Center, Ohda, Shimane, 694-0013, Japan; mooe@affrc.go.jp

Skeletal muscles are composed of different types of muscle fibers, which are determined by the expression of myosin heavy chain (MyHC) isoforms. In bovine muscles, three MyHC isoforms (MyHC-slow, MyHC-2a and MyHC-2x) are expressed and these isoforms correspond to muscle fiber type I, IIA and IIB. Muscle fiber type affects sensory quality of meat, thus it is important to classify muscle fiber type accurately. In our previous study, the composition of MyHC isoforms in ten different bovine muscles was determined by the expression of MyHC mRNA. However, this method is not sufficient to clarify the accurate expression of MyHC isoforms in individual muscle fibers. The objective of this study is to reveal mRNA expression level of each MyHC isoform in isolated single muscle fibers. A total of 192 single fibers were isolated from three muscles (masseter, semispinalis and semitendinosus), and then total RNA was prepared from each fiber. The expression levels of MyHC isoforms were analyzed by real-time PCR. Our results showed that the fibers isolated from masseter expressed only MyHC-slow isoform. Most of the fibers isolated from semispinalis expressed either MyHC-slow or MyHC-2a. In semitendinosus, the fibers containing either MyHC-2x or MyHC-2a were dominant. Furthermore the hybrid fibers expressing both MyHC-2x and -2a in an identical single fiber were also detected in semitendinosus. Our mRNA analysis in single fibers may contribute to the accurate classification of muscle fiber type in individual muscle fibers.

Session 26b Poster 24

Correlation between GH gene polymorphisms and milk production trait in Holstein cattle

S. Abdel-Rahman[1], A. El-Nahas[2], S. Hemeda[2] and S. Nasr[2]
[1]Genetic Engineering and Biotechnology Research Institute, Department of Nucleic Acid Research, City of Scientific Research and Technological Applications, Alexandria, Egypt, [2]Faculty of Veterinary Medicine, Department of Animal Husbandry and Wealth Development, Alexandria University, Alexandria, Egypt; salahmaa@yahoo.com

PCR–RFLP technique was developed for the association between growth hormone (GH) gene polymorphisms and milk production trait in Holstein cattle. Forty-eight female Holstein cattle reared under Egyptian conditions were selected precisely according to their milk productivity, and DNA from blood samples of these animals wer extracted to amplify 329-bp of the gene encoding GH. Based on the breeding value, the 48 animals were ordered from the highest to the lowest milk productivity levels. Restriction analysis of PCR-RFLP- HapaII of the GH gene (329-bp) showed three different genotypes MM, MN and NN with frequencies 0.04, 0.25 and 0.71, respectively. The frequencies of the M and N alleles were 0.17 and 0.83, respectively. The results indicated that the MN cows yielded more milk than MM and NN cows. Sequencing (GenBank JF826521) revealed that six mutations (115C→T, 249C→T, 251C→A, 261T→C, 264T→C and 269T→C) occurred in the genotype NN of Holstein cattle. These findings can be used as marker-assisted selection (MAS) for high milk production traits in Holstein cattle.

Session 27 Theatre 1

QTL mapping for performances in sport horses

S. Brard[1] and A. Ricard[2,3]
[1]Institut National de la Recherche Agronomique, UR 631, Auzeville BP 52627, 31326 Castanet Tolosan Cedex, France, [2]Institut Français du Cheval et de l'Equitation, Recherche et innovation, Jumenterie du Pin, 61310 Exmes, France, [3]Institut National de la Recherche Agronomique, UMR 1313, Allée de Vilvert, 78352 Jouy-en-Josas, France; sophie.brard@toulouse.inra.fr

QTL mapping was performed for performances of sport horses in jumping. The 1,010 horses included in the study were Selle Français (70%), Foreign Sport Horses (17%) and Anglo-Arabians (13%). Two samples were used: a sample with all horses, and another one with SF and FSH only, because AA were proved to be genetically different from these breeds. 93% of horses were stallions. Phenotypes were got by de-regressing EBVs, which were corrected for fixed effects and preferential mating of stallions. Horses were genotyped using Illumina EquineSNP50 BeadChip. 54,602 SNP were available and 44,424 SNP were retained after quality tests. Two models were used: a single-SNP mixed model and a haplotype-based mixed model. They included a polygenic effect to take into account familial structures. Because of the high number of independent tests, a Bonferroni correction was applied to significance and suggestive thresholds, which were set at 10^{-5} and 10^{-4}. In the sample with all horses, the single-SNP mixed model detected a suggestive QTL on chromosome 1. The haplotype-based mixed model detected it, and other suggestive QTLs on chromosomes 11 and 16. In the sample without AA, the haplotype based model detected the suggestive QTLs of chromosome 1 and 16. Suggestive QTL of chromosome 16 was found with only one model, so it could be a false positive. Suggestive QTL of chromosome 11 was detected by both models, but in only one sample. It could be a marker of difference between breeds, but the aim of our study was to find QTLs for jumping shared by the three types of horses. QTL on chromosome 1 (BIEC2 31196) got low P-values in all cases. Performance in jumping is a complex trait, but investigations could be done around the suggestive QTL of chromosome 1 to find potential genes influencing this trait.

Session 27 Theatre 2

Opportunities for joint genetic evaluations of Danish and Swedish sport horses

Å. Viklund[1], S. Furre[2], O. Vangen[2] and J. Philipsson[1]
[1]Swedish University of Agricultural Sciences, Animal Breeding and Genetics, P.O. Box 7023, 75007 Uppsala, Sweden, [2]Norwegian University of Life Sciences, Department of Animal and Aquacultural Sciences, P.O. Box 5003, 1432 Ås, Norway; asa.viklund@slu.se

The Nordic Interstallion project aims at a joint genetic evaluation for the Nordic sport horse populations. The populations use foreign stallions of similar origin to a great extent, but a stallion may have too few offspring in each country for a reliable national breeding value. All Nordic countries should therefore benefit from a joint genetic evaluation. The present study included data from show jumping and dressage competitions in Sweden and Denmark through 2011. Show jumping data consisted of 28,000 and 19,500 competing horses in Sweden (SWB) and Denmark (DWB), respectively. For dressage the number of horses was 15,200 SWB and 20,800 DWB. Lifetime performance in each discipline was defined as lifetime accumulated points. The joint pedigree file was traced 7 generations back ending up with 174,000 horses. For the competition horses the pedigree completeness index was on average above 80%. In total 1074 stallions had competing offspring in both countries. The genetic similarity (GS) between SWB and DWB was calculated to 60%. Both countries contributed almost equally to GS, 52% from SWB and 48% from DWB. Genetic analyses were performed with a bivariate animal model with performance in the different countries considered as different traits. The genetic correlations between performances in the two countries were close to unity, 0.99 for show jumping and 0.98 for dressage. Heritabilities for dressage were estimated to 0.23 for both Swedish and Danish data. For show jumping the heritabilities were estimated to 0.34 and 0.30 for Swedish and Danish data, respectively. The high genetic correlations between performance traits, and high GS between the both populations, show that the joint data can be used to estimate common breeding values. Next step in the project is to include competition data from Norway and Finland.

Session 27 Theatre 3

The inclusion of international showjumping results in the genetic evaluation of Irish Sport Horses

K.M. Quinn-Brady, D. Harty and A. Corbally
Horse Sport Ireland, Beech House, Millennium Park, Osberstown, Naas, Co. Kildare, Ireland; dharty@horsesportireland.ie

The breeding objective of the Irish Sport Horse studbook is to produce a performance horse that is sound, athletic with good paces and suitable temperament and capable of winning at the highest international level in FEI disciplines. Eventing and showjumping are the main disciplines prioritised by breeders. Breeding values for showjumping ability have been estimated with a multi-trait animal model since 1995 but no international performance information was included. A study was carried out to investigate the inclusion of international performances of Irish Sport Horses in the genetic evaluation. Data used in the study included pedigree data on Irish Sport Horses (n=57,136) and a database of national and international performances constructed from FEI data, manually retrieved international performances and national performance data (n=1,119,770). The Lifetime Performance Rating (LPR) of 16,727 horses was assessed as the 'highest level' successfully achieved. Success was defined as two double clear rounds at 11 possible levels determined by fence height and competition level, ranging from national 1.00 m competitions to CSI major championships. Genetic parameters were estimated in uni- and multivariate animal models. The heritability of LRP was estimated at 0.28. The genetic correlation between LPR and national performances at 1.40 m/1.50 m level was found to be 0.96. The correlation between breeding values (BVs) for LPR and BVs for national performances (including all levels from 0.90 m to 1.50 m in a multi-variate model) was 0.77 showing a relatively large effect on BVs and ranking of stallions. This reflected a decrease in emphasis on young horse results. The LPR methodology results in a BVs that better reflect the studbook's breeding objective and are now available.

Session 27 Theatre 4

Health status and conformation in young sport horses affect performance and longevity in competition

L. Jönsson[1], A. Näsholm[1], L. Roepstorff[2], A. Egenvall[3], G. Dalin[2] and J. Philipsson[1]
[1]Swedish University of Agricultural Sciences, Department of Animal Breeding and Genetics, P.O. Box 7023, 75007 Uppsala, Sweden, [2]Swedish University of Agricultural Sciences, Department of Equine Sciences, P.O. Box 7043, 75007 Uppsala, Sweden, [3]Swedish University of Agricultural Sciences, Department of Clinical Sciences, P.O. Box 7054, 75007 Uppsala, Sweden; lina.jonsson@slu.se

Few studies have evaluated the influence of variation in conformation and health in young sport horses, on present and future performance. The objective of this study was to estimate phenotypic and genetic associations between conformation and health status in young riding horses on one hand, and performance at the day of testing and in future competitions, including number of years in competition (NYC), on the other hand. Analyses included 8,238 horses examined for conformation, health and talents for jumping and dressage by independent examiners, during the Swedish Riding Horse Quality Test (RHQT) of 4-5-year-old horses, and lifetime competition results. Single trait linear models and multi-trait animal models were used, for phenotypic effects and genetic correlations, respectively. Results indicated both health status and conformation to have significant phenotypic and genetic effects on gaits and jumping performance scores at day of testing. Favourable genetic correlations between health status and performance reached 0.23 for jumping scores and 0.37 for gait scores. Genetic correlations with NYC were favourable, 0.25-0.31, for all health traits. Conformation also showed positive correlations with NYC ~0.2. Lifetime performance showed genetic correlations of 0.26-0.36 to health traits and of 0.28 for the most important conformation traits (head-neck-body). Horses of intermediate size, withers height 164-171 cm, were found to be most durable in competitions. Results suggest that selection for improved health status and conformation of the young horse will also improve desired performance traits and longevity of competition horses.

Genetic analyses of linear conformation and performance traits in Warmblood horses

K.F. Stock[1], J. Duensing[2], W. Schulze-Schleppinghoff[3] and J. Krieter[2]

[1]Vereinigte Informationssysteme Tierhaltung w. V., Heideweg 1, 27283 Verden, Germany, [2]Christian-Albrechts-University of Kiel, Institute for Animal Breeding and Husbandry, Olshausenstrasse 40, 24098 Kiel, Germany, [3]Oldenburg horse breeding society, Grafenhorststrasse 5, 49377 Vechta, Germany; friederike.katharina.stock@vit.de

Linear descriptions of equine conformation and performance allow standardized recording of detailed information on traits of breeding relevance. With an innovative documentation approach, linear data for a broad spectrum of traits could be collected in connection with regular breeding events of the Oldenburg horse breeding societies in 2012. In this first broad praxis, test linear profiles were compiled in addition to the official evaluations for in total 832 foals, 441 mares and 271 stallions. To investigate the usability of linear traits for selection in the Warmblood, genetic parameters were estimated for 25 traits in juvenile horses (J) and 70 traits in adult horses (A) using VCE6. Pedigree information on at least 3 ancestral generations was considered, resulting in a relationship matrix with 7,731 animals. Uni- and bivariate analyses were performed for the 25 traits included for J and A, revealing mostly consistent and moderate heritabilities, with averages of 0.14 in J and 0.20 in A for conformation and 0.23 in J and 0.17 in A for movement. Among the movement traits, elasticity in free trot had the highest heritability (0.41 in J, 39 in A). Between analogous linear traits in J and A, we found mostly moderately to highly positive additive genetic correlations of ≥0.4. According to the results of this study, conformation and performance data collected with the new Oldenburg system of linear profiling are suitable for genetic analyses, regardless of age and evaluation conditions. Compared to the traditional system of valuating scoring, linear trait definitions are much more specific, implying opportunities for targeted improvement of breeding goal characteristics in Warmblood riding horses.

Genetic parameters of type traits scored at adult age in Italian Heavy Draught Horse

F. Folla[1], C. Sartori[1], G. Pigozzi[2] and R. Mantovani[1]

[1]Dept. of Agronomy Food Natural Resources Animals and Environment, Viale dell'Universita', 16, 35020 Legnaro (PD), Italy, [2]Italian Heavy Draught Horse Breeders Association, Via Verona, 90, 37068 Vigasio (VR), Italy; roberto.mantovani@unipd.it

The aim of this study was to estimate genetic parameters for type traits scored at adult age in the Italian Heavy Draught Horse and analyze genetic correlation with the same traits scored at younger age (i.e. on 6 month-old foals), and used for selection. The initial databases consisted in 7,133 evaluations of adults and 15,945 of foals scored by 35 classifiers in 20 subsequent years (i.e. from 1992 to 2011) on 14 linear type traits with a 9 point scale system (from 1 to 5 including half points). After editing, 4,205 scores on adults and 8,490 scores on foals were retained for further analysis. Data were merged and 2,180 animals resulted scored at both ages. Bi-traits Gibbs Sampling analyses were carried out on a total of 10,515 records related to 15,649 animals in the pedigree. The model considered the following fixed effects: herd-year-classifier (1,855 classes for foals and 1,297 for adults), sex (2 classes), age at evaluations (9 classes for foals and 5 for adults), and age of mare at foaling (5 classes, for foal data only). The heritability estimates ranged from 0.10 to 0.37 for traits scored at adult age, and were similar to those obtained on foals. Small differences were observed on traits under selection, with the exception of fleshiness, where heritability was reduced from 0.35 for scores on foals to 0.23 for scores on adults. The genetic correlations between traits measured in young and adult age ranged from 0.55 to 0.97, and the phenotypic correlations ranged from 0.15 to 0.51. The lowest genetic correlations between foal and adult scores were observed on thorax depth and fleshiness (i.e. 0.55 and 0.63, respectively). Greater genetic correlations were estimated for the other traits under selection, i.e. fore and rear diameters (0.85 and 0.87, respectively), head-size and expression (0.88) or temperament (0.82).

Inbreeding status and conservation possibilities of the endangered Faroese Horse

P. Berg[1], A. Præbel[1] and D. Joensen[2]
[1]NordGen, P.O. Box 115, 1431 Ås, Norway, [2]Búnaðarstovan, Frammi i Dal 166, 410 Kollafjørður, Faroer Islands; peer.berg@nordgen.org

The Faroese Horse was brought by the first settlers to the Faroe Islands 1,200 years ago. Traditionally, horses have been living free in the mountains and gathered only to work. As a result, the Faroese Horse has adapted to the harsh environment on the Faroe Islands. Current interest is in sustainable management and conservation of a purebred Faroese Horse population. Our objective was to describe the current Faroese Horse population relative to the level and development of inbreeding. Additionally, using optimal contribution selection, mating scenarios were examined to test possibilities for sustainable management of this endangered population and give recommendations to Faroese horse breeders. The present Faroese Horse population comprises only 63 individuals, from a bottleneck of one stallion and four mares in 1960's. The stallion has contributed 48% of genes to foals born in 2011 and 2012. Pedigree consists of a total of 146 individuals. In the period from 1980 and onwards the effective population size has been 34. In 2012, 38 individuals (19 females, 19 males) were available for breeding. Average inbreeding coefficient for the breeding candidates was 25.6%. Optimal contribution selection (OCS) of parents (females mated once, males mated 1-4 times) were studied for 6 to 12 matings in the next cohort. Repetitive use of stallions resulted in the lowest level of average relationship, although differences between scenarios were low. With the high additive relationships between breeding candidates (within and across male and female groups 55-62%), levels of inbreeding will be high. OCS restricted rate of inbreeding. Maximum avoidance of inbreeding amongst selected individuals resulted in 0.04 lower inbreeding coefficients compared to random mating. Due to the high level of inbreeding careful registration of traits such as conception rate, foaling difficulties, birth defects and mortality is highly recommended.

Investigations into genetic variability in Holstein Horse breed using pedigree data

L. Roos[1], D. Hinrichs[1], T. Nissen[2] and J. Krieter[1]
[1]Institute of Animal Breeding and Husbandry, Christian-Albrechts-University, Hermann-Rodewald-Straße 6, 24118 Kiel, Germany, [2]Verband der Züchter des Holsteiner Pferdes e.V., Abteilung Zucht, Steenbeker Weg 151, 24106 Kiel, Germany; lroos@tierzucht.uni-kiel.de

A pedigree data set including 129,923 Holstein warmblood horses was analyzed to determine genetic diversity, coefficients of inbreeding, the age of inbreeding and the genetic contributions of founder animals and foreign breeds. The reference population includes all horses which were born between 1990 and 2010. The average Pedigree Completeness Index for the reference population was determined as 0.88 and the average complete generation equivalent was computed at 5.62. The mean coefficient of inbreeding for the reference population (inbred and non-inbred horses) was 2.27%. The proportion of inbred animals increased from 79% in 1990 up to 98% in 2010. Most of the inbreeding was defined as 'new' inbreeding, which had evolved during recent generations. The effective population size and the effective number of founders were calculated to be 55.3 and 50.2 effective individuals, respectively. The most influential foreign breed was the English Thoroughbred with a contribution of 25.98%, followed by Anglo Normans (16.38%) and Anglo Arabians (3.27%). At 2.75%, Hanoverian warmblood horses were determined to be the most contributing German horse breed. The stallions Cor de la bryere, Ladykiller xx and Cottage son xx were found to be the most important male ancestors. The mare Warthburg was defined as the most affecting female. It was possible to detect the occurrence of the loss of genetic variability within the Holstein horse breed, related to unequal founder contributions caused by the intensive use of particular sire lines. However, a slight increase in the effective population size and a stagnation of inbreeding during the last generation might show the impact of more open access given to foreign stallions in the recent past.

Session 27 Theatre 9

Effect of energy supplementation on grass intake, performances and parasitism in lactating mares

C. Collas[1,2], G. Fleurance[1,2], W. Martin-Rosset[2], J. Cabaret[3], L. Wimel[1] and B. Dumont[2]
[1]IFCE, Direction des Connaissances et de l'Innovation, 49411 Saumur, France, [2]INRA, UMR1213 Herbivores, 63122 Saint-Genès-Champanelle, France, [3]INRA and Université François Rabelais Tours, UMR1282 IASP, 37380 Nouzilly, France; claire.collas@clermont.inra.fr

Horse farming systems have to cope with high feeding costs, so that a major challenge is reducing the inputs required for production, e.g. by feeding horses on grasslands. Few studies have so far recorded daily intake of lactating saddle mares at pasture, and assessed the need to providing them supplements. Sixteen lactating saddle mares, eight receiving a daily barley supplement (60% of energy requirements for lactation) and the remaining eight animals being non-supplemented, were rotationally-grazed on permanent pastures from June to September 2012. Each mare was experimentally infested with 5,000 nematode larvae (cyathostomes) at the start of the experiment. Stocking rate was 3.1 LU/ha during the first cycle characterized by active grass growth and 1.5 LU/ha in the second and third cycles. Sward nutritive value remained relatively stable around 11.1%MAT and 44.7%NDF. Data were analyzed using the Mixed Procedure of SAS for repeated measurements and initial conditions were considered as covariates. There was no effect of energy supplementation on the evolution of mare liveweight (on average 597.4±3.4 kg), body condition (3.5±0.1), and foal growth (1st cycle: 1,175±45 g/d, 2nd cycle: 1,020±10 g/d, 3rd cycle: 520±30 g/d). The level of parasitic excretion after mares had been infested with nematode larvae was also similar, which suggest that barley supplementation did not increase mare resistance. Grass daily intake of supplemented and non-supplemented mares did not significantly differ during the first and second cycles. In the third cycle, grass intake of non-supplemented mares was higher (15.7±0.5 vs. 12.2±0.4 g DDM/kgLW/d; $P<0.001$). This behavioural flexibility allowed them ensuring a good foal growth under our grazing conditions.

Session 27 Theatre 10

Haematology and plasma metabolites in horses fed linseed oil over a 4 months period

S. Patoux, C. Fabry, O. Dotreppe, G. Haeghens, J.-L. Hornick and L. Istasse
Université de Liege, Nutrition Unit, Bd. de Colonster 20, 4000 Liège, Belgium; spatoux@ulg.ac.be

Sport horses are offered diets high in concentrate during the training and racing seasons,. Fat has been suggested as an alternative compound to cereals to provide energy and to decrease the level of starch and associated disturbances. Oils increase the energy but also the essential fatty acids supplies. In this study, eight exercised adult horses were used during four months. The diet was made of 50% grass hay and 50% compound feedstuff. The control concentrate was composed of 48% of whole spelt, 48% of rolled barley, 3% of molasses and 1% of a mineral mixture. In the treated grou with oil, 8% of barley was substituted by 8% of first pressure linseed oil. The horses remained healthy over the four months of the experiment. The hay and the compound feedstuffs were completely eaten within one hour after being offered. The inclusion of linseed oil did not affect the plasma concentrations of glucose and of insulin. By contrast, there were reductions in plasma concentrations of urea (4.68 vs. 5.54 mmol/l, $P<0.001$) and triacylglycerol (0.20 vs. 0.26 mmol/l, $P<0.001$). An increase in total cholesterol concentration (2.69 vs. 2.41 mmol/l, $P<0.01$) was also observed. There were period effects on concentrations of plasma glucose ($P<0.001$), total cholesterol ($P<0.01$) and insulin ($P<0.01$) with a large increase in plasma insulin when month 4 was compared with month 1 (39.7 vs. 65.6 UI/ml). In terms of haematology, the linseed oil inclusion significantly ($P<0.05$) reduced the erythrocyte counts (6.8 vs. 7.6×10^{12} cells/l), the haemoglobin content (11.6 vs. 13.0 g/dl) and the haematocrit (0.32 vs. 0.36 l/l) but there were no period effects on haematology. Since linseed oil supplementation did not affect intakes but improved metabolic pathways, linseed oil supplementation could be of interest for racing horses.

Session 27 Theatre 11

Innovation in dairy products in France: preliminary results on donkey milk production

M. Peter, H. Tonglet, G. Jard, H. Tormo and D. Ali Haimoud-Lekhal
INP EI Purpan, Sciences Agronomiques et Agroalimentaires, 75 voie du TOEC, 31076 Toulouse, France; magali.peter@purpan.fr

Research interests on the nutritional properties of donkey's milk have increased. Authors have highlighted its resemblance to human milk. Furthermore, studies have suggested that it can be useful in the prevention of atherosclerosis or in the stimulation of the immune system. Recently, donkey's milk whey proteins have been observed to exert *in vitro* anti-tumor activity. Donkey's milk production could be very dynamic, but consumption of milk is marginal and its use remains essentially cosmetic. Considering its interesting nutrient profile the main goal of this study was to explore opportunities for development of donkey's milk in France. Available data on donkey's milk was taken on Italian and Chinese farms and shows variability depending on the management conditions. The first objective of our study was to characterize the composition of milk and to observe its variability. Samples were collected manually from four donkeys on two different farms (organic and traditional) 2 times per month over 4 months. The composition of milk was determined with a focus on the fatty acid composition (determined by GC) and the lysozyme content (determined by RP-HPLC). Analysis of variance was used to compare the two management conditions. Our results were consistent with data in literature: high levels of lactose, low levels of dry and mineral matter, proteins and lipids content. Lipids were characterized by high levels of linoleic (17%) and linolenic acid (9%). Proteins contained high levels of lysozyme (3.6±1.3 mg/ml of milk). Management conditions had a significant statistical effect only on the fatty acid content, probably due to dietary factors. These preliminary results open up promising prospects for valuation of donkey milk as a functional food but more in depth studies are required. Further complementary investigations are in progress.

Session 27 Poster 12

Conformation affects health of sport horses

L. Jönsson[1], A. Näsholm[1], L. Roepstorff[2], A. Egenvall[3], G. Dalin[2] and J. Philipsson[1]
[1]Swedish University of Agricultural Sciences, Department of Animal Breeding and Genetics, P.O. Box 7023, 75007 Uppsala, Sweden, [2]Swedish University of Agricultural Sciences, Department of Equine Sciences, P.O. Box 7043, 75007 Uppsala, Sweden, [3]Swedish University of Agricultural Sciences, Department of Clinical Sciences, P.O. Box 7054, 75007 Uppsala, Sweden; lina.jonsson@slu.se

Few population studies have been performed regarding associations between conformation and soundness of horses. The objective was to investigate phenotypic and genetic associations between conformation and health status in young riding horses. Analyses included 8,187 horses examined for conformation and health by independent examiners during the Swedish Riding Horse Quality Test of 4-5-year-old horses. Both descriptive conformation traits and assessment scores of overall conformation were studied. Single trait linear models and multi-trait animal models were used, for phenotypic effects and genetic correlations. Four major health indicators were constructed from the detailed veterinary examinations. Conformation assessment scores were all favourably related to at least one of four studied health traits, both phenotypically and genetically. Best health status was found for intermediate sized horses, with well-positioned neck, light front, and no limb deviations. Correct movements in trot were also important. Among limb deviations stiff pasterns, toe-out/toe-in forelimbs, and small/large hock angles had negative effects on health. Genetic correlations reached 0.75 between conformation scores for head-neck-body and health and 0.43 for limb scores and health. Heritabilities of withers height, cannon bone circumference and overall scores (except for limbs), were medium to high, 0.20-0.67. For specific limb deviations present in ≥2% of examined horses, heritabilities were 0.01-0.15. Considerable heritabilities of conformation and favourable correlations to health indicate good opportunities for genetic improvements of conformation, and to some extent health, except for some limb deviations.

Genetic parameters for chronic progressive lymphedema in Belgian draught horses

K. De Keyser[1], S. Janssens[1], M. Oosterlinck[2], F. Gasthuys[2] and N. Buys[1]
[1]KU Leuven, Department of Biosystems, Livestock Genetics, Kasteelpark Arenberg 30, 3001 Leuven, Belgium, [2]UGent, Department of Surgery and Anaesthesiology of Domestic Animals, Faculty of Veterinary Medicine, Salisburylaan 133, 9820 Merelbeke, Belgium; steven.janssens@biw.kuleuven.be

Belgian draught horses (BDH) are susceptible to chronic progressive lymphedema (CPL). The main cause is a failure of the lymphatic system of skin and subcutis, resulting in excessive protein rich interstitial fluid (lymphedema). Symptoms are mainly restricted to skin anomalies and deformations of the lower limbs. Initial signs (from 1.5 yr) are often missed in younger horses, however, with disease progression, increased disability frequently justifies early euthanasia in older horses (from 6 yr). A genetic background is suggested and as disease prevalence in BDH approaches 70%, urgent measures are needed. The aim of this study was to estimate genetic parameters for CPL in a BDH population. Analyses were performed on two datasets, comprising clinical examinations (CPL-scores), collected at official horse contests and stable visits throughout Belgium. The first dataset comprised all collected data, whereas in the second data set, scores for horses younger than 3 years were omitted. Age (yr), coat color and sex were included as fixed effects, whereas date of recording was a random effect. Variance component estimates were obtained using the REML method (VCE4 and 6) by means of an animal model. Heritability coefficients (standard error, SE) for CPL ranged from 0.25 (0.07) to 0.46 (0.09). Phenotypic and genetic correlations (SE) for CPL-scores between fore -and hind limbs were high and respectively counted 0.85 and 0.96 (0.02). Date of recording was a factor of importance, where the proportion of variance (SE) ranged from 0.16 (0.05) to 0.22 (0.04). To improve reliability of estimated breeding values, continued collection of CPL-scores is advised. This study indicates the potential to reduce chronic progressive lymphedema susceptibility and prevalence in BDH through selection.

Preliminary analysis of genetic variability for heart rate in endurance horses

I. Cervantes[1], M.J. Sánchez[1], M. Valera[2], A. Molina[3] and J.P. Gutiérrez[1]
[1]Department of Animal Production, University Complutense of Madrid, 28040, Spain, [2]Department of Agro-forestry Sciences, University of Sevilla, 41013, Spain, [3]Department of Genetics, University of Córdoba, 14071, Spain; icervantes@vet.ucm.es

The endurance competition requires a strong physical condition of the horses. Nowadays many breeds are using genetic evaluation for performance traits as rank, speed or time to select animals. Horses can be eliminated by veterinary controls when they exceed a determined heart rate during the endurance exercise. The aim of this study was to explore the genetic variability for heart rate in horses participating in this competition in order to determine the possibility of including it as an additional selection goal. A total of 647 records from 346 animals (71% Arab horses) from competitions held between 2007 and 2011 were used. The analysed traits were basal heart rate, heart rate at phase I, II, III, and the increases of heart rate between basal value and at the different phases. The model included sex, age, competition and the speed covariate as fixed effects. Except in the basal heart rate, rider effect was fitted as random effect joint to the animal and the residual. Heritabilities ranged between 0.00 and 0.14. Highest heritabilities were found for basal value (0.05), heart rate at phase III (0.09) and increase between both values (0.14). The genetic correlation between basal value with phase III value and with the increment was -0.69 and -0.93, respectively. The rider component was important mainly in the values related to phase I. The results revealed that genetic variability of the heart rate was not negligible. The genetic correlation could suggest that selecting animals with low basal heart rate would be accompanied by higher heart rate during exercise but, as the number of records was low, further analyses are needed to confirm the results.

Transformed variables for the genetic evaluation of the competition performance of jumping horses

J. Posta, A. Rudiné Mezei and S. Mihók
University of Debrecen, Department of Animal Husbandry, Böszörményi str. 138, 4032 Debrecen, Hungary; postaj@agr.unideb.hu

The aim of the study was to compare different transformation of competition performance of show-jumping horses and the estimation of genetic parameters based on these measurement variables. Show jumping competition results collected between 1996 and 2011 were analysed. The database contained 358,342 starts of 10,199 horses. The results were gathered from Hungary and other European countries. Identity number, name and sex of the horse, rider, competition year, the level and location of the competition, fault points and ranks were recorded in the database. Competitions were categorized into five groups based on their difficulty levels. The transformed variables were differently weighted with the difficulty level. The transformed variables were the following: Variable 1: (10-log2(rank)) × sqr(level); Variable 2: (15-sqrt(rank)) × level; Variable 3: (3-log10(rank)) × sqr(level); Variable 4: (6-ln(rank)+3 × sqr(level); Variable 5: (Blom-normalised ranks + 3) × level; Variable 6: (Blom-normalised ranks + 3)+3 × level; Variable 7: Exponential scale based on difficulty level. The goodness-of-fit of the transformed variables was evaluated with the determination coefficients of the models and the distribution of the residuals. The used repeatability animal model included fixed effects for age, gender, competition place, year of competition (and competition level in case of non-weighted measurement variables), and random effects for rider, animal and permanent environment effect. Variance component were estimated with VCE-6 software package. The goodness-of-fit of the models was moderate and varied between 0.446 and 0.519. Heritability values were low for each measurement variables between 0.054 and 0.09. The best goodness-of-fit model was the Variable 7, and the highest heritability value was estimated for Variable 6, respectively.

Joint genetic evaluation of Nordic sporthorses

S. Furre[1], A. Viklund[2], J. Philipsson[2], B. Heringstad[1] and O. Vangen[1]
[1]Norwegian University of Life Sciences, Department of Animal and Aquacultural Sciences, P.O. Box 5003, 1432 Ås, Norway, [2]Swedish University of Agricultural Sciences, Animal Breeding and Genetics, P.O. Box 7023, 75007 Uppsala, Sweden; siri.furre@umb.no

The Nordic Interstallion project aims at providing tools for estimating breeding values for sport horses across the Nordic countries. The project consists of two parts, joint analyses of young horse tests and joint analyses of competition results, the present study deals with the first part. The four Nordic Warmblood populations consists of two smaller studbooks; Norwegian (NWB) and Finnish Warmblood (FWB), and two larger studbooks, Swedish (SWB) and Danish Warmblood (DWB). For SWB and DWB the criteria for publishing official breeding values (BV) is that a stallion has at least 15 offspring with test results within country. If test results from the other countries can be utilized a stallion receives a more reliable BV in shorter time. A previous study in the NWB and SWB has shown an increase in accuracy of BVs for common stallions when performing joint genetic evaluations, especially for the NWB. Data included records from 63,396 young horses evaluated at young horse performance tests in one of the four studbooks from 1981 to 2010. Out of a total of 2,837 sires, 631 have been used in two or more of the populations and sired around 50% of the tested horses. The genetic similarity between the countries was as follows: FWB-NWB: 12%, DWB-FWB: 16%, FWB-SWB: 22%, DWB-NWB: 27%, SWB-NWB: 31%, and DWB-SWB: 43%. All studbooks would increase the number of stallions they could provide official BVs for by performing joint genetic evaluations; DWB from 180 to 270, SWB 179 to 359, NWB 3 to 97 and FWB 1 to 167. Use of either the de-regressed proofs of national BVs or raw data for the joint genetic evaluation will be tested and compared. Differences in the accuracy of BVs as well as cross-validation of the models will be presented.

Genetic correlations between young horse performance and later results in show-jumping

I. Cervantes[1], *E. Bartolomé*[2], *M.J. Sánchez*[1], *M.D. Gómez*[2], *M. Solé*[3], *A. Molina*[3] *and M. Valera*[2]
[1]*Department of Animal Production, University Complutense of Madrid, 28040, Spain,* [2]*Department of Agro-forestry Sciences, University of Sevilla, 41013, Spain,* [3]*Department of Genetics, University of Córdoba, 14071, Spain; icervantes@vet.ucm.es*

The Young Horse Selection Tests (YHST) are held in Spain since 2004. These competitions are aimed to evaluate the sport performance of young horses (4 to 6 years old), in order to make an early genetic pre-selection of the elite sport horses and future breeding stock. Likewise, the breeding values (EBVs) predicted from YHST data are useful to assess the efficiency in selection for later performance. The aim of this study was to analyze the genetic relation between sport results obtained earlier at Young Horse Selection Tests and later in life, in Show-Jumping competitions. For this purpose, the genetic correlation between performance evaluated at YHST and their later results (more than 6 years old) in show-jumping competitions were estimated. Data included 7,266 records from 1,381 horse participants at YHST and 48,498 Spanish Royal Equestrian Federation (SREF) competition performance records from 3089 horses. The genetic parameters for rankings in both types of competitions were obtained using a bivariate animal model. The genetic model included competition, sex, breed, competition score type, competition level and kind of exercise as fixed effects, with age as a covariate. The rider was included as a random effect joint to the animal and the residual. VCE v.6.1 software was used to estimate this model. Heritability estimates for YHST results and for SREF show-jumping competitions were 0.23 and 0.13, respectively. The genetic correlation between the YHST and later results obtained in SREF show-jumping competitions was 0.77. This relatively high correlation supported genetic relationship of young horse selection test sport performance with later competition results for show jumping.

Genetic and environmental effects on 60-days performance tests for mares

N. Cieklińska[1], *J. Wejer*[1] *and D. Lewczuk*[2]
[1]*Uniwersity of Warmia and Mazury in Olsztyn, ul. Oczapowskiego 2, 10-719 Olsztyn, Poland,* [2]*IGAB PAS Jastrzębiec, ul. Postepu 36a, 05-552 Magdalenka, Poland; d.lewczuk@ighz.pl*

The aim of the study is to investigate the influence of genetic (sire and breed) and environmental (year, training centre, data of evaluation, kind of breeder) factors influencing the results of 60-days performance tests for mares. Investigated material consists of results of 878 mares (bred as Wielkopolska, Polish Halfbred Horse, Małopolska and foreign) that were officially tested during the years 2007-2011. Following traits were evaluated on performance test: character, temperament, trainability, free jumping, walk, trot, gallop and rideability in the scale 0-10. Most of these traits (except rideability) were characterised by low coefficient of variance. The influence of the sire was statistically significant for most investigated traits except character and temperament. The effect of the breed was not significant. The effects of the year and training centre were statistically significant for five from eight investigated traits. The date of evaluation was statistically significant for interior traits. The type of breeder (state or private) was not statistically significant for performance results of mares. The average age of mares was 1,326 days (SD-290 days). The effect of the age (in days) considered in the statistical model as regression was not statistically significant. Most elements of results of mares performance test could be used for breeding value estimation, but the evaluation of character and temperament has to be changed.

Session 27 Poster 19

The genetic analysis of Sztumski and Sokolski mares enrolled in horses conservation programs

G.M. Polak
National Research Institute of Animal Production, Animal Genetic Resources Conservation, National Focal Point, Wspolna, 30, 00-930 Warsaw, Poland; grazyna.polak@izoo.krakow.pl

The horse genetic resources conservation programs given an opportunity for restoration of the native cold-blooded horse breeds: Sokolski and Sztumski, created in the middle of 19th century. In the end of 20th century two populations were intensively crossed with limited group of Ardennes and Belgian stallions to improve certain features. The aim of this study was estimate the coefficient of inbreeding, kinship and the most common founders of 672 Sztumski and 823 Sokolski mares, which were enrolled in the conservation programme between 2008-2011. The results shows that over 97% of Sztumski and Sokolski mares were inbred. The average inbreeding coefficient were: for Sokolski population 1.64% and increased by 0.2% in comparison with 2008; for Sztumski mares 1.84% and decreased compared to 2008 (1.93%). The level of meankinship, remained at 2.74% for Sokolski mares and 2.58% for Sztumski mares. The most common founders were a group of 17 stallions found in the pedigrees of more than 70% of the mares.

Session 27 Poster 20

The analysis of the distribution and origin of the Sztumski and Sokolski horses

G.M. Polak
National Research Institute of Animal Production, Animal Genetic Resources Conservation, National Focal Point, Wspolna, 30, 00-930 Warsaw, Poland; grazyna.polak@izoo.krakow.pl

The local varieties of Polish cold-blooded horses were created in the middle of 19th century. Changes in breeding goals and importation of stallions from abroad influenced the phenotypic and genetic transformation. In 2012 the population of cold-blooded horses under genetic resources conservation programs consists of 661 Sokolski and 698 Sztumski mares. The aim of analyses covered the effective size, distribution in Poland, number and the origin of common ancestors (founders of the population). Distribution of the herds does not fully reflect the historic region where the breed originated, while the population is kept in many herds, most of which have 2 or 3 mares. The Sztumski population was found to be descended from 767 founders, including 545 mares and 222 stallions; Sokolski population from 787 founders: 543 mares and 244 stallions. Sokolski type founders belonging to 10 foreign races and domestic horses; Sztumski type founders belonging to 13 breeds; including Thoroughbred horses, half bred and domestic horses.

Session 27 Poster 21

Empirical percentile growth curves considering compensatory growth for Japanese Thoroughbred horses

T. Onoda[1], R. Yamamoto[2], K. Sawamura[3], H. Murase[4], Y. Nambo[4], Y. Inoue[4], A. Matsui[5], T. Miyake[1] and N. Hirai[1]

[1]Graduate School of Agriculture, Kyoto University, Kyoto, 606-8502, Japan, [2]The Japan Bloodhorse Breeders' Association, Tokyo, 105-0004, Japan, [3]JRA Facilities Co. Ltd., Tokyo, 105-0004, Japan, [4]Hidaka Training and Research Center, Japan Racing Association, Hokkaido, 057-0171, Japan, [5]Equine Research Institute, Japan Racing Association, Tochigi, 320-0856, Japan; tonoda@kais.kyoto-u.ac.jp

Percentile growth curves are often used for a clinical indicator to evaluate variations of children's growth status. In this study, we propose an empirical percentile growth curves using Z-scores adapted for Japanese Thoroughbred horses, with considerations of the seasonal compensatory growth (CG) that is typical characteristic in seasonal breeding animals. Thoroughbred foals born in spring generally show a CG pattern, in which their growth rate decline in winter and increase in next spring. Formerly, we developed a new growth curve equations for Japanese Thoroughbreds adjusting the CG. Based on Richards' growth curve equation, a sigmoid sub-function adjusting the CG was developed and adapted to the biological parameters responsible for maturity. Individual horses and residual were included as random effects in the growth curve equation model and their variance components were estimated by SAS NLMIXED procedure. Based on the Z-scores of the estimated variance components, the empirical percentile growth curves were constructed. For the analyses, we used a total of 3,961 and 4,341 body weight and age measurements for male and female Thoroughbreds, respectively, collected from Hidaka Training Farm, Japan Racing Association (JRA) between 1999 and 2008. The developed empirical percentile growth curves using Z-scores are computationally feasible and useful for understanding of body weight distributions among individuals and for monitoring growing Thoroughbreds. It can be an alternative for the percentile growth curves of Thoroughbreds.

Session 27 Poster 22

Measuring neck angle of horses under different ceiling heights with an inertial measurement unit

S. Claar and P. Michanek

Flyinge AB, Kungsgården, 24032 Flyinge, Sweden; selma_claar@hotmail.com

It is important for the comfort and well-being of the horse that it is comfortable in the stable, and for example not prevented from standing in natural postures. This is especially important for horses, since they spend most of their time standing. Height of ceiling is one parameter that could affect the comfort of horses when standing. Legislation and recommendations for ceiling height in individual horse boxes varies considerably between countries, and to date there are no published scientific data for determining a limit when a low ceiling makes the horse assume an unnatural standing posture. In this pilot study, the neck angle of three horses of different sizes (height at withers: 158, 168 and 174 cm) was measured at ceiling heights down to 200 cm, using an inertial measurement unit, fastened to the ventral aspect of the horses necks with adhesive tape. The horses were kept in loose boxes where the ceiling height could be varied (200-250 cm) and the neck angle was registered 4 times per second between 16:00 and 07:00 during a total of 23 nights. The neck angel of all three horses was constant, irrespective of ceiling height, and no association between ceiling height and neck angel could be detected (regression analysis; r^2=0.0007)). It was concluded that a low ceiling probably has little effect on the natural standing posture of a horse as long as it does not physically interfere with the space normally occupied by the head. The measurements also generated interesting data about the horses behavior during the night, for example that most of the time was spent with the head close to the floor, probably searching for feed in the bedding, even though there was no feed vailable.

Session 27 Poster 23

Some morphological characteristics of mules raised in Van Province in Turkey

F. Coskun[1], O. Yilmaz[2] and M. Ertugrul[3]
[1]Ahi Evran University, Technical Vocational High School, Department of Vegetable and Livestock, Bagbasi Mah, Bagbasi Yerleskesi, 40100 Kirsehir, Turkey, [2]Canakkale Onsekiz Mart University, Faculty of Agriculture, Department of Animal Science, Terzioglu Yerleskesi, 17100 Canakkale, Turkey, [3]Ankara University, Faculty of Agriculture, Department of Animal Science, Diskapi, 06110 Ankara, Turkey; fusuncoskun40@gmail.com

The aim of this research study is to define some morphological characteristics of mules raised in Van province. A total of 62 mules, 35 males and 27 females, in three age groups (3-6, 7-8 and 9-15 years) were examined. Descriptive statistics of morphologic traits were as follow: Withers height=129.9±0.87 cm; height at rump 128.3±1.64 cm; body length 134.2±0.83 cm; heart girth circumference 148.5±0.84 cm; chest depth 59.8±0.54 cm;cannon circumference 16.2±0.16 cm; and head length 54.9±0.53 cm. In this study the distributions of coat colour were 54.8% for bay colour, 24.3% for white, 4.8% for black, 4.8% for brown, 8.1% for mouse gray and 3.2% for chestnut. It can be concluded that body development continues until 3 years of age and thereafter only slight increases can be seen in this trait. The present data also showed that mules raised in province of Van were slightly larger in body sizes than UK and Ordu Province of Turkey mules but nearly similar in body sizes with Eastern Anatolian and Turkish mules.

Session 27 Poster 24

Some aspects of horse vision

I. Majzlik, J. Ptacek, B. Hofmanova and L. Vostry
CULS in Prague, Animal Science and Ethology, Kamycka 129, Prague, 165 21, Czech Republic; majzlik@af.czu.cz

The quality of horse visual abilities is particularly relevant for ridden work esp. jumping. The horse does not appear to exhibit significant refractive errors, but there is a tendency towards hyperopia and therefore the horse should have excellent visual acuity for distant objects. Another authors reported in horses refractive errors ie. myopie and hyperopie. In our trial the prevalence of refractive errors in 103 horses (87 warmblooder and 16 coldblooder) was studied using halogen ophtalmoscop. In warmblooder group we found 168 emmertropic eyes, 1 eye blind, 2 myopic eyes and 2 hyperopic eyes. The coldblooder group showed 23 emmertropic, 7 blind and 2 hyperopic eyes resp. The difference of refractive errors between groups reached statistical signifikance, moreover, the coldblooder group showed significantly more errors for both right and left eye. There is also the incidence of tendency to hyperopia in warmblood horses under study which is in agreement with with literature. Second part of the paper deals with test of jumping abilities of 20 horses with emmertropia in free jumping (without rider) ie. the connection of eye quality and color design of two obstacle types (oxer, upright jump). A statistical significant differences were found among colors of upright jump-white color was found the worst one, the best color for this obstacle was proved as a green-white combination. There were found no differences on colors of oxers. Conclusions of the study are showing the benefit of testing the refractive eye errors in horses for jumping before the training.

Analysis of air quality in riding halls with special emphasis on fungal contamination

A. Blasse, K. Wicha, T. Lühe, R. Preissler and N. Kemper
Institute of Agricultural and Nutritional Science, Martin-Luther-University Halle-Wittenberg, Theodor-Lieser-Str. 11, 06120 Halle, Germany; nicole.kemper@landw.uni-halle.de

The exposure to fungi and their mycotoxins represent a predisposing factor for respiratory diseases in horses. Studies on air quality in riding halls are lacking, although horse and rider consume large amounts of air during training. This study investigates the air quality in riding halls by analysing the fungal and dust contamination before and after a standardized riding program. Air samples are collected weekly over the course of one year in four riding halls in Saxony-Anhalt, Germany. The measurements take place in the height of the horse's and the rider's nose (1.5 and 2.5 m) on 4 arena points using the HHPC-6 Airborne Particle Counter for determination of respirable dust particles, humidity and temperature. The MAS-100 Eco Air Sampler in combination with DG18-agar plates is used for the determination of fungal species (spp.) in colony forming units (cfu). Data analysis is performed using the statistical analysis software SAS 9.2. In the preliminary study (09/12-01/13), 285 air samples were analysed. A broad spectrum of fungi was found, with Aspergillus spp., Penicillium spp., Trichophyton spp. and Alternaria spp. as the most frequently. Mean number of different fungal species per sample was three (range 0-7, median 3) and varied significantly between the riding halls. On average, 80 cfu (range 0-400 cfu) were determined per plate. The number of cfu was significantly influenced by the fixed effect of riding hall and month of sampling, while height and arena point did not show significant influence. Similar findings were observed for the quantity of dust particles which was significantly related to the fixed effect of riding hall, month and time point of sampling. To identify further potential influencing factors on the fungal contamination of air, environmental parameters such as humidity, temperature and the coexistence of dust particles are investigated in further analysis.

Milk and blood serum rubidium and strontium concentrations in lactating donkeys

F. Fantuz[1], S. Ferraro[1], L. Todini[1], P. Mariani[1], R. Piloni[1], N. Zurlo[2] and E. Salimei[2]
[1]Univ. Camerino, Scienze ambientali, via Gentile da Varano, 62032 Camerino, Italy, [2]Univ. del Molise, Agricoltura, Ambiente e Alimenti, via DeSanctis, 86100 CB, Italy; francesco.fantuz@unicam.it

Donkey milk can be used as hypoallergenic food for infants suffering of cow milk protein allergy. Although a biological role has not been demonstrated, Rubidium (Rb) and Strontium (Sr) are potential essential elements in mammals nutrition. Aim of this study was to determine the concentrations of Rb and Sr in milk and serum of lactating donkeys. During the experimental period (3 month), individual milk and serum samples (n=112) were obtained from 16 lactating donkeys (averaging 205.4 kg body weight). Donkeys were fed meadow hay *ad libitum* and 2.5 kg of mixed feed daily. Based on Rb and Sr concentrations in feeds, and assuming dry matter intake at 3.2% body weight, the estimated dietary intake of Rb and Sr were approximately 70 mg/d and 350 mg/d, respectively. Feeds, milk and serum samples were analyzed for Rb and Sr concentrations by inductively coupled plasma-Mass Spectrometry. Data were elaborated by analysis of variance for repeated measures. Average (±SD) Rb and Sr concentrations in donkey milk were respectively 339.0±81.8 µg/l and 880.3±269.3 µg/l. The concentrations of Rb and Sr in donkey milk were respectively 4.6 and 3.5 times higher than those in blood serum (Rb 74.1±15.2 µg/l; Sr 254.8±56.0 µg/l). Milk Rb and Sr concentrations were strongly correlated ($P<0.001$) with their serum counterpart (Rb r=0.85; Sr r=0.75). Rubidium was correlated ($P<0.001$) with Sr both in milk (r=0.66) and serum (r=0.50). The effect of stage of lactation was significant for Rb and Sr concentrations both in milk and serum, and lower concentrations were observed during the second part of the trial. Current results suggest that the mammary gland plays an active role in transferring Rb and Sr from blood to milk. Considering that Rb and Sr are in the same chemical group of Potassium and Calcium, respectively, it is possible that such elements share transport mechanisms.

Minerals in blood serum of lactating donkeys: effect of dietary trace element supplementation

F. Fantuz[1], S. Ferraro[1], L. Todini[1], R. Piloni[1], P. Mariani[1], C. Amadoro[2] and E. Salimei[2]
[1]Univ. Camerino, Scienze Ambientali, via Gentile da Varano, 62032 Camerino, Italy, [2]Univ. Molise, Agricoltura, Ambiente, Alimenti, via DeSanctis, 86100 CB, Italy; francesco.fantuz@unicam.it

Aim of this study was to measure the concentrations of Calcium (Ca), Phosphorus (P), Potassium (K), Sodium (Na), Magnesium (Mg), Sulfur (S), Zinc (Zn), Iron (Fe), Copper (Cu), Manganese (Mn), Selenium (Se), Molybdenum (Mo), Cobalt (Co), and Iodine (I) in blood serum of lactating donkeys also considering the effect of dietary trace element supplementation. During the experimental period (3 month), individual blood serum samples (n=112) were obtained from 16 lactating donkeys, which were divided in two groups (control and treated). Donkeys were group fed meadow hay *ad libitum* and 2.5 kg of mixed feed daily. The mixed feed for the treated group was supplemented with a trace element premix providing 163 mg Zn, 185 mg Fe, 36 mg Cu, 216 mg Mn, 0.67 mg Se, 2.78 mg Co and 3.2 mg I/kg mixed feed. Feeds and serum samples were analyzed for Ca, P, K, Na, Mg, S, Zn, Fe, Cu, Mn, Se, Mo, Co, and I concentrations by inductively coupled plasma-Mass Spectrometry. Data were statistically analyzed by analysis of variance for repeated measures. Average (±SD) major mineral concentrations in serum were as follows: Ca 129.8±7.5 mg/l; P 102.2±20.8 mg/l; K 181.2±14.1 mg/l; Na 3,150.2±14.1 mg/l; Mg 25.7±4.3 mg/l; S 846.9±63.8 mg/l. The serum concentrations (mean±SD) of trace elements were as follows: Zn 631.0±80.1 µg/l; Fe 1,475.8±479.9 µg/l; Cu 1,112.8±137.4 µg/l; Mn 0.58±0.57 µg/l; Se 195.6±36.7 µg/l; Co 0.99±0.46 µg/l; Mo 30.1±18.5 µg/l; I 23.0±4.9 µg/l. Compared to the control group, donkeys from the treated group showed significantly higher concentrations of Mg (26.6 vs. 25.0 mg/l; $P<0.05$), Co (1.34 vs. 0.69 µg/l; $P<0.01$) and I (24.4 vs. 21.4 µg/l; $P<0.05$). Results from this study may be useful to define reference serum values for lactating donkeys and to support the assessment of dietary trace element requirements for lactating donkeys.

The grass height: a horses'grazing management tool

L. Wimel and P. Dupuy
IFCE, DCOI Station Expérimentale des Haras, Domaine de la Valade, 19370 Chamberet, France; laurence.wimel@ifce.fr

Grass is a fundamental alimentary resource for herbivore breeders whose herds are mainly fed with grazing. For these breeders, it is very important to optimize the new growth of grass, so that animals can take advantage of it: grass must have a good nutritional value, and must also be able to satisfy animals' food requirements. Several researches have been done about grazing management for cattle, but it still remains quite infrequent for horses. An experimental station owned by the French Institute for the Horse and Equitation conducts researches since 2007 about pastures and horses physiology (duration of grazing, horses weights, agronomic monitoring of pasture plots…). In order to establish a new technical reference and a decision-making tool for rotational grazing management, the station decided to consider the interest of grass height as an indicator for grazing management. Heights have been measured with a herbometre each time a herd was entering or leaving a pasture plot. Quantities of available grass for horses in the plot and grass consumption per horse have been calculated. A comparison with horses' nutritional requirements showed that grass consumption could sometimes insufficient to satisfy these needs, proving the interest of a grazing management tool. Theoretical days of grass reserve per plot can be calculated, considering data such as grass height, biomass, and plots acreage. They seem to be a helpful indicator to plan and adjust grazing duration, number of horses and size of the plot to better serve horses' nutritional needs.

Session 27 Poster 29

Chemical composition, biomass production and nutritive value of two Portuguese horse farms pastures

A.S. Santos[1,2], M.J. Fradinho[3], R. Fernandes[4], M.A.M. Rodrigues[2], R.M. Caldeira[3] and L.M.M. Ferreira[2]
[1]EUVG, Estrada da Conraria, 3040-714, Castelo Viegas, Portugal, [2]CECAV-UTAD, Animal Production, Po-Box 1013, 5001-901, Vila Real, Portugal, [3]CIISA, FMV-UTL, Av. Universidade Técnica, 1300-477 Lisboa, Portugal, [4]FMV-UTL, Av. Universidade Técnica, 1300-477 Lisboa, Portugal; assantos@utad.pt

Lusitano is a Portuguese horse breed produced in a pasture based system. Mares and foals are kept outdoors throughout the year and feeding practices including supplementation with concentrates and dry forages usually complement pasture availability. In spite of its importance to the breeding system, composition and nutritive value of pastures are very variable, and not well known. This study intended to characterize biomass production and nutritive value of irrigated and dry land pastures availability throughout 2012. For this purpose sampling was performed in Farm A (FA; dry land pasture) and Farm B (FB; irrigated pasture). Samples were taken in four periods: January, May, July and December. Samples were randomly collected at ground level using a 0.5 m^2 iron square. Pasture samples were immediately weighted for biomass production estimation. Samples were then dried and analyzed for dry matter (DM), crude protein (CP), crude fiber (CF), and minerals (Ca, Mg, Zn and Cu). Energetic value was estimated as UFC (/kg DM) and digestible protein was estimated as PDC (/kg DM). Differences observed are a reflection of water availability and climate. Higher and more constant pasture development was observed in FB whereas in FA the availability and nutritive value of pasture was lower. Zero pasture availability was only identified in the end of July for FA. From January to December, CF increased and CP decreased in the two pastures, and this had an influence in UFC and PDC. This study provides useful information to the breeders, in order to better adjust diets to offer the animals, according to the nutritive value and biomass production of pastures in their farms.

Session 27 Poster 30

Study on the use of long chain alcohols as diet composition and digestibility markers in equines fed

A.S. Santos[1,2], R. Celaya[3], R.W. Mayes[4], M.A.M. Rodrigues[2], K. Osoro[3] and L.M.M. Ferreira[2]
[1]EUVG, Estrada da Conraria, 3040-714 Castelo Viegas, Portugal, [2]CECAV-UTAD, Po-Box 1013, 5001-901, Vila Real, Portugal, [3]SERIDA Servicio Regional de Investigación y Desarrollo Agroalimentario, Apartado 13, 33300 Villaviciosa, Spain, [4]The James Hutton Institute, Craigiebuckler, AB15 8QH, Aberdeen, United Kingdom; assantos@utad.pt

Long-chain alcohols (LCOH) have been used with success as diet composition markers in different ruminant species. Nevertheless, to the authors' knowledge the use of these epicuticular wax compounds in equines wasn't yet evaluated. This study aimed to evaluate the use of LCOH to estimate diet composition and dry-matter digestibility (DMD) of equines fed with two different diets. Eight crossbreed mares (385±47 kg LW) divided in 2 groups were housed in metabolism pens and received daily a total of 1.0 kg of DM/100 kg LW of one the diets: 100% *Lolium perenne* (E1) and 70% *L. perenne* + 30% *Ulex gallii* (E2). Diet composition was estimates from LCOH concentrations (C_{20}, C_{22}, C_{24}, C_{26}, C_{28} and C_{30}) using the 'EatWhat?' software. C30 was used as internal marker for DMD estimation. Accuracy of diet composition estimates was calculated by the Kulczynski similarity index (KSI) and known and estimated values of DMD were compared. The effect of carbon-chain length and diet composition on the faecal recoveries (FR) of LCOH was examined by ANOVA. Plant species were clearly discriminated when using these markers. Faecal recovery of LCOH was not influenced by diet composition (P=0.452) and tended to increase (P<0.001) in a curvilinear fashion with carbon-chain length (CCL). DMD was significantly (P<0.001) overestimated by the C30 marker due to the bias in diet composition estimates as its faecal recovery did not differ from the unit (mean of 1.02 for both diets). Accuracy of diet composition estimates was high in both diets with mean KSI values of 92% suggesting the usefulness of LCOH as diet composition markers in equines.

Session 27 Poster 31

Heart rate of horses undergoing aerobic exercise and supplemented with gamma-oryzanol

A.A.O. Gobesso, I.V.F. Gonzaga, R. Françoso, F.M.P. Taran, T.N. Centini, J.R. Ferreira, F.P. Rodrigues, Y.N. Bortoletto and C.T. Marino

Sao Paulo University, Animal Science, Av. Duque de Caxias Norte, 225 Jd. Elite Pirassununga, SP, 13630-900, Brazil; gobesso.fmvz@usp.br

Heart rate has been used as an indicator of respiratory and circulatory functions of the horse and its measurement during exercise to monitor the fitness. This study aimed to evaluate the heart rate of horses submitted to aerobic exercise and supplemented with gamma-oryzanol. In 180 days we used 10 horses aged 35±8.15 months and body weight 375±22.78 kg, divided into two groups: control and treatment. The treatment group received 5.0 g of gamma-oryzanol crystalline, dissolved in 50 ml of linseed oil per meal. The control group received the same diet in the same manner, except for the addition of gamma-oryzanol. All animals were exercised five times a week for 60 minutes at maximum speed of 12 km/h in mechanical walker for horses. Measurements were made monthly heart rate (HR) of all animals. At each time, we obtained the HR: basal, maximum, end, and at 10 and 20 minutes after the end of the training. We used a completely randomized design with repeated measures, and adopted the procedure of mixed models of SAS computer program, with a significance level of 10%. The animals had lower basal HR over time, which can be explained by the effects of physical conditioning. HR reduction 20 minutes after the end of the exercise can be directly linked to training and supplementation with gamma-oryzanol, even justified by time and treatment interaction being observed, demonstrating that one or more of the various properties of the substance can postpone fatigue these animals. Supplementation with gamma-oryzanol of horses undergoing aerobic activity is beneficial and may reduce heart rate recovery of animals over time.

Session 27 Poster 32

Effect of dietary inclusion of omega-3 in concentration of immunoglobulin on mares colostrum

A.A.O. Gobesso, T.N. Centini, I.V.F. Gonzaga, F.M.P. Taran, R. Françoso, J.R. Ferreira, F.P. Rodrigues and A.C.R.C. Porto

Sao Paulo University, Animal Science, Av. Duque de Caxias Norte, 225 Jd. Elite Pirassununga, SP, 13630-900, Brazil; gobesso.fmvz@usp.br

Essential fatty acids comprise a class of molecules that cannot beefficiently synthetized by most mammals. In this group are linoleic and linolenic acid, also known as omega-6 and omega-3, respectively. The purpose of this study was to evaluate the effect of including dietary source rich in linolenic acid on the immunoglobulins concentration of colostrum mares. Eighteen pregnant mares, cross-breed, weighing 521±56 kg, were divided into groups: control, supplemented with 0.05% BW/day of soybean oil, source of omega-6, and supplemented with 0.05% BW/day of linseed oil, source of omega-3, individually added to the concentrate during the third trimester of gestation and 2 months of lactation. The colostrums samples were subjected to ELISA to determine the concentration of IgA, IgM, and IgGa IgGb. We used a completely randomized design, three treatments and six replicates per treatment, with repeated measures. It was observed that the isotope IgGb prevailing in the three groups of immunoglobulin analyzed, followed by IgGa. The average concentration of IgGa in colostrum of mares supplemented with linseed oil was almost 4 times higher than that found on control group (50.4 mg/ ml and 183.5 mg/ml) and almost 3 times higher than those supplemented with soybean oil (73.6 mg/ml and 183.5 mg/ml). The average concentration of IgGb in colostrums of mares supplemented with linseed oil was almost two times higher than that observed in the group supplemented with soybean oil (239.63 mg/ml and 155.3 mg/ml). We can conclude that mares supplemented with soybean oil and linseed had higher concentrations of immunoglobulin Ga and Gb on colostrum when compared to those without supplementation.

Microbiological characteristics of ass's milk in France

G. Jard[1], H. Tonglet[1], H. Tormo[1], M. Peter[1], C. Roques[2] and D. Ali-Haimoud Lekhal[1]
[1]INP, EI-Purpan, 75, voie du TOEC BP 57611, 31076 Toulouse cedex 3, France, [2]LGC, Campus INP-ENSIACET, 4 allée Emile Monso, 31432 Toulouse cedex 4, France; magali.peter@purpan.fr

Ass's milk provokes an increasing interest since its biochemical composition confers interesting nutritional potentialities. For now, the ass's milk is mainly used for cosmetics in France. To develop the direct use in human food, a good knowledge of this milk on a biochemical and a microbiological point of view and of management of dairy donkeys is necessary. Few studies have been performed on the microbiological composition of ass's milk but none of them in France. The objectives of this study were: (1) to assess the absence of pathogens in milk for direct human consumption; (2) to study the microbial composition in order to detect potential presence of alteration microorganisms and interest microbiota (lactic acid bacteria). The microbiological composition of raw ass's milk from 15 french farms (one sample per farm, mixed milk of between 2 and 11 animals) were studied in relation to the management practices (milking frequency, hygienic and milking practices). A high variability between farms was observed for total microbiota (1.5 et 4.5 log10 UFC/ml).The microbiota is thus lower than in other conventional milk. The relative high level in lactic acid bacteria, as shown in other studies suggests an interesting potential for health. However, no strict anaerobic microorganisms were found, suggesting an absence of probiotics such as Streptococci. Some groups was formed according to their microbial composition but no clear correlation was found between management practices and microbial composition of raw milk. Acidifying capacity of raw milks was performed and the pH decreased from 7.32 +-0.15 to 5.76 +-0.45.

Effect of specialization on genetic parameters in sport horses

G. Rovere[1,2], B.J. Ducro[1], P. Madsen[2], E. Norberg[2] and J.A.M. Van Arendonk[1]
[1]Animal Breeding and Genomic Centre, Wageningen University, De Elst 1, 6708 WD Wageningen, the Netherlands, [2]Department of Molecular Biology and Genetics, Aarhus University, Blichers Allé 20, 8830 Tjele, Denmark; gabriel.rovere@wur.nl

During the last decades the breeding practice within the Dutch Warmblood studbook (KWPN) has resulted in an increasing specialization of horses into dressage (DH) and show jumping (JH). Until now, breeding values for 32 common traits (related to conformation and movement) are estimated jointly for DH and JH. Even though it has been shown that the current populations of DH and JH are genetically linked, the increasing specialization could lead to differences in genetic parameters and makes joint evaluation suboptimal. The aim of this study was to investigate if the specialization has led to differences in genetic parameters for movement traits. Bivariate animal model analyses were used to estimate heritability and genetic correlation between each trait expressed in DH and JH. The material comprised 38,142 first inspections from 1998 through 2010. Average phenotypic scores significantly differed among DH and JH for some of the traits related to trot and walk. Estimated genetic correlations ranged from 0.86(±0.07) to 1(±0.01). Heritability estimates ranged from 0.21(±0.02) to 0.39(±0.03), in few cases differences between DH and JH were significant. These results indicate that the specialization until now has not lead to changes in genetic parameters that seriously affect the current genetic evaluations. Similar analyses might be extended to all traits. Subsequently, based on analysis of subsamples, inferences will be made as to what extent further divergence of traits in DH and JH might be anticipated in future generations, which might require re-evaluation of the genetic parameters.

Participatory identification of breeding objective traits of two goat breeds of Ethiopia

S. Abegaz, M. Wurzinger and J. Sölkner
BOKU-University of Natural Resources and Life Sciences, Vienna, Gregor-Mendel Strasse 33, 1180, Austria; soabgu96@yahoo.com

Goat production has a significant role in Ethiopian farming community. However goat production is more traditional and no systematic breeding programs are in place. The aim of this study was to identify breeding objective traits of Abergelle and Western Lowland goat. This information shell be used later on to develop breeding strategies. Preferred traits of selection of farmers were identified in a participatory approach. Phenotypic ranking of animals from the own flock of the owner was used to identify important traits. Sixty households from Western Lowland and thirty households from Abergelle were visited and household members were asked to pick their 1st best, 2nd best, 3rd best and one inferior doe in their flocks. The reasons of their selection, the life history of the ranked does were recorded and linear body measurements were taken. The preferred traits were almost similar in both study sites, but with marked difference in frequency (magnitudes). Milk yield (20.47%), drought resistance (14.96%), body size (14.17%), kid growth (11.81%) and multiple births (10.24%) were found as preferred traits for Abergelle goat breeders while multiple birth (20.26%), kid growth (18.50%), body size (15.42%) and mothering ability (15.42%) were preferred by Western Lowland goat breeders. The animals selected as the best animals had significantly higher value than the inferior animals for linear measurement and other traits considered. For instance, significantly ($P<0.001$) higher total number of weaned kids (9.78±0.41) were found for the 1st best does compared to the inferior does (1.61±0.41) of Western lowland goats. The average daily milk yields of the 1st best does (0.58±0.03 l) were significantly higher than the inferior does (0.16±0.03 l) of Abergelle goat. We conclude that in such type of area where no recording scheme is developed; own flock ranking method is an important tool to identify objective traits of farmers.

Genetic diversity and structure in Cameroon native goat populations using microsatellites

F. Meutchieye[1], P.J. Ngono[2], A. Djikeng[3], S. Ommeih[3], M. Agaba[4] and Y. Yacouba[1]
[1]University of Dschang, FASA, Dschang, P.O. Box 188, Cameroon, [2]University of Ngaoundéré, ESMV, Dan, N'déré, Cameroon, [3]ILRI, BecA Hub, Nairobi, Naivasha road, Kenya, [4]NM University of Technology, Biosciences, Arusha, TZ, Tanzania; fmeutchieye@gmail.com

Our study aimed to estimate genetic diversity in native goat population of Cameroun. Blood samples were randomly collected on 190 native goats belonging to 8 country agro ecological zones representing indigenous sub populations. They are morphologically known in 3 main groups, which are dwarf, medium and long legged. A total of 17 simple sequence repeats (SSR) markers selected from ISAG-FAO recommended list were applied. All 17 loci tested were polymorphic in each subpopulation, and the average number of alleles was 5.11. The average observed heterozygosity for all loci in individual subpopulation was in general less than expected heterozygosity. The population specific inbreeding coefficient (Fis) was ranging between 0.008 (dwarf coastal goat) to 0.25 (rain forest dwarf goat) showing a tendency of inbreeding. The Analysis of Molecular Variance (AMOVA) revealed that about 6% of genetic differences were due to subpopulations while 45% were explained by differences among individuals and 49% within individuals. The STRUCTURE analysis under Arlequin showed a tendency of admixture between various subpopulations. Our findings open ways for a better understanding of genetic makeup of native goats. These could give basis to investigate further for subsequent interesting traits, for better utilization and conservation of Cameroon native goat population.

Potential benefits of genomic selection on genetic gain of small ruminant breeding programs

F. Shumbusho[1], J. Raoul[1], J.M. Astruc[1], I. Palhiere[2] and J.M. Elsen[2]
[1]Institut de l'Elevage, 31321 Castanet-Tolosan, France, [2]INRA, UR631 SAGA, 31326 Castanet-Tolosan, France; fshumbusho@toulouse.inra.fr

In conventional small ruminant breeding programs, selection is based on pedigree and phenotypes but now there are prospects of including genomic information. The objective of this study was to predict the potential benefits of genomic selection on the genetic gain in today's French sheep and goat breeding designs. Classic and genomic scenarios were modeled for three breeding programs. The models included decisional variables that were optimized to maximize annual genetic gain (AGG) of: (1) a meat sheep breeding program that improved a meat trait of h^2=0.30 and a maternal trait of h^2=0.09; and (2) dairy sheep and goat breeding programs that improved a milk trait of h^2=0.30. The Bulmer effect was accounted for and the results presented are the averages of AGG after ten generations of selection. Results show that current breeding programs provide an AGG of 0.095 s.d. (genetic standard deviation) for meat and 0.061 s.d. for maternal traits in meat breed, 0.147 s.d. in dairy sheep and 0.120 s.d. in dairy goat breeds. By optimizing decisional variables, the AGG with classic selection methods increased to 0.139 s.d. for meat and 0.096 s.d. for maternal traits in meat breeding programs, and to 0.174 s.d. and 0.183 s.d. in dairy sheep and goat breeding programs, respectively. With a medium-sized reference population of 2,000 individuals (nref), the best genomic scenarios gave an AGG that was 17.9% higher than with traditional selection methods for combined meat and maternal traits in meat sheep, 51.7% in dairy sheep and 26.2% in dairy goats. These results were obtained when scenarios were compared at optimized decisional variables. The superiority of genomic schemes increased with nref and genomic selection gave the best results when nref >1000 individuals for dairy breeds and nref >2,000 individuals for meat breed. Correlation between meat and maternal traits had a large impact on the AGG of both traits.

Assessment of technical and economic efficiency of French dairy sheep genomic breeding programs

D. Buisson[1], J.M. Astruc[2], G. Baloche[1], X. Aguerre[3], P. Boulenc[4], F. Fidelle[3], B. Giral[4], P. Guibert[5], P. Panis[5] and F. Barillet[1]
[1]INRA, SAGA, 31326 Castanet-Tolosan, France, [2]Institut de l'Elevage, BP 42118, 31321 Castanet-Tolosan, France, [3]CDEO, Ordiarp, 64130 Ordiarp, France, [4]Ovitest, Rte d'Espalion, 12850 Onet-le-Château, France, [5]Confédération Générale de Roquefort, BP 348, 12100 Millau, France; jean-michel.astruc@toulouse.inra.fr

The Roquefort'in and Genomia programs aim at evaluating the feasibility of genomic selection in French dairy sheep breeds. Reference populations of nearly 2,500 (Lacaune) and 1,500 (red-faced Manech) progeny tested AI rams were set up. Genomic predictions exhibit an increase in accuracy compared to parent average, lower than in dairy cattle, but allowing selection of proven rams without progeny-test. The purpose of this work was to model different designs of genomic-based breeding schemes fitting the constraints of dairy sheep: high number of alive rams required to face a high number of fresh semen AI concentrated within a few weeks. The modeling considered a wide range of variations in the management of the rams in the AI center: genomic selection pressure at 3-month-old, semen production of rams, age at culling of rams, selection intensity in the AI centre. The outputs of the models are the annual genetic gain, the total number of alive AI rams and the number of rams to be genotyped. Costs of rams' maintenance are estimated. The technical and economical results are compared to the current situation based on progeny-test. With a genomic selection pressure of around one third, the genetic gain is slightly above the gain of the classical program, close to its optimum, without any extra costs. The sharp decrease in the number of rams in the AI center compensates the costs of the genotypings. A higher genetic gain would be obtained by increasing dramatically the number of genotypings, which would break the economic balance. The capacity of semen production and the cost of genotypings are the key points to make a genomic selection program successful in dairy sheep.

A model for predicting the impacts of gastrointestinal parasitism in dairy ewe: sensitivity analysis

F. Assenza[1], J.M. Elsen[1], S.C. Bishop[2], C. Carré[1] and C. Moreno[1]
[1]INRA, 24 Chemin de Borde Rouge, Auzeville, 31326 Castanet Tolosan, France, [2]The Roslin Institute and R(D)SVS, Easter Bush, EH25 9RG Midlothian, United Kingdom; fabrizio.assenza@toulouse.inra.fr

Gastrointestinal parasites are a major cause of economic loss in sheep production. Control has been operated largely through the use of anthelmintics, but the development of anthelmintic resistance in populations under treatment makes this strategy increasingly ineffective. Computer simulations for predicting the outcome of selection for resistance to gastrointestinal parasites have been published, however they have not investigated the effect of parasitism during lactation. The objective of this study was to enhance one of the existing models by adding a module for simulating the effect of parasitism on milk production. The model includes a description of lactation, host-parasite interactions and outputs in terms of worm burdens, egg counts and impacts on milk production. A global sensitivity analysis of the model was performed by radial design, previously applied to physics simulation models. This design allows both to identify the model's input parameters which have weak impact on the model's outputs (elementary effect sensitivity index) and to rank the parameters according their global impact, including their interactions with all other input parameters, on the variance of the model's outputs (total effect sensitivity index) simultaneously. The former index allowed us to simplify the model itself and to limit the factor prioritization to the influential parameters only. The latter index resulted both in an exhaustive mapping of the model's behaviour to the input space and in a guideline for an efficient investment of resources in data collection for the estimation of the input parameters. The optimally parameterized model will predict the efficacy of genetic selection in controlling dairy sheep's gastrointestinal parasites and help derive selection indices.

Housing and management risk factors affecting body condition and traits of animal health in ewes

G. Kern[1], I. Traulsen[1], N. Kemper[2] and J. Krieter[1]
[1]Institute of Animal Breeding and Husbandry, Christian-Albrechts-University Kiel, Olshausenstraße 40, 24098 Kiel, Germany, [2]Institute of Agriculture and Nutritional Science, University Halle-Wittenberg, Theodor-Lieser-Straße 11, 06120 Halle (Saale), Germany; gkern@tierzucht.uni-kiel.de

The aim of this investigation was to evaluate the animal health of ewes with regard to housing and management conditions and to analyze risk factors that affect the occurrence of health disorders. In total, 6,093 visual inspections of 1,562 ewes housed with three different primary purposes (meat, dairy and extensive) were performed on 20 organic farms in 2010 and 2011. The ewes were examined with respect to their body condition (BCS, Scale = -2 to +2), constitution of hoofs, respiratory symptoms and udder health. The effects 'primary purpose', 'age', 'year*production period' and 'grazing area' had a significant influence on BCS in fact of significantly higher values in 'extensive' ewes compared to ewes in 'dairy' and 'meat' systems. The odds ratio (OR) of 'constitution of hoofs' was better in 'meat' (OR=0.40) and 'dairy' (OR=0.47) than in 'extensive' (OR=1.00) breeds. With non-routine 'bedding' (OR=0.48), 'concentrate feeding' (OR=0.32) and 'medicinal treatment' (0.51) the risk of badly conditioned hoofs decreased. Risk of respiratory symptoms was higher in 'extensive' (OR=1.00) and 'dairy' (OR=0.96) than in 'meat' (OR=0.33) primary purpose. With routine 'bedding' and non-routine 'medicinal treatment', the risk of respiratory disorders increased three and four times. OR in udder health were lower in 'dairy' breeds (OR=0.25) than in 'meat' (OR=1.34) and 'extensive' (OR=1.00). In conclusion, the results indicate that an overview of the herd and the status of animal health should be optimized in order to detect problems in the on-farm management of sheep.

Session 28 Theatre 7

Amh secretion kinetic in doe kids

H. Guérin[1], A. Fatet[2], K. Boissard[3], D. Monniaux[2] and L. Briand-Amirat[1]
[1]Ecole Nationale Vétérinaire ONIRIS, Biotechnologie et Pathologie de la Reproduction, Route de Gachet, 44000 Nantes, France, [2]INRA, Physiologie de la Reproduction et des Comportements, UMR 85, 37380 Nouzilly, France, [3]INRA, Fourrage-Environnement Ruminants FERLUS, UE 1373, 86600 Lusignan, France; alice.fatet@tours.inra.fr

Anti-Müllerian hormone (AMH) is produced by granulosa cells on antral and pre-antral follicles and is used in human medicine as a marker of the ovarian reserve. In goats, AMH is a predictive endocrine marker of the response to a superovulation treatment and further embryo production. Doe kids fertility rates after artificial insemination are highly variable and low, around 40%. To be bred, doe kids should match two zootechnical criterias: age >5 months, weight >32 kg. There is no hormonal marker for sexual precocity. The main objective of this study is to monitor AMH secretion variations from birth to the first reproduction on doe kids and to confront AMH values at a given age to first reproduction fertility rates. The long term objective is to predict puberty and fertility with AMH plasma levels in young nanny goats. Two groups C1 (n=31) and C2 (n=13) of alpine doe kids, born from goats bred during the breeding season and during seasonal anoestrus respectively, are sampled once a month from birth to first reproduction at given ages (3, 30, 60, 90, 120, 150 and 180±3 days). At each sampling date, weight and chest width are recorded. A group P (n=110) of alpine and saanen doe kids born over a 16 days period during natural kidding season is sampled twice, at age 60 and 180±8 days. After a double centrifugation and freezing, the plasma concentration of AMH is determined by using the active MIS/AMH Elisa kit (Beckman Coulter France), as described for goat follicular fluid and plasma. Results are expected in March 2013.

Session 28 Poster 8

Production performance of Thamary and Tahami sheep breeds and their crossbreds in Yemen

T. Alkhewani, M. Momani Shaker, S.A. Al-Olofi and S. Al-Azazi
The Czech University of Life Sciences in Prague, Faculty of Tropical Agri-Sciences, Kamýcká 129 – Suchdol, 165 21 Praha 6, Czech Republic; alkhewani@its.czu.cz

The objective of this study was to evaluate the effect of crossbreeding of Thamary and Tahami sheep pure breeds and their crossbreds, on production performance as reflected by reproductive performance and growth ability of lambs. Including the effect of litter size, sex, dam age and year of rearing on lamb birth weight, weaning weight, daily weight gain, mortality rate, fertility rate and the rate of twins. Four hundred and forty-seven, 2- to 6-year-old multiparous ewes of two breeds were allocated to three groups [Thamary, (Th; n=100), Tahami, (T; n=110), F1 Thamary × Tahami (F1ThT; n=121), F1 Tahami × Thamary (F1TTh; n=126)]. The study was conducted at the Regional Research Station of Central Highlands in Yemen located in the northern part of Yemen at 32°34'N and an altitude of 510 m above sea level, (300-2000 mm rainfall). Results of this study indicated that the effect of genotype on production performance were highly significant($P \leq 0.01$). Lamb birth weight, weaning weight of lambs were greater ($P<0.05$) in the F1ThT and F1TTh group (2.74, 2.59 kg at birth resp. and 11.43, 12.28 kg at weaning resp.) and compared to the T group, which amounted to 2.21 kg at birth and 8.66 kg at weaning. In general, the results of this study demonstrated that F1 crossbreeds ThT and F1 crossbreds TTh had a positive effect on weight lambs at birth, at weaning and daily growth rate in compared with sheep Althammeh, while not so positive effect on the twins rate in the Community genetic different.

Session 28 Poster 9

Live weight and fertility characteristics of Karayaka sheep of Amasya Province in Turkey

E. Sirin[1], E. Soydan[1,2] and M. Kuran[2]
[1]Ahi Evran Universty, Faculty of Agriculture, Agricultural Biotechnology, Asikpasa Campus, 40000 Kirsehir, Turkey, [2]Ondokuzmayis Universty, Faculty of Agriculture, Agricultural Biotechnology, Kurupelit, 55200 Samsun, Turkey; emre.sirin@ahievran.edu.tr

Karayaka sheep is one of the native breeds of Turkey raised at the coastline of Black Sea Region. This sheep is well suited to the harsh climate, poor pasture and severe conditions that are the characteristics of the hills and uplands in the region. They are carpet-wool breed kept also for meat production. The body color of Karayaka sheep is white, and there might be black and brown plaque in head, ear, leg and body and occasionally black or brown animals are seen. In this study 4,180 data for Karayaka sheep were collected in 2011. Birth weight, 56 day weight and 140 day weight of Karayaka lambs were 3.69 kg, 17.67 kg and 26.02 kg respectively. Twinning rate was 6.40%. Infertility rate was 8.01%. These result showed lower live weight and fertility rate of Karayaka seep breed compared to some native breeds. Therefore, Karayaka sheep needs to be taken under the breeding programs in order to improve growth and fertility characteristics.

Session 28 Poster 10

Interactions between milk production and reproduction in Sicilo-Sarde dairy sheep

A. Meraï[1], N. Gengler[2], H. Hammami[2,3], M. Rekik[4] and C. Bastin[2]
[1]National Agronomic Institute of Tunisia, Animal Science Unit, Tunis, Tunis, Tunisia, [2]Gembloux Agro-Bio Tech, Animal Science Unit, Gembloux, Gembloux, Belgium, [3]National Fund for Scientific Research, Brussels, Brussels, Belgium, [4]National School of Veterinary Medicine, Tunis, Tunis, Tunisia; catherine.bastin@ulg.ac.be

This work aimed to investigate the association between productive and reproductive traits of the dairy Sicilo-Sarde ewes. After edits, a database containing 5,935 lactation records collected during 6 successive years in 8 dairy flocks in the North of Tunisia was used. Total milked milk in the milking-only period (TMM) was retained as productive trait. The interval from the start of the mating period to the subsequent lambing (IML) and the status of pregnancy (SP) was designed as reproductive traits. Sicilo-Sarde ewes had an average TMM of 60.93 L (±44.12) milked during 132.8 days (±46.6) after a suckling period of 100.4 days (±24.9). Average IML was 165.7 d. In a first step, the major factors influencing milk production and reproductive traits were determined. The significant sources of variation identified for TMM were: flock, month of lambing, year of lambing, parity, suckling length, litter size and milking-only length. Flock × month of mating, parity, year of mating and litter size were identified as significant factors of variation for IML. While, flock × month of mating, parity and year of mating were identified as significant sources of variation for PS. In a second step, variance components were estimated using a 3-traits threshold mixed model which combined SP as categorical trait and TMM and IML as continuous traits. Repeatability estimates were 0.21 (±0.03) for TMM, 0.09 (±0.02) for IML and 0.10 (±0.05) for SP. Moreover, TMM and IML were found to be favorably associated for the interval from the start of mating to the subsequent lambing (-0.45). This antagonism was largely explained by environmental factors especially flock × year of lambing effect (-0.45±0.18) but unfavorably associated for the animal effect (0.20±0.09).

Session 28 — Poster 11

Reproductive performance of Manchega dairy sheep farms (Spain): preliminary results

J. Rivas, J. Perea, C. Barba, E. Angón, P. Toro-Mujica and A. García
University of Cordoba, Animal Production, Campus Rabanales, 14071, Spain; pa1gamaa@uco.es

The Manchega sheep is an autochthonous breed used in traditional crop-dairy systems located in Castilla La Mancha region. Their milk production is entirely intended to cheese making and, since 1985, all is for the denomination of origin 'Queso Manchego', which allows for a differential pay for quality, contributing to the protection of this breed and the preservation of its production systems. Reproductive performance is a key element for the economic profitability of Manchega dairy sheep farms, but only few data exist. Therefore, information on reproductive performance and reproductive management variables was collected in 2012 through random sampling with proportional allocation by province. The sample of farms comprises 156 Manchega dairy farms. The results showed that the mean first lambing age was 10.2±1.8 month. Respective averages of replacement rate in females and males were 20.6±5.0% and 34.2±10.3%. Average lambing interval was 349.1±99.2 days, and average prolificacy was 1.3±0.1 lambs per lambing. Over 70%, 25% and 5% of farms have a reproductive objective of 1.5, 1.25 and 1 lambing per ewe and year respectively. Artificial insemination is used by 37% of farms, which are looking forward to obtained 1.5 lambing per ewe and year. Over 50% of farms do not use reproductive strategies such as male effect, flushing or hormone treatments. Over 75% of farms applied a three lambing season system, 10% of farms used a two lambing season system, while 15% of farms used a lambing continuous system. Farms with more lambing seasons significantly reached better economic results ($P<0.05$). This study has been development inside of the Research Fundamental Project oriented to agricultural resources and technologies in coordination with the Autonomous Communities of the Spanish National Institute of Research and Agricultural and Food Technology (RTA2011-00057-C02-02).

Session 28 — Poster 12

Performance recording for German Mutton Merino sheep in Saxony-Anhalt

R. Schafberg[1], G. Heckenberger[2], H.J. Rösler[3] and H.H. Swalve[1]
[1]University of Halle, Institute of Agricultural and Nutritional Sciences, Theodor-Lieser-Str. 11, 06120 Halle, Germany, [2]Landesanstalt für Landwirtschaft, Forsten und Gartenbau, Zentrum für Tierhaltung und Technik, Lindenstr. 18, 39606 Iden, Germany, [3]Landesschafzuchtverband Sachsen-Anhalt e. V., Angerstr. 6, 06118 Halle, Germany; renate.schafberg@landw.uni-halle.de

The German Mutton Merino is a sheep breed derived from fine-wool Merino sheep at the turn of the 19th and 20th century. Although this breed is still of the fine-wool type, under today's German conditions the main emphasis in the breeding goal is on lamb and meat production. Within Germany, the German Mutton Merino is only a small breed with around 7,700 breeding ewes of which 3,500 head are kept in the state of Saxony-Anhalt. Performance recording up to now mainly is carried out as a station test. With the present study, relationships between station performance test and a field test are examined. Data for this study consisted of results from 91 lambs that were performance recorded in the test station in the form of a progeny test and 590 lambs with progeny test records from a field test. From the station test, the traits recorded included fattening traits as well as slaughter data. In the field test, only lifetime daily gain (LDG) and ultra-sound measurements of muscle depth as well as fat depth were available. On station, LDG was 373 g while in the field test 317 g were achieved. Muscle depths were 2.97 cm and 2.32 cm, for station and field tests, respectively. For back fat thickness the corresponding values were 0.55 cm and 0.52 cm.

components of the female response to the ram effect: results of a study in Barbarine breed

S. Maatoug-Ouzini[1], G. Khaldi[1], D. François[2] and L. Bodin[2]
[1]INAT, 43, Avenue Charles Nicolle, Cité Mahrajène, 1082, Tunisia, [2]INRA, UR631 Station d'Amélioration Génétique des Animaux, 31320, Castanet-Tolosan, France; loys.bodin@toulouse.inra.fr

The objectives of this study were to (1) assess the reproductive performances of the Barbarine breed for cyclicity traits, and (2) estimate the phenotypic and genetic parameters of the female response to the 'ram effect'. A total of 16,150 matings of 4,201 ewes recorded during a period of 10 years in 7 flocks was used in this study. Four traits were analyzed: the response or not to the ram effect and the presence of spontaneous, normal or short sexual cycles after joining rams with ewe in spring. Mean performances of the ewes as well as averages of the four oestrus behaviour traits showed that approximately, 28% of the ewes presented oestrus behaviour during the first 14 days after introducing rams and were considered as already cycling. Among non-cyclic ewes before mating, 86% were responding to the ram effect, among them 47% had a normal cycle, while 53% started with a short cycle followed by a normal one. Flock, year and ageat lambinghad significant effects on the traits analysed. The spontaneous ovarian activity in spring was negatively correlated with the absence of response to a ram effect (r_g=-0.44) and positively correlated with the response by a normal cycle (r_g=0.63). For all variables, heritability values were ranging from 0.03 to 0.09 but significantly different from 0 and repeatability was slightly higher and varied from 0.06 to 0.1. These results can encourage the use of ram-effect technique, which allows increasing sheep productivity, reducing cost inputs in harsh conditions and making sheep breeding strategies more economical. Key-words: ram effect, cyclicity, genetic parameters, Barbarine.

Post-weaning growth of lambs grazing natural pastures supplemented with protein concentrates

M.J. Marichal[1], M.L. Del Pino[2], H. Deschenaux[2] and L. Piaggio[2]
[1]Facultad de Agronomía, Producción Animal y Pasturas, Garzón 780, 12900 Montevideo, Uruguay, [2]Secretariado Uruguayo de la Lana, Nutrición, Rbla. Baltasar Brum 3764, 11800 Montevideo, Uruguay; mariadejesus.marichal@fagro.edu.uy

Supplementing sunflower meal (SFM) or soybean meal (SBM) on post-weaning growth rate of lambs grazing natural pastures was investigated. The experiment was conducted at Research Center (S 33°52´, W 55°34´) of the Secretariado Uruguayo de la Lana (January 24 to April 28; 14 and 80 days of adaptation and measurements, respectively). Ninety 3 months old weaned Corriedale lambs were allocated to two blocks (45 lambs/block, 10 lambs/ha) of pastures (510 and 853 kg of initial available DM/ha, 74 and 86 g CP kg/DM and 54 and 58% IVOMD for blocks 1 and 2, respectively). In each block, lambs were randomly assigned to one of the following treatments:continuous grazing (T0), T0+SBM (100 g/lamb/d; 467 and 247 g CP and NDF kg/DM, respectively) (T1), T0+SBM (200 g/lamb/d) (T2); T0+SBM meal (300 g/lamb/d) (T3) or T0+SFM (270 g/lamb/d; 325 and 434 g CP and NDF kg/DM, respectively) providing CP equivalent to SBM in T2 (T4). Lambs initial weight was 20±1.6 kg. Animals were weighed every 15 days. Final body weights (FBW) and average daily gains (ADG) were analyzed in a completely randomized block design with sub sampling. Lambs in T3 and T0 registered the greatest and smallest ($P<0.01$) FBW (32.9 and 26.0 kg, respectively). The FBW of T4 and T1 were similar ($P=0.30$) but smaller ($P<0.04$) than T2 (28.2, 29.1 and 30.8 kg, respectively).Differences ($P<0.05$) in ADG were first registered on day 52; SBM supplementation resulted in greater ($P<0.03$) ADG than T0 and T4. Considering the overall experimental period, T0 presented the lowest ($P<0.04$) and T3 and T2 the greatest ($P<0.05$) ADG (58, 133 and 112 g/lamb/day, respectively); T4 resulted in greater ($P=0.02$) ADG (86 g/lamb/day) than T0 but similar ($P=0.14$) to T1 (104 g/lamb/day). SBM appeared as a more effective supplement than SFM to improve growth in lambs.

Session 28 Poster 15

Quantitative characteristics of carcasses of lambs weaned at different ages

N. Maria Brancacci Lopes Zeola, A. Garcia Da Silva Sobrinho, T. Henrique Borghi, C. Renato Viegas, L. Desidério Sena, V. Teixeira Santana, F. Alves De Almeida, A. Carolina Columbelli and L. Gabriel Alves Cirne

College of Agricultural and Veterinarian Sciences, São Paulo State University, Jaboticabal, Brazil., Department of Animal Science, Via de Acesso Prof. P.D. Castellane, 14884-900, Brazil; nivea.brancacci@ig.com.br

Twenty four mixed breed, ½ Lacaune ½ Ile de France, lambs weaned at 30, 45 and 60 days old were used in the experiment. During the weaning period through creep feeding, the lambs were fed '*ad libitum*' a concentrate with 20% crude protein. Subsequently, during the confinement period, housed in individual performance pens, the lambs were fed a diet with 18.0% crude protein, containing 50% sugarcane and 50% concentrate, which consisted of 36.0% ground maize, 61.0% soybean meal, 2.6% dicalcium phosphate and 0.4% mineral mixture. They were slaughtered when body weight reached 32 kg, following humane slaughter procedures. The objective of this study was to evaluate the quantitative characteristics of lamb carcasses with respect to weaning age. Experimental design was completely randomized with three treatments (30, 45 and 60 days) and eight repetitions. Means were compared by Tukey test at 5%. Weaning age did not affect significantly ($P>0.05$) any of the studied variables. Average body weight (BW) was 32.67 kg before fasting; 31.42 kg BW at slaughter (after fasting); 4.78% losses during fasting, 26.64 kg empty body weight; 14.31 kg hot carcass weight; 45.55% hot carcass yield; 13.84 kg cold carcass weight; 44.05% cold carcass yield; 53.72% biological yield and 3.30% chilling loss. Carcasses of lambs weaned at different ages displayed similar characteristics; therefore, it can be inferred that weaning at 30 days old is indicated, since ewes service period would be reduced, the new reproductive cycle of females would start, with consequent lamb production.

Session 28 Poster 16

Effects of crossing a local Romanian sheep with German Blackhead rams on growth and carcass traits

E. Ghita, R. Pelmus, C. Lazar, M. Gras and M. Ropota

National Research Development Institute for Animal Biology and Nutrition, Animal Biology, 1 Calea Bucuresti, 077015 Balotesti, Ilfov, Romania; gras_mihai@yahoo.com

The purpose of this study was to analyse the effects of crossing the Romanian local Teleorman Black Head (TBH) sheep with German Blackhead (GBH) meat-type rams. Because the Romanian consumers prefer the suckling lamb meat, we investigated the growth aptitudes of the lambs (29 TBH lambs and 27 hybrid lambs) up to weaning, when several lambs were slaughtered and evaluated. The weaned lambs were assigned to two groups, 19 TBH and 19 hybrid lambs and fattened for 100 days. At the end of the experiment 8 lambs from each group were slaughtered in order to determine the slaughter and commercial outputs, the proportion of the different carcass parts, the proportion of butcher parts, the meat to bone ratio (for parts/for entire carcass), specific measurements of the carcass, the chemical composition of the meat including the fatty acids and cholesterol level. During the suckling period, the hybrid lambs had a higher daily weight gain (0.286 kg) than the lambs of the local breed (0.253 kg) reaching a body weight of 21.16 kg versus 18.31 kg ($P<0.05$), when weaned at 60 days. During the fattening period, the daily weight gains were 0.232 kg (hybrid lambs) and 0.223 kg (local lambs), reaching a body weight of 43,850 kg and 41,836 kg ($P<0.05$) at the age of 160 days. Carcass conformation of the fattened hybrid lambs differed from that of the local lambs: carcass was shorter in the hybrid lambs, larger carcass width, larger breast width and larger upper thigh perimeter than in the local lambs. This produced a better meat to bone ratio in the hybrid lambs. The high quality butcher parts: rump, leg and rack had a higher output in the hybrid lambs than in the local lambs. In conclusion, the cross with German Blackhead meat rams improved the growth rate, the carcass conformation, the dressing with muscles and the meat to bone ratio.

Post-natal evaluation of local baladi capretto meat

S. Abi Saab[1], F. Esseily[2], M. Saliba[1] and P.Y. Aad[3]
[1]USEK, Agricultural Sciences, NA, Kaslik, Lebanon, [2]UL, Public Health, NA, Fanar, Lebanon, [3]NDU-Louaizeh, Sciences, 072 Zouk Mikael, Zouk Mosbeh, Lebanon; paad@ndu.edu.lb

The local Baladi breed is considered a common source of fresh meat in Lebanon. Kids meat, in particular milk-fed, are an expensive delicacy. However, few studies observed growth parameters, physico-chemical and sensory properties of meat in Baladi Kids. Therefore, 24 male Baladi kids were fed milk free choice for 24 days, then 6 randomly selected and fed free choice pasture grass and goat milk (Control) and 18 fed only milk (Capretto) twice per day. Animals were slaughtered at 4, 6 and 8 weeks of age and body organs compared. Feed intake was measured by weighing kids pre- and post- feeding and body weight measured twice per week. Meat chemical properties were total ether-extracted fat, kjeldhal protein, and minerals. Meat organoleptic properties were evaluated by consumers for color, taste, odor, richness, juiciness, tenderness and overall appreciation, as cooked by housewives or professionals. Data were analyzed as a CRD using SPSS 10.0 and presented as LSMeans ± SEM. Results show that Capretto consumed increasing levels of milk from 4 to 8 weeks, with ADG decreasing ($P<0.05$) from 4 to 8 weeks. Capretto showed 5% less ($P<0.05$) weight than Control kids. Carcass yield was higher ($P<0.05$) for Control (44.5±3.4%) than Capretto (36.3±3.3%), with Capretto showing smaller viscera and lower hemoglobin. Both blood and meat (protein, cholesterol) composition did not differ ($P>0.05$) between Control and Capretto but with lower meat fat content ($P<0.05$) in Capretto than Control Kids. Meat organoleptic characteristics were desirable at weeks 4 and 6 regardless for both Capretto and Control kids, whereas only Capretto meat stayed desirable at week 8. To conclude, Capretto kids have similar properties to their traditional counterparts, with added desirable properties for a longer growing period, in addition to the increased health benefit of lower fat. Project supported by Lebanese CNRS.

Fatty acids profile of intramuscular fat in light and heavy carcass lambs

M. Margetín[1,2], D. Apolen[2], M. Oravcová[2], O. Debreceni[1], D. Peškovičová[2], O. Bučko[1] and Z. Horečná[1]
[1]Slovak University of Agriculture Nitra, Tr. Andreja Hlinku 2, 949 76 Nitra, Slovak Republic, [2]Animal Production Research Centre, Hlohovecká 2, 951 41 Lužianky, Slovak Republic; milan.margetin@uniag.sk

The quality of 40 light carcass lambs from artificial and traditional rearing (LLAR, LLTR) and 40 heavy lambs from pasture rearing under mothers and traditional rearing in sheepfolds (HLPR, HLTR) was assessed on the basis of carcass characteristics and fatty acids profile (FAs) of intramuscular fat (IMF; Musculus longissimus lumborum et thoracis – MLLT). The average empty live weight of lambs before slaughter was 17.8, 17.6, 29.2 and 32.5 kg in LLAR, LLTR, HLPR and HLTR. The average age was 63.6, 55.3, 105.0 and 108.9 d. Profile of fatty acids (totally 69 FAs) was determined by gas chromatography. By means of ANOVA we detected significant differences among individual FAs or FAs groups in dependence on the lamb type (LLAR, LLTR, HLPR, HLTR) and lamb sex. We found significant effect of lamb type on SFA, MUFA, PUFA, cis-MUFA, trans-MUFA, CLA, n-6 PUFA, n-3 PUFA, essential FA, ratio of n-6/n-3 PUFA and ratio of LA/ALA ($P<0.001$). Ratio of n-6/n-3 in LLAR was several times higher than in LLTR, HLPR and HLTR lambs (14.08, 3.25, 1.86 and 2.21, $P<0.001$). The content of CLA in IMF in HLPR lambs (2.103 g/100 g FAME) was several times higher than in light lambs (0.193 in LLAR; 0.749 in LLTR) and HLTR lambs (0.645 g/100 g FAME, $P<0.001$). The content of trans-vaccenic acid (TVA), α-linolenic (ALA), rumenic (RA), eicosapentaenoic (EPA) docosapentaenoic (DPA) and docosahexaenoic acid (DHA) was the highest in HLPR lambs (4.05, 2.38, 1.82, 0.82, 0.92 and 0.29 g/100 g FAME). Content of SFA in heavy lambs was significantly higher (48.7 and 57.9) than in light lambs from artificial and traditional rearing (44.8 and 45.6 g/100 g FAME; $P<0.001$). Effect sex significantly influenced content of several FAs (OA, LA, ALA, RA, AA; $P<0.01$) and groups of fats (MUFA, PUFA, CLA, n-6 PUFA, essential FAs; $P<0.01$) in IMF of MLLT.

Fatty acids profile of extramuscular fat in light and heavy carcass lambs

M. Margetín[1,2], J. Margetínová[1], Z. Krupová[1], K. Vavrišinová[2], P. Polák[1] and L. Luptáková[2]
[1]Animal Production Research Centre, Hlohovecká 2, 951 41 Lužianky, Slovak Republic, [2]Slovak University of Agriculture Nitra, Tr. Andreja Hlinku 2,949 76 Nitra, Slovak Republic; milan.margetin@uniag.sk

Fatty acids profile (FAs) of extramuscular fat (EMF; fat samples were taken from root of tail) was assessed in 40 light carcass lambs from artificial and traditional rearing (LLAR, LLTR) and 40 heavy lambs from pasture rearing under mothers and traditional rearing in sheepfolds (HLPR, HLTR). The average empty live weight of lambs before slaughter was 17.8, 17.6, 29.2 and 32.5 kg in LLAR, LLTR, HLPR and HLTR. The average age was 63.6, 55.3, 105.0 and 108.9 d. Dressing percentage of lambs was in average 46.7, 47.8, 47.0 and 48.9%. Profile of fatty acids (totally 69 FAs) was determined by gas chromatography. By means of ANOVA we detected significant differences among individual FAs or FAs groups of EMF in dependence on the lamb type (LLAR, LLTR, HLPR, HLTR) and lamb sex. We found significant effect of lamb type on SFA, MUFA, PUFA, cis-MUFA, trans-MUFA, CLA, n-6 PUFA, n-3 PUFA, essential FA, ratio of n-6/n-3 PUFA and ratio of LA/ALA ($P<0.001$). Ratio of n-6/n-3 in LLAR was several times higher than in LLTR, HLPR and HLTR lambs (19.24, 3.52, 1.61 and 2.15, $P<0.001$). The content of CLA in EMF of HLPR lambs (2.205 g/100 g FAME) was several times higher than in light lambs (0.209 in LLAR; 0.815 in LLTR) and HLTR lambs (0.665 g/100 g FAME, $P<0.001$). The content of trans-vaccenic acid (TVA), α-linolenic (ALA), rumenic (RA), eicosapentaenoic (EPA) docosapentaenoic (DPA) and docosahexaenoic acid (DHA) was the highest in HLPR lambs (4.04, 1,38, 1.96, 0.084, 0.365 and 0.094 g/100 g FAME). Content of SFA in HLPR and HLTR was significantly higher (55.9 and 60.2 g/100 g FAME) than in light lambs from artificial and traditional rearing (46.3 and 49.1 g/100 g FAME; $P<0.001$). Effect sex significantly influenced content of saturated FAs (lauric; myristic, palmitic, $P<0.05$), and SFA ($P<0.01$) and MUFA ($P<0.05$).

Influence of Wiltshire Horn rams on Romanov ewes fertility and performance of crossbred lambs

R. Nainiene, B. Zapasnikiene, V. Juskiene and A. Siukscius
Institute of Animal Science, Lithuanian University of Health Sciences, R. Zebenkos 12, 82317 Baisogala, Radviliskis district, Lithuania; rasa@lgi.lt

According to the current demands of domestic and international markets, meat-producing sheep breeding is considered as having the best future. Although Romanov sheep ewes produce litters of 3 to 5 lambs at a time, but lambs show low growth intensity and the dressing percentage accounts for only 40%. The aim of our study was to determine the influence of Wiltshire Horn rams on the fertility of Romanov sheep, their progeny weight, meat percentage and wool length. Twelve Romanov ewes and two rams of Romanov and Wiltshire Horn breeds (2 groups of 6 ewes and 1 ram each) and their progeny (18 purebred and 14 crossbred) were used in the study. Ewe fertility was evaluated by the number of lambs born per litter. The lambs were weighed at day 1 and 7 months of age. Wool quality was assessed at 1 and 7 months of age. Lamb muscularity was determined visually on a 1 to 9 point scale. The mating fur bearing Romanov sheep with woolless mutton type Wiltshire Horn ram had a positive influence on lower fertility of ewes and higher weight of newborn lambs, their growth intensity and muscularity. Although the number of crossbred lambs born was by 26.50% lower, but they were by 1.04-1.25 kg heavier ($P<0.001$) and gained daily by 36.55-50.85 g more ($P<0.001$), and thus exceeded the purebreds by weight from 8.74 to 11.69 kg ($P<0.001$) and had 1.30-1.87 points higher muscularity than the purebred lambs. Crossbred lambs inherited white wool colour from Wiltshire Horn ram and low wool production as wool naturally moulted from 5-6.5 cm at one month of age to 2-3 cm at 7 months of age and did not require shearing. It is advisable to mate fertile and low weight Romanov ewes with mutton type woolless Wiltshire Horn rams in order to increase progeny meatiness and lower wool production when wool shearing three times a year is replaced by single shearing.

Session 28 Poster 21

Nutrient intake for lambs fed 'in nature' or hydrolyzed sugarcane under aerobic and anaerobic condit

V. Endo, A. Garcia Da Silva Sobrinho, N. Maria Brancacci Lopes Zeola, N.L. Lins Lima, F. Alves De Almeida and G. Milani Manzi
College of Agricultural and Veterinarian Sciences, São Paulo State University, Jaboticabal, SP, Braz, Department of Animal Science, Via de Acesso Prof. P.D. Castellane, 14884-900, Brazil; nivea.brancacci@ig.com.br

The objective of this study was to evaluate nutrient intake of dry matter (DM), ash, neutral detergent fiber (NDF) and acid detergent fiber (ADF) in lambs fed 'in nature' or hydrolyzed sugarcane. Ile de France lambs (n=24, from 15.0 to 32.0 kg BW) were subjected to one of three treatments: IN) 'in nature' sugarcane + concentrate; AER) hydrolyzed sugarcane with 0.6 kg of calcium oxide (CaO)/100 kg of sugarcane in aerobic environment + concentrate; ANA) hydrolyzed sugarcane with 0.6 kg of CaO/100 kg of sugarcane in anaerobic environment + concentrate. Feedlot lambs were raised individually and concentrate had 20.0% of CP. Animals were randomly assigned to treatments in a completely randomized design with eight animals per treatment. Data was analyzed by GLM procedure of SAS and treatment differences were detected at significance level of 5%. Treatments were compared by Tukey test at the same level of significance. Ash intake was lower (P=0.01) for lambs fed 'in nature' sugarcane (50.4 g/d) compared with lambs fed hydrolyzed sugarcane in anaerobic environment (69.7 g/d). While the condition of hydrolysis did not differ for ash intake. The lowest ash intake for lambs fed 'in nature' sugarcane was due to non-inclusion of calcium oxide. Average intakes were DM (951 g/d), NDF (246 g/d) and ADF (120 g/d) that were not difference between treatments. Lambs had daily gain of 240 g/d (P=0.3). The hydrolyzed sugarcane did not increase nutrient intake, particularly the NDF. The results showed that hydrolyzed sugarcane with alkaline agents did not improve intake.

Session 28 Poster 22

Performance, ingestive behavior and meat quality of lambs submitted to different feeding frequencies

E.L.A. Ribeiro, C.L. Sousa, F.A.B. Castro, I.Y. Mizubuti, B. Marson, F.A. Grandis, C. Constantino and N.A. Koritiaki
Universidade Estadual de Londrina, CCA, Zootecnia, Caixa Postal 10.011, CEP 86057-970, Londrina, PR, Brazil; elar@uel.br

This study had the objective of evaluating effects of feeding frequencies on performance, ingestive behavior and meat quality of feedlot lambs. Thirty-six crossbred Santa Ines ram lambs were used, with average age and live weight at the beginning of the experiment of 70 days and 21.9 kg, respectively. Lambs were randomly assigned to one of the three frequencies of feeding (1, 2 or 3 times a day). They were confined for 58 days in pens with two animals each. Diet consisted of sorghum silage and concentrate composed by soybean meal, ground corn and sunflower cake. The roughage/concentrate ratio was 50/50. Rations (20% CP and 68% TDN) were supplied *ad libitum*. Productive performance and nutrients intake were not affected (P>0.05) by feeding frequency. Averages for body weight at slaughter, daily weight gain, DM intake and feed conversion were 38.01 kg, 0.280 kg, 1.22 kg/day and 4.39, respectively. With regard to ingestive behavior, feeding frequency did not affect the evaluated activities. Lambs spent an average of 268 and 521 minutes/day, on ingestion and rumination of feed, respectively. Cold carcass weights were not different (P>0.05) among feeding frequencies and presented an average value of 16.73 kg. However, cold carcass yield were similar for lambs fed 1 and 2 times/day (44.82 e 44.26%, respectively) and superior (P<0.05) in relation to those fed 3 times/day (42.62%). Traits (pH and marbling) evaluated on lamb's meat and sensorial characteristics (tenderness, juiciness, overall acceptability and odor) evaluated by a trained panel, were not affected (P>0.05) by frequency of feeding. Meat was considered tender with a moderate odor and juiciness. Therefore, finishing lambs can be fed only once a day without compromising animal performance and meat quality.

Session 28 Poster 23

Grape pomace and grape seed extract in lamb diets: meat fatty acid profile and antioxidant activity

C. Guerra-Rivas[1], P. Lavín[2], B. Gallardo[1], A.R. Mantecón[2], C. Vieira[3] and T. Manso[1]
[1]Universidad de Valladolid, Ciencias Agroforestales, ETS Ingenierías Agrarias. Avd. Madrid s/n, 34004 Palencia, Spain, [2]Instituto de Ganadería de Montaña (CSIC-ULE), Finca Marzanas, 24346 Grulleros, León, Spain, [3] Instituto Tecnológico Agrario de Castilla y León, Estación Tecnológica de la Carne, C/ Filiberto Villalobos s/n, 37770 Guijuelo, Salamanca, Spain; cguerra@agro.uva.es

Forty-eight male Merino lambs housed in individual pens and fed *ad libitum* with concentrate (74% barley and 20% soya) and barley straw, were used to study the effect of feeding grape pomace, grape seed extract and other commonly used antioxidant (vitamin E), on M. Longissimus fatty acid profile and lipid oxidation at 5 days after slaughter (TBARS). Animals (14.3±2.05 kg initial body weight) were assigned according to the four dietary treatments: control (CTL, 50 IU of vitamin E per kg of concentrate), vitamin E (Vit E, 500 IU of vitamin E per kg of concentrate), grape seed extract (GSE, including 50 mg of grape seed extract per kg of concentrate) and grape pomace (GP, 5% of dry grape pomace from red wine production per kg of concentrate). Lambs were slaughtered when they reached 27 kg live weight. Data were subjected to ANOVA using the GLM procedure of SAS. Meat from GP lambs tended ($P<0.1$) to increase the polyunsaturated fatty acids (PUFAs) content compared to CTL and GSE treatments, and there were not differences ($P>0.05$) between Vit E and the other groups. The saturated and monounsaturated fatty acids meat content were not affected by the treatments. The lowest TBARS were observed in lambs from Vit E group ($P<0.05$). In spite of the greatest PUFAs content of GP, TBARS were not affected when they were compared with the others treatments. In conclusion, vitamin E was the most powerful antioxidant used in this study and grape pomace as ingredient in lamb diets tended to increase PUFAs content without effect on TBARS in meat.

Session 28 Poster 24

Effect of olive and fish oil calcium soaps fed to early lactating ewes on milk fatty acid profile

B. Gallardo[1], P. Gómez-Cortés[2], A.R. Mantecón[3], M. Juárez[2], M.A. De La Fuente[2] and T. Manso[1]
[1]Universidad de Valladolid, ETS Ingenierías Agrarias. Avd. Madrid s/n, 34004 Palencia, Spain, [2]Instituto de Investigación en Ciencias de la Alimentación (CSIC-UAM), C/ Nicolas Cabrera, 9.,28049 Madrid, Spain, [3]Instituto de Ganadería de Montaña (CSIC-ULE), Finca Marzanas, 24346 Grulleros, León, Spain; tmanso@agro.uva.es

The objective of this study was to evaluate the effects of different calcium soaps of fatty acids (CSFA) fed to early lactating ewes on milk performance and fatty acid (FA) profile. After lambing, thirty-six Churra ewes were fed daily 2.1 kg of a TMR containing lucerne and concentrate at a 40:60 ratio. Each ewe was assigned to one of the three dietary treatments, which received 3% (as-fed basis) of the corresponding CSFA: CSFA of palm oil (CTL), CSFA of olive oil (OLI, Olifat) and CSFA of fish oil (FO, StrataG). Milk yield and composition were recorded weekly during the first month of lactation. Data were evaluated by the MIXED procedure of SAS. Regardless the type of CSFA, milk yield and milk protein percentage were not different ($P>0.05$), but OLI and FO diet decreased milk fat content ($P<0.01$). Milk samples from FO had the lowest content of monounsaturated FA ($P<0.01$) and the greatest content of polyunsaturated FA (PUFA) ($P<0.001$). OLI and FO treatments caused an increase of 2-fold and 8-fold in trans-11 C18:1 and 2-fold and 6-fold in cis-9, trans-11 C18:2 in milk fat compared to the control diet respectively. Although there were no differences in trans-10, cis-12 C18:2, OLI and FO enhanced the content of trans-10 C18:1 ($P<0.05$). FO treatment caused an increase of 4-fold in n-3 PUFA in milk fat compared with the other two treatments ($P<0.001$), mainly C20:5, C22:5 and C22:6 while no differences ($P>0.05$) were observed between CTL and OLI. In conclusion, supplementing ewes' diet with CSFA of olive oil or fish oil could be a promising way to improve milk FA composition from a human health point of view, however fish oil CSFA would reduce the milk fat percentage.

Session 28 Poster 25

Linseed oil and natural or synthetic vitamin E in ewe diets: milk performance and fatty acid profile

B. Gallardo[1], M.G. Manca[2], A.R. Mantecón[3], A. Nudda[2] and T. Manso[1]

[1]Universidad de Valladolid, ETS Ingenierías Agrarias. Avd. Madrid s/n, 34004 Palencia, Spain, [2]Università degli Studi di Sassari, Science Zootechnique, Via E de Nicola 9, 07100 Sassari, Italy, [3]Instituto de Ganadería de Montaña (CSIC-ULE), Finca Marzanas, 24346 Grulleros, León, Spain; beatriz.gallardo.garcia@uva.es

The objective of this study was to evaluate the effects of dietary linseed oil and vitamin E, synthetic or natural, on milk performance and fatty acid (FA) profile in early lactating ewes. After lambing, forty-eight Churra ewes were fed daily 2.3 kg of TMR with lucerne and concentrate at a 40:60 ratio. The dietary treatments were: Control (CTL, without linseed oil), LO (with 3% linseed oil), LO+Syn E (LO plus 400 mg/kg TMR of synthetic vitamin E) and LO+Nat E (LO plus 400 mg/kg TMR of natural vitamin E). Data were subjected to ANOVA using the GLM and MIXED procedure of SAS. Milk yield and protein percentage were not affected by the three diets containing linseed oil compared to CTL ($P>0.05$), whereas milk fat percentage increased in dairy ewes fed LO with vitamin E ($P<0.05$). Milk from LO, LO+Syn E and LO+Nat E treatments had lower percentages of saturated FA and higher percentages of monounsaturated FA and polyunsaturated FA (PUFA) than ewes fed the CTL diet. Linseed oil supplementation, without vitamin E, caused an increase in trans-11 C18:1 (VA; $P<0.001$), trans-10 C18:1 ($P<0.05$), trans-10, cis-12 C18.2 ($P<0.05$), cis-9, trans-11 C18:2 (RA; $P<0.01$) and in C18:3 n-3 ($P<0.001$) in milk fat compared to the CLT. Addition of vitamin E to LO diets did not influence significantly ($P>0.05$) the majority of milk fatty acids compared to the LO diet. The LO+Syn E treatment resulted in higher percentage of RA ($P<0.01$) and trans-10, cis-12 C18:2 ($P<0.05$) than LO+Nat E and CTL treatments. In conclusion, feeding linseed oil to lactating ewes could be a way to increase VA, RA and PUFA n-3 in milk, whereas the type of vitamin E (natural or synthetic) added to the linseed oil diet could influence the content of some conjugated C18:2 isomers in milk.

Session 28 Poster 26

Digestibility of grass-based diets supplemented with four levels of *Gliricidia sepium* hay in lambs

S. Angeles-Campos[1], J.L. Valle-Cerdán[2], F. Castrejón-Pineda[1], J.N. Avilés-Nieto[1] and E. Vargas-Bello-Pérez[3]

[1]Universidad Nacional Autónoma de México, Departamento de Nutrición Animal y Bioquímica, Facultad de Medicina Veterinaria y Zootecnia, Ciudad Universitaria, 04510, Mexico, [2]Centro de Bachillerato Técnico Agropecuario 194, Miacatlán, Morelos, 62600, Mexico, [3]Pontificia Universidad Católica de Chile, Departamento de Ciencias Animales, Av. Vicuña Mackenna 4860, 6904411, Santiago, Chile; evargasb@uc.cl

In dry tropical areas of Mexico, *Gliricidia sepium* hay is readily available; therefore, the objective of the present study was to evaluate the effect of supplementing *G. sepium* hay at increasing levels of inclusion (0, 10, 20 and 30%) on digestibility of BG hay by hair sheep lambs. The objective of this experiment was to evaluate the effect of supplementing increasing levels of *G. sepium* with different levels of inclusion of Buffel grass (BG) hay on digestibility by hair sheep lambs. Eight male lambs were used in a replicated 4×4 Latin square design with 21-d experimental periods (n=4). Animals were fed BG with different levels of *G. sepium*: T1) 100% BG (control); T2) 90% BG + 10% *G. sepium*; T3) 80% BG + 20% *G. sepium*; and T4) 70% BG + 30% *G. sepium*. The intake of crude protein (CP), organic matter (OM) and gross energy (GE) was higher ($P<0.05$) in those lambs fed T4 diet than control. NDF and ADF digestibilities were higher ($P<0.05$) in T1 than in the other treatments. CP digestibility was higher ($P<0.05$) in T4 compared to T1, T2 and T3. The study showed that inclusion of *G. sepium* up to 30% with BG in forage based diets of sheep does not affect DM and fibre intake, nor influenced DM and OM digestibilities but increases CP intake and reduces NDF and ADF digestibilities.

Session 28 Poster 27

Impact of changing amount and frequency of concentrate feeding on growth and carcass of weaned lambs

A.S. Chaudhry[1] and M.I. Mustafa[1,2]
[1]Newcastle University, Agriculture, Food & Rural Development, Agriculture Building, NE1 7RU, United Kingdom, [2]University of Agriculture, Department of Livestock Management, Faisalabad, Pakistan; abdul.chaudhry@ncl.ac.uk

We examined the impact of two amounts and frequencies of concentrate (CON) feeding on daily dry matter intake (DDMI), live-weight gain (DLWG) and carcass of lambs consuming ryegrass hay *ad libitum* from weaning to slaughter. This factorial study involved 24 Texel × lambs by using 2 CON amounts each at 2 feeding frequencies for 2 genders each involving three lambs as replicates. The lambs received *ad libitum* grass hay but fixed daily amounts of either 500 or 250 g CON (11 MJ ME and 214 g CP/kg DM) as either single morning feeds or two equal feeds for 35 days. The lambs were slaughtered at 40 kg LW to assess their carcass and killing out (KO %). The data were statistically analysed for the effects of gender (G), CON amount (A) and frequency (F) of CON and their two way interactions on growth and carcass of these lambs for their significance at $P<0.05$. While the CON amount significantly affected DDMI ($P<0.05$), the feeding frequency and gender effects on DDMI were not significant ($P>0.05$). The A × G interactions were significant ($P<0.05$) for DDMI and KO% suggesting that the wether lambs had greater DDMI than the ewe lambs for higher CON amount but the ewe lambs had more DDMI on lower CON amount. The wether lambs had higher KO% for lower CON amount but the ewe lambs had higher KO% for higher CON amount ($P<0.05$). The amount and frequency of CON feeding had no appreciable effect on growth or carcass of weaned lambs. The absence of response may be partly due the relatively lower DDMI of low quality hay. However, the tendency of greater DDMI and KO % of weather than the ewe lambs suggest that the wether lambs tended to respond more to changes in the frequency and amount of CNN feeding in this study.

Session 28 Poster 28

Genetic parameters of growth and faecal egg counts obtained by using pedigree or SNP information

F. Assenza[1], J.M. Elsen[1], A. Legarra[1], C. Carré[1], G. Sallé[2], C. Robert-Granié[1] and C. Moreno[1]
[1]INRA, 24 Chemin de Borde Rouge, Auzeville, 31326 Castanet Tolosan, France, [2]Université François Rabelais, 10 boulevard Tonnellé, 37000 Tours, France; fabrizio.assenza@toulouse.inra.fr

Haemonchosis is a parasitosis causing severe losses in sheep production. Recently, the parasite's resistance to anthelmintics raised the need for alternative control strategies. Genetic selection is a promising candidate but its efficacy depends on the availability of genetic variation and on the occurrence of favourable genetic correlations between the traits under selection. The aim of this study was twofold: to estimate both the heritability of and the genetic correlations between growth traits and parasite resistance traits in two subsequent experimental infestations; to compare the precision of the estimates when using two different relationship matrices: including pedigree information only or including both pedigree and SNP information. The heritabilities of the growth traits and of the parasite resistance traits were weak and moderate, respectively. The estimates of the genetic correlation between the average daily gain before the infestation and faecal egg counts were different from the estimates of the genetic correlations between the average daily gain during infestation and faecal egg counts. The standard errors of the estimates obtained by including SNP information were smaller than those obtained without it. The use of molecular information results in more precise estimates than using pedigree only. The estimates of the genetic parameters suggest that is possible to select for resistance to *H. contortus* while keeping the growth performance in contaminated environments. Whereas, selection for growth in parasite free environment could result in more susceptible animals.

Genetic parameters for major fatty acids and proteins in French dairy sheep

H. Larroque[1], M. Ferrand[2], G. Miranda[3], J.M. Astruc[4], G. Lagriffoul[4], R. Rupp[1] and M. Brochard[2,3]
[1]INRA, UR631 SAGA, 31320 Castanet-Tolosan, France, [2]IDELE, MNE, 75012 Paris, France, [3]INRA, UMR1313 GABI, 78352 Jouy-en-Josas, France, [4]IDELE, SAGA, 31320 Castanet-Tolosan, France; helene.larroque@toulouse.inra.fr

The French PhénoFinlait project has been carried out to study milk composition in fatty acids (FA) and proteins of dairy ruminants. Profile in FA has been of increasing interest because of their importance in human health. Milk composition in protein is relevant to the dairy industry for cheese making process. The aim of this study was to evaluate feasibility of genetic selection to improve sheep milk quality. Genetic parameters were estimated by restricted maximum likelihood with an animal model using 67,013 test-day records from 11,747 Lacaune (LAC) and 8,159 Manech red faced (MRF) ewes in first and second lactation. Traits considered were milk yield, total fat content (FC) and protein content (PC), mono and poly-unsaturated FA (MUFA and PUFA) and saturated FA (SFA) contents in fat, and caseins (κ-CN, αS2-CN, αS1-CN, β-CN), total caseins (CN) and whey proteins (WP) contents in protein. Heritability estimates for FA were moderate (SFA: 0.26-0.28, MUFA: 0.23-0.25, PUFA: 0.25-0.25, for MRF and LAC resp.). FA contents in fat were not correlated with milk yield. FC was positively correlated to SFA (0.40 and 0.20 in MRF and LAC, resp.), and negatively with MUFA and PUFA (-0.27 and -0.41 with PUFA in MRF and LAC, resp.). Heritability estimates for proteins were moderate to high and ranged from 0.17 (for CN in MRF) to 0.46 (for WP in LAC), with larger differences between breeds than for FA. Genetic correlations between caseins were positive (0.13 to 0.47) except αS2-CN, showing null or negative correlations with others (-0.26 to -0.31 with αS1-CN). Selection on FC would involve an increase in SFA in milk fat and selection on major FA or protein profiles is possible in French dairy sheep. This program receives financial support from ANR, Apis-Gène, Ministry of Agriculture(CASDAR), CNIEL, FranceAgriMer and FGE.

Characteristics of Hair Goat in Turkey

E. Soydan[1,2], E. Sirin[1] and M. Kuran[2]
[1]Ahi Evran University, Faculty of Agricultural, Agricultural Biotechnology, Asikpasa Campus, 40000 Kirsehir, Turkey, [2]Ondokuz Mayis University, Faculty of Agricultural, Agricultural Biotechnology, Samsun, 55200 Kurupelit, Turkey; esoydan@ahievran.edu.tr

Hair Goat is one of the native breeds of Turkey. There are 7 million head of Hair Goat in Turkey. Hair goat is well suited in harsh climate or poor pasture conditions and forests. They are kept usually for meat production. The body color of Hair Goat ranges from white to black or brown. The male goats have more live weight than females. In this study, 4872 data of Hair Goats were collected in 2012. Birth weight, 56 day live weight and 140 day live weight of Hair Kids were 3.06 kg, 14.69 kg and 23.42 kg respectively. Twinning rate was 10.20%. Infertility rate was 10.81%. These results indicate low live weight and fertility rate of Hair Goats. Therefore, in order to improve growth and fertility characteristics of Hair Goat, the breeding programs should be carried out if the meat production is desired to be improved.

Session 28 Poster 31

EBVs and sire ranking: the impact of pedigree integrity on selection in South African Angora goats

C. Garritsen, C. Visser and E. Van Marle-Koster
University of Pretoria, Animal and Wildlife Sciences, Private Bag X 20, Hatfield, 0028, South Africa; carina.visser@up.ac.za

Pedigree integrity is vital for the success of any breeding selection programme. DNA marker-based parentage testing has become a useful tool in amending inaccuracies in on-farm records. Previous studies have determined that errors in pedigree records may have a negative effect of up to 15% in genetic improvement of livestock herds. This study quantified the extent of incorrect paternity records in seven South African Angora goat herds, amounting to 381 animals, using a panel of 12 microsatellite markers. A 6% misallocation rate was detected in the pedigree records of the breeder and these individuals were re-allocated. All offspring (40 animals) that did not have on-farm recorded parentage were assigned to sires using Cervus 3.0. Estimated Breeding Values for fleece traits (fibre diameter and fleece weight) as well as birth weight and weaning weight were calculated for 21 sires using ASREML, firstly using the breeder's recorded pedigree and secondly using the DNA marker-verified pedigree. The sires ranked lowest in the breeder's records for fibre diameter, birth weight and weaning weight were moved to the top three ranks in the DNA marker-verified pedigree. The ranking for fleece weight was not as severely affected, the lowest ranking sires in the breeder's pedigree remained in the lowest ranks in the DNA-verified pedigree, and similar effects realized in the top ranking sires. The significant change in sire ranking after DNA marker-based pedigree verification emphasised the importance of pedigree integrity in optimizing selection accuracy in the South African Angora goat industry.

Session 28 Poster 32

Pedigree analysis in goats: an example of White Shorthaired breed

M. Oravcová[1], M. Margetín[1,2] and J. Tomka[1]
[1]Animal Production Research Centre Nitra, Hlohovecká 2, 95141 Lužianky, Slovak Republic, [2]Slovak University of Agriculture, Tr. A. Hlinku 2, 94976 Nitra, Slovak Republic; tomka@cvzv.sk

Genealogical information of 1,682 animals of the White Shorthaired goat breed in Slovakia was analysed. The population under study (reference population) consisted of 670 animals born between 2008 and 2011 with at least one ancestor known in the second generation of ancestors. The numbers of founders, ancestors, effective founders, effective ancestors and founder genome equivalents were 286, 256, 73, 45 and 32, respectively. Fifty percent of genetic variability of the reference population was explained by fifteen ancestors. Marginal contributions of ten most influential ancestors were between 5.45% and 2.47%, and accounted for 39.8% of genetic variability of the reference population. The mean values of inbreeding and coancestry assessed for the reference population were 0.69% and 1.55%. The effective population size was assessed at 182 and 142 individuals, respectively. It was calculated either from the individual increase in inbreeding or from the individual increase in coancestry. The number of maximum generations traced, the number of fully traced generations and the number of equivalent complete generations traced were 5.62, 1.97 and 3.04, respectively. The first, second and third generation of ancestors of animals in the reference population was 100%, 83% and 71% complete. The completeness decreased to as low as 35% and 11% in the fourth and fifth generation of ancestors. To keep genetic links across generations, the amount of genealogical information should be increased. This is crucial for monitoring and management of genetic relations within the population.

Session 28 Poster 33

Importance of Zackel sheep for conservation of biodiversity at Sjenica-Pester plateau

M. Savic[1], M. Vegara[2], S. Vuckovic[3], S. Prodanovic[3] and B. Dimitrijevic[1]
[1]Faculty of Veterinary Medicine, Animal breeding and genetics, Bulevar oslobodjenja 18, 11000 Belgrade, Serbia, [2]Norwegian University of Life Sciences (UMB), Department of International Environment and Development Studies, Noragric, P.O. Box 5003, 1432 Aas, Norway, [3]Faculty of Agriculture, Nemanjina 6, 11000 Belgrade, Serbia; mij@beotel.net

The decreasing populations of various autochthonous breeds are being recognized as important elements to regional agro-biodiversity, more specifically in their relevance to the agro-ecosystems that encompass the natural and cultural heritage of all West Balkan countries. By recognizing that locally adapted animal breeds gained genetic resistance and adaptability through the evolutionary process, breeding strategies in sustainable farming practices today are far more attuned to the need for preserving and utilizing these autochthonous breeds. Multidisciplinary HERD Agriculture Project was applied in order to enable protection and rational utilization of natural resources and biodiversity in WB mountain eco-regions. Sjenica sheep is a transboundary type of Zackel breed. Sjenica sheep is traditionally reared at Sjenica-Pester plateau (900-1,200 m), wide pasture in southwestern Serbia. The uses and values of Sjenica sheep as AnGR have been evaluated. The estimation of production traits, health status included the analysis of disease frequency, metabolic profile and oxidative stress parameters have been performed. The special attention was laid on the parasitic infections monitoring, as the main health problem in sustainable sheep breeding. Botanical analysis has shown the presence of plant diversity as well as nutritious value sufficient to satisfy the requirements of low input sheep farming system. The positive interaction between extensive grazing system and ecosystem biodiversity should contribute to creating the sustainable farming of Zackel sheep which are bringing back the traditional landscape to the Sjenica Pester region.

Session 28 Poster 34

Milk consumption behavior of Romane lambs under artificial rearing

I. David[1], F. Bouvier[2], E. Ricard[1], J. Ruesche[1] and J.L. Weisbecker[3]
[1]INRA, UR631, SAGA, 31320 Castanet tolosan, France, [2]INRA, UE0332, Domaine de la sapinière, 18390 Osmoy, France, [3]INRA, UE0065, Domaine de Langlade, 31450 Montgiscard, France; ingrid.david@toulouse.inra.fr

To provide tools to improve the management of lambs artificially reared; the milk consumption behavior of 94 Romane lambs was studied. Lambs were reared with an automatic self-feed milk replacer from 2 to 28 days of age. The time, lamb identification and quantity of milk distributed at each visit to the teat were recorded. The meal criteria was estimated at 55 minutes. On average, a lamb made 10±3 meals per day with 1.5±0.7 visits to the teat per meal. The mean time interval between meals was 150 min±90. The mean quantity of milk drunk was 180±120 ml per meal and 1.7±0.7 l per day. The food conversion efficiency was estimated at 1.3±0.3. Factors influencing the feeding behavior were evaluated for each of the 6 previous traits using generalized linear mixed models. The following effects were tested: type of birth (single, twin …), sex of the lamb, year, type of sires (high or low direct effect on preweaning growth), weight at birth, pen, type of growth (normal if the weight of the lamb at 35 d. of age is higher than 6 kg, abnormal otherwise). Females had significantly (alpha risk 5%) more meals per day and more visits to the teat per meal and they drunk less milk per meal than males. We did not observe any differences in the feeding behavior of lambs depending on the type of their sires. Nonetheless, we noticed that the food conversion efficiency was slightly lower for lambs with high direct effects than lambs with low direct effects. The number of meals per day, the quantity of milk per meal and consequently the quantity of milk per day were significantly lower for lambs with an abnormal growth than for lamb with a normal growth. Conversely to the hypothesis claimed by some breeders that lambs with an abnormal growth had an erratic feeding behavior, we did not observe a lower repeatability of the traits for lambs with an abnormal growth in comparison with lambs having a normal one.

Mathematical modelling of lamb growth rate

J.M. Coyne[1], P. Creighton[2] and N. McHugh[1]
[1]Teagasc Animal & Grassland Research and Innovation Centre, Moorepark, Fermoy, Co. Cork, Ireland, [2]Teagasc Animal & Grassland Research and Innovation Centre, Athenry, Co. Galway, Ireland; jessica.coyne@teagasc.ie

Mathematical models have been extensively used in different species to summarise and predict live-weight. The objective of this study was to compare different mathematical models fitted to lamb live-weight data with particular emphasis on the ability to predict futuristic live-weight. A total of 6,594 live-weight records were collected on 556 lambs ranging from 0 to 205 days of age. Several growth functions and polynomial regressions models were fitted to the live-weight data of each animal individually. The R^2 and the root mean square error (RMSE) statistic were used to test goodness of fit. The accuracy of the Von Bertalanffy growth function and the random regression model to predict futuristic live-weight was determined by masking the live-weight data post 100 days of age in 25% of the animals. The Von Bertalanffy growth function with no random effect, the Von Bertalanffy with a random asymptotic weight, constant of integration or maturing rate, were compared to a linear random polynomial regression to predict the masked weights. The mean R^2 and RMSE values for the different models ranged from 0.90 to 0.99 and 1.39 kg to 11.04 kg, respectively. The Von Bertalanffy mixed model equation, with asymptotic mature weight included as a random effect had the strongest correlation between predicted live-weight and actual weights ($r=0.97$). The RMSE value for this model indicates 68% of predicted live-weight records were within ±1.53 kg of the actual weights. The correlation between actual and predicted weights for the first order random polynomial was 0.89. In conclusion, the Von Bertalanffy mixed model growth function with asymptotic mature weight included as a random effect had the highest accuracy when it came to predicting weights throughout all stages of growth. These results indicate that the prediction of live-weight is extremely promising in striving to produce a growth prediction tool for farmers.

Milk production and composition of Santa Ines ewes submitted to different levels of energy

E.L.A. Ribeiro, F.A.B. Castro, I.Y. Mizubuti, F. Fernandes Junior, B. Marson, F.A. Grandis, V.A.P. Guimarães and D.K.S. Tagliatella
Universidade Estadual de Londrina, CCA, Zootecnia, Caixa Postal 10.011, CEP 86057-970, Londrina, PR, Brazil; elar@uel.br

The objective of this work was to evaluate the influence of feeding energy levels during the last third of gestation and lactation on ewe's milk production and composition and on its lamb's performance. Fifty-five crossbred of Santa Ines ewes were used with 107 days of gestation, average live weight of 52.4 kg and average body condition score of 3.0 at the beginning of the experiment. Animals were confined in individual pens, distributed randomly in three diets: 2.0, 2.2 and 2.4 Mcal of ME/kg of DM. Rations were supplied *ad libitum*. Average daily milk production of 60 days of lactation was influenced ($P<0.05$) by diets, being 0.98, 1.32 e 1.80 liters for levels 2.0, 2.2 e 2.4, respectively. Energy levels did not exert any effect ($P>0.05$) on fat, protein, lactose or total solids in milk, being recorded means of 7.30, 4.66, 5.33 e 18.73%, respectively. However, the amounts of all milk nutritious compounds evaluated were increased linearly ($P<0.05$) by raising the energy level of the diet. Lamb weights at birth (3.77, 4.19 e 4.44 kg) and at weaning (11.57, 14.69 e 17.69 kg) were affected ($P<0.05$) by diets (2.0, 2.2 e 2.4, respectively). Lambs average daily gain from birth to weaning (0.125, 0.176 e 0.220 kg) was also affected ($P<0.05$) by diets (2.0, 2.2 e 2.4, respectively) and positively correlated ($r=0.67$; $P<0.01$) with the average milk production. Diet energy level did not influence ($P>0.05$) somatic cell count in milk and their mean value observed (258.20 × 1000/ml) can be considered low for sheep. Despite the fact that energy levels tested did not affect the percentage of milk compounds, lambs raised by females receiving more energetic feed showed higher growth rate due to the greater availability of nutrients provided by higher amounts of milk produced.

Session 28 Poster 37

Characterization of sheep and goats farming systems in Abu Dhabi, United Arab Emirates

S. Al-Shorepy[1], S. Sherif[1], A. Al-Juboori[2] and A. Al-Awad[1]

[1]United Arab Emirates University, College of food and Agriculture, P.O. Box 15551, Al Ain, United Arab Emirates, [2]Abu Dhabi Food Control Authority, P.O. Box 52150, Abu Dhabi, United Arab Emirates; salih.abdu@uaeu.ac.ae

Presently, sheep (1.4 million) and goat production (2.0 million) comprise 89% of the total livestock framings in the United Arab Emirates. This study aims to describe the sheep and goats framings in Abu Dhabi Emirate, with the goal of identifying areas where productivity could be improved. Six-hundred farms were surveyed in the three regions of Abu Dhabi Emirate (Al-Ain, Western Region, and Abu Dhabi). Data were collected using a structured direct questionnaire for in-depth interviews, including questions properly selected to obtain a general description of overall management practices. A multivariate analysis was used to determine the different farm characteristics. The number of clusters was decided using hierarchal cluster analysis, whereas, the farms were allocated in 3 clusters. The identified clusters were compared using one way ANOVA or a chi-squared test. The main differences were evident on flock size, breeds, facility and equipment availability, and application of preventive health programs. The average flock size for of clusters 1, 2 and 3 were 550, 250 and 100 head, respectively. 65% of the farms in cluster 1 practiced weaning on time compared with 37% in cluster 3. Almost all the farms in cluster 3 did not keep records. It can be cocnluded that although the numbers of sheep and goats have increased dramatically within the last 3 decades in United Arab Emirates, the management practices have not been improved accordingly.

Session 28 Poster 38

Development of a management information system for sheep

G. Kern, I. Traulsen and J. Krieter

Institute of Animal Breeding and Husbandry, Christian-Albrechts-University Kiel, Olshausenstraße 40, 24098 Kiel, Germany; gkern@tierzucht.uni-kiel.de

Within the project 'Development of preventive measures for the promotion of health and longevity in sheep on organic farms' a management information system (MIS) was developed. This MIS optimises the overview of the herd, the status of animal health and prolific performance because recorded information can be used for monitoring of the animals in an optimal way. For each sheep, individual information is available like pedigree, prolific performance or medicamental treatment. Additionally, continuous traceability of all ancestors is possible. This makes detecting genetic disorders easier for the farmer. In case of the performance, the MIS generates current management ratios automatically of fertility and performance traits of the flock or the individual ewe. For an easy overview the current situation is displayed in graphs. In addition, a comparison with the mean of all other farms is available. Sporadically recorded data like body condition can be entered as well. All records are checked for plausibility during the import process immediately. For a better detection of diseases, a decision-tree technique was implemented to support health-monitoring. Besides the functions related to animal health, the MIS provides an automatic documentation of changes in flock and medicamental treatment. Additionally, process sheets can be displayed. These functions relieve management of the flock. The on-line user interface of the MIS enables monitoring of herd performance at any time. The MIS can be used by any shepherd without charge after registration at sheep.tierzucht.uni-kiel.de. At the moment about 60 farmers with 6,000 sheep are registered.

Effect of feeding linseed or chia seed on fatty acid profile and lipogenic gene expression of lambs

O. Urrutia, A. Arana, J.A. Mendizabal, K. Insausti, B. Soret and A. Purroy
ETSIA. Universidad Pública de Navarra, Campus Arrosadia, 31006 Pamplona, Spain; olaia.urrutia@unavarra.es

The purpose of this work was to study the effect of the inclusion in the diet of linseed or chia seed, both rich in α-linolenic acid (ALA), on growth and carcass parameters, fatty acid composition and gene expression of key-lipogenic enzymes in subcutaneous (SC) and intramuscular (IM) adipose tissue. Thirty one Navarra breed lambs were distributed into three feeding groups: Control group (C, n=9) fed on barley and soya meal concentrate; Linseed group (L, n=11) receiving the same concentrate and 10% of whole linseed and Chia group (Chia, n=11) including 10% of chia seed. The fatty acid composition was determined and the expression of acetyl-CoA carboxylase (ACACA), lipoprotein lipase (LPL), stearoyl-CoA desaturase (SCD), fatty acid desaturase 1 (FADS1), fatty acid desaturase 2 (FADS2) and fatty acid elongase 5 (ELOVL5) were quantified. Lambs of the three groups had similar growth, carcass and fattening parameters. The inclusion in the diet of linseed or chia seed caused an increase in the ALA content in SC and IM adipose tissue (AT) ($P<0.001$). These diets also caused an inhibition in ACACA and SCD gene expression in IM AT ($P<0.01$) but not in SC AT. In the SC AT, FADS1, FADS2 and ELOVL5 genes expression were lower in L group than C group and only ELOVL5 mRNA was lower in Chia group than C group ($P<0.05$). In the IM AT, ALA significantly downregulated FADS2 gene in L and Chia and chia seed supplementation also downregulated FADS1 gene expression.

Reproductive characteristics and growth ability in lambs of Charollais breed in the Czech Republic

M. Momani Shaker, T. Alkhewani and S.A. Al-Olofi
The Czech University of Life Sciences in Prague, Faculty of Tropical AgriSciences, Department of Animal Science and Food Processing in Tropics and Subtropics, Kamycka 129, 165 21 Praha 6 – Suchdol, Czech Republic; alkhewani@its.czu.cz

The study was aimed at evaluation of selected reproduction characteristics in the sheep breed Charollais at pasture farming grazing. In addition the effect of ewes' age, sex of lambs and number of litters per live weight at birth and at the age of 30 and 70 days. Moreover, average daily live-weight gain has been observed. The trial comprised 55 ewes in total and they were distributed into two groups according to their age. Harem selection breeding was used for mating of ewes (from September 30, 2012 to October 30, 2012). In 85 born lambs live weight was determined at birth and in every 30 days. Based on the results obtained, selected reproduction characteristics were assessed (by the STATGRAPHIC program) and lambs' growth abilities were estimated as well (by the mathematico-statistical program – SAS). Pregnant lambing ewes received hay with an addition of grain by 0.20 kg per head/day more two weeks before the beginning of lambing. The lambs born were given hay and grain mix (*ad libitum*) until the common beginning of strip grazing on April 10, 2012. In the general assessment, the gradual increase in values of all reproduction indicators with rising age of ewes has been found out. The highest percentage of fertilization, percentage of fertility per ewe of basic stock and percentage of fertility per pregnant ewe was found in three-year old ewes (92.5; 170.2; 175.5). Age of ewes had a significant effect on live weight of lambs and weight gains at 30 days of age. In addition highly significant effect of sex on live weight at birth was found out ($P \leq 0.01$). Live weight at birth and at 30 and 70 days of age and average daily live-weight gain significant effect on the number of litters ($P \leq 0.01$).

Session 29 — Theatre 1

Impact of genetics and feeding on loin marbling levels of Canadian hogs

L. Maignel[1], J.P. Daigle[2], F. Fortin[2], S. Wyss[1] and B. Sullivan[1]
[1]Canadian Centre for Swine Improvement inc, Central Experimental Farm, Bldg #54, 960 Carling Avenue, Ottawa, ON, K1A0C6, Canada, [2]Centre de développement du porc du Québec, Place de la Cité, tour Belle Cour, 2590, boul. Laurier, bureau 450, Quebec City, QC, G1V 4M6, Canada; laurence@ccsi.ca

Intramuscular fat is known to be highly heritable but is also strongly influenced by feeding strategies. Since 2009, loin IMF predicted on live pigs using ultrasonic scanning has been included in the Canadian technician accreditation program and in the swine genetic evaluations for Duroc pigs. A research project involving 6,000 Duroc pigs scanned across Canada was designed to enlarge the live IMF database and confirm genetic parameters estimated in a previous study. Data collected in the project were also used for genetic evaluation and selection of Duroc boars with either low or high IMF EBVs to produce 1,350 commercial pigs tested in two research trials (in Western and Eastern Canada). Pigs were fed either a standard diet specific to the region, or a special diet formulated to increase marbling deposition. A total of 687 pigs were performance-tested from weaning to slaughter, and tracked at the packing plant for a full carcass and meat quality evaluation. Special feed increased loin IMF (+0.4 to +0.6%), backfat by 1 to 2 mm and decreased lean depth by 2 to 4 mm. It also decreased growth rate by 40 to 50 g/day. Using boars with high IMF EBVs allowed increasing loin IMF (by 0.3 to 0.5% in these trials) without adverse effects on growth and carcass quality. No effect on other meat quality traits (pH, drip loss or colour) was observed in sire lines or feeding programs. The project provided valuable information about the effect of genetics and management (especially feeding) on marbling level in pork loins. Moreover, Canadian pig producers now have new tools to customize marbling levels in their hogs using a combination of genetics and feeding.

Session 29 — Theatre 2

Effect of genetic type and dietary protein level on growth, carcass and ham traits of heavy pigs

L. Gallo, P. Carnier, A. Cecchinato, L. Carraro and S. Schiavon
University of Padova, DAFNAE, Viale Università 16, 35020 Legnaro, Italy; luigi.gallo@unipd.it

This study, supported by AGER (grant 2011-0280), investigated the effects of genetic type (GT) and dietary crude protein (CP) content on variation of growth performance, carcass and ham traits in finishing heavy pigs. Genetic types, selected according to their diffusion in commercial farms, involved Duroc × Large White ANAS (A), C21 Goland (G), Tempo × Topigs 40 (T) and DanBred (D) lines. A total of 184 pigs (gilts and barrows) were raised in 2 subsequent trials from 90 to 165 kg BW and were fed, under restricted feeding, conventional or low-protein diets providing, on average, 140 g CP and 6.45 g lysine/kg or 106 g CP and 4.85 g lysine/kg of diet, respectively. Pigs were housed in 8 pens equipped with feeding stations recording individual feed intake. Pigs were slaughtered at 9 months of age (average BW: 165±12.4 kg) and weights of carcass and typical cuts and backfat thickness were recorded for all carcasses. After 24-h chilling, hams were trimmed, weighed, scored and measured for several quality traits. A sample of trimmed fat was taken from hams to assess iodine number and linoleic acid content. All hams were cured according to the San Daniele procedure. Weight losses during curing and number of hams discarded were recorded. All traits were analysed by ANOVA. Genetic type affected variation of most growth and carcass traits: D pigs showed the highest ADG and gain to feed ratio and provided carcasses with the highest and lowest incidence of lean and fat cuts, respectively. Ham quality traits were mostly affected by GT, and hams from A and G had higher fat covering thickness, lower iodine number and a better overall evaluation than other genetic types. Incidence of hams not suitable for PDO production in D and T was twice as high as in A and G. Effects of diets were less important, but low dietary CP tended to increase ham subcutaneous fat covering and quality. No interaction effect between genetic type and diet was detected.

Session 29 Theatre 3

Within-litter variation in piglets' birth weight in reciprocal crosses

E. Sell-Kubiak[1], E.F. Knol[2], P. Bijma[1] and H.A. Mulder[1]
[1]Wageningen University, Animal Breeding and Genomics Centre, P.O. Box 338,6700 AH Wageningen, the Netherlands, [2]TOPIGS Research Center IPG B.V., P.O. Box 43, 6640 AA Beuningen, the Netherlands; ewa.sell-kubiak@wur.nl

The homogeneity of within-litter birth weight can increase piglet survival and ease of management. Therefore, decreasing variation in within-litter birth weight is of economic interest. The objective was to study genetic and non-genetic effects in within-litter standard deviation of birth weight (SD of BW) in two reciprocal crosses of pigs. Data on 3,891 litters came from TOPIGS commercial farms. Birth weight records were collected on 61,586 crossbred piglets (42,489 from AB-cross and 19,097 from BA-cross), which were offspring of 1,183 sows and 725 boars. Records included birth weight of 5,739 stillborn piglets. The pedigree had information over ~10 generations. SD of BW was analysed using linear dam model in ASReml 2.0. Fixed effects were: sow's parity, total number of piglets born in a litter and farm-year-season. Random components were: maternal genetic effect, permanent sow effect, and service sire effect. The maternal heritability for SD of BW was 0.11 in AB-cross and 0.14 in BA-cross. The permanent sow effects, as a proportion of the phenotypic variance, were rather small in both crosses: 0.0051 in AB and 0.0038 in BA-cross. The service sire effect was only estimable for AB-cross and small relative to the phenotypic variance: 0.009. The differences in heritability estimates are caused by higher maternal genetic variance in BA-cross than in AB-cross, whereas phenotypic variances being almost the same in both crosses. This result shows that A-line sows exhibit a larger genetic variance in SD of BW than B-line sows. The substantial maternal genetic variance for SD of BW shows that homogeneity of within-litter birth weight can be increased by selection. Subsequently, analyses of genetic heterogeneity of residual variance in BW and a GWAS will be performed to further unravel the genetics of homogeneity of within-litter birth weight.

Session 29 Theatre 4

Genomic selection in German Landrace

M. Gertz[1,2], J. Dodenhoff[3], C. Edel[3], K.-U. Götz[3] and G. Thaller[1]
[1]Institute of Animal Breeding and Husbandry, Christian-Albrechts-University, Institute of Animal Breeding, Olshausenstraße 40, 24098 Kiel, Germany, [2]Tierzuchtforschung e.V., Senator-Gerauer-Str. 23, 85586 Poing, Germany, [3]Bavarian State Research Centre for Agriculture, Institute of Animal Breeding, Prof.-Dürrwaechter-Platz 1, 85586 Poing, Germany; mgertz@tierzucht.uni-kiel.de

The objective of this study was to investigate the potential for an application of genomic selection in the breeding program for German Landrace in Bavaria. Samples for genotyping were available since birth year 1995 for boars, and since 2006 for sows. In total, 1,982 animals were genotyped with the Illumina PorcineSNP60 Bead-Chip. Traits analyzed were number of piglets born alive (NBA) and lean meat percentage (LMP). After editing of data, 1,868 pigs (314 boars, 1,554 sows) remained in the analysis. Year of birth was used as criterion to divide animals into a training data set (NBA: 1,529 animals; LMP: 947 animals) and a validation data set (NBA: 158 animals; LMP: 322 animals). Phenotypic records of validation animals were removed from genetic evaluation in order to create proper phenotypes for calibration. Direct genomic values (DGV) were calculated using the GBLUP approach with deregressed estimated breeding values (dEBV) as a response variable. Subsequently, genomic breeding values (GEBV) were calculated by combining direct genomic values with conventional parent averages (PA). For the validation set, squared correlations of dEBV with PA, DGV and GEBV were corrected for the reliability of dEBV and prior selection on PA to obtain 'realized genomic reliabilties'. For NBA, realized reliabilities were 0.38, 0.55 and 0.58 for PA, DGV and GEBV. For LMP, corresponding reliabilities were 0.41, 0.43 and 0.44, respectively. These results indicate that implementation of genomic selection can be beneficial for maternal traits. Especially for slaughter traits, realized reliabilities obtained from the current training data set may hardly justify the additional costs for genotyping.

Genetic parameters of number of piglets nursed

B. Nielsen, T. Ostersen, I. Velander and S.B. Bendtsen
Pig Research Centre, Breeding & Genetics, Axeltorv 3, 1609 Copenhagen V, Denmark; bni@lf.dk

In commercial pig production, breeding for increased litter size has been very successful the last decades. The increased litter size has entailed extensive use of cross-fostering and demands sows with increased nurse capacity. The aim of this study was to develop a new trait in the, which increases the genetic potential of the sows to nurse and wean more piglets. We need a trait that: (1) reflects the nurse capability of the sows; and (2) is heritable. An experiment was conducted in three large production herds, and at present time litters from more than 8000 crossbred gilts between Danish Landrace and Yorkshire are available. All matings were by AI using Duroc as terminal sire. At farrowing the total number of born (TNB) was recorded including the number of stillborn. During the first three days after farrowing piglets were exchanged between litters in order to obtain 14 piglets for each gilt within three days after farrowing. The exchanges between litters were kept as low as possible and many sows mostly had their own piglets. The following three weeks piglets could only be removed from the experimental gilts if the piglets were deemed too weak to stay. After three weeks the litter size were recorded as the number of piglet nursed (NN). Thereby NN ranged from zero to 14, and high values indicate high nurse capacity of the gilts, whereas low values indicate low nurse capacity. The results indicate that NN was heritable and had heritability about 0.07. The genetic correlation between NN and TNB was non-significant. In an additional study to consider a useable trait for selection of increased nurse capacity, we considered the number of nursed piglets until weaning in multiplier herds. Here with no limitations of litter equalization the heritability of this trait was also about 0.07, and the correlation to TNB was non-significant.

Session 29 Theatre 6

Estimation of variance components of sow longevity traits using discrete time model

M.-L. Sevón-Aimonen[1], R. Labouriau[2], R.P. Maia[2], S. Haltia[3] and P. Uimari[1]
[1]MTT Arifood Research Finland, Biotechnology and Food Research, Myllytie 1, 31600 Jokioinen, Finland, [2]Aarhus University, Center for Quantitative Genetics and Genomics, Blichers Allé 20, 8830 Tjele, Denmark, [3]Figen Ltd., P.O. Box 40, 01301 Vantaa, Finland; marja-liisa.sevon-aimonen@mtt.fi

Longevity of sows has a great impact on efficiency of piglet production but is also an animal welfare issue. Commonly used measures of longevity include stayability to certain age, parity or production level, length of productive life, total number of parities produced before culling, lifetime prolificacy, and pigs produced per day of life. Generally, the records of longevity are not available when selection decision of replacement AI-boars is taken place. However, genomic selection opens new possibilities to improve longevity of sows. The aim of the current study was to estimate variance components of sow longevity traits using a multivariate competing risk model with discrete time to describe the number of parities. Culling was treated as three traits according to culling reason: leg weakness (L), fertility problems (F), or other reason (O). The data was collected from the database of Finnish breeding company Figen Ltd and contained records from 31,742 Finnish Landrace (LR) and 31,252 Finnish Yorkshire (FY) sows that were born between 2000 and 2010. The data included both complete (the culling date was available) and censored observations and was analyzed with the DMU program package using a multi trait sire model. The fixed effects in the model were the parity number (baseline), year-season effect, size of the herd-year, age at first farrowing, and the litter size. Herd-year was treated as a permanent environmental effect and sire as a genetic random effect. Sire variances (SE) for L, F and O were 0.024 (0.003), 0.012 (0.002) and 0.010 (0.002) for LR and 0.012 (0.002), 0.015 (0.003) and 0.013 (0.003) for FY, respectively.

Session 29 Theatre 7

Does lameness in the sow influence her reproduction?

H.T. Le[1], N. Lundeheim[1] and E. Norberg[2]
[1]Swedish University of Agricultural Sciences, Department of Animal Breeding and Genetics, Gerda Nilssons väg 2, Ultuna, Box 7023, 75007 Uppsala, Sweden, [2]Aarhus University, Department of Molecular Biology and Genetics, Blichers Allé 20, P.O. Box 50, 8830 Tjele, Denmark; thu.le.hong@slu.se

Lameness is not only a welfare issue but also a major source of economic loss in pig production. Lame sows are assumed to suffer from pain and stress which might cause negative effects on her reproduction. Our aim was to estimate the genetic association between leg scoring (performed at performance testing, at approx. 100 kg live weight) and reproductive performance of the sows. Data analyzed consists of records on purebred Swedish Yorkshire pigs born and raised in nucleus herds. In total 119,345 pigs were tested for two lameness traits: movement score (MOV) and overall leg score (LEG). Both scores of MOV and LEG ranged from 1 (worst) to 3 (best) and were transformed using normal score transformation to obtain normality. Information on fertility in first parity (in total 13048 sows) included gestation length, number of total and alive born piglets, weaning to service interval (within 7 days after weaning; WSI7), and litter weight at 3 weeks of age. The DMU package using bi-variate and multitrait animal models was used to perform the genetic analyses. As expected, the two lameness traits (MOV and LEG) were highly correlated with a genetic correlation of 0.88±0.02. Due to relatively high standard error, correlation between lameness traits and reproductive traits in the first parity were not significantly different from 0. Lameness traits and WSI7 seem negatively correlated (r_g=-0.33±0.21), meaning that sows with worse score of lameness tend to have longer weaning to service interval. In order to have a better conclusion about genetic association between lameness of the sow and her reproduction performance, further studies, including information on later parities will be performed.

Session 29 Theatre 8

Accounting for variation among individual pigs in deterministic growth models

B. Vautier[1,2], N. Quiniou[1], J. Van Milgen[2] and L. Brossard[2]
[1]IFIP, BP 35104, 35651 Le Rheu cedex, France, [2]INRA-Agrocampus Ouest, UMR1348 PEGASE, 35590 Saint-Gilles, France; ludovic.brossard@rennes.inra.fr

Accounting for variation among pigs in deterministic growth models can be performed by repeating simulations using different realistic sets of parameters. However, model parameters are not independent and variation among them is therefore not random. The structure of relationships among parameters describing individuals has been considered only to a limited extent. In this study, the mean and covariance structure of model parameters describing daily feed intake (DFI) and growth were analyzed on data obtained from ten batches of growing pigs from different crossbreeds (dam: Large White × Landrace, sire: 10 different lines). Performance of group-housed crossbred gilts and barrows was tested from 70 days of age to approximately 110 kg body weight (BW). The DFI was recorded continuously and BW was measured at least every 3 weeks. Based on the observed DFI and BW, the InraPorc model was used to characterize each pig through a set of five model parameters, three of which described growth and two described *ad libitum* DFI. The data set included parameters for 1,288 pigs. Almost all parameters were affected by sex and batch, but not by crossbreed within batch. Covariance matrices computed according to batch, sex, crossbreed, or their combinations were all different as evaluated using the Flury hierarchy. The unit of covariance corresponded to the combination of batch, sex and crossbreed. As the variance structure is affected by batch, sex, and crossbreed, this information would be required for each situation, which is impossible from a practical point of view. The next-best solution is to use a generic covariance matrix structure based on the median matrix accounting for the size of subpopulations. This covariance matrix can be used, in combination with average parameters obtained on-farm to generate virtual populations of pigs and to evaluate different nutritional and management strategies on performance and variability of groups of pigs.

Repeatability of a behavioral play marker in piglets

W.M. Rauw
INIA, Departamento de Mejora Genética Animal, Crta. de la Coruña, km. 7.5, 28040 Madrid, Spain; rauw.wendy@inia.es

Behavioral markers in pigs, such as aggression, sociality, stereotypy, and fear, have been discussed by several authors. Assessment of play behaviour is a new and promising potential indicator of animal welfare. Evident emotions are associated with play such as joy and happiness. Animals play only if they are healthy, safe, well-fed and in a relaxed state, but not if they are under a stressful condition. Therefore, play behaviour can be used as an indicator of animal welfare. Repeatability of a play marker was investigated in 32 piglets (17 males and 15 females) from four litters at 37, 41, 44 and 48 days of age; animals had been weaned at 35 days of age. Piglets where released into a corridor measuring 1.1×5.3 m together with their litter mates for eight minutes. Since the animals had never been out of their pen before, the first test was considered as an adjustment period and was not included in the analysis. In the second to the fourth test, joyful brusque movements (jumping, turning and running) where recorded with a camera and number of movements and total time were estimated individually. In addition, body weight was recorded. Body weight was not significantly related to the number or time of joyful movements in any of the three tests. Since dominant pre-weaned piglets generally have higher growth rates than subordinates this may indicate that the joyful movements investigated in this study are relatively unrelated to their social position. Females had higher scores than males but this was significant for the number of movements in the third test only ($P<0.05$). The phenotypic correlations between total time in tests two to four, adjusted for the effect of litter and sex, were mostly positive but significant only between test three and four ($r=0.43$, $P<0.05$). The same was true for the total number of movements ($r=0.47$, $P<0.01$). Results indicate that the measurements were litter specific and day dependent ($P<0.05$). Further research will investigate the relationship between play markers and cortisol.

Individual behavioural pattern in pigs

K. Scheffler, I. Traulsen and J. Krieter
Christian-Albrechts-University Kiel, Institute of Animal Breeding and Husbandry, Olshausenstr.40, 24098 Kiel, Germany; kscheffler@tierzucht.uni-kiel.de

Stressful situations in pig life can affect the health and welfare as well as the production parameters. Dealing with such challenging conditions is called coping. To assess coping styles in pigs two behavioural tests (Backtest: BT, Human-Approach-Test: HAT) were performed in a single herd with the same animals at different age stages. The BT was performed twice with suckling piglets (n=1,382), the HAT twice with suckling piglets (n=1,318), four times with weaned pigs (n=1,317) and once with gilts (n=230). The data analysis was conducted with general linear models. The consistency across test situation and over time was calculated with Spearman-Rank-Correlation coefficients. Additionally, kappa coefficients as measurement of agreement between coping styles 'HR' (High reactive), 'LR' (low reactive) and 'D' (doubtful) were calculated for BT. Significant effects of batch, test day and birth weight were found for the analysis of number of escape attempts, duration of escape attempts and latency for escape attempts in BT ($P\leq0.05$). The correlation and kappa coefficients between first (rp: -0.43-0.73; κ: 0.38-0.49) and second BT (rp: -0.54-0.79; κ: 0.43-0.53) showed that the behaviour of the piglets was different in both tests. The effects of batch, test number and sex had significant influences on the trait of latency in the HAT ($P\leq0.05$). The correlation within the age stages showed moderate relations (rp: 0.20-0.52), between different ages (suckling piglets, weaned pigs, gilts) the correlations were small (rp: 0.01- 0.11). The BT and the HAT showed only a poor relation (rp: -0.07-0.17). Therefore both tests seemed to measure different behavioural pattern. The coping style is neither predictable with the BT nor with the HAT. In the remainder of the study the results of the behavioural tests will be linked with the agonistic behaviour of the pigs.

Session 29 Poster 11

Immunophenotype of LW pigs crossbred from parents selected for resistance to MPS and high immunity

T. Shimazu[1], Y. Katayama[1], L.J. Borjigin[1], R. Otsu[1], H. Kitazawa[1], H. Aso[1], K. Katoh[1], Y. Suda[2], A. Sakuma[3], M. Nakajoh[3] and K. Suzuki[1]
[1]Tohoku University, sendai, 981-8555, Japan, [2]Miyagi University, Sendai, 982-0215, Japan, [3]Miyagi Prefecture Animal Industry Experiment Station, oosaki, 989-6445, Japan; tomoyuki-shimazu@m.tohoku.ac.jp

Mycoplasma pneumonia of swine (MPS) results in considerable economic losses. We recently developed 2 potentially disease-resistant lines of pig by crossbreeding Landrace and Large White pigs (LW); the Landrace pigs were selected on the basis of reduced pulmonary MPS lesions and the Large White pigs were selected for the high immune activity in their peripheral blood. In this study, we compared the immunophenotypes and MPS scores of the crossbred LWs (hereafter selected-LW) and the non-immune-selected LWs (control-LW). Twelve castrated males each of selected-LW and control-LW pigs were used. Blood samples were collected at days -14, -7, 0, 2, 7 and 14, and MPS vaccine was injected intramuscularly twice on day -7 and day 0. The pigs were sacrificed on day14, and Peyer's patches, mesenteric lymph nodes, and hilar lymph nodes were collected. Blood samples were subjected to white blood cell (WBC) count and cell population analysis (B cells, T cells, etc.). Tissue samples were analyzed to determine the levels of cytokine expression. Statistical analyses were performed using the SAS MIXED or TTEST procedure. We did not find a significant difference in the WBC count and MPS score between the 2 lines. On the other hand, cell population analyses revealed some differences. For example, control-LW showed higher monocyte ratio at steady state and it decreased after vaccine sensitization, whereas the ratio did not change in selected-LW. In tissue sample analysis, we found that IL-10 mRNA expression was significantly decreased in all tissues of the selected-LW. Although MPS score was not different between lines, these results indicate that these 2 lines have different immunophenotypes. Further studies are required to understand the inheritance of the MPS-resistant phenotype.

Session 29 Poster 12

Expression of mRNA of MHC protein in l. dorsi muscle of the pig during pre and postnatal development

M. Katsumata[1], T. Yamaguchi[2], A. Ishida[1], A. Ashihara[1] and K. Nakashima[1]
[1]NARO Institute of Livestock and Grassland Science, 2 Ikeno-dai, Tsukuba, 305-0901, Japan, [2]Tokushima Agriculture, Forestry, and Fisheries Technology Support Center, Itano, Tokushima, 771-1310, Japan; masaya@affrc.go.jp

In order to manipulate the number and the type of myofiber in the pig, manipulation during fetal stage is required because myofiber formation occurs during fetal stage in the pig. However, our knowledge on the details of development of myosin heavy chain (MHC) protein isoforms during pre and postnatal development in the pig is limited. The aim of this study is to clarify the developmental changes in mRNA expression of the MHC protein isoforms during pre and postnatal development in the pig. We used in total 36 pigs in this study. Specimens of longissimus dorsi (l. dorsi) muscle were obtained from pigs of 90th day of fetal stage (90FS), 1, 12, 26, 45 and 75 days after birth. We measured the abundances of mRNAs of the type I, IIa, IIb, and IIx MHC proteins in the muscle with a real time RT-PCR method. The effects of growth stage were calculated with one-way ANOVA where the ages in days were the main factor. The abundances of mRNAs of all the isoforms measured changed during the development ($P<0.01$). The abundance of mRNA of the type I MHC protein on the day 26 increased to 9.8 times higher than that of 90FS and stayed stable thereafter. The abunadnce of mRNA of type IIa MHC protein was the highest on the day 1; it was 5.4 times higher than that of 90FS. It gradually declined thereafter and only 1.9 times higher on the day 75. The abundance of mRNA of type IIb MHC protein considerably up-regulated after birth; it reached to 412 times higher on the day 12 and to 704 times higher on the day 45 as compared with that of 90FS, respectively. The abundance of mRNA of type IIx MHC protein was the highest on day 12; it was 23 times higher than that of 90FS. The pattern of the developmental change in the expression was different among each isoform and the largest change was observed in type IIb MHC protein.

Session 29 — Poster 13

Heritability estimates evaluated as binary type scoring to linear scoring for leg weakness in swine

K. Ishii[1,2], K. Kadowaki[1] and M. Satoh[2]
[1]National Livestock Breeding Center, Fukushima, 9618511, Japan, [2]NARO Institute of Livestock and Grassland Science, Ibaraki, 3050901, Japan; kazishi@affrc.go.jp

Leg weakness is one of the most serious problems, because it is one of the most common reasons for involuntary culling of breeding pigs. In general, two types of scoring systems exist for leg weakness traits: (1) binary type of scoring, indicating whether the pig has a certain problem; and (2) a linear scoring system, where all pigs are evaluated for certain leg locomotions. In this study, Monte Carlo computer simulation was used to investigate the factors affecting heritability estimates in a closed strain herd of swine, when a linear scoring trait (LST) is evaluated as a binary scoring trait (BST). Records of LST in the range from 0 to 10 were generated by a computer with separate generations. Breeding values of LST for base population animals were sampled from normal distribution with mean of 0. Phenotypes of BST were made in the range from 0 to 5 by folding back the LST. A breeding herd of 20 sires and 100 dams in each generation was assumed; one selected sire was randomly mated to selected five dams. From each litter, 2 boars and 4 gilts were reared. Generation and sex constants were generated as a fixed effect. Selection of three schemes was either at random, on individual phenotype of BST, and on phenotype of BST within a litter. A hundred replications were simulated under various initial heritabilities (0.1, 0.3, 0.5) of LST, base population means (1, 3, 5) of LST, and the number of generations(3, 5, 9). The means of genetic parameter estimates were calculated for each condition of each replicate. Estimated heritabilities for LST were mostly the same as true values regardless of any simulated conditions. On the other hand, estimated heritabilities for BST were strongly affected by the base population means of LST. The influences of estimated heritabilities for BST owing to the selection methods and the number of generations were depended on the initial heritabilities for LST.

Session 29 — Poster 14

Relationship between birth order and birth weight of the pig

R. Charneca[1], A. Freitas[1], J.L.T. Nunes[1] and J. Le Dividich[2]
[1]Instituto de Ciências Agrárias e Ambientais Mediterrânicas, Universidade de Évora, Apartado 94, 7022-554 Évora, Portugal, [2]INRA, 32, Avenue Kennedy, 35160 Breteil, France; rmcc@uevora.pt

The objective of this study was to determine whether birth weight of the pig is related to its birth order. The study involved 292 sows from 2 genotypes (Large White × Landrace crossbred (LL), n=247 and Alentejano (AL), n=45) of mixed parity and their piglets. Most sows farrowed naturally. Each piglet was identified, weighed (±1 g) (mummies excepted) and its birth order (BO) recorded within 2 min of birth. A total of 3418 LL and 375 AL piglets were born of which 43 and 7 were mummified, and 205 and 6 were stillborn, respectively. Number of total born (TB) and born alive piglets (BA) per litter ranged from 6 to 23 (mean, 13.85±0.19 (se)) and from 6 to 22 (mean, 12.84±0.17), and from 4 to 13 (mean, 8.33±0.31) and from 4 to 12 (mean, 8.04±0.31) in LL and AL litters, respectively. Within-litter regression was used to determine the relationship between BO and birth weight (PASW Statistics, version 18.0, 2009). To compare litters of different sizes, BO was expressed as relative BO (RBO) calculated as RBO = (BO-1) / (TB-1). The slopes of the regression lines relating birth weight of TB or BA piglets to RBO were positive and similar in both genotypes ($P>0.50$). The common slopes of the regression line relating birth weight of TB and BA piglets to RBO were: bTB=70±14 (g) RBO ($P<0.0001$, $R^2=0.007$) and bBA=81±14 (g) RBO ($P<0.0001$; $R^2=0.009$), respectively. Results indicated that birth weight of BA piglets increases by 0.81 g per one percent increase in RBO. It was concluded that RBO explained a small percentage of the total variability found in piglet birth weight.

Effect of HoxA11 gene polymorphisms on litter size in pigs

A. Mucha, K. Piórkowska and K. Ropka-Molik
National Research Institute of Animal Production, Sarego 2, 31-047 Krakow, Poland; aurelia.mucha@izoo.krakow.pl

The appropriate course of pregnancy is associated with the proper preparation of the uterus for implantation, as well as the development of the embryo. The HoxA genes control the development of different parts of the embryo along the anterior-posterior (head-tail) axis. The change in mRNA expression and protein levels of HoxA proteins affects uterine receptivity. There is a few research about the importance of HoxA11 gene in pigs breeding. Therefore, it is interesting to identify mutations in porcine HoxA11 gene and examining the effect of new polymorphisms on the level of selected breeding traits. In locus Hoxa11, we identified in 3'UTR region two substitution T3746G and A3786G (dbsnp ncbi 678251287 and 678251288). Analysis was performed on 480 sows from two Polish breeds: Large White and Landrace pigs. Investigated polymorphisms were identified by PCR-RFLP method using endonucleases: T3746G – BstAPI and A3786G – HpaII. After restriction digestion, the PCR products were separated on 4,5% agarose gel. Reproductive traits investigated were: number of piglets born alive (NBA) and the number of alive piglets in 21st day (N21) in parity 1 and parities 2 to 4. In all litters, homozygotes TT (T3746G) and AA (A3786G) had the highest values of the NBA and N21 parameters. The lowest NBA and N21 values were found for the GG genotype (T3746G) (except NBA in 2-4 litters) and for AG genotype (A3786G) (except N21 in 2-4 litters). Statistical analysis showed significant differences only between the NBA or N21 parameters in first litter for AA (A3786G) and AG, GG pigs ($P \leq 0,05$). The results obtained suggest that this researches should be continued on other pig populations or lines. This study was supported by the National Research Institute of Animal Production, statutory activity no. 01-4.06.1.

Towards improved boar fertility by genetic characterization and detection of traits important in spe

M. Van Son[1], A.H. Gaustad[1], I. Andersen-Ranberg[1], F. Myromslien[2] and E. Grindflek[1]
[1]Norsvin, P.B. 504, 2304 Hamar, Norway, [2]Hedmark University College, P.B. 400, 2418 Elverum, Norway; eli.grindflek@umb.no

The aim of this study is to establish key boar fertility parameters, study their underlying genetics, and to use this knowledge in practical breeding. AI sires are currently selected based on performance traits (EBVs) and only visual inspection of semen quality. Objective methods have therefore been developed to assess semen quality, including flow cytometry assays. Flow cytometry is too laborious and expensive for incorporation in breeding practices and it is therefore desirable to find genetic markers explaining the phenotypes. DNA Fragmentations Index (DFI%), in stored semen is found to be negatively correlated to number of piglets born alive in Norsvin Landrace and Duroc breeds. Boars with extreme levels of this parameter are therefore further investigated. So far, a total 330 of semen boars have been analyzed for % DFI and the testicle samples from extreme high and low Norsvin Landrace boars were selected for transcriptome sequencing. The sequences were aligned to version 10.2.68 of the published pig genome using the software TopHat, and approximately 75% of the sequences were successfully aligned. The software HTSeq was used to register the number of sequences aligned to every known transcript of the pig genome. The R package edgeR was used to test for gene expression differences between animals with high DFI and low DFI. The reads were aligned to a total of 19.334 different transcripts and 110 of the genes were found to be significantly differentially expressed (5% FDR) between samples with high/low % DFI. Several of the differentially expressed genes have previously been shown to be involved in traits such as fertility and semen quality in pigs or other species. The next step is to identify SNPs in differentially expressed genes to find putative genetic markers.

Session 29 Poster 17

Correlation of CTSD and CTSF genes expression and economically important traits in pigs in Poland

A. Bereta[1], K. Ropka-Molik[2], M. Różycki[1], K. Piórkowska[2], M. Tyra[1] and R. Eckert[1]
[1]National Research Institute of Animal Production, Department of Animal Genetics and Breeding, Sarego 2, 31-047 Krakow, Poland, [2]National Research Institute of Animal Production, Laboratory of Genomics, Sarego 2, 31-047 Krakow, Poland; anna.bereta@izoo.krakow.pl

The aspatic proteinase CTSD plays a role in the proteolysis of intra- and extracellular proteins, while cysteine protease cathepsin F is involved in proteolytic processing closely related with the remodeling of the intimal extracellular matrix. The aim of our study was to evaluation of expression profile of CTSD and CTSF genes in porcine skeletal muscles (m. semimembranosus and longissimus dorsi). Furthermore, we analyzed association of expression levels and economically important traits in pig. The analysis was performed on 164 animals belonged to four pig breed: Polish Landrace (n=50) and Polish Large White (n=50), Pietrain (n=16) and Puławska (n=48). Relative quantification of mRNA abundance was performed with 7500 Real-Time PCR System using TaqMan® MGB probes labeled with FAM, VIC or NED and with endogenous controls OAZ1 and RPL27. In both analyzed tissue, the lowest CTSD gene expression was obtained for Polish Landrace and Puławska breeds ($P<0.05$). The differences in CTSF expression between pig breeds were similar: the lowest mRNA level was observed in Puławska breed and the highest in Pietrain (statistically significant in longissimus dorsi, $P<0.05$). We obtained low correlation coefficient between expression of CTSF gene in m. semimembranosus and ham weight (0.24, $P=0.008$). Also, significant correlation we obtained for CTSD transcript level in m. semimembranosus and ham weight, average backfat thickness (cm), loin eye area (cm^2) and lean meat percentage (%), however correlation coefficients were low (0.15; 0.21; 0.15; and 0.19; respectively, $P<0.05$). The study was supported by the Polish Ministry of Science and Higher Education (project no. NN311349139).

Session 29 Poster 18

Longissimus and semimembranosus muscles transcriptome comparison in pig displays marked differences

F. Hérault[1], A. Vincent[1], O. Dameron[2], P. Le Roy[1], P. Cherel[3] and M. Damon[1]
[1]INRA, UMR1348 PEGASE, 35590 Saint-Gilles, France, [2]INSERM, UMR936, 35065 Rennes, France, [3]Hendrix-Genetics, RTC, 45808 Saint Jean de Brayes, France; frederic.herault@rennes.inra.fr

Longissimus lumborum (LM) and semimembranosus (SM) are used for different meat consumption. Both are classified as glycolytic muscles but have different myofiber composition and metabolic properties. Compare LM and SM transcriptome profiles may clarify the biological events which could explain their phenotypic differences. The 90 pigs used in this study were produced as an inter-cross between 2 commercial sire lines. Muscle samples were collected 20 minutes post-mortem, snap frozen and used for total RNA isolation. Transcriptome analysis was undertaken using a pig muscle microarray: the 15K Genmascqchip. Analyses were performed using R software. Raw data were submitted to quality filtration and normalization. Probes with the smallest expression variability were filtered out. Normalized data were analyzed using a linear model of variance taking into account fixed effects of slaughter date, sex, sire and muscle. Carcass weight was used as a covariate. Genes which were differentially expressed between muscles were clustered according to their semantic similarities. Semantic similarities were computed according to Wang's method using Gene Ontology (GO) Biological Process (BP) terms. Thus, functional characterizations of genes clusters were performed with WebGestalt using GO BP terms. A total of 3,867 genes were differentially expressed between the 2 muscles, out of which 1,729 and 2,138 were over-represented respectively in LM and in SM. A set of 1,047 differentially expressed genes with a muscle fold change ratio above 1.5 was used for functional characterization. Five clusters related to energy metabolism, cell cycle, gene expression, anatomical structure development and signal transduction/immune response were identified. These results shed light on differential transcriptome profiles between LM and SM. This variability could affect muscle development and hence meat quality.

Androstenone-levels in entire male pigs: a GWAS for direct and indirect genetic effects

N. Duijvesteijn[1,2], E.F. Knol[1] and P. Bijma[2]
[1]TOPIGS Research Center IPG B.V., P.O. Box 43, 6640 AA Beuningen, the Netherlands, [2]Wageningen University, Animal Breeding and Genomics Centre, P.O. Box 338, 6700 AH Wageningen, the Netherlands; piter.bijma@wur.nl

Androstenone is one of the compounds causing boar taint and is heritable (~0.6). Recently, indirect genetic effects (IGEs, also referred to as associative effects) were found for androstenone, meaning that other pen mates (only boars) also genetically influence the level of androstenone of a given pen mate. Similar to estimating variance components for a direct-indirect animal model, SNP effects for the DGE (direct genetic effect) and IGEs can be estimated for androstenone. With this new approach, this study aims to detect significant SNP associations for androstenone. The dataset consisted of 1,282 boars (993 boars genotyped) from 184 groups of pen mates. After quality control, 46,421 SNPs were included in the model. Two models were fitted: the first model included the direct effect of the SNP of the individual and the second model included the indirect effect of the SNP effects of its pen mates. None of the SNPs (direct or indirect) were found to be genome-wide significant, but two QTL on SSC6 and SSC13 were detected chromosome-wide significant for the direct SNPs. A single association on SSC9 and a QTL on SSC14 were found for the indirect SNPs. A backwards elimination method and haplotypes were used to determine the variance explained by the SNPs. Using both methods, the QTL on SSC6 explained 2.5% and 2.1% of the phenotypic variance, and the QTL on SSC13 explained 6.4% and 3% of the phenotypic variance respectively. Using both methods, the QTL on SSC14 explained 3% and 2% of the phenotypic variance respectively. All QTL found explain a considerable part of the phenotypic variance which together would account for ~14% of the phenotypic variance. Besides the newly discovered QTL, this study also shows a new methodology to model SNPs for indirect genetic effects.

Growth performance and risk of boar taint for non-castrated Walloon pigs

S. Dufourny, V. Servais, J.-M. Romnee, V. Planchon and J. Wavreille
Walloon agricultural Research Centre, rue de Liroux 9, 5030 Gembloux, Belgium; s.dufourny@cra.wallonie.be

The aim of the study was to determine the effect of the non-castration on growth performance, carcass quality and risk of boar taint. The study achieved results in particular fattening conditions (Belgian genetic lines, *ad libitum* feeding, straw bedding). Entire males (EM), as well as vaccinated males (VM), were included in the experiment in comparison with castrated males (CM) and females (F). Five series of pigs (363 crossbred animals Landrace K+ × Piétrain) were fattened by group of 8 per treatment (EM, VM, CM or F) and per pen (8 m^2), to reach an average body weight of 120 kg. There were all slaughtered at the same time. Samples of fat from all treatments were collected and stored at -24 °C. They were smelled and scored (scale from 0 to 100) for boar taint by a jury of 3 experts. The results were statistically analyzed using a general linear model (1 fixed and 1 random effect factors). Body weight and average daily gain of EM (not different from F) were lower than those of VM (not different from CM): -6 kg body weight, -58 g/d. Daily feed intake of EM (not different from F) was lower than VM (-0.2 kg/d) and CM (-0.3 kg/d). Feed conversion ratio was better for EM, VM and F than for CM (-0.2 kg/d). A lower carcass yield was obtained for EM and VM in comparison with CM and F (-2.1%). Lean meat percentage of EM and F was higher than VM and CM (+2.9%). Concerning boar taint, a moderate odor (score between 33.6/100 and 67/100) was detected in 15% of the EM fat samples. An average risk of 4.6% moderate odor inacceptable for human consumption is expected. No carcass with strong boar taint (average odor >67/100) was identified by the jury.

Session 29 Poster 21

Comparison of texture parameters in loin and ham of Polish Landrace, Duroc and Pietrain pigs

G. Żak[1], M. Pieszka[1] and W. Migdał[2]
[1]National Research Institute of Animal Production, Sarego 2, 31-047 Krakow, Poland, [2]University of Agriculture, Balicka 122, 31-149 Krakow, Poland; grzegorz.zak@izoo.krakow.pl

The level of hardness, cohesiveness, springiness, resilience, chewiness, and shear force was analysed in m.l.d. and m. semimembranosus of 3 pig breeds: Polish Landrace, Duroc and Pietrain. Springiness of m.l.d. ranged between 0.4990 cm (Duroc) and 0.5220 cm (Pietrain). Differences between Pietrain and the other breeds were statistically no significant. No significant differences were found between the analysed muscles. Cohesiveness measurements ranged from 0.410 for Duroc to 0.444 for Polish Landrace. The differences between the breeds were statistically significant. The differences between m.l.d. and m. semimembranosus were highly significant. Resilience was almost identical in the meat from all three breeds (around 0.15 cm), although significant differences were found between m.l.d. and m. semimembranosus. Analysis of hardness and chewiness showed that the analysed samples of meat from Pietrain pigs differ highly significantly from the samples of meat obtained from Polish Landrace and Duroc pigs. The meat of Pietrain pigs was the hardest (131.43 N) and showed the highest chewiness (30.01 N). Shear force was highest for the meat from Polish Landrace pigs (7.73 kg/cm^2), and the lowest for the meat from Pietrain pigs (5.88 kg/cm^2). Statistically significant differences were found in muscle shear force values between the breeds and between the muscle types studied. It is concluded from this preliminary study that the meat of Pietrain pigs has slightly different, less favourable physical parameters compared to the meat of Polish Landrace and Duroc pigs. Supported by the State Committee for Scientific Research within project No N N311 1636 37.

Session 29 Poster 22

The production volume and economic efficiency in the piggery using hybrid hyperprolific sows

M. Sviben
Freelance consultant, Siget 22B, 10020 Zagreb, Croatia; marijan.sviben@zg.t-com.hr

The pig farm, having 8 farrowing rooms each equipped with 74 pens supposed to be occupied 56 days a shift, was built in Ihan, Slovenia, in 1962, when the production volume of 30,864 weaned piglets a year for the first so large pig unit in the world was considered as probable at the Veterinary Faculty of Zagreb, Croatia. As much as 99.59% of foretold value was registered at the Ihan farm on an average for 1963 and 1964. Twenty years later the review on hyotechnological estimations of the production volume, the proposals of intensification and the achievements in the pig industry of Yugoslavia ended with the conclusion that the farrowing room could be justly occupied 30 days a shift. The index of economic efficiency could be 97.33 weaned piglets/pen/year. In 2007, since achievements after the decades of the breeding work had been reviewed, the mean of 11.58 instead of 8 weaned piglets per litter was considered as probable to be attained at a large piggery or at a small one. Then at the farm having 6 farrowing rooms each equipped with 24 pens in Bratina, Croatia, hybrid hyperprolific sows became exploited. The production volume of 20,288 weaned piglets a year and the index of economic efficiency of 140.9 weaned piglets/pen/year could be considered as probable. Average annual production volume of 13,632 weaned piglets and the index of economic efficiency of 94.67 weaned piglets/pen/year were registered by Croatian Agricultural Agency at the piggery in Bratina for 5 years (2007-2011), since the farrowing rooms were occupied on an average 44.52 days a shift or 48.40% longer than 30 days. That resulted with 32.81% less production volume and economic efficiency than they could be expected, although the means of 11.54 weaned piglets per litter and 27.57 weaned piglets/sow/year were attained.

Session 29 Poster 23

Handling of supernumerary piglets in Flanders, a survey among pig breeders

J. Michiels[1], H. Vandenberghe[1], S. Van Gansbeke[2], J. Degroote[1], D. Fremaut[1] and B. Ampe[3]
[1]University College Ghent, Valentin Vaerwyckweg 1, 9000 Melle, Belgium, [2]Department of Agriculture and Fisheries, Burg. van Gansberghelaan 115A, 9820 Merelbeke, Belgium, [3]Institute for Agricultural and Fisheries Research, Scheldeweg 68, 9090 Melle, Belgium; jeroen.degroote@hogent.be

In previous decades, the number of piglets born alive per litter increased tremendously due to both genetic and management improvements. Supernumerary piglets refers to those piglets that exceed the number of available functional teats of the sow. Pig breeders need to implement management practices to cope with these challenges in order to assure a good technical performance and the well being of pig and sow. A questionnaire was sent to Flemish pig breeders and 217 surveys were returned correct and complete (response rate: 70%, survey sample size: 9%). The survey examined the application of management practices and looked at correlations with farm management and production characteristics. The most common practice was cross-fostering, which was applied on 95% of the farms. Cross-fostering is commonly performed before 48 h of age of the piglet and corresponds to 12±10.0% of the piglet population. Providing supplementary feed to piglets during lactation occurred on most farms as well (85%). More than half of the farmers (55%) performed euthanasia of disadvantaged piglets shortly after birth, though it accounted for only 1.6% of the piglets on these farms. The use of foster sows (44%), very early-weaning (<21 days of age) (31%) and alternating suckling (19%) occurred less often. Most farms used several practices simultaneously, with euthanasia + cross-fostering + supplementary feeding being the most popular combination (on 17% of the farms). In general, these practices were more frequently applied on larger farms, on farms with a higher number of weaned pigs per sow, on farms with batch farrowing systems and on farms with sows acquired externally. These practices were more often implemented by younger and higher educated pig breeders.

Session 29 Poster 24

Motherless rearing of piglets: effects on small intestinal morphology and digestion capacity

M. De Vos[1], V. Huygelen[1], S. Willemen[1], C. Casteleyn[1], S. Van Cruchten[1], J. Michiels[2] and C. Van Ginneken[1]
[1]Antwerp University, Department of Veterinary Sciences, Universiteitsplein 1, 2610 Wilrijk, Belgium, [2]University College Ghent, Faculty of Applied Biosciences, Valentin Vaerwyckweg 1, 9000 Gent, Belgium; maartje.devos@ua.ac.be

The use of hyper-prolific sows leads to an increasing number of supernumerary and underprivileged (e.g. low birth weight (LBW)) piglets. The effects of motherless rearing on the growth, small intestinal morphology and digestion capacity of these piglets remain unclear. Therefore, the aim of our study was to assess the effect of sow feeding versus artificial rearing on piglets' growth and function of the small intestine. For this purpose, pairs of LBW and normal birth weight (NBW) piglets (n=40) were allocated to four treatment groups. Groups 1 and 2 contained piglets that suckled naturally until either d 10 or 28 of age, respectively. Groups 3 and 4 contained animals that suckled naturally until 3 d of age and were then formula-fed until either d 10 or 28. Data were analyzed using a mixed model and all statistical calculations were performed in the software package R (version 2.13.1), in which P-values below 0.05 were considered significant. During d 3 to 10, formula-fed piglets showed reduced ADG and reduced lactase activities (P=0.01) compared to suckling piglets (P<0.001). In contrast, animals that were formula-fed until d 28 had a comparable ADG compared to sow-fed pigs. In addition, formula-fed piglets had a greater absorptive area (P=0.006), deeper crypts (P<0.03) and greater maltase and sucrase activities (P<0.045) compared to suckling piglets. In general, the differences in small intestinal morphology and digestion capacity between LBW and NBW piglets were scarce. These results suggest that the combination of *ad libitum* access to formulated milk and an increased capacity to absorb nutrients makes artificial rearing a good alternative to raise supernumerary and/or LBW piglets.

A starch binding agent decreases the rate of rumen fermentation of wheat in a dose-dependent manner

F.R. Dunshea, S.A. Pate, V.M. Russo and B.J. Leury
The University of Melbourne, Agriculture and Food Systems, Royal Parade, Parkville 3010, Australia; fdunshea@unimelb.edu.au

Rapid rumen fermentation of starch in wheat can result in sub acute ruminal acidosis (SARA) with resultant inhibition of rumen function. A starch binding agent (Bioprotect™, RealisticAgri, Ironbridge, UK) decreases the rate of *in vitro* gas production from wheat and the aim of this study was to determine the optimal dose. The active ingredient in Bioprotect™ is a stable non-volatile organic salt that complexes with the hydroxyl groups of starch at neutral or slightly acidic conditions, as observed in the rumen. These complexes decompose under more acidic conditions as in the abomasum and duodenum, making the starch available for enzymatic digestion. Soft and hard wheat and maize were ground and passed through a 1 mm sieve. The ground wheat was treated with various doses of Bioprotect™ (0, 4, 8 and 16 ml/kg) while maize was untreated. Samples (1.0 g) of the treated grain (n=16) were added to flasks containing buffered rumen fluid obtained from lactating dairy cows. The flasks were purged with CO_2 and maintained at 39 °C. Gas production was monitored every 5 min for 48 h using the ANKOMTM wireless *in vitro* gas production system. Gas production was modeled using a Gompetz equation to determine maximum amount of gas production (Rmax) and the rate constant (β). There was no effect of dose (P=0.21) or grain type (P=0.32) on Rmax. On the other hand, β was decreased (P<0.001) with increasing dose particularly in the soft wheat as indicated by an interaction (P<0.001) between wheat type and dose. The β was lower (P<0.001) for maize than for wheat even at the highest dose of Bioprotect™ (0.17 vs. 0.22 per min). These data demonstrate that wheat is fermented faster than maize and that Bioprotect™ can slow the fermentation of wheat, especially soft wheat. The optimal dose of Bioprotect™ appears to be 8 ml/kg. If more starch can pass through the rumen without being fermented it may reduce the incidence of SARA and allow for greater post-ruminal enzymatic digestion of starch.

Impact of concentrate supplementation on two Holstein cow strains in a pasture-based feeding system

C. Heublein[1,2], F. Dohme-Meier[2], K.-H. Südekum[1] and F. Schori[2]
[1]University of Bonn, Institute of Animal Science, Endenicher Allee 15, 53115 Bonn, Germany, [2]Agroscope Liebefeld-Posieux, Route de la Tioleyre 4, 1725 Posieux, Switzerland; carolin.heublein@alp.admin.ch

The aim of the present study was to examine the impact of concentrate supplementation on grazing behaviour, metabolic variables and milk production of two Holstein strains under organic farming conditions. In a crossover study, 12 Swiss Holstein cows (HCH) and 12 Holstein cows of New Zealand origin (HNZ) were used. They differed in body weight (HCH, 597 kg vs. HNZ, 554 kg, P<0.05) and were on average 91±18 d in milk. All cows grazed full time and were supplemented either with 6 kg/d of a commercial cereal grain mix or received no supplements. After an adaptation period of 21 d, feed intake was estimated using the n-alkane double indicator technique during the 7 d collection periods. Grazing behaviour was recorded using jaw movement recorders during 72 h. Milk yield was recorded daily and milk ingredients were analysed three times per collection period. Blood samples were taken from the jugular vein on two consecutive days at 07:00 and 14:00, respectively. The statistical analysis was carried out with a mixed model analysis using SYSTAT-13 and R. Supplemented cows had a lower (P<0.001) herbage intake than cows without supplementation, but total DM was higher (P<0.001) for supplemented cows (14.7 vs. 12.1 kg DM/d). Grazing time was reduced (P<0.001) for supplemented cows (448 vs. 558 min/d). Cow genetics (HCH vs. HNZ) had no effect on eating time, but HCH spent less time ruminating than HNZ (389 vs. 414 min/d, P<0.05). Supplementation caused a higher milk yield (30.0 vs. 24.9 kg/d, P<0.001) and HCH produced more (P<0.05) milk with supplementation than HNZ. Milk fat was lower (P<0.001) for supplemented cows (3.2%) than unsupplemented cows (3.8%). Results of blood analysis indicate that supplementation influenced concentration of glucose, urea, β-hydroxybutyrate, and non-esterified fatty acids.

Session 30 Theatre 3

Saccharomyces cerevisiae can alleviate the impact of subacute ruminal acidosis in dairy cattle

O. Alzahal[1], N. Walker[2] and B.W. McBride[1]
[1]University of Guelph, Guelph, N1G 2W1, Canada, [2]AB Vista, Marlborough, SN8 4AN, United Kingdom; oalzahal@uoguelph.ca

The objective of the study was to determine the effect of Saccharomyces cerevisiae supplementation on DMI, milk yield, milk components, and ruminal pH during a dietary regimen that leads to subacute ruminal acidosis (SARA). Sixteen multiparous, rumen-cannulated lactating Holstein cows were randomly assigned to 1 of 2 dietary treatments that included S. cerevisiae (4 g/d, AB Vista, Marlborough, UK) or placebo. During the first 6 weeks, all cows received a high-forage diet (HF; 77:23, F:C; NFC=35). Cows were then switched abruptly during week 7 to a high-grain diet (HG; 50:50, F:C; NFC=48) and remained on the HG until week 10. Feed intake and milk yields were recorded daily. Milk samples were collected 3 times per week and pooled by week. Ruminal pH was recorded continuously using an indwelling system during weeks 6, 7 and 10. Data were summarized by week and analyzed using Proc Mixed of SAS using treatment, week, and their interaction as fixed effects. The model accounted for repeated measurements. Orthogonal contrasts were utilized to compare treatment means within each week or dietary phase (HF vs. HG). Cows were considered to have SARA if the duration below pH 5.6 exceeded 300 min/d. Ruminal pH during week 6 was not different across treatments (15 ± 46 min/d <5.6, $P\geq0.05$). The dietary regimen successfully induced SARA during week 7, and ruminal pH (551 ± 46 min/d<5.6) was not different ($P\geq0.05$) across treatments. However, cows receiving S. cerevisiae treatment had an improved ruminal pH (122 ± 57 vs. 321 ± 53 min/d <5.6, $P<0.05$) compared with placebo during week 10. Additionally, cows receiving S. cerevisiae had higher ($P<0.05$) DMI (23.3 ± 0.66 vs. 21.6 ± 0.61 kg/d) and higher ($P<0.05$) 4%-fat-corrected milk yield (29.6 ± 1.2 vs. 26.5 ± 1.2 kg/d) compared with placebo during HG phase. In conclusion, S. cerevisiae can be used to minimize the occurrence of SARA caused by sudden dietary changes and consequently maintain the health and productivity of dairy cattle.

Session 30 Theatre 4

The effect of feeding level on goats milk and plasma fatty acids profile

E. Tsiplakou and G. Zervas
Agricultural University of Athens, Nutritional Physiology and Feeding, Iera Odos 75, Athens, 11855, Greece; eltsiplakou@aua.gr

Small ruminants usually face under/over-feeding situations under certain circumstances. The effect of long term under/over-feeding in ruminants has focused on milk chemical composition but not on milk fatty acids (FA) profile. Thus, the objective of this study was to determine the effects of long-term under/over-feeding on goats milk chemical composition and FA profile and on blood plasma insulin and leptin concentrations. Twenty-four dairy goats were used for the experiment. Three months post partum the goats were divided into three homogenous sub-groups (n=8). Each group fed the same ration, but in quantities which covered 70% (underfeeding), 100% (control) and 130% (overfeeding) of their energy and crude protein requirements. The data were analysed using a general linear model (GLM) for repeated measures analysis of variance (ANOVA). The results showed that the milk of underfed goats had significantly higher fat content compared to overfed, while the lactose content was significantly lower compared to that of control and overfed goats. The underfeeding reduced significantly the concentrations of C16:0 and long chain FA (LCFA) and increased that of C18:0 and the polyunsaturated FA (PUFA) in goats milk fat compared to the controls. The overfeeding increased significantly the cis-9, trans 11 C18:2 (CLA) milk fat content compared to the control, and the medium chain FA (MCFA) compared to underfeeding. The concentrations of non esterified FA (NEFA) and C18:1 in blood plasma was significantly higher in underfed, compared to overfed, goats while the opposite was observed for the C18:2n6c and C18:3n6 FA concentrations. No differences were observed in blood plasma leptin concentration between the dietary treatments, while the concentration of insulin was significantly higher in overfed compared to underfed goats. In conclusion, the long term under/over-feeding affected the goats milk chemical composition, milk and blood plasma FA profile and the insulin blood plasma concentration.

Intake rate evolution after a change in concentrate percentage in mid-lactation goats

S. Giger-Reverdin, A. Serment, C. Duvaux-Ponter and D. Sauvant
INRA-AgroParisTech, UMR791 MoSAR, 75005 Paris, France; sylvie.giger-reverdin@agroparistech.fr

High milk production is generally achieved by increasing the dietary concentrate percentage but can induce rumen subacidosis. Nevertheless, with the same diet, all animals do not suffer from acidosis, possibly because of different feeding behaviors. We studied the evolution of intake rate in mid-lactation goats by either increasing or decreasing the concentrate percentage. Twelve goats in individual crates with automatic measurement of the quantity of feed eaten every 2 min were fed *ad libitum* a control diet (52.5% of concentrate on a dry matter basis). After a transition week (5 days), they received *ad libitum* either a high-concentrate diet (70%) or a low one (35%) for two weeks. An index of intake rate was defined by Giger-Reverdin *et al.* as the proportion of dry matter eaten 90 min after the afternoon feeding which corresponded to two thirds of the daily feed allowance (P90). P90 of the control diet was averaged on 7 days per goat and varied from 0.43 to 0.83. For five out of the six goats of the L diet group, P90 decreased during the transition week but was not modified afterwards with the L diet and was of 0.65±0.160. For four of the goats of the H diet group, P90 remained stable during the transition period while it increased for two animals which had the highest intake rate with the control diet. With the H diet, P90 decreased linearly during the 15 days to a mean value of 0.27±0.069. In conclusion, feeding behavior estimated by the P90 index was quickly modified after the change in dietary concentrate proportion. With a lower percentage of concentrate, P90 decreased because forage intake rate is lower than concentrate one. With a higher concentrate percentage, some of the goats increased their intake rate, but this increase was followed by a greater decrease in intake rate probably in order to face a rumen subacidosis. More work is needed to better understand the destructuration of feeding behavior in goats when facing rumen subacidosis.

A new approach to measure methane production from *in vitro* rumen fermentation of concentrates

M. Cattani[1], L. Maccarana[1], F. Tagliapietra[2], S. Schiavon[2] and L. Bailoni[1]
[1]University of Padova, Department BCA, Viale dell'Università 16, 35020 Legnaro, Italy, [2]University of Padova, Department DAFNAE, Viale dell'Università 16, 35020 Legnaro, Italy; mirko.cattani@unipd.it

Analysis of methane (CH_4) produced from *in vitro* rumen fermentation is usually conducted on gas samples collected from headspace of bottles where gas is not vented. However, especially when concentrates are incubated, this procedure can lead to overestimation of CH_4 values, as no venting causes partial solubilization of carbon dioxide (CO_2) in fermentation fluid and an increased concentration of CH_4 in headspace of bottles. This study aimed to measure gas (GP) and CH_4 production using bottles where gas was vented into tight bags. Bottles (310 ml), filled with 60 ml of buffered rumen fluid (headspace=250 ml) and 0.040 g of feed sample, were not vented (NV) or vented (VE) at threshold pressure of 6.8 kPa by a valve connected to tight bag. Three concentrates, different for fermentative properties (corn meal, beet pulp, and flaxseed expeller), were incubated for 24 h at 39 °C. The experimental design was: 3 feeds × 3 replicates × 2 venting systems, plus 4 blanks, for a total of 22 bottles. A PC wireless-connected to bottles monitored GP, and GP was adjusted for solubilized CO_2. Gas samples (2 µl) were collected from headspace of bottles or from tight bags by a syringe and analyzed for CH_4 by gas-chromatography. Data were analyzed by ANOVA considering feed, venting system and their interaction as sources of variation. Compared to NV, VE system provided greater ($P<0.01$) GP values, for all feeds. Adjustment of GP values removed differences between systems for beet pulp and flaxseed expeller, but not for corn meal. Measured CH_4 concentrations were greater ($P<0.01$) for NV compared to VE, for all feeds. When CH_4 concentrations were adjusted for solubilized CO_2, differences between two systems tended to decrease but did not disappear. Results evidence that gas venting plays a key role to achieve reliable measurements of *in vitro* CH_4 production.

Comparison of prediction errors of published predictive equations of water dairy cow requirement

A. Boudon[1,2], H. Khelil-Arfa[1,2] and P. Faverdin[1,2]
[1]Agrocampus Ouest, UMR 1348 PEGASE, 65 rue de Saint-Brieuc, 35500 Rennes, France, [2]INRA, UMR 1348 PEGASE, Domaine de la Prise, 35590 St-Gilles, France; anne.boudon@rennes.inra.fr

Thirteen predictive equations of drinking water (DW) requirements of dairy cows were collected from the literature. These equations were established by multiple regression approach from datasets containing from 24 to 12821 individual measurements of DW intake on dry and lactating cows. Included predictive parameters were related to animal, diet composition or meteorological conditions. A dataset of 71 observations of DW intake obtained from groups of cows given a same treatment was also constituted from 18 studies collected from the CAB with the keywords 'water intake' and 'dairy cows'. All the data that could have been used to establish one of the 13 compared equations were previously removed. The average daily dry matter intake, DW intake and diet dry matter content in the dataset were 17.7 kg (±5.35), 66.5 l (±27.52) and 54.8% (±15.18) respectively. The dataset included 75 observations on lactating cows, with an average milk yield of 28.5 kg/d (±8.40) and 54 observations in condition of ambient temperature exceeding 15 °C. The root mean square prediction error (MSPE) of DW intake ranged between 15.4 and 25.5 l/d with an average of 23.0 l/d (±3.56) and a relative RMSPE of 35% (±5.4). For 2 equations, RMSPE were clearly lower, i.e. more than half of the initial RMSPE, when data obtained at ambient temperature exceeding 15 °C were removed, indicating that these equations should be preferred in thermoneutral conditions. For 4 of these equations, we observed a strong proportion of the MSPE explained by a bias on the average that could be explained by the difficulty to assess predictive parameters such as diet Na content or minimal daily temperature. This study shows that predictive performance of published equations is very variable. The choice of an equation should be driven by the relevance of the dataset from which it was obtained in relation with the type of variability that has to be predicted.

Interactions between milk amount and fat content of a starter feed on performance of Holstein calves

G. Araujo[1], M. Terré[1] and A. Bach[1,2]
[1]IRTA, Ruminant production, Torre Marimon, 08140 Caldes de Montbui, Spain, [2]ICREA, Institució de Recerca i Estudis Avançats, 08010 Barcelona, Spain; marta.terre@irta.cat

Sixty-eight Holstein calves (43±5.7 kg BW and 11±3.1 d age) were involved for 53 d in a study to evaluate the effects of dietary energy supply on performance. Calves were housed individually and allocated to one of 4 treatments following a 2×2 factorial design. Calves were fed either 4 or 6 L of milk replacer (MR)/d in 2 feedings (08:00 and 16:30) and offered either a low fat (LF) (19% CP and 3.5% fat) or a high fat (HF) (19% CP and 9.5% fat) starter feed *ad libitum* and weaned on d 42 of study. Calves were fed only the morning MR feeding for a wk before weaning. Individual MR and starter feed intake were recorded daily and BW was determined weekly. Data were analyzed using a mixed-effects model with repeated measures. Before weaning, starter feed intake was greater ($P<0.05$) for 4LF than for 4HF and 6LF calves (725 vs. 572±65.7 g of DM/d). However, total DM and protein intake in 6HF calves (1,309±70.3 and 264±8.3 g of DM/d, respectively) was greater ($P<0.05$) than for 4LF and 4HF calves (1,122±70.3 and 216±8.3 g of DM/d, respectively). Calves in 6HF treatment had the greatest ($P<0.05$) ADG during preweaning (0.7 vs. 0.5±0.05 kg/d). Gain-to-feed ratio was greater ($P<0.01$) for calves consuming 6 than for those consuming 4 L of MR/d (0.58 vs. 0.53±0.016, respectively). Beyond 6 d after weaning, calves consuming LF starter had ($P<0.05$) or tended to have ($P<0.1$) greater DM intakes than calves on HF. However, recently after weaning gain-to-feed ratio and ADG were ($P<0.05$) or tended to be ($P=0.06$) lowest for 6LF calves, respectively, but 1 wk after weaning such differences disappeared. In conclusion, increasing fat content of starter feed of calves receiving 6 L of MR/d resulted in a better growth performance before weaning and a reduced postweaning ADG and gain-to-feed ratio slump than those receiving a low fat starter feed.

Effect of feed form and particle size on diet digestibility in pigs

E. Magowan[1], M.E.E. Ball[1], V.E. Beattie[2] and A. Thompson[3]
[1]AFBI, Hillsborough, BT26 6DR, United Kingdom, [2]Devenish Nutrition, Belfast, BT1 3BG, United Kingdom, [3]John Thompson and Sons, Belfast, Bt15 3GW, United Kingdom; elizabeth.magowan@afbini.gov.uk

The effect of feed form and particle size on diet digestibility in pigs was investigated. Thirty two pigs were offered one of four dietary treatments. All diets were the same and based on barley, wheat and soya with 13.6 MJ/kg digestible energy (DE), 167 g/kg crude protein (CP) and 9.5 g/kg Lysine. In a 2×2 factorial design diets were in either meal or steam pellet form and were ground to attain a fine (13% 1.4-2.0 mm, 59% 0.5-1.4 mm, 28% <0.5 mm) or coarse (34% 1.4-2.0 mm, 41% 0.5-1.4 mm, 18% <0.5 mm) particle size profile. The average start weight of pigs was 44.8 kg and they were housed in metabolism crates for 14 days (7 days pre-feed + 7 days total collection period of faeces). Diet and faeces samples were analysed for digestibility of dry matter (DM), CP, ash and energy. Proximate analyses used the methods outlined by the Association of Official Analytical Chemists. Gross energy was determined using a bomb calorimeter. Data was analysed according to the experimental design using Analysis of Variance in Genstat V 10. There was no interaction (P>0.05) between particle size and feed form. Pelleting of the diet improved DM digestibility (85.1%) and DE content (15.4 MJ/kg DM) (both P<0.05) compared with feed in meal form (84.2% and 15.2 MJ/kg DM, respectively). CP digestibility was improved (P<0.05) when the diet with a fine particle size profile was offered (84.0%) compared with the coarse particle size diet (81.5%). DM digestibility (P=0.051) and DE content (P=0.094) also tended to be improved when the diet with a fine particle size profile was offered (85.1% and 15.4 MJ/kg DM respectively) compared with the coarse particle size diet (84.2% and 15.2 MJ/kg respectively). The cumulative effect of pelleting and fine grinding improved diet DE content by 0.4 MJ/kg DM compared to the same diet in meal form with a coarse particle size profile.

Comparative efficacy of L-methionine and DL-methionine in piglets

J. Van Milgen[1], J. Noblet[1], P. Looten[2], P. Fuertes[2] and C. Delporte[2]
[1]INRA-Agrocampus Ouest, UMR1348 PEGASE, Domaine de la Prise, 35590 Saint-Gilles, France, [2]Roquette Frères, Rue de la Haute Loge, 62136 Lestrem, France; jaap.vanmilgen@rennes.inra.fr

Methionine (Met) is an essential amino acid and a component of structural and functional body proteins. As a sulfur-containing amino acid, Met can transfer its sulfur group to serine to synthesize cysteine. Methionine is also implicated in methylation reactions. Methionine can be provided by dietary protein or by isomers (DL-Met) or analogues (DL-HMB) of L-Met. D-Met and DL-HMB have to be converted by the animal to L-Met to be biologically active. Although a large number of studies have been carried out comparing the efficiencies of DL-Met and DL-HMB, little information is available on the efficiency of L-Met. The objective of this study was to compare the efficiency of L-Met (obtained by fermentation) relative to that of DL-Met (obtained by chemical synthesis) in 77 individually-housed 12-25 kg piglets. The study was carried out with 7 treatments including a basal diet deficient in Met (based on barley, corn and pea protein) and 3 diets with 3 levels of additional L-Met or DL-Met. The standardized ileal digestible Met:Lys levels ranged from 20 to 35% and the (Met+Cys):Lys levels from 42 to 57%; the Lys supply was sublimiting (1.0%). Feed intake, daily gain and feed efficiency were used as response criteria. They increased curvilinearly with increasing levels of both sources of Met and attained a plateau at the highest levels of supplementation with no effect of the source of Met. A bent-stick model was used to estimate the efficiency of L-Met utilization. The model was parameterized to allow for different response trajectories between L-Met and DL-Met. The estimated efficiencies of L-Met relative to that of DL-Met were 1.15±0.12, 1.12±0.09 and 0.99±0.12 for feed intake, daily gain and feed efficiency, respectively. None of these values differed significantly from 1, indicating L-Met and DL-Met can be used equally efficiently as a Met source for growth in piglets.

Efficiency of a growth promoter additive for chickens and weaned piglets: synthesis of 63 trials

G. Benzoni[1], M.L. Le Ray[2], P. Pourtau[2], D. Marzin[2] and A. Guyonvarch[1]
[1]In vivo NSA, Research and Development, Talhouet, 56250 Saint Nolff, France, [2]Neovia, Talhouet, 56250 Saint Nolff, France; mleray@invivo-nsa.com

The ban of antibiotics growth promoters in Europe has dramatically increased demand for alternatives. In this context, a specific ion-exchanged clay, named B-Safe, was developed as such an alternative. Several trials illustrating its efficiency have already been published. Nevertheless, efficiency in independent trials is not enough to conclude on the quality of feed additive, and repeatability of response is a key factor. On this subject, Gordon Rosen published an evaluation method for growth promoter additives, based on seven criteria with quantified targets, that takes into account constancy of performance improvement. In order to properly evaluate the previously cited specific clay, results of the 36 broiler chickens and 27 piglets trials in which it was included were compiled and examined through this Rosen's screen. Broiler trials were conducted from 1 to 35 days of age. Piglet trials were conducted from weaning at 21 days to 67 days of age. All the trials included a contemporaneous negative control group in the same building than the group of animals treated with the tested additive. Some of the trials included also an antibiotic growth promoter as positive control. These trials were done in diversified stocking conditions, going from sanitized experimental farm environment to field conditions. Performances were improved in 76% of the trials and coefficient of variation of response was of 6% for both species. European efficiency factor was improved by 5% for chickens and by 6.7% for piglets. In average, additive effect was stronger on growth than on feed conversion in both species. Moreover, it was possible to modelize performance response to product dose. The optimal additive dose was modified by stocking conditions. All Rosen's screen targets were then achieved and this additive could therefore be considered as a reliable growth promoter for broiler chickens and weaned piglets.

Effect of grape pomace supplementation on broiler performance and eating quality

E.N. Sossidou[1], E. Kasapidou[2], V. Dotas[3], I. Ioannidis[3] and P. Mitlianga[2]
[1]Hellenic Agricultural Organization-Demeter, Veterinary Research Institute, Nagref Campus, 57001 Thermi, Thessaloniki, Greece, [2]Technological Educational institute of West Macedonia, Department of Agricultural Products Marketing and Quality Control, Terma Kontopoulou, 53100 Florina, Greece, [3]Aristotle University of Thessaloniki, Department of Animal Production, Faculty of Agriculture, 541 24 Thessaloniki, Greece; sossidou.arig@nagref.gr

The objectives of this study were to determine the effect of grape pomace, a major agro-industrial byproduct, on broiler performance and meat eating quality. 512 day old chicks were divided into four treatments replicated four times with 128 birds per treatment and they were fed on a standard commercial diet, containing either 0, 2.5, 5 or 10 g/kg feed ground and dried grape pomace, denoted as CON, DGP 2.5, DGP 5 and DGP 10 respectively, for 42 days before slaughter. Performance was assessed as average daily weight gain, live weight at the end of the feeding period and carcass weight. The oxidative stability was determined as thiobarbituric acid reacting substances (TBARS) on skinless breast samples on storage day. Eating quality was determined by a 10-member panel that used 5-point category scales for odour, taste, tenderness, juiciness, fibrousness and overall acceptability. Preliminary results from the study showed no significant differences between treatments ($P \geq 0.05$) for the performance parameters. Moreover, inclusion of increasing amounts of grape pomace in the diets had no effect ($P \geq 0.05$) on limiting the extent of lipid oxidation. Average lipid oxidation was 0.338 mg malonaldehyde/kg of sample and far below the threshold value for detection of rancidity. Sensory assessment showed no statistical differences ($P \geq 0.05$) between treatments in all examined traits. However, the highest scores for overall acceptability was recorded for the DGP 5 samples. Further analysis is required for the determination of optimum supplementation levels for improved bird performance and product eating quality.

Session 30 Poster 13

Effect of microwave irradiation on *in vitro* gas production parameters of linseed

J. Amini[1], M. Danesh Mesgaran[1], A.R. Vakili[1], A.R. Heravi Moussavi[1] and M.R. Ghaemi[2]
[1]Ferdowsi University of Mashhad, Department of Animal Science, 91775-1163, Mashhad, Iran, [2]Shafashir Toos Co., Research Department, Mashhad, Iran; ghaemi.mohammadreza@gmail.com

The effect of microwave irradiation on *in vitro* gas production kinetics of linseed was investigated. Samples were milled (1 mm), then exposed to a 2,450 MHz continuous microwave radiation by an average specific absorption rate (SAR) of 75.5 W/kg for 0.0 (CON), 10, 30, 60, 120 or 150 seconds. Approximately 500 mg of oven dried sample was incubated with 50 ml buffered rumen fluid (ratio of buffer to rumen fluid was 2:1) in a 125 ml glass bottle (n=9) at 38.6 °C using water bath for 2, 4, 6, 8, 12, 24, 48 and 72 h. Cumulative gas production values were corrected for blank incubation and fitted to an exponential equation $Y=b(1-e^{-(ct-l)})$, where b is the gas production from the fermentable fraction (ml), c is the gas production rate constant for b, t is the incubation time (h), l is lag time (h) and Y is the gas produced at time t. Data were analyzed using the GLM procedure of SAS 9.2 and the means were compared by the LSD test ($P<0.05$). The amount of gas produced from the fermentable fraction (b ± SE) was $53.3^{b}\pm0.83$, $48.4^{c}\pm0.88$, $56.4^{a}\pm1.59$, $42.9^{d}\pm0.97$, $31.8^{f}\pm1.08$ and $37.0^{e}\pm0.78$ ml; The fractional constant rate (c ± SE) was $0.058^{a}\pm0.0024$, $0.055^{a}\pm0.0025$, $0.044^{b}\pm0.0028$, $0.051^{ab}\pm0.0026$, $0.045^{b}\pm0.0034$, $0.046^{b}\pm0.002$ h; and the lag time (l ± SE) was $0.7^{c}\pm0.19$, $1.4^{b}\pm0.19$, $1.6^{b}\pm0.28$, $2.8^{a}\pm0.19$, $2.6^{a}\pm0.29$, $2.7^{a}\pm0.18$ h for 10, 30, 60, 120 and 150 s, respectively. Irradiation for 30 s caused to increase the amount of gas produced significantly ($P<0.05$) from the fermentable fraction (b) compared with CON. Irradiation for 30, 120 or 150 s decreased the fractional constant rate (c) of the samples than that of the CON. Furthermore, irradiation significantly increased fermentation lag time ($P<0.05$). Results indicated that the microwave irradiation for more than 60 s might significantly reduce both b and c parameters and could increase fermentation lag time.

Session 30 Poster 14

Comparison of the effects of three different diets on growth curves

H. Onder[1], B.Z. Sarıçiçek[2] and S.H. Abacı[1]
[1]Ondokuz Mayis University, Animal Science, Ondokuz Mayis University, Agricultural Faculty, Animal Science Dep., 55139 Samsun, Turkey, [2]Ankara University, Animal Science, Ankara University, Agricultural Faculty, Animal Science Dep., 06110 Dışkapı, Ankara, Turkey; hasanonder@gmail.com

The purpose of this study was to compare the effects of three different diets containing 0 mg/kg, 30 mg/kg and 45 mg/kg organic ZnO on growth curves. Different diets were given after weaning period. Growth curves of Karayaka lambs were analyzed using body weights measured at ages 1 (birth weight) to 174 d for each diet groups. Linear, quadratic and cubic curves were estimated. Analysis results showed that cubic growth curves has minimum error sum of squares and maximum coefficient of determination which were 0.38 and 0.995, 0.35 and 0.997, 0.21 and 0.997 for diets 0 mg/kg, 30 mg/kg and 45 mg/kg organic ZnO, respectively. The models for diets were estimated as $3.304 + 0.073t + 0.0011t^2 - 0.000022t^3$, $3.301 + 0.052t + 0.0014t^2 - 0.000003t^3$ and $3.447 + 0.112t + 0.00035t^2 + 0.000002t^3$, respectively (t: days). Effects of the diets were not statistically significant on growth curves. Results mean that parameters of the models could be admitted as same. Superiority of cubic models, it may be result of rapid growth after weaning than pre-weaning.

Intake and body gain of beef heifers as affected by concentrate content and fiber digestibility

B.S. Marques, M.V. Carvalho, D.O. Souza, B.S. Mesquita and L.F.P. Silva
Universidade de São Paulo, School of Veterinary Medicine, 355 Duque de Caxias Norte, Pirassununga, SP, 13635-900, Brazil; lfpsilva@usp.br

Effect of corn silage stalk fiber digestibility (NDFD) and level of concentrate on intake and body weight gain (ADG) was evaluated. Forty eight Nellore heifers were used in a randomized block design with a 2×2 factorial arrangement of treatments. Two corn hybrids with differing stalk NDFD were used: 30F90 with higher NDFD and 30S40 with lower NDFD. Treatment diets contained either 80 or 60% corn silage as roughage source. Animals were housed in collective pens (3 per pen) with free access to water and fed *ad libitum*. The experiment period comprised 10 days of adjustment followed by three periods of 21 days. Heifers were weighted at the beginning of each period and at the end of the experiment. Daily intake was recorded by weighting of the orts. Weekly samples of dietary ingredients and or orts were collected for chemical analysis. Main effects of corn hybrid (CORN), of concentrate content (DIET), and their interaction were tested by ANOVA. Dry matter intake as a percentage of BW was affected by CORN, by DIET and by the interaction (P=0.04). Corn silage with greater NDFD increased DM intake only in the diet with 20% concentrate (P<0.01), and not with 40% concentrate (P=0.33). The same was true for NDF intake, with greater intake for 30F90 only in the diet with 20% concentrate (P<0.01). The diet altered ADG, with greater ADG for the diet with 40% concentrate (P=0.05). There was only a tendency (0.86 vs. 0.73 kg/d, P=0.10) for greater ADG because of higher NDFD in the 20% concentrate diet. Feed efficiency was not affect by treatments. Using a corn hybrid with greater stalk NDFD increased intake and tended to increase ADG.

Effects of live yeast on performance of young Holstein calves

M. Terre[1], M. Gauthier[2], G. Maynou[1] and A. Bach[1,3]
[1]IRTA, Department of Ruminant Production, Caldes de Montbui, 08140, Spain, [2]Lallemand, Blagnac, 31702, France, [3]ICREA, Barcelona, 08007, Spain; marta.terre@irta.cat

The objective of this study was to determine the effects of Saccharomyces cerevisiae on performance of young calves. Ninety-six Holstein male calves (42±5.4 kg of BW and 9±5.8 d of age) were randomly distributed in two treatments: unsupplemented starter feed (CTR) or the same starter feed supplemented with 1.5×10^6 cfu of Saccharomyces cerevisiae CNCM-I-1077/g of starter (YEA). All animals were fed the same milk replacer (23% CP, 19.5% fat) at 12.5% DM and at the rate of 4 l/d from 1 to 7 d, 5 l/d from 8 to 14 d, 6 l/d from 15 to 21 d, 3 l/d from 22 d to preweaning (when calves consumed 900 g/d during 3 consecutives days) and 1.5 l/d during 7 d after the preweaning day. The study finished 56 d after the beginning of the study. Body weight was measured weekly, and individual calf starter intake was recorded daily from days 1 to 34, and in groups of 6 animals per pen afterwards. Twenty days after weaning, 34 calves were randomly selected to obtain a rumen sample via an esophageal tube, and pH was measured immediately. Data were analyzed using a mixed-effects model. During the first 5 wk of study, starter intake was similar between treatments (607±36.3 g/d), and from 6 to 8 wk of study, although not significant (P=0.13), starter intake of YEA was numerically greater than that of CTR calves (2,173 vs. 1,980±85.2 g/d, respectively). Furthermore, ADG tended (P=0.08) to be greater in YEA than in CTR calves at 5, 6 and 8 wk of study (0.69 vs. 0.58±0.04 kg/d, respectively). Although all calves were weaned at the same moment in both treatments (37.4±0.97 days), YEA calves tended (P=0.08) to weight 2.5 kg more than CTR calves at the end of the study. Rumen pH tended (P=0.06) to be greater in YEA than in CTR calves (5.5 vs. 5.2±0.12, respectively). In conclusion, the supplementation of Saccharomyces cerevisiae in weaned calf starter concentrates may improve calf performance, and rumen pH.

Session 30 Poster 17

Effect of dietary Lys levels on expression of cationic amino acid transporters and myosin in pigs

H. García, N. Arce, F. Grageola, V. Díaz, A. Araiza, M. Cervantes and A. Morales
Universidad Autónoma de Baja California, ICA, Ej. Nuevo León, Mexicali, BC, Mexico; adriana_morales@uabc.edu.mx

Lys is a cationic amino acid (AA), the first limiting AA in most feed ingredients for pigs and, along with Leu, is the most abundant AA in muscle proteins. Some AA seem to participate in the control of their own cellular uptake and protein synthesis process. An experiment was conducted with 21 pigs (20.1±1.63 kg BW) to analyse the effect of dietary Lys levels (deficient, adequate, or excess) on the expression of two cationic AA transporters ($b^{0,+}$ and CAT-1) and myosin in several tissues. A basal diet, Lys deficient (0.34% Lys), was formulated with wheat and added with Thr, Met, vitamins and minerals. Treatments were: 1, basal diet (DEF); 2, basal + 0.70% free L-Lys, adequate (ADE); 3, basal + 1.40% free L-Lys, excess (EXC). All pigs had free access to feed and water all the time. At the end of the 21-d study, five pigs per treatment were euthanized to collect samples from small intestine (jejunum) mucosa, liver, Longissimus (LM) and Semitendinosus (SM). Expression of $b^{0,+}$ was analysed in jejunum and liver, CAT-1 in all those tissues, and myosin in LM and SM. Expression values of DEF, ADE and EXC pigs (arbitrary units) for $b^{0,+}$ in jejunum: 7.99, 8.00, 12.97; liver, 0.56, 0.04, 0.28; for CAT-1 in jejunum, 0.25, 0.01, 0.06; liver, 0.43, 0.27, 0.43; LM, 0.006, 0.009, 0.004; SM, 0.095, 0.033, 0.037; for myosin in LM, 4.12, 3.96, 4.60; SM, 0.97, 6.63, 2.43, respectively. Expression of $b^{0,+}$ in jejunum was not affected by the Lys level, but in liver ADE pigs had the lowest value ($P<0.05$). CAT-1 expression value in jejunum was lower in ADE pigs as compared to DEF pigs ($P<0.05$), but it was not affected in liver, LM or SM. Myosin expression in SM was higher in ADE pigs as compared to DEF and EXC pigs ($P<0.01$), but it was not affected in LM. These data show a differential tissue effect of the dietary Lys level on the expression of its own transporters and myosin.

Session 30 Poster 18

Effect of free or protein-bound amino acids on amino acid transporters and myosin expression in pigs

F. Grageola, A. Morales, H. García, A. Araiza and M. Cervantes
Unoversidad Autónoma de Baja California, Instituto de Ciencias Agrícolas, Blvd. Delta, Mexicali, Mexico; miguel_cervantes@uabc.edu.mx

Free amino acids (AA) appear to be absorbed faster than protein-bound AA (PB-AA). We conducted an experiment to assess the effect of feeding pigs with a partially free (F-AA) or a totally PB-AA diet on the expression of the cationic AA transporters $b^{0,+}$ and CAT-1 in intestinal mucosa, liver, and Longissimus (LM) and Semitendinosus (SM) muscles, as well as that of myosin in LM and SM. Twelve pair-fed pigs (31.7±2.7 kg) were used. The F-AA diet was based on wheat, supplemented with 0.59% L-Lys, 0.33% L-Thr, and 0.10% DL-Met. The PB-AA diet was formulated with wheat and soybean meal to contain similar levels of Lys, Thr and Met as in F-AA. Average daily feed intake was restricted to 1.53 kg per pig. Expression of $b^{0,+}$ and CAT-1 were analyzed in jejunal and ileal mucosa, liver, LM and SM; myosin expression was also analyzed in both muscles. Expression values of F-AA and PB-AA pigs (arbitrary units) for $b^{0,+}$ were: jejunum 10.37, 4.05; ileum, 3.48, 5.71; liver, 0.29, 1.34; LM, 0.59, 1.34; SM, 0.18, 0.11; and for CAT-1: jejunum, 0.93, 2.47; ileum, 0.23, 0.24; liver, 0.52, 0.08; LM, 0.33, 0.87; SM, 0.33, 2.17; respectively. Myosin expression values of F-AA and PB-AA pigs: LM, 0.47, 0.60; SM, 0.45, 0.45. Expression of $b^{0,+}$ was higher ($P<0.01$) in jejunum but lower ($P<0.01$) in the liver of pigs fed the F-AA diet; CAT-1 tended to be lower in liver but higher in LM of PB-AA pigs. Myosin expression was not affected. Intestinal AA absorption appears to be faster in pigs fed the F-AA diet, but AA uptake by the liver seemed to be faster in pigs fed the PB-AA. These results indicate that the form (free or protein-bound) in which AA are provided in the diet affect the expression of their transporters and hence their intracellular availability for protein synthesis in pigs.

Nutritional evaluation of several extruded linseed product by *in vitro* gas production technique

M. Danesh Mesgaran[1], J. Amini[1], A.R. Vakili[1], A.R. Heravi Moussavi[1] and M.R. Ghaemi[2]
[1]Ferdowsi University of Mashhad, Department of Animal Science, 91775-1163, Mashhad, Iran, [2]Shafashir Toos Co., Research Department, Mashhad, Iran; ghaemi.mohammadreza@gmail.com

The present study was conducted to determine effect of different processing and extrusion method on *in vitro* digestibility of organic matter (DOM), metabolizable energy (ME) and net energy for lactation (NEL) values. Commercial products containing extruded linseed including Nutex Compact® (containing 56% extruded linseed) and LINOMAX®, and a pure extruded linseed (155-160 °C for 15-20 s) sample were evaluated. Samples grounded to pass through a 1-mm screen and subjected to *in vitro* gas production technique. Mixed rumen micro-biota obtained from four ruminally fistulated Holstein steers (420±13 kg, body weight) fed twice daily a diet containing 5.8 kg alfalfa hay and 3.0 kg concentrate mixture). Approximately 200 mg of each sample was weighed into a 125 ml flask (n=9) and then 50 ml rumen fluid-buffer mixture (in a 1:2 ratio) added into each bottle under CO_2 flush, followed by incubation in a water bath at 38.6 °C. Gas volume was recorded at 24 h of incubation. Metabolizable energy (ME), NE_L and DOM values of the samples were calculated using following equations: ME (MJ/kg DM) = 1.56 + 0.1390 GP + 0.0074 XP + 0.0178 XL; NE_L (MJ/kg DM) = 0.1010 GP + 0.0051 XP + 0.011 XL; DOM (g/kg DM) = 14.88 + 0.8893 GP + 0.0448 XP + 0.0651XA. Where GP is net gas produced after 24 h of incubation (ml/0.2 g DM), and XP, XL and XA are crude protein, crud fat and ash content of the feed (g/kg DM), respectively. The results showed that gas production for Nutex Compact®, LINOMAX® and pure extruded linseed samples at 24 hours were 21.4^a, 14.8^b and 12.7^c ml/0.2 g; respectively ($P<0.05$). The DOM of samples were 49.06^a, 39.6^b and $31.28\%^c$; respectively. The ME contents were 10.47^a, 10.62^a and 9.11^b MJ/kg DM, respectively and the NE_L contents were 5.97^a, 5.92^a and 4.87^b MJ/kg DM, respectively ($P<0.05$)

Use of electronic nose for corn silage screening

M. Moschini, A. Gallo, G. Giuberti, C. Cerioli, P. Fortunati and F. Masoero
Università Cattolica del Sacro Cuore, Feed and Food Science and Nutrition Institute, Via Emilia Parmense 84, 29122, Italy; maurizio.moschini@unicatt.it

Corn silages were randomly collected in the Po valley during the year 2012. Samples were taken from 18 concrete wall bunkers and from 3 different positions of freshly cut face: core or C, (1 meter high from the bottom), side or S (1.5 meter high from the bottom, 0.3 meter from the walls) and top or T (0.5 meter from the top). Collected samples were stored at 4 °C and subjected within 24 h to electronic nose analysis (Pen3 – Airsense AnalyticsGmbH, Schwerin, Germany) equipped with metal oxide semiconductor sensors (W1C, W3C, W6S, W5C, W1S, W1W, W2S, W2W, W3S). Each sample was weighed (20 g) into airtight glass jar, then jars were closed and let it stand at room temperature for 30 minutes to allow for headspace equilibrium. After reaching equilibrium, the headspace gas was pumped to sensors of the electronic nose (flow rate 400 ml/min). The measurement phase lasted 60 seconds with data collection interval of 1 second. A stand-by phase (320 seconds) was observed between each sample reading to allow for a cleaning of the system. Only one reading (at 59 second) for each sensor entered a data matrix of 54 rows (silage samples) and 9 columns (sensors). A correlation matrix was obtained from collected data and a principal component analysis (PCA) was performed using the FACTOR procedure of SAS. The PRIN method with Kaiser's criterion (eigenvalue≥1.00) and the orthogonal Varimax rotation were used to extract latent constructs and to produce loading vectors and sample scores. Three principal components (PC) were extracted: PC1 (W1C, W3C, W5C, W1S, W2S, W2W; eigenvalue=5.60), PC2 (W6S, W3S; eigenvalue=1.75), PC3 (W1W; eigenvalue=1.00). The PC1 allowed for clustering the silage samples into two populations being C and S+T, whereas the PC2 and PC3 tended to discriminate between S and T samples. Results suggest electronic nose could be a valuable laboratory tool for discriminating corn silages exposed to different preservation processes.

Effect of quebracho tannins on the biohydrogenation of linoleic and linolenic acid: *in vitro* study

M. Mele[1], A. Buccioni[2], A. Serra[1], G. Conte[1], S. Minieri[2], F. Mannelli[2], D. Benvenuti[2], A. Pezzati[2] and S. Rapaccini[2]
[1]University of Pisa, Dipartimento di Scienze Agrarie, Alimentari e Agro-Ambientali, via del Borghetto 80, 56124, Italy, [2]University of Firenze, Dipartimento di Biotecnologie Agrarie, via delle Cascine 5, 50100, Italy; mmele@agr.unipi.it

Tannins are phenolic compounds able to interfere with the biohydrogenation of polyunsaturated fatty acids. The aim of this trial was to study *in vitro* the effects of quebracho tannin extract (QT) on the rumen biohydrogenation (BH) of polyunsaturated fatty acids (PUFA) contained in soybean and linseed oils. Two control diets were prepared using grass hay (770 g/kg on dry matter, DM), soybean meal (55 g/kg DM), barley meal (135 g/kg DM), supplemented with 35 g/kg DM of soybean (SOC diet) or linseed oil (LOC diet). Other two diets (SOCT and LOCT) were obtained by the integration of SOC and LOC with QT (49 g/kg DM). Feeds (2 g DM) were incubated in triplicate with 200 ml of rumen liquor in a thermostatic chamber (39 °C) equipped with thirty-six 300 ml glass fermentation vessels. Incubation times were 6, 12 and 18 h. At each fermentation time, rumen liquor was fractionated in order to obtain liquid and solid associated bacteria (SAB). Only SAB were considered for the fatty acid analysis. QT induced a decrease of the BH of PUFA, favouring the accumulation of linoleic and alpha-linolenic acid. After 18 h of fermentation the content of linoleic acid was reduced by 76% and 42% in SOC and SOCT feed, respectively, whereas the content of linolenic acid was reduced by 85% and 73% in LOC and LOCT feed, respectively. Conversely, the accumulation in rumen liquor of vaccenic acid, total CLA and the other intermediates of BH process was lower when QT was added to both SOC and LOC feeds. The content of 18:0 was significantly lower in both SOCT (-33%) and LOCT (-17%) feeds, if compared with the same feeds without QT. In conclusion, QT was effective in the reduction of the BH of PUFA in rumen liquor, especially in the case of linoleic acid.

Blood features of lactating dairy cows fed sunflower cake as partial replacement of soybean meal

A. Dal Prà, L. Migliorati, M. Capelletti, F. Petrera, G. Pirlo and F. Abeni
Consiglio per la Ricerca e la Sperimentazione in Agricoltura, Centro di Ricerca per le Produzioni Foraggere e Lattiero-casearie, via Porcellasco 7, 26100 Cremona, Italy; giacomo.pirlo@entecra.it

The aim of this study was to assess the effect of partial substitution in the diet of dairy cows of sunflower cake for soybean on metabolic and haematological profile. Sixteen Italian Friesian cows (8 in early and 8 in mid lactation) were allotted in two groups and fed a control diet (C) based on soybean meal as main protein source, or a diet with 2 kg/head of sunflower cake (S). The experimental design was a change-over based on two period of 4 weeks (3 for adaptation + 1 sampling). Blood samples were drawn from jugular vein just before daily distribution of total mixed ration. The complete haematological profile was determined by an automated analyzer; erythrocytary osmotic fragility was assessed with a single point (0.6% buffered saline solution) method. Plasma metabolites related to energy and lipid (glucose, cholesterol, triglycerides, NEFA, and BHB), nitrogen (urea, creatinine, total protein, albumin), and enzymatic activities related to hepatobiliary damage and functionality (AST, ALT, GGT, LDH) were determined by an automated analyzer. Data were statistically processed by a GLM, with diet and stage of lactation as main factors. The only parameter which tended to be affected by diet was plasma urea, with lower values in S than C, probably as a consequence of a small difference in total protein supply. Stage of lactation affected plasma triglycerides, urea, creatinine, and protein concentrations. No effect of S diet was evidenced on erythrocyte osmotic fragility and on the main haematological features. The partial replacement of sunflower cake for soybean did not substantially affect metabolic and haematological profile of dairy cows in our experimental condition.

Session 30 Poster 23

Milk production and composition and blood and liver parameters of dairy cows fed various fat sources

R.M. Prado[1], I.N. Prado[1], G.T. Santos[1], C. Benchaar[2], M.-F. Palin[2] and H.V. Petit[2]
[1]Universidade Estadual de Maringá, Programa de Pós-Graduação em Zootecnia, Av. Colombo, 5790, 87020-900, Maringá, PR, Brazil, [2]Agriculture and Agri-Food Canada, Dairy and Swine R & D Centre, 2000 College St., J1M OC8, Sherbrooke, QC, Canada; rodolphoprado@hotmail.com

Twenty-nine Holstein cows were allotted 4 weeks before the expected calving date to 10 groups of 3 cows blocked within parity for similar calving date to compare the effects of feeding omega 3 and omega 6 fatty acids (FA) on milk production and composition, plasma metabolites and liver parameters. Cows were fed for *ad libitum* intake from 4 wk before until 12 wk after calving. Cows within blocks were assigned randomly to one of three isonitrogenous and isoenergetic total mixed rations with different fat sources: calcium salts of palm oil (MEG; 1.26% and 2.57% of the dry matter in prepartum and postpartum diets, respectively); omega 3 FA as whole flaxseed (WFL; 4.82% and 7.59% of the dry matter in prepartum and postpartum diets, respectively); or omega 6 FA as whole linola (WLO; 4.82% and 7.59% of the dry matter in prepartum and postpartum diets, respectively). Cow fed MEG had the highest milk production and lactose yield. Dietary fat had no effect on feed intake and blood concentrations of non-esterified fatty acids, β-hydroxybutyrate and glucose. There was no effect of dietary fat on glycogen concentration and activity of the enzymes superoxide dismutase and glutathione peroxidase in the liver. Hepatic concentration of lipids and triglycerides tended ($P=0.0961$ and 0.0834, respectively) to be lower for cows fed WFL and WLO than for those fed MEG on week 4 after calving. Cows fed WFL had greater catalase activity on week 4 after calving than those fed MEG and WLO. These results suggest that fat sources had little effect on hepatic fat metabolism.

Session 30 Poster 24

The nutritional value of corn grains for growing pigs: influence of the way of preservation

J. Danel[1], P. Callu[1], A. Samson[2], J.G. Cazaux[3] and M. Vilariño[1]
[1]ARVALIS, Institut du végétal, Pouline, 41100 Villerable, France, [2]INZO°, Rue de l'église, BP50019, 02407 Chierry Cedex, France, [3]FNPSMS, 21 Chemin de Pau, 64121 Montardon, France; j.danel@arvalisinstitutduvegetal.fr

In growing pig diets, the increase of the high-moisture corn grains utilization raises the question of their nutritional value compared with dry corn grains. In order to assess the impact of preservation practices of corn grains on ileal and faecal digestibility of various nutrients, two trials were performed in growing pigs. A batch of corn was divided at the harvest and stored in three forms: Dry Corn grains (DC), High-Moisture Whole Corn grains (HMWC) and High-Moisture Grounded Corn grains (HMGC). HMWC and HMGC were stored in natural anaerobic conditions. In the first study, 15 castrated male pigs were used for faecal measurements (3 diets × 5 pigs). In the second trial, 4 surgically modified pigs were used in a Latin square design for ileal measurements. The faecal energy digestibility of HMGC is significantly higher (88.4%) than those of HMWC (85.7%) or of DC (85.2%). The associated digestible energy values are: 4019, 3907 and 3854 kcal/kg DM respectively. In the same way, the ileal energy digestibility was of 83.8%, 80.8% and 77.3% for HMGC, HMWC and DC, respectively, but the differences between corn forms are more important. The ileal digestibility values for other nutrients are consistent with these results, particularly for fat (84.9%, 71.7% and 57.1%, respectively). The standardized ileal digestibility of the sum of amino acids is not significantly different between corn forms. Nevertheless, the hierarchy of the values is similar to those of other components (89.9%, 87.4% and 84.4% for HMGC, HMWC and DC, respectively). Considering all these results, a nutritional benefit of high-moisture corn grains for growing pigs can be highlighted, especially in the case of the grounded one, in comparison with dry corn grains.

Energy maintenance requirements of goats

A.K. Almeida, S.P. Silva, D.C. Soares, M.H.M.R. Fernandes, I.A.M.A. Teixeira and K.T. Resende
Univ Estadual Paulista/ Jaboticabal, São Paulo, Brazil, Department of Animal Science, Via de Acesso Prof. Paulo Donato Castellane s/n, 14884-900 Jaboticabal, SP, Brazil; almeida.amelia@gmail.com

We proposed to investigate the effect of gender on energy maintenance requirements of Saanen goat kids weighting from 30 to 45 kg of body weight (BW). We used 24 intact males, 23 females and 24 castrated males with initial BW of 30.0±1.09 kg and initial age of 258±53 days. At 30.0±1.09 kg, 6 intact male, 6 female and 6 castrated were slaughtered to estimate the initial body composition. The remaining animals were randomly allocated into 18 groups (blocks) of 3 animals of the same gender, subjected to 0, 25 or 50% of feed restriction. The restriction level was calculated for each group of three animals based on the *ad libitum* animal intake, when this animal reached 45 kg, it was slaughtered, dictating the number of days of the others two animals in the group. Heat production (HP, kcal/kg$^{0.75}$ of empty body weight (EBW)) was calculated as the difference between metabolizable energy (ME) intake and retained energy (RE). The antilog of the intercept of the linear regression between the log of HP and MEI was assumed to be the net energy requirement for maintenance (MRNE). The ME requirement for maintenance (MRME) was computed by iteratively solving the semilog linear regression equation until HP was equal to MEI. Data were analyzed using MIXED procedure of SAS (SAS Inst. Inc., Cary, NC). The MRNE of females and castrate males (357 KJ/kg$^{0.75}$EBW) was greater than that of males (345 KJ/kg$^{0.75}$ of EBW), which resulted in a greater MRME and a lower energy utilization for maintenance (k_m) for females and castrate males (601 KJ/kg$^{0.75}$ EBW; k_m=59.5%) compared to intact males (560 KJ/kg$^{0.75}$ EBW; k_m=61.6%). FAPESP process number: 2010/02482-4.

Do cows under chronic subacute ruminal acidosis attempt to self-medicate?

E. Hendriksen, O. Alzahal and B.W. McBride
University of Guelph, 50 Stone Road East, N1G 2W1 Guelph, Canada; ehendrik@uoguelph.ca

The objective of this study was to examine feed-sorting behaviour of dairy cows in response to a grain challenge that leads to subacute ruminal acidosis (SARA). Sixteen multiparous, rumen-cannulated lactating Holstein cows were used. During the first 49 days of the experiment, all cows received a high-forage diet (HF; 77:23, F:C; NFC=35). Cows were then transitioned to a high-grain (HG; 50:50, F:C; NFC=48) diet on days 50 and 51 and remained on the HG until day 72. Feed intake was recorded daily. Particle size distribution of feed and orts was analyzed using the Penn State Particle Separator (PSPS) on days 36 (HF) and 71 (HG). The extent of feed-sorting was evaluated by calculating sorting indices for each PSPS dietary fraction for each given diet. A sorting index of a given fraction of the PSPS was calculated as the actual DMI expressed as a percentage of the predicted DMI of that fraction. Ruminal pH was recorded continuously every minute using an indwelling system on days 36 (HF), 50, 51 (onset of SARA), and day 71 (chronic SARA). Data were analyzed using Proc Mixed of SAS with day as a fixed effect and accounting for repeated measurement. Orthogonal contrasts were utilized to compare days. Ruminal pH on day 36 was not different between treatments (16±46 min/d <5.6, P≥0.05). Ruminal pH recorded on days 51 and 71 (551±46 min/d <5.6 and 246±46 min/d <5.6, respectively) indicated an established and chronic SARA. Sorting assessment on day 36 showed that cows while on HF diet sorted against large particles (>19 mm, P<0.05) and concurrently sorted for the short and fine particles (<8 mm, P<0.05). While under SARA (HG), the cows have shown an altered preference by sorting for (P<0.05) long particles and against (P<0.05) short and fine particles and this effect was more pronounced during the evening. The results suggested that cows undergoing SARA may alter their feed-sorting behaviour, likely in an attempt to self-medicate their condition, by selecting long particles and avoiding short and fine particles.

Phosphorus excretion in dairy cows not affected by forage particle size or rumen degradable protein

L. Puggaard, P. Lund and J. Sehested
Aarhus University, Research Centre Foulum, Dep. of Animal Science, P.O. Box 50, 8830 Tjele, Denmark; jakob.sehested@agrsci.dk

Inevitable fecal loss (IL) of phosphorus (P) in dairy cows is estimated to 1 g P/kg DMI. It was hypothesised that IL of P would be reduced by: (1) reduced chewing activity (saliva secretion); or (2) reduced rumen microbial production. The effect of forage particle size (FPS), and rumen degradable protein (PBV) on fecal excretion of P was tested in 36 Holstein cows, 222±102 days from calving, 627±7 kg BW, 32±3 kg/d milk and allocated to 3 diets: CONTROL, SHORT and LOW-N. Diets contained 2.5 g P/kg DM and were based on the same feeds (% of DM): compound feed (30), corn silage (31), sugar beet molasses (19), grass hay (20); but varied in FPS of grass hay and urea content. PBV was optimized by supplmenting dietary urea for CONTROL and SHORT, whereas urea was excluded for LOW-N. FPS of grass hay was reduced to 3 mm for SHORT, as compared to 40-60 mm for CONTROL and LOW-N. Rations were fed *ad libitum* as TMR for 16 d, measurement of chewing activity were conducted on d 13 and 14, and balance trials were conducted on d 15 and 16. The internal marker INDF was used to estimate fecal output and total tract digestibilities. Jaw movements were recorded and analysed to estimate chewing time. Negative P balances (-9±1.3 g/d) and low plasma P (0.95±0.07 mmol/l) confirmed that P was fed below requirement, indicating that effects on fecal P excretion mainly originated from variations in IL of P. Daily chewing time tended to be lower (P=0.1) whereas daily DMI and fecal P excretion was increased with SHORT and none of these were affected by LOW-N, as compared to CONTROL. Digestibility of OM was reduced with SHORT and LOW-N as compared to CONTROL. Excretion of fecal P (g/kg DMI) was 1.77, 1.66 and 1.77±0.05 in SHORT, CONTROL and LOW-N, respectively. The results of the present study do not support the concept that fecal loss of endogenous P is affected by FPS or by rumen degradable protein supply and thus rumen microbial P incorporation.

Intake and ruminal kinetics of sugarcane as affected by fiber digestibility and conservation

D.O. Souza, B.S. Mesquita, J. Diniz-Magalhães, F.D. Rodriguez and L.F.P. Silva
Universidade de São Paulo, School of Veterinary Medicine, Pirassununga, SP, 13635900, Brazil; lfpsilva@usp.br

Effect of sugarcane fiber digestibility (NDFD) and method of conservation on intake and rumen kinetics was evaluated. Eight ruminally cannulated Nellore steers were used in a duplicated 4×4 Latin square design. Two sugarcane genotypes with differing stalk NDFD were used: IAC2480 with higher NDFD, and SP1049 with lower NDFD. Treatment diets contained 40% sugarcane as roughage source given as freshly-chopped or as silage. Animals were housed individually with free access to water and fed *ad libitum*. Periods lasted for 14 d, being 10 d for adaptation, and 4 d for sample collection. Dry matter intake was determined on days 10, 11 and 12, and ruminal contents were evacuated manually at 11:00 h (2 h after feeding) on d 12 and at 07:00 h (2 h before feeding) on d 13 of each period. Total ruminal content mass and volume were determined. Aliquots were squeezed through a nylon screen to separate into solid and liquid phases. Samples were taken from both phases for determination of nutrient pool size. Main effects of sugarcane (CANE), of method of conservation (CONS), and their interaction were tested by ANOVA. Dry matter and NDF intake were greater for steers consuming diets with higher sugarcane NDFD, however the interaction CANE × CONS was significant. The effect of greater NDFD on intake was only significant when feeding sugarcane as silage, having no effect on intake when sugarcane was offered as freshly-cut. Ruminal NDF passage rate was faster for steers fed silage, but only for the genopype with greater *in vitro* NDF digestibility. Ruminal NDF digestion rate was also faster for steers fed silage, and for steers consuming IAC2480, with no significant interaction. Total ruminal NDF digestibility was greater for steers receiving sugarcane as silage, with no effect of genotype. Increased *in vitro* NDFD improved intake and passage rate, but only when given as silage. Feeding sugarcane as silage increased intake and fiber passage and digestion rates.

Session 30 Poster 29

The use of pomegranate pulp silage in growing lamb rations

B. Kotsampasi[1], V. Christodoulou[1] and V.A. Bampidis[2]
[1]Hellenic Agricultural Organization Demeter, Animal Research Institute, Paralimni, 58100 Giannitsa, Greece, [2]Alexander Technological Educational Institute, Department of Animal Production, P.O. Box 141, 57400 Thessaloniki, Greece; bampidis@ap.teithe.gr

In an experiment with 24 male growing Florina (Pelagonia) lambs, effects of dietary pomegranate pulp silage (PS) on performance and carcass characteristics were determined. In the 9 week experiment, lambs were allocated to one of three dietary treatments (PS0, PS120, and PS240) of 8 lambs each. Lambs had an initial body weight (BW) of 18.8±2.28 kg, and were fed one of three isonitrogenous (crude protein 171 g/kg, dry matter – DM basis) and isoenergetic (net energy for gain 5.62 MJ/kg, DM basis) total mixed rations (TMRs) *ad libitum*. The PS was added to the TMR at inclusion levels (as mixed basis) of 0, 120, and 240 kg/t DM for treatments PS0, PS120, and PS240, respectively. No differences (P>0.05) occurred among PS treatments in final BW (34.3 kg), BW gain (0.24 kg/day), DM intake (0.94 kg/day), and feed conversion ratio (3.95 kg DM intake/kg BW gain). Moreover, carcass characteristics were not affected (P>0.05) with increased PS feeding, except for the fat color, fat firmness, wetness and overall acceptability of carcasses, which increased (P<0.05). Pomegranate pulp silage supplementation, at levels up to 240 kg/t DM of TMR, in isonitrogenous and iso (net energy) energetic diets for growing lambs did not affect their performance and carcass characteristics.

Session 30 Poster 30

How to use BMR sorghum silage to feed dairy cows and heifers?

B. Rouillé[1], J.M. Lamy[2], D. Plouzin[2] and P. Brunschwig[1]
[1]Institut de l'Elevage, Monvoisin, BP 85225, 35652 Le Rheu Cedex, France, [2]Chambre d'agriculture de Maine et Loire, Ferme expérimentale des Trinottières, 49140 Montreuil sur Loir, France; benoit.rouille@idele.fr

The high frequency of droughts pushes milk producers to use some alternatives to maize silage. Sorghum is a plant that requires less water than maize. This is why high disgestibility sorghum with BMR gene (brown mid rib) has been experimented during five years under French conditions in Pays de Loire. Eight trials on dairy cows and dairy heifers have been leaded on the experimental farm of Les Trinottières (Chambre d'agriculture de Maine et Loire) using three types of sorghum: Sweet Virginia, Elite and Choice. Sorghum silage represented between 50% and 100% of the forages of the total diets. Looking at dairy cows, standardised milk production was maintained while total intake significantly decreased. Fat content strongly and steady decreased in all experiments, between -2.6 to -4.4 g/kg of milk. Beyond 65% of of BMR sorghum silage in the forages, a significant decrease in global animal performances is observed. However, it allows to maintain a high level of productivity per cow. Looking at dairy heifers, three trials were made and led to the same conclusion. BMR sorghum silage associated with concentrates rich in proteins is well adapted to dairy heifer growth. The goal of 24-months calving has been reached in good conditions when using this forage. BMR sorghum silage is an interesting forage for dairy herds, specifically in drought conditions. It represents a good compromise between agronomical and animal issues. We are now able to deliver a precise message on how to use it for both dairy cows and heifers.

In vitro methane production of chicory and plantain collected at grazing heights

M.J. Marichal[1], L. Piaggio[2], O. Bentancur[1], G. Moreno[1] and S. Rosas[1]
[1]Facultad de Agronomía, Garzon 780, 12900 Montevideo, Uruguay, [2]Secretariado Uruguayo de la Lana, Baltasar Brum 3764, 11800 Montevideo, Uruguay; mariadejesus.marichal@gmail.com

The CH_4 generating potential of *Cichorium intybus* cv. INIA LE Lacerta and *Plantago lanceolata* cv. Ceres Tonic was investigated. At the Faculty of Agronomy (34°50'S; 56°13'W) forages were seeded in three blocks (two plots/block) in September 1 and plants of 15, 20 and 25 cm were harvested in December 15. The NDF, ADF and Lig were determined. In an *in vitro* gas production system 3 batches of 42 bottles were incubated. In cumulated gas at 8, 8 to 24 and 24 to 48 h CH_4 was measured. Results were analyzed in a split-plot design with main plots in randomized blocks. In fiber fractions, no interactions (P>0.20) were detected between effects evaluated. In chicory The NDF, ADF and Lig were 35 to 38, 25 to 27 and 12 to 14 g kg/DM in chicory and 33 to 36, 22 to 27 and 11 to 13 g kg/DM in plantain. In chicory NDF tended (P=0.10) to be greater than in plantain (35 vs. 33 g kg/DM). In both species, plants of 20 and 25 cm registered similar (P>0.20) NDF but greater (P<0.01) than 15 cm plants. Similar (P>0.20) ADF was observed in chicory. In plantain, plants of 20 and 25 cm presented similar (P=0.30) but greater (P=0.06) ADF than 15 cm. No differences (P>0.40) in Lig were observed. Interaction (P<0.01) between specie and plant height was observed in total CH_4 but not (P=0.43) in CH_4 production profile. Total CH_4 was 38 to 41 and 20 to 36 ml g/OM incubated of chicory and plantain. Both species presented similar (P=0.74) CH_4 production profile. In chicory, similar (P>0.30) total CH_4 was observed in plants of different heights. In plantain, plants of 20 and 25 cm presented similar (P=0.70) but greater (P<0.02) total CH_4 than 15 cm. The greatest (P<0.01) CH_4 was produced from 8 to 24 h. From 0 to 8 h and 24 to 48 h CH_4 production was similar (P>0.4) and accounted for 50% of overall production. Chicory and plantain presented similar methanogenic potential and CH_4 production profile.

Session 30 Poster 32

Changes in body content of magnesium in Iberian suckling piglets under different nutritional managements

R. Castellano, M.A. Aguinaga, R. Nieto, J.F. Aguilera, A. Haro and I. Seiquer
INAN-CSIC, C/Camino del jueves s/n, 18100 Armilla, Granada, Spain; rosa.castellano@rennes.inra.fr

Magnesium is associated with Ca and P in bone and it is also an important co-factor of numerous enzyme systems. Different nutritional strategies are usually applied to suckling piglets in order to improve performance and facilitate adaptation to solid feed consumption after weaning. Thirty-eight purebred Iberian sows were involved in two consecutive trials to determine the influence on Mg body content of different nutritional strategies applied to suckling piglets weaned at 35 d of age. Exclusively milk feeding (M), conventional suckling (CS) and intermittent suckling (IS) were studied. Only litters on CS and IS treatments had free access to creep feed from day 15 onwards. Piglets of the CS group had continuous access to their dams, and on IS treatment they were progressively separated from the sow during 6, 8 and 10 h on days 29-30, 31-32 and 33-34, respectively. Eight piglets at birth and one piglet per litter on day 35 of age were slaughtered and used to study whole-body content of Mg; mineral retention was calculated following the comparative slaughter procedure. Average content of Mg in Iberian sows' milk ranged 109-149 mg/kg during the suckling period, whereas in creep feed was 1,200 mg/kg (as fed). The nutritional treatment significantly affected Mg body content and concentration, as higher values were found in milk-fed piglets (P<0.001) than in the other groups (336, 271 and 257 mg/kg EBW for M, CS and IS groups, respectively). Mg retention during the whole period was also increased in M group, both expressed per day (62.7, 49.5 and 51.9 mg/d for M, CS and IS groups, respectively) and on a weight-gain basis (394, 314 and 292 mg/kg EBW gain for M, CS and IS groups, respectively) (P<0.001). The present results show that Mg bioavailability seems to be greater in sows' milk than in creep feed. Mg supply in piglet's nutrition deserves more attention, as the consumption of solid feed in the suckling piglet may affect negatively Mg body content at weaning.

Amino acid composition of muscle and intestine protein in pigs at different ages

J.A. Conde-Aguilera[1,2], C. Cobo-Ortega[1,2], S. Tesseraud[3], Y. Mercier[4] and J. Van Milgen[1,2]
[1]Agrocampus Ouest, UMR1348 PEGASE, Rennes, 35000, France, [2]INRA, UMR1348 PEGASE, Saint-Gilles, 35590, France, [3]INRA, UR83 Recherches Avicoles, Nouzilly, 37380, France, [4]Adisseo France, Antony, 92160, France; alberto.conde@rennes.inra.fr

Changes in the amino acid (AA) composition of whole body protein during growth have been reported. We hypothesize that part of this change is due to changes in the AA composition of different tissues. The objective of this study was to evaluate the AA composition of tissues in pigs at different ages. Two groups of 6 pigs each received diets formulated to meet nutrient requirements and were slaughtered at 10 and 23 wk of age, respectively. Samples of the longissimus dorsi (LM) and rhomboideus (RM) muscles, proximal (I1) and distal jejunum (I2), and ileum (I3) were taken for analysis. The relative weight of LM increased by 32% at 23 wks of age, but decreased for RM (-52%) and the small intestine (-39, -22, and -30% for I1, I2 and I3, respectively). Protein content increased with age by 13% and 7% in LM and RM, respectively. Total AA content (per 16 g N) increased with age from 85.5 to 90.7% on average for the tissues at 23 wk of age. For LM and RM, the increase in AA content was due to an increase in essential AA content (especially His, Lys, and Ile in LM, and His, Phe, and Tyr in RM), while the content of non-essential AA decreased with age (especially Gly and Pro). In I1, the contents of Thr, Ile, Phe, and His decreased with increasing age, while that of Gly and Pro increased. In I2, the Met and Asp content increased while that of Thr and Glu decreased. In I3, the content of Met, Cys, Ser, and Gly increased within increasing age and that of Lys, Phe, Tyr, Arg, and Asp decreased. The AA composition of tissue protein is affected by age (or body weight), and the changes in composition differs among tissues. This may contribute to the change in the whole body AA composition during growth.

Comparison of gastrointestinal tracts and pH values of Ross 308 broiler and indigenous Venda chicken

M. Mabelebele[1,2], O.J. Alabi[1], J.W. Ngambi[1] and D. Norris[1]
[1]University of Limpopo, Agricultural Economics and Animal Production, P/Bag X1106 Sovenga Turfloop, 0727 Polokwane, South Africa, [2]University of New England, Enviromental and Rural Science, University of New England, Armidale, NSW 2351, Australia, Australia, Australia; jones.ngambi@ul.ac.za

An experiment was conducted to determine and compare the pH values in the digestive tracts of Ross 308 broiler and indigenous Venda chickens. A complete randomized design (CRD) with two treatments (Ross 308 broiler and indigenous Venda chickens) having five replicates was used in this experiment. The chickens were both raised under intensive management system and fed a commercial diet until the age of 70 d. Five birds per treatment (breed) were taken, weighed and killed by cervical dislocation, then scalded and defeathered. Weights, lengths and pH values of different segments of the gastrointestinal tract were measured immediately by using a sensitive level electronic scale (RADWAG), millimeter graduate rule and a Model 191 pH meter equipped with a spear-like electrode (Knick, Berlin, Germany), respectively. Breed difference had effect ($P<0.05$) on live weight, crop, gizzard and large intestine weight. However, proventriculus, small intestine, caecum, and liver weights were not influenced ($P>0.05$) by the breed differences. All the length parameters measured were similar ($P>0.05$), however, the GIT of the broiler chickens was longer ($P<0.05$) than the indigenous Venda chickens. In addition, breed differences had effect ($P<0.05$) on the crop, gizzard and small intestine pH. However, the proventriculus, large intestine, caecum and pH at 2, 4, 6 and 24 h were not influenced by the breed differences ($P>0.05$). The differences observed in the present study could be attributed to genetic variation of the chickens.

Session 30 Poster 35

A sulphur amino acid deficiency changes the amino acid composition of tissue protein in growing pigs

J.A. Conde-Aguilera[1,2], C. Cobo-Ortega[1,2], S. Tesseraud[3], Y. Mercier[4] and J. Van Milgen[1,2]
[1]Agrocampus Ouest, UMR1348 PEGASE, 35000 Rennes, France, [2]INRA, UMR1348 PEGASE, 35590 Saint-Gilles, France, [3]INRA, UR83 Recherches Avicoles, 37380 Nouzilly, France, [4]Adisseo France, 92160 Antony, France; alberto.conde@rennes.inra.fr

Methionine and Cys are considered the second or third-limiting amino acid (AA) for most cereal-soybean based diets in growing pigs. The AA content of body protein is usually assumed constant. However, there are indications that this AA composition can be affected by the AA supply. The objective of this study is to evaluate the effect of feeding pigs diets with a deficient or sufficient total sulfur AA supply (TSAA; Met+Cys) on tissue composition and meat quality. Two groups of 6 piglets were selected 14 d after weaning and received a diet either deficient (TSAA–) or sufficient (TSAA+) in TSAA for 17 wk and were slaughtered thereafter. Samples of the longissimus dorsi (LM) and rhomboideus (RM) muscles, liver, intestines and skin section were taken for analysis. Diets were formulated to meet nutrient requirements except for diet TSAA– in which the supplies of Met and TSAA were 19 and 16% below requirements, respectively. The TSAA– pigs had a lower ADG, and weights of the LM, RM, and skin section ($P<0.05$). The protein content of LM and RM decreased ($P<0.05$) in TSAA– pigs, while this was not changed in other tissues. The TSAA supply affected the AA composition of protein in all tissues but the Met content was changed only in the liver ($P<0.05$), and the Cys content in RM, distal jejunum and ileum ($P<0.01$). We conclude that pigs use different mechanisms to cope with a prolonged AA deficiency as tissues respond differently in terms of weight, tissue protein and AA contents with a slight impact on meat quality. Muscles responded more to a TSAA deficiency than did other tissues. The observed changes in the tissue AA composition question the use of a constant AA profile of retained protein in swine nutrition.

Session 30 Poster 36

Use of phase feeding in combination with split gender grouping for pigs

E. Magowan
Agri-Food and Biosciences Institute, Hillsborough, BT26 6DR, United Kingdom; elizabeth.magowan@afbini.gov.uk

In a 2×3 factorial design, the performance of gilts and boars penned in single or mixed gender groups and offered either a single diet (single phase) or a two phase dietary regime during finish was compared. A total of 480 pigs (PIC337 × (Landrace/Large White)) were grouped in pens of 10. Eight replicates per treatment were completed (8 time periods). Diets were barley, wheat, soya based and formulated to contain 13.5 MJ/kg DE and an ideal protein ratio. Diet 1 (formulated to contain 180 g/kg crude protein (CP), 9.5 g/kg lysine) and was offered to pigs weighing between 45 and 120 kg in the single phase regime and between 45 and 80 kg in the two phase regime. Diet 2 (formulated to contain 167 g/kg CP, 8.0 g/kg lysine) was offered between 80 and 120 kg in the two phase dietary regime. Pigs were individually weighed and pen feed intake was taken to correspond with the change in diet (target 80 kg) and at slaughter. Data was analysed according to the experimental design using Analysis of Variance in Genstat V 10. Start weight was used as a covariate in the analysis of live pig performance and slaughter weight as a covariate for the analysis of carcass performance There was no interaction ($P<0.05$) between group gender and dietary regime. Dietary regime had no effect on pig performance (average daily gain (ADG) 905 g/day, feed conversion ratio (FCR) 2.56 between 44 and 120 kg, both $P>0.05$, SEM 12.9 and 0.029 respectively) or back fat depth at P_2 (average 12.2 mm, $P>0.05$, SEM 0.27). However, between 45 and 120 kg the performance of gilts was poorer (ADG 865 g/day, FCR 2.70; ($P<0.01$, SEM 13.7 and <0.001, SEM 0.025 respectively) than that of boars (ADG 942 g/day, FCR 2.46). Although the kill out percentage of gilts was better than boars (77.7 vs. 76.1%; $P<0.01$, SEM 0.31), their back fat depth (P_2) was greater (12.1 vs. 11.5 mm ($P<0.01$, SEM 0.31). The performance of the mixed gender pens was intermediate to that of pens of gilts and boars housed separately but their back fat depth at P_2 was greater (13.0 mm).

Session 30 Poster 37

Effect of dietary CP and rumen protected CLA on performance of lactating cows

F. Tagliapietra, S. Schiavon, G. Cesaro, A. Cecchinato and G. Bittante
University of Padova, DAFNAE, Viale dell'università 16, 35020 legnaro PD, Italy; franco.tagliapietra@unipd.it

The effects of dietary protein level (CP) and of rumen protected conjugated linoleic acid (rpCLA) on dry matter intake (DMI), milk yield, rumen activity, and nutrients digestibility were evaluated. Twenty Holstein-Friesian cows, homogeneous for parity, body weight, body condition score, DIM, and milk yield, were housed in 4 pens and fed 4 diets containing 15% (CP15) or 12% of CP (CP12), supplemented or not with 80 g/d of rpCLA, following a 4×4 Latin Square design. Each period lasted 3 weeks. The CP12 diet was formulated from CP15 by replacing soybean meals with barley, to contain similar NDF content (37% DM). DMI, rumen activity, milk yield and quality were daily monitored. Diet digestibility was evaluated using ADL as marker. Data were analyzed using GLM procedures in SAS. The reduction of dietary CP content did not influenced DMI, DM and NDF digestibility but slightly reduced true CP digestibility and increased rumination activity from 16.8 to 19.8 min/kg DMI (P=0.009). Reduction of dietary CP decreased milk yield only from 29.0 to 27.7 kg/d (P=0.072), milk protein yield from 1.00 to 0.91 kg/d (P=0.006), but N efficiency was improved (+14%, P<0.001). Supplementation of rpCLA reduced DMI from 21.4 to 19.6 kg/d (P=0.018), milk fat production from 0.99 to 0.93 kg/d (P=0.001), slightly increased rumination activity (P=0.06), without effect on diet digestibility and improved efficiency of N utilization (+7%, P=0.019). Results shown that a strong reduction of dietary CP influences milk yield less than it was expected from simulation using the NRC model which predicts for CP15 and CP12 milk yields of 23.8 and 13.9 kg/d, respectively. The effects of a reduction of diet CP on rumen degradation of feeds would be mitigated by a compensative increase of rumination activity. The significant effect of rpCLA on the efficiency of N utilization suggest that these bioactive molecules can exert influences on N metabolism.

Session 30 Poster 38

Influence of pre-grazing herbage mass on *in vivo* digestibility of perennial ryegrass in sheep

M. Beecher[1,2], R. Baumont[3], T.M. Boland[1], M. O'donovan[2], J. Aufrère[3] and E. Lewis[2]
[1]School of Agriculture & Food Science, University College Dublin, Belfield, Dublin 4, Ireland, [2]Teagasc, Animal & Grassland Research & Innovation Centre, Moorepark, Fermoy, Co. Cork, Ireland, [3]INRA, UMR 1213, Herbivores, 63122 St Genès Champanelle, France; rene.baumont@clermont.inra.fr

Pre-grazing herbage mass (HM) is a critical factor influencing herbage intake and animal performance. Low HM swards are perceived as being highly digestible leading to high animal performance. The aim of this study was to determine the effect of pre-grazing HM on the digestibility of perennial ryegrass in sheep. The study was run as a replicated Latin square design. Twelve Texel wether sheep were housed in individual stalls allowing for the total collection of urine and faeces. Animals were offered one of three treatments: low (1000 kg DM/ha), medium (2,000 kg DM/ha) and high (3,000 kg DM/ha) HM, fed *ad libitum* (10% refusal rate). Following a 6-day acclimatisation phase, a 6-day measurement phase (MP) began during which total daily intake and faeces produced were recorded. Three of these 12-day periods made up each Latin square. A representative sample of grass offered and faeces voided from each sheep was collected daily. The average dry matter digestibility (DMD) over the 6-day MP was determined for each sheep. Digestibility data were analysed using PROC MIXED in SAS, sheep within period was included as a random effect. Digestibility data for organic matter, neutral detergent fibre, acid detergent fibre and nitrogen are due shortly. There were no significant interactions. There was a significant effect of treatment on DMD (P<0.05). Average DMD for low HM was 0.725±0.0397, medium HM was 0.705±0.0185 and for high HM was 0.674±0.0231. There was no difference between the low and medium HM (P>0.05) but DMD was lower for sheep on the high HM compared to the low and medium HM (P<0.05). These results suggest that as HM increases above 2000 kg DM/ha there is a reduction DMD. This is consistent with previous studies which found that as HM increases DMD decreases.

Estimating degradability of free and fat-protected tryptophan using rumen *in vitro* gas production

G. Marín, G. Pichard, F. González and R. Larraín
Pontificia Universidad Católica de Chile, Facultad de Agronomía e Ingeniería Forestal, Departamento de Ciencias Animales, Vicuña Mackenna 4860, Macul, 6904411 Santiago, Chile; larrain@uc.cl

Tryptophan (Trp) fed to ruminants may be degraded in the rumen by microorganisms before it reaches the small intestine. The goal of this experiment was to estimate the availability to rumen microorganism of free tryptophan (L-Trp) and fat-protected Trp (FPT) using an *in vitro* gas production technique. An *in vitro* ruminal fermentation system was designed to minimize all foreign sources of nitrogen and let Trp be its main source. Tryptophan was protected by mixing with rumen-inert fat from hydrogenated fish oil with cold crystallization. The substrate provided 200 mg of fermentable polysaccharides and sugars. A standard curve was made using nine concentrations of casein hydrolysate (assumed to be 100% degradable), equivalent to 0, 1, 2, 3, 4, 5, 6, 7 and 8% w/w CP in the substrate. Availability of Trp was evaluated by estimating potential gas production by a logistic model and comparing it to the casein standard curve. Treatments were arranged in a 2×2 factorial with 2 sources of Trp (L-Trp and FPT) and Trp supplying 4 and 8% w/w CP in the substrate. Three replicates were used for each concentration of casein and 5 for each Trp treatment. The experiment was repeated 3 times and analyzed by ANOVA ($P<0.05$) using a complete block design. Degradability of Trp (±2.5%) was 16.3% for 4% CP from L-Trp, 9.3% for 8% CP from L-Trp, 18.9% for 4% CP from FPT and 12.2% for 8% CP from FPT. Results indicated no difference in availability between L-Trp and FPT (13.0% vs. 15.6±1.8%), but higher inclusion reduced Trp availability from 17.6 to 10.9±1.8%. We concluded that protection of L-Trp by fat was not effective, but its rumen availability was already low so it might be used directly as a source of rumen-escaped Trp.

Utilization of globe artichoke by-products in sheep feeding

F. Salman, Y.A.A. Elnomeary, A.A. Abedo, H.H. Abd El-Rahman, M.I. Mohmed and S. Ahmed
National Research Center, Animal Prod., Dokki 12622, Egypt; elnomeary@yahoo.com

This study was conducted to evaluate the effect of insert artichoke by-products in Rahmany sheep ration and its effect on animals performance. Twelve male growing Rahmany lambs aged 8 months with an average body weight 31.7±0.38 kg were fed three rations (4 animals/ration). The three groups were fed concentrate feed mixture (FCM) to cover 50% of requirements, plus kidney bean straw (control, 1st group), a 1:1 mixture of kidney bean straw and artichoke by-products (2nd group) or artichoke by- products (3rd group). The roughages were given *ad libitum*. Daily live weight gain, dry matter intake and feed conversion were measured. Digestibility trials were conducted to determine the nutrients digestibility coefficients and nutritive values of the three tested rations. Rumen parameters were also measured. Results obtained showed that artichoke by-products had higher value of CP (16.6%) compared with kidney bean straw and nearly similar with CFM, but the CF content was higher in artichoke by-products. The dry matter intake of group 3 was higher than the other tested rations. Nutrients digestibility and nutritive values in terms of TDN and DCP for group 3 were the highest followed by group 2 and group1, respectively. Concerning average daily gain, animal fed ration 3 recorded highest gain (263 g) followed by group 2 (256 g) and group 1 (247 g). The values for feed conversion were nearly the same for the three groups. All ruminal parameters values were in the normal range for all groups. In conclusion, insert the artichoke by- products in sheep ration improve animal performance without any adverse effect.

Impact of dietary protein and PUFA level on lipogenesis-related protein expression in porcine tissue

D. Dannenberger[1], K. Nuernberg[1], G. Nuernberg[1] and A. Priepke[2]
[1]Leibniz Institute of Farm Animal Biology, Institute of Muscle Biology and Growth, Wilhelm-Stahl-Allee 2, 18196 Dummerstorf, Germany, [2]State Institute for Agriculture and Fishing Research, Institute of Animal Production, Wilhelm-Stahl-Allee 2, 18196 Dummerstorf, Germany; dannenberger@fbn-dummerstorf.de

The clarification of mechanisms regulating de novo synthesis of FA and fat partitioning in tissues in pigs is an important step in designing the strategies for producing pigs with desirable FA composition and fat content. In total 40 male Landrace pigs were allocated into four experimental groups fed diets different in dietary protein (19.5% vs. 15.5%) level and oil supplementation (sunflower seed vs. linseed oil). The animals of the control group were fed a regular diet without vegetable oil supplementation. The pigs were fed from 60 kg to until 120 kg live weight. Diet effects were estimated by one-way analysis of variance using the GLM procedure of SAS software system. Protein expression levels of acetyl-CoA carboxylase, fatty acid synthase and stearoyl-CoA desaturase were not affected by different dietary protein as well as in muscle, back fat and liver, and resulted in unaffected fat contents and total SFA- and MUFA concentrations in the tissues. However, reduced protein diet combined with PUFA supplementation in muscle lead to a down-regulation of the mature form of mSREBP-1, compared to high protein diets. In back fat and liver, different diets caused no significant changes for protein expression of SREBP-1 forms, precursor (pSREBP-1) and mSREBP-1. The FA concentrations in the tissues were affected by the diets resulted in higher n-3 FA concentration of linseed oil containing diets compared to sunflower oil diets. Different dietary protein level and PUFA supplementation resulted in significantly lower total PUFA concentrations in all investigated tissues compared to control group.

Effects of probiotic protexin on the egg laying traits and gut bacterial load of Japanese quails

N. Vali, A. Ghaderi Samani and A. Doosti
Islamic Azad University, Shahrekord Branch, Shahrekord 8813733395, Iran; nasrollah.vali@gmail.com

One of the probiotics used in poultry feed is Protexin. This study was conducted to investigation effect of different levels of Probiotic protexin on the egg laying traits and bacterial load of Japanese quails gut. 120 Japanese quails (*Coturnix japonica*), 6 weeks of age were equally allocated to four treatments containing three pens (include 8 female and 2 male) in each. Birds were received diets which had been supplemented with 0, 0.250, 0.500, and 1.000 g Probiotic Protexin per kg diet as treatments 1 (control), 2, 3 and 4 respectively. During 63 days of experimental period, egg weight, albumen weight, yolk weight, and shell weight was measured. And also Samples of the small intestinal and ceca contents of 4 quails from each replicate were collected in polymerase chain reaction (PCR) tubes. Bacterial load of *Campylobacter*, *Closterdium perfringens* and *Lactobacillus* were measured using Real time PCR. Rotorgene 6000 software, version 1.7 used for variables and PCR setting. All analyses data were carried out with one way analysis of variance SPSS. Means were compared by using Duncan's test. Control group had the lower egg weight ($P<0.05$) than the other groups. Yolk weight of control group and 2 and 3 treatments did not show significant difference whereas they were different with 4th treatment ($P<0.05$). *Lactobacillus* bacteria concentration as beneficial digestive bacteria was the highest in treatment 2 and then treatments 4 and 3 had the high concentration. Concentration of *Closterdium perfringens* and *Campylobacter* in quails that fed whit treatment 2 had the highest concentration compared to the control group and then treated with 3 and 4, respectively, had the lowest concentration.

Diurnal variation of ruminal pH, and N-NH_3 concentration in dairy cows grazing two herbage allowance

L. Leichtle[1], M. Noro[2], R. Pulido[1] and R. Delagarde[3]
[1]University Austral of Chile, Valdivia, Chile, [2]Universidade de Passo Fundo, Rio Grande do Sul, Brazil, [3]INRA Agrocampus Ouest, Saint-Gilles, France; remy.delagarde@rennes.inra.fr

The aim of the trial was to determine diurnal variation of ruminal pH and N-NH_3 concentration of dairy cows grazing two herbage allowances and supplemented with high moisture corn (HMC) or cracked wheat (CW). The trial was conducted in Vista Alegre Experimental Station of the University Austral of Chile, for 56 days during the Spring of 2010. 4 Black Friesian ruminally cannulated dairy cows. (212±72 days in milk, live weight of 566±59 kg, body condition score of 2.97±0.1 (1 to 5), and milk production of 22.7±4.5 l). The cows were distributed in a Latin square design of 2×2, The 4 treatments resulting from the combination of the two herbage allowances(30 and 20 kg DM/cow/day; 18.8% DM, 20.7% CP, 2.78 Mcal ME/kg of DM) and supplemented with 3.5 kg DM/cow/day with HMC (69.2% DM, 7.7% CP and 3.30 Mcal ME/kg of DM) or CW (84% DM, 12.4% CP, 3.18 Mcal of ME/kg of DM), At day 14, samples of ruminal fluid were collected at fixed times to determine immediately pH values Another, 1.5 ml of ruminal fluid were acidified with 0,4 ml of trichloroacetic acid to determine N-NH_3 concentrations (Indophenol Method). Statistical analysis was performed using a linear mixed model on a repeated measures factorial design. Increasing pasture allowance produced similar changes in ruminal pH ($P>0.05$, 5.72 and 5.76 for high and low respectively). The ruminal N-NH_3 concentration is similarly affected by pasture allowance ($P>0.05$, 8.13 and 8.68 mmol/l for HMC and CW).Supplementation did not affect pH or N-NH_3 ($P>0.05$). Variations were observed associated with the sampling time, which would be associated with the grazing cycle. The ruminal pH showed its maximum at 6:30 h (6.4), and its lowest value (5.3) at 22:30 h ($P<0.05$). N-NH_3 concentrations remained constant during the morning, increasing sharply in the afternoon ($P<0.05$) with maximum values at 19:45 h (14 mmol/l) and then, in the evening, down to 4.4 mmol/l, $P>0.05$).

Ruminal and intestinal starch digestibility of rolled and NaOH treated rye in dairy cows

A.L.F. Hellwing, M. Larsen and M.R. Weisbjerg
Aarhus University, AU Foulum, P.O. Box 50, 8830 Tjele, Denmark; martin.weisbjerg@agrsci.dk

Wheat treated with NaOH is a rumen friendly starch source to dairy cows. Rye is a cost competitive alternative to wheat, however, knowledge on digestibility is scarce. The aim of the study was to examine starch digestibility of rolled and NaOH treated rye in different parts of the digestive tract in dairy cows. Four rumen, duodenal and ileal fistulated multiparous lactating Holstein cows were used in a cross over design and fed either rolled rye or NaOH rye grain (rye mixed with 10% water and 3% NaOH). Rations were composed of (% DM): grass-clover silage, 48.1; rye grain (rolled or NaOH), 43.8; soybean meal, 6.8; and minerals and vitamins, 1.3. Rations were fed *ad libitum* twice daily as total mixed rations. Chromic oxide was used as digesta flow marker. Samples of fluid from medial and ventral rumen contents, intestinal digesta, and faeces were obtained at 6 h intervals. Energy corrected milk yield was 31.0 and 32.5 kg/d ($P=0.80$) for rolled and NaOH rye, respectively. DM intake was 20.8 and 19.9 ($P=0.09$) and starch intake was 5.7 and 5.4 ($P=0.05$) kg/d for rolled and NaOH rye, respectively. Medial rumen pH decreased postprandial for rolled rye as compared with stable postprandial pH for NaOH rye (interaction, $P=0.01$). Starch digestibility (% of entering) in the rumen was 88.1 and 78.7 ($P=0.09$), in small intestine 76.3 and 62.1 ($P=0.30$), and total tract 99.2 and 96.6 ($P=0.04$) for rolled and NaOH rye, respectively. In conclusion, NaOH treatment of rye grain reduced ruminal digestibility of starch resulting in higher and more stable pH in comparison with rolled rye. The small intestinal starch digestibility remained relatively high in spite of the reduced ruminal digestibility. The total tract digestibility was reduced, however, to a limited extent without significance in practical feeding. Overall, the digestive profiles of starch from both rolled and NaOH treated rye were similar to observations with wheat.

Whole plant and cell wall digestibility in maize silage

J. Peyrat[1], A. Le Morvan[1], A. Férard[2], G. Cabon[2], P.V. Protin[2], P. Nozière[1] and R. Baumont[1]
[1]INRA de Clermont-Ferrand Theix, UMR-Herbivores, Saint-Genès Champanelle, 63122, France, [2]ARVALIS Institut du Végétal, SAEE, Station expérimentale de la Jaillière-La Chapelle Saint Sauveur, 44370, France; julie.peyrat@clermont.inra.fr

The evaluation of the energetic value of maize silage relies on organic matter digestibility (OMd) prediction. However, maize silage is composed of two energetic fractions, cell wall and starch, which vary in proportion among varieties and maturity stages, and have a different rate and extent of digestion. The aim of the present experiment was to analyse whole plant and cell wall digestibility of maize silage in order to improve evaluation of energetic value. The digestibility of sixteen maize silages (four varieties, A, B, C and D, which differ among type of grain and four stages of maturity harvested from 27% to 42% of dry matter) was measured on sixteen castrated sheep in individual metabolic crates. Measurements of digestibility were carried out according repeated latin square designs. Organic matter digestibility (OMd), and cell wall digestibility (NDFd) was calculated. Variety and maturity effects on digestibility parameters were analysed with the MIXED procedure of SAS. Starch and NDF contents of the silages varied from 29% to 47% of MS and from 29% to 40% of MS respectively. OMd of C-variety (42% of starch) is higher ($P<0.001$) than OMd of B-variety (36% of starch). Nevertheless, OMd did not vary with maturity stage. In contrast, NDFd was not significantly affected by variety but by maturity stage. NDFd of early stages were higher ($P<0.001$) than that of advanced stages. Moreover, a close negative relationship between digested OM and indigested cell wall (NDFnd) was observed ($r^2=0.78$). With maturity stage, the decrease in NDF content and the increase in starch content of the plant compensate the decrease in NDF digestibility, leading to a rather constant evolution of OMd. Advances in prediction of energetic value should be obtained by searching prediction criterion of the NDFnd fraction in maize silage.

Mineral requirements of pregnant dairy goats

C.J. Härter, A.R. Rivera, D.S. Castagnino, L.D. Lima, H.G.O. Silva, K.T. Resende and I.A.M.A. Teixeira
UNESP, Univ Estadual Paulista, Department of Animal Sciences, Jaboticabal, SP, 14884-900, Brazil; harter.carla@gmail.com

The gestation period of goats is marked by big gaps of information about minerals requirements. Therefore, the objective of this study was to determine the requirements for pregnancy of calcium (Ca), phosphorus (P), magnesium (Mg), sodium (Na) and potassium (K) in singleton and twin-pregnant dairy goats. The experiment was performed using comparative slaughter method with 51 multiparous goats (BW 49.5±7.8 kg). At the beginning of the experiment six goats (3 Oberhasli and 3 Saanen) were slaughtered to estimate the composition of the mammary gland and uterus in non-pregnant goats. After confirmation of pregnancy, the others animals were distributed in a completely randomized design in a 2×2×4 factorial arrangement, with two breeds (Oberhasli and Saanen), two types of pregnancy (single and twin) and slaughtered at different gestational age (50, 80, 110 and 140 days). Minerals retention was estimated in pregnant uterus and mammary gland during gestation. The requirements were obtained from the data fitted to Gompertz model. The goats with twin pregnancy had higher mineral requirements ($P<0.001$). Oberhasli goats had higher minerals requirements of pregnancy than Saanen goats ($P<0.001$). The requirements of Oberhasli ranged from 0.05 to 1.05 and 0.05 to 0.93 g/day of Ca, from 0.03 to 0.58 and 0.06 to 0.73 g/day of P; 0.006 to 0.05 and 0.005 to 0.03 g/day of Mg; 0.08 to 0.34 and 0.12 to 0.35 g/day of Na; 0.04 to 0.17 and 0.05 to 0.15 g/day of K at 50 and 140 days of gestation, in single and twin pregnancy, respectively. The requirements of Saanen ranged from 0.006 to 0.91 and 0.03 to 0.68 g/day of Ca, from 0.02 to 0.67 and 0.06 to 0.37 g/day of P; 0.002 to 0.05 and 0.006 to 0.03 g/day of Mg; 0.08 to 0.32 and 0.13 to 0.39 g/day of Na; 0.04 to 0.13 and 0.06 to 0.20 g/day of K at 50 and 140 days of gestation, in single and twin pregnancy, respectively.

Session 30 Poster 47

Utilization of sugar beet pulp in feeding sheep

A.Y. El-Badawy, S.S. Abdel-Magid, H.A.A. Omer, M.I. Mohamed and I.M. Awadalla
National Research Centre, Department of Animal Production, Dokki, Giza, 12622, Egypt;
soha_syd@yahoo.com

A total of 14 Ossimi lambs with average weight 30.20 +1.54 kg and aged 5-6 months were divided into two equal groups, each of 7 animals, to study the effect of using sugar beet pulp with 10% soy bean meal (SBP) instead of concentrate feed mixture (CFM). Feeding trial lasted 56 days and lambs were fed at 4% of live body weight and clover hay was offered *ad libitum*. The results showed that, dietary treatment had no significant effect (P>0.05) on all digestion coefficients (DM, OM, CP, CF, EE and NFE) and TDN value. The values of previous data of digestion coefficients and nutritive values were 74.56 and 76.49; 76.87 and 78.66; 69.64 and 64.34; 73.01 and 82.03; 82.27 and 64.02; 81.41 and 80.76; 73.96 and 73.95 and 9.54 & 8.49 for lambs fed CFM or SBP, respectively. Nitrogen retention was positive for two group animals. Ruminal fluid parameters of SBP (pH value, NH_3-N and TFV's concentrations) were significant decreased (P<0.05) in comparison with control diet. The values were 6.15, 21.14 and 9.03 for lambs fed CFM, vs. 6.83, 27.08 and 10.47 for group lambs fed SBP, respectively. Dietary treatment had no significant effect on ADG, feed intake and feed conversion. Values of ADG was 195 and 157 g; DM intake was 2.102 and 2.013 g/h/d and feed conversion ratio was 10.78 and 12.82 (kg intake/kg gain) for lambs fed CFM or SBS, respectively. Dietary treatments had no significant on blood parameters (total protein, albumin, globulin, glucose, ALT, AST, alkaline phosphates and creatinine). On the other hand, values of triglyceride, cholesterol and uric acid were significantly decreased (P<0.05). While, blood plasma urea was significantly increased (P<0.05). Carcass characteristic was not significant effected by replacing CFM by SBP in sheep diet.

Session 30 Poster 48

Growth performance of growing New Zealand White rabbit fed bean straw

S.S. Abdel_magid, I.M. Awadalla and M.I. Mohamed
National Research Centre, Department of Animal Production, Dokki, Giza, 12622, Egypt;
soha_syd@yahoo.com

This study aimed to search for an alternative feed sources instead of the formal ones to be used in rabbit diets. This was designed to study the effect of substitution of bean straw instead of clover hay at level of 0, 25, 50 and 100% in growing New Zealand White (NZW) rabbit diets during eight week fattening period from 6 to 14 weeks of age on growing rabbit performance, digestion coefficient. In this study, 36 weaned NZW rabbits, 6 weeks old, with an average initial weight of 690 g, were distributed randomly into four equally groups, each contained 3 replicates. Four experimental grower diets of iso-nutritive value but differ in their components. The results indicated a decrease in body weight of the rabbits fed 100% bean straw compared to the control group. While, group of rabbits fed 50% bean straw was the best one. The best feed conversion values were obtained by rabbits received 25% bean straw. The best nutritive value expressed as TDN was recorded by rabbits fed 50% bean straw. It could be noticed that all treatments supported positive nitrogen balance. In conclusion, the bean straw could be safely used in the commercial diet of growing rabbit diets instead of clover hay.

Session 31 — Theatre 1

The pig/pork production system as a complex adaptive system

K.H. De Greef[1] and J.S. Buurma[2]

[1]Wageningen UR Livestock Research, P.O. Box 65, 8200 AB Lelystad, the Netherlands, [2]LEI, Wageningen UR, Agricultural Institute, P.O. Box 29703, 2502 LS Den Haag, the Netherlands; karel.degreef@wur.nl

The pig production system is complex due to the many actors and the many interactions between technological development, market and society forces. Recently, the concept of Complex Adaptive Systems (CAS) was first applied here to enhance understanding of the developments. Elsewhere, a CAS-approach has proven its value for understanding dynamics, not in the least because of the availability of Agent Based Modelling (ABM) as a tool. In ABM, the separate actors (pigs, farmers, chain actors, consumers, etc.) are modelled as individuals, with each agent (person, animal) having its own agenda. The early results imply that the system (strictly spoken the European pig carcass commodity system) indeed acts as a Complex Adaptive System: individual actors conduct their own agenda's, it responds to the environment and especially: it shows emergence despite absence of a central driving force. The conclusion that the pig sector behaves along the criteria of a CAS has consequences as to how effects of external interventions other than legislation can be expected. For example: for enhancement of new system equilibriums, specific support to early initiatives is expected to be more effective than generic innovation support to all producers. Currently, the pig system is modelled from a CAS-perspective at three levels: the farm level: pigs as agents within the farm; the sector level: pig farmers acting as agents forming the sector; and at society level, where the interaction between representatives of the various stakeholder groups (NGO's, government, farmers organisations) interact. The case 'Towards cessation of castration' illustrates the relevance of a CAS multi-level analysis. Especially the upper systems level, the so called innovation arena where the chain and societal actors meet at representatives level turned out to be of major influence in the early stages of the systems change.

Session 31 — Theatre 2

Development of an animal model of ruminant production systems, from growing cattle to mature cows

E. Ruelle[1,2], L. Delaby[3], M. Wallace[1] and L. Shalloo[2]

[1]UCD, School of Agriculture, Food Science and Veterinary Medicine, Belfield, Dublin 4, Ireland, [2]Teagasc, Animal & Grassland Research and Innovation Centre, Moorepark, Fermoy, Co. Cork, Ireland, [3]INRA, UMR 1348, PEGASE, Domaine de la Prise, 35590 Saint Gilles, France; elodie.ruelle@teagasc.ie

Simulating production systems through mechanistic and dynamic models allows experimentation of key system components in a low cost, timely, and effective manor. With always limited research resources, models can be used to both test hypothesis and/or help guide the decision making process around experiments required for future systems. A new dynamic, mechanistic, stochastic model of dairy animals was developed. The model was developed with a daily time step using C/C++. Each animal is simulated from birth through the different stages of their life including growth, lactation and finally death. The model was evaluated using different 3 different animal groups around the genetic potential for milk production (LG: 25 kg/cow/day, MG: 35 kg/cow/day or HG: 45 kg/cow/day of milk at peak), 3 different herbage allowance (HA) groups (14 kg DM/cow/day, 18 kg DM/cow/day, 22 kg DM/cow/day above 4 cm) all with an allocation of 0 or 4 kg of concentrate per day throughout the whole lactation. The model outputs indicate that the model is capable of reacting in a sensible fashion to the variation in the genetic potential for milk production with a milk production difference of 5.2 kg/cow/day between the LG and MG and 4.45 kg/cow/day between the MG and HG. The effect of an increasing HA from 14 kg DM/cow/day to 18 kg DM/cow/day resulted in an increase in milk production by 0.75 kg/cow/day for the MG group. Supplementing with 4 kg of concentrate results in 3.0 kg/cow/day, 2.5 kg/cow/day and 2.1 kg/cow/day of milk at a low, medium and high HA for the MG group of animals. The model has been able to predict a higher milk response for HG compared to LG or MG cows to increased feed allowance and increased supplementation levels. All expected results are within previously published ranges.

Typology of Manchega dairy sheep farms in Castilla La Mancha (Spain): preliminary results

J. Rivas, J. Perea, E. Angón, P. Toro-Mujica, C. Barba and A. García
University of Cordoba, Animal Production, Campus Rabanales, 14071, Spain; pa1gamaa@uco.es

The region of Castilla La Mancha has a long tradition of sheep production, with a census which reaches 2.6 million heads, most of them Manchega breed (more than 1.5 million head). Their milk production is entirely intended to cheese making and, since 1985, all is for the denomination of origin 'Queso Manchego'. A survey was conducted on 156 farms in 2012, in order to establish the main technical, socio–economical and productive aspects of the Manchega dairy sheep farms. A principal component analysis showed four factors explaining 65.7% of the original variance. The systems differed in their intensification level, dimensionality, technological level and agricultural bias. The subsequent cluster analysis leads to characterize four different dairy sheep farm systems. Groups I and II are mixed crop–dairy farms with small size, which base their production on grazing. Group I (24.4% of farms) concentrates the most extensive farms with the lower productivity. Group II (51.9% of farms) is formed by farms with higher levels of technology and mechanization. Group III (14.1% of farms) are semi–intensive farms of medium size, mainly directed at cultivating cereals for sheep feeding. The food is based on conserved forages, produced on the farms and use of supplementary feed. Group IV (9.6% of farms) are large intensive farms specialized on dairy. This production model requires a higher contribution of external feedstuff, increasing feed costs despite of the intensive use of the land. This study has been development inside of the Research Fundamental Project oriented to agricultural resources and technologies in coordination with the Autonomous Communities of the Spanish National Institute of Research and Agricultural and Food Technology (RTA2011-00057-C02-02).

Session 31 Theatre 4

Developing a new indicator to assess nitrogen efficiency of various farming systems

O. Godinot, M. Carof, F. Vertès and P. Leterme
INRA, Agrocampus Ouest, UMR 1069 SAS, 65 rue de Saint Brieuc, 35000 Rennes, France; godinot@agrocampus-ouest.fr

Reactive nitrogen (N) fluxes are greatly increasing worldwide, mainly as a consequence of increased food production. They have major impacts on water, air and soil quality, biodiversity and human health. One of the main ways to increase food production while decreasing its detrimental effects is to increase nitrogen use efficiency (NUE). Two indicators are commonly used to evaluate N use and its efficiency at the farm scale: N farm gate balance (NB) and NUE. To take into account N fluxes linked to inputs, and to assess the whole farming system's nitrogen efficiency (SNE), we propose a new indicator combining NUE and a life cycle analysis (LCA) approach. SNE brings the following improvements to NUE: We consider useful products (crops, milk, animals) rather than total outputs. Manure is taken into account but not considered a product. Possible changes in soil organic matter status (storage / decrease of organic N) are estimated. We calculate a net input or a net production when similar products are both entering and going out of the farm. This prevents a mathematical bias occurring in the calculation of NUE. We use LCA references to estimate N emissions linked to input production and occurring outside the farm. We enlarge the farm perimeter to calculate global farming system efficiency; this avoids bias linked to farmer strategies relying on purchased inputs vs. self-sufficiency. The successive steps from NUE to SNE are calculated for a sample of 38 farms. SNE corrects some biases of NUE, and is a useful indicator to compare the N efficiency of different farming systems. However, a good value of SNE does not guarantee a low level of nitrogen losses in the environment, while a moderate excess of NB does not ensure an optimal use of resources for food production. Using both indicators gives a more detailed understanding of the productive efficiency and the potential environmental impacts of N in various farming systems.

Crop acreage allocation decisions on intensive mixed crop-livestock farms

B. Roche, C. Amiotte, H. Boussard, A. Joannon and G. Martel
INRA, SAD, UR 0980 SAD-Paysage, 35000 Rennes, France; broche@rennes.inra.fr

In agricultural area, the organization of crop and grassland acres at the landscape level is recognized as a main factor affecting natural resources. This organization results from individual decisions taken at the farm level. Agronomists have studied farmers' crop acreage allocation decisions, including crop rotation and crop spatial organization decisions. They have shown that farm territory characteristics and crop management are key parameters in crop acreage allocation decisions. But less attention has been paid to the links between livestock management and crop acreage allocation decisions on mixed crop-livestock farms. The aim of our study is to analyze, in an intensive livestock area, how livestock management interacts with farm territory characteristics and crop management in crop acreage allocation decisions. We carried out a comprehensive survey on 13 farms with various types of livestock in Brittany, France (cattle, pig, poultry). We analyzed the interviews as case studies. Our first results suggest that choices of animal type and feeding management create a minimum acreage required for each crop and their different uses but don't create maximum acreages (resulting from other minimum and farm area). These livestock choices also lead to compromises between minimum crop acreages involving a hierarchy of priorities: forage production if a farm has cattle, secure incomes (with wheat) if not. Our results also show a kind of homogeneity in agronomic decision-making rules. For instance, for all farmers interviewed, corn is grown with a two year waiting period, whatever its acreage and function; farmers also try to avoid mono-cropping. Lastly our results suggest that farm territory characteristics can prevent the implementation of certain livestock farming systems which can lead to adaptation of agronomic rules. We plan to carry out a complementary survey to confirm these first results. We are building a model of crop acreage allocation in mixed crop-livestock farms to test our hypothesis.

Effect of stocking rate and cow lactation stage on nitrogen balance of grazing dairy cows

A.I. Roca-Fernández, D. Báez-Bernal and A. González-Rodríguez
Agrarian Research Centre of Mabegondo, INGACAL, Abegondo 10, 15080 La Coruña, Spain; anairf@ciam.es

The effect of stocking rate (SR), low (L, 3.9 cows/ha) vs. high (H, 5.2 cows/ha), and cow lactation stage (LS), early (E, 31 DIM) vs. middle (M, 140 DIM), was investigated on nitrogen (N) balance of grazing Holstein-Friesian cows (n=72). Animals were randomly assigned to four treatments (LE, LM, HE and HM) in a 2×2 factorial design. ΣN inputs (grass, silage and concentrate) and ΣN outputs (milk and body weight gain) were evaluated in cows grazing rotationally on perennial ryegrass-white clover swards. Two periods of supplementation at pasture, with (P1, March-April) vs. without (P2, May-August), were considered. Pasture and silage intake were higher ($P<0.001$) in cows at middle LS (15.9±1.5 and 7.8±0.8 kg DM/cow/day) than in cows at early LS (12.6±0.2 and 6.2±0.2 kg DM/cow/day) while concentrate intake was higher ($P<0.001$) in cows at early LS (4.1±0.1 kg DM/cow/day) than in cows at middle LS (1.5±0.2 kg DM/cow/day). ΣN inputs and ΣN outputs were higher ($P<0.001$) in cows at early LS (189±14 and 121±9 g N/cow/day) than in cows at middle LS (163±2 and 105±5 g N/cow/day). Nevertheless, no differences were found between LS treatments for N excretion. Supplements intake was higher ($P<0.05$) in high SR groups (10.4±0.2 kg DM/cow/day) than in low SR groups (9.2±0.9 kg DM/cow/day). No differences were found between SR treatments for ΣN inputs, ΣN outputs and N excretion. ΣN inputs and ΣN outputs were higher ($P<0.001$) in P1 (237±8 and 140±10 g N/cow/day) than in P2 (114±19 and 86±8 g N/cow/day). N excretion was also higher ($P<0.001$) in P1 (479±67 g N/cow/day) than in P2 (68±24 g N/cow/day), but no differences were found between cows at both LS and SR treatments. The results pointed the interest of evaluating cow LS and SR at pasture on ∑N inputs and ∑N outputs to minimize N-losses and to validate N-balance as a tool for assessing grazing milk production systems. Supplementation highly increased N-excretion in grazing dairy cows.

Session 31 Theatre 7

Domestic cavies in Cameroon and eastern DR Congo for nutrition security and income diversification

F. Meutchieye[1], S. Bacigale[2], B. Wimba[3], W. Chiuri[4], A.T. Niba[1], G. Amzati[2], O.A. Mwai[5], D.E. Fon[1], B.L. Maass[4], A. Djikeng[6] and Y. Manjeli[1]

[1]University of Dschang, Cameroon, Dschang, Box 222, Cameroon, [2]UEA, Bukavu, Buk, Zaire, [3]INERA, Bukavu, MUL, Zaire, [4]CIAT, Nairobi, Box 823-00621, Kenya, [5]ILRI, Nairobi, Box 30709, Kenya, [6]BecA Hub ILRI, Nairobi, Box 30709, Kenya; fmeutchieye@gmail.com

Small livestock species in the tropics have received little attention by researchers, extension and development agents, despite their potentials in ensuring nutrition security and improving income in rural and poor urban households. The domestic cavy (i.e. Guinea pig) is an invisible animal in Sub-Saharan Africa within livestock censuses or research. One of the objectives of the AusAID-funded project 'Harnessing husbandry of domestic cavy for alternative and rapid access to food and income in Cameroon and the eastern Democratic Republic of Congo (DRC)' is to provide reliable statistics on cavy culture and make available an incipient estimate of nutrition and livelihood impacts of cavies to their owners in the two countries. A survey with 370 and 250 questionnaires has been conducted in Cameroon and DRC, respectively. Cavy populations are estimated to be at least 400,000 in southern Cameroon and 500,000 in Sud-Kivu Province of DRC. About 16 animals are kept per household in both countries, with maximum flocks of 200 or more. Smallholder flock sizes fluctuate highly, with significant decreases being observed when school fees are due. In western Cameroon main motivations for keeping cavies are income generation (66% of respondents), consumption (45%) and manure (30%); in the forest zone of Cameroon they were consumption (40%) and income (22%); and in Sud-Kivu consumption (65%) and income (20%). In Sud-Kivu, about 60% of both men and women consume cavies, while almost 100% of the children consume them. Cavies, thus, significantly contribute to nutrition security to rural households. An adult cavy fetches approximately 1.5 USD at farmgate, hence offers a source of income to their rural and peri-urban keepers.

Session 31 Theatre 8

Evaluation of soil intake in free ranged domestic animals to ensure food safety

S. Jurjanz[1], C. Jondreville[1], R. Delagarde[2], A. Travel[3], K. Germain[4], C. Feidt[1] and G. Rychen[1]

[1]Université de Lorraine, UR AFPA, TSA 40602, 54518 Vandoeuvre, France, [2]INRA, UMR PEGASE, Domaine de la Prise, 35590 Saint-Gilles, France, [3]ITAVI, UMT Bird, Centre de Tours, 37380 Nouzilly, France, [4]INRA, UE EASM, Le Magneraud, 17700 Surgères, France; stefan.jurjanz@univ-lorraine.fr

Consumer demand is for rearing systems with free ranged animals. Then, animals may be exposed to environmental pollutants when involuntary ingesting soil. Indeed, soil is considered as a sink able to store over long periods deposited organic pollutants or heavy metals. Because these compounds may be transferred to animal products, soil intake should be limited as far as possible to ensure the safety of the produced food. Studies on different species have been carried out to evaluate the soil intake in rearing conditions nowadays. Studies on grazing dairy cows showed daily intakes ranging from 0.15 to 0.85 kg dry soil but in disadvantageous conditions daily intakes increased up to 1.3 kg of dry soil. In order to control soil intake, sward height should not be <50 mm. The distribution of supplementary feed or restricted access time to pasture may limit soil intake on sparse plant cover. In free ranged poultry observed soil intakes were generally modest, i.e. <2 g in chicken and <5 g in laying hens. Contrarily soil intake increased rapidly, when rearing conditions became more difficult because of a sparse soil cover. Indeed up to 5 g daily have been observed in chicken exploring yards during winter on nearly bare soil. Moreover, hens fed with an unbalanced and coarse diet could ingest <30 g of soil. These extreme values show the need to control the rearing conditions to avoid the increase of soil intake. Preliminary work on free ranged sows indicates a dry soil intake of 0.2 kg (maximum 0.5 kg). Thus, soil intake of free ranged domestic animals is generally non negligible. Moreover, disadvantageous rearing conditions may dramatically increase the soil intake, affecting the food safety when these soils are contaminated with environmental pollutants.

Grassland management in large dairy farms

C. Gaillard[1], S. Mugnier[1], A. Destrez[1], A. Gérard[1] and S. Ingrand[2]
[1]AgroSup Dijon/UMR Métafort, BP 87099, 21079 Dijon Cedex, France, [2]INRA/UMR Métafort, Centre de Theix, 63122 Saint-Genès-Champanelle, France; c.gaillard@agrosupdijon.fr

Nowadays, the increase of dairy farms size is a major evolution that impacts on feeding practices. This evolution leads to much more conserved fodders in diets, especially maize silage, instead of grass and direct grazing. Such changes tend to decrease farming systems environmental sustainability. However, some dairy farmers still maintain the pasture of cows during all the grazing season. Our study focuses on the practices of these farmers in order to understand how they can keep a grass-based dairy system in two constraining situations: large scale farms and milk production under quality signs, requiring a strong link with soil, i.e. 'terroir'. Our study is situated in Franche-Comté area, producing PDO cheese (Comté). The specifications precise that: (1) grass has to be a main part of cow diet; and (2) grazing has to cover not less than 50% of daily cow diet between March and October. Thus, farmers have to use local grassland resources. We surveyed 29 farms with large dairy herds (84 cows on average) about farmers' practices: field pattern use, paddock management, forage and grazing schedule for a whole forage year. Four strategies of grassland functioning were identified (using multivariate analyses): (1) intensive grazing management in spring with high stocking rate and green feeding from the diminution of grass growth; (2) spring grazing with moderate stocking rate and increasing area at the beginning of summer with adding aftermaths; (3) only direct grazing which area increase with aftermaths in summer and low stocking rate during the whole season; and (4) complex adjustments of grassland areas to answer the requirements of cows by using fields regrowth, animal destocking at the beginning of July and late green feeding in August. Our results point out several possibilities to maintain the use of grassland resources in large-scaled dairy farms, partially depending on soil and climate capabilities, as also field-pattern structures.

Evaluation of the GrazeIn model of grass dry matter intake and milk yield prediction in NW Spain

A.I. Roca-Fernández[1,2,3], R. Delagarde[3], M.E. López-Mosquera[2] and A. González-Rodríguez[1]
[1]Agrarian Research Centre of Mabegondo, INGACAL, Abegondo 10, 15080 La Coruña, Spain, [2]University of Santiago de Compostela, IBADER, Campus Universitario s/n, 27002 Lugo, Spain, [3]INRA-Agrocampus Ouest, PEGASE, UMR 1348 Saint Gilles, 35590 Rennes, France; ana-isabel.roca-fernandez@rennes.inra.fr

GrazeIn model is a prediction model that simulates grass dry matter intake (GDMI) and milk yield (MY) of grazing dairy cows. Sward and animal data from a trial realized in Galicia (NW Spain) with Holstein-Friesian cows (n=72) managed at two lactation stages (days in milk), early (E, 29) vs. late (L, 167), and rotationally grazing in perennial ryegrass and white clover pastures at two daily herbage allowances (kg DM/cow/day), medium (M, 25) vs. high (H, 30), were used to evaluate the prediction accuracy of GDMI and MY predicted by the GrazeIn model from a randomized 2×2 factorial design (EM, EH, LM and LH). Sward variables (fill value, energy content, protein value, pre- and post-grazing grass heights, herbage mass and DHA), supplementation levels at pasture (concentrate and silage) and animal variables were introduced into the model. The mean actual GDMI of the database determined by differences between pre- and post-grazing grass heights (n=125) was 14.2 kg DM/cow/day and GrazeIn predicted a mean GDMI for the database of 13.8 kg DM/cow/day. The mean bias was -0.4 kg DM/cow/day. GrazeIn predicted GDMI for the total database with a relative prediction error (RPE) of 12.5% at herd level. The mean actual MY of the database (n=528) was 23.2 kg/cow/day and GrazeIn predicted a MY for the database of 21.8 kg/cow/day. The mean bias was -1.4 kg/cow/day. GrazeIn predicted MY for the total database with a mean RPE of 23.0% at cow level. From the evaluation, GrazeIn predicted MY of cows in late lactation (26.5%) with a larger error than in early lactation (19.5%). These errors could be reduced by adapting the persistency of the potential MY lactation curve for cows at different stage.

Session 31 Poster 11

Fatty acid profile of organic and conventional retail milk of Cyprus

O. Tzamaloukas, M. Orford, D. Miltiadou and C. Papachristoforou
Cyprus University of Technology, Department of Agricultural Sciences, Biotechnology and Food Science, P.O. Box 50329, 3603 Lemesos, Cyprus; c.papachristoforou@cut.ac.cy

Milk lipids have distinctive effects on nutritional, textural and organoleptic properties of milk and dairy products, while the fatty acid (FA) profile has been reported to be linked with a number of biological processes affecting human health. Unsaturated FAs have been associated with reduced risk of carcinogenesis, atherosclerosis, osteoporosis and obesity, while strengthening the immune system. The objective of the present work was to study the FA profile of retail cow milk produced under organic or conventional systems in Cyprus. A survey was conducted over a two-year period and samples of whole and semi skimmed fresh milk, either organic or conventional, were collected every two months from supermarkets and other retail outlets. All together, eighty-four samples from organic and conventional milk were collected and analysed by gas chromatography – mass spectrometry for fatty acid profile. The effect of production system on milk constituents was investigated by paired t-test analysis using the SPSS statistical package. The results obtained, were in agreement with results of earlier studies that compared organic and conventional milk lipids. Thus, under farm practices implemented in Cyprus, retail organic milk had by 40% higher concentrations of total polyunsaturated fatty acids than conventional milk (28.3 vs. 20.3 g/kg, respectively, of total FA; $P<0.001$). Further analysis of specific FAs showed increased linoleic acid (19.8 vs. 14.2 g/kg of total FA; $P<0.001$), α-linolenic acid (3.46 vs. 1.97 g/kg of total FA; $P<0.001$) and conjugated linoleic acid (cis- 9, trans-11 CLA, 3.36 vs. 2.38 g/kg of total FA; $P<0.001$) in organic compared to conventional milk. Most likely, these differences could be attributed to contrasting feeding practices applied in organic and conventional dairy farms in Cyprus.

Session 31 Poster 12

Characterization of production systems and marketing strategies of Iberian pig farms in SW Spain

A.F. Pulido, P. Gaspar, F.J. Mesías, A.J. Escribano, M. Escribano and F. Pulido
Universidad de Extremadura, Escuela de Ingenierías Agrarias, Av. Adolfo Suárez s/n, 06006 Badajoz, Spain; pgaspar@unex.es

Iberian pig farms in extensive production systems are very important for the economy of rural areas in SW Spain, as they represented about 13% of Final Agricultural Production in 2012. Nowadays, the sector is going through a structural crisis that is threatening the subsistence of many farms. The producers are taking decisions that may affect the extensive traditional farming systems with the risk that it entails both for environment and for rural areas. The aim of this study is to widen and to deepen the knowledge about pig farms located in Extremadura Region (SW Spain) using both, productive and commercial parameters. For this purpose, 55 randomly selected holdings have been surveyed during 2011 to obtain different indicators related to farm management and product marketing. Multivariate analyses of these indicators (Principal Component Analysis and Hierarchical cluster analysis), allowed the identification of the best strategies adopted by the farmers to adapt themselves to the current market situation. Best strategies were those adopted by farms that make an optimum grazing of rangelands keeping cattle and pigs at the same time in the farms. These farms obtained increased profits by livestock sales (pig and cattle) and by CAP subsidies linked to cattle. Another strategy was accomplished by farms that did not breed sows. They bought piglets from another farm and they only completed the pig fattening period called 'montanera' (by acorn feeding). This way, increased feeding cost affect lesser extent, because fodder is not required for breeding sows, and the number of piglets bought outside the farm for fattening may vary depending on the amount of acorns available every year on the farm.

Intramuscular fatty acid composition in organic and conventional beef, poultry and pork systems

D. Villalba, M. Tor, D. Cubiló, D. Babot and J. Álvarez-Rodríguez
University of Lleida, Animal Production, Lleida, 25198 Lleida, Spain; dvillalba@prodan.udl.es

There is still a lack of information about the differences in quality between organic and conventional livestock products. The objective of the study was to compare intramuscular fatty acid (FA) composition from meat obtained in organic (ORG) and conventional (CON) systems. Meat samples from animals reared under ORG (EU regulation) or CON systems were obtained in the same slaughter conditions from beef (L. dorsi, 59 vs. 37, 2 vs. 3 farms), poultry (Semimembranous, 24 vs. 12, 2 vs. 1 farms) and pork (L. dorsi, 31 vs. 29, 2 vs. 3 farms) and they were analysed for fatty acids profile by gas chromatography. PROC GLM of SAS was used to compare the within species husbandry system effect on individual FA composition. A multivariate analysis has been performed with PRINCOMP and DISCRIM procedures of SAS. For beef meat, 15 out of the 23 FA identified were different between systems. The percentage of saturated and monounsaturated FA was lower and the level of conjugated linoleic acid was higher ($P<0.05$) in ORG systems. For poultry meat, 13 out of the 16 FA identified were different between systems ($P<0.05$). The polyunsaturated to saturated ratio was higher ($P<0.05$) in ORG systems. For pork meat, 7 out of the 16 FA were different between systems ($P<0.05$). Meat from ORG system had a lower ratio n6/n3 than CON system ($P<0.05$). In the three species, the multivariate analysis of FA profile allowed a good discrimination between the ORG and CON systems. The estimated error in discriminating the system origin of an observation was less than 1% for beef and poultry and 3% for pork. Differences in FA profile, in general favourable to the ORG system, could be explained by differences in dietary feedstuffs, breed and age at slaughter, especially in poultry, but overall, the meat that arrive to consumers from the organic system could be considered different than the conventional in terms of FA.

Dairy cattle feeding system's analysis in a rainfed area in Morocco

S. Boumakrat[1], A. Araba[1] and T. M'chaar[2]
[1]Hassan II Agronomy and Veterinary Medicine Institute, Department of Animal Production and Biotechnology, BP 6202, Madinate Al Irfane, Rabat, 10080, Morocco, [2]Centrale laitière, Casablanca, 20320, Morocco; s.boumakrat@gmail.com

The main studies conducted in Morocco about dairy cattle feeding systems was concerning irrigated areas. However, the advent of the national agricultural strategy 'Plan Maroc Vert' allowed an expansion of dairy farming in rainfed areas. Therefore, the aim of this study is to analyze the feed system encountered in these regions, through surveys and analysis of forages samples collected from 20 dairy cattle farms located in rainfed region in Morocco, and a regular follow-up of their milk's chemical quality (fat content, protein content, dry defatted extract). Results showed that dairy cattle farms in the rainfed region use less green fodder which are replaced by straws, hays (oats) and silage (corn). More than that, this type of farms is characterized by a high intake of concentrates (0.55 UFL from concentrates per kilogram of produced milk). As a consequence, average milk fat content slightly exceed 35 g/kg in all of 20 studied farms, increasing also feeding costs and urging negative impacts on the health of dairy cows (frequent cases of acidosis). The forage analysis and prediction of nutrient value concerned 39 samples of prevalent forages in the region. Results were communicated to farmers in order to improve stock management and control the quality of silage conservation.

Session 31 Poster 15

Ammonia emissions from a naturally and a mechanically ventilated broiler house in Brazil

L.B. Mendes[1], I.F.F. Tinôco[1], N.W.M. Ogink[2] and M.S. Sousa[1]

[1]Federal University of Viçosa, Departent of Agricultural Engineering, Av. PH Rolfs, SN, Centro, 36570-000, Viçosa, MG, Brazil, [2]Wageningen University and research Centre, Livestock Research institute, Vijfde Polder 1, 6700 WC Wageningen, the Netherlands; mrmendes2010@gmail.com

In a global scale, Brazil is the 3rd biggest producer and 1st ranked exporter of broiler chicken. However, even with the considerable magnitude of animal production systems, very little effort has been given to estimate NH_3 emission factors (f_{NH3}) from poultry houses under the unique Brazilian conditions: tropical climate and non-insulated broiler houses, that can be either mechanically or naturally ventilated. Emissions of NH_3 is still not legislated in Brazil, but recent studies have evidenced changes in N sensitive ecosystems within states that have intense livestock activity, pointing out the urgent need for implementation of mitigation strategies. This study was conducted with the aim of monitoring NH_3 emissions from a mechanically and a naturally ventilated broiler house (MVB and NVB, respectively) in the southeastern state of Minas Gerais and calculate their f_{NH3}. Bird stocking density was 13.5 and 11.1 birds/m for the MVB and NVB, respectively. The marketing age was 43 d and bedding consisted of dried coffee husks in its first time of use. Ventilation rates were calculated with the metabolic carbon dioxide mass balance method as elaborated by CIGR (2002). Values of f_{NH3} were (0.38±0.12) g/bird/d and (0.32±0.08) g/bird/d for the MVB and NVB, respectively, and are in agreement to what was presented in other studies performed under similar conditions. Estimated f_{NH3} on yearly basis were 109 g/bird/year and 91 g/bird/year for the MVB and NVB, respectively. The results obtained with this study help providing reliable methodology for the determination of a solid database on NH_3 emission factors for tropical conditions that can be used for future inventories, when performed in a sufficient number of barns that is representative for Brazilian scenario.

Session 31 Poster 16

Adaptation of mixed sheep-crops systems to the socio-economic context in a high natural value area

A.M. Olaizola[1], F. Ameen[2], V. Silveira[3], A. Bernués[4] and E. Manrique[1]

[1]Zaragoza University, Ciencias Agrarias y del Medio Natural, Miguel Servet 177., 50013 Zaragoza, Spain, [2]Assiut University, P.O. Box 71526, Assiut, Egypt, [3]Universidad Federal de Santa María, DEAER UFSM, Avenida Roraima no. 1000, Cidade Universitária, Santa Maria, RS 97105-900, Brazil, [4]Norwegian University of Life Sciences (UMB), Dept. of Animal and Aquacultural Sciences, P.O. Box 5003, 1432 Ås, Norway; alberto.bernues.jal@gmail.com

Meat sheep farming systems play a central role in the management and conservation of large High Nature Value farmland areas, but have suffered a strong decline in the last few decades in most European Mediterranean regions. In order to evaluate the possible strategies of adaptation to different agricultural policies and off–farm labour scenarios, four representative mixed sheep-crops systems in the 'Sierra y Cañones de Guara' Natural Park (SCGNP) (Spain) were analysed. A mixed linear programming model for each farm system was developed representing the annual operation of the farm distributed into the twelve months of the year. The objective function maximized the farm Gross Margin and the decision variables were land size, land use, flock size, use of agricultural products, use of seasonal rented grazing areas and the possibility to start off-farm activities. Five scenarios combining CAP implementation (partial, total decoupling and absence of subsidies) and the possibility of starting off-farm activities (part-time farming) were considered in the analysis. Results revealed that under the current situation of high cereals prices, the cultivation of barley has economic interest for mixed sheep-crops farming systems of SCGNP. In farms with less labour availability this involved a slight decrease in flock size. Total decoupling of subsidies implied an increasing economic interest of part-time off-farm activities, reducing further flock size (from 21% to 33% or original size) and changing land use (reduction of forage crops). Under the hypothetical scenario of lack of subsidies this trend was strengthen.

Carbon footprint of heavy pig production in eight farms in Northern Italy

G. Pirlo[1], S. Carè[1], G. Ponzoni[2], V. Faeti[2], R. Marchetti[2] and G. Della Casa[2]
[1]CRA, FLC, Porcellasco, 26100 Cremona, Italy, [2]CRA, SUI, S. Cesario s/P, 41018 S. Cesario s/P, Italy; giacomo.pirlo@entecra.it

A preliminary LCA was performed for estimating the carbon footprint (CF) of the heavy pig production in eight fattening pig farms in Northern Italy. The average pig number was 1,967 (±1,379), the average number of pigs fattened every year was 4,123 (±3,349). Initial and final weight were kg 38.2 (±11.0) and 167.1 (±5.25) respectively. System boundaries of CF study comprised the growing and fattening phase in the pork production chain; they did not include the sow-piglet phase, slaughtering and feed production. Environmental burden of purchased feeds and piglets was estimated from database and was comprised in CF estimation as indirect carbon dioxide equivalent (CO_2eq) emissions. Functional unit (FU) was 1 kg of body weight at the farm gate. The global warming potential (GWP) was expressed as kilograms of CO_2eq; greenhouse gases (GHG) were CH_4, N_2O and CO_2 with a GWP of 25, 298 and 1 kg CO_2eq respectively. The average CF was 3.83 (±1.07) kg CO_2eq. The contribution of each emission source was: enteric CH_4 3.34%, manure CH_4 17.7%, manure N_2O 1.38%, direct soil N_2O 6.59%, indirect soil N_2O 6.39%, direct CO_2 1.00%, indirect CO_2 63.6% (comprising production of purchased feed and piglets, fossil fuel extraction and refing, electricity production, and transportations). In respect to other animal foods, such as milk, where the enteric fermentation is a main GHG source and the crop production is an important component of the system, this study showed that the main drivers of GHG production in growing and fattening phase of pork production chain are the off-farm feed production and transportation phase; on the contrary the activities within the system boundaries are of minor importance. Mitigation strategies should be addressed to feed production chains and manure management systems.

Milk performance of two dairy cow genotypes managed at two levels of supplementation at pasture

A.I. Roca-Fernández[1,2,3], L. Delaby[3], S. Leurent[4], Y. Gallard[4], M.E. López-Mosquera[2] and A. González-Rodríguez[1]
[1]Agrarian Research Centre of Mabegondo, INGACAL, Abegondo 10, 15080 La Coruña, Spain, [2]University of Santiago de Compostela, IBADER, Campus Universitario s/n, 27002 Lugo, Spain, [3]INRA-Agrocampus Ouest, PEGASE, UMR1348, Saint Gilles, 35590 Rennes, France, [4]INRA, PEGASE, UE326, Domaine du Pin-au-Haras, 61310 Exmes, France; ana-isabel.roca-fernandez@rennes.inra.fr

Milk performance of two cow genotypes, Holstein-Friesian (F, n=178) vs. Normande (N, n=174), supplemented at pasture with two levels of concentrate, low (0 kg DM/cow/day, n=174) vs. high (4 kg DM/cow/day, n=178), were studied in 2001-2005. Cows were randomly assigned to four treatments (F0, F4, N0 and N4) in a 2×2 factorial design under simplified rotational grazing system. Three paddocks per rotation were grazed in spring and 6 in summer-autumn for each group with a mean residence time per grazing cycle of 10-days. A maximum of milk yield (MY max.) was reached at 4-day followed by a drop of milk (Dm) at 10-day. MY and peak of MY per lactation were higher ($P<0.001$) in Holstein-Friesian (7,591 and 36.2 kg/cow) than in Normande (6,214 and 29.2 kg/cow), with lower ($P<0.001$) body weight (BW) and body condition score (BCS) in Holstein-Friesian (659 kg and 2.18) than in Normande (695 kg and 2.83). Supplemented cows had higher ($P<0.001$) MY and peak of MY (7,567 and 35.0 kg/cow) than cows without supplement (6,238 and 30.4 kg/cow), with higher ($P<0.001$) BW and BCS in supplemented cows (702 kg and 2.85) than in cows without it (652 kg and 2.15). At grazing, supplemented cows had higher ($P<0.001$) MY max. and lower ($P<0.001$) Dm in each cycle (24.9 and 5.3 kg/cow) than cows without supplement (21.2 and 6.1 kg/cow). MY max. and Dm were higher ($P<0.001$) in Holstein-Friesian (25.1 and 6.4 kg/cow) than in Normande (21.0 and 4.9 kg/cow). Milk performance was highly affected by cow genotype and level of supplementation at pasture. Good control of Dm in each grazing cycle was essential to keep milk reduction steady throughout lactation.

Impact of calving date and cowtype in a seasonal Alpine low-input dairy system

M. Horn[1], A. Steinwidder[2], W. Starz[2] and W. Zollitsch[1]
[1]University of Natural Resources and Live Sciences, Division of Livestock Sciences, Gregor Mendel Straße 33, 1180 Vienna, Austria, [2]Agricultural Research and Education Centre Raumberg-Gumpenstein, Institute of Organic Farming and Farm Animal Biodiversity, Trautenfels 15, 8951 Trautenfels, Austria; marco.horn@boku.ac.at

In order to reduce costs of production and dependence on external resources the implementation of a pasture-based milk production system might be an alternative for the near future in Alpine regions. Data for 73 lactations from a 4-year experiment were analysed in order to investigate the effects of calving date (CD) and cowtype (CT) on milk production, live weight and diet composition in an Alpine low-input dairy system. Calvings were assigned to one out of three groups (CD1, CD2 and CD3) and mean CD were Nov. 17th, Jan. 8th and Mar. 8th, resp. The two CT compared were Brown Swiss (BS) and a special strain of Holstein Friesian (HFL). BS represented the average Austrian BS population and was primarily selected for high milk yield. HFL was primarily selected for lifetime performance and fitness. The dataset was analysed using a mixed model, including animal within breed as a random effect. BS was superior for most milk performance traits, except for milk yield per unit of metabolic body weight. While late calving (CD3) sig. depressed the energy-corrected-milk yield of BS (6,473, 6,026, 5,371 for CD1, CD2, CD3, resp.), CD had no sig. effect on this trait for HFL (5,611, 5,656, 5,679 for CD1, CD2, CD3, resp.). Week of live weight nadir was sig. later for BS than for HFL, but CD only showed sig. differences for BS (34 and 15 weeks in CD1 and CD3, resp.). There was no CT effect for ration composition, but annual concentrate consumption decreased sig. from CD1 (600 kg) to CD3 (261 kg), while pasture ratio increased from 45% (CD1) to 55% (CD3). In conclusion, CD sig. influenced the ration composition. Comparing CT for milk production, the effect of CD was more pronounced for BS and differences between cowtypes decreased when delaying CD.

Effect of climatic conditions on milk yield and milking frequency of automatically milked dairy cows

S. Ammer, C. Sanker, C. Lambertz and M. Gauly
Livestock Production Systems Group, Department of Animal Science, Albrecht-Thaer-Weg 3, 37075 Göttingen, Germany; sammer1@gwdg.de

The objective of this study was to evaluate the influence of climatic conditions on milk yield and milking frequency of dairy cows, milked with an automatic milking system (AMS). Therefore, two Holstein-Friesian herds kept in an insulated (120 cows) and a non-insulated (130 cows) barn with two AMS each and free cow traffic were observed. From April to September 2009 the daily milk yield and the milking frequency of each cow were recorded. THI was calculated based on hourly air temperatures and relative humidity values recorded by meteorological stations. For the statistical analysis the 3-days average temperature values were divided into two classes (<20 °C, ≥ 20 °C) and THI values into three classes (<55, 55-60, ≥ 60). The mixed model included the stage of lactation, parity class, THI class, temperature class and milking frequency class and two-way interactions as fixed effects and the cow as random effect. The daily milk yield differed significantly ($P<0.001$) between the temperature (32.98±0.4 vs. 32.19±0.4 kg/d) and THI classes (33.75±0.4 vs. 33.16±0.4 vs. 32.01±0.4 kg/d), respectively. Furthermore, daily milking frequency per cow was significantly different between the temperature (2.70±0.04 vs. 2.78±0.04) and THI classes (2.63±0.03 vs. 2.70±0.03 vs. 2.76±0.03). The greatest effect of temperature was observed during mid-lactation (65 -185 DIM) while the most obvious effect of THI occurred in mid- and late-lactation (65-305 DIM). Milking frequency as well as milk yield were more affected in the non-insulated than in the insulated barn by increasing temperature and THI. Both THI and temperature showed negative effects on the level of milk yield under conditions of heat stress (THI$\geq$60 and temperature $\geq$20 °C). Nevertheless, the milking frequency increased with higher temperature and THI values.

Session 32 Theatre 1

Risk factors associated with reproductive tract status

T. Carthy[1,2], A. Fitzgerald[2,3], S. McParland[2], D.P. Berry[2], E.J. Williams[1], A.R. Cromie[4] and D. Ryan[3]
[1]UCD, School of Veterinary Medicine, Belfield, Dublin 5, Ireland, [2]Teagasc, AGRIC, Moorepark, Fermoy, Cork, Ireland, [3]Reprodoc Ltd, Fermoy, Cork, Ireland, [4]ICBF, Bandon, Cork, Ireland; tara.carthy@teagasc.ie

Ultrasound analysis is a non-invasive method to assess the reproductive tract. The aim of this study was to identify the risk factors associated with fertility phenotypes derived from ultrasound measurements in dairy and beef cattle. Data were available on 148,990 ultrasound records from 75,966 cows in 843 Irish herds over a 5 year period. Phenotypes derived from ultrasound measurements included: (1) cycling, the resumption of oestrus cycle post partum; (2) early cycling, the resumption of cycle within 15 days post partum; (3) cystic; (4) embryonic/foetal death; and (5) uterine score, level of inflammation of the uterine wall and fluid present in the uterine horn on a scale of 1 to 4. Factors associated with the logit of the probability of a positive outcome for each of the fertility traits with the exclusion of uterine score were determined using logistic regression; linear mixed model analyses was used for uterine score. Animal and herd-year-season were included in the model as random effects. The odds of cycling increased with parity (P=0.023), days since calving (P<0.001), and previous dystocia (P=0.049). Cows in dairy herds had increased odds of cycling (P=0.002) and becoming cystic (P=0.03) compared to cows in beef herds. The odds of becoming cystic increased with parity (P<0.001), days since calving (P<0.001), and month of examination (P<0.001). The odds of early cycling was greatest in cows that experienced no dystocia with the exception of those who had a caesarean (P<0.001). The odds of embryonic/foetal death increased with parity number (P<0.001) but decreased with days since calving (P<0.001). A higher (i.e. inferior) uterine score was observed in older animals and that had experienced dystocia but decreased with days since calving. Cows in beef herds had on average poorer uterine score than cows in dairy herds.

Session 32 Theatre 2

Evaluation of different devices for automated estrous detection in dairy cows

A. Chanvallon[1], S. Coyral-Castel[1], J. Gatien[2], J.M. Lamy[3], J.M. Philipot[4], J. Girardot[5], J.B. Davière[5], D. Ribaud[1] and P. Salvetti[2]
[1]Institut de l'Elevage, 149 r Bercy, 75595 Paris cedex 12, France, [2]UNCEIA, 13 r Jouët, 94704 Maisons-Alfort cedex, France, [3]Chambre d'Agriculture de Maine-et-Loire, 14 av Jean Joxé, 49006 Angers cedex 01, France, [4]CREAVIA, 69 r Motte Brûlon, 35702 Rennes cedex 7, France, [5]CLASEL, 141 bd des Loges, 53942 Saint Berthevin cedex, France; audrey.chanvallon@idele.fr

Considerable technological advances have been made in automated estrous detection in dairy cattle but few studies have focused on concomitant evaluations of their performances on same animals. Our study aimed: (1) to assess heat detection reliability of a pedometer Afitag® (P) and two activity recorders Heatime-Ruminact® (HT) and HeatPhone® (HP); (2) to analyze factors influencing their quality of detection. Sixty-three Holstein cows were housed in an experimental farm and fitted simultaneously with the three systems. Resumption of cyclicity occurred 27.7±9.8 days after calving and 3 cows (4.8%) did never resume cyclicity. A total of 214 follicular periods were identified to characterize 153 ovulatory cycles, 60% of the cows had cyclicity classified as normal. After removing periods of dysfunction of each tool, the sensitivities (SE) and positive predictive values (PPV) were respectively 71% and 71% for P, 62% and 84% for HT and 61% and 67% for HP. A new algorithm for HP was applied a posteriori on the collected data, SE and PPV were 62% and 87%, respectively. With this last result, performances of HT and HP were equivalent. The effects of estrus rank, interval from calving, cyclicity patterns, milk production, negative energy balance and parity of all monitors were estimated on SE and PPV. For all tools, detection performances were strongly lower for the first postpartum ovulation (P<0.05). This study shows that several environmental and physiological factors can affect efficiency of the estrus monitors. Further studies are needed to define new recommendations to help farmers in their choice, depending on production system.

Response of blood hormones and nutrients to an ACTH challenge and to a physical stressor in pigs

A. Prunier, N. Le Floc'h, C. Leclercq and E. Merlot
INRA, UMR1348 PEGASE, 35590 Saint-Gilles, France; armelle.prunier@rennes.inra.fr

The mobilization of body reserves is particularly important for an animal to face a stressor. However, precise data regarding the blood availability in metabolic substrates in response to a stressor are still missing in pigs. The present study aimed at describing the endocrine and metabolic response to a physical stressor (nose lasso applied during 3 min: NL) and to an i.v. injection of ACTH (5 µg/kg live weight) in 34 finishing male pigs. Concentrations of ACTH (only after NL), cortisol, glucose, lactate, free fatty acid (FFA) and amino acids (AA) were measured in blood samples collected serially through a jugular catheter before and after NL (3 to 95 min) or ACTH (3 to 240 min) applied 6 to 10 h after the last meal. Data were analyzed by ANOVA using SAS. After NL, ACTH increased from 3 to 60 min, cortisol from 10 to 30 min, glucose and lactate from 3 to 10 min and FFA from 3 to 30 min ($P<0.05$). AA were measured at -1 and 60 min. Several essential (Ile, Thr, Val) and non-essential AA (Asn, His, Orn, Pro, Ser, Tyr) decreased at 60 min whereas Glu increased ($P<0.05$). After the ACTH injection, cortisol increased from 3 to 120 min, FFA from 3 to 240 min, lactate from 10 to 240 min, glucose from 120 to 180 min whereas glucose decreased from 30 to 60 min ($P<0.05$). AA were measured at -1, 60, 120 and 240 min. Several essential (Ile, Leu, Lys, Met, Phe, Thr, Trp, Val) and non-essential ones (Ala, Arg, Asn, Cit, Glu, His, Orn, Pro, Ser, Tyr, Val) decreased at 60 min ($P<0.05$). Some of them were back to the pre-injection level at 120 min (e.g. Tyr) or 240 min (e.g. Gly). A few AA (Ala, Glu, Tyr) were even higher at 120 min than before ACTH ($P<0.05$). Present data suggest that a physical stressor induced a fast energetic demand supplied by the mobilization of all sources of body reserves (fat, glycogen, proteins) supported, at least in part, by the ACTH and cortisol release.

Response of blood nutrients to an LPS inflammatory challenge in pigs

E. Merlot, N. Le Floc'h, C. Leclercq and A. Prunier
INRA, UMR1348 PEGASE, 35590 Saint-Gilles, France; elodie.merlot@rennes.inra.fr

Understanding the nutrient metabolism during inflammation is a first step to develop feeding strategies that help sick animals to overcome or recover from an excessive inflammatory response. This study aimed at describing the response of blood nutrients to an LPS injection (O55:B5, 15 µg/kg) in 32 finishing male pigs. LPS was administered i.v., 6 h after the daily meal. Concentrations of cytokines, catecholamines, cortisol, plasma glucose, lactate, free fatty acids, urea and amino acids (AA) were measured in blood samples collected through a jugular catheter from -30 to +420 min and 24 h post LPS. Data were analyzed by ANOVA using SAS. The inflammatory response was revealed by the increase in TNF-alpha (at 60 and 90 min), IL-1 and -6 (from 180 to 300 min), adrenaline (at 60 and 240 min), noradrenaline (at 240 and 420 min), and cortisol (from 30 to 420 min) levels. Glycaemia continuously decreased from 120 to 420 min. Lactate concentrations were sharply increased between 30 and 420 min. Fatty acid and urea concentrations increased from 90 and 360 min respectively. After 24 h, glucose was back to basal levels but lactate and fatty acids were still elevated and urea was at its maximal level. Most of essential (His, Leu, Lys, Met, Phe, Thr, Trp) and non-essential AA (Arg, Asp, Gln, Gly, Orn, Pro, Ser, Tyr) had lower plasma levels at time points 60 or 180 min and were nearly back or even exceeded basal levels at 420 min. Plasma levels remained lower than the initial levels until 420 min for the two essential AA Val (-24%) and Ile (-27%) as well as Cit (-15%). Ala, Glu and Tau steadily increased from 60 min with maximal values at 420 min. Thus, an acute systemic inflammation generates a high and fast energetic demand, leading to anaerobic glucose utilization and release of fatty acids and AA from fat and protein for gluconeogenesis. The degradation of AA carbon skeleton for energy supply released amino groups that were probably used for the synthesis of Ala and Glu during the first hours and of urea at later stages.

Session 32 Theatre 5

Effects of high ambient temperature on responses of pigs to an inflammatory challenge

P.H.R.F. Campos, E. Merlot, M. Damon, J. Noblet and N. Le Floc'h
INRA, UMR1348 PEGASE, 35590 Saint-Gilles, France; paulo.campos@rennes.inra.fr

Although the effects of heat stress on pig performance are well described in the literature, little is known about how ambient temperature could affect their ability to resist, cope with or recover from an inflammatory challenge associated with heat stress. The aim of the study was to evaluate the effects of heat stress on performance and physiological responses of growing pigs submitted to repeated injections of *Escherichia coli* lipopolysaccharide (LPS). Catheterized and individually housed 48 kg growing pigs were assigned to thermo neutral (TN: 24 °C; n=16) or heat stress (HS: 30 °C; n=14) environmental conditions. Within each condition, pigs were submitted to 14 days of adaptation and then to a 17-d experimental period divided in a period without LPS (P1, from d -7 to -1) and a subsequent period with LPS administration (P2, from d 1 to 10). During P2, LPS was administered each 48 h (i.e. on d 1, 3, 5, 7 and 9) and the initial dosage of 30 μg/kg of BW was increased by 12% at each subsequent injection. Dry matter intake decreased during P2 and this decrease was greater in TN than in HS pigs (-28 vs. -19%, respectively; $P<0.05$). Average daily gain and feed efficiency were lower during P2 compared to P1 in TN pigs ($P<0.01$), while no differences ($P>0.05$) were observed in HS pigs. Rectal temperature and plasma haptoglobin and cortisol concentrations were higher ($P<0.001$) during P2 compared to P1. Total thyroxine levels were lower ($P<0.001$) during P2 than during P1. Plasma total triiodothyronine (T3) decreased ($P<0.05$) 24 h after the first LPS administration (LPS1) and then returned to pre LPS levels by d 6. A similar pattern of response was observed for free T3 but HS pigs returned to pre LPS levels earlier (d 6; $P<0.01$) than TN pigs (d 10). Plasma interleukin-6 levels were markedly increased ($P<0.01$) after LPS1 and this increase was twice higher ($P<0.01$) in TN than in HS pigs. In conclusion, the effects of an inflammatory challenge seem to be attenuated in heat-stressed pigs.

Session 32 Theatre 6

Increase in ambient temperature changes the relationship ruminal-core temperatures in lactating cows

A. Boudon[1,2], H. Khelil-Arfa[1,2] and P. Faverdin[1,2]
[1]Agrocampus Ouest, UMR1348 PEGASE, 35000 Rennes, France, [2]INRA, UMR1348 PEGASE, 35590 Saint-Gilles, France; anne.boudon@rennes.inra.fr

Automatic monitoring of ruminal temperature is a promising tool for identification of heat stress or other physiological disorders in dairy cows. However, the relationship between ruminal and body core temperature has to be established in situation of high ambient temperatures. Daily dynamics of ruminal temperature (RT) and vaginal temperatures (VT) were compared in 4 dry and 4 lactating Holstein cows submitted to 2 ambient temperatures (15 °C i.e. thermoneutrality (TN) or 28 °C i.e. high temperature (HT)) maintained constant throughout the day and 2 diet sodium contents (0.15 or 0.55 g/kg DM). RT and VT were recorded every 15 min with indwelling autonomous monitoring systems (MEDRIA S.A.). Drinking and feeding behaviors were continuously recorded. The treatments were compared during 4 periods of 15 d, according to a Latin square design for each physiological stage. Diet consisted on a total mixed ration based on maize silage. Dry cows were offered 14 kg DM/d while lactating cows were fed *ad libitum*. At TN, average RT was 38.9 °C (SEM=0.11) and VT was 38.4 °C (SEM=0.15), whatever the physiological stage ($P>0.15$). HT increased daily averaged RT and VT at both physiological stages ($P<0.0001$) but these increases were higher in lactating than in dry cows (+1.1 vs. 0.15 °C for RT in lactating and dry cows, and +1.05 vs. +0.1 °C for VT). In all situations, RT dropped at the times of the meals (-1.6 °C for the 08:30 h meal) and leveled off at a maximal value during the night. For lactating cows at HT, VT followed a parallel pattern to that of RT with post-prandial drop of -1 °C, whereas for lactating cows at TN and dry cows at both HT and TN, the hourly variation of VT was inferior to 0.3 °C. The huge infra-day variation of core temperature related to RT that was observed in lactating cows at HT clearly showed that the rumen can be a heat sink for the body when cow thermoregulation is impaired by high ambient temperature.

Plasma mineral metabolism and milk minerals in Modenese cows compared to Italian Friesian cows

F. Abeni[1], F. Petrera[1], A. Dal Prà[1], P. Franceschi[2], M. Malacarne[2], A. Summer[2], G. Catillo[3] and F. Napolitano[3]

[1]Consiglio per la Ricerca e la Sperimentazione in Agricoltura, Centro di Ricerca per le Produzioni Foraggere e Lattiero-casearie, via Porcellasco 7, 26100 Cremona, Italy, [2]Università di Parma, Dipartimento di Scienze degli Alimenti, via del Taglio 10, 43126 Parma, Italy, [3]Consiglio per la Ricerca e la Sperimentazione in Agricoltura, Centro di Ricerca per la Produzione delle Carni e il Miglioramento Genetico, via Salaria 31, 00015 Monterotondo (Roma), Italy; fabiopalmiro.abeni@entecra.it

The aim of this study was to compare the mineral metabolism and its relationships with milk macro-minerals in two different cattle breeds: Modenese (M) and Italian Friesian (IF). Milk samples from 12 M and 14 IF pluriparous cows were collected at 8 and 21 weeks of lactation (WOL). These cows were raised and managed together in the same herd, feeding the same total mixed ration. Blood plasma mineral profile (Ca, P, Mg, Na, K, and Cl), alkaline phosphatase (ALP, as indirect marker of bone deposition), and tartrate-resistant acid-phosphatase (TRAP, as indirect marker of bone mobilization) activities were assessed by an automated clinical biochemistry analyzer. Milk was analyzed for total contents of Ca and P and for their repartition into soluble and colloidal phases. Data were analysed by a mixed model (breed, WOL, breed×WOL, with the animal repeated in time). Plasma concentration of Na, K, ALP, and TRAP were higher ($P<0.05$), whereas Mg was lower ($P<0.05$) in M than in IF cows, and plasma Ca and P did not differ ($P>0.05$) between breeds. Total Ca content in milk was higher in M ($P<0.05$), but both the soluble and colloidal fractions were unaffected by breed ($P>0.05$). The contents of total and colloidal P were higher ($P<0.05$) in M than in IF milk. From these results, there is no apparent relation between plasma mineral status for Ca and P and their content in milk; however, the different plasma activities of ALP and TRAP need further studies for their explanation.

Influence of sampling procedure and skin contamination on skatole in fat of pigs

R. Wesoly, V. Stefanski and U. Weiler

Universität Hohenheim, FG Verhaltensphysiologie landwirtschaftlicher Nutztiere, Garbenstrasse 17, 70599 Stuttgart, Germany; wesoly@uni-hohenheim.de

Several breeding programs use measurements of androstenone and skatole (S) in biopsy material of young boars to estimate the individual breeding value for boar taint. In contrast to androstenone information about variability of S along the carcass of boars and the influence of biopsy technique are limited. Additionally, effects of soiling are discussed controversially. Thus in the present study S variability, the effects of sampling, and soiling on S in fat were analyzed. To determine the variability of S along the carcass of 8 boars, 36 fat samples/boar were measured (18/side: 3 neck, 6 loin, 4 ham, 5 belly). The effect of soiling was studied in 6 cannulated barrows after application of high S spiked feces (HSF; 565 µg/g) twice daily on a 5×5 cm area on the left shoulder. The contralateral area (COL) was cleaned daily. S levels before treatment were determined in bioptic fat from neck. Daily blood samples were used to monitor resorption. On day 7, pigs were euthanized and fat was collected from each area. To determine effects of sampling techniques, biopsies were collected from cannulated barrows with and without anesthesia. S and cortisol were determined in daily blood samples. S measurements along the carcass revealed low within animal variations and no differences between neck, loin and ham region, but higher levels ($P<0.05$) in the belly. Application of HSF for 7 days resulted in higher S levels close to the application area ($P<0.05$). This effect was restricted to the application side, as COL and blood concentrations were not affected. Biopsy techniques influenced S in blood ($P<0.05$). Anesthesia led to an increase in S within 7 h which was highly correlated with cortisol. Similar effects were not observed after biopsy in non-anesthetized animals. It is concluded, that samples for S determinations should be taken preferentially without anesthesia from the dorsal side of pigs. Soiling on the ventral side is not likely to influence levels in the neck region.

Difference in microRNA expression profiles between bovine masseter and semitendinosus muscles

S. Mruoya[1], M. Taniguchi[2], M. Shibata[3], M. Oe[1], K. Ojima[1], I. Nakajima[1] and K. Chikuni[1]
[1]NARO Institute of Livestock and Grassland Science, Tsukuba, Ibaraki, 305-0901, Japan, [2]National Institute of Agrobiological Sciences, Tsukuba, Ibaraki, 305-0901, Japan, [3]NARO Agricultural Research Center for Western Region, Ohda, Shimane, 694-0013, Japan; muros@affrc.go.jp

MicroRNA (miRNA) is noncoding small RNA involved in post-transcriptional gene regulation in various biological processes. The objective of this study was to determine differentially expressed miRNAs between the semitendinosus (ST) and masseter (MS) muscles from three Japanese Black cattle using deep sequencing technique. Differential gene expression of myosin heavy chain (MyHC) isoforms confirmed that ST and MS were MyHC-2x- and MyHC-1-abundant muscles, respectively. In total, 192 known miRNAs and 20 potential new bovine miRNAs were obtained from the sequencing. The differentially expressed miRNAs with more than 2-fold difference in each muscle were identified. In particular, miR-196a and miR-885 were exclusively expressed in ST muscle, which was validated by qRT-PCR ($P=0.045$ and $P<0.001$, respectively), whereas a slow type-directing miR-208b was highly expressed in MS compared with ST (false discovery rate <0.05). The results of functional annotation combined with in silico target analysis showed that the predicted target genes of miR-196a/b and miR-885 were enriched in gene ontology (GO) terms related to skeletal system development and regulation of transcription, respectively. Moreover, the predicted targets of ST-abundant miRNAs were enriched in GO terms of the embryonic skeletal system, regulation of cell motion, chordate embryonic system, and embryonic development ending in birth or egg hatching. On the other hand, GO terms of the pattern specification process, reorganization, and sensory organ development were obtained from predicted MS-abundant miRNA targets. The distinct miRNA distribution between the muscles suggests that differentially expressed miRNAs are associated with muscle type-specific tissue formation and/or maintenance.

Calving monitoring in dairy cattle

A. Chanvallon[1], S. Coyral-Castel[1], A. Leblay[2], J. Girardot[3], J.B. Davière[3] and J.M. Lamy[2]
[1]Institut de l'Elevage, 149 r Bercy, 75595 Paris cedex 12, France, [2]Chambre d'Agriculture de Maine-et-Loire, 14 av Jean Joxé, 49006 Angers cedex 01, France, [3]Clasel, 141 bd des Loges, 53942 Saint Berthevin cedex, France; audrey.chanvallon@idele.fr

Calving monitoring is essential to ensure breeding progress. With increasing size of herd, monitoring tools are in expansion. Various devices, placed on the tail, in the vagina or around the abdomen of the animal, can alert the farmer of an imminent calving. The Vel'Phone® system is based on the analysis of temperature variation using a vaginal thermometer. Three alerts can be transmitted to the farmer from the Vel'Phone®: probable calving within 48 h (PRO), calving expected within 48 h (WAIT) or thermometer expulsion with the breaking of the foetal membranes (EXP). Three assays were conducted in an experimental farm to assess this tool. A first assay was conducted on 3 heifers and 32 dairy cows that calved between November 2010 and February 2011, a second assay was carried out on 33 heifers and 53 dairy cows that calved between August 2011 and February 2012 and a third assay concerned 19 heifers and 66 dairy cows that calved between August 2012 and December 2012. On average, calving took place 4.9±4.8 days before expected full-term in heifers (n=55) and 3±5 days in cows (n=151). Interval between the introduction of the thermometer and calving was 5.4±3.3 days in heifers (n=44) and 4±3.2 days in cows (n=130). Sensibilities of detection were for heifers 27%, 79% and 100% for alerts PRO, WAIT and EXP, and for cows 54%, 62% and 97% for alerts PRO, WAIT and EXP. Interval between alert EXP and calving was on average 148±68 minutes for heifers and 85±49 minutes for cows. To conclude, optimal time to place the thermometer is difficult to determine because of the delay between expected calving and real calving. The alert expulsion of the thermometer is a good calving predictor. The Vel'Phone® is an interesting device to reduce the constraint of supervision and thus to improve the comfort of the farmer.

Session 32 Poster 11

Changes of the resorption surface area of the small intestinal villi of ostriches in ontogenesis

L. Mancevica and A. Mugurevics
Faculty of Veterinary Medicine, Preclinical Department, Kr. Helmana 8, 3004 Jelgava, Latvia; lauma.mancevica@llu.lv

The small intestine has a physiological importance in digestion of the consumed feed and absorption of nutrients in the blood, which is directly affected by the intestinal epithelial villi area. The aim of the research was to find out the changes of the length, width and longitudinal section area of the small intestinal villi in 18 ostriches raised in Latvia at the age of 4, 6, 8 and 12 months. For microscopic analysis of the intestinal mucous membrane, tissue sections were stained with hematoxylin and eosin. In each sample, the length (μm), width (μm) in the middle of the villus, and longitudinal area (μm^2) were measured in 10 villi. The data were statistically processed with SPSS 20.0 program using ANOVA Post Hoch test and Pearson correlation analysis. The size of the duodenum villi had a tendency to increase regularly during the postnatal ontogenesis period. A close correlation was observed between the duodenum villi length and area ($r=0.73$; $P<0.01$). The jejunum villi length and area were increased ($P<0.05$) during the experimental period of ontogenesis, but more rapid increase ($P<0.05$) of these parameters was observed in animals from 6 to 8 months of age with a close correlation ($r=0.73$; $P<0.01$). The ileum villi area increased throughout the observed period of ontogenesis ($P<0.001$), but the length and area increased significantly between 6 and 8 months of age ($P<0.05$) with a medium close correlation ($r=0.68$; $P<0.01$). The villi width of the duodenum, jejunum and ileum increased regularly during the observed period of ontogenesis. Up to 6 months of age, the longitudinal section area was the largest in the duodenum, while from 8 months of age – in the jejunum. The obtained results, possibly, indicate that the functional activity of some small intestinal segments takes place between 6 and 8 months of age that should be taken into account when selecting an optimal nutritional regime.

Session 32 Poster 12

The effect of short term nutrition on the estrus characteristics in ewes with synchronized estrus

M. Kuran[1], U. Sen[1], E. Sirin[2], Y. Aksoy[3] and Z. Ulutas[3]
[1]Ondokuz Mayis University, Faculty of Agriculture, Department of Animal Science, 55139, Turkey, [2]Ahi Evran University, Faculty of Agriculture, Department of Animal Science, 40100, Turkey, [3]Gaziosmanpasa University, Faculty of Agriculture, Department of Animal Science, 60250, Turkey; ugur.sen@omu.edu.tr

The aim of this study was to determine the effect of short term nutrition during pre-mating period on the estrus characteristics in Karayaka ewes with estrus synchronized. Intravaginal sponges containing 30 mg flugestone acetate were inserted into the ewes for 12 days followed by an intramuscular injection of 1 ml of PGF2α. Eight days after the insertion, the ewes (average body weight of 48.9±0.7 kg, loin thickness of 2.10±0.03 mm, fat thickness of 0.50±0.02 mm) were allocated randomly into three groups and were fed until mating (6 to 8 days) as follow; daily requirement for maintenance (control group, C; n=58) or 0.5×maintenance (undernutrition group, UN; n=33) or *ad libitum* (overnutrition, ON; n=27, consumed 1.6×maintenance). The percentage of ewes in estrus in ON, C and UN groups were 44, 41, and 42% within 24 to 48 h, 49, 47, 55% within 49 to 72 h and 7, 12, 3% within 73 to 96 h after the removal of sponge. There were no significant differences between nutrition groups in terms of the time span from the sponge removal to the estrus ($P>0.05$). A reduced proportion of ewes were in estrus in ON (70.6%) and UN (67.3%) groups compared to those in C (89.4%) group ($P<0.01$; $\chi 2=8.97$). These results show that a short period of over and under nutrition before estrus in ewes with synchronized estrus does not influence the time of estrus but reduces the number of ewes in heat.

Session 32 Poster 13

Effects of GH SNPs on endocrine status and glucose metabolism in Japanese Black calves

R. Kobayashi[1], H. Dian[2], H. Sano[2], T. Hirayama[3], K. Katoh[1] and S.-G. Roh[1]
[1]Tohoku University, GSAS, Sendai, 981-8555, Japan, [2]Iwate University, Morioka, 020-8550, Japan, [3]University of the Ryukyus, Naha, 903-0213, Japan; kato@bios.tohoku.ac.jp

Single nucleotide polymorphism (SNP) in the gene of growth hormone (GH) significantly affects the performance of meet quality of Japanese Black cattle. On GH SNP, cattle with alleles A and B have larger body weight and greater GH release induced by GHRH stimulation, but lower intramuscular oleic acid %, than those with allele C. The aims of the present study were to analyze the effect of GH SNPs on endocrine status and glucose metabolism in young Japanese Back calves (genotypes AA and BB, n=4 and 5, respectively) at age of 9 months under glucose-clamped condition. The clamp technique is employed to measure glucose utilization (insulin sensitivity) and also to investigate the effects of high insulin condition on stress hormone levels. Glucose-clamp was successfully established for 2 hrs by continuous infusion of insulin (6 mU/metabolic body size/min) and 20% glucose solution at every 5 min: plasma glucose and NEFA concentrations were not changed during the clamp period. The clamp condition significantly reduced plasma GH concentrations (t-test, $P<0.05$), mimicking postprandial reduction in GH levels. However, there were no significant changes in plasma ACTH ($P=0.76$) and cortisol ($P=0.83$) concentrations, showing no stress even under the clamp condition. There was no difference in these parameters between both genotypes (P values for ACTH and cortisol were 0.28 and 0.22, respectively). Finally, glucose availability (infusion rate) in calves with genotype BB was significantly smaller ($P<0.05$) than that with genotype AA, although the body weight for BB was smaller than that for AA ($P<0.05$).

Session 32 Poster 14

Effect of season feeding on the activity aminopeptidases and protein fraction in milk

A. Jóźwik, N. Strzałkowska, M. Markiewicz-Kęszycka, P. Lipińska, W. Grzybek, E. Poławska, E. Bagnicka, J. Krzyżewski and J.O. Horbańczuk
Institute of Genetics and Animal Breeding PAS, Postępu str. 36a Jastrzębiec, 05-552, Poland; aa.jozwik@ighz.pl

The aminopeptidases activity and protein fraction profile effects milk nutritional quality and consumer's health. The aim of the study was to estimate the influence of the feeding season on the activity of aminopepdidases and the composition of the protein fraction in cows' milk. Forty Montbeliarde cows were kept in the different feeding seasons: during autumn/winter (A/W) season cows were maintained on the typical winter ration (TMR-Total Mixed Ration), during spring/summer (S/S) season a typical summer feeding was applied (grazing on grass pasture+concentrates with mineral-vitamin premix). At the end of each season, milk samples were taken (at the end of A/W and S/S in May and in September, respectively). The activities of the alanine, leucine and arginyl aminopeptidases, total protein (TP) content and casein fraction (CF) of whole milk were determined. The results showed significantly higher activity of aminopeptidases in milk of cows maintained on pasture as compared to cows kept indoors during winter ($P<0.05$). The negative Pearson's correlation between activity of all aminopeptidases and TP content and CF in milk (from -0.38 to -0.67) only after S/S was observed ($P<0.05$). In conclusion, milk obtained during summer season may have better nutritional parameters than that yielded in winter. Thus, milk obtained during summer feeding season may positively influence the consumer's health. BIOFOOD–innovative, functional products of animal origin no. POIG.01.01.02-014-090/09 co-financed by the European Union from the European Regional Development Fund within the Innovative Economy Operational Programme 2007-2013.

Session 32 Poster 15

Effect of acute heat stress on thermoregulatory responses in water buffaloes

C.G. Titto[1], R.A. Vilela[1], E.A.L. Titto[1], A.M.F. Pereira[2], T.M.C. Leme[1], A.C.A.P.M. Geraldo[1], J.C.C. Balieiro[1], E.H. Birgel Junior[1] and L. Martello[1]

[1]Faculdade de Zootecnia e Engenharia de Alimentos, USP, Av. Duque de Caxias Norte 225, 13635-900, Pirassununga, SP, Brazil, [2]Instituto de Ciências Agrárias e Ambientais Mediterrânicas, Universidade de Évora, Núcleo da Mitra, Ap 94, 7002-554, Évora, Portugal; martello@usp.br

The study aimed to assess the adaptive thermoregulatory and hematological responses of buffaloes towards to acute heat stress, and monitor the ability to maintain thermal balance in absence of solar radiation. Twelve 18-month old heifers were analysed during 2 periods of 3 days inside a climatic chamber: 3 days (P1) with air temperature (Ta) and relative humidity (rH) kept at 26 °C and 78%; and 3 days (P2) simulating the cycle of daily temperatures under heat stressful conditions (Ta:36 °C, rH: 78% from 6:00 to 19:00 and 27 °C and 80% from 19:00 to 6:00). Animals were subjected to a switchback design. For rectal temperature (RT) and respiratory rate (RR) period (P1, P2) and sampling time (7, 10, 13, 16, 19 and 22) were considered as fixed effects, and their interaction and animal as random effects with repeated measures in the same animal across time. For sweating rates (SR), haematological parameters and white blood cells, period was included as fixed effect and animal as random effect with repeated measures in the same animal across time. Significant different means were submitted to Tukey test (P=0.05). In P2 an increase in RR was observed ($P<0.05$), which quadrupled in relation to P1. The SR were quite high in both periods, however P2 (1,146 $g/m^2/h$) was higher than P1 (567 $g/m^2/h$; $P<0.05$). A slight increase was observed in RT along the day ($P<0.05$). The heat stress did not influence the results of erythrogram and leucogram or Na^+, but originated a significant reduction in K^+. The moderate heat storage and the subsequent fast recovery of the thermal equilibrium seem to be associated with high SR. The SR shown is of importance to understand the sudoresis in the maintenance of thermal equilibrium of the buffalo.

Session 33 Theatre 1

Fallen livestock: animal welfare concerns and consequences

J. Baumgartner, M. Mlak, M. Klager, A. Geier and C. Hofer-Kasztler

Vetmeduni Vienna, Veterinaerpl. 1, 1210, Austria; johannes.baumgartner@vetmeduni.ac.at

A considerable number of farm animals destined for food production end as fallen animals in the animal waste processing system. In 2009 the proportion of fallen stock in Austria was approximately 11% for young cattle, 2% for older cattle and 5% for pigs >50 kg. The causes of death of fallen animals remain unclear within the standard procedure. There are indications that some of these animals die miserably due to a lack of appropriate intervention by the animal keeper. In our study we wanted to highlight the concealed fate of fallen animals from an animal welfare point of view. The cadavers of 1,070 cattle and 987 pigs delivered to an Austrian animal waste processing plant during 20 work days were inspected for pathological changes by means of a predefined protocol. 134 fallen cattle (12.5%) with obvious pathological changes were registered. 75% of these cadavers showed decubitus ulcers and 17% an excessive loss of weight, 20% were extremely dirty, 13% had pathological claws and 7% were found with swollen joints or abscesses. In pigs 203 fallen cadavers (21%) with external pathological changes were registered. Bite injuries at tail and ears were most frequent (62%) followed by severe decubitus ulcers (39%) and tissue swellings (14%). The majority of the cadavers with findings showed a combination of different pathological signs. Only in a few of fallen cattle and pigs signs of appropriate emergency killing could be found. (Too) many inspected cadavers did substantiate suspicion of unnecessary pain and prolonged suffering of the animals prior to death. Particularly carcasses with long lasting severe pathological changes such as multiple decubitus ulcers and cachexia point to either a lack of appropriate medical treatment or to missing euthanasia. Both from an economical and ethical point of view the number of fallen livestock with obvious signs of unnecessary suffering has to be reduced. Transfer of knowledge and the installation of a feasible monitoring system could lead to specific counteractive measures.

Effects of journey time to slaughter on the welfare of pigs

M.A. Mitchell[1], P.J. Kettlewell[1], M. Villarroel[2], M. Farish[1], K. Stoddart[3] and H. Van de Weerd[4]
[1]SRUC, AVS, the Roslin Institute Building, Easter Bush, Midlothian, EH25 9RG, United Kingdom, [2]Polytechnic University of Madrid, Department of Animal Science, School of Agricultural Engineering, Madrid, Spain, [3]BPEX, n/a, Stoneleigh Park, Kenilworth, Warwickshire. CV8 2TL, United Kingdom, [4]Cerebrus Associates, n/a, The White House, 2 Meadrow, Godalming, Surrey GU7 3HN, United Kingdom; malcolm.mitchell@sruc.ac.uk

There is much debate concerning the maximum permitted journey times for livestock, however, there is little research that can be directly applied to the determination of acceptable limits for journey times. The current study examined commercial pig transportation to a single UK slaughterhouse on journeys of durations ranging from less than two hours to over 10 hours. Post transport measures including mortality, injury, lesions, casualty slaughter rates, lameness, behaviours in lairage and meat quality parameters were employed to assess welfare status of the pigs upon arrival and to estimate the degree of pre-slaughter stress imposed. The data were analysed to determine if journey time had any direct effects upon stress and animal welfare and to determine if any specific upper limit of journey time for safe transportation could be identified. There were no statistically significant indications that journey time had a detrimental effect of the pigs transported under commercial conditions in this study. The results suggest that other factors are more important for the welfare of pigs in transit. These are thermal conditions on the vehicle, the nature of the system in which the pigs were produced prior to transport and interruptions to the journey i.e. stationary periods the most important of which may be the 'standing time' at the abattoir. Whilst the effects of all these factors may be exacerbated by excessive journey times the journey time per se may not be the most important issue when attempting to optimise welfare in transit.

Development of a force plate scale to measure weight distribution and lameness in gestating sows

S. Conte[1], R. Bergeron[2], H. Gonyou[3], J. Brown[3], M.L. Connor[4] and N. Devillers[1]
[1]Agriculture and Agri-Food Canada, Dairy and Swine R&D Centre, Sherbrooke, QC, J1M 0C8, Canada, [2]University of Guelph, Alfred Campus, Alfred, ON, K0B 1A0, Canada, [3]Prairie Swine Centre, 2105, 8th Street East, Saskatoon, SK, S7H 5N9, Canada, [4]University of Manitoba, Animal Science Building, Winnipeg, MB, R3T 2N2, Canada; nicolas.devillers@agr.gc.ca

Lameness is the second cause of culling and the first cause for euthanasia in sows and needs to be assessed using reliable and quantitative methods. A scale with 4 platforms was developed to measure weight distribution between limbs was validated for repeatability using 10 sows measured 4 times on 2 different days. The mean and SD of % body weight (BW) on each limb, and the average ratio between weights on right and left limbs (RATIO) were calculated. The weight shifting (at least 2.5% of BW above or below average weight on a leg) was expressed in terms of frequency, amplitude and % time above or below average weight. All variables had good to very good intra-sow repeatability (CV≤16%). Data from the force plate scale were compared using PROC GLIMMIX (RATIO) and PROC MIXED of SAS between 60 sows visually scored for lameness (non-lame (NL), n=24; mildly lame (ML), n=19; lame (L), n=17). On average (mean ± S.E.), sows put 28.8±0.14% of their BW on an anterior limb and 21.2±0.14% on a posterior limb. L sows had a lower RATIO (L: 0.64 [0.61-0.67]; ML: 0.69 [0.66-0.72]; NL: 0.70 [0.68-0.73]; P<0.01), a higher frequency of weight shifting (L: 32.3±1.4; ML: 23.3±1.4; NL: 21.5±1.2 shifts per min; P<0.001) and tended to have a higher % time above or below average weight (L: 64.8±1.9%; ML: 60.0±1.8%; NL: 59.7±1.6%; P=0.09) per leg than ML and NL sows. SD of %BW on each limb and amplitude of weight shifting were not different between lameness scores. In conclusion, the force plate scale is efficient in discriminating lame sows which on average showed a higher variation in the weight applied on their limbs. Further analyses are needed to enable detection of the affected leg(s).

Milk FA profile measured by MIRS as an indicator of ketosis status in dairy cow: a preliminary study

A. Théry[1,2], G. Chesneau[1,2], E. Besnier[2] and S. Couvreur[1,3]
[1]FCI, La Messayais, 35210 Combourtillé, France, [2]Valorex, La Messayais, 35210 combourtillé, France, [3]LUNAM Université, Groupe ESA, Unité Recherches sur les Sysèmes d'Elevage, 55 rue Rabelais, 49007 Angers, France; a.thery@valorex.com

Ketosis diagnosis is not possible without blood parameter measurements. The recent development of FA composition measurement by Mid-InfraRed Spectroscopy (MIRS) may help in diagnosing ketosis, as it is well known milk FA profile is linked to ketosis metabolic status. Our objective was to assess if FA composition measured by MIRS may help in diagnosing ketosis in dairy cows. 27 dairy cows at the beginning of lactation ($\leq$3 months) were selected and split in 2 groups by two vets in 6 farms based on their metabolic status: healthy cows (HC; n=16) vs. ketosis cows (KC; n=11) (0.45 vs. 0.30 mg glucose/dl et 0.88 vs. 3.15 mmol BOH/L, respectively; P>0,001). Individual blood and milk samples were taken out to measure out plasma and milk composition (MIRS). The effect of ketosis on plasma and milk compositions was assessed using the general linear model. A decision tree was created to assess if milk composition (MIRS) is a good indicator of ketosis status. In comparison to HC, KC had a more negative energy balance and a shorter lactation stage. Fat content, fat/protein ratio, mono-unsaturated FA, C18:0, C18:1 levels were higher and C16:0 and saturated FA levels lower in KC milk than HC milk. Zn and Cu contents, and acetate/propionate ratio were higher and Ca, urea and thyroxin contents were lower in KC blood than HC blood. Milk composition (MIRS) was, in our population, a good indicator of ketosis status: a decision tree using C16:0 then fat/protein ratio makes possible a good prediction of the metabolic status of 26 cows. These results must be confirmed with bigger databases.

Minimizing negative effects of poultry red mite in layer farms using an automated monitoring device

M.F. Mul[1], T.G.C.M. Van Niekerk[1], B.G. Meerburg[1] and P.W.G. Groot Koerkamp[1,2]
[1]Wageningen UR Livstock Research, environment, Edelhertweg 15, 8219 PH Lelystad, the Netherlands, [2]Wageningen University, Farm Technology Group, Droevendaalsesteeg 1, 6708 PW Wageningen, the Netherlands; monique.mul@wur.nl

The poultry red mite (*Dermanyssus gallinae*) is a common ectoparasite in poultry farms worldwide, feeding on blood of poultry and sometimes humans in order to development to the adult stage and reproduce. A poultry red mite (PRM) infestation may result in high economic losses, veterinary risks and allergic reactions among farm workers. Control of PRM has become more difficult due to development of resistance to some acaricides and a ban on others. Therefore recent research has been focused on more environmentally friendly control methods such as the use of natural enemies, attract and kill methods using fungi and development of a vaccine. For an effective, timely, durable and place specific application of the control methods, monitoring of the size, place and the development of the poultry red mite population is necessary. Therefore, we aim to develop an automated monitoring device for PRM in layer farms which is composed of an automated counter for PRM and a dynamic adaptive model. This monitoring device enables to assess the PRM population in: (1) the actual situation; (2) after a treatment (effect); and (3) in future situations (necessary to indicate timely treatment). A preliminary PRM farm model was made in which the interactions are determined of a PRM population on feed intake, egg production, flock size and farm economics. At the beginning of 2013 the first prototype of the automated mite counter will be validated under laboratory conditions. Therefor the counted number of mites by the automated poultry mite counter will be compared with the number of mites present in the test facility. The model and results of the test with the automated mite counter will be presented.

Session 33 Theatre 6

Estimate of the impacts of the first outbreak of Schmallenberg virus in French sheep herds in 2012

B. Mounaix[1], L. Sagot[1], C. Delvalle[1], L.M. Cailleau[1], K. Gache[2] and F. Dion[3]
[1]Institut de l'Elevage, 149 rue de Bercy, 75595 Paris, France, [2]GDS France, 149 rue de Bercy, 75596 Paris, France, [3]Races de France, 149 rue de Bercy, 75595 Paris, France; beatrice.mounaix@idele.fr

As a first attempt to assess the consequences of the outbreak of the Schmallenberg virus in the French sheep farms, the variability of the impacts of the virus which were observed in 348 herds during the winter 2012 was analyzed using multivariate analysis. The resulting 3 class typology showed that more than 70% of these herds suffered moderate impacts: 6.3% of average mortality in winter lambs and 0.9% of average mortality in ewes. The two classes of highly and very highly impacted herds showed higher mortalities of winter lambs, respectively 21% and 41%, but limited impacts in ewes. Detailed interviews were realized in 20 farms. They indicated that ewe mortality was mainly due to the difficulties in the first lambing of malformed newborns and was rapidly corrected by special care to females after lambing. The variability in the mortality of lambs was explained by the differences in the date of breeding; no other risk factor could be demonstrated in this study. The average mortalities of lambs and ewes in the 3 classes of herds were affected to two models of farming systems to estimate the economical consequences of the impacts of the virus at the farm level. The economical losses were comparable in both systems, from 2% of losses in the gross margin per ewe for the least impacted farms, to 19% of losses in the very highly impacted ones. This estimate only took into account the congenital effects of the virus and did not include the possible impacts of the acute infection. This study was coordinated with the main national and regional federations of livestock technicians and veterinarians and was released to farmers and technicians with recommendations to take a special care of lambing ewes to reduce the long term consequences of the virus in the herd.

Session 33 Theatre 7

Contributions of neighborhood and animal movements to *Coxiella burnetii* infection in cattle herds

S. Nusinovici[1], T. Hoch[1], A. Joly[1], J. Frössling[2], S. Widgren[2], A. Lindberg[2] and F. Beaudeau[1]
[1]Oniris, UMR1300 Biology, epidemiology and risk analysis in animal health, Atlanpole, La Chantrerie, BP 40706 Nantes, France, [2]National Veterinary Institute, Travvägen 20, 751 89 Uppsala, Sweden; simon.nusinovici@oniris-nantes.fr

Coxiella burnetii infection occurs mainly after inhalation of aerosols contaminated by shedder ruminants. The propagation of *C. burnetii* between ruminant herds may result from a contiguity transmission between neighbours herds and/or the introduction of infected shedder animals in free herds. The objective of this study was to quantify and compare the relative impacts of the local cattle density and the introduction of possibly infected animals on *C. burnetii* spread between herds. Dairy herds located in France (n=2,829) and Sweden (n=117) were included. Herds were tested using ELISA kits in the bulk tank milk in May 2012 and June 2011 respectively. The total number of cattle introduced (ID) in herds was calculated during a period of 2 years before the date of ELISA test. The local cattle density was calculated using the total number of cattle located in a 5 km radius circle around herds. Logistic regression was used to assess the risk for a herd to be tested positive associated with an increase in the local cattle density and in the ID. The risk for a herd to be tested positive was higher for herds with a higher local cattle density (e.g. OR=4.05, CI95% 1.5-10.9, for herds with a local density between 100 and 120 compared to herds with a local density below 20). The risk was higher for herds with higher ID (e.g. OR=2.4, CI95% 1.7-3.4, for herds with ID over 3 compared to herds with no animal introduction). The variable country was not significant, indicating that results are not country specific. Population attributable fraction was 64% for the local density and 16% for ID. Results highlight the main role of contiguity in the dispersion process of *C. burnetii* between herds, suggesting the main role of wind. These results will enlighten the type of control measures to be implemented in herds at the regional level.

Effects of maternal selenium supply during gestation on colostrum quality and IgG transfer in lambs

F. Corbiere[1,2], J.M. Gautier[2,3] and L. Sagot[3,4]
[1]National Veterinary School of Toulouse, 23 chemin des Capelles, 31076 Toulouse, France, [2]UMT Small Ruminants Heard Health, 23 chemin des Capelles, 31076 Toulouse, France, [3]Institut de L'élevage, BP 42118, 31321 Castanet Tolosan, France, [4]CIIRPO, Le Mourier, 87800 St Priest Ligoure, France; f.corbiere@envt.fr

It is the aim to investigate effects of Selenium (Se) supply during late gestation on colostrum quality and passive transfer of immunity in neonatal lambs. 80 Vendéen ewes were allocated to 2 treatment regimens. The treated group received a mineral preparation containing inorganic Na-selenite and organic Se yeast during the last 5 weeks of gestation providing a daily selenium supply of 15 µg/kg BW. The control group was left with low daily Se allowances (0.7 µg/kg BW). After birth lambs were left to suckle their dam naturally. The butterfat (BT) and Immunoglobulin G1 (IgG1) concentration in colostrum was assessed at parturition. The plasma Se and IgG1 concentration was assessed in lambs 24 h after birth, and their Se status evaluated further at 20 days of life. At allocation the ewes Se plasma concentration was low (26.3±5.4 µg/l) in both ewe groups. One week after parturition the Se plasma concentration and the glutathione peroxidise (GPx) activity in the treated group were higher than in the control group ($P<10^{-4}$). Similarly the plasma Se concentration and GPx activity in lambs were higher ($P<10^{-4}$) in the treated than in the control group at each evaluation point. No effect of Se supply could be evidenced for colostrum BT (P=0.94) and IgG1 concentrations (P=0.50) and for lamb serum IgG1 concentration at 24 h of life (P=0.35). This study did not show any effect of selenium supply during late gestation on colostrum quality and passive transfer of immunity in neonatal lamb. However the selenium supply during late gestation enabled to restore the selenium status of deficient ewes and to provide their lambs with a satisfactory selenium status at birth.

A model to quantify individual tolerance and resistance to infectious diseases

J. Detilleux
University of Liege, B43 Boulevard de Colonster, 4000 Liège, Belgium; jdetilleux@ulg.ac.be

A system of ordinary equations embedded in a structural equation model is proposed to estimate direct and indirect levels of resistance (i.e. reduced pathogen transmission at contact and pathogen growth rate once infection has occurred) and tolerance (i.e. limited damage inflicted by the pathogen load or caused by the immune response) to infectious diseases. The system describes how within-host pathogens and immune cells interact after infection and consists of two differential equations, one for pathogen load and the other for an index of immunity. According to parameter values, the equations generate realistic outcomes in response to infection: healthy response, recurrent infection, persistent infectious and non-infectious inflammation, and severe immunodeficiency. Solutions of the differential equation model (random effects) are included in the structural equation model for individual host performances. Results of the study are tested with simulated and real data.

Session 33 Theatre 10

Efficient fragmentation of animal trade networks by targeted removal of central farms

K. Büttner[1,2], J. Krieter[2], A. Traulsen[1] and I. Traulsen[2]
[1]Max Planck Institute for Evolutionary Biology, August-Thienemann-Str. 2, 24306 Plön, Germany, [2]Institute of Animal Breeding and Husbandry, Christian-Albrechts-University, Olshausenstr. 40, 24098 Kiel, Germany; kbuettner@tierzucht.uni-kiel.de

Centrality parameters in most animal trade networks have right-skewed distributions, meaning networks are highly resistant regarding the random removal of farms but vulnerable to targeted removal of the most central farms. The aim of this study is to understand how the targeted removal affects animal trade networks and which parameter is an appropriate measure for this procedure. Furthermore, the optimal combination of three farms regardless of their centrality is identified and compared to the targeted removal. Contact data from a producer group in Northern Germany (2006-2009) were analysed in the three-year network, the yearly and the monthly networks. The data contain information of 4,635 animal movements between 483 farms. Results showed that centrality parameters regarding ingoing contacts (in-degree, ingoing infection chain, ingoing closeness) were not suitable for a rapid fragmentation. More efficient was the removal by parameters considering outgoing contacts or betweenness. In the three-year network (yearly and monthly networks) about 7% (5%) of farms had to be removed to reduce the size of the largest component by more than 75%. For the three-year network the smallest difference to the optimal combination was the removal by out-degree with 1%, followed by outgoing infection chain and outgoing closeness (8%) and betweenness (10%). In the yearly networks the removal by out-degree and outgoing closeness were located 2% above the optimal strategy, followed by outgoing infection chain (6%) and betweenness (11%). For the monthly networks the smallest difference to the optimal combination was the removal by out-degree, outgoing infection chain and outgoing closeness centrality (2%), followed by betweenness (10%). Most efficient interruption of the infection chain was obtained by using targeted removal by out-degree.

Session 33 Poster 11

Promoting high quality control posts in the European Union

K. De Roest, P. Rossi, P. Ferrari and A. Porcelluzzi
Research Center for Animal Production, Agricultural Economics and Engineering, Viale Timavo 43/2, 42121 Reggio Emilia, Italy; k.de.roest@crpa.it

The transport of live animals is a crucial step in the animal production process, involving farmers, traders, slaughterhouses, control posts (CPs) and transport companies and potentially affecting the environment, animal health and animal welfare. CPs are establishments where animals transported by road over long journeys must be unloaded, fed and watered and be rested for at least 24 h, according to Regulation (EC) No 1/2005. Due to the lack of enforcement of animal transport rules across Europe and to reduce the spread of animal diseases, the European Commission has been called by Member States to adopt measures to secure full and uniform monitoring of adherence to the transport conditions. To this end DG SANCO has financed a preparatory action to renovate 15 CPs located at the cross roads of important flows of animals transported over Europe according to highest quality standards set up by a previous feasibility study of a certification scheme for high quality CPs. The involved CPs partners have been visited by a team of experts to plan and design the interventions taking into account the requirements for high levels of animal welfare, sound systems of bio-security, work safety and efficiency of personnel and environmental sustainability. Co-financing procedures for CPs have been adopted by using analytical cost calculations based on official tariffs and price lists of equipment and building materials. To this end eight national databanks (IT, FR, ES, DE, PL, GR, HU, RO) of prices and tariffs of equipment have been constructed which serve as objective reference values for the final investment plans. An important achievement of the project has been also the production of a certification scheme, for CPs developed with an adequate involvement of relevant stakeholders represented by an appointed advisory board. The system enables the CPs to certify their facilities according to high quality standards of animal welfare standards and bio security.

Session 33 Poster 12

The effects of improved welfare on quality characteristics of pork

A. Gastaldo, A. Rossi, E. Bortolazzo, M. Borciani, A. Bertolini, E. Gorlani and P. Ferrari
Research Center For Animal Production (CRPA), Viale Timavo 43/2, 42121 Reggio Emilia, Italy; p.ferrari@crpa.it

The aim of this study was to assess the effect of improved animal welfare on farm, during transport and pre-slaughter operations on the quality of pork of heavy pigs in Italy. 11 fattening pigs' farms and 44 pigs' lots were evaluated and classified according to two animal welfare assessment systems: Farm Welfare Index (FWI) for farms, and Transport and Pre-slaughtering Welfare Index (TPWI) for transport and pre-slaughtering operations. Physiological parameters (aldolase, creatine kinase, cortisol levels) as well as quality and biochemical indicators (lean meat percentage, water holding capacity – WHC, pH) were measured and at the same time pork sensory quality (appearance, olfactory, taste and flavour-related and textural) was evaluated. All data collected were organized and processed in order to perform the statistical analysis using the SPSS. Two-way Anova with interaction analysis was performed using the Duncan mean separation test (alfa=0.05) to identify differences between the factors. The following results have been obtained: (1) the increase of animal welfare level on farms (higher FWI score) improves the sensory characteristics of pork, mainly those related to texture and appearance (intramuscular fat, tenderness, juiciness); (2) when the number of pigs that slip, climb over or fall during loading/unloading operations, the incidence of green hams haematomas and the level of aldolase in blood increases; (3) longer transport times increase the creatine kinase levels in blood; and (4) at slaughterhouse, longer lairage times improve the qualitative parameters of the pork and green hams (colour, juiciness and chewiness). In conclusion improved animal welfare conditions on farm, duringtransport and in the lairage influence positively the quality characteristics of pork and consequently the farmer and slaughterer income.

Session 33 Poster 13

The mortality of calves in French beef cattle farms: a national review

B. Mounaix[1], P. Roussel[1] and S. Assie[2]
[1]Institut de l'elevage, UMT Maitrise de la santé des troupeaux bovins, 149 rue de Bercy, 75595 Paris, France, [2]ONIRIS, INRA, UMR1300, UMT Maitrise de la santé des troupeaux bovins, CS 40706, 44307 Nantes, France; beatrice.mounaix@idele.fr

The mortality of calves (up to 6 months old) in French beef cattle farms was analyzed from the national database of bovine identification to better understand its variability. Survival models (Proc Lifetest, SAS®) were elaborated from the individual dates of birth and death of 8.6 million calves which were born in 33,982 beef cattle herds between 2005 and 2009. The average yearly mortality of calves within this period of time was 8.36%, but it varied a lot between herds: 75% of them showed 10% or less calf mortality; about 10% of herds showed 20% or more calf mortality. Although a significant increase in the mortality of calves was observed from 2007, probably due to the bluetongue outbreak, the survival curves showed similar patterns each year. Most of the calf mortality was observed during the first week (6.57%) then the slope of the survival curves stabilized in the following weeks. Males suffered a significantly higher mortality (9.54%) than females (7.04%). Greatest variations in the average calf mortality were observed between breeds: from 5.04% in Aubrac and Salers cattle to 13.8% in Rouge des près. The survival of crossed breed calves varied according to the crossing. Calves from first- and second-parity cattle showed a significantly higher mortality -respectively 9.72% and 7.58%- than those from following parities (6.58%). The seasonal calving pattern also impacted the survival of the newborn calf. No significant effect of the herd size was observed in our data. Most of the variability in the survival of calves was explained by differences in calving difficulties and stillbirth. These results can now be used to elaborate supporting methods to help the farmers to reduce the mortality of calves and to improve the productivity of beef cattle farms.

The influence of feeding levels before and after 10 weeks of age on osteochondrosis in growing gilts

D.B. De Koning[1], E.M. Van Grevenhof[1], B.F.A. Laurenssen[1], P.R. Van Weeren[2], W. Hazeleger[1] and B. Kemp[1]
[1]Wageningen University and Research Centre, Animal Sciences, P.O. Box 338, 6700 AH Wageningen, the Netherlands, [2]Veterinary Faculty, Utrecht University, Equine Sciences, P.O. Box 80.163, 3508 TD Utrecht, the Netherlands; danny.dekoning@wur.nl

Osteochondrosis (OC) in the epiphyseal growth cartilage develops in a short time frame in young growing gilts, which can cause lameness and reduced longevity. Feeding levels may be associated with OC. The aim of this study is to investigate age dependent effects of feeding levels, *ad libitum* versus restricted (80% of *ad libitum*), on OC prevalence in gilts at slaughter (26 weeks of age). At weaning (4 weeks of age), 211 gilts were subjected to 4 treatments. Gilts were administered either *ad libitum* feeding from weaning until slaughter (AA); restricted feeding from weaning until slaughter (RR); *ad libitum* feeding from weaning until 10 weeks of age after which feeding levels were switched to restricted feeding (AR); or restricted feeding from weaning until 10 weeks of age after which feeding levels were switched to *ad libitum* feeding (RA). At slaughter, the elbow joints, hock joints, and knee joints were collected. Joints were scored macroscopically for articular surface deformations indicative of OC and were analysed for treatment effects using binary logistic regression. Results show that gilts in the RA treatment have significantly higher odds to be affected with OC than gilts in the RR and AR treatments in the hock joint (OR=3.3, P=0.04 and OR=8.5, P=0.002, respectively), and at animal level (OR=2.5, P=0.001 and OR=1.9, P=0.01, respectively). Gilts in the AA treatment have higher odds to be affected with OC than gilts in the AR treatment in the hock joint (OR=5.3, P=0.01). In conclusion, switching to a higher feeding level after 10 weeks of age increases OC prevalence as opposed to a restricted feeding level. Therefore, feeding levels appear to have an age dependent effect on OC prevalence in gilts.

Using meta-analysis to aid reduction of the use of farm and pet animals in research

M. Speroni[1], B.H. Sontas[2], L. Migliorati[1], C. Stelletta[3], S. Romagnoli[3] and G. Pirlo[1]
[1]Consiglio per la ricerca e la sperimentazione in agricoltura, Centro di Ricerca per le Produzioni Foraggere e Lattiero-Casearie, via Porcellasco 7, 26100 Cremona, Italy, [2]Faculty of Veterinary Medicine, Department of Obstetrics and Gynaecology, Avcilar 34320, 34320 Istanbul, Turkey, [3]University of Padova, Department of Animal Medicine, viale dell'Università 16, 35020 Legnaro, Italy; marisanna.speroni@entecra.it

Optimal allocation of limited resources such as animals is both a scientific and an ethical issue. The study reports two cases showing how meta-analysis aids reduction of the use of animals in research. The first case is a meta-analysis on data obtained from four experiments carried out, over the time, at the same experimental barn and measuring individual dry matter intake (DMI) of a total of 75 dairy cows fed 13 diets. The meta-analysis wanted to estimate the effect of dietary factors on N excreted by dairy cows; DMI, live weight, protein content of the diet, milk yield, protein and urea nitrogen content of milk were all used to estimate the individual balance between ingested and excreted N. Data of individual DMI of farm animals are valuables in animal research; here is an example of how a meta-analysis, getting new results from existing data, reduces the need of dedicated experiments. Second case is about progesterone (P4) in bitches. Measurement of serum P4 is useful to manage risk pregnancies in the bitch; guidelines in the decision as to whether to supplement P4 or not in cases of impending abortion due to luteal insufficiency are lacking. Range of P4 during the canine pregnancy as reported by a number of studies varies widely due to the sources of variation (breeds, age, lab, sampling). A meta-analysis was used to combine results of 18 studies involving a total of 148 bitches; an average weekly pattern with confidence intervals of serum P4 was estimated. The results can impact on the reduction of use of animals in further studies providing information about variability of P4 during canine pregnancy.

Session 33 Poster 16

Air quality for meat quails during initial life cycle under different thermal environments

M.S. Sousa[1], I.F.F. Tinôco[1], L.B. Mendes[1], H. Savastano Júnior[2] and R.S. Reis[1]
[1]Federal University of Viçosa, Department of Agricultural Engineering, Av. P.H. Rolfs, S/N, Centro, 36570-000, Viçosa, MG, Brazil, [2]Universdity of São Paulo, College of Animal Sciences and Food Technology, Department of Agricultural Engineering, Av. Duque de Caxias Norte, 225, Jardim Elite, 13635-900, Pirassununga, SP, Brazil; mrmendes2010@gmail.com

The theme 'air quality' in rearing meat quail is still widely discussed in tropical and subtropical countries, such as Brazil, due to the lack of a database that relates productivity, thermal environment and air quality. For this reason, the aim of this research study was to evaluate the effect of different thermal environments on the quality of air inhaled by meat quails in the initial life cycle (1-21 d age). Groups of 180 meat quails were randomly placed in 5 climate chambers at a density of 60 bird m^{-2}. Five treatments were applied to each chamber and consisted of controlled levels of dry bulb temperature, with average temperatures (°C) for the first 3 weeks of age, of: Severe Cold Stress (SCS): (30.1±0.6), (33.0±0.8) and (23.8±0.8); Moderate Cold Stress (MCS): (33.0±0.8), (30.2±0.9) and (26.8±0.7); Thermo Neutrality (TNT): (36.0±0.6), (33.3±0.7) and (30.0±0.9); Moderate Heat Stress (MHS): (39.1±0.6), (35.8±1.3) and (32.9±0.7) and Severe Heat Stress (SHS): (41.9±0.8), (39.1±0.6) and (35.8±0.5). Air quality was evaluated through concentrations of ammonia (NH_3) and carbon dioxide (CO_2). Data analysis indicated that the levels of NH_3 were below the detection limit of the sensor. Regarding CO_2 levels, the following average concentrations (ppm) were observed: (915±176), (1,235±316) and (1,578±184); (844±65), (1,005±43) and (1,542±96); (1,001±63), (1,104±163) and (1,374±116); (991±150), (1,335±276) and (1,322±101), and finally, (1,138±180), (1,488±118) and (1,730±318) for treatments SCS, MCS, TNT, MHS and SHS, at weeks 1, 2 and 3, respectively. Results remained within acceptable limits (>3,000 ppm continuous exposure) for the proper development of birds.

Session 33 Poster 17

Effect of oral natural vitamin e supplementation to sows on the immunoglobulin levels in newborn pig

A.I. Rey[1], D. Amazan[1], G. Cordero[2], C. Piñeiro[2] and C.J. Lopez-Bote[1]
[1]Facultad de Veterinaria. Universidad Complutense, Producción Animal, Ciudad universitaria s/n, 28040, Spain, [2]PigChamp pro Europe, 40006 Segovia, Spain; anarey@vet.ucm.es

The influence of natural vitamin E supplementation (D-α-tocopherol) in drinking water vs. the synthetic form (Dl-α-tocopheryl acetate) in feed to sows during the gestation and lactation periods on tocopherol concentration of colostrum and milk and the immunoglobulin levels in piglets serum at 2, 14 and 28 days of age was studied. From day 107 of gestation to 28 of lactation sows (n=36) were divided in three groups that received: (1) 30 ppm of dl-α-tocopherol in feed; (2) ½ of the natural form in drinking water respect to the synthetic form; (3) 1/3 of the natural form in drinking water respect to the synthetic form. Consequently, sows received 150, 75 and 50 mg/d respectively. Extraction of α-tocopherol in colostrum and milk samples was carried out by saponification and it was analysed by reverse phase HPLC. Immunoglobulins A and M were determined in serum piglet using a pig ELISA quantification kit. The statistical analysis of data was carried out using the statistical program SAS v. 9.2. Vitamin E source and dose were evaluated in the statistical model. A repeated measurement test was used to study time and treatment effects and its interactions. α-Tocopherol in colostrum tended to be higher (P=0.06) in sows supplemented with the natural form in water, but no differences were found in milk. Ig A concentration of piglet serum at day 2 of age was higher (P=0.0031) when sows were supplemented with 75 mg/d of the natural form in water compare to the other groups and the lowest values were found when sows were supplemented with 50 mg/d of the natural form (P=0.0009). Differences were not detected for Ig M concentration or Ig A at days 14 or 28 of age. In conclusion, the form and dose of vitamin E supplementation to sows during gestation and lactation affects on the α-tocopherol concentration of colostrum and modifies the ig A concentration in piglets serum at day 2 of age.

Session 33 — Poster 18

Evaluation of the level of mycotoxin contamination in European feedstuffs from July to December 2012

H.V.L.N. Swamy[1], P. Caramona[2] and A. Yiannikouris[3]
[1]Alltech India, Karnataka, Bangalore, India, [2]Alltech UK Ltd, Rhyall Road, Stamford, United Kingdom, [3]Alltech Research Headquarters, Nicholasville, Kentucky, USA; pcaramona@alltech.com

Mycotoxins are produced both in the field and during storage. A previous European survey (Jan. to June 2012) indicated 100% contamination of samples tested, which has negative implications for animal health. Hence, the purpose of this study was to monitor levels of mycotoxin contamination in feeds/feed ingredients using a highly sensitive and validated technique. Ninety two European feedstuffs from the 2012 harvest (July to Dec) were subjected to ultra performance liquid chromatography coupled to tandem mass spectrometry (UPLC-MS/MS) (37+ Programme, Alltech Inc., KY) allowing simultaneous detection of multiple mycotoxins. Criteria for mycotoxin selection included prevalence in the field and established toxicological impact. To aid interpretation of potential total toxicity, toxins were categorised according to chemical properties and effects: aflatoxins, ochratoxins, Type A trichothecenes/T-2, Type B trichothecenes/DON, fumonisins, zearalenone, Penicillium mycotoxins, ergot mycotoxins and Alternaria mycotoxins. Ninety six % of the samples tested contained one or more mycotoxins, lower than the 100% contamination from the previous survey. Fumonisins (66%) and Type B trichothecenes (65%) were the predominant mycotoxins detected, followed by *Penicillium* mycotoxins (35%), Type A trichothecenes (33%) and zearalenone mycotoxins (27%). Remaining toxin groups were detected at levels between 4 and 18% of the total mycotoxin level. Fumonisin contamination was lower compared with the previous survey (94%) possibly due to variation in environmental temperature as higher temperatures favour fumonisin production in corn in the field. Mycotoxin contamination appears unavoidable, highlighting the importance of being able to detect multiple mycotoxins accurately and rapidly in order to take steps to minimize their impact on animal health.

Session 33 — Poster 19

***Neospora* spp. and *Toxoplasma gondii* antibodies in equine from Southern Italy**

T. Machačová and E. Bártová
University of Veterinary and Pharmaceutical Sciences, Department of Biology and Wildlife Diseases, Palackého 1/3, 612 42, Brno, Czech Republic; h11003@vfu.cz

Donkeys and horses are farm animals used for their meat or milk in many European countries. The aim of this study is to screen the hygienic risk of *Toxoplasma gondii* and *Neospora caninum* in equines in Southern Italy. The sera of 238 donkeys and 155 horses from southern Italy were tested for *T. gondii* antibodies by Latex Agglutination Test (LAT) and by the Indirect Fluorescent Antibody Test (IFAT); a titre ≥50 was considered positive. The same sera were tested for *N. caninum* antibodies by a Competitive-Inhibition Enzyme-linked Immunosorbent Assay (cELISA); samples with ≥30% inhibition were considered positive. In the case of donkeys, antibodies against *T. gondii* were found in 12 (5%) and 19 (8%) donkeys by LAT and IFAT, respectively. Antibodies against *Neospora* spp. were found in 28 (11.8%) donkeys with inhibition ranging from 30% to 44%. In case of both *T. gondii* and *N. caninum*, no statistical difference (P>0.05) was found between genders, age, use and their seropositivity. We found statistical difference (P>0.05) in breeds (18%) compared to crossbreeds (5%) for *N. caninum* with different seroprevalence in individual breeds; however no statistical difference was found for *T. gondii*. The present study describes for the first time the presence of *Neospora* spp. and *T. gondii* in donkeys from Italy. In case of horses we found antibodies against *T. gondii* in 8 (5.2%) horses by IFAT. Antibodies against *Neospora* spp. were found in 2 (1.3%) horses with low inhibition 3.2% and 301% by cELISA and in 2 (1.3%) horses by IFAT. In case of both *T. gondii* and *N. caninum*, no statistical difference (P>0.05) was found between genders, age, use and their Consumption of meat and milk from donkeys and horses does not represent important hygienic factor for transmission of *T. gondii* and *N. caninum* infection. This study was funded by the grant from the Ministry of Education, Youth and Sports of the Czech Republic, from IGA VFU Brno, Czech Republic.

Session 33 Poster 20

Do mirrors reduce stress in captive fish?

W.M. Rauw[1], L. Gomez-Raya[1], J. Fernandez[1], L.A. Garcia-Cortes[1], M.A. Toro[2], A.M. Larran[3] and M. Villarroel[2]
[1]Instituto Nacional de Investigación y Tecnología Agraria y Alimentaria, Departamento de Mejora Genética Animal, Carretera de La Coruña km 7,5, 28040 Madrid, Spain, [2] Universidad Politécnica de Madrid, Departamento de Producción Animal, Senda del Rey s/n., 28040 Madrid, Spain, Spain, [3]Instituto Tecnológico Agrario, Centro de Investigación en Acuicultura, Ctra. Arévalo, s/n, 40196 Zamarramala, Spain; jmj@inia.es

It is not well known how mirrors can affect stress in fish. An open field test has been carried out with 30 trout (*Oncorhynchus mykiss*), weighing between 11 and 26 g, challenged for 10 min in an aquarium (50×24×30 cm) with a mirror covering all walls and another 10 min in an aquarium without mirrors, covered in black. After the second test, blood was collected for analyses of cortisol and glucose. Behavior was recorded individually with an overhead camera. The software Panlab Harvard Apparatus was used to estimate the total distance for each trout in each of the tests. The correlation between the total distance for the same fish with and without the mirror was 0.34 but non-significant ($P=0.16$). Preliminary results suggest an effect of fish and further that the mirror might disturb fish activity. The correlation between body weight and total distance with the mirror was -0.64 ($P<0.01$) but this was not significant when tested without the mirror ($r=0.10$, $P=0.67$). There was a positive and significant correlation between distance in the test without the mirror and cortisol levels in blood ($r=0.75$; $P<0.05$), but it was not significant for the test with the mirror. These preliminary results suggest that activity and cortisol levels increase in the absence of mirrors. This is not in agreement with previous observations that a fish might act aggressively towards their mirror image. However, social behavior is perhaps related to the age of the fish. If the predominant behavior of fish at young ages is shoaling or schooling, then the image of his reflection in the mirror might help to reduce stress by reducing activity and by triggering shoaling behavior.

Session 33 Poster 21

Development of a multicriteria evaluation system to assess animal welfare

P. Martin[1], C. Buxade[2] and J. Krieter[1]
[1]Institut of Animal Breeding and Husbandry, Christian-Albrechts University, Olshausenstr. 40, 24098 Kiel, Germany, [2]ETSIA, UPM, Av. Complutense, 28040 Madrid, Spain; pfernandez@tierzucht.uni-kiel.de

It is a general accepted fact that animal welfare is a multidimensional concept. The aim of this study is to develop a multicriteria evaluation system to assess animal welfare. In a first step the Welfare Quality® assessment protocol for fattening pigs was implemented. Welfare Quality® identified 4 main animal welfare criteria (Good feeding, Good housing, Good health and Appropriate behaviour), each one of these criteria is defined by several subcriteria (12 in total) which are in turn assessed by one or various measures. Multiattribute utility theory was used to aggregate the 32 welfare measures into the corresponding 12 subcriteria. The utility functions and the aggregation functions were constructed in two separated steps. Firstly, utility functions for each measure were determined using the MACBETH method. In the second step, measures were aggregated. As aggregation functions, the weighted sum (WS) and the Choquet Integral (CI) were used. Ten simulated farms were used as an example to draw conclusions about the preferences of the decision maker (DM). A progressive interactive approach was used in the CI determination to define the relative importance of the measures and the allowance or not of compensation between them. Results showed that for some subcriteria (absence of prolonged thirst, thermal comfort and absence of pain induced by management procedures) measures were independent, being the results obtained by the WS and the CI equal, but for the rest of the subcriteria (except of those assessed just by one measure) there were interactions between the measures. In conclusion, the use of the CI is suitable to aggregate those measures that are considered by the DM to have interactions between them, in order to limit the compensation, but, for the other measures, a simpler aggregator can be used, for instance the WS or the decision trees used in the Welfare Quality® protocol.

Session 34 Theatre 1

Computing strategies for a single step SNP model with an across country reference population

Z. Liu[1], M.E. Goddard[2], F. Reinhardt[1] and R. Reents[1]
[1]VIT, Heideweg 1, 27283 Verden, Germany, [2]Univ of Melbourne, Parkville, 3010, Australia; zengting.liu@vit.de

In comparison to currently widely used multi-step genomic models, single step genomic models provide more accurate genomic evaluation by jointly analysing phenotypes and genotypes of all animals. An alternative model to the single step genomic model based on genomic relationship matrix includes an additional step for estimating SNP effects. The single step SNP model allows flexible modelling of SNP effects in terms of the number and variance of SNP markers. In addition, the single step SNP model contains a residual polygenic effect with a trait-specific variance. For genomic evaluations of Holstein dairy breeds, countries usually have set up an across country genomic reference population via exchange of genotypes. The objective of this study was to develop efficient computing algorithms for solving equations of the single step SNP model applied to an across country reference population. A core calculation of the SNP model involves repeated multiplications of the inverse of average relationship matrix of genotyped animals with a vector, thus we proposed a new algorithm for the multiplication by decomposing the relationship matrix as well as its inverse. For estimating SNP effects a special updating algorithm was proposed to separate residual polygenic effects from the SNP effects. A general prediction formula was derived for candidates without phenotypes, which can be used for frequent, interim genomic evaluations without new phenotypes. As a result of foreign bulls included in reference population, international MACE evaluation of bulls had to be used as phenotypes in addition to domestic cow phenotypic data. A random regression test-day model for milk production traits was chosen as an example to illustrate the integration of genomic prediction into a multiple trait conventional evaluation. The issue of different trait definitions between national and international evaluations was discussed.

Session 34 Theatre 2

Combination of genotype, pedigree, local and foreign information

J. Vandenplas[1,2], F.G. Colinet[2] and N. Gengler[2]
[1]National Fund for Scientific Research, FNRS, 1000, Brussels, Belgium, [2]University of Liege, Gembloux Agro-Bio Tech, 5030, Gembloux, Belgium; nicolas.gengler@ulg.ac.be

Simultaneous use of all data by Best Linear Unbiased Prediction is a condition to predict unbiased estimated breeding values (EBV). However, this condition is not always fully met. For example, small scale local populations lead to evaluations based only on local data while foreign bulls are used. Although these bulls were strongly selected, foreign data used to select them is unavailable leading to potential biases in local evaluations. Local EBV will be also less accurate because only incomplete data (i.e. foreign data not included) is available. Genomic selection could increase these problems. Initial implementations of genomic prediction used Multiple Across Country Evaluation (MACE) results which mitigated these issues for sires. Single step genomic evaluations (ssGBLUP) could reduce potential biases by the optimal combination of local genomic, pedigree and phenotype information. However, foreign information, like MACE EBV and associated reliabilities (REL), are usually not integrated into ssGBLUP. Therefore, the aim of this study was to assess the potential of a Bayesian approach, based on ssGBLUP, to simultaneously combine all available genotype, pedigree, local and foreign information in a local evaluation by considering a correct propagation of external information and no multiple considerations of contributions due to relationships and due to records. Local information refers here to local EBV and associated REL estimated from all available local data. The Bayesian approach has the advantage to directly combine EBV and REL without any deregression step. The approach was tested using a pedigree of 27,376 Holstein animals including 11,550 animals with a Walloon EBV and 1345 bulls with a MACE EBV. A total of 1351 cows and bulls were genotyped. For bulls with MACE EBV, correlations between MACE EBV and combined genomic EBV were 0.985 to 0.989 for yield traits. This approach has the potential to improve current genomic prediction strategies.

Efficiency of BLUP genomic prediction models that use pre-computed SNP variances

M.P.L. Calus[1], C. Schrooten[2] and R.F. Veerkamp[1]
[1]Wageningen UR Livestock Research, Animal Breeding and Genomics Centre, P.O. Box 65, 8200 AB Lelystad, the Netherlands, [2]CRV BV, P.O. Box 454, 6800 AL Arnhem, the Netherlands; mario.calus@wur.nl

Rapidly expanding dimensions of SNP data used in breeding programs imply that the processing time of genomic prediction models is rapidly increasing as well. Our objectives were to develop and compare fast BLUP models that can be applied at high frequency for routine estimation of breeding values, using SNP variances that are re-estimated at much lower frequency. Those objectives were investigated using both a simulated dataset and real dairy cattle data. The simulated data contained 2,000 reference and 3 generations of each 500 validation animals. The dairy cattle data contained 16,663 reference and 724 validation animals. The models used to estimate SNP variances were Bayesian Stochastic Search Variable Selection (BSSVS) and BayesC. The BLUP models were used to estimate breeding values, either using variances obtained from BSSVS (BLUP-SSVS) or BayesC (BLUP-C), or assuming that each SNP explained the same amount of genetic variance (RR-BLUP). Animals used for estimation of variance components were subsets of the animals used to estimate breeding values, comprising: (1) all animals; (2) a random subset; (3) the best 50% of the animals; or (4) the worst 50% of the animals. Accuracies of BLUP-C and BLUP-SSVS in the simulated data decreased by 0.03-0.06 when SNP variances were estimated from a non-random subset of the data, but were unchanged when SNP variances were estimated from a random subset. All BLUP models converged very fast in the simulated data (i.e. within <10 iterations vs. ≥10,000 when SNP variances were estimated). These results will be compared with those from the Dutch cattle population. In conclusion, it appears that the BLUP models can be used to efficiently analyse rapidly growing datasets, while combined with separated estimation of SNP variances in a Bayesian model they are able to capture the same features as the variable selection methods used to estimate the SNP variances.

Session 34 Theatre 4

Strategies for computation and inversion of the additive relationship matrix among genotyped animals

P. Faux and N. Gengler
University of Liège, Gembloux Agro-Bio Tech, Animal Science Unit, 2 Passage des Déportés, 5030 Gembloux, Belgium; pierre.faux@ulg.ac.be

Large-scale genetic evaluations became only possible with the direct construction of the inverse of the numerator relationship matrix (A). For genomic evaluations, even if computing the inverse of the genomic relationship matrix might not be needed depending on the implementation, computation of a subpart of the numerator relationship matrix (A), the sub-matrix (A_{22}), and its subsequent inversion is still required. Also, for genomic evaluations, one can expect that it will include two sets of genotyped animals: animals already involved in a previous evaluation and newly genotyped animals. Therefore, strategies that reuse already performed computations could be valuable. Two strategies for computation and inversion of A_{22} were compared. In the first one, currently mostly used strategy, A_{22} is computed, inverted and used at each evaluation but not stored on disk. In the second one, alternative strategy, only relationships between newly genotyped animals and all others genotyped animals are computed. Then, the stored inverted A_{22} of the previous evaluation is updated for newly genotyped animals for the new evaluation. Comparisons between strategies were performed using two criterions: required computing time and memory for both computation and inversion. Tests were conducted using the Walloon pedigree for routine genetic evaluations with different simulated scenarios: (1) number of already genotyped animals and (2) number of newly genotyped animals added in current evaluation. Depending on factors (1) and (2), it may be shown that first strategy was efficient for small number of genotyped animals whereas the second one prevails for greater numbers. As for large-scale genetic evaluations, large-scale genomic evaluation may be greatly facilitated with advanced algorithms allowing the successive construction of the inverse of the sub-matrix (A_{22}) or even the genomic relationship matrix.

Session 34 Theatre 5

An inversion-free method to compute genomic predictions using an animal model approach

J. Ødegård[1] and T.H.E. Meuwissen[2]
[1]Aqua Gen AS, Norwegian University of Life Sciences, P.O. Box 5003, 1432 Aas, Norway, [2]Dept. of Animal and Aquacultural Science, Norwegian University of Life Sciences, P.O. Box 5003, 1432 Aas, Norway; jorgen.odegard@umb.no

The GBLUP approach described in Meuwissen *et al.* can be fitted using an equivalent animal model, where the numerator relationship matrix is replaced by a genomic relationship matrix. Solving Henderson's mixed model equations requires the inverse of the relationship matrix, but no algorithm currently exists for direct calculation of the inverse genomic relationship matrix which is typically completely dense. Thus, 'brute force' inversion of such matrices can be computationally challenging for large datasets. The aim of this study was therefore to develop an inversion-free method for solving genomic animal models. The mixed model used was: y = Xb + Zu + e, where b are the fixed effects, u~N(0, G) are the genomic breeding values, G is the genomic (co)variance matrix, e~N(0, R) and V = G + R. We then define: s = inv(V)(y – Xb), i.e. Xb + Vs = y. The latter can replace the lower (animal) part of Henderson's mixed model equations, allowing computation of s without involving the inverse of G. Still, computing an exact solution to the reformulated equations can be computationally demanding, but the equation system can be iteratively solved by Gauss-Seidel. Finally, it can be shown that genomic breeding values can be predicted from s by: u = GZ's. The proposed algorithm was used to analyze a simulated dataset with a dense G for 11,200 animals, of which 10,000 had phenotypes. The analysis was done in Matlab on a Linux computer with a 64-bit Intel Xeon 2.60 GHz processor. Gauss-Seidel iterations and prediction of u were finalized in less than 16 seconds. As expected, the solutions converged towards the exact solutions obtained by solving the standard Henderson's mixed model equations. The proposed method represents an efficient computation method for genomic predictions in large data sets without the need for inverting the genomic relationship matrix.

Session 34 Theatre 6

Haplotype-assisted genomic evaluations in Nordic Red dairy cattle

T. Knürr[1], I. Strandén[1], M. Koivula[1], G.P. Aamand[2] and E.A. Mäntysaari[1]
[1]MTT Agrifood Research Finland, Myllytie 1, 31600 Jokioinen, Finland, [2]NAV Nordic Cattle Genetic Evaluation, Agro Food Park 15, 8200 Århus N, Denmark; timo.knurr@mtt.fi

In admixed populations originating from different base breeds, such as the Nordic Red dairy cattle, allele identity by descent constitutes a more important source of information for genomic predictions than in homogenous populations. In this study, we aimed at increasing the validation reliabilities in genomic evaluations by exploiting haplotype information in phased SNP data. In a first step, the genome was scanned for QTL signals: (1) via fitting all SNPs simultaneously by BayesB; or (2) by estimating variances of haplotypes of single chromosomal segments one-by-one jointly with a polygenic effect. Based on these results, the chromosomal segments showing the strongest QTL signals were then pre-selected. In a second step, we estimated relative variances for the pre-selected segments in a multi-locus model without a pedigree-based polygenic component. Finally, these segments were included into the actual prediction model, in which a pre-defined proportion (w) of the genetic variance was assigned to pedigree and the rest (1-w) to the segments. The accuracies of alternative approaches were assessed using the Interbull GEBV validation test. Validation test reliability (R^2) was higher for models with pre-selection based on BayesB. For protein, milk and fat, respectively, the highest R^2s observed were 0.45, 0.51 and 0.54, when 500 segments (each spanning a block of 5 neighbouring SNPs) were pre-selected; the improvement over G-BLUP was 9, 16 and 4 percentage points. The estimation of relative variances of segments was not critical, and even a constant variance over segments yielded satisfactory results. R^2s were best when w was set to 0.2-0.3 for the three traits. The variance consistency measure (b_1) was close to 1 for w=0.5, without much sacrifices in R^2.

Session 34 Theatre 7

Comparison of genetically and genomically estimated variance

A. Loberg[1], L. Crooks[2], W.F. Fikse[1], J.W. Dürr[1,3] and H. Jorjani[1,3]
[1]SLU, Animal Breeding and Genetics, Box 7023, 75007 Uppsala, Sweden, [2]Wellcome Trust, Sanger Institute, Cambridge, United Kingdom, [3]Interbull Centre, Box 7023, 75007 Uppsala, Sweden; anne.loberg@slu.se

To reach desired goals in animal breeding, the right breeding animals have to be selected. To find superior animals, estimation of genetic variance for the trait in question is fundamental, on national as well as international level. In this study estimation methods have been compared, by comparing the amount of genetic variance estimated. Data for 7,038 Brown Swiss bulls have been included in this study. The phenotypic records included nationally and international estimated predicted genetic merits (NPGM, IPGM) for a production, a health, a conformation and a fertility trait. The genomic records included 44,826 SNPs (after different quality checks). The data represent seven geographical regions (countries). Five ways of estimating genetic variance were included in this study: (1) estimates from each geographic region using phenotypic records on cows (using various REML implementations); (2) Estimates from simple variance of the available NPGM values; (3) estimates from an international EM-REML implementation; (4) estimates using allele frequencies and allele effects in an international genomic evaluation model with a non-linear, heavy-tailed prior for marker effects analogous to Bayes A); and (5) genomic estimation of genetic variance with an AI-REML implementation. Results from method (1) were used as a reference base. The amount of genetic variance estimated with different methods showed much variation within and across traits, also did the depth of pedigree affect results and will be investigated further.

Session 34 Theatre 8

Allele frequency changes due to hitch-hiking in genomic selection programs

H. Liu[1], A.C. Sørensen[1], T.H.E. Meuwissen[2] and P. Berg[1,3]
[1]Aarhus University, Department of Molecular Biology and Genetics, Blichers Allé 20, 8830 Tjele, Denmark, [2]Norwegian University of Life Sciences, Department of Animal and Aquacultural Sciences, Arboretveien 6, 1432 Ås, Norway, [3]Nordic Genetic Resource Center, Raveien 9, 1431 Ås, Norway; lhmsai007@yahoo.com

Genomic selection was recognized to result in lower pedigree inbreeding compared to best linear unbiased prediction (BLUP). However, pedigree inbreeding might not well reflect the true level of inbreeding and the loss of genetic variation as a result of allele frequency changes and hitch-hiking. This study aimed at understanding the consequence of long-term genomic selections in terms of allele frequency changes and inbreeding. The simulated genome of each animal in the population consisted of five chromosomes, with defined genetic architecture by varying h2 and the number of QTL. The population consists of 400 animals and selections were performed for 25 continuous generations, using the following selection criteria: phenotypic selection (PS), BLUP, Genomic BLUP (GBLUP) and Bayesian Lasso (BL). The investigation was performed in terms of allele frequency changes, loss of favorable alleles and level of inbreeding measured by pedigree and runs of homozygosity (ROH). The results showed that the 'linkage drag' due to hitch-hiking was maximal at the location of QTL and was reduced as the distance of linked neutral loci to the QTL increased. Hitch-hiking in the vicinity of QTL became stronger with a higher accuracy and a lower number of QTL. For hitch-hiking the results showed that GBLUP > BL > BLUP > PS. Hitch-hiking was substantial, which resulted in a higher rate of inbreeding measured by ROH than by pedigree. We concluded that genomic selection can reduce pedigree inbreeding in relative to BLUP, but it also gradually reduces the genetic diversity at QTL as well as its surrounding region. With limited number of QTL, BL is superior to GBLUP in maintaining favorable alleles, reducing the loss of heterozygosity and controlling the level of inbreeding.

Session 34 — Theatre 9

Artificial neural networks for genome enabled predictions of milk traits in Simmental cattle

A. Ehret[1], D. Hochstuhl[2] and G. Thaller[1]
[1]Institute of Animal Breeding and Husbandry, Christian-Albrechts-University, Olshausenstr. 40, 24098 Kiel, Germany, [2]Institute for Theoretical Physics and Astrophysics, Christian-Albrechts-University, Leibnizstrasse 15, 24098 Kiel, Germany; aehret@tierzucht.uni-kiel.de

Artificial neural networks (ANNs) are proposed to be universal approximators of complex functions without explicitly defining a fixed model. This concept is important for high dimensional noisy data, as in genetic marker based approaches in genome enabled predictions of complex traits, especially when a high amount of markers is available. An issue which should be considered in using such a method is how the marker information can be incorporated into models to enhance the predictive ability of future phenotypes and to reduce the computational costs. The data in this study represented daughter yield deviations (DYD) of three milk traits of 3,341 genotyped Simmental cattle bulls. To train the networks an early stopping back propagation algorithm and a non-linear transformation function in the hidden layer were used. Architectures with 1 to 10 hidden neurons were examined. The average correlation between the true and the predicted phenotypes in a 10-fold cross-validation was used to assess the predictive ability. In the different models the additive genomic relationship matrix (G), the marker genotypes (X), or their principal component score (UD) of a 50k marker panel were employed. The recent study shows that predictive abilities of the different models varied markedly. Dimension reduction methods can yield in a higher prediction performance of the ANNs as well as the choice of the optimal architecture. A network with 10 neurons in the hidden layer achieves a correlation coefficient of r=0.482 for the trait milk yield when X was used. The predictive ability was better with using G as input matrix (r=0.608) and best (r=0.654) when the UD matrix was used. Similar results can be shown for the other milk traits. In a following study we will proof these results with a higher marker panel (777k).

Session 34 — Theatre 10

Genome wide associations study for fertility and longevity in cattle

P. Waldmann[1], G. Mészáros[2] and J. Sölkner[2]
[1]Linköping University, Division of Statistics, Department of University, 581 83 Linköping, Sweden, [2]University of Natural Resources and Life Sciences, Vienna, Division of Livestock Sciences, Gregor-Mendel-Str. 33, 1180 Vienna, Austria; gabor.meszaros@boku.ac.at

Genome wide association studies (GWAS) are frequently used to pinpoint connections between SNPs and the trait of interest, utilizing diverse methodological approaches. Since the focus of GWAS is to map regions of interest in the genome, the ideal methodology should find all associated SNPs (no false negatives) and do not catch any additional non-associated SNP (no false positives). In our study we have tested lasso and elastic net for their ability to identify associations between genotypes and phenotypes, using two simulated data sets of various complexities and one real data set. The first simulated set consisted from 50,000 predictors from which 25 were significantly associated with QTL. The second simulation (QTLMAS2010) was biologically more complex, with total of 10,000 SNPs and 37 QTLs, each QTL being surrounded by 19-47 polymorphic SNPs located within 1Mb distance. The real data consisted of 41,008 SNPs from 1,907 Fleckvieh bulls from Austria with DYDs for longevity as phenotype. The elastic net selected all important SNPs and had good false positive rates in the simple simulation and seemed to be performing better compared to the lasso in the QTLMAS data. Based on the real data set, the lasso did not select any SNPs associated with longevity. The elastic net selected 143 SNPs for the same trait, but after correction for population structure this was reduced to zero as well.

Interpretation of dominant and additive variances from genomic models

Z.G. Vitezica[1], L. Varona[2] and A. Legarra[3]
[1]INRA, INPT, UMR 1289 TANDEM, Avenue de l'Agrobiopole, PB 32607, F31326 Castanet-Tolosan, France, [2]Facultad de Veterinaria, Universidad de Zaragoza, Miguel Servet, 177, 50013 Zaragoza, Spain, [3]INRA, UR 631 SAGA, 24 Chemin de Borde Rouge, 31326 Castanet-Tolosan, France; zulma.vitezica@ensat.fr

Genomic evaluation models typically fit only SNP additive effects. However, it is possible to include also dominant SNP effects. Under quantitative genetics theory, breeding values (A) of individuals are generated by substitution effects, which involve both additive and dominant effects, whereas the dominance deviations (D) only include dominant effects. From the genotypic value, we can also define (A*) and (D*) as the parts attributable to the additive and dominant effect of the markers. Note that this A* is not a breeding value. We show that the variance of genotypic values due to additive effects of markers (σ^2_{A*}), is not equal to additive genetic variance (among breeding values) and that the variance of genotypic dominant effects (σ^2_{D*}) is not the dominance variance. In fact, only when the allele frequencies are equal to 0.5 all variances are identical. In addition, dominant relationship matrices constructed from markers depend on which of the two decompositions is used. As the total genetic variance can be partitioned into additive and dominance components, we know that $\sigma^2_A + \sigma^2_D$ is equal to $\sigma^2_{A*} + \sigma^2_{D*}$. In our study, we show that it is easy to define a one on one relation between 'breeding' and 'genotypic' additive and dominance variance and that even being equivalent, the estimated variance components will be different. The 'genotypic' model overestimates the dominance variance and, consequently, underestimates additive variance. We illustrate these results with parameter estimations in mice data. The differences between 'breeding' and 'genotypic' model must be taken into account in the interpretation and use of genomic predictions and genomic estimates of variance components.

Genomic prediction of heterosis for egg production traits in White Leghorn crosses

E.N. Amuzu-Aweh[1,2], P. Bijma[1], B.P. Kinghorn[3], A. Vereijken[4], J. Visscher[4], J.A.M. Van Arendonk[1] and H. Bovenhuis[1]
[1]Wageningen University and Research Centre, Animal Breeding and Genomics Centre, P.O. Box 338, 6700 AH Wageningen, the Netherlands, [2]Swedish University of Agricultural Sciences, Department of Animal Breeding and Genetics, P.O Box 7023, 750 07 Uppsala, Sweden, [3]University of New England, School of Environmental and Rural Science, NSW 2351, Armidale, Australia, [4]Hendrix Genetics, Institut de Sélection Animale B.V., P.O. Box 30, 5830 AE Boxmeer, the Netherlands; esinam.amuzu@wur.nl

The genetic basis of heterosis has puzzled geneticists for decades. Accurate prediction of heterosis would benefit animal and plant breeding by identifying parental lines suitable for crossbreeding. Prediction of heterosis has a long history with mixed success, partly due to low numbers of genetic markers and/or small data sets. We investigated prediction of heterosis for egg number, egg weight and survival time in domestic White leghorns, using ~400,000 individuals from 47 crosses and allele frequencies on ~53,000 genome-wide SNPs. For a single locus, heterosis is solely due to dominance and proportional to the squared difference in allele frequency between parental lines (SDAF). We, therefore, used linear mixed models where phenotypes of crossbreds were regressed on the SDAF between parental lines. Accuracy of prediction was determined using leave-one-out cross-validation. SDAF predicted heterosis for egg number and weight with an accuracy of ~0.5, but not for survival time. Heterosis predictions allowed pre-selection of pure lines prior to field-testing, saving ~50% of field-testing costs with only 4% loss in heterosis. Accuracies from cross-validation were lower than those from the model-fit, indicating that values in the literature may be overestimated. Cross-validation also indicated dominance cannot fully explain heterosis. Nevertheless, the dominance model yielded a considerable accuracy, clearly greater than that of a general-specific combining-ability model. Our results show that SDAF can be used to predict heterosis in commercial layer breeding.

Session 34 Theatre 13

Unknown-parent groups and incomplete pedigrees in single-step genomic evaluation

I. Misztal[1], Z.G. Vitezica[2], A. Legarra[3], I. Aguilar[4] and A.A. Swan[5]
[1]University of Georgia, Athens, GA 30605, USA, [2]Université de Toulouse, UMR 129, Castanet-Tolosan 31326, France, [3]INRA, Station d'Amelioration, Castanet-Tolosan 31326, France, [4]INIA, Las, Brujas, Uruguay, [5]AGBU, University of New England, Armidale, Australia; ignacy@uga.edu

In single-step genomic evaluation using best linear unbiased prediction (ssGBLUP), genomic predictions are calculated with a relationship matrix that combines pedigree and genomic information. For missing pedigrees, unknown selection processes, or inclusion of several populations, a BLUP model can include unknown-parent groups (UPG) in the animal effect. For ssGBLUP, UPG equations also involve contributions from genomic relationships. When those contributions are ignored, UPG solutions and genetic predictions can be biased. Several options exist to eliminate or reduce such biases. First, mixed model equations can be modified to include contributions to UPG elements from genomic relationships (greater software complexity). Second, UPG can be implemented as separate effects (higher cost of computing and data processing). Third, contributions can be ignored when they are relatively small but they may be small only after refinements to UPG definitions. Fourth, contributions may approximately cancel out when genomic and pedigree relationships are constructed for compatibility; however, different construction steps are required for unknown parents from the same or different populations. Finally, an additional polygenic effect that also includes UPG can be added to the model (slower convergence rate). Chosen options need to reflect different origins of UPGs: missing pedigrees in a closely selected population, multiple breeds, external lines or combinations of origins. Incomplete pedigrees may also cause biases and convergence problems even when UPGs are not in the model. In such cases, choices include restoration or truncation of pedigrees. Severity of problems with UPG and incomplete pedigrees greatly depends on the population structure.

Session 35 Theatre 1

Effect of project 'YoungTrain' on participants` career

P. Polák
Animal Production Research Centre Nitra, Department of Animal Breeding and Product Quality, Hlohovecká 2, 951 41 Lužianky, Slovak Republic; polak@cvzv.sk

The main aim of the project 'YoungTrain' was to train a group of 40 early carrier scientists from new Member States and countries in Southern and Central Europe, the Newly Independent States (former Soviet Republics) and the Mediterranean rim in transparency of the food chain and quality in meat products. The animal scientists were supported by mentors and a new network that includes regional experts. By means of e-learning packages and workshops, they learned to pass on the safe and high-quality food message and to develop RTD projects in the field. The group received full attention from specialists in the field of meat quality and quality of meat products as well as about procedures of preparing a proposal in framework of RTD projects. One very important issue was the selection of participants. Project leaders made balanced selection according gender, age, regions, ability to communicate and willingness to cooperate. Support action YT enabled young scientists to meet not only specialists in the field of meat and product quality, but also persons actively involved in support managing and creating ideas of new call for RTD projects. Participants of YT project started or continued their scientific career in field of animal science during or after the project with support of skills and contacts obtained as result of the project. Lot of them finished PhD study, became assistant professor or heads of department in home universities or institutes. One of the more important outputs from the project is networking and dissemination of knowledge among participants and their working places. It is not surprising that quite a visible part of participants have started to work actively in structures of EAAP, in Scientific Commissions, Working Groups or on organising annual meetings. Nowadays participants communicate and cooperate on creating international projects using skills and personal contacts established on the base of the YT project.

Potential of using leftovers to reduce the environmental impact in animal production

H.H.E. Van Zanten[1,2], *H. Mollenhorst*[2], *P. Bikker*[1], *T.V. Vellinga*[1] *and I.J.M. De Boer*[2]
[1]*Wageningen University, Livestock Research, Vijfde Polder 1, 6708 WC Wageningen, the Netherlands,* [2]*Wageningen University, Animal Production Systems Group, De Elst 1, 6708 WD Wageningen, the Netherlands; hannah.vanzanten@wur.nl*

The livestock sector has a major impact on the environment. This impact may be reduced by feeding co-products (e.g. beet tails) to livestock, as this transforms inedible products for humans into edible products, e.g. pork or beef. Co-products, however, have different applications, such as bio-energy production. The aim of this PhD project is to explore the potential of using co-products to reduce environmental emissions in the livestock sector taking the alternative use of co-products into account. To accomplish this, we first developed a framework, which we verified with two case studies. We used consequential life cycle assessment to illustrate the overall consequences for the two cases, regarding land use and greenhouse gas emissions (including land use change). In the first case, we analysed the consequences of increasing the use of wheat middlings in dairy cattle feed at the expense of using it in pigs feed, whereas in the second case we analysed the consequences of increasing the use of beet tails in dairy cattle feed at the expense of using it to produce bio-energy. Results show that increasing wheat middlings in dairy cattle feed decreased land use by 169 m^2 and greenhouse gas emissions by 329 kg CO_2-eq per ton, whereas increasing beet tails decreased land use by154 m^2 and greenhouse gas emissions by 239 kg CO_2-eq per ton. We concluded that feeding co-products has the potential to reduce the environmental impact of the livestock sector, and, therefore, we want to explore other possibilities. We identified two important aspects we would like to discuss during the EAAP conference: (1) which livestock sector has most potential to reduce the environmental impact by feeding co-products?; (2) can we use insects, reared on waste only, as alternative protein source to reduce the environmental impact of the livestock sector?

Survey study of economic parameters in Holstein and Slovak dual purpose breeds

M. Michaličková, J. Huba, P. Polák and D. Peškovičová
Animal Production Research Centre Nitra, Institute for Animal Breeding and Product Quality, Hlohovecká 2, 951 41 Lužianky, Slovak Republic; michalickova@cvzv.sk

The aim of this study was to evaluate economic efficiency for Holstein cattle in lowland areas (9 farms, A) and Slovak dual purpose cattle (Simmental and Slovak Pinzgau) and crosses in mountain and foothill areas (13 farms, B) for the years 2007-2011 (database of APRC Nitra). In cost evaluation formula for farms A as well for farms B the most important items were costs for feed (46% and 43%) and other direct costs (19% for both). Depreciation as well was a big proportion of costs (16% and 18%), however they are implicit cost items. The labour costs represented 7% and 8% from costs totally in farms A and B, respectively. Profit in milk production was achieved in the years 2008 and 2011 in farms A. The highest profit (0.012 € per kg of milk) in 2008 was determined mainly by the level of milk price (0.348 € per kg of milk) and by the level of milk yield (18.89 kg of milk per day). On contrary, profit in milk production was not achieved in farms B over the analyzed period. The lowest loss (-0.007 € per kg of milk) was observed in 2008 due to level of milk price (0.355 € per kg) along with the low value of total costs (4.623 € per day). Sharp drop in milk prices in 2009 (-18% farms A and -20% farms B) and in 2010 (-15% farms A and -18% farms B) compared to the year 2008, caused highest losses on farms A in 2009 (-0.021 € per kg of milk). On the opposite, the highest loss (-0.074 € per kg of milk) for farms B was in 2010. Except for price of milk (0.292 € per kg of milk) it was determined also by the higher value of total costs (5.014 € per day), which were partly eliminated with milk yield (13.67 kg of milk per day). Based on the prognoses of costs trend on dairy farms and prices of the cattle products in 2013 it is recommended to achieve milk yield 23.8 kg, 19.4 kg and 15.7 kg per cow and day for Holstein, Simmental and Slovak Pinzgau breeds, respectively.

Relation between farm size and sustainability of dairy farms in Wallonia

T. Lebacq[1,2], D. Stilmant[1] and P.V. Baret[2]
[1]Centre wallon de Recherches Agronomiques, Rue de Serpont, 100, 6800 Libramont, Belgium, [2]Université catholique de Louvain, Croix du Sud, 2, L7.05.14, 1348 Louvain-la-Neuve, Belgium; theresa.lebacq@uclouvain.be

In the context of transition towards sustainable farming systems, there is a need to consider the environmental, economic and social effects of development pathways. The Belgian dairy sector has undergone structural changes during the past 30 years, with a decrease in the number of farms and an increase of their size. Semi-directive interviews with stakeholders highlighted contrary perceptions about the sustainability of this evolution. Therefore, this study aimed to assess the role of farm size, i.e. the total milk production, on the sustainability of dairy farms in Wallonia, the southern part of Belgium. Environmental, economic, social and structural indicators were quantified for 478 specialized dairy farms in 2008 and 2009, based on accounting data. A clustering approach consisted of analyzing farm diversity by grouping farms having similar characteristics of sustainability and structure. This analysis led to identify five farm groups. Bivariate analyses were performed to analyze the effect of farm size on economic and environmental performances, between and within these groups. Our results showed that one group included farms that were significantly larger, with in average 79 hectares of agricultural area, a production of 756,218 liters, 2.1 labor units and 110 dairy cows. Larger farms tended to be more intensive in terms of production and stocking rate. They had a higher gross operating surplus per labor unit and a higher environmental impact per hectare, in terms of energy consumption, nitrogen surplus and pesticide costs. The number of cows per labor unit increased with the farm size. In addition, we observed a great diversity within the group including the largest farms, notably for the pesticide costs, nitrogen surplus and gross operating surplus. This diversity paves the way for identifying farms combining a large size with good environmental and economic performances.

Natural antibodies measured in blood plasma and milk of Dutch dairy cattle

B. De Klerk[1], J.J. Van Der Poel[1], B.J. Ducro[1], H.C.M. Heuven[1,2] and H.K. Parmentier[3]
[1]Wageningen University, Animal Breeding and Genetics, P.O. Box 338, 6700 AH Wageningen, the Netherlands, [2]Utrecht University, P.O. Box 80163, 3508 TD Utrecht, the Netherlands, [3]Wageningen University, Adaptation Physiology, P.O Box 338, 6700 AH Wageningen, the Netherlands; britt.deklerk@wur.nl

Increased longevity will be beneficial from an economical as well as an ethical viewpoint, as it is related to lower incidence of (painful) diseases. To improve the ability to resist diseases (resilience) of dairy cows parameters such as natural antibodies (NAbs) can be used. Natural antibodies (NAb) are a part of the innate immune system and it is anticipated that NAb titres may reflect the ability of an animal to stay healthy. NAbs were measured in blood plasma and in milk collected on 3,000 cows from 30 farms. The farms were chosen based on average herd life, either high or average. Milk samples were collected from all lactating cows, while blood samples were taken from ca. 60 cows per farm (lactating and non-lactating). NAb titres binding Keyhole limpet hemocyanin (KLH)-antigen were obtained using ELISA tests. Preliminary analysis revealed that there were significant farm differences, but no significant effect of farm type (average of high herd life). Furthermore, parity had a significant effect on milk and plasma Nab levels and lactation stage had a significant effect on milk Nabs. Somatic cell score (SCS) was related to milk NAbs ($P<0.0001$) and to IgM in plasma ($P=0.047$) indicating that there is a correlation between a cows health status, since SCS is high correlated with (subclinical) mastitis. From the relation to SCS it can be concluded that NAbs are involved in the immune response of the cows. A moderate correlation between IgG and IgM in milk and in plasma (0.4 and 0.34 respectively) was observed. The correlations between NAb levels from milk and blood indicate that NAbs from milk and blood might reflect different aspects of the health status of a dairy cow.

Veterinary practices, control, animal health and food security: the weak points and perspectives

I. Stokovic
University of Zagreb Faculty of Veterinary Medicine, Heinzelova ulica 55, 10000 Zagreb, Croatia; igor.stokovic@vef.hr

Animal health and food security issues are indicated as very important ones in new EU Common agricultural policy. That is not surprising since we are recently bombarded with outbreaks of various diseases and food safety issues common to animals and humans. Since live animals and their products could be a source of threat to human health, animal feed safety, animal health care and animal products safety control mechanisms are combined in a Veterinary Public Health concept. Core domains of VPH, as defined in Teramo in 1999, are: diagnosis, surveillance, epidemiology, control, prevention and elimination of zoonoses; food protection; management of health aspects of laboratory animal facilities and diagnostic laboratories; biomedical research; health education and extension; and production and control of biological products and medical devices. Other VPH core domains may include management of domestic and wild animal populations, protection of drinking-water and the environment, and management of public health emergencies. Antibiotic use in animals is another emerging issue. The veterinary sector is aware of before mentioned threats and actions are in place to prevent possible outbreaks, nevertheless there are weak points to be considered and prevention measures developed. Conclusions of the VET 2020 project are that Veterinary Public Health is a field of special care in the future. In this presentation focus will be on future points of attention and vision developing.

Organochlorine compounds residues in muscle of wild boar and red deer in the Czech Republic

L. Zelníčková, P. Maršálek, J. Mikuláštíková, Z. Svobodová and Z. Hutařová
University of Veterinary and Pharmaceutical Sciences Brno, Department of Veterinary Public Health and Toxicology, Palackeho 1-3, Brno, 61242, Czech Republic; h10026@vfu.cz

The aim of the present research was to investigate the levels of contamination by twenty-six organochlorine pesticides and seven polychlorinated biphenyls in muscle (Musculus latissimus dorsi) of wild boar (*Sus scrofa* L.) and red deer (*Cervus elaphus* L.) from the Czech Republic. OCPs and PCBs are known to be very resistant in the environment and their lipophilic nature is the reason for their concentration and bioaccumulation in the food chain. The content of DDT and its metabolites was higher ($P<0.01$) in wild boar than in red deer, while PCBs and HCH were higher ($P<0.01$) in red deer than in wild boar. The concentrations of DDT and its metabolites, hexachlorobenzene and hexachlorocyclohexane isomers were higher ($P<0.05$) in juvenile wild boar than in adults. The most abundant pesticide and polychlorinated biphenyl were metabolite p, p'-DDE and PCB 153, respectively. Therefore, these results show that the Czech Republic is not at contamination risk from organochlorines and polychlorinated biphenyls and moreover is free from health problems for the customer of boar and deer meat.

Consumption of fish from important rivers of the Czech Republic: evaluation of health risk

L. Sedláčková, K. Kružíková and Z. Svobodová
University of Veterinary and Pharmaceutical Sciences Brno, Department of Veterinary Public Health and Toxicology, Palackého tř. 1/3, 612 00 Brno, Czech Republic; h10023@vfu.cz

The aim of the work was to evaluate hazard risk resulting from consumption of fish from major rivers of the Czech Republic from the perspective of mercury content. For the determination of mercury indicator species of fish chub (*Leuciscus cephalus* L.) was chosen. In this study a number of fish meat servings that are allowed for weekly consumption was determined and hazard index for moderate consumer and fisher's family was assessed. For the calculations leading to evaluation of health risk it was necessary to know the content of total mercury and methylmercury. In total, 130 fish were caught at 12 locations on 11 rivers. Total mercury (THg) concentrations were determined by cold vapour atomic absorption spectrometry using AMA 254 (Altech Ltd., Czech Republic) analyser. Methyl mercury (MeHg) concentrations were determined by gas chromatography with the electron capture detector GC 2010A. The hazard index was calculated according to Kannan *et al.* using a reference dose (RfD) for THg (0.3 mg/kg body weight per day) set forth by US EPA. To determine the maximum possible consumption of fish meat, the provisional tolerable weekly intake limit (PTWI) of 1.6 μg MeHg per kg body weight per week was used. The lowest number of fish meat servings that can be eaten was calculated for the Vltava – Vraňany (3.6) and the Labe – Obříství (2.8) and the highest number of fish meat servings that can be eaten was calculated for the Lužnice – Bechyně (10.1) and the Berounka – Srbsko (9.6). The value of Hazard Index 1 was not exceeded for fisher's family members and for moderate consumer. The results of this study helped evaluate contamination levels of rivers that flow through the Czech Republic.

What is that horse doing: the need for a working horse ethogram

M. Pierard
KU Leuven, Centre for Animal Husbandry, Bijzondere Weg 12, 3360 Lovenjoel, Belgium; marc.pierard@biw.kuleuven.be

The behaviour of domestic horses is the most visible and accessible tool to evaluate their welfare, performance or emotions. Interpretation of equine behaviour is fundamental for researchers, veterinarians, breeders, trainers, riders and grooms. Consistent and objective discussion of horse behaviour depends on clearly defined and generally accepted descriptions of behaviour units. A working horse ethogram, defining all behaviours of horses during interactions with humans, does not exist at this time. An ethogram is defined as a set of comprehensive descriptions of the characteristic behaviour patterns of a species. The first step in the development of an ethogram is to observe the behaviours and describe them correctly. It starts with looking at and describing the range of body movements and the coordination of different body parts. This is the first necessary phase in the process of developing a functional ethogram. Unequivocally defined behaviour units allow structural research into function, causation and ontogeny of equine behaviour. Interpretation of behaviour can only be validated if it is based on purely descriptive definitions of behaviour patterns. The use of a descriptive ethogram would benefit all stakeholders. Research becomes comparable and compatible, allowing meta-analyses. It would also facilitate communication between researchers, veterinarians, riders, trainers, welfare inspectors and other officials of government or equestrian institutions. To serve its purpose, an ethogram needs to be tested for a high intra- and interobserver reliability. It could be a powerful tool to convince the horse industry to evolve towards more evidence based practices. This would improve horse welfare, decrease injury rates amongst people working with horses and optimize performance, whatever the horses are used for. A working horse ethogram would be a significant progress for equitation science and the study of human-horse relations.

Session 36 Theatre 2

Stress in riding horses

J.W. Christensen
Aarhus University, P.O. Box 50, 8830 Tjele, Denmark; jannewinther.christensen@agrsci.dk

Several aspects of training and management can influence the welfare of riding horses. For instance, management factors such as reduced social contact, feeding, and repeated mixing can lead to increased stress levels and decreased welfare. Recent studies have documented high levels of glandular ulceration in horses (55.2%), which is linked to stress reactivity. Although the affected horses showed few obvious external signs, they had an increased concentration of cortisol metabolites in faeces (FCM) in response to a stressor (26% higher FCM in ulcer horses compared to controls; P=0.018). Besides, the amount of starch in the feed and riding discipline (dressage vs. jumping) affected both type and severity of ulceration in different regions of the stomach. Furthermore, training methods may affect acute stress levels in horses. We recently investigated salivary cortisol, heart rate variability (HRV) and behavioural reactions of dressage horses (medium to Grand Prix level) ridden in a long frame (LF); competition frame (CF) and hyperflexion (Low-Deep-and-Round; LDR) in a balanced order across three different test days. The horses had the highest cortisol concentrations in saliva directly after LDR (P=0.005). Furthermore, the LF treatment resulted in significantly less conflict behaviour with the head (P=0.028) and mouth (P<0.001), whereas HRV did not differ between treatments. In other studies, we have investigated the link between stress, social rank, fearfulness and learning. We failed to demonstrate an effect of baseline stress on learning capacity, possibly due to a general low stress level in the participating young horses, kept 24 h on pasture in a stable group. However, we demonstrated a negative correlation between fearfulness and performance in a learning test in a novel environment (P=0.040) and also that lower ranking horses had higher FCM levels (P=0.035), indicative of more stress. In conclusion, understanding of stress responses in riding horses is relevant to optimize horse welfare and performance and may contribute to increase safety in horse-human interactions.

Session 36 Theatre 3

More than '9 to 5': thinking about horse working in the 21st century

R. Evans[1] and N. Ekholm Fry[2]
[1]Hogskulen for Landbruk og Bygdeutvikling, 213 Postvegen, 4353, Klepp Stasjon, Norway, [2]Prescott College, Counseling Psychology, 220 Grove Avenue, Prescott, AZ 86301, USA; rhys@hlb.no

In the Scandinavian languages the term arbeidshest, or a close variant, refers to a working horse, contributing to agriculture and industry by pulling and carrying both people and things. An element of leisure was also present in horse keeping. For centuries the two roles were separate and distinctive for horses and well as for people. There has been a shift in how horses are both viewed and used in current society. Horses help us counter the stresses of urban life, horses provide us a connection to nature and an ineffable relationship with another species. For some, horses represent continuous and open-ended learning opportunities. This is the era of the 'Leisure Horse'. The distinction between leisure and work is becoming increasingly blurred, and this blurring is becoming increasingly formalized. New areas, such as equine-assisted therapy and equine tourism where the 'work horse' term is not being used, are growing steadily. In both industries, we expect horses to cooperate with us, to interact with strangers, and to deliver benefits through our interaction with them. Is this not work? We propose that we think of horse work similarly to the way we think of human work. Providing therapy is work for people, as is catering to tourists. Accepting this, it raises important questions about the way we think about horses and work. In this paper we raise the question and ask, how must we re-think our relationship with horses and what we ask them to do? What can we do to help our companions to cope with the stresses and strains of these new forms of labor? From our research in equine-assisted therapy in the USA and in equine tourism in the Nordic countries, we propose some tentative conclusions about how we must re-think our ethical relationship to our horses, and make a few suggestions on how we can, in the least, perceive our horses needs and respond to them as they work hard to help us make our lives better.

Session 36 Theatre 4

The role of human-horse relationship in a pilot project about autism and therapeutic riding

S. Cerino[1], N. Miraglia[2] and F. Cirulli[3]

[1] A.I. Sant'Elia Horses, Therapeutic Riding Center, Via Terre dei Consoli 1/20b, 01030 Monterosi, Italy, [2]University of Molise, Via F. De Sanctis, Campobasso, Italy, [3]Istituto Superiore di Sanità, Neuroscienze comportamentali, Viale Regina Elena 64, Roma, Italy; s.cerino@alice.it

Many experiences and empirical researches have pointed out how autistic patients are among the most involved in Therapeutic Riding Sessions. On the contrary there are really only few scientific studies to demonstrate the possible positive effects of this activity. In this paper the authors would like to discuss a pilot research project about Therapeutic Riding and autistic children, organized in Italy in 2011-2012. Aims of the project are the analysis of the level of participation and relationship, motors ability development and cognitive functions in autistic children. Particularly the authors point out the role of human-horse relationship in this activity, to underline how the horse behavior can influence the therapeutic setting and how important is to study it together with the interactions with the patients and the proposed methodology of equestrian work.35 children, aged 6-12, have been involved into this project, divided in two cross mode groups. They had been diagnosed as autistic (clinical and test diagnosis), with neither mental retard nor previous experiences in Therapeutic Riding. The test battery used (and submitted at T/0, T/6, T/F) has been: Ados Scale, Leiter Scale, ICF, Vineland, London Tower and ABC movement.T The applied Therapeutic Riding methodology and the preliminary evaluation of the results will be examined closely, together with the analysis of the behavior of the horses involved into the project and its influence on the relationship with the patients.

Session 36 Theatre 5

Repeatabilities and genetic parameters for behavioural and physiological parameters in ridden horses

U. König V. Borstel, C. Glißmann, V. Peinemann, S. Euent, P. Graf and M. Gauly

U. of Göttingen, DNTW, Albrecht-Thaer Weg 3, 37075 Göttingen, Germany; koenigvb@gwdg.de

When considering novel parameters e.g. for use in breeding programmes or for evaluation of animal welfare, prior assessment of suitability of such parameters is of paramount importance. Besides questions regarding the practicality of data collection, one component that determines suitability is the repeatability of a parameter. In the case of breeding purposes, genetic parameters are also of importance. Therefore, the present paper gives an overview of repeatabilities and, where available, genetic parameters for physiological parameters as well as different aspects of behaviour horses show under a rider. While repeatabilities calculated from variance components of mixed models for heart rate (HR) (e.g. baseline HR during riding: r=0.88), HR variability (e.g. RMSSD during novel object tests: r=0.75) as well as rein tension parameters (maximum (r=0.64±0.11), mean (r=0.50±0.14) and variability (r=0.52±0.14) of rein tension during dressage training) were generally in an acceptable range, repeatabilities for behavioural parameters vary considerably, which is likely a result of varying rider influences. For example, repeatabilities for refusal to jump (r=0.05) and bucking (r=0.19) in cross-country training were rather low, while they varied for dressage training (e.g. head-tossing: r=0.56; tail-swishing: r=0.07) and novel object tests (emotionality: r=0.40; stops: r=0.15). Preliminary results for heritabilities from an animal model for behaviour related to fear reactivity assessed in 575 horses in a ridden novel object test revealed values in a low to moderate range (e.g. reactivity: h^2=0.46±0.19; emotionality: h^2=0.26±0.21; intensity of rider's aids: h^2=0.17±0.29; interest in stimuli: h^2=0.02±0.16). Genetic correlations to performance traits were generally positive, indicating that consideration of such parameters in breeding programmes could aid in improving equine personality traits and therefore ridden horse behaviour and welfare in the long term.

Session 36 Theatre 6

May work alter horse's welfare: a review

M. Hausberger, C. Lesimple, C. Lunel and C. Fureix
CNRS/Université de Rennes1, Ethologie animale et humaine, Campus de Beaulieu, Avennue du Général Leclerc, 35042 Rennes Cedex, France; martine.hausberger@univ-rennes1.fr

From early on in the domestication history of horses, their relation to humans has been largely centered around a working relation through harnessing and riding. Archeological data trying to estimate when riding first occurred are largely based on examination of teeth and spine, both showing traces of bit actions and rider's weight on the back respectively. Thus, work does affect the horse's body but the question remains of whether this is so that horses' welfare may be altered both during the working sessions and/or in a more chronic way, with consequences outside the working sessions. In this presentation, we will review the existing evidence of the potential effect of work on the overall welfare of horses and try and identify the behavioural indicators of discomfort at work as well as the indicators of work related problems outside the working situation. Studies converge to show that increased emotionality, behavioural disorders and aggressiveness outside work may in some cases result from the type of work the horses are used for or the way it is performed. Different approaches lead to a high prevalence of back disorders in sport horses that could explain for some part some of these welfare problems. We will try and disentangle the potential mechanisms relating welfare issues to work in horses and propose manageable solutions of improvement that would not only avoid altering horses'well being but also may lead horses to enjoy working, as seen in certain facilities.

Session 36 Theatre 7

Temperament of school horses: relation with sport discipline and level of riders

M. Vidament[1], H. Schwarz[1], M. Le Bon[2], O. Puls[2] and L. Lansade[1]
[1]INRA, CNRS, Univ. Tours, IFCE, PRC, 37380 Nouzilly, France, [2]IFCE, IFCE, 49411 Saumur, France; marianne.vidament@tours.inra.fr

Temperament is an important factor when using horses. Behavioral tests have been developed to measure certain dimensions in horses. The aims of this study were to evaluate the relationship between temperament and: (1) sport discipline; (2) the ease to be ridden. Five dimensions of temperament have been measured in 56 horses belonging to National Riding School of Saumur (mainly French Saddle Horses, mean age: 11,5 years). They were specialized in one discipline (dressage (n=25), show jumping (n=12), eventing (n=19)) and ridden only by riders having the national basic level for competition, for education. Horses temperament was tested as described by Lansade: fearfulness/curiosity (tests: crossing a novel surface (2.7×2.7 m), suddenly opening an umbrella, novel object), gregariousness (test: isolation), locomotor activity (during the other tests), reactivity/curiosity to a non familiar human (tests: passive and active human), tactile sensitivity. Jumping horses had a lower level of fearfulness: they presented a lower intensity of the flight after umbrella opening than eventing horses (P=0.03) and a quieter manner to cross alone the novel surface than dressage horses (P=0.04). Horses of all disciplines were divided into 2 groups depending on the level of riders that could ride them safely, according to a questionnaire addressed to the riding teachers. Horses that could be ridden by intermediate level riders (n=26) were less fearful (P=0.02), less gregarious (P=0.03) and less active (P=0.01) than horses that were safely ridden only by pre-national and national competition level riders (n=28). In conclusion, there are relations between temperament dimensions and discipline, and between temperament dimensions and rideability.

Session 36 Theatre 8

Using appropriate reinforcement to trigger attention: the example of horse training

C. Rochais[1], S. Henry[1], S. Brajon[1], C. Sankey[1], A. Górecka-Bruzda[2] and M. Hausberger[1]
[1]UMR 6552 CNRS Université Rennes 1, Laboratoire Ethologie Animale et Humaine, EthoS, Station biologique de Paimpont, 35380 Paimpont, France, [2]Polish Academy of Sciences, Institute of Genetics and Animal Breeding, Jastrzebiec, 05-552 Magdalenka, Poland; celine.rochais@univ-rennes1.fr

The emotional valence of a situation can modulate selective attention. The valence of training experience is crucial for horses, the use of a positive reinforcement as a food reward has been shown to enhance learning performances and to promote a positive relation to the trainer. Here, we investigated the impact of different reinforcement (no reinforcement, primary positive reinforcement) or action (scratch the wither) on the horses' attentional state in a learning task. In a first study, 15 young males Angloarabian were trained to remain still in response to a vocal command. Immobility was rewarded either with a piece of food (n=8) or with no reward (n=7). In a second study, 15 young polish horses (6 females, 9 males) were trained to the same task, but here, immobility was rewarded either with a piece of food (n=8) or with a tactile contact (scratching on the withers during 5 seconds, n=7). For both studies, monitoring, gazes and behaviours directed towards the trainer revealed that the use of a food reward as positive reinforcement increased horses' selective attention towards their trainer (MW, $P<0.05$ in all cases). These studies suggest that attention may well be the process underlying the efficiency of appropriate reinforcements leading to better learning performances. This also explains increased bonding in human-animal training situations when positive emotions are involved. Horse attentional state is a key for better interaction, better security and better performances in a working situation. Nevertheless, only individuals can tell what an appropriate reward is and which reward induces positive emotions and hence triggers attention.

Session 36 Poster 9

How to ensure a simple, secure and efficient training of the young horse?

S. Henry, C. Sankey and M. Hausberger
UMR 6552 é 1, Station Biologique de Paimpont, 35380 Paimpont, France; severine.henry@univ-rennes1.fr

Horses' education requires repeated interactions with a trainer and training sessions may be special moments during which the horse builds up its own positive or negative representation of humans, depending on the quality of the interaction. Previous works shed light on the beneficial effects of using the maternal influence in order to establish a trustful human-foal relation and facilitate the first education (haltering for instance) of the foal. But, how to ensure more efficiently the learning of more complex tasks such as saddle-breaking? In a series of studies, we demonstrate that: (1) the use of a positive reinforcement-based training improves not only speed of training (learning to remain immobile and undergo several handling procedures, accepting the first steps of saddle-breaking), but also attitude of young horses during training (i.e. less agressive behaviours) and horse-trainer relationship in contexts other than training, with long-term and generalized effects; (2) a negative reinforcement-based training induces conversely an increased emotional state in the horse leading to a less positive relationship; (3) only food reward is efficient as a primary positive reinforcement, as scratching the withers (commonly used as a classical reward in horse training) do not allow real progress in training, nor do it impact on the human-horse relationship during the tests. In conclusion, creating a positive learning situation appeared to benefit both learning and behaviour during the training sessions. It is also essential in order to ensure secure training in horses.

Equitherapy and autism: a pilot study about visual attention

E. Dubois[1], M. Hausberger[1] and M. Grandgeorge[2]
[1]UMR6552, Campus Beaulieu, 35000 Rennes, France, [2]Centre Ressources Autisme, Rte de Ploudalmézeau, 29820 Bohars, France; dubois.elo@live.fr

Equitherapy is widely practiced and its positive effects are quite well known. However, dynamics of communication between the horse, the user and the therapist is little described. Here, we proposed a pilot study that focused on visual attention of these three partners at two moments: before and during horseback riding (HR). The triad had been observed during the equitherapy session (around 45 min) of four boys with autism (6-9 yo). Ten-second scan samplings recorded the gaze direction of each subject. Data were converted in percentage of time. Horses had mostly gazed at physical environment both before (76.2±6.6%) and during (91.6±1.6%) HR. Human they mostly gazed was the therapist before HR (8.7±3.3%). Similarly, boys mostly gazed at physical environment throughout the session (61.7±17.1% and 46.9±15.3%, respectively). Focus on social partner showed that boys mostly gazed at horse especially during HR (8.7±4.7% and 13.2±9.6% respectively). Likewise their visual attention to humans was enhanced during HR (9.0±1.5% to 19.3±3.1%). At last, before HR, therapists mostly gazed at boy (33.8±9.4%) and horse (38.4±16.7%). During HR, their attention on boy increased slightly (36.7±10.0%) but surprisingly, their attention on horse decreased extremely (7.2±0.9%) in favor of the physical environment (26.1±11.5% to 50.3±12.7%). This pilot study on visual attention during equitherapy suggested that children with autism, whose attentional skills are impaired, seemed to be more attentive to social partners during HR. In this situation, horses seemed few attentive to humans that might be explain by either apathy or working situation. Finally, the reduction of therapist's visual attention to the horse during HR suggests a decrease in vigilance. This was an example of routine where human pays less attention to animal, which constituted one of the most accidental situations. Thus, therapists must be greatly attentive when horses are used in such therapeutic programs where safety is indispensable.

Are horses sensitive to humans' emotional state during a leading task?

V. André[1], A.S. Vallet[1], S. Henry[1], M. Hausberger[1] and C. Fureix[2]
[1]Université de Rennes 1, UMR CNRS 6552 EthoS, 35042 Rennes, France, [2]University of Guelph, Animal & Poultry Science, Guelph, Canada; vanessa.andre@univ-rennes1.fr

Both humans and animals appear to be sensitive to cues displayed by each other while interacting and adapt their behaviours accordingly. However, very little is still known about the relevant elements that have to be considered when humans interact with horses. Here we investigated whether humans' emotional state had an impact on horses' heart rate and level of obedience in a simple leading task. Professionals (6 women, 2 men) and non-professionals (3 women, 3 men) were asked to lead a horse along a given path. Experiment 1 was performed on 8 professionals and 3 horses kept in natural conditions (site A); experiment 2 on 6 non-professionals and 13 horses, of which 5 from the site A and 8 from a riding school (site B). Humans' and horses' heart rates were recorded during the interaction. People also reported a posteriori on their positive (e.g. pleasure) and negative (e.g. fear) emotional states (questionnaire). Women reported on more satisfaction and more intense emotions than men ($P<0.05$). Interestingly, horses' emotional states might be related to humans' emotional states in women/horses dyads: heart rates of horses and women were higher in the site A than in the site B ($P<0.05$). Level of expertise also seems to be at stake in humans' emotional states: non-professionals had lower heart rates and expressed less negative feelings (i.e. misunderstanding) than professionals ($P<0.05$). Horses however showed higher heart rates with non-professionals ($P<0.001$). These results suggest that horses are able to perceive humans' emotional states. On-going analysis performed on horses' behaviours will allow us to explore further the relationship between heart rates variations and the valence of the horses' emotional states. The results reinforce the idea that knowledge of factors which might impact on humans' emotions should be promoted, as appropriated emotions might be key elements to prevent accidents.

Evaluation of bedding material for horses in a practice situation

E. Søndergaard[1], J.W. Christensen[2] and P. Kai[1]
[1]AgroTech, AgroFoodPark 15, 8200 Aarhus N, Denmark, [2]Aarhus University, Blichers Alle 20, 8830 Tjele, Denmark; evs@agrotech.dk

In Scandinavia, horses are stabled many hours per day for at least 6 months per year, most commonly on concrete floor with a bedding material. Straw was the most common bedding material for horses but in recent years it has been replaced by materials like peat, wood shavings, etc. Quality of a bedding material relates to its influence on horse welfare and on stable climate, economy, disposal, work load, etc. In the future also climatic and environmental aspects may be important. An all-round evaluation of bedding materials was performed in a stable at an agricultural school. 17 horses housed in single boxes and 14 horse owners were included in the study lasting 8 weeks. Four bedding materials: (1) straw; (2) peat with wood shavings; (3) wood shavings; and (4) wood pellets, were evaluated in two-week periods in a semi-randomised design. At the end of each period, surface temperature and ammonia concentration were measured in each box and pH was measured in hoof material. Each box was categorised in relation to amount and distribution of the bedding material and the owners filled in a questionnaire in relation to cleaning, dust, etc. Temperature and humidity were monitored continuously. Volume, elasticity and water binding capacity of the bedding material were measured in the laboratory. Horses were recorded on video during the 8-week experimental period. Water binding capacity was lower in peat than in the other materials. Volume and elasticity was highest in straw and lowest in wood pellets. Overall, horse owners ranked wood pellets highest and straw lowest. There were no difference in surface temperature and ammonia concentration. Data on horse behaviour remains to be analysed. Evaluating bedding material in relation to all relevant parameters is a costly and time consuming task and it could be useful to have some indicators of quality of bedding materials. This will be further explored in future projects.

Session 37 Theatre 1

Is genomic selection compatible with organic values?

K.K. Jensen
University of Copenhagen, Department of Food and Resource Economics, Rolighedsvej 25, 1958 Frederiksberg C., Denmark; kkj@foi.ku.dk

In a subproject within the EU project LowInputBreeding, it is planned to examine the potential for using genomic selection in breeding of organic dairy cows. This plan has been subject of some controversy within the organic movement. Taking these discussions as its point of departure, this paper aims at analyzing the question whether or not genomic selection, when used to promote organic production, still must be considered incompatible with basic organic values. Applying the IFOAM principles on breeding lead me to the following statements: Breeding should ensure that animals are well adapted to their conditions in the ecosystem made up by the farm. Organic agriculture should maintain local breeds continually over time. It is wrong to breed animals to live in conditions not in accord with their 'physiology, natural behavior and well-being', and breeding should not involve serious risk of adverse effects on future health and well-being of humans and/or animals. How should one assess the use of genomic selection from this perspective? The problem is that the continuous adaptation has been broken. Many years breeding for higher productivity have made many breeds less well adapted to organic conditions. Genomic selection could have a potential in breeding for functional traits to make animals better adapted to organic conditions. Would this violate organic values? Perhaps the most important concerns is that genomic selection leads to use of unacceptable reproduction techniques. Concerning the first, I suggest that organic values are already violated in terms of many cases of poorly adapted animals. In this situation, a strict interpretation could imply that production must be stopped. But a more pragmatic view would imply, I argue, that better adapted animals should have more weight than using organically acceptable reproduction techniques. So if genomic selection serves this goal better than traditional breeding (and if certain other concerns are met), it should be favored.

Ethical aspects of two alternatives to the killing of male chicks

M.R.N. Bruijnis[1], H.G.J. Gremmen[1], V. Blok[2] and E.N. Stassen[1]
[1]Wageningen University, Adaptation Physiology Group, Department of Animal Sciences, P.O. Box 338, 6700 AH Wageningen, the Netherlands, [2]Wageningen University, Management Studies, Department of Social Sciences, P.O. Box 8130, 6700 EW Wageningen, the Netherlands; bart.gremmen@wur.nl

In the Netherlands annually over 45 million, globally billions, male chicks are killed after hatching. In the laying hen industry there is no use for male chicks as they do not lay eggs and are not suitable for meat production. This mass killing raises discussions in society, shown by different newspaper articles and the position of animal rights activists that this practice is unethical. Several years ago the Dutch minister of agriculture commissioned research projects to identify alternatives with the best prospects to the killing of male chicks, resulting in two technically possible alternatives: (1) Genetic modification (GM): by developing a line of chickens where the male eggs can be recognized optically by the expression of enhanced Green Fluorescent Protein (eGFP) which is transmitted only to the males. These male eggs do not have to be hatched. An important part of the innovation is that the laying hens, and also their eggs, will not be GM. In practice this alternative will mainly have impact on the processing of parent stock and the hatcheries but will not change production processes of eggs or chicken meat. (2) Dual purpose chickens: in this case the males can be used for meat production. These dual purpose chickens do need more time to reach the desired weight than the current broilers and the hens produce fewer eggs than specialized laying hens. Both alternatives may be problematic from an ethical perspective. The aim of this paper is to analyse the ethical aspects in order to evaluate the alternatives compared to the current situation. We will present an ethical framework based on a combination of a deontological- and a consequentialist perspective, and describe the ethical arguments in the debate about the alternatives. Finally we will compare and evaluate the two alternatives to the current situation.

Killing new born animals for efficiency reasons; genetic selection as a cause for a dilemma

F. Leenstra[1], V. Maurer[2], M. Bruijnis[3] and H. Woelders[1]
[1]Wageningen UR Livestock Research, P.O. Box 65, 8200 AB Lelystad, the Netherlands, [2]FiBL, Postfach 219, 5070 Frick, Switzerland, [3]Wageningen University, Adaptation Physiology, P.O. Box 338, 6700 AH Wageningen, the Netherlands; ferry.leenstra@wur.nl

In commercial egg production, male chicks are killed immediately after hatch as they are not profitable for meat production. Some of them are utilised as feed for zoo or pet animals, or snack for humans, but they do not have a life of significance. In many countries people have objections against this practice. The origin of this problem is the development and use of specialised breeds for specific purposes, to obtain increased production efficiency and low-priced animal products. Specialization can overcome the opposite requirements for high efficiency in the production of meat and eggs (milk), respectively. For efficient meat production, a high growth rate is essential. In contrast, for efficient production of eggs or milk, low animal maintenance costs, i.e. a high production rate per kg body mass, is most important. This dichotomy is most clearly seen in modern industrialized poultry production. Egg type males require 3 times more time and 2-4 times more feed than meat type birds to reach an acceptable slaughter weight, while meat type hens require much feed for growth and maintenance which makes them inefficient for egg production. Selection of layer type birds for improved growth rate could make it more attractive to rear the males for meat production, but would strongly compromise efficiency of egg production by the females. A similar situation, albeit less extreme (for now?) can be found in dairy goats and cattle. Male offspring of dairy goat and some typical dairy cattle breeds do not have an economic value for meat production and may be killed at birth. In terms of economics, resource efficiency, or animal welfare (provided killing is carried out in a humane way), this may not be a problem but ethically it is. We discuss this ethical dilemma and explore technological and niche market alternatives as possible solutions.

Breeding on polled genetics in Holsteins: chances and limitations

H. Täubert[1], D. Segelke[1], F. Reinhardt[1] and G. Thaller[2]
[1]Vereinigte Informationssysteme Tierhaltung w.V., Heideweg 1, 27283 Verden, Germany, [2]Institute of Animal Breeding and Husbandry Christian-Albrechts-University, Olshausenstraße 40, 24098 Kiel, Germany; dierck.segelke@vit.de

Goal of this study was to investigate chances and limits of breeding polled cattle. Based on the new innovative method of imputation there is a possibility to identify additional polled animals in the herdbook with a low error rate. This can be used to extend the genetic base and additional animals for selection are available. Analyzing SNP-genotypes gives information on breeding values and recessive gene-defects, which can be considered in the different selection steps. The biggest limitation for an intensive breeding on polled animals at the moment is the use of two polled sires and their direct offspring with high breeding values in performance and functional traits. Different simulations were performed to analyze the consequences of different breeding strategies on the frequency of the desired polled allele, the genetic gain in male and female animals as well as the development of inbreeding. It can be shown how the selection of the homozygote polled genotype for already SNP-genotyped animals is the most effective way to increase the polled-allele in the population with simultaneous increase of genetic gain. For non-SNP-genotyped selection on the phenotype is sufficient, genotyping the polled status is not mandatory.

Session 37 Theatre 5

Animal breeding and ethical values

M. Marie[1,2] and G. Gandini[3]
[1]INRA, ASTER-Mirecourt, 662 Avenue Louis Buffet, 88500 Mirecourt, France, [2]Université de Lorraine, ENSAIA, 2 Avenue de la Forêt de Haye, TSA 40602, 54518 Vandoeuvre lès Nancy cedex, France, [3]Università degli Studi di Milano, Dipartimento VSA, Facoltà di Medicina Veterinaria, Via Celoria 10, 20133 Milano, Italy; michel.marie@mirecourt.inra.fr

The rapid evolution of scientific knowledge and technics in the field of genetic which occurred from the second half of the 20th century impacted the farm animal breeds and individuals and questions the geneticists, the farmers, the agricultural profession as well as the society. Among other cases, we can cite the choice of selection criteria, the management of domestic biodiversity, at the level of the species, with the loss of small size local breeds, or intra-breed with restricted selection bases, the negative health side-effects demonstrated in meat chicken, swine or beef such as the Blanc Bleu Belge breed, or the consequences of transgenesis on health as for swine, or on environment as for transgenic fish. This paper will review the literature describing such cases and identify the moral values which can be referred to, such as animal welfare, naturalness, integrity, intrinsic value, dignity, responsibility. We will explore how the reflexive equilibrium method can be used in order to form a moral position at the individual or collective level.

Session 37 Poster 6

Ethical aspects of breeding: may be limited to genetics alone?

G. Bertoni[1], N. Soriani[1], A. Chatel[2], G. Turille[2] and L. Calamari[1]
[1]Università Cattolica del Sacro Cuore, Via E. Parmense, PC, Italy, [2]Institut Agricole Régional, Rég. La Rochère, AO, Italy; giuseppe.bertoni@unicatt.it

There is no doubt that the genetic improvement of animals, either under ignorance in the past as well as with the utilization of genetic markers today, have a closed relationship with the available environment for animal life or with the possibility to adapt it to the improved individuals. For instance, Valdostana has been selected for his adaptability to mountain pastures, while Holstein Friesian has been selected within more 'sophisticated' conditions: good hygiene, nutrition, monitoring, skillness, etc. Therefore, in our view, the ethical aspects of animal breeding cannot consider the genetic improvement detached from the available environment, natural or specially adapted. Aim of the paper is to show that breeds with extremely different milk yield, when kept in proper conditions, can experience a good welfare. Two herds, one of Italian Friesian (F) with high milk yield (36 kg/d) and a second of Valdostana (V) with lower milk yield (18 kg/d) were used. Physiological indices in early lactating cows, main milk characteristics, clinical diseases prevalence and fertility parameters were measured; in the F herd the welfare was assessed by using our model and the score was above the acceptable value considered in our model (75/100), while V cows were often on pastures. In both herds the blood parameters were within the reference range suggesting good health conditions. Namely, the acute phase proteins (haptoglobin and albumins), indicators of inflammations were good and similar as well as milk fat and protein content (3.5-3.7% and 3.3-3.4%, respectively). Somatic cells were slightly different, but again good: 111 and 160 n/μl in F and V respectively. The clinical diseases prevalence was within acceptable range, and the fertility was near to optimal also in F (137 days open). It can be therefore suggested that the judgement of breeding effects can only be expressed if the 'new' animals are kept in suitable conditions.

Session 38 Theatre 1

Sow group housing: an introduction

A. Velarde
Animal Welfare Subprogram, IRTA, Finca Camps i Armet s/n, 17121 Monells, Girona, Spain; antonio.velarde@irta.es

From 1 January 2013 sows and gilts in the European Union should be kept in groups from four weeks after the service to one week before the expected time of farrowing. A lack of knowledge might be the major obstacle to implementation the transition from stall to group housing. Producers, veterinarians and animal scientists may not be sufficiently familiar with the range of options available for housing pregnant sows in groups. Even more importantly, they may lack the expertise needed to manage such systems effectively. Moreover, although it is widely accepted that group-housed sows in well managed systems have higher standards of welfare than sows kept in crates, poor management of groups can increase aggression, lameness and culling rates. In turn, these welfare problems will reduce productivity, product quality and profitability. This session aims to share strategies to overcome difficulties in implementing the Council Directive 2008/120/EC as it refers to group-housing of pregnant sows.

How does Canada deal with the issues of animal welfare and sow group housing?

N. Devillers
Agriculture and Agri-Food Canada, Dairy and Swine R&D Centre, 2000 rue College, Sherbrooke, QC, J1M 0C8, Canada; nicolas.devillers@agr.gc.ca

Pork is the 4th agricultural production in Canada representing 8% of Canadian agriculture income and 27 millions of pigs produced in 2010. More than 50% of the pork meat produced is exported mainly to USA (29%), Japan (18%), Russia (12%) and China (10%). In 2012, there were 1,293,400 breeding sows and gilts in Canada, mainly in Québec (28%), Ontario (26%), and Manitoba (25%). In Québec, only 5% of them were housed in groups in 2012. Canadian pig producers are facing the issue of converting their farms to group housing of gestating sows because of the move occurring in the EU, the USA and other countries and the increasing pressure of consumers and animal rights associations towards the ban of gestation stalls. However, farm practices in Canada are not much regulated and adoption of animal friendly practices is more the result of the supply and demand process. Since 2007, several North-American meat processing and agri-food companies, such as Smithfield Foods or Maple Leaf, intended to ask their suppliers to progressively convert gestation stalls to group-housing within a 10 years period. Canadian pig producers associations have also adopted a common HACCP assurance quality program which now includes an Animal Care Assessment, but it does not request group-housing for gestating sows. Finally, the voluntary Code of Practice for the care and handling of Pigs is currently under revision by the National Farm Animal Care Council and should include updated recommendations on the housing of gestating sows. To date, there is very little information on the number of farms that already convert to group-housing in Canada. In Québec, 75% of farms having group-housing use floor feeding. However, new facilities are usually built for large static groups with electronic sow feeders. In the near future, one of the major issues that producers will have to face is how to fund the conversion to group housing in a context of depressed market.

Session 38 Theatre 3

Effects of type of floor and feeding system during pregnancy on performance and behaviour of sows

F. Paboeuf[1,2], M. Gautier[2], R. Cariolet[1], J. Lossouarn[3], M.C. Meunier-Salaün[4] and J.Y. Dourmad[4]
[1]ANSES, SPPA, 22440 Ploufragan, France, [2]Chambre d'Agriculture de Bretagne, Av. Ch Sans Pitié, 22195 Plérin, France, [3]AgroParisTech, 16 rue C. Bernard, 75231 Paris, France, [4]INRA, Agrocampus Ouest, UMR Pegase, 35590 Saint-Gilles, France; jean-yves.dourmad@rennes.inra.fr

The aim of this study was to compare contrasted group housing systems for gestating sows, differing by the type of floor, either fully slatted floor (SF) or straw bedding (SB), combined with two types of pen arrangement, either small groups of 6-8 sows with individual feeding stalls (FS) or larger groups of 24 sows with an electronic feeding station (EF). Data were collected on sows from different parities (a total of 540 reproductive cycles). All sows received the same diet and were fed according to their body condition at insemination. The numbers of piglets born (13.2) or weaned (10.9) per litter were not affected by the type of floor during gestation, neither by the type of pen. However, the interaction between type of floor and type of pen tended to be significant. SB sows tended to wean more piglets in EF than in FS pens, whereas the opposite was found for SF sows. This was mainly related to a reduced piglets mortality during lactation in SB-EF litters. The weight of sows before farrowing and at weaning was lower for EF than for FS sows, by 16 and 12 kg, respectively, whereas no effect of type of floor was observed. During pregnancy, the frequency of stereotypies was higher in SF than in SB sows, and in FS than in EF sows, the lowest occurrence being observed in SB-EF sows (44%) and the highest in SF-FS sows (85%). The occurrence of body lesions was twice as high in SF compared to LP sows, whereas vulva lesions were significantly higher in EF compared to SF sows, independently of the type of floor. This study suggests that the type of floor and the feeding system during gestation have rather limited impacts on reproductive performance of sows, whereas they significantly affect the occurrence of body lesions and the behaviour of pregnant sows.

Gestation features of group-housed sows affecting growth rate and feed intake in finishers

E. Sell-Kubiak, E.H. Van der Waaij and P. Bijma
Wageningen University, Animal Breeding and Genomics Centre, P.O. Box 338,6700 AH Wageningen, the Netherlands; ewa.sell-kubiak@wur.nl

The focus of this study was to identify sow gestation features that underlay permanent sow and common litter effects, estimated as fraction of phenotypic variance, of growth rate (GR) and feed intake (FI) in sows' offspring as finishers. Data of 17,743 finishers, coming from 604 sires and 681 crossbred sows, were obtained from TOPIGS Research Center IPG. Average size of gestation group was 22.5. Sow gestation features were collected during multiple gestations and divided into three clusters describing: (1) sow body-condition, i.e. weight, gestation length; (2) group-housed sows' feed refusals (FR), i.e. difference between offered and eaten feed during three periods of gestation: 1-28, 25-50, 45-80 days; (3). sow group features, i.e. number of sows, average parity in gestation group. Sow features were added to the basic model one at a time to study their effect on GR and FI. Significant sow features ($P<0.1$) were fitted simultaneously in an animal model to evaluate their effect on common litter and permanent sow effects. Gestation length had effect on GR (1.4 [g/day]/day) and FI (6.8 [g/day]/day). Sow's weights at insemination (0.07 [g/day]/kg) and at farrowing (0.14 [g/day]/kg). Days with FR during 25-50 (-1.1 [g/day]/day) and 45-80 days of gestation (-1.2 [g/day]/day) and average FR during 45-80 days of gestation (-24 [g/day]/kg) had negative effect on GR. Sow FR from 1-28 days of gestation were not significant. Size of gestation group had effect on FI (-9 [g/day]/group_member) and day sow entered group had an effect on GR (-0.9 [g/day]/day). Effect of sow FR on GR in finishers suggests that process of muscle development (myogenesis) in offspring might have been disturbed by FR during gestation. This results in slower growing offspring. Sow gestation features explained 1-3% of the total phenotypic variance. Gestation features did explain phenotypic variance due to permanent sow and part of phenotypic variance due to common litter effects for FI, but not for GR.

Effect of group housing on the productive efficiency of sow farms in Flanders

J. Hamerlinck[1], J. Van Meensel[1] and M. Meul[2]
[1]Institute for Agricultural and Fisheries Research, Social Sciences Unit, Burg. van Gansberghelaan 115 bus 2, 9820 Merelbeke, Belgium, [2]University College Ghent, Department of Animal Production, Faculty of Applied Bioscience Engineering, Valentin Vaerwyckweg 1, 9000 Ghent, Belgium; marijke.meul@hogent.be

The aim of this study is to analyze the effect of group housing on the productive efficiency of sow farms in Flanders and on the underlying parameters that determine the efficiency levels. The frontier method DEA (Data Envelopment Analysis) is used to assess efficiency scores for 48 sow herds (21 with individual housing and 27 with group housing of sows) using farm accountancy data from 2011. Efficiency levels reflect the transformation of five inputs into one output, being the 'total weight of piglets produced in one year'. The selection of inputs is based on their significant effect on the economic and environmental performance of sow farms. They comprise (1) number of sows, (2) amount of feed used, (3) replacement rate, (4) health costs (including fertility costs) and (5) fixed costs. Preliminary results show that the technical efficiency is not significantly different between farms with individual housing and farms with group housing of sows. Nevertheless, sow farms with group housing use a significantly ($P<0.05$) higher amount of feed per sow, have higher health and fixed costs per sow and a higher replacement rate. Cluster analysis using efficiency results of the 27 sow herds with group housing reveals that sow herds applying the same group housing system are situated among the different clusters. This indicates that the type of group housing system has no influence on the technical efficiency. An extended analysis using a larger dataset and considering more inputs (e.g. energy use) is planned to validate these findings.

Group housing of sows in Europe: looking for strategies to transfer knowledge

D. Temple, E. Mainau and X. Manteca
UAB, Facultat de veterinària, 08193, Spain; deborah.temple@uab.cat

In the European Union, from the 1st of January 2013, pregnant sows should be kept in groups during a period starting from four weeks after the service to one week before the expected time of farrowing. Still, many farmers started the transition too late and will have difficulties meeting the new legal requirements. Part of the difficulties in implementing the legislation have been associated to a lack of knowledge. Stakeholders may face difficulties in interpreting the legislation or may not be sufficiently familiar with the range of options available for housing pregnant sows in groups. The adaptation to group housing system has also be delayed by the scepticism of vets, farmers and scientists on the efficiency of this new management system. Group housing of sows is one specific example of the EU legislation on animal welfare that illustrates the existing gasp between competent authorities and stakeholders when implementing a new regulation. To fill this gap, the EuWelNet® project aims to develop strategies based on the transfer of knowledge to competent authorities and stakeholders in order to improve the level of understanding and implementation of the legislation. Several research groups through Europe are evaluating knowledge-based educational material (fact sheets and digital information) to assist pig producers and competent authorities to assess and improve compliance with the council directive regarding the group housing of pregnant sows.

Session 38 Theatre 7

Inter- and intra-observer repeatability of three locomotion scoring scales for sows

E. Nalon[1], S. Van Dongen[2], D. Maes[1], M.M.J. Van Riet[3,4], G.P.J. Janssens[3], S. Millet[4] and F.A.M. Tuyttens[3,4]
[1]Ghent University, Veterinary Medicine, Dept. of Obstetrics, Reproduction and Herd Health, Salisburylaan 133, 9820 Merelbeke, Belgium, [2]University of Antwerp, Dept. of Biology, Evolutionary Ecology Group, Groenenborgerlaan 171, 2020 Antwerp, Belgium, [3]Ghent University, Veterinary Medicine, Dept. of Nutrition, Genetics and Ethology, Heidestraat 19, 9820 Merelbeke, Belgium, [4]Institute for Agricultural and Fisheries Research (ILVO), Animal Sciences Unit, Scheldeweg 68, 9090 Melle, Belgium; elena.nalon@ugent.be

We compared the inter- and intra-observer repeatabilities (interOR, intraOR) of three locomotion scoring scales for sows: a tagged visual analogue scale (tVAS), a 5-point (5P) and a 2-point (2P) ordinal scale with identical descriptors. Veterinary medicine students (n=108) were trained to use the scales and asked to score 90 videos (30 with each scale) of sows with normal and abnormal gait. Thirty-six videos were shown once and 18 were shown 3 times, of which one mirrored horizontally. Inter- and intraOR were estimated from the variance components of a mixed model with sow, student and their interaction as random factors using Monte Carlo Markov Chains in a Bayesian framework. The mean student score of each sow was regressed against the experts' scores for all scales and the correlation coefficient (CC) between the individual students' and the experts' scores was calculated. Inter- and intraOR were higher with the tVAS than the 2P scale and lower for sows with a lameness score of <45 mm on the tVAS. Other variables (order of video in the sequence, original vs. mirrored video, previous experience with locomotion scoring) did not affect repeatabilities. Correlation coefficients between students' and experts' scores were generally high but lower for the 2P scale (both for individual sows and across all sows). Using a 5-point or even a 2-point ordinal scale did not improve repeatabilities or correlations with experts' scores compared to using a tVAS. Repeatabilities were higher for lame versus non/mildly lame sows.

Session 38 Theatre 8

Addressing lameness in group housed sows

L. Boyle[1], A. Quinn[1,2], J. Calderon-Diaz[1,3], P. Lawlor[1], A. Kilbride[2] and L. Green[2]
[1]Teagasc, Pig Development Department, Moorepark, Fermoy, Co. Cork, Ireland, [2]University of Warwick, Life Sciences, Coventry, CV4 7AL, United Kingdom, [3]University College Dublin, Department of Animal Science, Belfield, Dublin 4, Ireland; laura.boyle@teagasc.ie

Lameness in sows is a welfare concern and contributes to poor longevity. The aim of our project was to evaluate risk factors for lameness and establish environmental and nutritional means of addressing it. Lameness and claw lesions were compared in 85 sows on transfer from stalls (ST) or a dynamic group housing (GH) system to farrowing crates. 100% of sows had claw lesions but while differences in claw health between the housing treatments were minimal, GH sows had a higher risk of poor locomotory ability suggesting that lameness will increase with the switch to GH. This was supported by findings of a survey of lameness in 70 Irish pig units with ST and a variety of GH systems. Preliminary results also suggest that slats wider than the minimum width (80 mm) specified in Directive (2008/120/EC), are associated with less lameness. Replacement gilts are often fed diets formulated for sows or finisher pigs which do not meet their requirements for bone/claw strength and fat deposition. Gilt diets reduced lameness, claw abnormalities and joint lesions. No improvement was detected in bone mineral density. A longitudinal study was conducted on a commercial farm which followed 160 GH gilts from entry to the breeding herd through two parities on either rubber covered or uncovered concrete slats. Rubber flooring reduced the risk of lameness and limb lesions. These benefits were likely mediated by improved comfort and sows showed a preference for lying on rubber. Nevertheless the risk of toe overgrowth, heel-sole crack and white line disease was higher on rubber possibly because the floor was dirtier. The findings of this project illustrate that in GH with slatted floors sows are at high risk of lameness. The effects can be ameliorated by improvements in gilt nutrition and to the flooring environment.

Session 38 Theatre 9

Preventing lameness in group housed sows

H.M. Vermeer and I. Vermeij
Wageningen UR Livestock Research, P.O. Box 65, 8200 AB Lelystad, the Netherlands; herman.vermeer@wur.nl

Lameness is a fundamental problem on sow farms. In group housing systems sows are even more dependent on healthy claws than in individual housing systems. Lameness is both a threat for animal welfare as for production. A review on sow claw health in the last 20 years learns how to prevent lameness in group housed sows. Although lameness can be scored relatively easy, the incidence and repeatability is lower than of claw lesions. An inspection of 15,000 claws showed that hind claws had three times more lesions that front claws and that the outer digits had six times more lesions than the inner digits. Scoring the hind outer digits turned out to be the most efficient way to assess sow claws. The most important floor characteristics effecting claw health are abrasiveness, friction, softness, slot/slat size ratio, and moisture. Aggressive interactions also play an important role in the aetiology of lameness. Abrasiveness of good quality concrete floor gives symmetric wear of the hooves resulting in the right gait. Plastic and smooth metal floors are not suitable for sows. A lack of friction like on dirty floors leads to slippery floors with risks on claw and leg injuries. Especially in a situation with aggressive interactions a high level of friction is important. Softness is important to prevent erosion of the heel and cracks in sole and side wall. Slatted floors should be designed on claw dimensions with slots narrow enough to prevent injuries but wide enough to prevent dirty and slippery floors. Openings of 18-22 mm and slats of 80-90 mm seem to be the optimum. Claws become fragile in moisty conditions with manure and urine. A dry floor is necessary for good claw health. In a social environment with aggressive interactions the risk on claw lesions and lameness is high and highly correlated with skin lesions. Minimizing mixing moments is a key factor for good claw health in group housing. The decrease of lameness in Dutch group housing systems in the last 20 years learned that claw health can be positively affected by housing and management.

Willingness to walk for a food reward in lame and non-lame sows

E. Bos[1,2], E. Nalon[1], M.M.J. Van Riet[1,2], S. Millet[2], G.P.J. Janssens[1], D. Maes[1] and F.A.M. Tuyttens[1,2]
[1]Ghent University, Veterinary Medicine, Salisburylaan 133, 9820 Merelbeke, Belgium, [2]Institute for Agricultural and Fisheries Research (ILVO), Animal Sciences Unit, Scheldeweg 68, 9090 Melle, Belgium; emiliejulie.bos@ilvo.vlaanderen.be

Lameness causes welfare and production related problems. Research is needed to detect mild cases of lameness and its impact on welfare and production. Lameness could also influence the willingness to walk for feed rewards. Motivation tests are used to measure animals' willingness to work for a reward, e.g. feed. The objective of this study was to investigate the impact of lameness on the ability to cover distances for feed rewards. Ten non-lame and 10 mildly to moderately lame hybrid sows were selected by using gait score. All sows were fed daily with 2.6 kg of a commercial gestation diet. After habituation to the experimental area, sows had to walk around a Y-shaped barrier, placed between two feeders, to receive a new reward. Sows were trained daily for 10 minutes. Training was considered successful when sows received at least 4 feed rewards during a session of 10 minutes. Feed rewards (apples and raisins) were presented after using both light and sound as a signal of an available reward. Sows were tested 3 times individually, once per day on 3 non-consecutive days. The effect of lameness on willingness to work for a feed reward was investigated using a Poisson model with gestation stage, parity range and body weight as fixed effects. The willingness to walk for feed rewards did not differ significantly between lame and non-lame sows ($P=0.164$). These results suggest that either the locomotary ability is not strongly reduced in sows with a low to moderate degree of lameness, or that the present motivation test is not sufficiently sensitive to detect these sows. Sensitivity of the test could possibly be improved by reducing the attractiveness of the feed rewards or increasing the work load during the test.

Session 38 Theatre 11

Sow group housing: general discussion

A. Velarde
Animal Welfare Subprogram, IRTA, Finca Camps i Armet s/n, 17121 Monells, Girona, Spain; antonio.velarde@irta.es

Sow housing is a current debate topic around the world. In the EU, group housing of sows is compulsory since 1 January 2013. Following this trend, other pork meat producer countries, as Canada, USA and Brazil are moving also to eliminate gestation crates. In these countries, the conversion to group housing is not regulated by legislation as in the UE, but is more a results of the supply and demand process. Although it is widely accepted that group-housed sows in well managed systems have higher standards of welfare than sows kept in crates, some housing and management factors might impair animal welfare and reduce productive and reproductive performance. The type of floor and the feeding system significantly affect the occurrence of lameness, body lesions and behaviour, but have a limited impact on reproductive performance of sows. Sows in slatted floors are at high risk of lameness. However, dirties floors, as rubber flooring increases the risk of some toe diseases. To overcome these problems, the education of farmer and veterinaries might play a relevant role. The development of transfer program, will allow the pig producers to identify risk factors and set up remedial solutions.

Session 39a Theatre 1

Effect of a patented combination of plant extracts on piglets performance

G. Benzoni[1], J.M. Laurent[1], D. Coquil[2], A. Morel[2] and M.L. Le Ray[3]
[1]INVIVO NSA, R&D, Talhouet, 56250 Saint-Nolff, France, Metropolitan, [2]LPA LE NIVOT, 29590 Loperec, France, Metropolitan, [3]NEOVIA, BP 394, 56009 Vannes Cedex, France, Metropolitan; mleray@invivo-nsa.com

Feed transition, housing change, separation from mother, enzymatic and immune adaptation are some of the great changes of weaning. To help animals coping with this stressful period a combination of plant extracts has been tested in the farm of an agricultural school. These three plant extracts have been selected for their anti-inflammatory and anti-oxidant properties. A trial was conducted on 77 sows and 624 piglets during three batches. Half of sows received plant extracts before farrowing and during lactation. Half of piglets received plant extracts from 20 to 67 days of age. Data were analysed with the analysis of variance except number of dead piglets in maternity treated with the Chi^2 method. Results in maternity have shown a reduction of dead piglet/sow between 24 h after birth and weaning in the group of sows fed with plant extracts (0.46) compared to the negative control group (0.17). After weaning, average daily gain was significantly improved in the group of piglets fed with plant extracts and coming from sows fed with plant extracts compared to control group (+11.4%, $P<0.001$). When plant extracts were distributed only in maternity or only in post-weaning, results were better than those of control group and lower than those of the group fed with plant extracts both in maternity and in post-weaning. Additional weight of piglets was due to a higher feed intake (+7% compared to negative control) and a better feed conversion ratio (-4.3% compared to negative control group). This trial demonstrated a cumulative effect of plant extracts distribution both in maternity and in post-weaning on piglets performance after weaning. This highlighted maternal impact on weaning preparation. Inflammatory and oxidative processes can be reduced during lactation and a right sows' management has indirect positive impacts on piglets' performance after weaning.

Session 39a Theatre 2

Feed additives may play a role on animal Welfare?

J. Brufau;, R. Lizardo and B. Vilà
IRTA, Monogastric Nutrition and Animal Welfare, Mas Bover, Ctra., Reus-El Morell, Km 3.8, 43210 Constanti, Spain; joaquim.brufau@irta.cat

The EU regulation 1831/2003 on feed additives has adopted animal as a functional category. Welfare criteria involved is discussed in an opinion published by EFSA in 2008. It describes how feed additives may be involved in the future of feed manufacturing with respect to animal welfare. However, welfare encounters unexpected difficulties to be implemented in animal husbandry by farmers and technicians. Recently the EU produced an impact assessment on the EU strategy for the Protection and Welfare of animals 2012-2015, which recognizes the difficulties for its implementation. This presentation describes potential actions of the application of feed additives to improve animal performance as well as welfare in animal husbandry. In practice, we aim to answer questions such as: Can zootechnical additives improve animal performance through improved gut health? Are these health improvements easily measurable? What are the main indicators to consider? Can these indicators connected with animal performance? How is immunity involved in animal performance? Some examples of feed additives that contribute to animal welfare conditions will be discussed.

Session 39a Theatre 3

Feed additives regulation and assessment in the EU: past, present and future

D. Jans
FEFANA, Av. Louise, 130A, 1050 Brussels, Belgium; info@fefana.org

The EU legislation is today a pivotal aspect for successfully bringing feed additives and other specialty ingredients on the market. Over the years, and through a deep evolution of the legislative context, these considerations have progressively taken a prominent place in companies' product strategies, from the earliest stage of product conception and positioning, down the final authorisation, through the demanding authorization process. A wrong legislative strategy applied to the best product could have dramatic consequences. Through its most recent developments, the EU legislation entered in the delicate area of functionality and claims in general. This triggered opportunities, including the recognition of new feed additives functionalities, but is also accompanied by a wide legal uncertainty. This lecture will bring an overview of the state regulation in the EU, focusing on assessment procedures and the role of EFSA in the assessment. It will explore the notion of functionality and its relationship with claims as well as initiative taken by industry in this area. It is anticipated to be a good basis of reflection for innovative projects. Didier Jans Secretary General, FEFANA (EU Association of Specialty Feed Ingredients and their mixtures)

Session 39a Theatre 4

Modification of gut microflora in rainbow trout using live yeast

P. Tacon[1], M. McLoughlin[2], S. Doherty[3], K. Maxwell[3], P. Savage[3], E. Pinloche[1] and E. Auclair[1]
[1]Lesaffre Feed Additives, 137 rue Gabriel Péri, 59703 Marcq en Baroeul, France, [2]Aquatic Veterinary Services, 35 Cherryvalley, BT5 6PN Belfast, United Kingdom, [3]Agri-Food & Biosciences Institute, Stormont, BT4 3SD Belfast, United Kingdom; marian@aquatic-veterinary.co.uk

Bacterial probiotics are becoming more used in aquaculture, however there is little work done on the effect of live yeast as a probiotic in fish. The objective of the present feasibility study was to evaluate the impact of live yeast on growth and gut bacterial changes in rainbow trout (*Oncorhynchus mykiss*). 100 rainbow trout, weighing 150 g, were stocked in each of four 750 litre tanks, supplied with fresh water (pH 7.0, temperature 10 °C). The fish were fed with Skretting LA30 feed at the ratio of 200 g/day. After a 7 day acclimatization period, 2 tanks received the same diet (control), and 2 tanks received the feed top dressed with 0.1% Actisaf live yeast. After 4 weeks a sample of fish from each tank was anaesthetized and weighed to determine the biomass and growth rates for each tank. After 8 weeks all of the fish from each tank were euthanised. Their weight and fork length were recorded. In addition mid-gut from 8 fish from each tank was sampled before being frozen at -70 °C. Feed top dressed with Actisaf was analysed for the presence of live yeast. 16s rDNA profiles from the gut samples underwent T-RFLP to analyse the bacterial population in each group. Growth and length data were analysed using One Way ANOVA and T-RFLP data with Multivariate ANOVA. Overall, it was possible to differentiate the fish with and without probiotic live yeast diets by analysing their gut microbiota. T-RFLP showed that there were major differences in the bacterial population. However there were no differences in growth and length, probably due to the use of a high performance diet, masking differences. Further studies are required to determine any beneficial effects. This study also highlighted the need to optimize the addition of the live yeast to the feed in order to provide the correct dosage to the fish.

Session 39a — Theatre 5

Use of new molecular biology techniques for the evaluation of zootechnical additives

C.J. Newbold
Aberystwyth University, nstitute of Biological, Environmental and Rural Sciences, Cledwyn Building, Aberystwyth University, SY23 3DD, Aberystwyth, United Kingdom; cjn@aber.ac.uk

Microbial biomass produced in the rumen and the resultant fermentation end-products provide the host animal with a substantial proportion of the nutrients required for the production of meat and milk but also result in the production of methane, an important greenhouse gas, and the loss of excess nitrogen in excreta. Given the importance of the rumen fermentation both in terms of the nutrition of the host and in a wider environmental context, it is perhaps not surprising that a great deal of effort has been devoted to investigating methods for manipulating this complex ecosystem. Traditional studies on rumen microbiology have relied on our ability to culture and characterize microorganisms from the rumen. However whilst significant progress has been made using these techniques over the years, it has been recognised that only a relatively small proportion of the microbes within the rumen are recovered by such techniques leaving us ignorant about the roles and activities of the vast majority of the rumen microbial ecosystem. Molecular techniques based on amplification of ribosomal genes have allowed both quantitative and qualitative studies on microbial populations in the rumen to be carried out. I believe that the use of such techniques to help further elucidate on the mode of action zootechnical additives in the rumen. By way of example recent we used a model of sub-acidosis to demonstrated that a supplementation of probiotic yeast could restore a healthy fermentation in the rumen of lactating cow (VFAs, pH, Eh and lactate) and that these improvements were accompanied with a shift in the main fibrolytic group (Fibrobacter and Ruminococcus) and lactate utilising bacteria (Megasphaera and Selenomonas).

Session 39a — Poster 6

Effect of feed additives on rumen pH and protozoa count of cattle fed abruptly high concentrate diet

C.A. Zotti, J.C.M. Nogueira Filho, R. Carvalho, A.P. Silva, T. Brochado and P.R. Leme
Faculdade de Zootecnia e Engenharia de Alimentos, USP, Zootecnia, Av. Duque de Caxias Norte, 225, 13635900 Pirassununga, SP, Brazil; prleme@usp.br

Essential oils from plant extracts and monensin have antimicrobial activity and may decrease the risk of acidosis when the animals suffer an abrupt transition from forage to high concentrate. The purpose of this trial was to verify the rumen pH and protozoa count of twelve Nellore steers with 532±14 kg of BW submitted to abrupt diet change from hay to a high concentrate diet (92% grains) fed *ad libitum* during two periods of 21 days. This diet had no additives added (CON) or with the inclusion of a blend of castor oil acid and cashew oil fed at 400 mg/kg DM (COC) or monensin at 30 mg/kg DM (M30) or monensin at 40 mg/kg DM (M40). Rumen fluid samples were taken six hours post-feeding and rumen pH was recorded every 15 minutes by one indwelling pH probe, both on d5 and d21. Steers fed with CON diet had higher DMI (1.58%BW, P=0.03) than M40 (1.33%BW), however there were no difference between COC and M30 (1.51%BW, P=0.56). The Entodinium was the main protozoa, consisting of 81, 88, 93 and 94% of the protozoa population of CON, COC, M30 and M40, respectively. Animals fed CON diet had the higher protozoa counts (P<0.001) of entodinium, diplodinium, epidinium, isotricha and dasytricha than the others diets. Both monensin diets decreased all the ciliate protozoa counts more than in COC and CON, without difference between the monensin levels (P>0.05). Despite lower protozoa count of the diets with the feed additives, there was no difference in mean ruminal pH (5.81±0.1), in the time below pH 5.8 (732.7±107 min) and below pH 5.2 (121.6±66) among the treatments but, although not statistically significant, time below 5.2 on d5, considered a critical day, was lower for M40 and M30 than CON and COC, respectively 59, 88, 162 and 176 minutes. Feed additives in high concentrate diet fed abruptly suppress the ciliate protozoa and may decrease the DMI, mainly in those fed M40.

Studying effect of adding different level of turmeric to NRC on performance & broilers chick quality

K.H. Ghazvinian[1], A. Mahdavi[1], M.S. Ghodrati[2], B. Roozbehan[2], M.A. Reisdanai[2] and P. Kazeminejad[2]
[1]University, Semnan, Iran, [2]University, Semnan, Iran; khosroghazvinian@yahoo.com

The effect of adding different levels of turmeric to Ross 308 species of broilers chicken NRC was studied on performance and the quality of their carcass. In this experiment 500 one-day-old broilers chicken in completely randomized project include 5 treatments and 4 experimental replicate was used meanwhile in each experimental replicate 25 birds exist. Experimental chickens have been fed with standard NRC according to the balanced 1994 NRC tables. Experimental treatments included levels 0, 250, 500, 750, 1000 / diet turmeric. They were free to use water and food and they received the same amount of energy and protein. In order to study the weight change of carcass organs including liver, pancreas, cecum and length of intestine and the length of cecum and examine abdominal fat, meat of breast and thigh, a chicken out of each replicate has been slaughtered. The result showed significant decrease in cecum and intestine length (P<0.05), in pancreas and liver's weight and also amount of feed intake and percentage of abdominal fat (P<0.05). Experimental treatment did not have any prominent impact on the weight of the spleen and cecum. Also adding turmeric to chicken diet firstly caused a decrease in amount of food intake and increase in feed conversion ratio (FCR) (P<0.05) and also significantly reduced the percentage of breast, thigh and fat (P<0.05).As the comparative item, control treatment has the highest percentage rate breast, thigh and the carcass fat. Results indicate that the addition of turmeric to the diet reduced the amount of feed intake which consequently can result in reducing body weight, pancreas, liver, intestine and the cecum length as well as the performance got reduced. It goes without saying the reduction of abdominal fat percentage was caused by reduction feed intake. In conclusion addition of turmeric to broiler diets did not improved performance and carcass quality.

Acid Buf as natural alternative to monensin in beef feedlot diets

L.J. Erasmus[1], F.M. Hagg[2], R.H. Van der Veen[2], E. Haasbroek[1] and S. Taylor[3]
[1]University of Pretoria, Animal and wildlife Sciences, Lynnwood Rd, 0001 Pretoria, South Africa, [2]Allied Nutrition, Doringkloof, 0001 Pretoria, South Africa, [3]Celtic Sea Minerals, Carrigaline, Cork, Ireland; lourens.erasmus@up.ac.za

Public concern over the emergence of antibiotic resistant bacteria and consumers demand for safe nutritious food has stimulated the search for natural alternatives to replace antibiotic ionophores in ruminant diets. The objective of this study was to determine whether a commercial buffer (Acid Buf, Celtic Sea Minerals) , originating from calcified marine algae, could replace monensin in beef cattle feedlot diets. In Trial 1, 120 newly weaned Bonsmara type male beef cattle (av 225 kg) were randomly allocated to one of two treatments: (1) Monensin (21-33 mg/kg DM); or (2) Acid Buf supplemented at 6 g/kg DM. The same basal diets containing 10.6-11.7 MJ ME/kg DM and 30.0-34.5% starch (DM) were fed during the starter, grower and two finisher feeding phases, the only difference being the supplemental additive. Six pens, each standing 10 animals were randomly allocated to each treatment for a period of 119 days. In a second trial, simulating commercial conditions, 780 recently weaned male beef cattle (av 225 kg) were randomly allocated to 3 pens per treatment (130 animals/pen) for a period of 115 days. Growth, feed intake and health parameters were monitored. Statiscal analyses were performed using ANOVA. In Trial 1 there were no differences (P<0.05) in DMI, FCR or ADG between treatments. In Trial 2, ADG tended (P=0.09) to increase when comparing Acid Buf supplemeted cattle (1.74 kg/d) to monensin supplemented cattle (1.70 kg/d). The FCR (DMI/weight gain) was 5.26 and 5.22 respectively for monensin and Acid Buf supplemented cattle (P>0.10). Percentage healthy rumens, showing no rumen damage, was improved (P<0.01) when Acid Buf (49.2%) replaced monensin (27.1%). Results suggest that a rumen buffer (Acid Buf), could be used as a natural alternative to the ionophore antibiotic monensin, in commercial beef feedlots without impairing animal performance.

Effect of level of Natuzyme® on methane production in diets with varoius forage sources

M. Danesh Mesgaran, E. Parand, A. Faramarzi Garmroodi and A. Vakili
Ferdowsi University of Mashhad, Mashhad, College of Agriculture, Department of Animal Science, 0098, Iran; danesh@um.ac.ir

This experiment aimed to investigate the effects of level of Natuzyme® (Bioproton Co.) on *in vitro* gas production (GP, ml/200 mg DM), dry matter disappearance (IVDMD) methane production (MET, ml/200 mg DM) and fermentation efficiency as mg IVDMD per ml MET (FE) of various ruminant diets containing wheat straw (D1: 6.2% wheat straw, 39.4% corn silage) or alfalfa hay (D2: 21.3% alfalfa, 37.2% corn silage) at half time (t1/2) of gas production. Approximately, 200 mg (DM) of each diet was weighted in a 125 ml serum bottle, while 24 h prior to incubation each bottle received 0.84, 1.68 and 2.52 g/kg DM of the enzyme (E1, E2 and E3, respectively) in an aqueous suspension to maintain same moisture content (40%), run=3 and n=3. The gas production procedure was followed by pipetting buffered rumen fluid into the bottles and incubated at 38.6 °C for desired intervals. In a pre-trail, pressure of gas was recorded at 2, 4, 6, 8, 10, 12, 24, 48, 72 and 96 h of incubation. Pressure data was converted to volume using an experimental curve and was modeled to estimate t1/2. Main trail incubation was continued until t1/2 and volumes of GP and ME, and residual DM was measured. Data were analyzed as 3×2 factorial arrangement in a completely randomized design. Results showed that D1 compared with D2 had higher FE (6.77 vs. 5.59), less GP (33.68 vs. 35.71) and MET (9.20 vs. 10.87), ($P<0.05$). In addition, both E2 and E3 compared with E1 had significantly ($P<0.05$) higher GP (36.70 and 36.75 vs. 30.63) and MET (10.81 and 10.39 vs. 8.91). The IVDMD was significantly ($P<0.05$) higher in E3 than those of the E1 and E2 (33.90 vs. 27.90 and 28.40, respectively). It seems that combination of improved IVDMD and FE using E3 comparing with E1 and E2 could be advantageous but the outcome can vary considering type and forage content of diet.

Effect of a combination of plant extracts on milk persistency and somatic cell counts of dairy cows

C. Gerard[1] and M.L. Le Ray[2]
[1]Invivo NSA, R&D Department, Talhouet, 56250 Saint Nolff, France, [2]NEOVIA feed additives, Talhouet, 56250 Saint Nolff, France; cgerard@invivo-nsa.com

For dairy cows, total milk production per lactation is largely dependent on the shape of the lactation curve, which can be described through 2 main parameters: peak yield and milk persistency after the peak. These parameters are mostly negatively correlated. Hence, finding nutritional strategies improving milk persistency could be a way to enhance productivity of dairy herds. In this context, the specific supply of a plant extracts combination was tested on 24 (control) + 24 (supplemented) dairy Holstein cows (average milk production =30 kg/day) fed with a diet composed of 48% corn silage, 35% pasture, and 17% complete feed. Most of the cows had passed the lactation peak at the beginning of the trial (average Days in Milk =101). Comparisons of milk production data were done through ANOVA, data of milk somatic cell counts (SCC) were analysed through the Chi^2 method, after classification of the samples in 4 groups according to their SCC level. The results showed a statistically relevant higher average milk production for the supplemented group (+ 0.6 kg/day), essentially linked to a strong higher milk production (+2.3 kg/day) observed for the highest producing cows (initial milk production over 30 kg). Even if milk fat and milk protein contents were slightly lower for the supplemented group, total milk protein and fat exportations were not affected by the plant extracts supply. When only the highest producing cows were considered, milk protein production was even slightly enhanced (+ 4%). In terms of SCC, the proportion of milk samples containing more than 250,000 SCC was significantly lower for the supplemented group (9 vs. 30% for the control group) during the trial period. This trial showed that the use of specific plant extracts could improve milk production through an enhancement of milk persistency after the peak, especially for high producing cows, and could have beneficial effects on milk SCC levels.

Assessment of microorganisms as zootechnical feed additives in the European Union

R. Brozzi, C. Roncancio Peña and J. Galobart
European Food Safety Authority, Feed Unit, Via Carlo Magno 1/A, 43121 Parma, Italy; rosella.brozzi@efsa.europa.eu

In the European Union (EU) feed additives must undergo an authorization procedure as provided by Regulation EC (No) 1831/2003 before being placed on the market. The European Food Safety Authority (EFSA) is responsible for assessing the safety and efficacy of feed additives. Based on EFSA's opinion, EU risk managers will grant or deny the authorization for use of the product. The assessment is based on the technical dossier prepared by the applicant and focuses on the identification and characterisation of the additive, the safety and the efficacy of its use. For microorganism based-additives, the origin and the means to identify them must be detailed. They should not produce antibiotics nor contribute to the spread of antibiotic resistance and the absence of toxin production and/or virulence determinants should also be demonstrated. The additive must be safe for the target animals, consumers of food derived from animals fed the additive, users and environment. The Qualified Presumption of Safety (QPS) approach can be applied to certain microorganisms and it requires the full identification and any qualification set to be met. QPS microorganisms are presumed safe for the target animals, consumers of food from animals fed the additive and the environment. In the remaining cases specific safety studies are requested. Tolerance studies with the target species/categories are needed to demonstrate the absence of adverse effects at use level. The safety for consumers is assessed by means of genotoxicity studies and a subchronic oral toxicity study. For user safety information on the potential for skin and eye irritation and skin sensitisation should be provided. Microorganisms are generally considered as respiratory sensitizers. No risk for the environment is envisaged if the active agent is a natural component of the environment. Efficacy has to be demonstrated in each target species/categories by at least three trials showing positive effects relevant to the claim made.

Addition of selenium and vitamin E in diet increases NK cell cytotoxicity in cattle

G. Greghi, A. Saran Netto, A. Latorre, E. Raspantini, M. Zanetti and L. Correa
FZEA/USP, 225, Duque de Caxias Norte st, 13635-900 Pirassununga, Brazil; saranetto@usp.br

NK cells play an important role in innate immunity and in the shaping of adaptive cellular immune responses. The study is not intended to prevent cancer in Nellore, because this is not a common disease, but the objective of this study was to evaluate the effect of including selenium, vitamin E and canola oil in the diet of feedlot cattle on the cytotoxic activity of peripheral NK cells, 48 Nelore bulls of approximately 2 years of age, received the following diets for 84 days: (1) control (Co); (2) Selenium + Vit. E (SE) – addition of 2.5 mg of Se and 500 UI of Vit. E/kg of total DM; (3) Selenium + Vit. E + Canola (SEC) – addition of 3% of canola oil, 2.5 mg of Se and 500 UI of Vit. E/kg of total DM; (4) Canola (C) – addition of 3% of canola oil/kg of total DM. After 70 days of supplementation, blood samples were collected from each animal in heparinized tubes. The leukocytes were separated by the Ficoll-Paque density gradient after centrifugation at 400×g/40 min. at 18 °C. These effector cells were adjusted and incubated with target cells BL3 tumor-1, previously marked by the fluorophore CFSE in a 100:1 (Effector:Target) ratio in triplicate. After 3 h incubation in a humidified stove at 37 °C and 5% CO_2, 40µl of propidium iodide (PI) [12.5 ug/ml] was added to identify the dead target cells by flow cytometry. Data were analyzed by FlowJo and presented as percentage. The percentage of cells killed by BL3-1 cytotoxicity of peripheral NK cells was 2.18±1.04 for (Co), 3.35±0.58 for (SE), 3.10±1.77 for (SEC) and 3.35±1.77 for (C). Supplementation with Se + Vit.E increased the cytotoxicity of peripheral NK cells (P=0.0091, Co vs. SE, Mann Whitney test). Thus, it may be concluded that Nellore beef cattle supplemented with Se + Vit.E are less susceptible to bacterial, viral and parasitic infections, because supplementation with selenium and vitamin E above the required minimum values increases the cytotoxicity of peripheral NK cells.

Concerns of different stakeholder groups about pig husbandry

T. Bergstra, H. Hogeveen and E.N. Stassen
Wageningen University, Hollandseweg 1, 6706 KN Wageningen, the Netherlands; tamara.bergstra@wur.nl

Concerns about pig husbandry differ between stakeholder groups, e.g. citizens and pig farmers. Differences in these concerns result in friction between stakeholder groups and their demands for pig husbandry. This friction results in little support from citizens for measures in pig husbandry implemented by pig farmers. To find a balance in support for pig husbandry of different stakeholder groups, concerns about pig husbandry of these groups have to be defined. The objective of this study was to determine concerns about pig husbandry of citizens, conventional pig farmers, organic pig farmers, pig veterinarians and pig yard entrants, e.g. feed advisers and accountants, and compare these concerns. A questionnaire was distributed with questions about different entities of pig husbandry, i.e. animal, human (animal keeper and consumer) and environment. For each entity, respondents could indicate how much additional care, i.e. extra attention compared to the current situation, they found necessary for aspects related to pig husbandry. Data of the questionnaire was analyzed with ordered multinomial regression. Results showed that for several aspects different stakeholder groups indicated different levels of additional care. For almost all aspects citizens agreed significantly with organic pig farmers in that additional care or maximal additional care was necessary. For aspects of the animal these stakeholder groups almost always disagreed significantly with conventional pig farmers, pig veterinarians and yard entrants. Conventional pig farmers and yard entrants often agreed significantly, but not always agreed significantly with veterinarians in the level of additional care. The indicated level of additional care can be representative for levels of concern, as it is likely that respondents with higher concerns give higher additional care levels. It can be concluded that citizens and organic pig farmers have different concerns about pig husbandry than conventional pig farmers, yard entrants and veterinarians.

Societal conformity of European pig production systems

K.H. De Greef[1], H.W.J. Houwers[1] and M. Bonneau[2]
[1]Wageningen UR Livestock Research, P.O. Box 65, 8200 AB Lelystad, the Netherlands, [2]INRA UMR1079, SENAH, 35590 Saint Gilles, France; karel.degreef@wur.nl

A tool was developed to assess the degree of social acceptance of pork production systems. Societal conformity was defined as the degree to which production systems meet the demands and expectations of society. The tool assesses judgements of informed professionals on nine sustainability themes: Animal health, Animal welfare, Economic sustainability, Environmental Impact, Genetic diversity, Human working conditions, Meat safety, Meat quality, and the overall term Public image. An explicit distinction is made between stakeholders involved in the production chain (Insiders) and those around the chain (NGO-representants, ' Outsiders'). The tool was applied in 5 countries, assessing the stakeholder views of the conventional pork production system (farm level) and two diversified systems in each country. Both Outsiders and Insiders judged the overall sustainability of the conventional pork system as worrisome. Both stakeholder groups expected considerable improvements in sustainability in diversified systems. Virtually no undesired effects of systems changes were predicted. Outsiders were more outspoken in their views, both for the level of sustainability of the conventional system and for the degree of improvement of the diversified systems. Results showed consistency between the sustainability themes and the parameter Public Image, although the results suggest that the underlying views of Outsiders is of another nature than those of the Insiders. From the study, it can be concluded that there is a deficit in social acceptance of European conventional pork production systems, which can be met with available contrasting systems, virtually irrespective of their contrast to the conventional system.

Long-lasting effects of early life factors on immune competence of pigs

D. Schokker[1], J. Zhang[2], H. Smidt[2], J.M.J. Rebel[3] and M.A. Smits[1,3]
[1]Wageningen UR Livestock Research, Genomics, Edelhertweg 15, 8200 AB Lelystad, the Netherlands, [2]Wageningen University, Laboratory of Microbiology, Dreijenplein 10, 6703 HB Wageningen, the Netherlands, [3]Central veterinary Institute, Infection Biology, Edelhertweg 15, 8200 AB Lelystad, the Netherlands; dirkjan.schokker@wur.nl

Colonization of microbiota as well as competent immune development are affected by early-life environmental variation. Previously we reported on the immediate effects of an antibiotic treatment and an antibiotic treatment in combination with stress on day 4 after birth on microbiota composition and immune development in 8 day-old piglets. Here we report on the long-lasting effects (day 55 (~8 weeks) and 176 (~25 weeks) after birth) of this treatment on microbiota composition and immune competence. To investigate the long-lasting effect of early-life treatment with antibiotics and/or stress in piglets, the following experiment was performed. Piglets were divided into three different groups receiving the following treatments: (1) no antibiotics and no stress; (2) no antibiotics and stress due to handling; and (3) antibiotics and stress due to handling, all these differences in the management conditions were applied at day 4 after birth, which is frequently used in intensive pig husbandry systems. Sampling was performed at day 55 and day 176 after birth. The results indicate that antibiotic treatment and/or stress due to handling at day four after birth affect the expression immune related genes in intestinal tissue later in life. However, no effect is observed in the microbiota composition at later age. This suggests that both treatments may have an effect on immune capacity, most probably due to the differences in early colonization of the gut by microbiota and therefore the programming of the immune system.

Vitamin A and colour parameters in pig's fat as possible biomarkers of feeding traceability

R. Álvarez[1], A.J. Meléndez-Martínez[2], I.M. Vicario[2] and M.J. Alcalde[1]
[1]University of Seville, Dept. Agricultural and Forestry Science, 41013-Sevilla, Spain, [2]University of Seville, Food Colour & Quality Laboratory, Dept. of Food Science and Nutrition, 41012-Sevilla, Spain; ralvarez1@us.es

The aim of this study was to assess the validity of different carotenoids, vitamin A (retinol) and colour parameters (L*, a*, b*, C* and h°) in pig's fat as biomarkers of feeding traceability. Forty-five animals divided into three groups were considered: Iberian breed pigs reared in a Montanera extensive system, with a diet based on acorns and pasture (group 1), Iberian breed pigs fed indoors with a diet based on concentrate (group 2) and pigs from a commercial crossbred receiving, indoors, a diet based on concentrate. Carotenoids were analyzed, by HPLC, both in the diets an in the renal fat. In addition, retinol and colour parameters were measured in fat. Carotenoids were not detected in acorns, in all the other feedstuffs, lutein was detected, being its concentrations higher ($P<0.01$) in pasture than in concentrate samples. In addition, β-carotene, (9Z)-β-carotene, violaxanthin and (9'Z)-neoxanthin were also detected in pasture samples (group 1). The concentrate samples (groups 2 and 3) showed a similar carotenoid profile. In the fat samples carotenoids were not detected but retinol was. Significant differences ($P<0.001$) were found in retinol concentration for group 3 (6.02±0.28 mg/g fat) relative to the other two groups (4.09±0.36, group 1; 3.69±0.22, group 2). Thus, retinol seemed to differentiate the pigs according to their genetic but not according to their diet. On the other hand, all the colour parameters showed significant differences ($P<0.001$, $P<0.01$) for the different groups of animals studied. Additionally, considering all these colour parameters in a discriminant analysis, 78.9% of the animals were correctly classified according to their diet. Therefore, it can be claimed that colour of fat might be useful to differentiate pigs according to their diet from a traceability point of view.

siMMin™: on line software tool to simulate zinc balance in feeding programs of growing pigs

S. Durosoy[1] and J.-Y. Dourmad[2]

[1]Animine, 335 chemin du noyer, 74330 Sillingy, France, [2]INRA-Agrocampus Ouest, UMR1348 Pegase, Domaine de la Prise, 35590 Saint-Gilles, France; sdurosoy@animine.eu

Copper and zinc are essential nutrients that are usually supplied above nutritional requirements in pig diets. As a result, high Zn and Cu concentrations are found in animal wastes, which poses a risk of soil accumulation when pig manure is spread on arable land. Moreover, technological treatments of pig slurry concentrate zinc in the solid fraction and this by-product may exceed maximal authorized Zn content when used as organic fertilizer. The mass balance approach has been recently updated to measure the excretion of heavy metals (Cu, Zn) from pig production. siMMin™ has been developed with the support of INRA with the following objectives: (1) the software should be intuitive and easily usable in any pig farm whatever the conditions; (2) it focuses on the pig growing life, from the weaning until slaughter; (3) calculation can be adapted to each user, taking into account farm variables of pig growth performance and feeding program; (4) the software enables to simulate changes in each variable compared to the existing situation, in order to measure the rate of improvement in the total reduction of zinc in the life of the growing pig and the change in Zn concentration of animal waste; (5) the calculator tool benchmarks any situation to existing EU regulation. The software is on line since January 2013 at www.animine.eu/simmin and can be utilized by all stakeholders involved in pig production.

Effects of diet microbial phytase, vitamin C and copper levels on cadmium retention in pigs

E. Royer[1] and N. Lebas[2]

[1]Ifip-institut du porc, 34 bd de la gare, 31500 Toulouse, France, [2]Ifip-institut du porc, Les Cabrières, 12200 Villefranche de Rouergue, France; eric.royer@ifip.asso.fr

Recent calculations of a limited safety margin between the cadmium dietary exposure and health-based guidance values have heightened the need to reduce exposure to cadmium in European food products. An experiment was undertaken to determine the effects of microbial phytase, vitamin C and copper dietary contents on cadmium accumulation in pig kidneys. From 13.5 kg live weight, 36 female pigs (LW×LD × LWxPiétrain) were assigned to control diets or to experimental diets containing cadmium-contaminated wheat and sunflower meal without phytase (PHOS), or with 1000 FTU/kg phytase (PHYT), or with 1000/kg FTU phytase, 1000 then 700 mg/kg vitamin C and low copper content (44 then 17 mg/kg) (CuVitC). Experimental diets had Cd concentrations ranging from 0.54 to 0.72 mg/kg and were given *ad libitum* for the phase 2 period (27 days of exposure) or the phase 2 and growing periods (69 days) before returning to the control diets, or for the whole fattening period (132 days). All pigs were slaughtered on the same day at an average body weight of 113.1 kg. Cadmium content in the kidney was significantly increased by the contaminated diets ($P<0.001$) and by the duration of exposure ($P<0.001$). However, a significant variability was found as the kidney cadmium levels varied on average by a factor of two between individuals given the same treatment. Kidney cadmium concentration was slightly but not significantly lower in PHOS pigs than in PHYT pigs ($P=0.14$). Pigs fed the CuVitC diets had lower cadmium level in kidney in comparison to the pigs fed the PHYT diets after 69 days ($P<0.05$) or 132 days of exposure ($P<0.05$). Management of calcium, phosphorus and phytase levels, reduction in copper content and supplementation with vitamin C are factors that could help to limit cadmium accumulation in the kidneys of exposed pigs.

Influence of dietary calcium on growth performance and mineral status in weaned piglets

P. Schlegel and A. Gutzwiller
Agroscope, Tioleyre 4, 1725 Posieux, Switzerland; patrick.schlegel@alp.admin.ch

Calcium (Ca) and phosphorus (P) are essential minerals, closely linked in their digestive processes and metabolism. With the generalized use of low P diets containing exogenous phytase, the interest in the optimal dietary Ca level has increased. Two experiments were conducted to study the dose response effect of Ca with, respectively 6 or 8 g and 4, 7 or 10 g Ca/kg diet using, respectively 34 and 36 28-day old weaned piglets for respectively 6 and 5 weeks. In each experiment, piglets were blocked by BW, gender and litter. The piglets were housed in 4, respectively 3 pens, each equipped with a computer controlled feeding station. The basal diets contained, per kg, 14 MJ DE and per experiment, respectively, 170 and 186 g CP, 4.1 and 4.2 g P, 1,300 and 650 FTU microbial phytase and 3.0 and 2.9 g digestible P. The data was submitted to an analysis of variance $Y = a + a_{block} + Ca + Ca*Ca + e$, whereas the a_{block} is the effect of the animal block on the intercept a. The piglet was considered as the experimental unit. With increasing dietary Ca, body weight gain (linear, P=0.05, R^2=0.51) and feed efficiency were deteriorated (linear, P<0.001, R^2=0.57); Urinary (mol/mol creatinine) Ca increased (linear, P<0.001, R^2=0.84) and urinary P decreased (quadratic, P<0.001, R^2=0.69); serum Ca increased (linear, P<0.001, R^2=0.75) and serum P decreased (linear, P<0.001, R^2=0.70); bone ash (quadratic, P<0.10, R^2=0.41) and bone breaking strength (quadratic, R^2=0.71) both increased to reach a plateau at 6.5 g Ca/kg diet; bone density was not affected (P>0.05). Finally, the present data indicate that: (1) the lowest dietary Ca level of 4 g/kg was insufficient as urinary P was lost to a high extent over urine; (2) at a dietary Ca >8 g/kg, the serum P level fell below the lower critical concentration of (2.5 mmol P/l); (3) for a maximal growth performance and for optimal bone strength, dietary Ca was respectively<6.0 and 6.5 g Ca/kg.

Cereal extrusion affects the volatile fatty acid concentration in the caecum of piglets and poultry

D. Torrallardona, E.A. Rodrigues and M. Francesch
IRTA, Monogastric Nutrition, Ctra. Reus, El Morell km 3.8, 43120 Constantí, Spain;
david.torrallardona@irta.es

Thirty two newly weaned piglets (6.9±0.83 kg; 26 days of age) and 160 newly hatched chickens were offered four dietary treatments arranged according to a 2×2 factorial design, with cereal nature (rice or barley) and cereal extrusion (with or without) as main factors. Diets contained 55% of the cereal under study and they were offered for 27-28 days (piglets) or 24-25 days (chickens). At the end of the trial, the animals were humanely killed and samples were obtained for the determination of volatile fatty acid concentration (VFA) in the caecal digestive contents. For piglets, individual samples of caecal digesta were considered, whereas for chicken pools of caecal digesta from 5 animals were used. Caecal VFA concentrations (all values in μmol/g) in chickens were higher (P<0.05) than in piglets for formic (0.8 vs. 0.1), butyric (14.3 vs. 7.7), lactic (3.4 vs. 0.5) and succinic (4.2 vs. 0.2) acids, whereas they were higher (P<0.05) in piglets than in chickens for propionic (16.6 vs. 6.9), isobutyric (2.9 vs. 1.4), isovaleric (1.6 vs. 0.3) and valeric (2.0 vs. 0.9) acids. In piglets, no effects of cereal source on VFA were observed, but cereal extrusion reduced (P<0.05) the concentrations of propionic (20.0 vs. 13.2) and butyric (11.0 vs. 4.4) acids and tended to reduce total VFA (77.2 vs. 51.0; P=0.06). In chickens, significant differences (P<0.05) were observed between rice and barley for acetic (55.8 vs. 39.7), propionic (8.6 vs. 5.6), isobutyric (1.9 vs. 1.0), isovaleric, (0.5 vs. 0.2), valeric (1.2 vs. 0.7), lactic (1.2 vs. 5.6) and succinic (2.2 vs. 6.2) acids. On the other hand, extrusion reduced (P<0.05) the concentrations of acetic acid (56.8 vs. 38.8), butyric (17.6 vs. 11.3) and total VFA (93.8 vs. 66.9). It is concluded that VFA concentration i the caecum are affected by the animal species, the nature of cereal (only in chickens) and the process of cereal extrusion.

Effects of supplying extra milk on production in suckling piglets on a high production farm

A. Bulens[1,2], S. Van Beirendonck[2], J. Van Thielen[2], N. Buys[1] and B. Driessen[2]
[1]KU Leuven, Department of Biosystems, Kasteelpark Arenberg 30, 3001 Heverlee (Leuven), Belgium, [2]KU Leuven|Thomas More Kempen, Kleinhoefstraat 4, 2440 Geel, Belgium; anneleen.bulens@khk.be

Optimising production is an important aspect in pig farming. Despite the fact that nowadays sows are able to reach a high litter size born alive, postnatal piglet mortality remains a problem in many pig farms. Due to the increased litter size, the birth weight of individual piglets is lower and as a consequence, the chances of survival of certain piglets are very low. It is therefore important that every piglet can consume enough milk. To ensure this, additional milk can be supplied in the farrowing crate, for example by introducing Rescue Cups. This means an extra cost for the pig farmer and the effects on production parameters of the piglets must be considered. To this end, 1,710 piglets from 127 sows were followed during a suckling period of 22 days, from which 854 piglets (from 63 sows) had access to extra milk using Rescue Cups (treatment group) and 856 piglets (from 64 sows) had no access to extra milk (control group). The piglets in the treatment group had access to extra milk from the age of 4 days until the age of 15 days. Birth weight, weaning weight and piglet mortality were recorded as production parameters. Data were analysed using the generalized linear mixed model. No significant difference was found in piglet mortality. The mean mortality rate however was already very low (7.53%). This shows that the management in general at the pig farm in this study was already good. There was no difference found in daily growth rate. The uniformity within the litters does not seem to increase when extra milk was provided. This might be due to the fact that not only small piglets drink milk from the cups, but also heavier piglets. There was also no difference found in weaning weight between treatment and control group within the group of piglets with the lowest weaning weight.

Comparative study to analyse effects on sows and piglets performance by providing supplemental milk

J. Pustal[1], I. Traulsen[2], R. Preissler[1], K. Müller[3], T. Große Beilage[4], U. Börries[4] and N. Kemper[1]
[1]Institute of Agricultural and Nutritional Sciences, Martin-Luther-University Halle-Wittenberg, Theodor-Lieser-Str.11, 06120 Halle, Germany, [2]Institute of Animal Breeding and Husbandry, Christian-Albrechts-University Kiel, Olshausenstr. 40, 24098 Kiel, Germany, [3]Education and Research Centre Futterkamp of the Chamber of Agriculture Schleswig-Holstein, Futterkamp, 24327 Blekendorf, Germany, [4]Boerries GmbH&Co. KG, Mühlenberg 17, 49699 Lindern, Germany; nicole.kemper@landw.uni-halle.de

Providing supplemental milk in addition to sow's milk in the farrowing crate is one practice to support rearing large litters. The aim of this study was to compare a supplemented group (n=60 sows), where piglets had free access to supplemental milk, provided by special cups, from the 2nd day of life on, with a non-supplemented control group (n=60 sows). In accordance with animal welfare requirements, sows of the supplemented group retained as many piglets as they had functional teats, whereas sows of the control group retained one piglet less than they had functional teats. Body weight (BW) of both the sows and the piglets, backfat thickness (BT) and body-condition-score (BCS) of the sows were analysed using a generalized linear mixed model involving fixed effects (group, batch, parity number), random effects (sow), and covariates (duration of suckling period/week). In the supplemented group, 13.5 piglets, and in the control group, 12.4 piglets were weaned. Piglets had the same weaning weight in both groups (7.8 kg), even though one more piglet had to be fed in the supplemented group. Piglets consumed averagely 0.2 l supplemental milk per day. Regarding the total weaning weight of the litter, significant differences were apparent (supplemented group=104.9 kg, control group=96.7 kg). The decreases of BW, BT and BCS were not significantly different between sows of the control and the supplemented group. Feed intake of sows was equal in both groups. In summary, supplementing milk supports fostering large litters.

Fattening and slaughter performance as related to meat quality in Polish Landrace pigs

G. Żak and M. Tyra
National Research Institute of Animal Production, Department of Animal Genetics and Breeding, Sarego 2, 31-047 Krakow, Poland; grzegorz.zak@izoo.krakow.pl

The high share of pork in total meat consumption in Poland (around 42 kg per capita) and consumer demands regarding its quality represent a major challenge to pig breeders. The principal goal is to find ways of producing pork characterized by good taste, culinary, technological and dietetic qualities while maintaining high lean meat percentage. These targets can be achieved mainly through efficient breeding work and based on the knowledge of relationships between traits being improved. The experiment with 130 Polish Landrace pigs sired by 21 males was carried out in a pig testing station. Animals were slaughtered at 100 kg of body weight. Slaughter was followed by dissection of carcasses and measurement of the analysed traits according to the testing station method. Analysis of the results showed that coefficients of correlation between fattening traits and meat quality parameters were rather low. The highest coefficient was obtained between feed consumption and water holding capacity (r=-0.22) and between daily gain and intramuscular fat (IMF) (r=0.21). A statistically significant correlation was also observed between feed conversion (kg/kg gain) and meat colour lightness L* (r=0.19). Analysis of the relationships between slaughter traits and meat quality showed that loin eye area and carcass meat content correlate the highest with a* and b* colour values of meat (r=0.36-0.43). These correlations are highly significant. Slightly lower but still significant correlations were noted between loin eye area and: pH45 of loin (r=0.24), water holding capacity (r=-0.21), and pH24 of loin (r=0.20). Carcass lean meat content was significantly correlated to IMF content (r=-0.21), L* colour (r=0.17) and pH45 of loin (r=0.15). These correlations, however, were not high. It is also worth noting a very low IMF level (1.34%) in meat (m. longissimus dorsi) from the Polish Landrace pigs studied. Supported by the National Centre for Research and Development, Grant No. N R12 0059 10.

Level of fatty acids in meat from Polish Landrace pigs and their association with slaughter traits

G. Żak[1], M. Pieszka[1] and W. Migdał[2]
[1]National Research Institute of Animal Production, Sarego 2, 31-047 Krakow, Poland, [2]University of Agriculture, Department of Animal Products Technology, Balicka 122, 31-149 Krakow, Poland; grzegorz.zak@izoo.krakow.pl

Research concerning the effect of fatty acid composition on meat flavour showed that it correlates positively with saturated fatty acids, and negatively with unsaturated fatty acids. Polyunsaturated fatty acids are highly desirable from a consumer perspective because they improve the dietetic value of meat, but their excessive amounts in animal fat compromise the flavour and aroma of meat and its storability. This study analysed meat from 100 Polish Landrace pigs. The level of fatty acids in m. longissimus dorsi and m. semimembranosus was investigated. The level of saturated fatty acids (SFA) was found to be significantly lower in m. semimembranosus (34.6%) compared to m.l.d. (35.89%). In m.l.d., the level of unsaturated fatty acids (UFA) was 63.91%, with n-6 PUFA of 21.83% and n-3 PUFA of 1.31%. In m. semimembranosus, the respective values were higher at 65.09%, 25.95% and 1.60%. Differences between the muscles were statistically significant. The level of fatty acids in m. semimembranosus was slightly more favourable than in m.l.d. The n-6 to n-3 PUFA ratio was 15:1 in m.l.d. and 16:1 in m. semimembranosus. According to WHO recommendations, the optimal n-6/n-3 fatty acid ratio in the human diet should average 5:1. Correlations were also analysed between the level of fatty acids and slaughter traits. In loin, SFA and UFA correlated the highest with loin eye area (r=0.17-0.18) and n-3 PUFA with backfat thickness (r=-0.13), but these correlations were not significant. In ham, the highest correlations were between n-3 PUFA and backfat thickness (r=-0.32), and between UFA and loin eye area (r=0.28). These correlations were statistically significant. No correlation was found between the level of fatty acids and daily weight gain. Supported by the State Committee for Scientific Research within project No N N311 1636 37.

Session 39b Poster 13

Analysis of relationships between intramuscular fat level and selected fattening, slaughter and meat

M. Tyra and G. Żak
National Research Institute of Animal Production, Department of Animal Genetics and Breeding, Sarego 2, 31-047 Krakow, Poland; miroslaw.tyra@izoo.krakow.pl

The aim of the study was to estimate coefficients of heritability for intramuscular fat (IMF) content and other fattening, slaughter and meat quality traits of the pig breeds raised in Poland. In addition, genetic correlations were estimated between IMF content and a group of fattening, slaughter and meat quality traits, which enables this parameter to be included in the BLUP estimation of breeding value. The experiment used Polish Landrace (PL), Polish Large White (PLW), Puławska, Hampshire, Duroc, Pietrain and line 990 animals. A total of 4430 gilts of these breeds, tested at Pig Performance Testing Stations (SKURTCh), were investigated. Heritability of IMF was at intermediate level for the two most common breeds raised in Poland (h2=.318 for PLW, h2=.291 for PL). In the group of meat quality traits, high heritability was noted for meat colour lightness (L*) measured by Minolta (from h2=.453 to h2=.572). No relationships were found between IMF level and indicators of fattening performance. The highest value observed in this group of traits concerned the genetic relationship with daily feed intake (rG=.227) for the entire group of animals. For the PLW and PL breeds, these relationships were with feed conversion (kg/kg gain) (rG=.151 and rG=.167, respectively). One of the higher relationships observed were genetic correlations with water holding capacity (above rG=-.3) and, for the PLW and PL breeds, with meat redness (a*), which amounted to rG=.155 and rG=.143, respectively. Supported by the National Centre for Research and Development, Grant No. N R12 0059 10.

Session 39b Poster 14

Plasma oxidative status in piglets changes upon weaning

J. Degroote[1,2], W. Wang[1], C. Van Ginneken[3], S. De Smet[1] and J. Michiels[1,2]
[1]Ghent University, Proefhoevestraat 10, 9090 Melle, Belize, [2]University College Ghent, Valentin Vaerwyckweg 1, 9000 Gent, Belgium, [3]University of Antwerp, Universiteitsplein 1, 2610 Wilrijk, Belgium; jeroen.degroote@hogent.be

Weaning is a critical process which results in an increased susceptibility to diseases. Some authors indicate that weaning also influences the oxidative status of piglets. Nevertheless, factors altering the piglet oxidative status are poorly described. The aim of this study was to assess the effect of birth weight, weaning treatments and days post-weaning on the plasma oxidative status in piglets. Therefore, newborns were weighed within 24 h after birth. Ninety pairs of low birth weight (LBW; 0.85±0.091 kg) and normal birth weight (NBW; 1.34±0.177 kg) sex-matched littermates were selected and assigned to one of three weaning treatments; i.e. weaning at 3 weeks of age (19.6±0.50 d), at 4 weeks of age (26.5±0.50 d) and removal from the sow at 3 days of age and fed a milk replacer until weaning at 3 weeks of age (19.8±0.38 d). After weaning, piglets were fed a starter diet *ad libitum* and were sampled at 0, 2, 5, 12 and 28 days post-weaning. Plasma samples were collected to analyze glutathion peroxidase (GSH-Px) activity, malondialdehyde (MDA) concentration, and the ferric reducing antioxidant power (FRAP) as markers of oxidative stress. Data were analyzed by linear models. Results showed significant effects on plasma GSH-Px activities of weaning treatment (P=0.021), days post-weaning (P<0.001) and the interaction term of these two main factors (P=0.039). GSH-Px activities peaked on day 5 post-weaning whereby the smallest increase was found in the 3w treatment. MDA concentrations were also altered by days post-weaning (P=0.001) and followed an opposite pattern as GSH-Px. FRAP values were significantly different between weaning treatments (P=0.01), with lower values for the 4w treatment. Unlike weaning treatment and age post-weaning, birth weight had minor effects on these parameters. These findings provide new insights into the complex changes in response to weaning.

Session 39b Poster 15

Labour time required for piglet castration with isoflurane anaesthesia

S. Weber[1], G. Daş[1], K.H. Waldmann[2] and M. Gauly[1]
[1]University of Göttingen, Department of Animal Sciences, Albrecht-Thaer-Weg 3, 37075, Göttingen, Germany, [2]University of Veterinary Medicine Hannover, Clinic for Swine and Small Ruminants, Bischofsholer Damm 15, 30173, Hannover, Germany; gdas@gwdg.de

Isoflurane anaesthesia combined with an analgesic represents a welfare-friendly method for castration of piglets. However, it requires an equipped inhaler device, which is unprofitable for small farms. Thus sharing a device among several farms may be an economical option if the shared use does not increase labour time and resulting costs. Therefore, this study aimed at investigating the amount and components of labour time required for isoflurane anaesthesia performed with stationary and shared devices. Piglets (n=1,579) from 12 organic farms were anaesthetised with isoflurane and castrated using either stationary or shared devices. The stationary devices were used in a group (n=5) of larger farms (84 sows), whereas smaller farms (n=7; 32 sows) shared one device. Each farm was visited four times and labour time required for each defined process step was recorded. The complete process included the machine set-up, anaesthesia and castration by a practitioner, and preparation, collection and transport of piglets by the farmer. Labour time required for the complete process was increased (P=0.012) on farms sharing a device (266 s/piglet) compared to farms using stationary devices (177 s/piglet). The increase was due to elevated time spent for preparation (P=0.055), castration (P=0.026) and packing (P=0.010) when sharing a device. However, on a percentage base, components of the total time budgets of farms using stationary or shared devices did not differ significantly (P>0.05). Cost arising from the time spent by farmer did not differ considerably between the use of stationary (0.26€ per piglet) and shared (0.28€) devices. It is concluded that costs arising from the increased labour time of the shared use of the device can be considered marginal, since the high expenses that originate from purchasing an inhaler device are shared among several farms.

Session 39b Poster 16

Influence of gonadal status on nutrient mobilization during an inflammatory challenge in pigs

E. Merlot, C. Leclercq, N. Le Floc'h and A. Prunier
INRA, UMR1348 PEGASE, 35590 Saint-Gilles, France; elodie.merlot@rennes.inra.fr

Male pigs are surgically neonatally castrated or immunocastrated before puberty to prevent boar taint in the meat. However, sex hormones modulate muscle and adipose tissue metabolism, but also immune function. Thus castration methods might modify nutrient partitioning during an energetically costly inflammation, with a possible impact on animal resilience. In this study, the inflammatory response of entire (E, n=7) pigs, neonatally surgically castrated (SC, n=9) pigs, pigs immunized against GnRH at 82 and 117 days (immunocastration, IC, n=9), and neonatally surgically castrated pigs immunized against GnRH (SIC, n=7) were compared. LPS (O55:B5, 15 µg/kg) was administered i.v. at 149 days of age, 6 h after the daily meal. The cytokine (IL-1, IL-6, TNF-α), cortisol, catecholamines (CAs) and metabolic (plasma glucose, lactate, free fatty acid (FFA), amino acids (AA) and urea) responses were investigated in blood samples collected through a jugular catheter from -30 to +1,440 min post LPS. Cytokine, CAs and lactate rose similarly in the 4 groups after LPS. In neonatally castrated pigs, cortisol increased more from 60 to 420 min and Lys, Val and Thr (at all time points), Leu (from -1 to 180 min) and Ile (at 60 and 180 min) was higher and Asp and Hypro was lower (at all time points) than in neonatally intact pigs (SC+SIC vs. E+IC). Immunocastration delayed the rise in FFA (240-360 min) after LPS, increased plasma levels of Lys and His, and decreased those of Hypro (IC+SIC vs. E+SC). Entire pigs had a more pronounced hypoglycaemia at 300 min and lower urea levels from -30 to 420 min relatively to the 3 other groups. Thus non-surgically castrated pigs had more difficulties to face the metabolic demand due to the inflammatory reaction, maybe because of their lower cortisol release. The effects of immunocastration that were not observed after surgical castration might result from a long term effect of immune stimulation by vaccination rather than from sex steroid suppression itself.

The analysis of technological pork quality by using the diet with the addition of organic chromium

O. Bučko, A. Lehotayová, J. Petrák, O. Debrecéni and M. Margetín
Slovak University of Agriculture in Nitra, Tr. Andreja Hlinku 2, 949 76 Nitra, Slovak Republic; ondrej.bucko@uniag.sk

The aim of the experiment was to determine the influence of the addition of organic chromium on the technological quality of pork. DNA tests have detected genetic marker RYR 1 (malignant hyperthermia syndrome) and all pigs corresponded to the NN genotype. The pigs were divided in a control group of 20 pigs (10 barrows, 10 gilts) and the experimental group of 20 pigs (10 barrows, 10 gilts). The control group was fed by a standard feed ration, it consisted of 3 feed mixtures, which were used at the different growth stages. The experimental group was fed by the same feed mixtures in the same growth stages as the control group, whereby the mineral-protein premix was used in all 3 feed mixtures and it was enriched by 750 μg/kg chromium nicotinate by the inactivated yeast Saccharomyces cerevisiae fermented on the substrate, which was from the natural sources with the higher content of trivalent chromium. From the indicators of the technological quality was evaluated the actual acidity 45 min p.m. and 24 h p.m., drip loss 24-48 h p.m., the colour of meat 24 h p.m. and 7 days p.m. in CIE L*, a*, b* and the Warner-Bratzler shear force. The results showed that there were no significant differences between the experimental and the control group in the parameter pH_1 and pH_{24}. It was found that the chromium did not have significant effect on the drip loss, the colour 24 h p.m. and the shear force. The differences between the indicators of the colour CIE a * and CIE b * were statistically significant ($P \leq 0.05$). In conclusion obtained results showed that the diet with the addition of organic chromium did not have a significant effect on the parameters of technological quality of meat with exception of the colour 7 days p.m. which may be related to oxidative stability. However, this requires further study. This work was supported by projects VEGA 1/0493/12, VEGA 1/2717/12, ECACB – ITMS 26220120015 and ECACB Plus – ITMS: 26220120032.

The analysis of technological pork quality by using the diet with the addition of organic zinc

O. Bučko, A. Lehotayová, J. Petrák, O. Debrecéni and J. Mlynek
Slovak University of Agriculture in Nitra, Department of Special Animal Husbandry, Tr. Andreja Hlinku 2, 949 76 Nitra, Slovak Republic; Ondrej.Bucko@uniag.sk

The aim of the experiment was to analyse the influence of the addition of organic zinc on the selected technological parameters of the pork quality. DNA tests have detected genetic marker RYR 1 (malignant hyperthermia syndrome) and all experimental pigs corresponded to the NN genotype. The pigs were divided in a control group of 18 pigs (9 barrows and 9 gilts) and the experimental group of 18 pigs (9 barrows and 9 gilts). The control group was fed by a standard feed ration consisted of the three feed mixtures, which were used at the different growth stages. The experimental group was fed by the same feed mixtures in the same growth stages as the control group, whereby the mineral-protein premix used in all three feed mixtures was enriched by 66 mg/kg organic zinc in the form of chelate zinc and amino acids, hydrate (optimin-zinc 15% LL101711). From the indicators of the technological quality of meat was evaluated the actual acidity 45 minutes post mortem and 24 h post mortem, electric conductivity 45 minutes and 24 h post mortem, drip loss 24-48 h post mortem, colour of meat 24 h post mortem in the values CIE L*, a*, b*. The results showed that there were significant differences ($P \leq 0.05$) between the experimental and the control group in the parameter pH_1 and EC_{24}. It was found that there was a statistically significant difference ($P \leq 0.01$) in the parameters pH_{24}, colour CIE L* and CIE b*. The differences between the genders of the indicators CIE a* a CIE b* were statistically significant ($P \leq 0.05$). In conclusion obtained results showed that the organic zinc influenced positively the technological parameters of pork quality. This work was supported by projects VEGA 1/0493/12, VEGA 1/2717/12, ECACB – ITMS 26220120015 and ECACB Plus – ITMS: 26220120032.

Session 39b Poster 19

The effect of storage time of slurry on the concentration of odorous compounds in manure

O. Hwang[1], S.B. Cho[1], K.Y. Park[1], S.H. Yang[1], D.Y. Choi[1], JH. Cho[2] and I.H. Kim[2]
[1]National Institute of Animal Science, RDA, 77 Chuksan-gil, Gwonsungu, Suwon, Republic of Korea, 441-706, Korea, South, [2]Dankook University, 29 Anseodong, Cheonan, Choognam, 330-714, Korea, South; hoh1027@korea.kr

This study was conducted to investigate the effect of italian rye grass (IRG) supplementation in feed and storage time of slurry on the concentration of odorous compounds in manure. Ten pigs weighed 80~120 kg were fed basal diet (control) or IRG diet (basal diet + IRG powder). Manure was collected after 3 wk feeding trial. Levels of odorous compounds were analyzed from the manure incubated at 20 °C in the pilot chamber which is similar to slurry pit at 0, 2, and 4 wk period. Levels of odorous compounds at incubation times of 0, 2 and 4 wk were as follows: phenol compounds were 193.15, 172.01, 146.23 ppm, indole compounds were 6.57, 5.54, 8.36 ppm, VFA was 9,594, 8,354 and 6,775 ppm, I-VFA was 1,030, 1,060 and 1,356 ppm, NH4-N level was 1,230, 1,574 and 1,083 ppm, respectively. Levels of phenol compounds and VFA were decreased ($P<0.01$) as incubation period increased. Levels of phenol compounds were lower ($P<0.01$) in IRG group compared to control group: 215.06, 187.40, 153.79 ppm in basal diet group and 171.23, 156.62, 138.68 ppm in IRG treatment group at 0, 2 and 4 wk, respectively. However, concentration of indole compounds was higher in IRG than control group: 3.78, 4.11 and 7.29 ppm in control group and 9.37, 6.98, 9.43 ppm in IRG group at 0, 2 and 4 wk, respectively. VFA levels were lower ($P<0.01$) in IRG than control group: 10,423, 9,185, 7,516 ppm in control group and 8,766, 7,523, 6,033 ppm in IRG group at 0, 2, and 4 wk, respectively. In conclusion, levels of VFA and phenol compounds were decreased as slurry was stored for longer periods as well as when pigs were fed with diet containing IRG. More studies are needed to reduce the indole level in the manure.

Session 40 Theatre 1

Enhancing diversity in livestock farming system to strengthen their resilience: a review of evidence

M. Tichit[1], L. Fortun-Lamothe[2], E. Gonzalez-Garcia[3], M. Jouven[3], M. Thomas[4] and B. Dumont[5]
[1]INRA, UMR1048 SADAPT, 75231 Paris, France, [2]INRA, UMR1289 Tandem, 31326 Castanet Tolosan, France, [3]SupAgroM, UMR868 Selmet, 34060 Montpellier, France, [4]INRA, USC0340 AFPA, 54505 Vandoeuvre-les-Nancy, France, [5]INRA, UMR1213 Herbivores, 63122 Saint-Genès-Champanelle, France; muriel.tichit@agroparistech.fr

Agricultural intensification has reduced the diversity of both plant and animal species and the variety of management practices. Recent empirical evidence suggests that we have underestimated the potential for diversity in livestock farming systems (LFS) to strengthen their resilience. Here, we review the major results from research that has analyzed the issue of how and why diversity in LFS could increase their resilience. At herd level, diversity in animal species, genetic strains, physiological status and management practices offers a risk-spreading strategy against droughts, disease outbreaks and market price fluctuations. Combining several species at pasture or in fish ponds enables higher overall resource capture, liveweight gain and production per unit area because contrasted feeding behaviors enable the use of multiple spatial niches and food resources. Managing diversity over time becomes a central issue in large herds where management strategies targeted at different herd segments are expected to increase overall herd performance. Diversity in lifetime performance emerged from complex interactions between herd management practices and individual biological responses. A diversity of forage resources also helps secure the feeding system against seasonal and long-term climatic variability. Finally, recent work has emphasized that a diversity of grazing management practices, i.e. in terms of stocking rate and grazing periods, can enhance the ability of grassland-based systems to overcome drought events. We conclude on research needs that could reduce knowledge gap and better serve strategies for the empowerment of farmers in resilience management.

Sensitivity of beef cattle and sheep farms to weather hazards according to their forage systems

C. Mosnier[1], A. Boutry[1], M. Lherm[1] and J. Devun[2]
[1]INRA, UMR1213 Herbivore, Theix, 63122 St Genès champanelle, France, [2]Livestock Institute, 9 allée Pierre de Fermat, 63170 Aubière, France; cmosnier@clermont.inra.fr

Designing livestock systems both profitable and resilient to weather risks and climate change is at the heart of many reflections. Diversifying forage productions is usually considered as a way to decrease farm exposure to weather risks. This study aims to verify whether the beef cattle and sheep farms with forage crops and temporary grassland perform better under risks than forage systems where permanent grassland prevails. To do so, we analyse technical and economic observations from a panel data of 504 farms specialized in beef or sheep production over the period 2000-2009, totalling 3,214 entries spread over the French territory. To compare forage systems, observations are divided into four groups according to the importance of forage crops, temporary and permanent grassland into forage area. We focus on between year variability. This is measured by the difference between annual records and farm long term average over at least 5 years. Overall variability is approximated by standard deviation. To assess sensitivity to weather variability, we analyse annual technical and economic variations at the light of weather hazards measured by variation of grass harvested by livestock unit. Our results demonstrate that most farmers adjust the area harvested and extra feed purchase rather proportionally to grass yield variations. Although positive variations of grass harvested have almost no effect on economic results, economic losses are more than proportional to negative ones. Forage crops are effective in reducing the variability of fodder harvested and extra feed purchase in case of variation of grass harvested. However, the presence of forage crops and temporary grass don't reduce the overall variability of production and economic results. It is likely that these farms have increased risk exposure of their animal production systems (e.g. higher stocking rate) while decreasing risk exposure of their forage system.

Relationships between trajectories and vulnerability on smallholder dairy farms in Brazil

M.N. Oliveira[1,2,3], B. Triomphe[2], N. Cialdella[2] and S. Ingrand[1]
[1]INRA, UMR Métafort, Clermont Ferrand, Theix, 63122 Saint-Genes-Champanelle, France, [2]CIRAD, Agricultural Research for Development, UMR Innovation, Bâtiment 15, 73 rue Jean-François Breton, 34398 Montpellier, Cedex 5, France, [3]EMBRAPA, Brazilian Agriculture Research Corporation, Embrapa Cerrados, Rodovia BR 020, km 18, Caixa Postal 08223, Planaltina, DF, Brazil; marcelo.nascimento-de-oliveira@clermont.inra.fr

In a context of limited supply of family agriculture products to the local market, mainly due to problems in delivery quantity and regularity, smallholder farmers adopt different strategies to adapt their system in order to enter into an organized and competitive supply chain. We interviewed 24 smallholder farmers in the municipality of Unaí (Brazilian Cerrados), which deliver different amounts of milk to the local cooperative. We first described the different trajectories of each farmer since inception (a few years ago). Then, we built some indicators to compare the vulnerability of each system face to different events (identified by the farmers themselves). The variability observed in the trajectories of milk production systems is linked to different technical choices and management practices, and leads to a greater or lesser capacity of the systems to cope with unpredictable events. We propose to assess this capacity by using a set of 21 indicators based on the concept of vulnerability, i.e. the capacity to adapt to exogenous and endogenous shocks. Three components of vulnerability are considered by the different indicators: exposure, sensibility and adaptive capacity. We show that the same perturbation has not the same effect on the farming system according to its previous trajectory (including workforce, money and technical management).

Session 40 Theatre 4

A modelling framework to evaluate benefits of animal adaptive capacity for livestock farming systems

L. Puillet[1], O. Martin[1], M. Tichit[2] and D. Réale[3]
[1]INRA, UMR 791 MoSAR AgroParisTech, 75005 Paris, France, [2]INRA, UMR 1048 SAD-APT AgroParisTech, 75005 Paris, France, [3]UQAM, Chaire de Recherche du Canada en Ecologie Comportementale, H3C 3P8 Montréal, Canada; laurence.puillet@agroparistech.fr

Livestock farming systems (LFS) are facing the challenge of producing more with less resource in a context of increasing uncertainty. Individual variability in the adaptive capacity of animals can be seen as a potential lever by which to improve LFS resilience. However, evaluation methods of such lever are currently lacking. The objective of this study was to develop a modelling framework to evaluate the effects of management strategy and environmental perturbation on the biological responses of animals within a herd. The framework is centred on the animal level, seen as: (1) an integrative level for biological functions, within which trade-offs in energy allocation are expressed; and (2) an elementary component of a population level, within which emergent properties such as resilience are expressed. At the animal level, adaptation is formalised through a dynamic pattern of energy allocation among life functions, allowing the characterization of adaptation in a multidimensional perspective. This representation allows a fuller description of adaptive capacities by representing different forms of adaptation. At the population level, adaptation is formalised as a trait evolving over several generation cycles, under the artificial selection achieved by the farmer through culling and breeding decisions. The originality of the animal model is to integrate life functions usually considered in animal nutrition models (growth, maintenance, lactation, gestation and reserves) with other traits like immunity, thermoregulation or behaviour. We finally apply the modelling framework to contrasted case studies of adaptation, for various environmental perturbations (thermal stress, pathogen pressure, resource shortage) and species.

Session 40 Theatre 5

Identification of strategies increasing the trade-off between N balance and income in dairy farms

A. Grignard[1], D. Stilmant[1] and J. Boonen[2]
[1]Centre wallon de Recherches agronomiques – CRA-W, Département Agriculture et milieu naturel, Rue de Serpont, 100, 6800 Libramont, Belgium, [2]Lycée Technique Agricole d'Ettelbruck, 72, avenue Salentiny, 9080 Ettelbruck, Luxembourg; d.stilmant@cra.wallonie.be

The DAIRYMAN INTERREG project aims to enhance the environmental and economic sustainability of dairy sector in 10 regions from North West Europe by improving the competitiveness and ecological performance of dairy farming. Accordingly, between 2009 and 2011, economic and environmental performances (e.g. N balances) were recorded and compared for a network of 76 farms specialized in dairy production. Thanks to a hierarchical clustering on principle components based on 2009 data (year of milk price crisis), three groups of farms were identified according to their economical (Income per family Annual Work Unit (fAWU)) and environmental (N balances per hectare and per ton of milk) performances. The first group P+ (n=14) is characterized by the best performances [48,578±14,705 €/fAWU, 98±54 kg of N/ha and 8.6±2.3 kg of N/ton of milk], the second group P- (n=13) is characterized by the worst performances [-21,888±37,280 €/fAWU, 241±54 kg of N/ha and 17±4.8 kg of N/ton of milk] while the last group P (n=49) got performances close to the average. P+ group strategy is mainly based on feed autonomy (less kg of concentrate) and lower productive level (less milk per cow, less milk per ha and less milk per fAWU). Furthermore, their milk is valorized in a better way (31.3±6.3 €/100 kg of milk vs. 26.0±1.7 €/100 kg of milk). The evolution of the performances from the farms of the P+ group in 2010 (year economically favorable), shows two different tendencies: stable farms (n=9) that maintain similar and good performances and opportunist farms (n=5) that have strongly intensified their production to increase their income (up to 3 times higher) but with a degradation of their N balances (up to 2.5 times higher loses).

Breeding can make sheep farming systems more resilient to climate uncertainty

G. Rose and H.A. Mulder
Animal Breeding and Genomics Centre, Wageningen University, P.O. Box 338, 6700 AH Wageningen, the Netherlands; gus.rose@wur.nl

Sheep are difficult to manage when the climate and pasture growth are uncertain, such as in Mediterranean climates. In these regions the length and severity of the annual periods of drought in summer and autumn are becoming harder to predict. During these periods sheep need to be fed grain which is expensive. Our objective was to compare the impact of breeding for reproduction or wool on the resilience of meat and wool sheep farming systems to climate variability. We also investigated if sheep that lose less live weight during summer droughts and need to gain less weight before lambing in winter reduces grain requirements and increases profit. We modeled the monthly energy and protein requirements of sheep when wool weight, number of lambs weaned and live weight loss during summer are changed by 1 genetic standard deviation. We then maximised profit per hectare (ha) by optimising stocking rate and grain feeding over a range of years, varying pasture growth from 1,700-5,200 kg dry matter per year. The model included interactions between the sheep and pasture, most importantly, the effects of pasture on pasture intake. Profit increased most when sheep weaned more lambs (average €18.9 per ha) followed by growing more wool (€8.1 per ha) and losing less summer weight (-€2.1 per ha). The relative importance of traits were 74.6% for number of lambs weaned, 17.3% for wool and 8.1% for body weight loss in the worst year and 58.1% for number of lambs weaned, 40.2% for wool and 1.7% for body weight loss in the best year. This is because most of the energy requirements for reproduction are in winter, aligned with the period of pasture growth compared to wool which requires energy and protein during the whole year. Selecting sheep to lose less weight during summer is only profitable in years with severe drought. We concluded that breeding for reproduction will contribute most to making sheep farming systems more resilient to climate variability.

Roles of summer mountain pastures for the adaptation of livestock farms to climate variability

C. Rigolot[1], S. Roturier[1,2], B. Dedieu[1] and S. Ingrand[1]
[1]Inra, Site de Theix, 63122 Saint Genes Champanelle, France, [2]AgroParisTech, Université Paris Sud, 91405 Orsay, France; cyrille.rigolot@clermont.inra.fr

In mountain regions like Auvergne in France, Summer Mountain Pastures (SMPs) correspond to high altitude pastures, characterized by a specific use for summer grazing and in some cases collective management. SMPs have assets to contribute to the resilience of livestock farms faced with climate variability: They are exposed to a fresh and rainier climate and have a higher resistance to drought due to their high botanical diversity. However, long distance to farms and collective rules of SMPs may be constrains for farmers to adapt their use to climatic conditions. To better characterize the contributions of SMPs to farm resilience, a survey was conducted in 2012 in Auvergne. The managers of 7 collective units (1 large cattle unit (500 users), 2 sheep and 4 mixed cattle/sheep units with 12 or less users), 7 cattle and 12 sheep farmers using individual and/or collective SMPs have been interviewed. A special focus was put on the practices associated to SMPs (animals, beginning and ending dates) and the way they were changing to cope with particular climatic events. In the long run, farmers' use of SMP follow diverse trajectories: In some cases, practices have remained the same from the beginning, while in other, practices have evolved, either progressively, or suddenly, or slightly each year to cope with circumstances (including climate events). During a given year, the use of SMP by farmers can be strictly fixed in advance, or only constrained by individual factors, or negotiated with other farmers. Finally, SMPs can contribute to the resilience of farms faced with climate variability in two complementary and concurrent ways: enhancing adaptive capacities and decreasing system sensitivity. We identify 'flexibility profiles' corresponding to trade-offs between these two ways at different time scales, at the interface between individual farming systems and collective management practices.

Session 40 — Theatre 8

Sensitivity of beef cow reproduction to body lipid dynamics: a modeling approach

E. Recoules[1,2], J. Agabriel[1,2], A. De La Torre[1,2], N.C. Friggens[3,4], O. Martin[3,4], D. Krauss[5] and F. Blanc[1,2]
[1]Clermont Université, VetAgro Sup, UMR 1213 Herbivores, 63000 Clermont-Ferrand, France, [2]INRA, UMR 1213 Herbivores, 63122 Saint Genès Champanelle, France, [3]AgroParisTech, UMR 791 MOSAR, 16 rue Claude Bernard, 75005 Paris, France, [4]INRA, UMR 791 MOSAR, 16 rue Claude Bernard, 75005 Paris, France, [5]INRA, La Sapinière, 18390 Osmoy, France; fabienne.blanc@vetagro-sup.fr

Reproductive performance of beef cows is influenced by their nutritional status and body condition. Due to a changing context nutritional constraints may occur at any time over the cow's production cycle. To simulate interrelations between nutritional status and reproduction a dynamic model was developed relying on a framework proposed in the dairy cow. Reproductive performance is described as successive events: parturition, ovulation, conception. The cow's physiological status influences its ability to mobilize or gain body lipids. These dynamics lead to a genetically driven body lipid trajectory. Under nutritional constraints, body lipids deviate from this trajectory towards an adaptive one that may affect the reproductive performance. Empirical laws were derived to represent the effects of difference between both trajectories on the interval from calving to resumption of cyclicity, estrus expression and probability of conception. These laws were adjusted using experimental data from Charolais cows reared with or without nutritional constraints. Calibration has to be improved but the model allows testing of how body lipid dynamics influence reproductive performance and vice versa over the productive career of the cow. For example, the predicted effect of a postpartum underfeeding period on the interval from calving to conception differs between a slight but long nutritional constraint compared to a short but intense one. We will present of how the variability of the reproductive performance at the herd level can be tested by taking into account the respective sensitivity of individual cows to changes in feeding levels occurring at any time of their productive cycle.

Session 40 — Theatre 9

How to assess the diversity of dairy cows adaptive capacities?

E. Ollion[1,2], S. Ingrand[2], C. Espinasse[3], J.M. Trommenschlager[4] and F. Blanc[1]
[1]Clermont Université, Vetagro-Sup, UMR 1213 Herbivores, Lempdes, 63000 Clermont-Ferrand, France, [2]Inra, UMR 1273 Métafort, Theix, 63122 Saint-Genes Champanelle, France, [3]Inra, UMR 1213 Herbivores, Theix, 63122 Saint-Genes Champanelle, France, [4]Inra, UR ASTER, Av L. Buffet, 88500 Mirecourt, France; emilie.ollion@vetagro-sup.fr

Uncertainty is one of the current challenge dairy farmers have to face. In low input livestock systems it mainly concerns time variations in feed availability and quality. The ability of the herd to recover or absorb such disturbances and to maintain its level of productivity relies both on the adaptive capacity of reproductive females and on farmers practices. In this study we focus on the adaptive capacities of the animals and we assume that increasing the diversity of individual adaptive capacities between cows within the herd contributes to improve the herd resilience. Therefore, we suggest that a better assessment of cows' adaptive capacities could broaden their use as a lever to face uncertainties, like feed supply instability. On the literature basis, we hypothesize that there is not only one pattern of adaptive capacity, but several, varying according to the physiological stage of the cow and to some animal characteristics such as its breed and the level of expression of its genetic merit. As unsuccessful reproduction is a major cause of culling in dairy herds, we establish that adaptive capacity of dairy cow can not only be appreciated by taking into account the ability of the cow to produce milk. Thus, we propose different patterns of adaptive capacity, characterized by the analysis of co-dynamics between body reserves, milk production and the reproductive success of cows experiencing a disturbance. This approach considers that priorities exist among physiological functions (maintenance, lactation and reproduction).The study was achieved using data from Inra experimental dairy farms from 1999 to 2012.

Session 40 Theatre 10

Dynamilk: a farming system model to explore the trade-offs between grassland and milk productions

A.L. Jacquot[1,2], G. Brunschwig[1,2], L. Delaby[3], D. Pomies[1,2] and R. Baumont[1,2]
[1]INRA, UMR1213 Herbivores, 63122 Saint-Genes-Champanelle, France, [2]VetAgro Sup, Clermont Universités, UMR Herbivores, BP 10448, 63000 Clermont-Ferrand, France, [3]INRA, UMR1348 INRA-AgroCampus PEGASE, Domaine de la Prise, 35590 Saint-Gilles, France; anne-lise.jacquot@vetagro-sup.fr

A model at farm-scale, Dynamilk, has been designed and implemented to explore grass-based dairy systems with contrasting production strategies in order to understand what are the possible trade-offs between animal production, forage self-sufficiency and grasslands use. Dynamilk mimics the dynamic relationships among dairy cattle, climate, forage resources and farmer management. The main output of Dynamilk, among others, is milk production according to herbage and feed supply depending on farmer's management, cattle and grasslands potential production and characteristics along with weather data. A validation of dairy cattle model and Dynamilk as a whole has been carried out by comparison against experimental data. From the initial assumption that matching animal needs with feed and herbage offer enables the farming system to lean towards a better forage self-sufficiency and to be more resilient to changes, two contrasted systems have been simulated. The first one is based on autumn calving distribution (AC), and the second one on spring calving distribution (SC). Their performances have been analyzed on a long climate time series (1995-2011). Then, several simulations have evaluated effects of stocking rate increase and concentrate decrease on farming system performances. Results have shown that: (1) for similar production performances at low stocking rate, SC system is less sensitive to climatic variations than AC; (2) under-utilization of grass at low stocking rate allows a positive performance responses in case of moderate stocking rate increase at farm-level; (3) effects of concentrate feed reductions on milk yields are softened by a better use of grass. Dynamilk is a relevant tool to test a wide-ranging of grass-based dairy systems.

Session 40 Theatre 11

Animal and farming system crossed approaches to reveal the goat production resilience in Guadeloupe

G. Alexandre and N. Mandonnet
INRA-URZ, Domaine Duclos, 97170 Petit-Bourg, Guadeloupe; nathalie.mandonnet@antilles.inra.fr

In the tropics, the capacities of the Creole goat genotype to living and producing under the biotic and abiotic constraints are frequently assigned to their adaptation. The numerous crossbreeding that occurred within the population, the natural selection to which it was submitted and its multiple bio-economic use, have assigned very original traits. It has stored up many alleles that allow the species survival. Qualified as hardy breed it present different capacities of resistance and flexibility. It goes too for the systems of production known as very diverse and multi-purpose. The hazards and dysfunctions of the markets (distancing between local vs. imported, informal vs. formal) the official frames of governance (legal constraints, models of development) and more than any, the natural disasters have traced the history of the animal production in Guadeloupe. The traditional system, an inheritance of a past colonial, continues in different forms that offer a panel of variable solutions. At the opposite, the intensive modern system, passed through many crises and is 'perennial' only owing to regular public subsidies. The adaptative and productive capacities of the Creole goat are known as necessary for sustaining the diverse husbandry activities. Thus, its resilience towards parasitism has been included in its breeding program. The goat farming systems, described in their great diversity, contain hidden sources of functioning that could be valorized. Thus it could be valuable for the future to build adaptative trajectories and suggest intermediary models of development in order to reach sustainability in a hard environment. The multifunctionality of the animal and of the system allows passing through the failures of the maximization of the productivity under limiting tropical conditions: for example there is allocation of nutrients between adaptative and productive animal physiological functions or there is a repartition of risks between the diverse bio-economic purposes of the system.

Effect of climate conditions on fat and protein yields in small dairy ruminants

M. Ramón[1], H.M. Abo-Shady[2], C. Díaz[3], A. Molina[2], M.D. Pérez-Guzmán[1], J.M. Serradilla[2], M. Serrano[3] and M.J. Carabaño[3]
[1]Centro Regional de Selección y Reproducción Animal, Avenida del Vino 10, 13300 Valdepeñas, Spain, [2]Universidad de Córdoba, Producción Animal, Campus de Rabanales, Ctra. Madrid-Cádiz km 3, 14071 Córdoba, Spain, [3]INIA, Mejora Genética Animal, Ctra. de La Coruña km 7.5, 28040 Madrid, Spain

The model initially proposed for dairy cattle to evaluate the impact of heat stress (HS) on milk yield assumes a comfort zone and a zone above a threshold (To) with a linear decrease of production. Studies with this or other models are scarce for small dairy ruminants. Our aim was to examine the shape of response on production associated with increasing values of temperature and a temperature and humidity index (THI) in two breeds. A total of 1,675,886 and 116,258 daily fat and protein yields from the official milk recording of 191,641 Manchega ewes and 11,259 Florida goats, respectively, were used. Average (Tave) and maximum daily values for temperature and THI at the test day were obtained from the Spanish Meteorological Agency. Two data sets, all and only high producing animals (above 1.5 SD) were analyzed per breed. Two types of models were used, one fitting splines (SP) with one knot at To and a linear slope b afterwards, and another fitting Legendre polynomials (LP) of varying order (up to cubic). Models included herd-year of test day, number and stage of lactation, age of the animal at recording, prolificacy, milking time and animal apart from the temperature/THI functions. For Tave, To was found at 29 °C for sheep and at unexpectedly small values around 10 °C for goats. Values of b were low in general, but larger for highly productive ewes, with declines of up 16 g/°C of protein. LP models resulted in better fits in both species, lower HS thresholds for sheep and similar for goats. It was also interesting the assessment of thresholds for cold stress in sheep. The latter might suggest the existence of a zone of comfort with two thresholds, outside of which yields decrease.

Genetic effects of heat stress on milk yield and MIR predicted methane emissions of Holstein cows

M.-L. Vanrobays[1], N. Gengler[1], P.B. Kandel[1], H. Soyeurt[1] and H. Hammami[1,2]
[1]University of Liege, Gembloux Agro-Bio Tech, 5030 Gembloux, Belgium, [2]National Fund for Scientific Research, 1000 Brussels, Belgium; mlvanrobays@ulg.ac.be

Dairy cows both contribute to and are affected by climate change. Breeding for heat tolerance and reduced methane (CH_4) emissions is a key requirement to mitigate interactions between dairy cows and climate change. This study was aimed to estimate genetic variation of milk yield and CH_4 emissions over the whole trajectory of temperature humidity index (THI) using a reaction norm approach. A total of 257,635 milk test-day (TD) records and milk mid-infrared (MIR) spectra from 51,782 Holstein cows were used. Data were collected between January 2007 and December 2010 in 983 herds by the Walloon Breeding Association (Ciney, Belgium). The calibration equation developed by Vanlierde *et al.* (R^2 of cross-validation=0.70) was applied on the spectral data in order to predict CH_4 emissions values (g CH_4/d). These values were divided by fat and protein corrected milk yield (FPCM) defining a new CH_4 trait (g CH_4/kg of FPCM). Daily THI values were calculated using the mean of daily values of dry bulb temperature and relative humidity from meteorological data. Mean daily THI of the previous 3 days before each TD record was used as the THI of reference for that TD. Bivariate (milk yield and a CH_4 trait) random regression TD mixed models with random linear regressions on THI values were used. Estimated average daily heritability for milk yield was 0.17 and decreased slightly at extreme THI values. However, heritabilities of MIR CH_4 traits increased as THI values increase: from 0.10 (THI=28) to 0.14 (THI=75) for MIR CH_4 (g/d) and from 0.14 (THI=28) to 0.21 (THI=75) for MIR CH_4 (g/kg of FCPM). Genetic correlations between milk yield and MIR CH_4 (g/d) ranged from -0.09 (THI=28) to -0.12 (THI=75) and those between milk yield and MIR CH_4 (g/kg of FPCM) from -0.75 (THI=28) to -0.71 (THI=75). These results showed that milk production and CH_4 emissions of dairy cows seemed to be influenced by THI.

Session 40 — Poster 14

Physiological adaptations and fertility of Holstein and Montbeliarde cows under low-input systems

J. Pires, Y. Chilliard, C. Delavaud, J. Rouel, K. Vazeille, D. Pomiès and F. Blanc
INRA, VetAgro Sup, Clermont Université, UEMA1296, UMRH1213, 63122 Saint-Genès-Champanelle, France; jose.pires@clermont.inra.fr

The objective was to study production, metabolic and hormonal profiles during the periparturient period, and fertility of 24 Holstein-Friesian (HO) and 22 Montbeliarde (MO) cows, under two low-input semi-mountain production systems with seasonal spring calving. An extensive (EXT) system was based on permanent grassland and zero concentrate supplementation (n=12 HO and 12 MO), and a semi-extensive (SEMI) system was based on temporary grassland and up to 4 kg/d of concentrate postpartum (12 HO and 10 MO). Pasture turnout occurred at 13±8 and -34±16 days in milk for cows calving prior and after (n=30) the beginning of grazing season, respectively. Individual measurements and sampling were performed from week -4 to 12 relative to calving. Ovarian cyclicity was assessed via milk progesterone from 10 DIM until the end of reproduction season, and estrous behavior was monitored twice daily. Milk yield was 22.1 and 24.4 kg/d during the first 12 wk of lactation for EXT and SEMI ($P<0.01$), respectively. HO produced more milk (24.7 vs. 21.8 kg/d; $P<0.01$), and reached a lower nadir of BCS (1.2 vs. 1.5 in wk 12 of lactation, a 0 to 5 scale; $P<0.01$) than MO. The IA success rate was particularly low in cows under EXT (27%) and low for SEMI (42%), which can be explained by cyclicity abnormalities and a low rate of estrus detection (63 and 42% for EXT and SEMI, respectively). HO tended to have lower plasma glucose than MO (59.6 vs. 61.4 mg/dl, $P=0.08$), which is in agreement with greater milk yield of HO. Plasma glucose was lower in EXT than SEMI (59 vs. 62 mg/dl, $P<0.05$), despite the lower milk yield in EXT. Plasma IGF-1 was lower in EXT (62 vs. 71 ng/ml, with significant differences between week 5 and 12 postpartum), probably reflecting a longer period of uncoupling of GH : IGF-1 axis. Glucose and IGF-1 may explain the interactions between nutritional status and fertility in the two systems.

Session 40 — Poster 15

Effect of different feeding strategies on GHG emissions and sustainability in dairy sheep

C. Pineda-Quiroga[1], N. Mandaluniz[1], A. García-Rodríguez[1], S. Marijuán[2] and R. Ruiz[1]
[1]NEIKER, P.O. Box 46, 01080 Vitoria-Gasteiz, Spain, [2]SERGAL, P.O. Box 46, 01080 Vitoria-Gasteiz, Spain; rruiz@neiker.net

Dairy sheep production in the Basque Country has been traditionally based on the management of the Latxa sheep breed through pasture based farming systems. However, there has been an intensification process during the last years even with the introduction of higher productive foreign breeds managed permanently kept indoors. The assessment of the quality of the diets employed throughout the production cycle, in terms of nutritional value, digestibility, and kinetics of fermentation metabolic pathways, are critical to improve their efficiency and reduce the emissions. The objective of this study was to characterise the diets provided during prelambing and lactation periods in different dairy sheep production systems, and to assess the potential impact on GHG emissions. Feed samples (concentrates and forages) of 15 flocks were collected during autumn-winter of 2012-2013. The nutrient content of every feed was assessed, and the ingestibility and digestibility of the forages were calculated according to Calsamiglia. Fermentation kinetics of the diets, organic matter digestibility (IVOMD), volatile fatty acids and methane were monitored *in vitro*. According to the results, the concentrate:forage ratio was 30:70 during prelambing and 40:60 during lactating period. Regarding forages, flocks were fed with a high variability of forage resources with acceptable-good nutritive values. Concentrates provided most of the protein in Assaf systems, whereas forages were the most important protein source in Latxa systems. The effect of all these different diets in fermentation kinetics, IVOMD, methane emissions and volatile fatty acids' contents of the diets will be discussed in the paper to propose alternatives to reduce emissions and improve the sustainability of the system. The authors want to gratitude to INIA for the financial support (RTA-2011-00133-C02-01), and to the farmers that have participated.

High growth breeding values increase weight change in adult ewes

S.E. Blumer[1,2], M.B. Ferguson[1,2], G.E. Gardner[1,2] and A.N. Thompson[1,2]
[1]Australian Cooperative Research Centre for Sheep Industry Innovation, Armidale, 2351, Australia, [2]Murdoch University, School of Veterinary and Life Sciences, Murdoch, 6150, Australia; sarah.john@agric.wa.gov.au

Ewes that lose less weight under restricted nutrition are potentially more profitable as less supplementary feeding is required or stocking rate can be increased. Heavier strains of Merino sheep have been shown to lose less weight when grazed on dry, poor pasture. Given that sire estimated breeding values (EBVs) for weight positively correlate with mature size, we hypothesise that adult ewes from sires with high EBVs for weight will have reduced annual fluctuation in weight. Spline functions were fitted to liveweight data for ewes from 8 Information Nucleus sites to determine annual weight change (max-min-max) for each ewe. The 2 to 4 year olds were born between 2007 and 2009 and there were 5,242 records for 2,783 animals. Weights were corrected for conceptus and greasy fleece. Weight gain and loss were analysed using linear mixed effects models with fixed effects for site, breed, year, age, lamb birth type and rear type, and sire of the ewe was included as a random term. Sire EBVs for muscle, fat and growth and, ewe average annual liveweight (frame size) were included simultaneously as covariates. Ewes from sires with low EBVs for growth had no significant change in weight gain across a range of frame size (40-70 kg). In contrast ewes from high growth sires demonstrated similar weight gain at frame sizes of 40 kg, but increased in weight gain by 2.4 kg across the range of frame size. Contrary to our hypothesis, it was sires with less genetic potential for growth that produced progeny with reduced changes in liveweight. In both cases the magnitude of weight gain represented a diminishing proportion of frame size as it increased indicating phenotypically larger animals are more resilient to weight change. Ewes from high growth sires may require more careful management to minimise weight change, particularly when maintained at phenotypically higher weights.

Session 41 Theatre 1

Cellular physiology of secretory processes

C.H. Knight
University of Copenhagen, IVKH, Grønnegårdsvej 7, 1970 Frederiksberg, Denmark; chkn@sund.ku.dk

The lactating mammary cell orchestrates numerous and complex secretory processes that culminate in the production of milk, an intracellular-like fluid containing significant quantities of protein, fat and carbohydrate. These processes were intensively studied at the cellular level over several decades, but in recent years the research focus has moved to the molecular level. Partly as a consequence, a number of significant questions remain unanswered, for example: How is glucose uptake and trafficking to the Golgi regulated? How does water pass across the Golgi membrane? How is docking of secretory granules to the apical membrane achieved? How is regulation of tight junction functionality achieved in such a dynamic cell? The objective of the session that this short presentation will introduce is to combine knowledge from mammary gland researchers and others working in different secretory cells to review recent developments in the cellular physiology of endocytosis, membrane flux, exocytosis and junction biology.

Session 41 Theatre 2

New developments in membrane channel physiology, with focus on ion and water flux

D.A. Klærke
University of Copenhagen, IVKH, Grønnegårdsvej 7, 1870 Frederiksberg, Denmark; dk@sund.ku.dk

Secretory cells are dynamic structures that undergo significant changes in volume as fluid and solutes are taken up and subsequently secreted. In some cases these volume changes may be accompanied by alterations in cell shape. Changes in volume and shape impose mechanical stress on the cell membranes. However, the view that increased volume necessarily results in membrane stretch is probably overly-simplistic. Cell membranes are also dynamic structures, undergoing their own fluxes as a consequence of exocytotic and other events. The question that arises is, what are the consequences of changes in size/shape/stretch for the functionality of membrane channels, the essential transporters of ions, water and solutes without which secretion could not happen? We have examined this question with particular reference to potassium channels and aquaporins, and have found that certain ion channels are highly sensitive to stretch, while others are can be considered 'sensors of cell volume' provided they are coexpressed with aquaporins. This functional interaction between ion channels and water channels may be especially relevant to the mammary secretory cell, which undergoes more radical changes in size and, probably, shape than most other secretory cells.

Session 41 Theatre 3

Regulation of glucose uptake and trafficking by the mammary secretory cell

F.-Q. Zhao
University of Vermont, Department of Animal Science, 211 Terrill Hall, 570 Main Street, Burlington, VT 05405, USA; fenq-qi.zhao@uvm.edu

Glucose is the major and an essential precursor of lactose synthesis in the Golgi vesicle of the lactating mammary secretory cell. Its mammary uptake is mainly mediated by facilitative glucose transporters (GLUTs), of which there are 14 known isoforms. Mammary cells mainly express GLUT1, GLUT8 and GLUT 12 with GLUT1 being the predominant isoform with a K_m of 9.8 mM. Mammary glucose transport activity increases approx. 40 fold from virgin state to the midlactaion state, and there is a concomitant increase in GLUT expression. We originally hypothesized that the accepted lactogenic hormones are responsible to stimulate GLUT expression during lactogenesis. However, our recent study rejected this hypothesis because the lactogenic hormones have no effect on GLUT1 and GLUT8 expression in mammary explants and primary epithelial cells although they are able to dramatically stimulate expression of milk protein and lipogenic genes. Our new evidence indicates that low oxygen tension resulted from increasing metabolic rate and oxygen consumption during lactogenesis may play a major role of stimulation of glucose uptake and GLUT1 expression in mammary secretory cells. Hypoxia treatment of mammary epithelial cells dramatically increases glucose uptake and GLUT1 expression in these cells and these effects are hypoxia-inducing-factor (HIF)-1α-dependent. In addition to its expression on the plasma membrane, mammary GLUT is also expressed on the Golgi membrane and is likely responsible for facilitating uptake of glucose and galactose to the site of lactose synthesis. The way(s) in which GLUT and glucose itself are trafficked between the plasma and Golgi membranes are unknown. Since lactose synthesis dictates milk volume, regulation of GLUT expression and trafficking represent potentially fruitful areas for further research.

Session 41 Theatre 4

Regulation of exocytosis in the mammary secreotry cell, and the role of SNARE proteins

S. Truchet
INRA, UR1196 Genomique et Physiologie de la Lactation, 78352 Jouy-en-Josas Cedex, France; sandrine.truchet@jouy.inra.fr

Lactating mammary epithelial cells (MECs) secrete a huge amount of milk, an aqueous fluid containing proteins, milk fat globules (MFGs) and elements such as lactose and minerals. These nutrients have two origins: some are produced by the MEC, while others are transferred from blood to milk by transcytosis. The MEC can thus be seen as a crossroad for both the uptake and the vectorial secretion of the milk constituents. These processes are likely to involve a cross-talk between the endocytic/exocytic compartments in order to regulate spatio-temporally the secretion of milk products. The molecular mechanisms underlying the secretion of milk products are still poorly characterized. The major milk proteins, caseins, are secreted by exocytosis while the MFGs are released by budding at the apical plasma membrane. Casein exocytosis thus provides membrane which may be reused to enwrap the budding MFG. Although milk secretion appears to be mostly constitutive, prolactin was shown to activate a phospholipase A2 which, produces arachidonic acid, leading to the acceleration of casein transport and/or secretion. Thus, MECs may possess both constitutive and regulated secretory pathways. Whatever their secretory mode, intracellular trafficking and exocytosis of the caseins probably involve SNARE (Soluble NSF Attachment Protein (SNAP) Receptor) proteins. Due to their ability to form highly stable four-helix bundle complexes bridging donor and acceptor membranes, SNAREs promote membrane fusion in a targeted manner. Moreover, SNAREs bind arachidonic acid, thus facilitating exocytosis. In MECs, certain SNAREs are associated with both casein-containing vesicles and intracellular lipid droplets. By orchestrating the intracellular trafficking of milk components in a hormonally responsive manner, SNAREs may contribute as a key point for the regulation of both the coupling and the coordination of milk product secretion at time of suckling.

Session 41 Theatre 5

New developments in tight junction functionality; endocytic recycling of junctional proteins

P. Whitley, C. Bryant, J. Caunt, A. Chalmers, J. Dukes and L. Fish
University of Bath, Department of Biology and Biochemistry, Centre for Regenerative Medicine, Bath, BA2 7AY, United Kingdom; p.r.whitley@bath.ac.uk

Tight junctions consist of many proteins, including transmembrane and associated cytoplasmic proteins. They act as selectively permeable intercellular barriers that regulate diffusion of small molecules between epithelial cells. They also contribute to cell polarity by maintaining the asymmetric distribution of proteins and lipids within the plasma membrane of epithelial cells. Tight junctions are highly dynamic structures and are regulated by various extracellular signals, but the mechanisms underpinning this regulation are still obscure. We have shown that the tight junction protein, claudin-1, is constitutively endocytosed and rapidly recycled back to the plasma membrane in unstimulated epithelial monolayers of kidney, colon, and lung epithelial cells. We propose that the regulation of endosomal trafficking of junction and polarity proteins is important in controlling tight junction properties and function. We are currently investigating whether oncogenic signalling pathways promote tissue instability by altering the recycling of junction and polarity proteins.

Session 41 Theatre 6

Exfoliation of mammary epithelial cells in milk is linked with lactation persistency in dairy cows

M. Boutinaud[1,2], L. Yart[1,2], P. Debournoux[1,2], S. Wiart[1,2], L. Finot[1,2], E. Le Guennec[1,2], P.-G. Marnet[1,2], F. Dessauge[1,2] and V. Lollivier[1,2]

[1]Agrocampus Ouest, UMR 1348 PEGASE, 35000, Rennes, France, [2]INRA, UMR 1348 PEGASE, Domaine de la prise, 35590 Saint Gilles, France; marion.boutinaud@rennes.inra.fr

In ruminants, milk yield is gradually reduced after the peak of lactation and ovariectomy has been recently shown to limit this decline and thus improving the lactation persistency. These effects on milk yield are partly controlled by the regulation of the number of mammary epithelial cells (MEC) in the gland, which results of apoptosis/proliferation balance. Moreover, MEC are shed into milk during the lactation process. In order to characterize exfoliation and apoptosis of milk MEC related to lactation persistency, 14 multiparous Holstein cows were either ovariectomized (Ovx, n=7) or sham-operated (Sham, n=7) around 60 days in milk. Milk was collected at 5, 21, 37, 47 and 52 weeks of lactation to purify MEC from milk after centrifugation and immunocytochemical sorting. MEC exfoliation was evaluated using the determination of MEC concentration in milk. The percentage of apoptotic MEC was determined by flow cytometry after TUNEL labelling. RNA was extracted from milk-purified MEC and analyzed by RT-PCR. As expected daily milk yield was decreased as the stage of lactation advanced ($P<0.001$) whereas ovariectomy limited the decline in milk yield ($P<0.05$). MEC has the tendency to be more exfoliated in milk during the advanced stage of lactation and ovariectomy decreased it at 47 weeks ($P<0.05$). The stage of lactation significantly affected the percentage of apoptotic milk purified MEC and ovariectomy decreased it at 47 weeks ($P<0.05$). The mRNA level of α-lactalbumin was reduced and the ones of the pro and anti apoptotic bax and blc-2 rose as the stage of lactation advanced ($P<0.05$). Ovariectomy did not affect transcript content in milk purified MEC. Taken together these results suggest that the MEC exfoliation in milk and apoptosis are negatively linked to the lactation persistency.

Session 41 Poster 7

The efficiency of melamine absorption in the mammary gland of lactating dairy cows

T. Calitz and C.W. Cruywagen

Stellenbosch University, Animal Sciences, Private Bag X1, Matieland, Stellenbosch 7602, South Africa; cwc@sun.ac.za

Five Holstein cows producing 39±3.3 (SE) kg milk/d were used in a trial to determine melamine absorption by the mammary gland. Cows received 10 g of melamine daily via treatment boluses (5 g of melamine/bolus) twice daily for three consecutive days. On the morning of day 3, catheters were inserted into the caudal superficial epigastric vein (milk vein) and the caudal auricular artery, following administration of a local anaesthetic. Arterial and venous blood samples were collected hourly for the following 9 h. Cows had access to fresh water, lucerne hay and semi-complete dairy pellets throughout the 9 hour period. Catheter patency was maintained by flushing the catheters with heparinised saline solution between blood collections. After the final blood collection, the catheters were carefully removed and cows were milked immediately thereafter. After each blood collection, samples were centrifuged (15 min, 1,800×g) and plasma was decanted and stored at -20 °C. Milk yield was recorded and milk samples were collected for milk content and melamine analyses. Plasma samples were analysed for melamine and amino acid content. Phenylalanine and tyrosine contents of milk and plasma were used to calculate mammary blood flow and. Blood and milk samples were collected on day 3 to determine melamine absorption by the mammary gland through arterio-venous (A-V) difference. Melamine in milk and plasma was determined by LC/MSMS. Because only one treatment was applied (10 g of melamine/cow daily), only standard errors were determined to indicate the amount of variation. A net positive melamine flux was observed, indicating net absorption of melamine by the mammary gland. Melamine absorption efficiency by the mammary gland was 0.29% and melamine excretion efficiency (melamine excreted as percent of intake) was 1.47%. It was concluded that melamine ingested by cows results in net absorption of melamine by the mammary gland, but that absorption efficiency is low.

Organic and free range egg production systems: effects of genotype and management

F. Leenstra[1], V. Maurer[2], M. Bestman[3] and F. Van Sambeek[4]
[1]Wageningen UR Livestock Research, P.O. Box 65, 8200 AB Lelystad, the Netherlands, [2]FiBL, P.O. Box 219, 5070 Frick, the Netherlands, [3]Louis Bolk Institute, Hoofdstraat 24, 3972 LA Driebergen, the Netherlands, [4]Institut Sélection Animales, P.O. Box 114, 5830 AC Boxmeer, the Netherlands; ferry.leenstra@wur.nl

Within the EC FP7 project LowInputBreeds, researchers from the Netherlands (NL), France (F) and Switzerland (CH) search for the ideal combination of genotype and management for free range egg production systems. In total 257 farmers with free range layers (organic and conventional) with 273 flocks were interviewed to determine the relationships between genotype of the hens, management and performance. Almost 20 different genotypes (brands) were present on the farms. In F, all birds were brown feathered. In CH and NL, there were brown, white, and silver hens. In CH, mixed flocks (brown/white) were also present. Overall performance in organic and conventional systems differed significantly (higher mortality and lower egg production among organic hens). The difference was highly significant in NL, and showed a non-significant tendency in the same direction in CH and F. White hens tended to perform better than brown hens. Silver hens appeared to have a higher mortality and lower production. There were no significant relationships between production, mortality, feather condition and use of outside run or with flock size. There was more variation in mortality and egg production among small than among large flocks. As a second step, 40 farms each were visited in NL and CH to find possible reasons for these differences and to look at management as well as animal health and welfare into more detail. First results indicate that in NL free range hens scored better on plumage condition and wounds than organic hens, while in Switzerland organic hens scored better on plumage condition and keel bones than free range hens. Effects of management and genotype are currently analysed. Furthermore we examine egg quality and application of prolonged laying periods or moulting in the visited farms.

Evaluating the need for organic breeding programmes and assessing possible implementation strategies

S. König, T. Yin and K. Brügemann
University of Kassel, Department of Animal Breeding, Nordbahnhofstr. 1a, 37213, Germany; sven.koenig@uni-kassel.de

This paper outlines motivations for implementing independent dairy cattle breeding programs for low input or organic production systems. Subsequently, we suggest and evaluate possible breeding strategies. From a scientific perspective, motivations for implementing organic breeding programs are based on additional or new breeding goals with a focus on animal health and welfare, possible genotype by environment interactions, and limitations in the use of biotechnologies. Hence, we will give a general overview of existing organic breeding programs along with their breeding goals, we will present results from own studies related to genotype by environment interactions and from gene expressions in harsh environments (genetic studies on heat stress), and we discuss the potential and limitations of reproductive and molecular technologies. A special focus is on aspects of genomic selection for new phenotypes using calibration groups of cows, and including imputing strategies in a designed experiment. A stochastic simulation was conducted to evaluate different breeding program designs by including aspects of genotype by environment interactions, accuracies of genomic breeding values, and various mating designs (e.g. natural service sires versus artificial insemination). Overall evaluation criteria were true breeding values of selected sires and their offspring, and the development of inbreeding and relationships in the low input population.

Session 42 Theatre 3

Genetic basis of functional traits in low input dairy cattle

A. Bieber[1], M. Kramer[2], M. Erbe[2], B. Bapst[3], A. Isensee[1], V. Maurer[1] and H. Simianer[2]
[1]Research Institute of Organic Agriculture, Ackerstr. 21, 5070 Frick, Switzerland, [2]Georg-August-Universität Göttingen, Department of Animal Science, Albrecht-Thaer-Weg 3, 37075 Göttingen, Germany, [3]QUALITAS AG, Chamerstr. 56, 6300 Zug, Switzerland; anna.bieber@fibl.org

Phenotypic data of Brown Swiss cows collected from 40 Swiss dairy farms within the EU-funded project LowInputBreeds were analyzed with two aims: (1) testing methods to describe functional traits; and (2) estimating genetic parameters and accuracies of breeding values for novel functional and conformation traits. (1) A data set of 1112 cows was analyzed to evaluate a commonly used Body Condition Scoring (BCS) system regarding its ability to assess the back fat thickness (BFT) and to generate a more objective scoring method. The results of multiple regression models showed that the BCS system, which takes the overall condition of the animal into account when scoring, was not only able to explain the BFT best, but also did outperform other apparently more objective scoring systems. (2) Estimation of genetic parameters and prediction of EBVs on 1799 Brown Swiss cows with ASReml revealed heritabilities for milking speed, udder depth, position of labia, rank order in herd, general temperament, aggressiveness, milking temperament and days to first heat in similar ranges as reported in literature. Values on some traits (e.g. udder depth h^2=0.42±0.06) were at the high end, whereas estimates for others (e.g. days to first heat h^2=0.04±0.05) showed low heritability. Position of labia, genetically analyzed for the first time, showed a moderate heritability. Moreover, genetic parameters and accuracies of breeding values for milk content traits of individual udder quarters revealed significant systematic differences in fat, protein and lactose content between front and rear udder quarters, while content of urea, SCS and hyperkeratosis did not. Our findings suggest that the front and the rear udder could be considered as partly genetically different organs.

Session 42 Theatre 4

Effect of season and management system on 'Sfakion' sheep milk fatty acid profile

N. Voutzourakis[1,2], N. Tzanidakis[2], I. Atsali[1], E. Franceschin[3], A. Stefanakis[2], S. Sotiraki[2], C. Leifert[1], S. Stergiadis[1], M.D. Eyre[1], G. Cozzi[3] and G. Butler[1]
[1]Newcastle University, School of Agriculture Food and Rural Development, Newcastle upon Tyne, NE1 7RU, United Kingdom, [2]Hellenic Agriculture Organization-DEMETER, Veterinary Research Institute, NAGREF Campus, 57001 Thermi, Thessaloniki, Greece, [3]University of Padua, Department of Animal Science, Viale dell' Università 16, Agripolis, 35020 Legnaro, Italy; nvoutz@hotmail.com

Recent research has demonstrated possible beneficial effects of several milk fatty acids (FA) on human health. However, in contrast to cows, little is known about factors affecting the milk FA profile of small ruminants. Our study investigates seasonal variation of FA profile of sheep milk from two management systems. Ten extensive and 10 semi-intensive 'Sfakion' sheep flocks on Crete, Greece, were monitored for two consecutive lactations, collecting monthly bulk milk samples and managerial records. Milk FA profiling was carried out by gas chromatography. Analysis of variance was performed by linear mixed effects models in R, using 'management', 'month' and 'year' as fixed factors and 'flock' as a random factor. Significant variations of the FA profile were found between sampling months, especially comparing January with July; saturated FA were 6.8% lower and monounsaturated and omega-3 FA were 17.4% and 31.7% higher ($P<0.001$) respectively in the later. Differences were greater in extensive flocks, which had higher concentrations ($P<0.001$) of monounsaturated (+6.4%) and omega-3 FA (+21.7%) and lower concentration of saturated FA (-2.8%) compared to semi-intensive flocks. Differences were also identified between the years of this study; milk in year 2 had higher ($P<0,001$) concentration of monounsaturated (+18.2%), polyunsaturated (+28.8%) and omega-3 FA (73.1%) and lower concentration of saturated FA (-8.3%). Sheep milk FA profile highly varies within and between lactations, but these changes can be modified by managerial practices.

Effects of different proportions of sainfoin pellets combined with hazel nut peels on infected lambs

M. Girard[1], S. Gaid[2], C. Mathieu[3,4], G. Vilarem[3,4], V. Gerfault[5], P. Gombault[5], F. Manolaraki[1] and H. Hoste[1]
[1] INRA, UMR 1225 Interactions Hôte Agents Pathogènes, 23 chemin des Capelles, 31076 Toulouse Cedex, France, [2]Université F. Rabelais, Fac. Sciences et Techniques, Parc de Grandmont, 37000 Tours, France, [3]Université de Toulouse, INP-ENSIACET, LCA (Laboratoire de Chimie Agro-industrielle), 31030 Toulouse, France, [4]UMR 1010 INRA-ENSIACET, LCA (Laboratoire de Chimie Agro-industrielle), 31030 Toulouse, France, [5]SARL Multifolia, Viapres le Petit, 10380 Viapres le Petit, France; h.hoste@envt.fr

Tannin-rich plants are nutraceuticals helping to control GIN infections in ruminants. The aim of the study was to evaluate the anthelmintic activity of pellets of sainfoin completed by agro industrial by-products in *H. contortus* infected lambs. The study lasted for 7 weeks (D0 to D42). On D0, 24 lambs were individually infected with 4000 L3 and composed 4 groups (G1, G2, G3, G4), fed first *ad libitum* on hay plus 500 g lucerne pellets. On D21 post infection (PI), the G2, G3, G4 groups were offered sainfoin dehydrated pellets (i.e. 33; 66, 100% of the concentrate diet). G1 remained fed on lucerne pellets (control group). Moreover, from D35 to D42PI, G2, G3 and G4 received a daily individual supplementation of 500 g of hazelnut peels (HZP). The mean overall refusals of concentrate and HZP were measured from D21 to D34PI; and from D35 to D41PI. Packed cell volume (PCV) and faecal egg counts (FEC) were measured weekly. Last, worm counts were measured after necropsy (D42PI). There were no refusals of concentrate for the 2 experimental periods. The mean refusals of HZP from D35-D41 PI ranged from 68 to 82%. A constant decrease in PCV values was found but without any differences between groups. The reductions in FEC in the treated groups reached a maximum value of 60%. The differences between treated and control groups showed a trend ($P<0,09$) after HZP addition. The worm counts showed establishment rates ranging from 30 to 44% but with no differences between groups.

Improving low input pig production systems

J.I. Leenhouwers
TOPIGS Research Center IPG, Schoenaker 6, 6641 SZ Beuningen, the Netherlands; jascha.leenhouwers@topigs.com

In comparison with conventional pig production systems, low input systems are characterised by smaller herd size, more space per animal, lower capital investment, often outdoor management, greater labour requirements and focus on animal welfare. In order to improve production efficiency in low input pig production systems, an extensive research program was set up, aiming at developments in the areas of breeding, management and product quality. The program included research on breeding infrastructures and strategies in order to design dedicated breeding solutions for the low input sector. Key breeding goal traits, such as pig survival, sow longevity and heat stress resistance of sows, were evaluated for optimal inclusion in specifically designed breeding programmes. Breed choice for low input systems was investigated by experimental studies and surveys to compare reproductive performance and carcass and meat quality of modern versus traditional pig breeds. Various gilt rearing and lactation environments were compared for their effects on mothering ability and piglet health and welfare. So far, research highlights and key results of the project include the implementation of an economically viable replacement breeding strategy for organic pig production in The Netherlands. This concept is designed in such a way that it easily can be adapted and transferred to other low input systems across Europe. Another highlight is the definition and design of a sow robustness concept that will be implemented in the breeding goal of a newly developed genetically robust sow line. In conclusion, results from this project contribute to improvements in production efficiency, animal health and welfare and product quality in low input pig production systems. This will underpin consumer perceptions about added value quality characteristics of pork products from these systems and thus may help to maintain economic sustainability of such systems.

Can pig breeding contribute to the sustainability of low input production systems?

L. Rydhmer[1] and J.-L. Gourdine[2]
[1]Swedish University of Agricultural Sciences, Dept. Animal Breeding Genetics, Box 7023, 75007 Uppsala, Sweden, [2]French National Institute for Agricultural Research, Tropical Animal Science Unit, UR 143, 97170 Petit-Bourg Guadeloupe, France; lotta.rydhmer@slu.se

Low input systems (LIS) are often based on specific values such as cultural traditions or principles for organic production. Low amounts of external inputs imply a closed nutrient cycle. Climate change, growing world population and loss of biodiversity put high demands on all systems; conventional as well as LIS have to be efficient. A LIS breeding goal typically includes pigs' ability to efficiently use local feed (preferably waste and by-products), thrive in their climate (heating and cooling are energy consuming), stay healthy (limited use of chemotherapy), and maternal ability (piglet mortality decreases efficiency and sow milk alternatives are external inputs). With grazing, strong legs are needed. Systems based on internal inputs are exposed to larger variation in feed quality than conventional systems where inputs come from a global market. Thus, low environmental sensitivity is an additional goal trait. Traits listed above are relevant also for conventional production, but economic weights differ between systems. Socio-economic impact and acceptance of goal traits must be considered for each LIS. Organic producers in Sweden want higher weight on disease and parasite resistance. In a EU project, 15 production systems were studied. Many alternative systems used animals bred for conventional production. The claimed added values of the products were therefor not reflected in the breeding. Some systems with local breeds were studied. Pig population size and human and technical resources were limiting factors for their breeding work. This illustrates that the small scale of LIS (related to their local nature) is problematic, since breeding is more efficient for large populations. Choosing animals suitable for LIS from a conventional breeding programme can be a more realistic strategy than specific LIS breeding.

Comparison of growth intensity and muscle thickness in Pinzgau and Limousine heifers

P. Polák, J. Tomka, M. Oravcová, D. Peškovičová, M. Michaličková and J. Huba
Animal Production Research Centre Nitra, Department of Animal Breeding and Product Quality, Hlohovecká 2, 951 41 Lužianky, Slovak Republic; polak@cvzv.sk

A decreasing population of Pinzgau cattle in Slovakia is changing production from dairy to beef in system of suckler cows. The aim of our investigation was to compare growth intensity and muscle thickness measured by ultrasound in Pinzgau (39 heads) and Limousine (37 heads) heifers kept in low-input production system of mountain region of Upper Orava. Live weight at birth (WB), live weights before pasture season at an average age of 21 and 33 months (W1 and W3, respectively), and live weight after pasture season at an average age of 26 months (W2) were weighed. Ultrasound measurements of musculus longissimus thoracis at lumborum at loins and musculus gluteus at rump were measured at an age of 21 months. Growth intensity was calculated for period between each weight (WB – W1; WB – W2; WB – W3, W1 – W2; W1 – W3 and W2 – W3). The average muscle thickness in Pinzgau heifers were 49.03 mm at loins and 91.74 mm at rump. Both ultrasound measurements were significantly higher in Limousine heifers (by 12 mm at loins and 13 mm at rump). Because of a higher fat layer, a lower muscle layer in Pinzgau heifers was found. W1, W2 and W3 were higher in Limousine heifers; the differences were 50.66, 55.33 and 37.37 kg, respectively. The average daily gains from birth to W1, W2 and W3 were significantly higher in Limousine heifers. Because of a smaller increase of weight in Limousine heifers during last two controlled periods, the average daily gains between WB and W2, and between W1 and W2 were higher in Pinzgau heifers. This pilot study proved the hypothesis that Pinzgau cattle is of less musculature in hind part of body. More research is needed for good characterization of important beef production traits in suckler cows production system and to create an enhanced breeding protocol for Pinzgau breed.

Heterosis and combining ability for body weight in a diallel crossing of three chicken gnotypes

D. Norris and J.W. Ngambi
University of Limpopo, Animal Science, Private Bag X1106, Sovenga 0727, South Africa; david.norris@ul.ac.za

The aim of the study was to evaluate heterotic and combining ability effects for growth in nine chicken genotypes. A 3×3 complete diallel mating system involving two indigenous breeds named Venda (V) and Naked Neck (N) and one commercial broiler breed named Ross 308 (R) were used to produce three purebred (V × V, N × N, R × R), three crossbreds (R × V, R × N, V × N) and three reciprocals (V × R, N × R, N × V). The nine genetic groups of crosses were reared up from hatch to 13 weeks of age in deep litter open house. Body weights of 180 chicks (20 chicks per genetic group), recorded at 0, 3, 5, 7, 9, 11, and 13 weeks of age, were used to estimate heterosis, general (GCA) and specific (SCA) combining abilities. Results showed that the Ross 308 had the heaviest body weight at all weeks of measurement except for hatch. With respect to crosses, the V × R and its reciprocal cross, R × V had the heaviest body weights at 13 weeks. Heterosis estimates for body weight were higher in the Venda male X Ross 308 female and Venda male and Naked Neck female crosses. GCA was significant ($P \leq 0.01$) for body weight from hatch to 13 weeks of age while SCA and RE were both significant ($P \leq 0.05$) for body weight at all ages of measurement except for hatch. The Ross 308 gave the highest positive effect of GCA for body weight except for hatch. V × N gave the highest and positive effects of SCA for body weight. Results indicate that it may be important to consider developing a composite chicken breed based on the estimates of heterosis.

Effect of laying stage on egg characteristics and yolk fatty acid profile from different age geese

V. Razmaitė and R. Šveistienė
Lithuanian University of Health Sciences, Institute of Animal Science, R. Žebenkos 12, 82317 Baisogala, Radviliškis district, Lithuania; ruta@lgi.lt

The objective of the study was to determine some egg characteristics and yolk fatty acid profile in the early and full lay by Lithuanian Vishtinės geese of different age. The geese were kept in large open indoor pens with free access to outdoor area and were fed the same diet. The study was conducted on randomly collected twenty eggs within 24 h of laying at the beginning and after seven weeks at full lay by the first and third year geese. Chemical composition of egg yolk was determined by AOAC (1990) methods. The FAMEs were analysed using a gas liquid chromatograph (GC – 2010 SHIMADZU). The data were subjected to analysis of variance (ANOVA) with Tukey's tests to determine the significance of differences of least square means between the groups. All statistical analysis was performed using MINITAB 15. In early lay, eggs from young geese had lower yolk and higher albumen ratios, respectively, than those from older geese ($P<0.01$), however, in full lay there were no differences between the geese from different age groups. The age of geese and the laying stage did not appear to affect the proportions of total saturated fatty acids and monounsaturated fatty acids. The laying stage of the third year geese tended to show effect on the proportion of total polyunsaturated fatty acids (PUFA). However, the laying stage showed effects on the proportions of separate n-3 PUFA and other fatty acids. The yolk from all the geese in full lay had more than twice higher contents of n-3 PUFA ($P<0.001$), including increase of EPA (C20:5n-3), DHA (C22:6n-3) and DPA (C22:5n-3). The laying stage did not influence the atherogenic index and hypocholesterolemic/hypercholesterolemic ratio of yolk lipids. The age of geese appeared to affect fatty acid composition only in early lay. In full lay, there were no differences in the fatty acid composition of the yolk lipids between the first year and third laying year geese.

Session 42 Poster 11

PATUCHEV: an experimental device to assess high-performance and sustainable goat breeding systems

H. Caillat[1], A. Bonnes[2] and P. Guillouet[1]
[1]INRA, UE1373 FER, 86660 Lusignan, France, [2]REDCap-BRILAC, BP 50002, 86550 Mignaloux-Beauvoir, France; hugues.caillat@lusignan.inra.fr

In western France, which concentrates over half of the country's capacity in terms of dairy goat production, goat farms have gradually turned into intensive farming over the last 10 years, thereby significantly increasing their need for purchased input. In this context, the Patuchev platform is aimed at assessing and proposing innovative goat farming systems in order to lead to low input and sustainable goat farms. In 2012, a goats shed with a solar-heated air hay dryer has been built especially. This device is based on comparing 3 types of systems with 60 goats each: two grazing herds, one kidding at the end of the winter and the other one in autumn, and the last one fed hay indoors all year round and kidding in autumn. 10 hectares divided between multi-specific cultivated grassland and a cereal-protein crops mixture are allocated to each system. Evaluation and comparison are based on multi-criteria approach with data collected throughout lactation, dairy goats' careers and crop rotations. Since December 2012, the feed intake and milk yield are measured weekly and body condition scores and weight monthly. Health data are recorded and individual goat parasitism will be controlled to evaluate the kinetics of infestation in natural conditions. The production of grassland and botanical composition are evaluated every week for the grazed pastures and before each harvest for the 3 systems. The input and output flows are handled separately for each group and recorded weekly. For a better knowledge diffusion, professional goat farming organizations have implemented a coordinated Research and Development scheme called REDCap. This network includes 34 volunteer goat farms with the aim to improve and promote grass-based dairy goats systems and feed self-sufficiency. The first associated experience between these two projects is an evaluation of a multi-species grassland mixture sown on Patuchev platform and in 10 farms in autumn 2012.

Session 43 Theatre 1

Fibre genetics on alpaca

J.P. Gutiérrez
Universidad Complutense, Avda/ Puerta de Hierro s/n, 28040 Madrid, Spain; gutgar@vet.ucm.es

Pacomarca S.A. runs a genetic improvement program for alpaca fibre. Mating is carried out individually and breeding values are used for selection and embryo transfer. It wills extend its advances to the small rural communities. The aim of these studies was analysing the genetics of the fibre. Traits such as fibre diameter (FD), coefficient of variation of FD (CV), greasy fleece weight (GFW), staple length (SL), shearing interval (SI) and textile value index (TV) were analysed to estimate genetic parameters. Results allowed concluding that expected selection response for TV was higher when FD was considered as selection goal instead of TV itself. FD at different ages was after considered as different traits and analysed by a multitrait animal model. Shearing at two years of age was shown to be the best showing the genetic value of the animal. Genetic parameters were after estimated for the traits in the selection criteria, (FD; CV, comfort factor (CF); and standard deviation (SD) of FD) jointly with type traits (fleece density, crimp, lock structure, head, coverage, and balance). Heritabilities for fibre traits were moderate to high and fibre and type traits were, in general, genetically poorly correlated. Afterwards, the optimal weighting of those traits was analysed under a selection index to conclude that the weight applied to CF should be surprisingly negative, and that morphological traits might be penalized if all the weight of the objective was on the fibre traits. Trying to model the relationship between FD and its variability, FD was studied under two innovative procedures. The results suggested that a genetic selection program is plausible to modify the evolution of the fibre diameter along time together with a favourable correlated decrease in the fibre diameter. A final study focused on the search of major genes in fibre traits. Significant segregating major genes were found associated with decreased FD, SD, CV values and increased CF values. The major gene variance was larger than the polygenic variance for all traits.

Session 43 — Theatre 2

Design of a community-based llama breeding program in Peru: a multi-stakeholder process

M. Wurzinger[1], A. Rodriguez[2,3] and G. Gutierrez[3]
[1]BOKU-University of Natural Resources and Life Sciences, Gregor-Mendel-Strasse 33, 1180 Vienna, Austria, [2]UNDAC-Universidad Nacional Daniel Alcides Carrion, Edificio Estatal N° 03, Pasco, Peru, [3]UNALM-Universidad Nacional Agraria La Molina, Av. La Universidad s/n. La Molina, 12 Lima, Peru; maria.wurzinger@boku.ac.at

The Peruvian llama population counts about 1 million animals and the sale of meat and breeding animals is of economic importance for many smallholders. Nevertheless, there are a number of factors hindering a higher productivity, one of them are well established breeding programs. The llamas of the central highlands (Department of Junin and Pasco) are well-known by llama producers as these animals are very tall and heavy and breeding stock is sold every year to other regions of Peru. Farmers raised their concern of losing potentially good genetics, when there is no concerted breeding management in place. Therefore the aim of this study was the design of a community-based breeding program and a multi-stakeholder consultation process was started. This involved personal interviews with farmers, but also a series of workshops with farmers, representatives of local government, an NGO and universities. These platforms were used to distribute information on breeding programs, but also to discuss and agree on level of involvement, roles and responsibilities of different stakeholders in a breeding program. In a first step, 70 farmers of 20 communities agreed to form a breeders association. At the same time a phenotypic characterisation of the llama population was performed. In addition, a preliminary market analysis was carried out to get a better understanding of the complete value chain of llama meat. Alternative breeding strategies, such as central versus dispersed nucleus, were presented and discussed with farmers. This participatory approach with the involvement of different actors ensures commitment and ownership of all parties which is a pre-requisite for the long-term sustainability of a breeding program.

Session 43 — Theatre 3

Performance of alpacas from a dispersed open nucleus in Pasco region, Peru

G. Gutierrez, J. Candio, J. Ruiz, A. Corredor and E. Flores
Universidad Nacional Agraria La Molina, Animal Production, Av. La Molina s/n. Lima 36, Peru; gustavogr@lamolina.edu.pe

The alpaca (*Vicugna pacos*) is a native animal from the High Andes of Peru. The main product is a special fiber of high value for the textile industry. In 2010 a genetic improvement program has started in Pasco Region of Peru in order to improve fiber quality and quantity. Six alpaca production units agreed to form a dispersed open nucleus with the technical support of UNALM. Alpacas were selected by visual appraisal and grouped in the following categories S, A, B, C and R. Fleece weight (FW) and body weight (BW) were recorded at shearing time in 2011 and 2012. Also fiber samples were taken to measure fiber diameter (FD) and coefficient variation of FD (CV) using IWTO-12 regulation. A model that includes effects such as category, age group, sex and their corresponding interactions was used to analyze the data by using SAS. Alpacas from different age group and categories differed in FD and BW but not in CV and FW. Alpacas from category S had lowest DF. Also, DF and BW tended to increase with age. Interaction sex by age group was significant for FW. Next step is to build a selection index for simultaneous genetic improvement of FD, CV and FW.

Session 43 Theatre 4

Characterization of llama (*Lama glama*) milk proteins

B. Saadaoui[1], L. Bianchi[2], C. Henry[3], G. Miranda[2], P. Martin[2] and C. Cebo[2]
[1]Faculté des Sciences de Gabès, Cité Erriadh Zrig, 6072 Gabès, Tunisia, [2]INRA, UMR 1313 Unité Génétique Animale et Biologie Intégrative, 78350 Jouy en Josas, France, [3]INRA, UMR 1319 MICALIS, Plateforme PAPSSO (Plateforme d'Analyse Protéomique Paris Sud Ouest), 78350 Jouy en Josas, France; besma_sadawi@yahoo.fr

Llamas belong to the *Camelidae* family along with camels. While camel milk has been broadly characterized, data on llama milk proteins are scarce. Previously released studies were only limited to the analysis of gross composition of milk (i.e. total fat, protein, or lactose content). The objective of this study was thus to investigate the protein composition of llama milk. Data were compared with those from dromedary milk, a closely related species. First, the protein concentration of llama and dromedary milk was determined. Surprisingly, the average value of protein concentration was roughly twice higher in llama milk compared with dromedary milk. Skimmed llama milk proteins were further characterized by a two-dimensional separation technique coupling Reverse Phase High Pressure Liquid Chromatography (RP-HPLC) in the first dimension with sodium dodecyl sulphate-polyacrylamide gel electrophoresis (SDS-PAGE) in the second dimension. Identification of proteins was achieved using peptide mass fingerprinting. This proven methodological approach allowed us to identify the major proteins in llama milk, namely caseins (αs1-, αs2-, β- and κ-caseins), α-lactalbumin, lactoferrin, lactophorin and serum albumin. Significant quantitative and qualitative differences were observed between camel and lama milk samples. Finally, we characterized proteins of the Milk Fat Globule Membrane (MFGM), the membrane surrounding fat in milk, in the llama species. The MFGM protein profile from llama was found to be highly similar to the MFGM protein profile from camel milk. Taken together, these data provide for the first time a thorough description of the milk protein fraction from llama, a new-world camelid.

Session 43 Theatre 5

Energy requirements during lactation in llamas (*Lama glama*)

A. Riek[1], A. Klinkert[1], M. Gerken[1], J. Hummel[1,2], E. Moors[1] and K.-H. Südekum[2]
[1]University of Göttingen, Department of Animal Sciences, Albrecht-Thaer-Weg 3, 37075 Göttingen, Germany, [2]University of Bonn, Institute of Animal Science, Endenicher Allee 15, 53115 Bonn, Germany; mgerken@gwdg.de

Llamas and alpacas have become popular as companion and farm animals in Europe and North America. However, scientific knowledge on the nutrient requirements especially for lactating and suckling llamas is sparse. Thus, most of the nutrient recommendations for New World Camelids available today have been derived from other livestock species, such as goats and sheep. Therefore we aimed to investigate the energy needs for llamas during lactation. For this purpose we measured daily milk output in 5 llama dams at different stages of lactation (i.e. at week 3,10,18 and 26 pp) using an isotope dilution technique (IDT). The method involved the application of the stable hydrogen isotope Deuterium (^{2}H) to the lactating dam. We also related estimated milk outputs to daily energy intakes. Furthermore we validated the IDT by measuring total water turnover (TWT) directly and compared it with values estimated by the IDT. Water intake and TWT decreased significantly with lactation stage, whether estimated by the isotope dilution technique or by calculation from drinking water and water ingested from feeds. Results from measured and estimated TWT revealed that the IDT estimated TWT with high accuracy with only small variations. Calculated ME intakes during lactation decreased with lactation stage but remained constant per kg milk output. However, lactation stage had no effect on the milk water fraction, i.e. the ratio between milk water and TWT. Although recommendations for energy requirements in lactating llamas so far have been based on extrapolations from sheep and goat data, the comparison with our measured data shows that these extrapolations seem to be fairly appropriate. However, our more detailed and measured data on ME intakes in lactating llamas could serve as a more accurate basis for further recommendations for the energy requirements in New World Camelids.

Genes involved in hair follicle cycle of cashmere goat

W. Zhang[1,2,3], D. Allain[4], R. Su[1,3], J. Li[3] and W. Wang[2]
[1]Institute of ATCG, Nei Mongol Bio-Information, Hohhot, 010018 Hohot, China, P.R., [2]Chinese Academy of Sciences, Kunming Institute of Zoology, No.32 East Jiaochang Road, 650223 Kunming Yunnan, China, P.R., [3]Inner Mongolia Agricultural University, College of Animal Science, No. 306, Zhaowuda Road, 010018 Hohot, China, P.R., [4]INRA, UR631, SAGA, Auzeville, CS52627, 31326 Castanet Tolosan, France; atcgnmbi@aliyun.com

On behalf of International Goat Genome Consortium(IGGC) The hair follicle (HF) is central to most economically important fiber growth in livestock. However, the changes in expression of genes that drive these processes remain incompletely characterised. As model animal of hair biology, cashmere goat might help deciphering genes involved in primary and secondary HF of skin. We used RNA-seq to study gene expression profiles of HF including anagen, catagen and telogen in Nei Mongol Cashmere Goat (NMCG), which will increase our understanding of HF biology and contribute to the development of strategies to improve cashmere. Exome data from 7 secondary HFs and 11 primary HFs in NMCGs was mined. 22,176 annotated genes in reference genome were evaluated, 7,481 of which were shown to be expressed in HFs. Join together, 1,923 of 7,481 were expressed in all HFs, 3,219 and 2,339 of which were co-expressed in primary and secondary HFs exclusively. GO analysis showed that GO: (0044464, 0003824, 0032501, 0032502) were associated gene sets. High expressed gene in secondary and primary HFs are mainly involved in 14 GO categories which included in above mentioned GO items. High expressed genes in two type HFs of 1,923 are mainly located in CHI 7, 18 and 19 (P<0.01). Genes located in 4.6~4.9 Mb of CHI1 might be one gene cluster which has important function in HF development. Six SNPs related to hair follicle based on ~30K polymorphism loci different from reference, were functionally identified in CHI 1 and 19 respectively. Compare to secondary HF, primary HF expresses more genes to deal with stimulations from all-environment. Additionally, 13 HF specific expressed genes were identified according to comparative transcriptome analysis.

SNP mapping of QTL affecting wool traits in a sheep backcross Sarda × Lacaune resource population

D. Allain[1], S. Miari[2], M.G. Usai[2], F. Barillet[1], T. Sechi[2], R. Rupp[1], S. Casu[2] and A. Carta[2]
[1]INRA, UR631, SAGA, Auzeville, CS52627, 31326 Castanet Tolosan, France, [2]AGRIS-Sardegna, Settore genetica e biotecnologie, DiRPA, 07704 Olmedo, Italy; daniel.allain@toulouse.inra.fr

A QTL detection experiment was organized in a backcross Sarda × Lacaune sheep resource population. The aim of this experiment was primarily the search for loci influencing milk production and several other traits including wool traits through a whole genome scan on 968 females from 10 sire families. Ten fleece characteristics: greasy fleece weight, length of long and fine wool as well as fibre diameter (mean and CV), fibre curvature (mean and CV) and medullation content (objectionable, flat and medullated fibres) using OFDA methodology were measured on 892 6-months old females. For QTL detection, the Illumina OvineSNP50 beadchip that provided 44,859 SNP markers after quality control was used. Within and across families analyses were performed with the QTLMAP software. The statistical techniques used were linkage analysis, linkage disequilibrium analysis and joint linkage and association analysis (LDLA) using interval mapping. High significant QTL (P<0.001 at genome wide) affecting 7 wool traits: greasy fleece weight, fibre diameter (mean and CV), Fibre length, medullation content and CV of fibre curvature were found on chromosome 25 within a 2 cM interval suggestting that one or some genes with major effect on fleece characteristics are located on this chromosomal segment. Other high significant QTL's (P<0.001 at genome wide) influencing medullation content and, fibre length on chromosome 20 and 15 respectively were also detected. Other putative QTL's (P<0.01 at chromosome wide) were also observed on chromosome 3 and 6, 13 and 18, and 14 for greasy fleece weight, fibre length, and fibre curvature respectively. The linkage disequilibrium analysis and joint LDLA analysis confirmed the locations of the QTL mapped on OAR25, 20 15, 13 and 6 and all QTL found were discussed.

The agouti gene in black and brown alpaca

C. Bathrachalaman, C. Renieri, V. La Manna and A. La Terza
University of Camerino, School of Environmental Sciences, Via Gentile III da Varano s/n, 62032 Camerino, Italy; carlo.renieri@unicam.it

The Agouti gene encodes agouti signaling protein which regulates pheomelanin and eumelanin synthesis in mammals. To investigate the role of Agouti in coat color variation of alpaca, we characterize the agouti gene on 27 black and 12 brown alpaca. The exon-4 hosts three loss of function recessive mutations, g.3836C>T, g.3896G>A and g.3866_3923del57, involved in eumelanin synthesis. The deletion at the position p.C109_Rdel19 eliminates the two beta sheets and the R-F-F- motif from the agouti functional domain, which are essential agaist alfa-MSH. Therefore, the deleted allele appears to lose function. The other SNPs observed at the amino acid position 98 and 118 change the conserved R to C and the R-F-F- motif into H-F-F-. The R-F-F- motif is important for functioning at MCRs; the disruption in this motif may result in a non functional agouti protein since the alteration of residues in and around R-F-F- causes a decrease in agouti protein inhibition of alfa-MSH binding to MCRs during signal trasduction. The three mutations are randomly distributed among the black alpaca. In our sample, we observed two genotypes: g/3836C>T / g.3896G>A (10 animals) and g.3836C>T / g.3866_3923del57 (17 animals). Among the brown alpaca, 2 are homozygous for the wild allele, twelve are heterozygous for g.3896G>A mutation, carriers for black phenotype.

Fine mapping of birthcoat type in the Romane breed sheep

M. Cano[1,2], D. Allain[3], D. Foulquié[4], C.R. Moreno[3], P. Mulsant[2], D. François[3], J. Demars[2] and G. Tosser-Klopp[2]
[1]INTA, Instituto de genetica, CICVyA, CC25, CP 1712 Castelar, Argentina, [2]INRA, UMR444, Génétique Cellulaire, Auzeville, CS52627, 31326 Castanet Tolosan, France, [3]INRA, UR631, SAGA, Auzeville, CS52627, 31326 Castanet Tolosan, France, [4]INRA, UE 321, Domaine de La Fage, Saint Jean et Saint Paul, 12250 Roquefort, France; julie.demars@toulouse.inra.fr

Birthcoat type is an important component of lamb survival for sheep raised under harsh environment. At birth two types of coat were observed: a long hairy coat or a short woolly one. It was shown that hairy coat lambs are more adapted to survive around lambing time due to a better coat protection with less heat losses at coat surface and show better growth performances up to the age of 10 days than woolly coat lambs. Birthcoat type was estimated to be a highly heritable trait and it was reported that its determinism seems to be under the control a few major genes. A QTL detection design was initiated in a Romane breed population to search for loci influencing adaptive traits including birthcoat type through a whole genome scan with the OvineSNP50 beadchip on 824 lambs issued from 8 sire families. A highly significant ($P<0.1\%$ Genome Wise (GW)) and a putative QTL affecting birthcoat type were found on chromosomes 25 and 13 respectively. Fine mapping with additional markers, comparative mapping and sequencing of the QTL segment on OAR25 revealed the presence of a 2 kb DNA deletion segment. All animals from the experimental design were genotyped for the presence or absence (ins/del) of this segment on OAR25. Homozygous del/del animals were all bearing a hairy coat at birth but not all the homozygous ins/ins animals were bearing a woolly coat. When including the ins/del genotype on OAR25 as a fixed effect within the linkage analysis model, a highly significant ($P<0.1\%$ GW) QTL was found on OAR13 with a significant interaction ($P<0.001$) between QTL on OAR13 and the ins/del genotype fixed effect of OAR25. It was suggested that both QTL on OAR13 and OAR25 are involved as major genes in the determinism of birthcoat type.

Session 43 Theatre 10

The DNA-Project of the llama and alpaca registry Europe (LAREU)

C. Kiesling
LAREU, P.O. Box 666, 3900 Brig-Glis, Switzerland; christian.kiesling@cern.ch

As a truly European registry, the llama and alpaca registry Europe (LAREU) is providing an online registration system for breeders and owners of South American camelids (SACs), free of charge. Founded in the year 2005, over 10000 animals from more than 10 European countries are stored in LAREU's database up to now, with a yearly growth of about 20%. The database itself is professionally maintained by the animal welfare organization TASSO, based in Germany, hosting data from a few million animals, mostly dogs and cats. The data of the animals registered with LAREU are entirely separated from the TASSO database, the registration software has been established independently. The transparent borders within the European Union are offering breeders of alpacas SACs great opportunities for improving the genetics of their animals, both with respect to fiber quality as well as to conformation and health. An essential factor in a successful breeding program is the detailed knowledge of the genetic information which is the only reliable way to prove parentage. While DNA testing of alpacas and llamas is done in several national camelid associations, usually working with different DNA marker sets, there is obviously no way to obtain reliable pedigrees for animals across associations and borders. LAREU has recognized from the beginning the importance of a Europe-wide registry, and, for the reasons given, of a world-wide standard of the DNA markers used for SACs. In 2007 LAREU has contacted the 'International Society of Animal Genetics (ISAG)', responsible for the standardization of animal DNA typing, and suggested a research program with the aim of finding a suitable set of markers for SACs. In this paper we will present statistical material from the LAREU database, shortly review the basic genetics leading to the ISAG set of markers, give the details of the new DNA marker set, and then describe the procedures developed together with internationally accredited DNA laboratories to provide a transparent and reliable system for the camelid owner to obtain and evaluate the DNA data of his animals.

Session 43 Theatre 11

Animal welfare problems in alpacas and llamas in Europe

I. Gunsser
AELAS e.V., Römerstr. 23, 80801 München, Germany; ilona.gunsser@t-online.de

Alpacas and llamas (South American camelids, 'SACs') are becoming more and more popular in Europe for two different reasons. One group of owners is buying these animals as a business investment to produce fiber animals and to sell their fiber and their offspring to the market. Another group of owners are buying SACs as companion animals with the intension to train them for trekking excursions or animal assisted activity and therapy. Very often both groups of persons become owners of these animals with very little or no experience how to handle and care about agricultural animals. The results are problems which may interfere with the health and the animal welfare of SACs. The lack of knowledge about the typal behavior and the needs of alpacas and llamas can lead to excessive stress and anxiousness, or even aggression. In the case of companion animals, which are mostly castrated males or non-pregnant females, one reason for stress can be wrong training techniques with incorrect material and without recognizing and reacting to the typical signs of stress. Other reasons may be the use of untrained animals for certain techniques of human-animal interactions, or owner-induced problems of mal-imprinting of young animals. In the case of animals being kept for breeding and fiber production – mostly alpacas – problems may arise from wool–blindness or lack of regular sheering, or sheering with fixation in an un-physiological stretching technique. Also the keeping of intact males in groups may lead to stress, dangerous fights and injuries. Grass and hay eating ruminating animals have a digestion system which is quite different to other animal species. Even if told not to do so, quite a number of owners feed SACs with grain and vegetables. This can lead to acidity of the stomach and severe illness or death. As well as insufficient understanding of the problems caused by internal and external parasites. In this paper examples of the above-mentioned problems will be presented in detail and methods for avoiding them will be discussed.

Animal fibers in Argentina: production and research

J.P. Mueller[1], M.E. Elvira[2] and D.M. Sacchero[1]
[1]National Institute for Agricultural Technology, CC 277, 8400 Bariloche, Rio Negro, Argentina, [2]National Institute for Agricultural Technology, Avda 25 de Mayo 87, 9103 Rawson, Chubut, Argentina; mueller.joaquin@inta.gob.ar

Argentina is amongst world's largest producers of wool and mohair in addition to llama and small amounts of cashmere, silk, vicuña and guanaco fiber. Most wool and all mohair are produced in the extensive cold desert of Patagonia. More than 50% of the wool is merino which averages 20.5 mic and 1% vegetable content. Above 90% of the wool is exported mainly to the EU and China, 80% of it in tops and garments. Argentinean mohair is white and in general medullated (2-6%). Adult mohair averages 30 mic and kid mohair 24 mic. Mohair is exported mainly to the EU. Only about a third of the llamas are shorn. Fleece weights are 0.8-1.2 kg, 40% of llamas are single coated and about 30% are white and fiber diameter averages 22 mic. Following strict regulations, vicuñas and guanacos are captured in the wild, shorn and released. Wild guanaco fleeces weigh 0.4 kg and average 15.5 mic. Vicuña fleeces weigh 0.3 kg and average 13 mic. Three animal fiber testing laboratories analyze wool bale core test samples representing 95% of first sale of wool. These labs also process individual animal samples aiding selection programs and research. Research has been on management issues like optimum stocking rates and shearing dates. Shearing prepartum instead of the conventional postpartum date has increased wool clean yield and marking percentage by close to 10 percentage points. Research on GxE interaction and fiber diameter profiles showed that it is possible to produce 18 mic merino wool as long as shearing is prepartum, thus avoiding mid breaks of staples. Central progeny testing results convinced many breeders of the usefulness of objective measurements for management and breeding. Extensive research went into the development of the national sheep genetic evaluation scheme: Provino. Other fibers underwent quality description surveys and research including the search QTL's responsible for wool and mohair quality traits.

Animal fiber production in Turkey: present situation and future

G. Dellal[1], F. Söylemezoğlu[2], Z. Erdogan[2], E. Pehlivan[1] and Ö. Köksal[3]
[1]Ankara University Agricultural Faculty, Animal Science, Ankara University Agricultural Faculty Department of Animal Science, 06110 Ankara, Turkey, [2]Ankara University School of Home Economics, Handicrafts, Ankara University School of Home Economics Department of Handicrafts, 06110 Ankara, Turkey, [3]Ankara University Agricultural Faculty, Agricultural Economy, Ankara University Agricultural Faculty Department of Agricultural Economy, 06110 Ankara, Turkey; gdellal@agri.ankara.edu.tr

In today's world almost nine different animal kinds are used to produce fiber for trading purposes. All of these animals are mammals except silkworm. Mainly in Turkey wool, mohair, silk, goat coarse hair production is made and cashmere fiber and Angora rabbit wool production is made in very low levels. Between the years 1991 and 2011; the production of wool, mohair, goat coarse hair and silk decreased significantly. Also there is not enough data about production levels of cashmere fibers and Angora rabbit wool. The reasons of the decrease in animal fiber production in Turkey are rapid increase in the use of chemical fibers, changes in fashion which have negative impacts on the consumption of fibers especially mohair, the import of wool, and mohair in low prices and systematic problems on production of sheep, Angora goat, hair goat and silkworm. In contrast to the situation in Turkey, in recent years EU countries have attempted to increase both industrial production and income of small agricultural establishments which are on non-agricultural lands with different animal fiber production systems. However, Turkey has significant potential in animal fiber production. Therefore, in order to utilize this potential effectively, the development of different models, especially increasing the consumption of fiber products, will regularly provide raw materials needed for the textile industry and it will also considerably contribute to the rural development and conservation of the native animal's genetic resources and national culture at an important level.

Session 43 Theatre 14

Genetic trends for fleece traits in South African Angora goats

M.A. Snyman[1], E. Van Marle-Koster[2] and C. Visser[2]
[1]Grootfontein Agricultural Development Institute, Private Bag X529, Middelburg, 5900, South Africa, [2]University of Pretoria, Department of Animal and Wildlifes Sciences, Private bag X20, Hatfield 0028, South Africa; evm.koster@up.ac.za

South Africa is considered as a major contributor of a high quality mohair clip to the world market. Mohair traits included in selection programs consist of both production and quality traits with fibre diameter being the price-determining trait. Fiber diameter is moderate heritable, with estimates varying between 0.30 and 0.45 for the South African Angora goat population, and fleece weight being lowly (0.19) to moderately (0.24) heritable. Unfavourable positive correlations between fibre diameter and fleece weight ranging between medium (0.35) and strong (0.55) have been confirmed in South African studies. Records from 7 breeders over twelve years (2000-2011) were analysed to estimate genetic trends for fleece weight, fibre diameter, coefficient of variation of fibre diameter and staple length. Average fleece traits recorded at the third shearing at eighteen months of age were 1.5 kg for fleece weight, 28.8 μm for fibre diameter, 24.2% for coefficient of variation of fibre diameter and 129 mm for staple length. Although a selection index (SI) for increasing body weight, maintaining fleece weight and decreasing fibre diameter has been used by many breeders (SI = (13 × Body Weight) + (4 × Fleece Weight) – (23 × Fibre Diameter)), fleece weight has decreased over the years, in part due to the emphasis placed on decreasing fibre diameter. Currently, breeders are putting more emphasis on fleece weight and staple length, while maintaining the low fibre diameter. A continuous evaluation of breeding objectives and selection criteria to maintain a high quality mohair clip and efficient production of South African mohair is necessary.

Session 43 Theatre 15

Selection and genetic variability of the French Angora goat breed: 30 years of a breeding program

C. Danchin-Burge[1] and D. Allain[2]
[1]Institut de l'Elevage, Genetic & Phenotypes, 149 rue de Bercy, 75595 Paris Cedex 12, France, [2]INRA, UR631 Station d'amélioration génétique des animaux, BP 52627, 31326 Castanet-Tolosan, France; coralie.danchin@idele.fr

The French Angora goat breeding program was based on animals imported at the beginning of the eighties mainly from North America, South Africa, Australia and New Zealand. The whole selection program was based on these founders' animals and their offspring as no foreign animals were imported since then. The main selection objectives were to improve the quality and the quantity of the mohair produced. This article will present the genetic progress on the three traits selected by the program (fineness, fleece weight and kemp medullation score) over the last decade. The genetic variability of the breed was also assessed by using its pedigree information, thanks to the PEDIG software (INRA). The pedigree depth can be considered to be good (5.5 equivalent generations). Various indicators of genetic variability were calculated, such as the effective number of founders and ancestors, average inbreeding and kinship and effective population size. The evolution of the various countries of origin was also assessed and showed a clear selection of the North American genes. Based on the average genetic trends, the breeding program could be assessed as clearly successful. One of its downside would be the rather narrow genetic basis of the breed, which could also be explained by the small number of founders used at the start of the selection scheme. In order to preserve the future of the breed, specific measures such as cryoconservation of semen were decided.

Processing and commercial opportunities for European animal fibre producers

N. Thompson
Biella The woolcompany, Via Vittorio Veneto 2, 13816 Miagliano, Italy; nathompson@biellathewoolcompany.it

The challenge was to offer top quality woolen processing systems developed over many decades in the Biella textile district directly to the producers themselves. Apart from the more evident difficulties involved due to the necessity that the operating base is able to communicate in several languages, the cultural differences between a variety of producers from a predominantly agricultural background and the processors in the textile industry, which in most cases, they have never had any direct contact with. New consumer trends are offering opportunities to the European fibre grower to provide products which fully match todays up and coming demand for sustainable fashion. This demand can be satisfied by encouraging the producer to learn more about his fibre, to aid him in his promotion using by indirect social network approaches and providing him with a traceability system from farm to factory to farm.

Regional projects valuing wool in Europe

M.T. Chaupin
Association Atelier/Wools of Europe, c/o Filature de Chantemerle, 05330 Saint Chaffrey, France; atelier5@orange.fr

After decades of neglect of the wool industry in Europe, we see in the past few years the development of some regional projects valuing wool. The impetus for these projects comes rarely from the 'traditional' breeders who are unfortunately convinced of the insignificance of this raw material and mostly concerned about its rapid elimination. But this natural and renewable resource find a new interest among consumers concerned by the origin of raw materials for their clothing and home textiles (local production, non-use of non-renewable resources, specific identity, ...). The initiators of these projects are from the wool sector, the 'atypical' sheep breeders, some professional sectors away from these two worlds or even from the social economy. But we had to build up again a complex chain whose parts have gone away far from Europe due to delocalization, to give farmers the pride of this raw material and the basic knowledge necessary for its recovery, to recreate collective structures which only allow a efficiency. Coordination and exchange of experiences between regional groups is a major challenge for the coming years. As well as thoughts on the technical tools needed in the sector. Keep to each one its specific identity while sharing experience and skills is a long and difficult way but necessary for solid reconstruction and long-term industry. Four examples will be presented: (1) APPAM in the Alpes Maritimes, France valorisation of wool from 4 breeds: Brigasques into carpet, Mourerous, Préalpes and Mérinos d'Arles into knitting wool and knitting garments (2) Consorzio Biella the Wool Company, Italy Valorisation of wools from Piedmont and Abruzzo (3) Vielfacht der Kollektion, Germany Valorisation of the rare breeds Coburger Steinschaf and Alpines Steinschaf (4) FIWO, Switzerland Valorisation of Swiss wool in a wide range of products: insulation, bedding,... The different projects will be presented from several points of view: breeding, local economy, socio-economic conditions and environmental consequences.

Intra-chromosomal recombination of Agouti gene in white alpaca

C. Bathrachalaman, C. Renieri, V. La Manna and A. La Terza
University of Camerino, School of Environmental Sciences, Via Gentile III da Varano s/n, 62032 Camerino, Italy; carlo.renieri@unicam.it

The role of the agouti gene in white phenotype was explored in mice. The agouti signaling protein (ASP) can inhibit the differentiation of melanoblasts through the inhibition of the alfa-MSH-induced expression of microphtalmia (MITF) and its binding to a M box regulatory element. The level of microphtalmia in the cells is reduced. To investigate the role of agouti in this phenotype we characterised the agouti genomic and transcript structures and relative mRNA expression levels in 13 white alpaca. The reverse transcription analysis of mRNA purified from skin biopsies revealed the presence of three transcripts with different 5'untranslated regions (UTRs) and color specific expression. One of the transcripts, possibly originating from a duplication event (intra-chromosomal recombination) of the agouti gene is characterised by a 5'UTR containing 142 bp of the NCPOA6 gene sequence. Furthermore, the relative level expression analysis of mRNA demonstrates that the agouti gene has up-regulated expression in white skin, suggesting a pleiotropic effect of agouti gene in the white phenotype.

Session 43 Poster 19

High degree of variability within αs1-casein in llama (*Lama glama*) identified by isoelectric focusin

I.J. Giambra[1], S. Jäger[1], M. Gauly[2], C. Pouillon[1] and G. Erhardt[1]
[1]Justus-Liebig-University, Department of Animal Breeding and Genetics, Ludwigstrasse 21B, 35390 Giessen, Germany, [2]Georg-August-University, Department of Animal Science, Division of Livestock Production, Albrecht-Thaer-Weg 3, 37075 Goettingen, Germany; georg.erhardt@agrar.uni-giessen.de

Studies concerning milk yield and composition of llamas, as one of the main species of the New World Camelids, are until now scarce. First analyses concerning electrophoretic separation of milk proteins were made, but to our knowledge no milk protein variability was described for this species until now. In a first step we analysed protein variability of alpha$_{s1}$-casein (α_{s1}-CN), coded by CSN1S1, as one of the main milk proteins in llama milk. Therefore, milk samples of 45 llamas (*Lama glama*), hold in different flocks in Germany, the Netherlands, Switzerland, and Chile, were analysed by isoelectric focusing in ultrathin layer polyacrylamide gels using carrier ampholytes. Llama milk samples showed a high variability within α_{s1}-CN with the simultaneous identification of four α_{s1}-CN alleles, preliminary named with 1 to 4, on the basis of increasing isoelectric point, whereas allele 2 showed highest frequency (0.79). Homozygous phenotypes are characterized by four bands with different intensity and could be demonstrated for the alleles 1 to 3. The results show a higher degree of variability within *L. glama* in comparison to *Camelus dromedarius*, where, until now, two alleles are described. Analyses of further llama milk samples and DNA-based studies concerning CSN1S1 are in progress to get a more complete picture about α_{s1}-CN/CSN1S1 variability within llama and to use them in evolutionary and population studies.

Effect of linseed (flax) ingestion and oil skin-application on hair growth in rabbits

K.B. Beroual[1], Z.M. Maameri[1], B.B. Benleksira[1], A.A. Agabou[2] and Y.H.P. Hamdi Pacha[1]
[1]Laboratory pharmacologytoxicology, Institute Veterinary University Mentouri Constantine, Route Batna, El khroub, Constantine 25100, Algeria, [2]Laboratory PADESCA, Institute Veterinary University Mentouri Constantine, Route batna el khroub Constantine 25100, Algeria; beroualk@yahoo.fr

However, it was shown that flaxseed chutney diet doesn't affect γ-glutamyl transpeptidase load. This microsomal enzyme is an indicator of hair growth (associated to alkaline phosphatase). The following study aims to assess the effects of linseed (flaxseed) (*Linum usitatissimum*) on hair growth in rabbits (quantitatively and qualitatively) and to study also its safety. Two trials have been conducted on adult New-Zealand rabbits (divided into four groups according to the route of administration: ingestion of flax seed and topical application of its oil (with control group for each route). Weekly rabbits were weighed and each month hair was shaved from a same delimited area on the back of each rabbit and blood samples were taken. Results showed a slight increase in mean weight (+3%) and a significant decrease in glycemia (-9%) and cholesterolemia (-22%) in the group fed daily with the ground seed, compared to control one. These findings are similar to those reported in the literature. Nevertheless, the results related to the trichogen effects are original. Hair growth recorded no significant difference during the first month, but a better and motivating growth was recorded later (second and third months). An increase in hair length (+26%) was observed in the third month (+2.04±0.27 cm) with a slight positive effect (+7%) on hair diameter (4.03±0.48 µm). Weigh of locks harvested from the oil topical application group, has increased (over than 53%) compared to the control one. The results of this study motivate more investigations to determine with exactness doses and schedule intake/application while taking into account the seasonal variations in hair growth.

Session 43 Poster 21

EAAP Animal Fibre Working Group (AFWG): experience of funding applications to the EC COST Framework

C. Renieri, M. Gerken, D. Allain, J.P. Gutierez, R. Niznikowski, A. Rosati and H. Galbraith
EAAP Animal Fibre Working Group, c/o University of Camerino, Environmental and Natural Sciences, via Pontini 5, 62032 Camerino, Italy; h.galbraith@abdn.ac.uk

The EAAP AFWG was constituted in 2007 with a view to enhancing the role of animal fibre in EU27, utilising an approach based on science and technology. Although an unrecorded and neglected product, annual production of wool from 62 m breeding sheep (Eurostat) alone is substantial at an estimated 186,000 tonnes (FAO). Recent outputs include organised symposia and publications defining current knowledge. In recognising the need for better networks of scientists and technologists, a total of 5 applications for financial support has been made to the EC COST Framework since 2010. Such applications, by initial pre-proposal, are assessed in 6 categories with a maximum score of 6 for each, giving a maximum score of 36. Pre-proposals scoring most highly, on average, are invited to submit a full proposal. Evaluation has been characterised by large variation in scores of individual assessors. For one example, scores of 36, 33, 32, 32, 28, 21,16 were awarded by 7 assessors, giving mean value=28.3; SD=7.23; CV%=25.5. The divergence of the median value of 32 from the mean (28.3) shows a skewed distribution. Removal of the two lowest outliers, gives a mean of 32.2; SD=2.86; CV%=8.9 and median=32 and removes the skew. The use in ranking, of such a simple average of means, is clearly unreliable. Another example, with a mean score of 31.25, gave rise to an invitation to submit a full proposal. This was done, involving 14 EU partner, and 4 international 'reciprocal agreement', countries. The outcome of this application was a score of 53, and below the cut-off score of 55, for further progression. The consensus conclusion of evaluation was that 'the expected benefits are likely to be non-European'. This conclusion is surprising and essentially without explanation. The selection of evaluators remains a concern.

Issues and challenges for animal fibre in Europe: a view of EAAP AFWG (Animal Fibre Working Group)

C. Renieri, M. Gerken, D. Allain, M. Antonini, J.P. Gutierez, R. Niznikowski, A. Rosati and H. Galbraith
EAAP Animal Fibre Working Group (AFWG), c/o University of Camerino, Environmental and Natural Sciences, via Pontoni 5, 62032 Camerino, Italy; h.galbraith@abdn.ac.uk

Animal fibre, in the form of the natural product, 'wool', is a neglected and socially under-recognised renewable natural resource in EU27. Its production is substantial at 186,000 tonnes of raw wool (FAO) for 62 m breeding sheep alone (Eurostat). To this may be added smaller amounts from goats (mohair, cashmere) and South American camelids (alpaca, vicuña). Relevant biological issues include (1) environmental costs from enteric methane and nitrogenous greenhouse gas emissions; and (2) requirement for dietary nutrients, calculated at 45 million tonnes annually (2 kg/animal/day). Wool has potential to improve returns to offset these costs. Sociological issues concern lack of social recognition, where wool is also not recognised in the Common Agricultural Policy, low prices and frequent unprofitability. These are associated generally, with increasing urbanisation and competition with petrocarbon-based artificial fibres. Technical issues relate to frequent poor quality and lack of uniformity in post-harvest collection and sorting. Challenges include enhancing quality by improving understanding of wool biology, animal genetics and husbandry, and technology of processing in the context of development of existing and novel products. An additional challenge is for the EAAP AFWG to expand its current successful scientific connections both within and outwith Europe. The aim will be to benefit from knowledge and expertise across scientific disciplines, in a global environment, for transfer to animal fibre production in Europe. This process is ongoing.

Improving the reliability of fertility breeding values through better data capture

J.E. Pryce
Department of Primary Industries, Victoria, Biosciences Research Division, 5 Ring Road, Bundoora, 3083, Australia; jennie.pryce@dpi.vic.gov.au

Over the last 10 to 20 years, a decline in fertility in dairy cattle has been observed around the world. In Australia the genetic trend of calving interval has deteriorated by +0.5 days per year over the period 1986 to 2001. Fertility breeding values were implemented in Australia in 2003 using a single trait model and since then the genetic trend in fertility ABVs appears to have stabilised. In 2013 a new multi-trait fertility ABV was launched that includes the following predictors of 6 week in-calf rate: calving interval, lactation length, days to first service, non-return rate and pregnancy rate. The new multi-trait fertility model has been calculated to increase average first proof bull reliabilities (i.e. on around 30 daughters) from 0.33 to 0.38. The limitation to realising the full potential of this model is the capture of data. Among cows that calved in 2009 the proportion of cows with mating and pregnancy data was 18% and 13% respectively. The Dairy Futures CRC in conjunction with ADHIS have recently embarked on a co-ordinated effort to capture many more mating and pregnancy records that are electronically recorded on-farm but currently do not contribute to fertility ABVs. The success of this project rests on how much extra phenotypic data is collected; we have estimated that the impact could be as much as increasing the response to selection in 6 week in calf rate by 13% after 10 years of selection. This is through increasing the reliability and the intensity of selection or proportion of 1st proof bulls selected (i.e. bulls graduating progeny-testing with first crop daughters). Finally the contribution of genomic data has been shown to increase the reliability of fertility ABVs of first proof bulls by a further 10%. The collection of extra phenotypic data in addition to genomics is expected to increase the response to selection in fertility in the Australian dairy herd.

Session 44 Theatre 2

Genomic predictions of clinical mastitis within and between environments

K. Haugaard[1], L. Tusell[2], P. Perez[2], D. Gianola[2], A.C. Whist[3] and B. Heringstad[1]
[1]Norwegian University of Life Sciences, P.O. Box 5003, 1432 Ås, Norway, [2]University of Wisconsin, Madison, WI 53706, USA, [3]The Norwegian School of Veterinary Science, P.O. Box 8146 Dep, 0033 Oslo, Norway; katrine.haugaard@umb.no

Genome-based predictions of daughter-yield-deviations for clinical mastitis within and between environments were performed for 1,126 Norwegian Red bulls using Bayesian Ridge Regression. The objective was to investigate whether predictability was different within and between environments, and if the statistical model trained with observations in one environment could be used to predict breeding values in another environment. Better predictability within than between environments would indicate genotype by environment interactions for clinical mastitis. The environments were defined based on the herds' prevalence of the contagious mastitis pathogens *Staphylococcus aureus*, *Streptococcus dysgalactiae* and *Streptococcus agalactiae* in milk samples analyzed by the mastitis laboratories. Herd-5-year classes were categorized as follows: <50% (L50) and ≥50% (H50) contagious pathogens and ≤25% (L75), 25-75% (M75) and ≥75% (H75) contagious pathogens. Predictive ability was evaluated using a 10-fold cross validation. For comparison, predictions based on all data across environment groups were also calculated. Predictive correlations ranged from 0.04 (L75) to 0.15 (H75). There was little difference whether the predictions were done within or between environments. The predictive correlation for the full dataset was 0.19. The variation in predictive ability were large within each dataset (even some folds yielding negative correlations) and small between datasets. Rank correlations of the SNP-effects from each environment ranged from 0.15 to 0.92. This may indicate that SNP effects could be specific for some mastitis pathogens or pathogen groups. The study did not reveal any difference in predictive ability between environments, but indicated that pathogen-specific SNP-effects may exist.

Session 44 Theatre 3

Physiology of cows with divergent genetic merit for fertility traits during the transition period

S.G. Moore[1,2], P. Lonergan[1], T. Fair[1] and S.T. Butler[2]
[1]School of Agriculture and Food Science, UCD, Dublin 4, Ireland, [2]Teagasc Moorepark, Fermoy, Co. Cork, Ireland; stephen.moore@teagasc.ie

Cows with similar genetic merit for milk production, but with extremes of good (Fert+; n=15) or poor (Fert-; n=10) genetic merit for fertility traits were monitored. Dry matter intake (DMI) was recorded daily from wk -2 to 5 relative to calving. Blood metabolites and metabolic hormones were measured from wk -2 to 8 relative to calving. Vaginal mucus (VM) was scored weekly on a scale 0 (no pus) to 3 (≥50% pus) from parturition to wk 6 and uterine polymorphonuclear neutrophil (PMN) count was measured at wk 3 and 6. Continuous data were analyzed using mixed model procedures. PROC NPAR1WAY was used to analyse VM score data. The proportion of each genotype classified as having endometritis (PMN>6%) was analysed using Fishers exact test. Fert+ cows tended to have greater DMI than Fert- cows (17.9±0.6 vs. 16.3±0.8 kg DM/day, P=0.1). Mean BCS at calving was similar for both genotypes (3.1 vs. 3.0 units, P=0.1). During the first 35 wk of lactation, the Fert+ cows had greater milk solids production (1.9±0.05 vs. 1.7±0.04 kg/d, P<0.001). Mean BCS and BCS nadir were greater in the Fert+ cows (3.01 vs. 2.75 units, P<0.001 and 2.71 vs. 2.44, P=0.007, respectively). Mean glucose (P=0.03) and IGF-1 (P<0.001) concentrations were greater in the Fert+ cows but NEFA (P=0.6) and BHB (P=0.9) were not affected by genotype. Fert- cows had greater VM score than Fert+ cows on wk 1 (3.0 vs. 2.0, P=0.06), 3 (2.3 vs. 1.1, P=0.02) and 6 (1.0 vs. 0.0, P=0.01) postpartum. A similar proportion (0.77) of each genotype was classified as having endometritis on wk 3 postpartum. By wk 6, however, a greater proportion of the Fert- cows was classified as having endometritis than the Fert+ cows (0.75 vs. 0.22, P=0.06). These results indicate that good genetic merit for fertility traits is associated with a more favourable bioenergetic and uterine health status and greater BCS after parturition, without antagonizing milk production.

Genetic evaluation of mastitis liability and recovery through longitudinal models of simulated SCC

B.G. Welderufael[1,2], D.J. De Koning[2], W.F. Fikse[2], E. Strandberg[2], J. Franzén[3] and O.F. Christensen[1]
[1]Aarhus University, Department of Molecular Biology and Genetics, Building K23, 8830 Tjele, Denmark, [2]Swedish University of Agricultural Sciences, Department of Animal Breeding and Genetics, Box 231, 750 07 Uppsala, Sweden, [3]Stockholm University, Department of Statistics, 106 91 Stockholm, Sweden; berihu.welderufael@slu.se

Genetic evaluation of mastitis is performed either with cross-sectional or longitudinal models. In this study we aim to develop better longitudinal models using simulated Somatic Cell Count (SCC) which usually is used as a proxy to label clinical mastitis. Data was simulated for mastitis liability and recovery for two scenarios (28 and 95% mastitis cases/lactation) and two daughter groups of 60 and 150 per sire in 1,200 herds. Weekly observations for SCC were simulated assuming a baseline curve for non-mastitic cows and deviations in case of a mastitis event. Binary data was created to define presence or absence of mastitis as 1 if the simulated SCC was above pre-specified boundary and 0 otherwise. The boundary was allowed to vary along the lactation curve modelled by a spline function with a multiple of 10 or 15. The dynamic nature of the SCC was taken in to consideration with the longitudinal approach and the patterns were captured by modelling transition probability of moving across the boundary. Thus, a transition from below to above the boundary is an indicator of the probability to contract mastitis, and a transition from above to below the boundary is an indicator of the recovery process. Estimated breeding values for mastitis liabilities and recovery were calculated in DMU. Our preliminary results showed the correlation between true and estimated breeding value for the simulated mastitis liability was 0.72 which is as good as the estimations based on clinical mastitis. Though the estimation accuracy for recovery (0.42) was not as high as for mastitis liability the transition probability model enables us to generate breeding values for mastitis recovery process.

Metabolic disorders and reproduction in dairy cows receiving folic acid and vitamin B12 supplement

M. Duplessis[1,2], C.L. Girard[1], D.E. Santschi[3], J. Durocher[3] and D. Pellerin[2]
[1]Agriculture and Agri-Food Canada, Sherbrooke, QC, Canada, [2]Université Laval, Département des sciences animales, Québec, QC, Canada, [3]Valacta, Ste-Anne-de-Bellevue, QC, Canada; melissa.duplessis.1@ulaval.ca

Previous studies showed that a combined supplement of folic acid and vitamin B_{12} enhances energy metabolism in dairy cows in early lactation. Therefore, the aim of the current project was to determine if this supplement given from 3 wk before the expected calving date until 8 wk after parturition would reduce metabolic disorders and improve reproduction in dairy herds. A total of 805 cows in 15 commercial dairy herds were enrolled according to parity, predicted 305 d milk production, and calving interval. Treatments consisted of weekly intramuscular injections (5 ml) of: (1) saline 0.9% NaCl (C); or (2) 10 mg of vitamin B_{12} + 320 mg of folic acid (V). Data were obtained from producers, DHI agency (Valacta), and veterinarians. Ketosis incidence was estimated by measuring β-hydroxybutyrate (BHB) in milk between 3 and 21 days in milk (DIM) using Keto-test strips (Elanco Animal Health, ON, Canada). MIXED and GLIMMIX procedures of SAS were used to analyze data. The vitamin supplement had no effect on DIM at first breeding for primiparous cows (P=0.44) but reduced it by 3.8 d for multiparous cows (P=0.05). Conception rates at first and second breedings were not affected by treatments (P>0.15). For V, 38.2±4.5% of cows had a BHB level over 100 µmol/l compared to 42.2±4.6% for C (P=0.31). No treatment effect was found on incidence of displaced abomasum, metritis, retained placentae, mastitis, and milk fever (P>0.37). The earlier first breeding date in multiparous cows receiving folic acid and vitamin B_{12} supplement could be explained by the supplement minimizing negative energy balance in early lactation. This is supported by the reduction in body weight loss and milk fat concentration without effect on milk production during the first 60 d of lactation reported previously for these cows.

Session 44 Theatre 6

Genetics of mastitis in the Walloon Region of Belgium

C. Bastin[1], J. Vandenplas[1,2], A. Lainé[1] and N. Gengler[1]
[1]University of Liège, Gembloux Agro-Bio Tech, Animal Science Unit, 5030, Gembloux, Belgium, [2]National Fund for Scientific Research (F.R.S.-FNRS), 1000 Brussels, Belgium; catherine.bastin@ulg.ac.be

The objective of this study was to estimate genetic parameters of mastitis traits for Walloon dairy cows. Data included a total of 4,489 lactations from 3,277 cows in 38 herds. The following mastitis traits were investigated: mastitis as a binary trait based on whether or not the cow had at least 1 mastitis case either from 10 days before calving till 365 days after calving (MAS) or from 10 days before calving till 50 days after calving (MAS50) or from 50 days after calving till 365 days after calving (MAS365). The number of mastitis cases from 10 days before calving till 365 days after calving (NMAS) was also studied. The frequency was 22% for MAS, 10% for MAS50, and 15% for MAS365. Variance components were estimated by Gibbs sampling using a 4-trait threshold mixed model that combined binary (i.e. MAS, MAS50, and MAS365) and continuous (i.e. NMAS) traits. The model included herd, year × season of calving, and age at calving × lactation as fixed effects and genetic and permanent environment as random effects. Heritability was 0.08 (±0.03) for MAS, 0.09 (±0.04) for MAS50, 0.08 (±0.03) for MAS365, and 0.05 (±0.02) for NMAS. Genetic correlations were 0.78 between MAS and MAS50, 0.88 between MAS and MAS365, 0.99 between MAS and NMAS, 0.44 between MAS50 and MAS365, 0.86 between MAS50 and NMAS, and 0.83 between MAS365 and NMAS. Standard error of correlation estimates ranged from 0.01 to 0.25. Genetic parameters from this study were in line with the literature. Further studies will investigate the genetic correlations between mastitis traits and somatic cell score and other production traits. This research is a part of a larger project which aims to develop selection tools to allow dairy farmers to prevent mastitis in their herd.

Session 44 Theatre 7

Potential use of mid-infrared milk spectrum in pregnancy diagnosis of dairy cows

A. Laine[1], A. Goubau[1], L.M. Dale[1], H. Bel Mabrouk[1], H. Hammami[1,2] and N. Gengler[1]
[1]University of Liege, Gembloux Agro-Bio Tech, Animal Science Unit, 5030 Gembloux, Belgium, [2]National Fund for Scientific Research, FNRS, 1000 Brussels, Belgium; aurelie.laine@ulg.ac.be

Fertility issues are a large part of economic losses for the dairy farmers. Early identification of pregnant and non-pregnant cows is a key element to improve reproductive performances and reduce costs for the farmer. The mid-infrared (MIR) spectrum obtained from milk recording routines is an inexpensive and quick method to obtain a fingerprint of the milk composition. This study was conducted in the context of the European project OptiMIR (INTERREG IVB North West Europe Program). The objective was to investigate the potential use of the entire milk spectrum to identify if a cow is pregnant or not. Investigation was based on 7,840 spectral records linked to confirmed pregnancy status coming from Luxembourg milk recording. The method was based on comparing a given spectrum to the expected spectrum if the cow would have been non-pregnant. The expected spectra were obtained from solutions of a mixed model (fixed effects: parity, herd, milking moment and days in milk; random effects: animal across lactations) applied to MIR spectra from a subset of non-pregnant cow. Therefore the solutions obtained in the model were used on the whole dataset to obtain predicted MIR spectral values for all test-days and prediction errors (residuals) representing the factors not present in the model (reproductive status, unaccounted factors, and error). A predictive quadratic discriminant function was then constructed on the residual spectra to predict the pregnancy status. Leave one out cross-validation showed promising results with an error rate equal to 1.8% and 6.8% for non-pregnant cow and for pregnant cow respectively. Results have shown that MIR milk spectra might be used as a pregnancy diagnosis tool. Therefore, this kind of diagnosis could be made routinely and at a low cost for farmers.

Relationships between udder health, milking speed and udder conformation in Austrian Fleckvieh

C. Fuerst[1] and B. Fuerst-Waltl[2]
[1]ZuchtData EDV-Dienstleistungen GmbH, Dresdner Str. 89/19, 1200 Vienna, Austria, [2]Univ. of Nat. Res. and Life Sci. Vienna, Sust. Agric. Syst., Gregor Mendel-Str. 33, 1180 Vienna, Austria; birgit.fuerst-waltl@boku.ac.at

Breeding for udder health is even more important when considering the enormously increased level of milk production in many populations worldwide. Apart from costs for mastitis treatments and reduced income for milk, udder diseases are among the most important disposal reasons for dairy cows. For Austrian Fleckvieh cows, routine genetic evaluations are not only available for SCS and udder conformation, but also for the direct health trait mastitis based on veterinary diagnoses. Mastitis was defined as a binary trait (0 = healthy or 1 = mastitis diagnosis) within the interval from -10 to 150 d after calving. The mean frequency across all lactations was 10%. Phenotypic relationships as well as genetic correlations between mastitis, milking speed, and udder conformation traits were analysed. While the phenotypic relationship between milking speed and mastitis was only weakly pronounced, genetic correlations revealed increased mastitis occurrence in case of higher milking speed. Udder score, which is scored subjectively from 1 (worst) to 9 (best), showed a curvilinear relationship to mastitis. Cows with udder scores lower than 7 showed progressively increased mastitis frequencies; the genetic correlation (r_a) was -0.38. Similar phenotypic relationships were found for the linear type traits udder depth (deep-high), fore and rear teat placement (outwards-inwards), fore udder attachment (loose-strong), and suspensory ligament (weak-strong). With r_a=-0.74 the highest genetic correlation was found for udder depth. Teat length and thickness had intermediate optima with regard to mastitis occurrence. Results suggest that cows with higher udders that are more tightly attached and have slightly inwards placed teats which are of average length and thickness have less mastitis treatments. Mastitis, SCS, selected conformation traits and milking speed may be included in a future udder health index.

Are milk content traits adequate ketosis indicators?

B. Fuerst-Waltl[1], H. Manzenreiter[1], C. Egger-Danner[2] and W. Zollitsch[1]
[1]Univ. of Natural Resources and Life Sciences, Vienna, Sust. Agric. Syst., Gregor Mendel-Str. 33, 1180 Vienna, Austria, [2]Zuchtdata EDV-Dienstleistungen GmbH, Dresdner Str. 89/19, 1200 Vienna, Austria; birgit.fuerst-waltl@boku.ac.at

The mobilization of fat and protein reserves in phases of energy deficiency during early lactation results in a formation of metabolites. If further metabolisation is made impossible, which will mainly be due to a lack of sufficient amounts of glucose, ketosis will occur. The aim of this study was to analyse the relationship between the milk constituents recorded during routine milk performance testing, and veterinarian ketosis diagnoses. The latter were collected within a nation-wide health monitoring system. Ketosis mainly (80%) occurs during the first 50 days of lactation, and about 35% of the diagnoses were made during the first 10 days of lactation. A significant difference was found between dairy cows with and without a ketosis diagnosis in terms of the content of milk constituents. However, it is not possible to sufficiently differentiate dairy cows with and without ketosis based on a defined threshold value for any of these traits. The commonly used fat-protein ratio threshold of 1.5 has to be questioned. Apart from significant breed differences that became obvious, 58% of the cows of the Austrian main breed, Fleckvieh, with a ketosis diagnosis had a fat-protein ratio smaller than or equal to 1.5. The practical utilization of information from milk performance testing is further hampered by the fact that for 49% of the positively diagnosed dairy cows no performance testing was conducted within a relevant time period before ketosis was diagnosed. To support the further development of early lactation ketosis, indicators on the basis of traits recorded during milk performance testing, a differentiation according to breeds, the critical assessment of the suitability of the fat-lactose ratio in comparison to the fat-protein ratio and the adaptation of threshold values need to be considered in order to detect a greater proportion of cows with ketosis.

Genetic trends for fertility in Swedish Red cattle using different models

S. Eriksson[1], W.F. Fikse[1], H. Hansen Axelsson[1] and K. Johansson[2]
[1]Swedish University of Agricultural Sciences, Dept. of Animal Breeding and Genetics, P.O. Box 7023, 75007 Uppsala, Sweden, [2]Växa Sverige, P.O. Box 7023, 75007 Uppsala, Sweden; susanne.eriksson@slu.se

Unfavourable genetic trends in female fertility have been observed in dairy cattle, explained by unfavourable genetic correlations with milk production. Scandinavian Red breeds have long been selected for functional traits, and less deterioration of fertility has been observed. The aim of this study was to estimate genetic trends in Swedish Red dairy cattle for two female fertility traits, udder health traits and protein yield, using different models. The fertility traits were number of inseminations per service period (NINS), and interval from calving to first insemination (CFI). Data for virgin heifers from 1998 and cows in first and second lactation from 1990 until 2007 was included. The number of observations per trait varied from 141,003 (udder conformation) to 1,046,206 (clinical mastitis). (Co)-variance components were estimated prior to the prediction of breeding values. Multi trait (MT) animal and sire models were used for estimation of variance and covariance components. For prediction of breeding values MT-animal models were used. Traits were analysed both within trait groups (e.g. fertility traits), as in the current Nordic evaluation, and using a full MT model with all functional traits and protein yield. Analyses were made with and without heifer data. The resulting genetic trends for Swedish AI bulls were neutral or favourable for udder health traits, CFI, and for NINS in virgin heifers. Unfavourable trends were seen for NINS in first and second lactation. In cows, the genetic trend seemed unfavourable for somatic cell score, clinical mastitis and NINS in lactating cows. These estimated trends were less unfavourable when evaluations were done within trait groups compared with using the full MT-model. Excluding heifer data had a smaller than expected effect on the estimated genetic trends.

Early pregnancy diagnosis in cattle using blood or milk samples

L. Commun[1], K. Velek[2], J.B. Barbry[1], S. Pun[3], A. Rice[2], C. Egli[3] and S. Leterme[2]
[1]INRA, UMR1213 Herbivores, Vet school of Lyon, 69280 Marcy l'Etoile, France, [2]IDEXX Laboratories Inc, Westbrook, ME-04092, USA, [3]IDEXX Switzerland AG, Liebefeld-Bern, 3097, Switzerland; loic.commun@vetagro-sup.fr

Early pregnancy diagnosis is a key point of reproductive management programs in cattle. Detection of open cows permits to increase economic performances. Our objective was to test an ELISA based on detection of pregnancy associated glycoproteins in blood (Bovine Pregnancy Test, IDEXX Laboratories) and milk samples from cows at several days post artificial insemination (AI). 102 dairy cows were tested at 16, 30 and 41 days post AI using plasma, serum and milk samples, and another milk sample at day 53. Samples were analyzed by the ELISA test, requiring adaptation for use in milk. Results obtained were compared to a transrectal ultrasound (TU) performed at day 41, 100% reliable at this stage. For each session and each type of sample, we calculated the sensitivity (SE) and specificity (SP) of the ELISA. The test characteristics were compared using the Chi-square or Fisher's exact test. We searched the optimum cut-off value to interpret ELISA on milk using a ROC curve method. The optimum cut-off for milk was 20% lower than for blood. With this adjustment, test performances were the same independent from the samples. At 16 days post AI, all results were negative. However, from 30 days post AI, SE reached 100% on serum and plasma samples and was 98% for milk. At days 41 and 53, SE was between 98 and 100%. At day 30, SP was 89 and 90% for blood and milk samples, and stayed between 92 and 100% at days 41 and 53. Four cows showed evidence of abortion with positive ELISA result at days 30 and 41 but negative TU at day 41, which explains the lower SP cited above. This ELISA test provides a reliable and simple tool to diagnose early pregnancy in cattle. The milk data described here was obtained off-label with a particular adaptation (Milk Pregnancy Test, IDEXX Laboratories Inc.). This is a breakthrough for pregnancy diagnosis in cattle.

Indirect online detection of udder health with an automated California Mastitis Test

A.-C. Neitzel, W. Junge and G. Thaller
Institute of Animal Breeding and Husbandry, Christian-Albrechts-University, Olshausenstr. 40, 24098 Kiel, Germany; aneitzel@tierzucht.uni-kiel.de

Somatic cell count changes in milk are one of the most important parameters to assist monitoring mastitis. With more detailed knowledge about SCC changes, mastitis can be predicted and possibly prevented. At bail SCC measurement can be done with several sensors, which are commercially available. The measurement principle behind the sensor used in this study is based on the automated California Mastitis Test. The viscosity of the formed gel (milk mixed with detergent-based chemical reagent) is called drain time: the time the gel needs to flow through a standardized bore. From the drain time of the gel the SCC of the milk sample is obtained. The aim of this study is to analyse the information content of the drain time and the obtained SCC. Data used were collected on the dairy research farm Karkendamm with 7 sensors in the milking rotary (in total 28 bails) from April 2011 to December 2012. Daily milk yield and weekly analysed data of the milk control association were measured additional to the drain time. The drain time is not normally distributed. For further assessment the logarithmic drain time was used. First results consider data from 272 Holstein-Friesian cows, each captured approximately three times a week. The mean logarithmic drain time is 1.18 s with a standard deviation of 0.07 s. The initial repeatability of the logarithmic drain time was 27%. The low value indicates a multifactorial influenced drain time similar to the SCC in milk. The repeatability of SCC measured by the milk control association was 52%. Effects on the drain time can be f. e. milk ingredients, like protein content (r=0.25) and fat content (r=0.12), milk quantity (r=-1.19) and milk flow (r=-0.15) (P<0.0001). Further effects will be analysed. Although several sensors detect real time information about udder health for management decisions, for breeding decisions it is important to implement inexpensive and adequately reliable sensor systems.

Reproductive performance in primiparous beef cows showing different growth patterns

A. Sanz, J.A. Rodríguez-Sánchez and I. Casasús
CITA de Aragón, Avda. Montañana 930, 50059 Zaragoza, Spain; asanz@aragon.es

An experiment was conducted to analyse the influence of nutrition levels from heifers' birth to first mating on their subsequent onset of puberty, fertility rate and performance in primiparous cows calving at two years. Twenty-nine Parda de Montaña (Brown Swiss for beef production) heifers, born in autumn, were assigned to two nutrition levels during Lactation (0-6 months: L-High vs. L-Low) and Rearing (6-15 months: R-High vs. R-Low) periods. At 15.5 months heifers were treated with an intravaginal progesterone device (PRID, CEVA, Spain) and Ovsynch protocol, being inseminated 14 days later. A second IA at heat detection was performed in non-pregnant heifers. Blood samples were collected weekly during rearing and postpartum periods for progesterone analysis (Ridgeway Science, UK). Productive parameters were controlled from heifers' birth until weaning of their first calves (30 months). Understandably, both lactation and rearing nutrition levels influenced on average daily gains in the different phases, heifers being able to compensate the lower growth rates in previous phases, depending on the food availability. The age at onset of puberty was affected by the nutrition level offered during lactation (10.3 vs. 12.0 months, in L-High and L-Low, P<0.01) and rearing (9.8 vs. 12.5 months, in R-High and R-Low, P<0.001) periods. However, no differences were found in live-weights at onset of puberty (327 kg, corresponding to 56% adult live-weight in this breed), conception age (16.4 months) or fertility rate (89%). Primiparous cows' performance was not affected by the growth patterns registered. Only weight at calving (495.8 vs. 454.4 kg, in R-High and R-Low, P<0.01) and postpartum anoestrus (77.7 vs. 106.5 days, in R-High and R-Low, P<0.05) were influenced by the rearing nutrition level. These preliminary results would confirm the feasibility of advancing the first service from 21 to 15 months of age in beef cattle, provided that growth rates close to 1 kg/d in the rearing period are guaranteed.

Session 44 Poster 14

Potential of probiotics and their glycopeptides in treatment of bacterial mastitis in dairy cows

G. Gulbe[1] and A. Valdovska[2]

[1]Research Institute of Biotechnology and Veterinary Medicine 'Sigra', Instituta 1, 2150 Sigulda, Latvia, [2]Latvia University of Agriculture, Faculty of Veterinary Medicine, K. Helmana 8, 3001 Jelgava, Latvia; anda.valdovska@llu.lv

A treatment of mastitis in dairy farms frequently results in the use of antibiotics. With bacterial resistance becoming potential threats to humans and animals, development of alternative treatment methods like a probiotic therapy and antimicrobial peptides is essentially important. Objectives of this study were to examine some probiotics and their glycopeptides for the inhibition of causative agents of bovine mastitis and to test haemolytic activity of glycopeptides. *Lactobacillus reuteri*, *Pediococcus pentosaceus*, glycopeptides of *Lactobacillus helveticus* in titre 109 cfu/g (GP1), in titre 107 cfu/g added casein (GP2) and GP1 glycopeptides in titre 107 cfu/g with β-glucans (GP3) were tested *in vitro* to examine their for the inhibition of *Staphylococcus aureus*, *Staphylococcus saprophyticus*, *Kocuria kristinae*, *Streptococcus uberi*, *Escherichia coli* and mixed culture of above mentioned bacteria isolated from cows with subclinical mastitis. The inhibition of bacterial isolates was tested by well and disk diffusion assays. Results obtained that GP1 had a greatest inhibitory effect (8.83±3.02 mm) against bacterial cultures while GP2 and GP3 as well worked effectively (6.67±1.73 and 5.67±1.52 mm, resp.) in dilution 20 mg/ml. GP1 had a greatest inhibitory effect also in dilution 20 mg/0.5 ml (15.42±0.31 mm), whereas GP3 and GP2 showed lower growth inhibition (4.14±0.21 mm and 2.67±0.00 mm, resp.). All bacterial strains were resistant to test solutions containing probiotics *L. reuteri* and *P. pentosaceus*. Glycopeptides displayed no haemolysis when tested with sheep blood. This study demonstrates that *L. helveticus* glycopeptides have a great potential for the treatment of bacterial mastitis in dairy cows.

Session 44 Poster 15

Late embyronic losses identified by ultrasonagraphy on four dairy farms from Trinidad

K. Beharry, A. Mohammed and N. Siew

The University of Trinidad and Tobago, Biosciences; Agriculture and Food Technology, Caroni North Bank Road off Arima, 01 and Centeno, Trinidad and Tobago; aphzal.mohammed@utt.edu.tt

Embryonic mortality in dairy cattle results in lower pregnancy rates, slower genetic improvement and financial losses to the dairy farmer. The incidence of late embryonic mortality (day 30-60) was investigated at four dairy cattle farms in Trinidad. Eighty seven animals exhibiting either natural oestrus or induced oestrus were inseminated either artificially (AI) (n=25) in the former or using a Fixed time AI protocol (n=62) in the latter. Pregnancy diagnosis (PD) was carried out using real time ultrasonography at day 25-30 and subsequently at day 60. At day 25-30, and at day 60, respectively, and using AI, 83.3% (21 of 25) and 100% (21of 21) animals from natural estrus were diagnosed as pregnant. At day 25-30 and at day 60, respectively, using FTAI, 67.3% (40 of 62) and 90% (36 of 40) animals from induced estrus were diagnosed as pregnant. Of the 61 animals diagnosed pregnant at day 25-30, only 6.5% (4 of 61) of late embryonic mortality was recorded at day 60. The incidence of embryonic mortality was independent of farm, oestrus type and physiological stage (chi square P>0.05). The presence of low embryonic mortality existing on these farms was probably due to good management.

Effect of genetic merit for energy balance on postpartum reproductive function in dairy cows

N. Buttchereit[1], R. Von Leesen[1], J. Tetens[1], E. Stamer[2], W. Junge[1] and G. Thaller[1]
[1]CAU, Institute of Animal Breeding and Husbandry, Olshausenstr. 40, 24098 Kiel, Germany, [2]TiDa GmbH, Bosseer Str. 4c, 24259 Westensee, Germany; nbuttchereit@tierzucht.uni-kiel.de

On the phenotypic level, it is well documented that postpartum energy status is critically important to the reproductive performance of dairy cows. However, little is known about the genetic association between energy balance (EB) and fertility. In this study, we investigated the impact of genetic merit for EB on postpartum reproductive function of primiparous Holstein Friesian cows. Feed intake, milk yield and live weight were recorded between day 11 and 180 in milk and EB was estimated on a daily basis. In total, 824 cows with an average number of 83 EB estimates per animal were studied. Daily EB breeding values were calculated using a random regression model. For a subset of cows (n=334), milk progesterone concentration was repeatedly determined and used as an indicator for luteal activity. Different traits describing the genetic merit for EB were defined to evaluate their relationship with fertility. Special emphasis was given to the breeding values obtained for day 11 to 55 in milk, since cows were, on average, in a negative EB until lactation day 55. It was found that genetic merit for early lactation EB has a significant effect on the interval from calving to first luteal activity (CLA). Cows with a low genetic merit for EB had significantly longer anovulatory periods after parturition compared to cows with a high genetic merit for EB. The sum of negative EB breeding values from lactation day 11 to 180 also had an impact on CLA. Cows genetically disposed to suffer from a severe negative EB had the longest CLA. Moreover, the results suggest that at least part of the phenotypic variation in days open can be related to the genetic predisposition for a prolonged negative EB. We concluded that an inclusion of some indicator for EB into future breeding programs would help to offset negative side effects of selection for milk yield on fertility.

Effect of enzymatic supplement on some reproductive and biological parameters of Ossimi sheep

A. Al-Momani[1], E.B. Abdallah[2], F.A. Khalil[2], H.M. Gado[2] and F.S. Al-Barakeh[3]
[1]The Ministry of Agriculture, Amman Directorate of Agriculture, P.O. Box 426002, 11140 Jabal Al-Naser -Amman, Jordan, [2]Animal Production Dept., Fac. of Agric., Ain Shams Univ., Shoubra El-Kheima, Cairo, Egypt, [3]National Center For Agricultural Research and Extension, Al- Baq'a, Al-Balqa, Jordan; esmat54@hotmail.com

Two experiments were conducted to evaluate the effect of ZADO on reproductive performance of Ossimi ewes. Exp.I. twenty eight, multiparous, adult Ossimi ewes were randomly assigned to two groups, group 1 fed ration supplemented ZADO (G1; n=14) while group 2 fed ration with no additives (G2; n=14). All ewes were fed maintenance ration for one month (during June) followed by flushing ration for five weeks (two weeks before and three weeks after ram introduction). ZADO (15 g/h/d) started to be added with flushing rations. Fertile rams were introduced for two successive estrous cycles. Fecundity was lower ($P<0.01$) in G2 (control group) than (G1). Exp. II twenty recently lambed Ossimi ewes were used with their offsprings in this experiment. Ewes were equally divided into two groups (ten ewes in each). Group 1 was supplied with 10 g/head/day ZADO (powder enzymatic supplement) mixed with total ration, while group 2 had the same ration with no additives and served as control. The average daily milk yield (DMY) was 907 vs. 787 g and weaning weight was 17.46 vs. 15.20 kg for ZADO supplemented and control groups, respectively. ZADO supplementation could improve fecundity, DMY and lamb weight at weaning.

Session 44 Poster 18

Rapid identification of *Corynebacterium* spp. in bovine milk by MALDI/TOF-MS

J.L. Gonçalves, P.A.C. Braga, J.R. Barreiro, T. Tomazi, C.R. Ferreira, S.H.I. Lee, J.P. Araujo jr., M.N. Eberlin and M.V. Santos
University of São Paulo, Rua Duque de Caxias, 225, Pirassununga, SP 13635900, Brazil; mveiga@usp.br

Corynebacterium spp. is one of the most frequently isolated microorganism causing bovine mastitis, however, there are no simple methodologies to differentiate species within the genus *Corynebacterium*. The objective of this study was to evaluate the matrix-assisted laser desorption/ionization mass spectrometry (MALDI/TOF-MS) to identification to the species level of *Corynebacterium* spp. isolated from mammary quarters. Milk samples were collected from 1,242 dairy cows on 21 farms, totaling 2,382 milk samples which were submitted to microbiological culture. For MALDI/TOF-MS analysis, the bacterial isolate was thawed and cultured for 72 h. Bacterial lysis and extraction was performed by the tube method using formic acid/acetonitrile. Obtained spectra were analyzed using MALDI Biotyper 3.0 software at default settings and score values higher than 2.3 were considered reliable for species identification. Isolates identified by MALDI/TOF-MS were subjected to gene sequencing of 16S rRNA. A total of 222 isolates were identified by MALDI/TOF-MS as *Corynebacterium* spp. of which, *Corynebacetium bovis* were the most isolated (n=208). Species no-lipophilic represented 4.5% of isolates (n=10) of these *C. auriscanis* (n=3), *C. xerosis* (n=3), C. amycolatum (n=1), *C. casei* (n=1), *C. efficiens* (n=1), *C. pseudotuberculosis* (n=1). Only four samples (1.8%) were not identified to the specie level. There were equivalence results of 95% between the two identification techniques used. The technique of MALDI/TOF-MS was applied with 91.57% sensitivity for identifying the species of *Corynebacterium* on cases of bovine mastitis subclinical when compared to the results of genes sequencing 16S rRNA. In conclusion, MALDI-MS was successfully applied for the identification of *Corynebacterium* species from bovine subclinical mastitis.

Session 44 Poster 19

Improving repeat breeder cows fertility by synchronizing ovulation and timed insemination

M. Kaim, H. Gasitua and U. Moallem
ARO, Department of Ruminants Science, P.O. Box 6, 50250 Beit Dagan, Israel; uzim@volcani.agri.gov.il

Improving repeat breeder cows fertility by synchronizing ovulation and timed insemination The prevalence of cows with 4 or more artificial inseminations (AI) out of the total in the Israeli dairy herd during 2011 was 30%. The conception rate (CR) of the repeat breeder cows (RBC) were 20-30% lower than CR at first AI. The objective of the current study was to improve RBC's conception rate by synchronizing ovulation followed by timed insemination (TAI). The concept behind the suggested method was to skip the current estrus, synchronize a new follicle wave, induce ovulation, and inseminate at optimal timing. This protocol might also increase the endogenous progesterone secretion during the luteal phase preceding TAI. Cows with 3 or more AI that did not conceive were defined as RBC and were randomly assigned into 2 treatment groups: (1) Control – were inseminated after a behavioral estrus; (2) TRT – Cows that returned in estrus (day 0) were not inseminated. Seven days later cows were treated with a GnRH injection, followed by a PG injection at d 14, and a second GnRH injection 50-60 h later. Cows were inseminated14-16 h after the second GnRH. The study was conducted in 5 large commercial herds and data included 2064 AIs. Data were analyzed using the logistic regression procedure of SAS, and the model included the effects of treatment, season (winter-spring or summer-autumn), parity, AI number (3+4 or ≥5), health disorder, milk yield and days in milk. The overall CR in TRT and control cows were 39.9 and 24.4%, respectively (P<0.0003). The CR in TRT cows with 3-4 or ≥5 AI were 38.7 and 40.7%, compared to 27.1 and 22.4% in the control cows, respectively. The CR in TRT cows in winter-spring and summer-fall were 43.7 and 31.1%, compared to 26.8 and 19.5% in the control cows, respectively. In conclusion, the treatment suggested in this study improved the CR of RBC cows by 15.5 percent units, and this procedure could be implemented in commercial dairy herds to improve fertility of RBC.

Session 44 Poster 20

Genetic analysis of atypical progesterone profiles in Holstein cows from experimental research herds

S. Nyman[1], K. Johansson[2], D.P. Berry[3], D.J. De Koning[1], R.F. Veerkamp[4], E. Wall[5] and B. Berglund[1]
[1]Swedish University of Agricultural Science, Animal Breeding and Genetics, P.O Box 7023, 750 07 Uppsala, Sweden, [2]Vaxa Sverige, P.O. Box 7023, 750 07 Uppsala, Sweden, [3]Teagasc, Animal & Grassland Research and Innovation Centre, Moorepark, Fermoy, Co. Cork, Ireland, [4]Wageningen UR Livestock Research, Animal Breeding and Genomics Centre, P.O. Box 65, 8200 AB Lelystad, the Netherlands, [5]Scottish Agricultural College, Sustainable Livestock System Group, Easter Bush, EH25 9RG Midlothian, United Kingdom; britt.berglund@slu.se

The objective of this study was to investigate the genetic parameters for atypical progesterone profiles in Holstein cows. Atypical profiles have been correlated with inferior fertility results in our earlier studies. Data were collected from approximately 1,200 Holstein cows from experimental research herds in Ireland, the Netherlands, Sweden and United Kingdom Genetic variances and correlations were estimated for the four progesterone profiles; normal cycle, delayed cyclicity (first progesterone peak later than 56 days post-partum), prolonged luteal phase (high progesterone level for more than 20 days) and cessation of cyclicity (interrupted cycle with low progesterone level for at least 14 days). In a first analysis of the genetic variance components data from Sweden and the Netherlands were used. In total 552 progesterone profiles from 239 Swedish Holstein cows and 675 profiles from 588 Dutch Holstein were analyzed. The proportion of delayed cyclicity and prolonged luteal phase decreased with increasing lactation number while cessation of cyclicity increased in later lactations. A mixed linear sire model including the fixed effects of lactation, calving season and calving year within country was used. There were 99 bulls in the Dutch data set and 102 bulls in the Swedish data set, of which 10 bulls were in common. Preliminary results showed heritability estimates for progesterone profiles that varied from 0.00-0.20. This indicates that progesterone profiles can offer potential for improved selection for fertility.

Session 45 Theatre 1

Physiological aspects of feed efficiency in ruminants

A. Bannink[1], C. Reynolds[2], L. Crompton[2], K. Hammond[2] and J. Dijkstra[1]
[1]Wageningen University Research Centre, Division of Animal Production, P.O. Box 65, 8200 AB Lelystad, the Netherlands, [2]The University of Reading, School of Agriculture, Policy and Development, Earley Gate RG6 6AR Berkshire UK, P.O. Box 237, Reading, United Kingdom; andre.bannink@wur.nl

Ruminants have the important role of turning human-inedible (fibrous) products into human-edible products. The efficiency of feed conversion (FCE) largely determines productivity, profitability and environmental impact of animal production, justifying efforts to improve FCE. Next to the profound impact of level of ruminant productivity on FCE, at a specific productivity level the variation in feed digestion contributes most to variation in FCE, and hence both are main targets for improvement. Nutritional and digestive factors directly affect FCE by their influence on rumen fermentation, site of digestion, feed digestibility, and nutrient supply and utilisation in ruminant metabolism, alongside on-going efforts to improve ruminant productivity through genetic and technological developments. Differences in nutrient post-absorptive metabolism contribute to variation in feed efficiency, although there are indications it has undergone limited changes with genetic selection. Enteric methane is a loss of feed energy for the ruminant and thus variation in methane per unit feed (methane yield) may contribute to variation in FCE. Mitigating methane does not necessarily increase net energy supply and FCE however. The beneficial effect depends on the mode of action of the mitigation, but seems less than expected. The latter illustrates the danger of making assumptions in relation to FCE, e.g. a reduction in methane yield increases FCE, without considering underlying mechanisms and effects on other metabolic pathways.

Session 45 Theatre 2

Genetics of feed efficiency in ruminant and non-ruminant animals

D.P. Berry
Moorepark Dairy Production Research Center, Fermoy, Co Cork, Ireland; donagh.berry@teagasc.ie

The growing demand for food, coupled with ever-increasing competition by non-food industries for land as well as societal pressures for cognisance of environmental footprint, imply that more food will need to be produced from a decreasing land base available to agriculture without compromising the environment. A meta-analysis of scientific publications in ruminants and non-ruminants clearly showed that genetic variation in feed efficiency exists. Meta-analysis of the literature indicates a mean heritability (range in parenthesis) for residual feed intake (RFI) in growing cattle, mature cattle, poultry and pigs of 0.33 (0.07 to 0.62), 0.04 (0.01 to 0.23) 0.34 (0.21 to 0.49), and 0.24 (0.10 to 0.45), respectively. The heritability (range in parenthesis) of gross feed efficiency (e.g. FCR) in growing cattle, mature cattle, poultry and pigs was 0.23 (0.06 to 0.46), 0.06 (0.05 to 0.32), 0.32 (0.11 to 0.67), and 0.26 (0.06 to 0.45), respectively. Of less certainty is how best to incorporate feed efficiency into breeding goals. Precise estimates of genetic correlations with other traits, especially low heritability traits like reproduction and health are, nonetheless, lacking to facilitate the evaluation of the expected responses to selection for alternative breeding goals. Moreover, the marginal benefit of collecting feed intake data over and above the exploitation of information from correlated traits using selection index theory has not been fully explored. Analysis of growing beef cattle and lactating dairy cows indicate that most of the genetic variation in feed intake can be explained by information on routinely measured traits.

Session 45 Theatre 3

Joint estimation for curves for weight, feed intake, rate of gain, and residual feed intake

J. Jensen
Aarhus University, Molecular Biology and Genetics, AU Foulum, 8830 Tjele, Denmark; just.jensen@agrsci.dk

Rate of gain and feed efficiency are very important traits in most breeding programs for growing farm animals. Rate of gain (GAIN) is usually expressed over a certain age period and feed efficiency is often expressed as residual feed intake (RFI), defined as observed feed intake (FI) minus expected feed intake based on live weight (WGT) and GAIN. In any case the basic traits recorded are always WGT and FI and other the traits are derived from these records. A bivariate longitudinal random regression model were employed on 9284 individual longitudinal records of WGT and FI from 2827 bulls of six different beef breeds that were performance tested on a central test station in Denmark. Genetic and permanent environmental covariance functions for curves of WGT and FI were estimated using Gibbs sampling. The covariance function were based on fourth order Legendre polynomials but other functions were also possible. Genetic and permanent covariance functions for curves of GAIN were estimated from the derivative of the function for WGT and finally the covariance functions were extended to curves for RFI, based on the conditional distribution of FI given WGT and GAIN. Furthermore, the covariance functions were extended to include GAIN and RFI defined over different periods of the performance test. These periods included the whole test period as normally used when predicting breeding values for GAIN and RFI for beef bulls. Breeding values and genetic parameters for derived traits such as GAIN and RFI defined both longitudinally or integrated over (parts of) of the test period can be obtained from a joint analysis of the basic records which consists of WGT and FI. The resulting covariance functions for WGT, FI, GAIN and RFI are usually singular but the procedure presented here do not suffer from the estimation problems associated with trying to define these traits individually before the genetic analysis. All results are thus estimated simultaneously and the set of parameters are consistent.

Comparison of energetic efficiency of Holstein-Friesian and other dairy cow genotypes

L.F. Dong[1,2], T. Yan[2], C.P. Ferris[2], A.F. Carson[2] and D.A. Mc Dowell[1]
[1]University of Ulster, Co. Antrim, BT37 0QB, United Kingdom, [2]Agri-Food and Biosciences Institute, Co. Down, BT26 6DR, United Kingdom; lifeng.dong@afbini.gov.uk

Data from 32 respiration calorimeter studies undertaken at this institute since 1992 were used to compare energy expenditure and energetic efficiency of Holstein-Friesian (HF; n=823) dairy cattle and other dairy cow genotypes (non-HF: 50 Norwegian Red, 16 Norwegian Red × HF and 46 Jersey × HF) during lactation. A mixed REML model was used to examine effects of cow genotype on the relationship between ME intake (MEI) and ME requirement for maintenance (ME_m), or milk energy output adjusted to zero energy retention ($E_{l(0)}$), with the effects of variation between individual studies removed. The ME_m was calculated from heat production minus energy losses from the inefficiencies of ME use for lactation, tissue change and pregnancy, with these based on measured milk energy output and energy change and predicted energy for pregnancy. There was a significant relationship between ME_m or $E_{l(0)}$ and MEI ($P<0.001$). However, the cow genotype group had no significant effect on the calculated ME_m ($MJ/kg^{0.75}$). With a common constant (0.483 $MJ/kg^{0.75}$), increasing MEI ($MJ/kg^{0.75}$) by one unit increased ME_m by 0.118 and 0.123 $MJ/kg^{0.75}$ with the HF and non-HF cows, respectively. A similar result was also obtained in the relationships between $E_{l(0)}$ and MEI ($MJ/kg^{0.75}$) with HF ($E_{l(0)} = 0.631\ MEI - 0.431$) and non-HF ($E_{l(0)} = 0.628\ MEI - 0.431$) cows. There was no significant difference in the derived efficiency of ME use for lactation (k_l) between the two groups of cows (0.631 vs. 0.628). Thus within the current data set dairy cow genotype had no significant effect on ME_m or k_l. However, the maintenance energy requirements obtained in the present study are considerably higher than those used within energy feeding systems for dairy cows currently adopted in West Europe and North America. In addition, ME_m is not a fixed value as currently used within these feeding systems, but increases with increasing MEI.

Does cow genetic merit influence maintenance energy requirement and energetic efficiency?

L.F. Dong[1,2], T. Yan[2], C.P. Ferris[2], A.F. Carson[2] and D.A. Mc Dowell[1]
[1]University of Ulster, Co. Antrim, BT37 0QB, United Kingdom, [2]Agri-Food and Biosciences Institute, Co. Down, BT26 6DR, United Kingdom; lifeng.dong@afbini.gov.uk

A dataset from this Institute comprising 31 calorimeter studies involving Holstein cows was used to examine the effects of genetic merit on ME requirement for maintenance (ME_m) or the efficiency of ME use for lactation (k_l). Within the UK cow genetic merit is defined using two economic indexes, Profit Index (PIN) and Profitable Lifetime Index (PLI). The former is based on milk production traits, while the latter includes both milk production and functional traits. A mixed REML model was used to examine if there was a relationship between PIN (n=736, -£54 to 63) or PLI (n=407, -£131 to 145) and ME_m or k_l. The ME_m was calculated from heat production and the efficiencies of ME use for production and k_l estimated from milk energy output adjusted to zero energy retention ($E_{l(0)}$) divided by ME available for production. There was no significant relationship between k_l and PIN or PLI, or between ME_m ($MJ/kg^{0.75}$) and PIN or PLI. A further analysis was carried out to evaluate effects of cow genetic merit on the relationship between $E_{l(0)}$ and ME intake (MEI), by dividing the whole PIN (<£3, £3-15 and >£15) or PLI (<£23, £23-67 and >£67) dataset into 3 subsets, categorised as low, medium and high genetic merit. With a common constant of 0.442 $MJ/kg^{0.75}$ (i.e. net energy requirement for maintenance (NE_m)), the slope increased only marginally with increasing PIN (i.e. k_l=0.631, 0.638 and 0.643, respectively). Similarly, the constants were similar between the 3 PIN groups (NE_m=0.449, 0.434 and 0.441 $MJ/kg^{0.75}$, respectively) with a common slope of 0.632. A similar result was also obtained between the 3 PLI groups. These results indicate that cow genetic merit (PIN and PLI) has no significant effect on NE_m, ME_m or k_l. Thus the high milk production potential of high genetic merit cows is likely due to higher nutrient intakes, and their ability to partition more energy for milk production.

Feed efficiency across species: aquaculture, monogastrics and ruminants

S. Avendano
Aviagen, Newbridge, EH28 8SZ, Scotland, United Kingdom; savendano@aviagen.com

Globally food demand is increasing rapidly to satisfy population growth. The majority of this predicted rise in meat consumption is attributed to the increase in poultry consumption worldwide. Critically, this expansion occurs at a time when there will be competition for land, water and energy; where the effects of climate change are expected to become more evident (Foresight Report 2011). High prices and volatility of feed and raw materials represent around 70% of the costs of global poultry production. Consequently, there is an industry wide focus on increasing feed utilisation efficiency, which also has a significant effect on environmental impact. Over the last ten years selection candidates have been evaluated on feed efficiency records from contrasting environments along with the characterisation feeding behaviour and its genetic basis. Simultaneously, a balanced breeding strategy with broad breeding goals is implemented, including not only traits associated with production but also traits related to liveability, skeletal and metabolic support, reproductive fitness and welfare, translating benefits to the whole industry. Here we will describe Aviagen's R&D strategy for developing broiler products which are able to perform optimally and efficiently in a wide range of environments to satisfy the current and future demand for healthy and affordable poultry meat.

Session 45 Theatre 7

The relationship between residual feed intake in growing heifers and cows

J.E. Pryce[1], L. Marett[1], W.J. Wales[1], Y.J. Williams[1] and B.J. Hayes[1,2]
[1]Department of Primary Industries (Victoria), Computational Biology, 5 Ring Road, Bundoora, 3083, Australia, [2]La Trobe University, 5 Ring Road, Bundoora, 3083, Australia; jennie.pryce@dpi.vic.gov.au

Research indicates that feed conversion efficiency (measured by residual feed intake; RFI), is heritable in growing heifer calves. In feeding trials conducted in Australia 108 heifers that were divergent for RFI were retested to determine if the difference was maintained during their first lactation. The difference in RFI between the high and low efficiency groups in growing heifers, tested at approximately 6 months of age, was 1.43 kg/d. As first lactation cows the difference between the two RFI groups was 0.43 kg/d ($P<0.05$). The correlations between calf and lactating cow RFI were $r=0.34$ (efficient group; $n=47$; $P<0.01$) and $r=0.17$ (inefficient group; $n=57$; $P=0.10$). This demonstrates that RFI in calves can be used as a predictor of RFI in lactating cows. However, more data are required to understand the genetic consequences of selecting on growing heifer RFI. This important result indicates that DNA markers for RFI derived from growing heifers can be used to predict genetic merit for RFI during lactation.

Selection for eye muscle depth breeding value increases dressing percentage in lambs

G.E. Gardner[1,2], A. Williams[1,2], A.J. Ball[1,2], R.H. Jacob[1,2] and D.W. Pethick[1,2]
[1]Cooperative Research Centre for Sheep Industry Innovation, University of New England, University of New England, Armidale, NSW 2351, Australia, [2]Murdoch University, School of Veterinary and Life Sciences, South Street, Murdoch, WA, 6150 Murdoch, WA, Australia; g.gardner@murdoch.edu.au

Pre-slaughter live weight taken directly off pasture and hot standard carcase weight data was collected from 7,849 lambs produced at 8 sites across Australia over a 4 year period (2007-2010) as part of the Sheep Cooperative Research Centre's information nucleus flock experiment. These lambs were the progeny of 363 Terminal, Maternal, and Merino sires divergent for post-weaning eye muscle depth breeding values (PEMD). Dressing percentage was calculated by dividing hot standard carcase weight by pre-slaughter weight expressed as a percentage, and all three of these terms were analysed using a linear mixed effects models. The base model included fixed effects for site, year of birth, kill group, sex, birth-rear type, sire type and dam breed, with sire and dam identification included as random terms. Sire post-weaning fat depth, weight, and PEMD breeding values were included as covariates. The PEMD breeding value was found to have no effect on pre-slaughter weight, however it did increase ($P<0.01$) hot standard carcase weight by 1 kg across the 7 unit PEMD range. This increase was attributable to the impact of sire PEMD on dressing percentage which increased ($P<0.01$) by 1.7 dressing percentage units across the PEMD range. This represents a clear production advantage for producers using high PEMD sires, the progeny of which will have a similar live weight to low PEMD sires, but will deliver heavier carcase weights on the basis of their improved dressing percentage. Although live growth rate does not appear to be impacted, the implications for feed efficiency due to reduced fat deposition are currently being explored.

Genome wide association analysis for residual feed intake in Danish Duroc boars

D.N. Do[1], T. Ostersen[2], A.B. Strathe[1,2], J. Jensen[3], T. Mark[1] and H.N. Kadarmideen[1]
[1]University of Copenhagen, Department of Clinical Veterinary and Animal Sciences, Faculty of Health and Medical Sciences, Groennegaardsvej 7, 1870 Frederiksberg C, Denmark, [2]Danish Agriculture & Food Council, Pig Research Centre, Axeltorv 3, 1609 Copenhagen V, Denmark, [3]Aarhus University, Department of Molecular Biology and Genetics, Blichers Alle 20, 8830 Tjele, Denmark; ddo@sund.ku.dk

This aim of this study was to identify genomic regions controlling feed efficiency defined by two estimated residual feed intake (RFI) measures in Danish Duroc boars. RFI1 was calculated based on regression of individual average daily feed intake (30-100 kg) on initial test weight and average daily gain (30-100 kg). RFI2 was the same as RFI1 except that it was also regressed on backfat (BF). A total of 868 boars had phenotypic and genotype records. A total of 33,945 SNPs were available for genome wide association studies (GWAS) after quality control. Deregressed EBVs were used as response variables in GWAS implemented in the R-GenABEL package. A total of 16 and 17 loci were significantly associated ($P<5\times10^{-4}$) with RFI1 and RFI2, respectively. 12 SNPs were in common implying the existence of common genetic mechanisms. A mutation (A/G) at locus ALGA0098358 on CPVL gene on SSC18 was highly associated with both RFI1 ($P=1.8\times10^{-4}$) and RFI2 ($P=4.5\times10^{-5}$). Moreover, the genomic region 30.5-31.5Mb on SSC1 contained high numbers of significant SNPs (8 loci) for both RFIs. The SNPs within region of MAP3K5 on SSC1, GTF2IRD2 on SSC3, and WDR70 on SSC16 may be interesting markers for both RFIs. The ANKRD6 on SSC1, CADM1 on SSC9 and RHBDD1 on SSC15 were distinct markers for each RFI1 and RFI2, respectively. Further systems genetics investigations are being carried out to reveal functional importance of genes and haplotypes on candidate chromosomes with high numbers of significant SNPs. This study enhanced our biological knowledge of the genes and variants controlling RFIs in pigs.

Session 45 Poster 10

Fatty acid and transcriptome profile of loin muscle and liver in pigs fed with PUFA enriched diet

M. Pierzchała, E. Poławska, P. Urbański, M. Ogłuszka, A. Szostak, A. Jóźwik, J.O. Horbańczuk, D. Goluch and T. Blicharski

Institute of Genetics and Animal Breeding, Polish Academy of Sciences , Jastrzębiec, 05-552, Poland; m.pierzchala@ighz.pl

The aim of the study was to characterize the transcriptome profile in pigs' liver and loin muscle in relation to PUFA n-6/n-3 ratio. The study was carried out on 90 pigs fed on standard, linseed and rapeseed enriched diets (different level of PUFA in diets). Twenty Polish Landrace pigs showing divergent ratios of n-6/n-3 fatty acid (group H and L – high and low ratio) in liver and muscle were used to identify differentially-expressed genes. The muscle and liver samples were sequenced using RNA-Seq and analyzed with microarrays. The high-throughput technologies such as next genome sequencing and microarray analysis were used for determining overall gene expression profile at transcriptome level. Pathways analysis revealed that genes belong to biological functions related to lipid and fatty acids metabolism had different expression profile. In concordance with the n-6/n-3 pig group classification, the pathways analysis inferred that linolenic and linoleic acids metabolism was altered between divergent individuals. The expression in liver of ca. 170 (L) and ca. 250 (H) lncRNAs was also detected. The highest repeatability of the RNA-Seq data and porcine expression microarrays were validated by RT-qPCR, showing a strong correlation between RT-qPCR and RNA-Seq data (ranking from ca. 0.8 to 0.9), as well as between microarrays and RNA-Seq (ca. r=0.7). A differential expression analysis between H and L pigs identified ca. 62 genes differentially regulated. This study showed the potential gene networks, which affect lipid and fatty acid metabolism. These results may help in the design of strategies to improve the nutritional quality of pork meat. Realised within project 'BIOFOOD–innovative, functional products of animal origin' no. POIG.01.01.02-014-090/09 co-financed by the European Union from the European Regional Development Fund within the Innovative Economy Operational Programme 2007-2013.

Session 45 Poster 11

Zip4-like zinc transporter variability is widespread in different commercial pig crossings

G. Erhardt, S. Welp and I.J. Giambra

Justus-Liebig-University, Department of Animal Breeding and Genetics, Ludwigstrasse 21B, 35390 Giessen, Germany; georg.erhardt@agrar.uni-giessen.de

Associations between a single nucleotide polymorphism (SNP) within exon 9 of porcine zip4like zinc transporter (Zip4like) and pancreatic zinc (Zn) concentration as well as apparent Zn absorption of piglets have recently been described. The A>C-SNP leading to an amino acid substitution (p.Glu477Ala) was demonstrated in a small number of pure- and crossbred Piétrain (Pi), while Deutsche Landrasse and Deutsches Edelschwein were homozygous for the probable wild-type allele C. Data about the occurrence of this SNP in commercial pig populations are missing. Due to the essentiality of Zn in several metabolic pathways, its low availability for monogastric animals from food, and restricted addition of Zn in the diets, an effective exploitation of the provided Zn is necessary. Therefore, the occurrence of genetic variability within the Zn metabolism will open a better understanding of bioavailability of Zn. Within this study screening for the Zip4like A>C-SNP was done in random samples (n=148) of crossings between sows from the breeding companies Hypor, JSR, and TOPIGS and Pi boars, originating from two farms in Germany. Typing was done by sequencing a polymerase chain reaction product including exon 9. Therefore, DNA was isolated from pig tails with a commercial DNA extraction kit. The study was done according to the German Animal Welfare Act. A high variability within the samples analysed could be demonstrated. In JSR × Pi and TOPIGS × Pi allele C was predominant (0.66, respectively, 0.71), whereas in Hypor × Pi allele A showed highest frequency with 0.63. All possible genotypes occurred within the animal material including the, until now, not described genotype AA. Due to the recently identified superior effect of the AC genotype in comparison to CC in the apparent Zn absorption, the high frequency of the A allele in the three commercial crossings, offers new promising possibilities in the field of animal breeding and nutrition.

Session 46 Theatre 1

Farmed animal genomes: status and where next

A.L. Archibald[1], D.W. Burt[1] and B.P. Dalrymple[2]
[1]The Roslin Institute and R(D)SVS, University of Edinburgh, Easter Bush, Midlothian, EH25 9PS, United Kingdom, [2]CSIRO, 306 Carmody Road, St Lucia 4067, Australia; alan.archibald@roslin.ed.ac.uk

High quality annotated reference genome sequences for organisms of interest are critical to contemporary biological research. Reference genome sequences are available for the major farmed animal species and several commercially important aquatic species, but these reference sequences are only of draft quality. Moreover, establishing its sequence is only the first step in characterizing a genome. Identifying the functional elements within the genome sequence is essential for understanding the phenotypic consequences encoded in the genome. Annotation of farmed animal genomes is currently limited to gene models deduced from alignments with expressed sequences (cDNA, ESTs, RNAseq) and some sequence variation (SNPs, CNVs). Sequencing of multiple genetically diverse individuals from each species is revealing a wealth of genetic variation, incl. millions of single nucleotide polymorphisms (SNPs). Platforms for genotyping 3-750K SNPs per individual have been developed and exploited in GWA studies and for genomic selection. Thus, genomics tools such as the SNP chips together with new statistical tools are exploited to dissect the genetic control of complex and economically important traits. As a result the accuracy with which the phenotypic and performance consequences encoded in genome sequences can be predicted has improved. Identifying the underlying causal molecular genetic variation in coding sequence is tractable with the current tools and resources. However, much of the causal genetic variation responsible for performance trait variation is expected to lie in regulatory sequences. Coordinated international collaborative research will facilitate the improvements in the functional annotation of farmed animal genomes required to enable identification of causal variation in regulatory sequences and the next step forward in predicting consequence from sequence. This research was funded by many sources, including EC-FP7's Quantomics-222664 and 3SR-245140.

Session 46 Theatre 2

The experience of managing cutting-edge science in Quantomics

A. Bagnato[1], C.C. Warkup[2] and A.W.M. Roozen[2]
[1]University of Milan, Via Celoria 10, 20133 Milan, Italy, [2]Biosciences KTN, Easter Bush, Midlothian, EH25 9RG, Roslin, United Kingdom; alessandro.bagnato@unimi.it

Animal genomics-related science and technologies are developing rapidly. In order to maximize scientific outcomes and economic benefits it is fundamental that researchers and funders anticipate these developments, and design and manage projects such that future scientific developments can be applied in the project. The Quantomics project has developed tools which are expected to have wide application in all farmed species, as it aimed to: (1) develop new tools for application in the project and by industry; (2) extend these tools to deal with new types of information, such as Copy Number; and (3) provide new ways for managing biodiversity using molecular DNA. Quantomics was supported by the EC in adopting a strategy in which it could include cutting-edge science in order to extend beyond the originally envisaged deliverables. Those included strategies on genome sequencing, SNP calling and CNV mapping in cattle; whole genome analysis in cattle; validation steps. Collaboration with, and the availability of information from, third parties and other collaborative projects (whether EC- or otherwise funded, completed or ongoing) (i.e 'SABRE', '1000 bull genome project') leveraged the financial investment and extended scientific excellence beyond the project boundaries. The experience of managing science in Quantomics underlines that future projects will need, during the advance of the project, great flexibility in the use of technologies, tools and resources and in the composition of its consortium. A project's managerial capability and its funders' flexibility to rapidly adapt will be crucial for having a world leading scientific community in Europe. 'Quantomics' ('From Sequence to Consequence – Tools for the Exploitation of Livestock Genomes') is an EC-funded FP7 Large Collaborative research project (contract no. 222664). This 4½-year (1 June 2009-30 November 2013), €8.14 million project involves 17 leading research groups and businesses: www.quantomics.eu.

Identification and conservation of novel long noncoding RNAs in cattle using RNASeq data

D. Kedra[1], G. Bussotti[1], P. Prieto[1], P. Sørensen[2], A. Bagnato[3], S.D. McKay[4], R. Schnabel[4], J.F. Taylor[4], R. Guigo[1] and C. Notredame[1]

[1]Centre for Genomic Regulation (CRG), Doctor Aiguader, 88, R440.04, 08003 Barcelona, Spain, [2]Agroecology Research Centre Foulum, Blichers Allé P.O. Box 50, 8830 Tjele, Denmark, [3]University of Milan, Via Celoria 10, 20133 Milano, Italy, [4]University of Missouri, 920 East Campus Drive, Columbia, MO 65211-5300, USA; darek.kedra@gmail.com

Long noncoding RNAs (lncRNAs) are a group of RNAs that are not protein-coding and are longer than 200 nucleotide (nt). At present, the Ensembl bovine annotation lacks bovine lncRNA genes and information about bovine orthologues of even well known human lncRNAs. We used bovine RNASeq data from multiple tissues (liver, skeletal muscle, ovary, small intestine, udder) for annotating novel lncRNAs. The Illumina reads were mapped by GRAPE using Ensembl reference genome assembly and annotation, gene models were called using cufflinks. The predicted spliced gene models were filtered using known transcripts in public databases including Ensembl and GenBank to identify novel transcripts. These were further analyzed using Coding Potential Calculator. These predicted to be non coding and located outside of known genes were cross-mapped to several mammalian genomes (human mouse, dog, horse, pig, sheep, water buffalo, and yak) using PIPER pipeline. We discovered hundreds of putative novel lncRNAs conserved among other mammalian species. This research was funded by many sources, including the EC-funded FP7 Project Quantomics-222664.

Snpredict: an integrated tool to assess the functional impact of SNPs

H. Tafer and P.F. Stadler

Universität Leipzig, Institut fur Informatik, 04107 Leipzig, Germany; studla@bioinf.uni-leipzig.de

High-throughput sequencing platforms are generating massive amounts of genetic variation data for diverse genomes. It has remained a challenge, however, to pinpoint the small subset of variants that have a direct functional effect. We describe here a new computation tool, Snpredict, that categorizes SNPs and scores the predicted effects of variants at genome-wide genome scales. Snpredict identifies gains and losses of start and codons codons, as well as the disruption of splice site and the creation of possible new splice sites. At the level of protein sequences, Snpredict combines the predictions of PolyPhen and PROVEAN to assess the effect of amino acid changes. Snpredict integrates RNAsnp to quantify the deleterious effect of variations in mRNAs and ncRNAs on RNA secondary structure. If differential expression data are available, Snpredict furthermore can weight the functional effects of SNPs on this basis. As part of the EC FP7 Quantomics project, Snpredict was used to annotate SNPs affecting mastitis. The results provide additional evidence on which of the polymorphisms are most likely causative, and provide a basis for further functional studies. Snpredict is an open source program that can easily be extented to consider additional SNP effects by adding further analysis modules. The tool can be applied with ease to newly sequenced genomes. Research supported by EC-FP7 'Quantomics', agreement no. 222664.

Validation of QTL affecting mastitis in dairy cattle

J. Vilkki[1], M. Dolezal[2], G. Sahana[3], A. Bagnato[2], H. Tafer[4], T. Iso-Touru[1], F. Panitz[3], E. Santus[5] and M. Soller[6]
[1]MTT Agrifood Research Finland, BEL, 31600 Jokioinen, Finland, [2]Universita degli Studi di Milano, Dept VESPA, 20133 Milano, Italy, [3]Aarhus University, Foulum, 8330 Tjele, Denmark, [4]Universität Leipzig, Institut fur Informatik, 04107 Leipzig, Germany, [5]ANARB, Loc Ferlina, Bussolengo, Italy, [6]Hebrew University of Jerusalem, Dept Genetics, Jerusalem 91904, Israel; johanna.vilkki@mtt.fi

Mastitis is the most costly and common disease in dairy cattle. As part of the EC FP7 project Quantomics (agreement n° 222664), we have used most recent genomic tools to characterize QTL affecting mastitis in dairy cattle. By high-density GWAS in the Finnish Ayrshire and Brown Swiss breeds we identified eight chromosome regions with major effects on resistance to mastitis. These genomic regions were analyzed for polymorphisms from 20X whole-genome sequences of 38 ancestral bulls of the two populations. The SNPs called within the regions were ranked according to their estimated effect using a pipeline developed within the project. A set of highly ranked SNPs were tested for their effect on mastitis traits in new samples from the breeds Finnish Ayrshire (SCC, clinical mastitis) and Valdostana (SCC, milk bacteriological results). In addition to the genotyped SNP set, independent SNP genotypes (50K) outside the chosen regions were used for correction of population stratification. Simultaneously, association analysis across the regions (RWAS) was performed with mixed model using imputed genotypes in the Danish Red breed. The sequences of the regions from Holstein, Danish Red, and Danish Jersey were used together with high-density genotype data from the sequenced individuals (as reference for imputation) to impute the sequence for approximately 1000 Danish Red individuals with nine mastitis phenotypes. The results provide additional evidence on which of the polymorphisms are most likely causative, and provide a basis for further functional studies. The information can be used to place prior weighting on genomic segments in Gene-Assisted-GWS.

Genomic information assisted selection

T.H.E. Meuwissen
Norwegian University of Life Sciences, Box 5003, 1432 Ås, Norway; theo.meuwissen@umb.no

Currently genomic selection is increasingly used to estimate breeding values using increasingly dense SNP-chips. With the advent of second-generation-sequencing technology, the use of whole-genome sequence data is becoming within reach. However, improvements in accuracy when moving from 50,000 to 700,000 SNP-chips were very modest in dairy cattle. Based on this one may expect an even worse cost/benefit ratio for the move towards sequence data. Although whole genome-sequence data contains all causative polymorphisms, it has a low signal to noise ratio in that relatively few causative effects are hidden amongst millions of polymorphisms. With some success, BayesB-type of methods attempt to eliminate the neutral polymorphisms based on the SNP effect. Here we develop a method for Genomic Information Assisted Selection where additional genomic information sources, e.g. is the SNP near a gene, is used to a priori differentiate the importance of the SNPs (for the trait). The actual genomic information used may be investigated, and the use of combinations of genomic information sources may prove most effective. We extended the BayesCpi method towards a hierarchical model where the indicator variables I_i, which indicates whether SNP_i is having an effect or not, are themselves modeled by a threshold model where their liabilities for having an effect is modeled by a regression on a (combination of) genomic information source(s) on the SNPs. Research supported by EC-FP7 'Quantomics', agreement 222664.

Partitioning of genomic variance using prior biological information

S.M. Edwards, L. Janss, P. Madsen and P. Sørensen
Aarhus University, Center for Quantitative Genetics and Genomics, Department of Molecular Biology and Genetics, Blichers Alle 20, 8830 Tjele, Denmark; stefan.hoj-edwards@agrsci.dk

Most complex diseases such as susceptibility to mammary gland infections (mastitis) have a complex inheritance and may result from variants in many genes, each contributing only a small effect to the trait. Genome-wide association studies have successfully identified numerous loci at which common variants influence complex diseases. Despite the successes, the variants identified as being statistically significant have generally explained only a small fraction of the heritable component of the trait, the so-called problem of missing heritability. Insufficient modelling of the underlying genetic architecture may in part explain this missing heritability. Evidence collected across genome-wide association studies in human provides insight into the genetic architecture of complex traits. Although many genetic variants with small or moderate effects contribute to the overall genetic variation, it appears that the associated genetic variants are enriched for genes that are connected in biological pathways or for likely functional effects on genes. These biological findings provide valuable insight for developing better genomic models. These are statistical models for predicting complex trait phenotypes on the basis of single nucleotide polymorphism (SNP) data and trait phenotypes and can account for a much larger fraction of the heritable component of the trait. A disadvantage is that this 'black box' modelling approach does not provide any insight into the biological mechanisms underlying the trait. We propose to open the 'black box' by building SNP set genomic models that evaluate the collective action of multiple SNPs in genes, biological pathways or other external biological findings on the trait phenotype. As a proof of concept we have tested the modelling framework on susceptibility to mastitis in dairy cattle. Research supported by EC-FP7 'Quantomics', agreement n° 222664.

Session 46 Theatre 8

The weighted interaction SNP hub (Wish) network method for systems genetics analyses

L.J.A. Kogelman and H.N. Kadarmideen
University of Copenhagen, Dep. of Veterinary Clinical and Animal Sciences, Grønnegårdsvej 7, 1870, Denmark; hajak@sund.ku.dk

Several network approaches have been used to detect pathways and underlying causal genes to unravel the biological and genetic background of complex diseases, e.g. Weighted Gene Co-expression Network Analysis (WGCNA). WGCNA uses gene expression microarray data to create a network based on their correlations. This results in the detection of gene modules and hub genes, which have been shown to be of biologic importance. Our main objective was to create an algorithm and R-package to develop scale-free weighted interaction networks using the estimated effects of SNP markers from high density genotype data, based on the WGCNA-principles. The new Weighted Interaction SNP Hub (WISH) network method is based on using the output of genome wide association studies (GWAS): the estimated effects of SNP markers and P-values. Data reduction is achieved by selecting the most varying and significant SNPs for network construction. To detect interaction patterns between the SNPs, the adjacency matrix (A) was created by calculating the correlations between the allelic substitution effects of selected SNPs. To create the WISH network, a similarity measure is created, where we use the Topological Overlap Measure (TOM) matrix to reflect the degree of overlap in shared neighbors between pairs of SNPs. Modules of closely connected SNPs were defined using a tree cutting algorithm on the dendrogram created from the dissimilarity TOM. Modules are selected for functional annotation based on their correlation with the trait of interest, and hub SNPs are detected based on their connectivity. The WISH network method has been tested on GWAS results from obesity related traits in a F2 pig resource population. Annotation and enrichment of detected modules showed key functional and biological pathways related to obesity. In conclusion, the WISH network method is a novel 'systems genetics' approach and is able to detect various SNP modules of high biological importance for the trait of interest.

High resolution copy number variable regions in Brown Swiss dairy cattle and their value as markers

M.A. Dolezal[1,2,3], K. Schlangen[3], F. Panitz[4], L. Pellegrino[3], M. Soller[5], E. Santus[6], M. Jaritz[2] and A. Bagnato[3]
[1]Vetmeduni Vienna, Josef Baumanngasse 1, 1210 Vienna, Austria, [2]FH Campus Vienna, Favoritenstraße 226, 1100 Vienna, Austria, [3]University of Milan, Via Celoria 10, 20133 Milan, Italy, [4]Aarhus University, Nordre Ringgade 1, 8000 Aarhus, Denmark, [5]Hebrew University Jerusalem, Edmond J. Safra Campus Givat Ram, 91904 Jerusalem, Israel, [6]ANARB, Associazione Nazionale Allevatori Razza Bruna, Loc Ferlina 204, 37012 Bussolengo, Italy; marlies.dolezal@gmail.com

This study presents a high resolution map of CNV regions (CNVRs) in Brown Swiss dairy cattle and characterizes identified CNVs as markers for quantitative and population genetic analysis. After careful quality filtering, CNVs were called in a set of 164 proven sires with PennCNV, CNAM (Goldenhelix) and genoCN at each of 735,238 loci (Illumina's bovine HD SNP chip) anchored to bovine autosomes on the UMD3.1 assembly. Overlapping CNVs were summarized at the population level into CNVRs. PennCNV identified 2,377 CNVRs comprising 1,162 and 1,131 gain and loss events, respectively and 84 regions of complex nature. CNAM identified 370 CNVRs comprising 23 and 240 gain and loss events, respectively, and 107 comprising both. The concordance for low frequency and short CNV calls between methods was low, emphasizing the need to establish consensus calls employing several algorithms, to avoid false positives. Consensus CNVRs are enriched for protein coding-, pseudo- and retroposed genes and biological functions involved in immunity. We then used total allele counts called by genoCN in consensus CNVRs between PennCNV and genoCN and phased them with polyHap v2 to disentangle the alleles sitting on each chromosome. Characterisation of LD between CNVs and SNPs in and around CNVRs is ongoing. This study was supported by the EC funded FP7 project 'Quantomics' contract n. 222664-2.

Analysis of next-generation sequence data of Swiss dairy populations

C.F. Baes[1,2], S. Jansen[3], B. Bapst[2], M. Dolezal[4], J. Moll[2], R. Fries[3], B. Gredler[2] and U. Schnyder[2]
[1]Swiss Federal Institute of Technology, Institute of Agricultural Sciences, Tannstrasse 1, 8006 Zürich, Switzerland, [2]Qualitas AG, Deparment of Breeding Value Estimation, Chamerstrasse 56, 6300 Zug, Switzerland, [3]Technische Universität München, Deparment of Animal Breeding, Liesel-Beckmann-Straße (Hochfeldweg) 1, 85354 Freising-Weihenstephan, Switzerland, [4]Università degli Studi di Milano, Dept. VESPA, Via Celoria 10, Milano 20133, Italy; christine.baes@qualitasag.ch

Next-generation sequencing (NGS) technology is the driving force behind genomics research and its applications. Switzerland's dairy cattle industry is characterised by many small populations of different breeds (Simmental, Swiss Fleckvieh, Red Holstein, Original Braunvieh, Swiss Braunvieh, Brown Swiss). The implementation of whole genome sequence (WGS) data will improve accuracy of breeding values and increase genetic gain, especially for functional traits with low-heritability in these breeds. WGS data of 65 bulls are available for analysis; these animals explain 74-79% of the genetic diversity in the corresponding Swiss dairy populations. Average read depth over all animals was 11x, whereby individual read depths per base varied from 0 to approx. 52,000. PCR duplicates were marked; it is likely that other technical factors, such as assembly errors, attributed to the excessive read depth at a few loci. All distributions of read depth showed positive skewness. Multi-sample variant calling by breed resulted in more variants; inclusion of all animals across breeds in a multi sample calling analysis resulted in the highest number of variants, the highest mean minor allele count, the lowest mean minor allele frequency and the lowest number of singletons. Here we present the systematic assessment of various preliminary data analysis steps on variant calling results. We provide an overview of useful quality parameters and compare single and multi-variant calling techniques, as well as single breed and multi breed variants.

Session 46 Theatre 11

Genomics and metabolomics approaches to identify markers associated with economic traits in pigs

L. Fontanesi[1], S. Dall'olio[1], F. Fanelli[2], E. Scotti[1], G. Schiavo[1], F. Bertolini[1], F. Tassone[1], A.B. Samorè[1], G.L. Mazzoni[1], S. Bovo[1], M. Gallo[3], L. Buttazzoni[4], G. Galimberti[5], D.G. Calò[5], P.L. Martelli[6], R. Casadio[6], U. Pagotto[2] and V. Russo[1]
[1]University of Bologna, Department of Agricultural and Food Sciences, Division of Animal Production, Viale Fanin 46, 40127 Bologna, Italy, [2]University of Bologna, Department of Medical and Surgical Sciences, Endocrinology and Metabolism Unit, Via Massarenti 9, 40138 Bologna, Italy, [3]Associazione Nazionale Allevatori Suini (ANAS), Via L. Spallanzani 4, 00161 Roma, Italy, [4]Consiglio per la Ricerca e la Sperimentazione in Agricoltura, Via Salaria 31, 00016 Monterotondo Scalo (Roma), Italy, [5]University of Bologna, Department of Statistical Sciences Paolo Fortunati, Via delle Belle Arti 41, 40126 Bologna, Italy, [6]University of Bologna, Biocomputing Group (BiGeA Dept), Via San Giacomo 9/2, 40126 Bologna, Italy; luca.fontanesi@unibo.it

Application of genomics (next generation sequencing and high throughput genotyping) and metabolomics technologies in farm animals are opening new opportunities for the identification of genetic factors affecting traits of economic relevance. In this study we combined several resources and data with the final aim to identify DNA polymorphisms and metabolites associated with production traits in Italian heavy pigs. Two genome wide association studies were carried out using a selective genotyping approach in Italian Large White pigs based on extreme and divergent estimated breeding values (EBVs) for average daily gain (ADG) and back fat thickness (BFT). Next generation sequencing was carried out using the Ion Torrent technology to identify single nucleotide polymorphisms (SNPs) from two reduced representation libraries constructed from pigs with extreme BFT EBVs. Metabolomics information was obtained using a mass spectrometry (MS/MS) analytical pipeline in a performance tested population. Integration of these data made it possible to identify markers (SNPs, copy number variation and metabolites) associated with ADG, BFT and several other correlated traits.

Session 46 Theatre 12

Search for pleiotropic effects in genome regions affecting number of teats in pigs

B. Harlizius, N. Duijvesteijn, M.S. Lopes and E.F. Knol
TOPIGS Research Center IPG B.V., Schoenaker 6, 6641 SZ, Beuningen, the Netherlands; barbara.harlizius@topigs.com

Breeding for female fertility in piglet production requires a balanced selection to prevent negative effects of increased litter size on vitality and growth of piglets. Deeper knowledge of the biological background of the main regulators of genetic antagonisms should result in improved selection strategies to avoid negative side-effects of increasing prolificacy. Genome-wide association analyses were performed separately for several female reproduction traits and screened for regions with pleiotropic effects. De-regressed breeding values were calculated for 1,470 animals of a commercial Large White line for the following traits: number of teats (NTE), age at first insemination (AFI), total number of piglets born (TNB), litter birth weight (LBW), litter variation (LVR), farrowing survival (FSL) and lactation survival (LSL). For GWAS, a Bayesian variable selection model was applied. Five major QTL (Bayes Factor BF>100) were found at high resolution for NTE, on SSC7, 9, 10, 12, and 14 of which the QTL on SSC9 and SSC14 are the first ones to be reported on these chromosomes. Three of these regions on SSC7, 12, and 14 show also significant effects (BF >30) on LBW and TNB. A more effective search for pleiotropic effects might come from a multi-trait GWAS approach or from a deeper knowledge of the biological background. Knowledge of the underlying functional mutation would enable to differentiate between pleiotropy and linkage of two different genes closely located together. Different attempts to disentangle the genetic correlation between traits on a molecular level as well as possibilities to handle risks and opportunities of pleiotropic effects in genomic selection schemes will be discussed.

Session 46 Theatre 13

Investigation of inter-individual epigenetic variability in bovine clones: a high throughput study

H. Kiefer[1], L. Jouneau[1], E. Campion[1], M.-L. Martin-Magniette[2], S. Balzergue[2], P. Chavatte-Palmer[1], C. Richard[1], D. Le Bourhis[1,3], J.-P. Renard[1] and H. Jammes[1]

[1]INRA, BDR, 78350 Jouy-en-Josas, France, [2]INRA, URGV, 91057 Evry, France, [3]UNCEIA, UN, 94704 Maisons-Alfort, France; helene.kiefer@jouy.inra.fr

Reprogramming the somatic cell to totipotency can be obtained following the transfert of its nucleus into an enucleated oocyte (cloning). We are using cattle clones as a model to assess the interindividual epigenetic variability and its consequences on phenotypes, including agronomical relevant traits and developmental pathologies. Indeed, the developmental defects frequently associated with cloning could be related to the insufficient extent of reprogramming, leading to long term consequences on phenotypes. To identify epigenomic patterns affected by incomplete reprogramming, we used immunoprecipitation of methylated DNA followed by hybridization on a new bovine promoter microarray (MeDIP-chip). The microarray targets the upstream region (-2,000 to +1,360 bp) of 21,416 genes (UMD3.1 assembly). We first focused on liver, because overgrowth of this organ is often observed in clones. The microarray has been hybridized with MeDIP samples from livers of normal Holstein animals (4 perinatal controls and 8 adults, obtained by artificial insemination and all healthy) and livers of Holstein clones (7 perinatal clones, either stillborn or suffering from severe pathologies, and 7 adult clones, with normal to pathological phenotypes). After normalization of the data, enriched probes were identified using ChIPmix. Results of exploratory analysis, including correlation clustering, Principal Component Analysis (PCA) and Independent Component Analysis (ICA), will be presented. A statistical test based on differences in the spatial distribution of the enriched probes revealed that most of the promoters exhibit a clustered distribution of the enriched probes. This local enrichment shows interindividual variability for some of the promoters, which are currently being identified and validated.

Session 46 Poster 14

Developing a sheep gene expression atlas

A.L. Archibald[1], A. Joshi[1], L. Eory[1], S. Searle[2], T. Ait-Ali[1], S.C. Bishop[1], F. Turner[1], K. Troup[1], P. Fuentes-Utrilla[1] and R. Talbot[1]

[1]The Roslin Institute and R(D)SVS, University of Edinburgh, Easter Bush, Midlothian, EH25 9PS, United Kingdom, [2]The Wellcome Trust Sanger Institute, Hinxton, Cambridge, CB10 1SA, United Kingdom; alan.archibald@roslin.ed.ac.uk

The International Sheep Genomics Consortium (ISGC) has assembled a high quality reference genome sequence for the domestic sheep (Ovis aries). We have generated >1 Terrabases of RNAseq data in order to characterise the sheep transcriptome and for annotation of the sheep genome sequence. Tissues were taken from a trio of individuals (ram, ewe + offspring) from the Texel breed plus a day 16 embryo from a mating of the same ram-ewe pair. 'Stranded' RNAseq libraries were generated from ribo-depleted RNA. Paired-end reads (2×150 bp) were sequenced from at least 30 million fragments per tissue sample. RNAseq data were produced from thirty-seven different tissues, including immune, reproductive, gastrointestinal tract, endocrine, skeleton-muscular and brain tissues. The RNAseq data will be assembled into transcripts. A gene expression atlas describing the expression levels of different transcripts in the sampled tissues will be developed. The availability of transcriptomics data from related individuals will facilitate analyses for evidence of allele-specific expression and imprinting. The RNAseq data have been provided to the Ensembl group for annotation of the sheep genome sequence. These results are obtained through the EC-funded FP7 Project 3SR-245140 and with support from the Roslin Foundation and a BBSRC Institute Strategic Programme Grant.

SNP detection workflow for deep re-sequencing data in cattle

L.-E. Holm[1], M.A. Dolezal[2], A. Bagnato[2], J. Vilkki[3] and F. Panitz[1]
[1]Aarhus University, Molecular Biology and Genetics, Blichers Allé 20, 8830 Tjele, Denmark, [2]Universitá degli Studi di Milano, VESPA, Via Celoria 10, 20133 Milano, Italy, [3]MTT Agrifood Research, Biotechnology and Food Research, Myllytie 1, 31600 Jokioinen, Finland; larserik.holm@agrsci.dk

One objective of Next Generation Sequencing (NGS) is to provide direct measurements of genetic variation to help answering biological questions – thus the quality of variant detection and genotypes directly impacts the scientific interpretation of an experiment. Within the Quantomics project (research suppported by EC-FP7 'Quantomics', agreement no. 222664) we performed deep whole-genome resequencing of 38 Brown Swiss and Finnish Ayrshire bulls for variant calling to characterize QTLs affecting mastitis in dairy cattle. Our workflow is based on the Genome Analysis Toolkit or GATK: NGS data processing starts by mapping raw reads to the genome reference (BWA; samtools) followed by deduplication (Picard). Base Quality Recalibration and Local Realignment steps generate analysis-ready reads for SNP and indel calling using the UnifiedGenotyper. The performance and scalability of the variant discovery step can be increased by multi-sample calling and reducing BAM files by data compression. The resulting data set of raw variants has been shown to contain a number of false positives, which impedes ranking and prioritization of candidates for validation studies. Improvements of the workflow will include further (quality) filtering, Variant Quality Score Recalibration and integrative analysis by annotating raw variants using external data (e.g. known genotypes/variation; population structure; functional effects).

A case control GWAS for APEC in a commercial broiler cross

M.A. Dolezal[1], T.H.E. Meuwissen[2], F. Schiavini[1], K.A. Watson[3], J. Vilkki[4], D.W. Burt[5], M. Soller[6] and A. Bagnato[1]
[1]University of Milan, Via Celoria 10, 20133 Milan, Italy, [2]Norwegian University of Life Sciences, Box 5025, 1432 As, Norway, [3]Aviagen Ltd, 11 Lothend Rd, EH28 8SZ Midlothian, United Kingdom, [4]MTT Agrifood Research Finland, Myllytie 1, 31600 Jokioinen, Finland, [5]University of Edinburgh, Easter Bush, EH25 9RG Midlothian, United Kingdom, [6]Hebrew University of Jerusalem, Edmond J. Safra Campus Givat Ram, 91904 Jerusalem, Israel; marlies.dolezal@gmail.com

We performed GWAS for susceptibility to Avian Pathogenic *Escherichia coli* (APEC) in 1119 case and 982 control birds from a commercial broiler cross, collected at a Danish abattoir. Cases were identified by visual inspection by trained veterinarians as animals showing APEC cellulitis and bacteriologically verified. Birds were genotyped with an Illumina 41K chicken chip. After filtering for MAF>0.02 and call rate >0.95, 26,614 SNPs remained on gga1-gga28 for analysis. Haplotypes coming from each of the parental lines in windows of 20 markers were identified using a custom written hidden Markov model. 26,105, 24,346 and 24,014 SNPs with MAF>2% within lines remained for association testing in each of the parental lines, respectively. Association testing of deconvoluted haplotypes within each of the parental lines and for convoluted genotypes across lines was performed with SVS7.6.4. Several regions met genome wide significance after multiple testing corrections using haplotype based per block and per haplotype analysis of 3-SNP-haplotypes. Across line mapping with convoluted genotypes did not yield genome significant regions after multiple testing corrections, but the regions identified in single line analyses could be confirmed as regions of elevated test statistic signals. The most promising regions on GGA1, 3, 5, 6 and 11 are being further investigated to refine and validate identified signals. To our knowledge this is the first attempt to map QTLs segregating in the parental lines using crossbred commercial offspring. This study was supported by the EC funded FP7 project 'Quantomics', contract n. 222664-2.

Session 46 Poster 17

Marker to marker LD in relation to the history of the site

M. Soller[1], M.A. Dolezal[2], A. Bagnato[2], N. O'sullivan[3], J. Fulton[3], E. Santus[4] and E. Lipkin[1]
[1]The Hebrew University of Jerusalem, Genetics, Edmund Safra Givat Ram Campus, 91904 Jerusalem, Israel, [2]University of Milan, Veterinary Medicine, Via Caloria 10, 20133 Milan, Italy, [3]HyLine International, 2581 240th St., Dallas Center, IA 50063, USA, [4]ANARB Italian Brown Cattle Breeders' Association, via Ferlina 204, 37012 Bussolongo, Italy; soller@mail.huji.ac.il

Marker to marker LD, measured by r^2, was examined using Illumina Bovine SNP50 v1 chip in the Italian Brown Swiss dairy cattle population, and a custom Illumina 36K microarray SNP chip in a chicken brown-egg layer line. In both populations, maximum LD showed a strong inverse relation to the distance between the markers of the pair (dS); and, independently of this, to the difference between them in minor allele frequency (dMAF). In both cases, LD values occupied the entire range between the maximum LD for the given dS or dMAF, and zero. Maximum LD values for given dMAF were sharply and precisely defined (i.e. there was no 'scatter' about the best curve fitted to the points representing maximum LD values against dMAF); while maximum LD values for given dS were 'fuzzy', i.e. points representing LD against dS were dispersed about the best fit curve to these points. It is shown that maximum LD is set deterministically for dMAF, but stochastically for dS. These relationships are further explored in terms of the number and frequency of haplotypes of the two markers as these develop historically, taking into account that the mutations that created the polymorphic sites at the two markers were necessarily separated in time. This study funded by EC (FP7/2007-2013), agreement n° 222664, 'Quantomics' (this Publication reflects only the authors' views and the EC is not liable for any use that may be made of the information contained herein). Hy-Line International and Italian Brown Cattle Breeders' Association provided data.

Session 46 Poster 18

Linkage disequilibrium in Brown Swiss cattle with Illumina 50K and 777K arrays

E. Lipkin[1], A. Bagnato[2], M.A. Dolezal[2], E. Santos[3] and M. Soller[1]
[1]The Hebrew University of Jerusalem, Genetics, Edmund Safra Givat Ram Campus, 91904 Jerusalem, Israel, [2]University of Milan, Veterinary Science, Via Caloria 10, 20133 Milan, Italy, [3]ANARB Italian Brown Cattle Breeders' Association, Via Ferlina 204, 37012 Bussolongo, Italy; soller@mail.huji.ac.il

Linkage disequilibrium (LD) was compared in 1131 Brown Swiss bulls genotyped by the Illumina 50K array; 147 were also genotyped by the 777K array. Marker order was highly conserved on the two arrays. 777K genotypes were imputed to 50K-genotyped individuals. Comparing imputed and actual haplotypes showed that short HT stretches were imputed accurately, but HT stretches originating from the two chromosomes were intermixed across the entire chromosome, with an average of 12.6 intermixed stretches/100 Mb. Hence imputed HTs were not used further. Considering pairs of adjacent markers on the 'diagonal' of the LD matrix, 10.9% (50K) and 52.2% (777K) presented LD≥0.70. Considering haplotype blocks (HB) on the diagonal defined by LD≥0.70: with the 50K array there were a total of 2,769 HBs mean size 60.1 kb, covering 166.4 Mb (4.8% of the bovine genome); with the 777K array there were 83,949 blocks, mean size 11.1 kb, covering 933.8 Mb (26.7% of the bovine genome). Thus, although the 777K array gave a dramatic improvement relative to 50K with respect to LD between adjacent markers, there is still considerable room for improvement. From analysis of LD between marker pairs within the 777K array that were separated by ≤1,500 bp, we project that the anticipated 2.2M array will increase the proportion of marker pairs presenting LD≥0.7 by 35%, from 0.52 to 0.70. This is an appreciable improvement, but it is not clear to what extent it will translate into increased power for GWAS or for GEBV estimation. Unexpectedly, complete-matrix HBs obtained by all possible marker pairs showing LD≥0.70, were fragmented and interdigitated in both arrays. (Supported by FP7 agreement n° 222664, 'Quantomics'. The Italian Brown Cattle Breeders' Association provided data).

A medium resolution SNP array based CNV scan in Italian Brown Swiss dairy cattle

L. Pellegrino[1], M.A. Dolezal[1], C. Maltecca[2], D. Velayutham[1], M.G. Strillacci[1], E. Frigo[1], K. Schlangen[1], A.B. Samoré[1], F. Schiavini[1], E. Santus[3], C. Warkup[4] and A. Bagnato[1]
[1]University of Milan, Via Celoria 10, 20133 Milano, Italy, [2]North Carolina State University, Box 7621, NC 27695-7621, Raleigh, USA, [3]ANARB, Loc. Ferlina, 37012, Bussolengo (VR), Italy, [4]Biosciences KTN, Easter Bush, Midlothian, EH25 9RG, Roslin, United Kingdom; alessandro.bagnato@unimi.it

Recent reports indicate copy number variations (CNVs) to be functionally significant. This study presents a medium resolution map of CNV regions (CNVRs) in the Italian Brown Swiss dairy cattle, from – to this day – the largest CNV genome scan in any cattle breed. We genotyped 1,489 bulls and after quality filtering on males we called CNVs with PennCNV and with 'Copy Number Analysis Module' (CNAM) of SVS7 software (Goldenhelix) for a total of 46,728 loci anchored on the UMD3.1 assembly on 983 and 561 bulls respectively. We corrected for sequence composition flanking each SNP and employed principal component analysis for CNAM to correct for technical background noise to reduce false positive calls. PennCNV and SVS7 identified a total of 4,501 and 1,289 CNVs segregating in 983 and 559 bulls respectively. These were summarized at the population level into 483 (401 losses, 61 gains, 21 complex) and 277 (185 losses, 56 gains and 36 complex) CNVRs, covering 114 Mb (4.59%) and 33.7 Mb (1.35%) of the autosome, respectively. We then obtained the consensus between the two CNV scans using the approaches suggested by Redon *et al.* (2006), union set, and by Wain *et al.*, intersection, covering 37.7 Mb (1.51%.) and 13.4 Mb (0.54%), respectively. CNVRs were annotated with the bovine Ensembl gene set v69 and tested for enrichment of GO terms using DAVID database. Consensus CNVRs are enriched for protein-coding genes and genes involved in MHC class II protein complex and VEGF signaling pathway. This study funded by EC-FP7, agreement n°222664, 'Quantomics'.

RN gene polymorphisms effect in a family based-structure scheme in French purebred pig populations

M.J. Mercat[1], K. Feve[2], N. Muller[3], C. Larzul[4] and J. Riquet[2]
[1]IFIP, BP 35104, 35651 Le Rheu Cedex, France, [2]INRA, UMR444, 31326 Castanet-Tolosan, France, [3]INRA, UETP, 35653 Le Rheu Cedex, France, [4]INRA, UMR1313, 78350 Jouy en Josas, France; marie-jose.mercat@ifip.asso.fr

The study was based on performance recording (with 23 meat quality traits) in progeny testing station of half-sib families composed of 50 offspring (castrates and females) from purebred sires. The aim of this scheme was to estimate, in French purebred pig populations, the effect of polymorphisms. Data will be presented for 1,740 genotyped animals belonging to 4 groups of breeds: LW (3 Large-White type populations), LF (French Landrace), D (3 Duroc populations) and CH (4 Chinese-European lines). Eight polymorphisms in RN (PRKAG3) gene were analyzed: R200Q, V199I, G52S, K131R, P134L, T30N, V41I, L53P. No polymorphism was found for R200Q and L53P. Six haplotypes were defined with the remaining mutations. Effect of haplotypes was estimated with MIXED procedure (SAS software) with sex and slaughter date as fixed effects, mother and father as random effects and carcass weight as covariate. Most significant results were observed for pH_{SM} (semi-membraneous pH 24 h post mortem) and MQI (Meat Quality Index combining pH_{SM}, Minolta L* and water holding capacity both on gluteus superficialis) in LW, LF, and D, for drip loss in LW and LF and for color traits (Minolta L*, a*, b*) in LW, LF and CH. Results will be illustrated focusing on 2 meat quality traits (MQI, and drip loss) and 2 haplotypes. Haplotypic frequencies estimated on parents are 26% and 10% in LW, 64% and 20% in LF, 40% and 47% in D and 12% and 21% in CH for haplotype 1 and 2 respectively. Haplotype 1 is favorable for the 2 considered traits. Regarding combinations of haplotypes, 11 is significantly better than 16 in LW, LF and D and even more than 66 in D. Estimated effects between 11 and 16 are between 0.2 and 0.8 phenotypic standard deviation; the highest being observed in LW. In D, differences between 11 and 66 animals are estimated to be 0.9 phenotypic standard deviation for MQI.

A genome wide association study for additive and dominance effects of number of teats in pigs

M.S. Lopes[1,2], J.W.M. Bastiaansen[1], E.F. Knol[2] and H. Bovenhuis[1]
[1]Wageningen University, Animal Breeding and Genomics Centre, De Elst 1, Building 122, 6708 WD, Wageningen, the Netherlands, [2]TOPIGS Research Center IPG B.V., Schoenaker 6, 6641 SZ, Beuningen, the Netherlands; marcos.lopes@topigs.com

Associations between genetic markers and different phenotypes have been extensively studied in several livestock species. So far the main focus of this research has been towards additive genetic effects. Identification of genes associated with non-additive effects such as dominance could further improve performance of especially crossbred animals. The objective of this study was to simultaneously identify genomic regions with additive and dominance effects related to number of teats in pigs. A total of 1,797 animals from a Landrace based line were genotyped using the 60K SNP chip. After editing for call rate, minor allele frequency, deviations from Hardy-Weinberg equilibrium and Mendelian inconsistences, 1,707 animals and 39,855 SNPs were included in a single SNP animal model. Additive and dominance effects were fitted and a false discovery rate (FDR) threshold of 0.10 was applied. A total of 7 regions (chromosomes 4, 6, 7, 10, 11, 12 and 14) with significant additive effects were identified. The highest significance ($-\log_{10}(P)$ values=8.4) was found for a SNP on chromosome 7 with an additive effect of 0.23 teats. No genomic regions were found with significant (FDR<0.10) dominance effects. However, the regions associated with additive effects on chromosomes 4 and 11 showed suggestive evidence for dominance effects ($-\log_{10}(P)$ values >4). The additive effect of the most significant SNP on chromosome 4 was -0.38 and the dominance effect was 0.43 teats. The results showed clear evidence for seven QTLs with additive effects of number of teats as well as two suggestive regions with both additive and dominance effects. Further analyses for non-additive gene action will include testing for imprinting effects on number of teats in pigs.

Mapping QTL for fatty acids in Italian Brown Swiss breed using a selective DNA pooling

M.G. Strillacci[1], E. Frigo[1], F. Canavesi[1], F. Schiavini[1], M. Soller[2], E. Lipkin[2], R. Tal-Stein[2], Y. Kashi[3], E. Shimoni[3], Y. Ungar[3] and A. Bagnato[1]
[1]Facoltà di Medicina Veterinaria, Università di Milano, Dipartimento di Scienze Veterinarie per la Salute, la Produzione Animale e Sicurezza Alimentare, Via Celoria 10, 20133 Milano, Italy, [2]The Hebrew University of Jerusalem, Department of Genetics, 91904 Jerusalem, Israel, [3]Faculty of Biotechnology & Food Engineering, Technion-Israel, Institute of Technology, Technion City, 32000 Haifa, Israel; alessandro.bagnato@unimi.it

A selective DNA pooling approach in a daughter design was applied to perform a GWA in Italian Brown Swiss cattle, to identify QTLs for Δ9desaturase (D^9D), conjugated linoleic and vaccenic (CLA and VA) acids. A total of 120 daughters for each of five selected families (60 animals with higher residual values and 60 with lower residual values) were pooled. DNA samples, extracted from sire's semen and milk pools were genotyped using Illumina Bovine SNP50 BeadChip. Statistical analysis was performed with respect to SNPs for which sires were heterozygous. Using the R software a procedure was implemented to perform a single and multiple marker sire test. A multiple testing correction was applied using the proportion of false positives among all positive test results. Association tests were carried out to identify genes with an important role in pathways for milk fat and fatty acids metabolism. Above all, BTA 19 showed a highly significant association with CLA, VA and D^9D. A large number of regions were significantly associated with the studied traits. Some of these regions harboring genes known to be involved in fat synthesis as reported in literature. The feasibility and the effectiveness of a selective DNA pooling approach using Bovine SNP50 BeadChip for the identification of QTLs was underlined in this study. This study was part of QuaLAT project financially supported by Regione Lombardia.

Session 46 Poster 23

Comparison of the Polish and German Holstein-Friesian dairy cattle populations using SNP microarrays

J. Szyda[1], T. Suchocki[1], S. Qanbari[2] and H. Simianer[2]
[1]Wrocław University of Environmental and Life Sciences, Department of Animal Genetics, Biostatistics Group, Kożuchowska 7, 51-631 Wrocław, Poland, [2]Georg-August-University, Department of Animal Sciences, Albrecht-Thaer-Weg 3, 37-075 Göttingen, Germany; joanna.szyda@up.wroc.pl

Genomic evaluation of dairy cattle has been conducted in many countries, but because of the commercial character of genetic evaluations the data sets are not available for research. The study provides a unique opportunity to exchange data on bulls subjected to genomic evaluation in the two countries and thus to compare genetic composition of both national populations representing the same breed. In particular, the aim of the project is to use estimated breeding values and genotypes of single nucleotide polymorphisms (SNP) to compare the pattern of genetic diversity and linkage disequilibrium (LD) as well as SNP effects calculated using exactly the same set of SNPs and the same statistical model. Two data sets will be used for the comparison: 2,294 German Holstein-Friesian bulls and 2,243 Polish Holstein-Friesian bulls. For each individual four different pseudo-phenotypes in form of the estimated deregressed breeding values for milk-, fat- and protein yields as well as for somatic cell score were available. Each bull was genotyped using the Illumina BovineSNP50 Genotyping BeadChip consisting of 54,001 SNPs. The first step of the analysis was to compare the pattern of linkage disequilibrium between German and Polish populations. Further, SNP effects were estimated based on the mixed model used in the Polish genomic selection project. Additionally, using a bivariate mixed model, covariances between genetic and residual effects in these populations will be calculated. The LD decay with increasing distance was significantly faster in the German, than in the Polish populations which also has generally lower LD. Differences between estimated SNP effects (based on a sliding window average) revealed 1, 1, 2, and 12 regions with different effects for milk-, fat-, protein yields and somatic cell score respectively.

Session 46 Poster 24

Validation of a new rabbit microarray: transcriptome variation in PBMCs after *in vitro* stimulation

V. Jacquier[1,2,3], J. Estellé[3], B. Panneau[4], J. Lecardonnel[3], M. Moroldo[3], G. Lemonnier[3], T. Gidenne[2], I. Oswald[1], V. Duranthon[4] and C. Rogel-Gaillard[3]
[1]INRA, UR66 TOXALIM, Toulouse, France, [2]INRA, UMR1289 TANDEM, Castanet Tolosan, France, [3]INRA, UMR1313 GABI, Jouy en Josas, France, [4]INRA, UMR1198 BDR, Jouy en Josas, France; vincent.jacquier@toulouse.inra.fr

Due to both physiological and economic reasons, rabbits have proven very useful as biomedical models for humans. However, genome-wide expression studies targeting the immune response are still limited in this species. We have designed a long oligonucleotide-based DNA microarray by combining a generic Agilent rabbit gene expression microarray with missing well-annotated genes known to be involved in immune and inflammatory responses. Rabbit peripheral blood mononuclear cells (PBMCs) were isolated and either mock-stimulated or stimulated with lipopolysaccharide (LPS) or a mixture of phorbol myristate acetate (PMA) and ionomycin for 4 (T4) and 24 (T24) h. After statistical analyses using LIMMA package in R, we have shown that the number of differentially expressed genes after PMA/ionomycin stimulation was 20 and 24 times higher at T4 and T24, respectively, compared to the LPS stimulation at similar time points. At T4, LPS stimulation induced an over-expression (FDR<0.05) of pro-inflammatory (IL23A) or regulatory (IL12A) cytokines and chemokine molecules (CXCL11), but a down-regulation of the LPS receptor CD14. At T24, IL1B, IL10, IL23A and IL36G were found over-expressed. At T4, PMA/ionomycin induced both Th1 (IL2 and IL10) and Th2 responses (IL4 and IL13). NFKBID and TBX21 were over-expressed whereas IL16, LTB and CD79B were found significantly down-regulated. At T24, IL2 and IL4 were still found over-expressed after PMA/ionomycin stimulation whereas IL1R2 and IL1B were down-regulated. Part of our results agrees with a similar previous report on the transcriptome of pig PBMCs. A microarray enriched in immunity-related genes is now available and can be used as a new tool for genome-wide expression studies of immunity in the rabbit species.

Productivity improvement in sheep associating polygenic selection and diffusion of the FecXR allele

J. Folch[1], B. Lahoz[1], J.L. Alabart[1], J.H. Calvo[1], E. Fantova[2] and L. Pardos[3]
[1]CITA de Aragón, Av. Montañana 930, 50059, Spain, [2]Oviaragón-Grupo Pastores, Crta Cogullada s/n, 50014, Spain, [3]EPS Huesca, Univ. of Zaragoza, Ctra Cuarte s/n, 22071, Spain; jfolch@aragon.es

The Cooperative Oviaragón carries out since 1994 a selection program for prolificacy in Rasa Aragonesa sheep, with 205000 sheep at present. Within this program, a mutation in the BMP15 gene located in the X chromosome (FecX^R allele, ROA®) was discovered in 2007. This polymorphism increases prolificacy in heterozygous carriers (R+), resulting in 0.35 extra lambs/lambing ewe, and produces sterility in homozygosis. Due to the productive interest, a controlled program for the outreach of the FecX^R allele by AI has been developed (7,500 AI/year). The program includes: (1) Electronic identification and computerized control to avoid sterility; (2) research on the effects of the FecX^R allele on reproductive traits: ovulation rate (+0.44 in ewe lambs and +0.63 in adults); response to eCG (increased response to the standard dose); preovulatory LH surge (no differences); fertility after AI (trend to increased fertility: +11.6%). 3) Studies on the offspring (no effect on birth weight, growth rate and meat characteristics). 4) Technical-economic studies on 47 farms classified in 3 groups: 'R+' (polygenic selection and at least 5% of R+ ewes), 'Selection' (polygenic selection without R+) and 'Non-selection'. The number of lambings/year were 1.28, 1.19 and 1.03 ($P<0.01$), and prolificacy was 1.57, 1.36 and 1.29 ($P<0.001$), respectively, with no differences in triplets or lambs mortality. The 'R+' group sold 0.34 and 0.55 extra lambs/ewe/year compared to 'Selection' and 'Non-selection' groups. Although production costs were higher in 'R+' and 'Selection' groups, they obtained greater gross margins (25,153€ and 25141€ vs. 15,644€). As result, the population of R+ has been increasing up to 8,900 recorded ewes. The selection program for prolificacy goes on with a combined polygenic selection and FecX^R allele dissemination. Financed by MEC and INIA (B. Lahoz grant)

Investigation of bovine leukocyte adhesion deficiency genetic defect in Iranian Holstein cows

B. Hemati and M. Ranji
Islamic Azad University, Animal Science, Azadi Blvd., Eram Blvd., Mehr-Shahr, Karaj, 3187644511, Iran; bzdhmt@gmail.com

In the present research, molecular detection of Bovine leukocyte adhesion deficiency (BLAD) in Holstein cows has been done using milk somatic cells. BLAD is a monogenic and autosomal recessive heredity lethal syndrome in Holstein dairy cattle which characterized by affecting the haematopoietic system via reduced expression of the heterodimeric β2integrin that may cause many defects in leukocyte function. The molecular cause of BLAD is a single point mutation with substitution of Adenin (A) by Guanine (G) at position 383 in the CD18 gene. Tank milk samples from 50 herds in the provinces of Tehran and Alborz were collected in three separated and different times. Samples from each herd were transferred to the research laboratory of the faculty of agriculture and natural resources at Islamic Azad University, Karaj Branch, Iran. After DNA extraction, polymerase chain reaction (PCR) was amplified using specific primers for 136 bp DNA,(CD18 gene).This mutation results in a substitution of glycine by an aspartic acid at position 128 in the D128G protein. Taq I enzyme was used to identify both BLAD alleles of the CD18 gene by digestion of PCR products. Restricted products were analyzed by electrophoresis in 1% agarose high resolution gel stained with DNA safe Stain. In this herd, the total number of dominant homozygote (AA), heterozygote (Aa) and recessive homozygote (aa) genotypes were 100, 0 and 0% respectively. In the present study we showed that, detection for all genetic defects using tank milk samples is more economic than the other methods. For example if we assume the average of 100 dairy cows for each non-infected herd, in fact we have done 50 tests instead of 5,000.

Session 47 Theatre 1

Microbiomics of farm animals: elucidating the interplay of microbiota with its mammalian host

J. Zhang, O. Perez-Gutierrez and H. Smidt
Wageningen University, Laboratory of Microbiology, Dreijenplein 10, 6703 HB Wageningen, the Netherlands; hauke.smidt@wur.nl

Soon after birth the gastrointestinal (GI) tract of humans, pigs and other monogastric animals is colonized by a myriad of microbes, referred to as microbiota, which plays an important role in host health and nutrition. The intricate interplay between microbial colonization in early life and the development of the mammalian host is addressed in the INTERPLAY project, an EU-FP7-funded collaborative project integrating the expertise of 11 partners from Europe and China. Knowledge generated in the project will be exploited for the identification of innovative management strategies that address host genotype as well as nutritional means to provide a framework for sustainable animal production at improved animal health and welfare. The GI tract microbiota is characterized by its wide phylogenetic and functional diversity. Despite all efforts in improving cultivation methods, a large fraction remains uncultured. The introduction of molecular approaches, mainly those based on 16S rRNA and its encoding gene, provided a more comprehensive insight into temporal, spatial and inter-individual dynamics of GI microbiota. Novel technologies, such as barcoded pyrosequencing of 16S rRNA genes, as well as DNA microarrays including the Human and Pig Intestinal Tract Chips have recently been described, and provide the necessary throughput that is needed to identify potential correlations between microbiota signatures and host health. In addition, meta-omics approaches have been developed to study the genetic potential and in situ activity of the microbiota. The application of these approaches to understanding the interplay of microbiota and animal health, also in response to the rearing environment as well as dietary additives, can provide the basis for the development of innovative nutritional strategies towards more sustainable animal production. Examples will be provided from current research on the pig, and an overview of the current state of the art of microbiomics will be given.

Session 47 Theatre 2

Microbiota, metabolism and immunity: the role of early-life in determining piglet performance

C.R. Stokes, M.C. Lewis, Z. Christoforidou and M. Bailey
University of Bristol, School of Veterinary Science, Langford House, Langford, Bristol, BS40 5DU, United Kingdom; chris.stokes@bristol.ac.uk

Although neonatal piglets are born capable of much greater independent movement and foraging than human infants, their immune systems are still poorly developed at birth. Immunological organisation develops over a period of weeks, and adult levels are not reached until six weeks old. Experiments using germ-free animals have demonstrated that this development is dependent on colonisation by environmental bacteria. It is now clear that very early events which affect microbial colonisation (maternal microbiota, weaning age, diet, environmental 'hygiene') have long-term effects on human health and disease. In contrast, the idea that early-life events can have long-term effects on piglet health and performance is not widely accepted. However, the immune and metabolic systems are key determinants of pig performance, governing resistance to infectious disease and conversion of feed to body mass. Early life events which have persistent, long term effects on these systems will be key targets for improving pig health, welfare and performance in future. Our recent studies have examined the effects of manipulation of colonisation using piglets reared under a range of management conditions. Rearing environment has very early effects on establishment of intestinal microbiota, and this is associated with changes in recruitment of antigen-presenting cells to the intestinal mucosa. Later in life, rearing environment affects recruitment of effector and regulatory CD4+ T-cells and functional responses to antigen. We suggest that there is considerable potential for intervention at birth and at weaning aimed at improving pig health and performance throughout the growing and finishing periods.

Session 47 Theatre 3

Effect of sow antibiotics and offspring diet on microbiota and barrier function throughout life

J.P. Lallès[1], M.E. Arnal[1], G. Boudry[1], S. Ferret-Bernard[1], I. Le Huërou-Luron[1], J. Zhang[2] and H. Smidt[2]
[1]INRA, ADNC, 35590 Saint-Gilles, France, [2]Wageningen University, Dreijenplein 10, 6703 HB Wageningen, the Netherlands; jean-paul.lalles@rennes.inra.fr

Neonatal microbial colonization of the gut participates in anatomical and functional development of this organ. Perinatal use of antibiotics (AB) is suspected to contribute to later development of diet-related diseases in human, through early alterations in microbial colonization impacting gut and distant-organ function. In that respect, gut barrier function, including permeability, detoxification, immune and cytoprotection systems are important aspects. AB treatment around farrowing also happens in pig production. However, the consequences on gut barrier throughout life in offspring are unknown. During the Interplay EU project, we developed a pig model of AB-induced early microbiota disturbances for investigating short- and longer-term consequences on offspring gut barrier function. Sows were given orally a broad spectrum AB around farrowing (day -10 to +21). Offspring were sacrificed during suckling, after weaning and at 6 months of age. In the long-term protocol, offspring were fed a high fat (vs. low fat) diet for 4 weeks as a model of diet-induced low grade inflammation. AB treatment of sows affected their faecal microbiota and that of their offspring. We observed age-dependent alterations in offspring ileal and colonic permeability, ileal epithelial transcriptome, mucosal inflammation traits, digestive enzymes, ileal and colonic protective heat shock proteins and in digesta concentrations of pro-inflammatory bacterial components. Interactions between perinatal antibiotic treatment of mothers and offspring adult diet was often significant for gut parameters. Data indicate that both early gut microbial colonization disturbances and late nutritional environment condition gut function in adulthood according to specific spatio-temporal patterns. Underlying molecular mechanisms and correlations between physiological and microbial traits are under investigation.

Session 47 Theatre 4

***In vitro* models to analyse nutritional and microbial antigens at the intestinal mucosal surface**

E. Mengheri[1], I.P. Oswald[2] and H.J. Rothkötter[3]
[1]INRAN Rome, Via Ardeatina 546, 00178 Roma, Italy, [2]INRA Toulouse, 180, Chemin de Tomefeuille, 31027 Toulouse, France, [3]Institute of Anatomy, Medical Faculty, Otto-von-Guericke-University, Leipziger Strasse 44, 39120 Magdeburg, Germany; hermann-josef.rothkoetter@med.ovgu.de

The health status of the animals is mainly influenced by the integrity of the intestinal barrier. Modern rearing conditions use a well-defined diet including the use of pre- and probiotics. *In vitro* systems are a first step to understand the benefit of the food additives. We have developed systems to analyse the effects of nutrients and probiotics on the cellular regulation and transport function of the epithelium. The main tool in our studies are epithelial cell monolayers, providing a tight barrier. They are cultures on transwell filters, thus the luminal (apical) and the body-side (basolateral) are accessible in the same culture. For this method especially the intestinal porcine epithelial cell lines 1 and J2 and the human-based gut epithelial cell line Caco-2 are used. These cell lines are analysed after stimulation with probiotics or food compounds, using microscopic examination, epithelial resistance measurements, protein biochemistry and gene expression as well as cellular transport. Our results demonstrate that bacteria affect the barrier via humoral mediators or via direct contact. There is evidence that the probiotic *Lactobacillus amylovorus* has a specific close contact to the epithelial surface. This is currently under examination. Pre-incubation of epithlial monolayers using probiotic diminishes damage to the cell layer and prevents alteration of the cytoskeleton during a challenge with pathogens. In addition, our *in vitro* test systems enable the analysis of cytokine expression after probiotic or pathogen stimulation and are used to study cell-cell-contacts in the epithelium – structures essential for the intestinal tightness. In summary, the *in vitro* test systems used in our studies can serve as an assesment of the benefit of pre- and probiotics prior to their use in *in vivo*.

Early microbiota association and dietary interventions affect gut development and function in pigs

A.J.M. Jansman[1], S.J. Koopmans[2], J. Zhang[3] and H. Smidt[3]
[1]Wageningen UR Livestock Research, P.O. Box 65, 8200 AB Lelystad, the Netherlands, [2]Wageningen University, Adaptation Physiology Group, Department of Animal Sciences, P.O. Box 338, 6700 AH, Wageningen, the Netherlands, [3]Wageningen University, Laboratory of Microbiology, Dreijenplein 10, 6703 HB Wageningen, the Netherlands; alfons.jansman@wur.nl

The development of intestinal microbiota in the gut of pigs in the immediate post-natal period may have profound effects on the functional development of the gut. The establishment of the intestinal microbiota in piglets starts at birth and is influenced by microbiota originating from the sow, the immediate environment of the piglet, the piglet's genetics and by the composition of diet in the pre- and post-weaning period. In the INTERPLAY project a model was developed in piglets that allows investigation of the effects of the postnatal association with a simple or a complex microbiota on gut health and development at a later stage. The model uses caesarean delivered piglets which are either associated with a mixture of three microbial species on days 1, 2 and 3 after birth or, in addition, on days 3 and 4, with a complex microbiota in the form of a faecal inoculant of an adult sow. In the first 4 days the piglets receive a milk replacer diet and in the subsequent period a moist diet. Microbiota composition showed consistent differences between treatments in microbiota composition of digesta in the small and large intestine and in the faeces up to at least 28 d of age. The further effects of dietary fat source (coconut oil as a medium chain triglyceride vs. soya/palm oil) in the post natal period have been investigated on gut function and microbiota development. It is concluded that manipulation of intestinal microbiota development by association treatments and/or dietary interventions in the postnatal period has long lasting effects on gut development and function in young piglets and opens new perspectives to support gut and animal health in sustainable pig production systems.

A mucin-enriched fermentation model to assess prebiotic potential of new indigestible carbohydrates

C. Boudry[1], T.H.T. Tran[1,2], Y. Blaise[1], A. Thewis[1] and J. Bindelle[1]
[1]ULG, Gembloux Agro-Bio Tech, Animal Science Unit, Passage des deportes 2, 5030 Gembloux, Belgium, [2]Wallonie-Bruxelles International, place Sainctelette 2, 1080 Brussels, Belgium; yblaise@ulg.ac.be

Screening the prebiotic potential of novel indigestible carbohydrates (ICH) is a challenge for feed and food industry and *in vitro* models are increasingly used for such purposes. Recently extracellular binding proteins responsible for the adherence to intestinal mucus were described for several *Lactobacillus* species. As this genus is known for its beneficial effect on gut health, we enriched the *in vitro* gas fermentation model with mucin in order to evaluate the prebiotic potential of 5 ICH. Mucin-coated microcosms (MCM) were prepared as described by Van den Abbeele *et al.* and introduced in the fermentation bottles with an inoculum prepared from fresh faeces of 3 sows mixed with a nutritive buffer solution. Fermentation was performed at 39 °C, using 200 mg of substrate, 30 ml of inoculum and 6 MCM, yielding approx. 20 mg mucin each, in 140 ml glass bottles. A first study was performed with inulin and cellulose as substrates, with and without mucus in the bottles. A second study was performed with 5 substrates (inulin, IMO, beet pulp POS, cellobiose and gluconate) in presence of mucus. After 8 and 72 h, SCFA and the microflora of fermentation broth was analysed as well as the microflora on the MCM. The comparison of the microflora evolution with and without mucus showed a better development of the *Lactobacillus* in the fermentation broth with MCM, mainly in presence of inulin. The development of the *Lactobacillus* genus allowed the classification of the 5 substrates tested in the second study (Inulin > IMO > Gluconate > Cellobiose > POS) ($P<0.05$) which was not possible without mucus ($P>0.05$). Inulin and IMO showed also the highest development of Bifidobacteria ($P<0.05$) and the highest levels of butyrate production ($P<0.05$) compared to the three other substrates, indicating a high prebiotic potential.

Responses in gastrointestinal ecosystem of pigs fed liquid diets containing fermented whey permeates

S. Sugiharto, C. Lauridsen and B.B. Jensen
Depart. of Animal Science, Aarhus University, Blichers Allé 20, 8830 Tjele, Denmark; sugiharto@agrsci.dk

The objective of the present study was to investigate the effect of feeding liquid diet containing whey permeate (WP) fermented with different lactic acid bacteria (LAB) on microbial population of gastrointestinal tract (GIT) and mucosal immune responses of pigs. The study consisted of 52 weaned pigs, which were allotted to: (1) no infection and fed no WP (INF–WP–); (2) infection and fed no WP (INF+WP–); (3) infection and fed non-fermented WP (INF+WP+); (4) infection and fed WP fermented with *S. thermophilus*/*L. bulgaricus* (INF+WP+LAB1); (5) infection and fed WP fermented with *L. plantarum* (INF+WP+LAB2); and (6) infection and fed WP fermented with W. viridescens (INF+WP+LAB3). Pigs in group 2 to 6 were inoculated with *E. coli* F4 on d2 and 3. Faeces were collected on d2, 3, 4, 5, 6 and 8. Immediately after killing, digesta from GIT, intestinal segments and mucosa were collected. MIXED procedure for repeated measurement was used to analyze the faecal microbial populations, and the data generated from different sites of intestine or GIT segments were analyzed based on nested design. Posterior incorporation of the fermented WP into liquid diet immediately prior to feeding did not impair the growth of pigs. Species of LAB used to ferment WP affected ($P<0.05$) the feed intake of pigs. Faecal populations of LAB and Clostridium spp. were lower ($P<0.05$) in INF+WP– pigs during the first 6 d, and in INF–WP– pigs during the first 5 d of the experiment, respectively, than the other pigs. Feeding fermented WP lowered ($P<0.05$) enterobacteriaceae population in the GIT digesta, resulting in less ($P<0.05$) stimulation of IgA production in the ileum and bile. The wet weight of intestine was higher ($P<0.05$) in INF–WP– pigs than the infected pigs. In conclusion, feeding liquid diet containing fermented WP maintained intestinal microbial ecosystem and modulate the mucosal immune responses of *E. coli* F4 infected weaned pigs.

Session 47 Theatre 8

A maternal scFOS supplementation modulates maturation of the immune system of piglets

C. Le Bourgot[1], S. Ferret-Bernard[1], E. Apper[2], S. Blat[1], F. Respondek[2] and I. Le Huërou-Luron[1]
[1]INRA, Domaine de la Prise, 35 590 Saint-Gilles, France, [2]Tereos Syral, Z.I. Portuaire, 67 390 Marckolsheim, France; cindy.lebourgot@gmail.com

Peri-partum nutrition is essential for immune development of neonatal piglets. Short chain fructo-oligosaccharides (scFOS) are defined as prebiotics. A scFOS supplementation of sows and piglets might modulate gut immune maturation by modulating the composition of the microbiota. Effects of maternal scFOS supplementation on immune transfer to the offspring and on their mucosal immune system maturation were investigated. Twenty-six sows received a control or a scFOS diet (10 g/d of maltodextrin or scFOS, Profeed®) for the last 4 weeks of gestation and during lactation. Immune quality of colostrum was determined. Twenty-six piglets were slaughtered on d21. Intestinal contents were sampled to analyse short chain fatty acids (SCFA) concentration. Morphometry of Peyer's Patches (PP) was determined. Mononuclear cells were isolated from PP and mesenteric lymph nodes (MLN). PP and MLN cell density was determined. Their cytokine patterns and secretory IgA (sIgA) were studied following culture. Flow cytometry was performed to quantify immune cell populations. Maternal scFOS supplementation increased concentrations of IgA and TGFβ1 in colostrum ($P<0.05$). It did not modify the morphometry of the PP but promoted the development of cellular immunity in piglets, as suggested by the increased secretion of IFNγ and sIgA by PP cells ($P<0.10$), a higher cell density in MLN cells with a greater IFNγ/IL-10 ratio ($P<0.05$) and an improved secretion of TNFalpha ($P<0.10$). Concentrations of SCFA and flow cytometry are being done. Maternal supplementation with scFOS enhances immune functions in a crucial period of maturation of the mucosal immune system of piglets. This effect is probably related to the modification of the microbiota of the sows, which in turn, impacts piglet's microbiota. Such results underline the key role of the maternal diet in supporting the development of mucosal immunity in neonates.

Session 47 Theatre 9

Perinatal antibiotic treatment of sows affects intestinal barrier and immune system in offspring

S. Ferret-Bernard, G. Boudry, L. Le Normand, V. Romé, G. Savary, C. Perrier, J.P. Lallès and I. Le Huërou-Luron

INRA, UR1341 ADNC, Domaine de la prise, 35590 Saint-Gilles, France; stephanie.ferret@rennes.inra.fr

Maternal environment during pregnancy and lactation influences health of the offspring and its microbiota. Immune tolerance to dietary and commensal microbiota antigens develops early in life. Our aim was to investigate intestinal barrier and local immune response to LPS in piglets deriving from sows whose microbiota was manipulated by antibiotics. Long-term consequences on gut adaptation to a high fat diet were examined in growing offspring. Sows were given amoxicillin (AMX; 40 mg/kg/d) per os (n=11 AMX) or not (n=12 CTL) from 10 d before to 21 d after parturition. One piglet per litter was sacrificed at post-natal day (PND) 14, 21 and after weaning at PND42. The remaining ones were given either a low fat (LF) or a high fat (HF) diet (to enhance LPS epithelial passage) from PND140 to 169. At PND14, ileal permeability was greater in AMX than CTL piglets ($P<0.05$). At PND21, no increase in proliferation of lamina propria mononuclear cells (LPMCs) in response to LPS was noticed, demonstrating LPMC tolerance in both groups. Analysis of laser-captured enterocyte transcriptome indicated specific changes in immune response pathways. Especially differential expression of BFIFB1, HPSE and FSTL4 genes demonstrated stimulation of proinflammatory activity, corroborating the increased TNFα secretion in response to LPS observed in LPMCs of AMX piglets. At PND42 gene analysis suggested a repression of the immune response towards weaning-induced stress in AMX piglets. At PND169, HF diet increased ileal permeability in CTL offspring only. By contrast, ileal explant sensitivity to LPS was altered in HF pigs, especially in AMX-HF pigs. Manipulating maternal microbiota modified offspring ileal barrier and immune response to LPS during the neonatal period. It also influenced the intestinal adaptive response to a HF diet in adult offspring, suggesting microbiota-related imprinting.

Session 47 Poster 10

Single and combined effects of *Bacillus subtilis* and inulin on performance of late-phase laying hens

A. Abdelqader[1], A. Al-Fataftah[1] and G. Daş[2]

[1]The University of Jordan, Department of Animal Production, Faculty of Agriculture, the University of Jordan, Amman 11942, Jordan, [2]University of Göttingen, Department of Animal Sciences, Albrecht-Thaer-Weg 3, 37075 Göttingen, Germany; a.abdelqader@ju.edu.jo

This study investigated whether performance, eggshell quality, Ca retention and intestinal health and morphology of late-phase laying hens can be improved by single and combined inclusions of *Bacillus subtilis* and inulin in the diet. Eighty hens (64 wk old) were randomly distributed into 4 treatment groups each consisting of 5 replicates of 4 hens. Birds in the groups were fed a basal diet (control) or basal diet plus 1 g/kg *B. subtilis* (2.3×10^8 cfu/g) or basal diet plus 1 g/kg inulin, or basal diet plus a synbiotic combination of *B. subtilis* and inulin for 12 wk. Dietary supplementation of *B. subtilis*, inulin or synbiotic improved feed conversion, egg performance, eggshell quality and calcium retention compared with the control ($P<0.05$). Inulin and synbiotic exhibited the highest increase in eggshell thickness and eggshell Ca content, and the lowest eggshell deformations ($P<0.05$). Unmarketable eggs were reduced from 8.4% in control group to 3.5%, 1.7%, and 1.5% in *B. subtilis*, inulin and synbiotic groups, respectively ($P<0.05$). Tibia density, ash, and Ca content increased by inulin and synbiotic inclusions ($P<0.05$). *B. subtilis*, inulin and their synbiotic combination increased villus height and crypt depth in all intestinal segments, compared with the control ($P<0.05$). *B. subtilis* and inulin modulated the ileal and caecal microflora composition by decreasing numbers of *Clostridium* and Coliforms and increasing bifidobacteria and lactobacilli ($P<0.05$). Colonization of the beneficial microflora along with increasing the villi-crypts absorptive area were directly associated with the improvements in performance and eggshell quality. It is concluded that egg production and eggshell quality of late-phase laying hens are improved by inclusions of *B. subtilis* and inulin in the diet.

Session 47 Poster 11

A nutrigenomic approach to identify pathways involved in stress response in gut weaning piglets

L. Bomba[1], E. Eufemi[1], N. Bacciu[2], S. Capomaccio[1], A. Minuti[1], E. Trevisi[1], M. Lizier[3], F. Lucchini[4], M. Rzepus[5], A. Prandini[5], F. Rossi[5], P. Ajmone Marsan[1] and G. Bertoni[1]
[1]Università Cattolica del Sacro Cuore, Ist. Zootecnica, Via E. Parmense 84, 29122 Piacenza, Italy, [2]Associazione italiana allevatori, via tomassetti, 9, 0161 Roma, Italy, [3]CNR, Ist. Tecnologie Biomediche, via Cervi 93, 20090 Milano, Italy, [4]Università Cattolica del Sacro Cuore, CRB, via Milano 24, 26100 Cremona, Italy, [5]Università Cattolica del Sacro Cuore, Ist. Scienze degli alimenti e della nutrizione, via Emilia Parmense 84, 29122 Piacenza, Italy; giuseppe.bertoni@unicatt.it

During weaning, piglets deal with stressors such as loss of mothering, mixing with other litters, end of lactational immunity, change in their environment and gut microbiota. Antibiotics are no more allowed to overcome weaning disorders, while organic acids are becoming widespread. Here we assess the effect of weaning on the transcriptional state of ileum and investigate the effect of a dietary acidification with sorbic acid on the expression of genes involved in intestinal metabolism. Ileum from 4 piglets at weaning day (T0) and 12 piglets five days after weaning (T5) were collected. Two diets were administered to the T5 group: a standard diet with 5 g/kg of sorbic acid, and a standard diet with 5 g/kg barley flour. Gene expression was assessed through Combimatrix whole transcriptome microarrays. The combination of the short period of treatment and the criticality of the weaning stage hide the diet effect on gene expression, metabolism and histology. Conversely, comparing piglets after weaning (T5) and pre weaning (T0), we identify 205 genes differentially expressed in response to the metabolic and environmental adaptation during weaning. The activation of pathways related to immune and inflammatory response correlates with a general atrophy of the small intestine. Our approach well described the intestinal piglets metabolism during weaning and may be useful to test nutrients or additives to control stress response and improve animal welfare.

Session 48 Theatre 1

French livestock systems facing the new context of market and policy

C. Perrot and P. Sarzeaud
French Livestock Institute, Economy, 149 Rue de Bercy, 75595 Paris cedex 12, France; christophe.perrot@idele.fr

In Europe, France appears in the three main countries as far as livestock of ruminant is concerned. The recent treatment of the 2010 census provides a new picture of the hudge diversity of the farming systems and its sensitiveness to changes in market and politic contexts. Led by the French Livestock Institute, this analysis is based on a farm classification considering the main breeding activities (dairy, beef, sheep and goat) and the combination of production from the last 2010 general census. The methodology gives the opportunity to describe the level of specialization and implication in the production. It states also with the means of production and management such as land and labor use. In regard of those figures and according to the new CAP negotiations, different issues seemed to highly impact French livestock systems: With the milk quota system removal, the dairy sector seems opened to follow up its restructuration leading to bigger size and more intensive farms, especially in plains areas and for farms impacted by the single farm-payment convergence. But what will be the consequence for dairy producers in mountains areas and for beef activities coupled with dairy production such as steers or bulls? France detains the biggest suckler cow's herd in Europe. This production is characterized by the rather old age of the breeders that announced difficulties in the replacement of farms owners. Main part of the herd is located in the mountains and the grasslands areas of Massif Central where others activities are not available. But in the plain areas, the concurrency of the cereals speculation appears to weaken the relative stability of the production. This competition between crop and livestock affected already in the last 10 years the sheep flock, now concentrated in mountains areas. With the rising up of price at the European level, the lamb production improved its rentability but it seems to be deficient in order to help the replacement of breeders.

The impact of the CAP reform on the profitability of dairy and beef farms in Italy until 2020

K. De Roest, C. Montanari and A. Menghi
Research Center for Animal Production, Agricultural Economics and Engineering, Viale Timavo 43/2, 42121 Reggio Emilia, Italy; k.de.roest@crpa.it

The profitability of calf fattening in Italy primarily depends on numerous factors such as the price of bullocks and feed, the interest rate on invested capital and the technical efficiency of the production process. Beef farms specialized in the fattening of imported bullocks from primarily France enjoy a significant income support from decoupled CAP premia. The new proposals for the CAP reform for the period 2014-2020 which include a scaling down and a flat hectare rate of direct income premia, the introduction of greening and the possibility of coupled premia for quality production will have a significant impact on the incomes and profitability of beef farms in Italy. In the presentation the effect of the CAP reform on representative samples of beef farms in Veneto and Piedmont will be illustrated by using full production costs analyses and partial budgeting simulation techniques until the year 2020. Economic analysis of farm income of dairy farms always reveal a strong variability between farms which is due to differences in the technical efficiency, the dairy farm structure and the quality of management. Dairy farms receive direct income support based on their milk quota detained at the 31st of March 2006. Decoupling of the CAP premia occurred three years later. The CAP reform for the period 2014-2020 and the abolishment of milk quota will have a significant impact for the profitability of dairy farms in Italy. By means of production cost analyses and partial budgeting techniques the impact of the CAP reform will be simulated on samples of dairy farms in Veneto, Lombardy and Emilia-Romagna until the year 2020. In this last region a sub sample of dairy farms is dedicated to the production of milk for Parmigiano-Reggiano cheese.

Beef production in Europe and Germany: status and perspectives

C. Deblitz
Thünen Institute of Farm Economics, agri benchmark, Bundesallee 50, 38116 Braunschweig, Germany; claus.deblitz@ti.bund.de

Global cattle and beef production are undergoing changes which are driven by increasing global demand, rising beef but also rising feed prices and concerns about sustainability and animal welfare. In many countries significant productivity increases as well as intensification of production can be observed while in the EU there seems to be a (policy) trend towards extensification which is expressed via policy measures like greening and increasing animal welfare standards. At the same time, the private sector is asking for 'greener' and more transparent production through traceability of animals. Production does not always manage to cope with the increasing requirements. Recent years have shown increasing beef prices globally but also increasing feed and livestock prices, sometimes with different patterns and having different impacts on profitability. Coming from a global background, the contribution highlights the latest price and market developments as well as farm level performance in selected European countries. It also looks at the impact that changes in feed, livestock and other prices had on the profitability. The role of subsidies in the economic equation of beef production is also considered. In addition, the latest developments in German dairy, suckler-cow and beef production and markets are outlined. Taking the previous information and the likely policy development into account, the contribution closes with an assessment of the expected mid-term developments of beef production in Europe.

CAP impact on cattle sector, markets and prices in Central and Eastern European countries

I. Stokovic[1], P. Polak[2], M. Klopcic[3] and A. Svitojus[4]
[1]University of Zagreb Faculty of Veterinary Medicine, Heinzelova 55, 10000 Zagreb, Croatia, [2]Animal Production Research Centre Nitra, Hlohovecká 2, 951 41 Lužianky, Slovak Republic, [3]University of Ljubljana Biotechnical Faculty, Jamnikarjeva 101, 1000 Ljubljana, Slovenia, [4]Heifer Baltic Foundation, S. Konarskio 49, 03123 Vilnius, Lithuania; igor.stokovic@vef.hr

Common agricultural policy of the EU since 1962 had huge impact on the whole European agricultural sector. The diversity of these impacts is frequently stressed. The same could be said for CEE countries which entered the EU after 2004. Special emphasis should be put on the fact that most of those countries came from different political system and just before or even during EU accession went through the period of transition. This means that those countries went from collectively owned farms transformed to private establishments. During that period lots of farms disappeared, land ownership became a problem and that had huge impact on number of cattle, crop and feed prices and the market in whole. When looking at the prices of milk, meat and maize there is obvious trend of price stabilization in the period from 2004 till 2006 in all CEE countries. The same was not observed in cattle number. A major problem for the new EU countries was also the adaptation on for them new EU rules. It is often criticized that those rules favour old EU countries or vice versa that old member states are better prepared for implementation of those rules. Changes in CAP which are expected to be active in near future will impose new ways which have to be taken into account when thinking about cattle sector future in CEE countries.

Economic performance and greenhouse gas emissions of Irish beef cattle production

P. Crosson, E.G. O'riordan and M. McGee
Teagasc, Animal & Grassland Research and Innovation Centre, Grange, Dunsany, Co. Meath, Ireland; paul.crosson@teagasc.ie

There are 2.2 million cows in Ireland with approximately half belonging to the suckler beef and dairy herds, respectively. Similarly, the 0.5 mt of beef carcass produced annually is sourced in approximately equal proportions from the suckler beef and dairy herds. Beef exports in 2011 were worth €1.8bn accounting for almost 35% of gross primary agricultural output. However, farm profitability is extremely low, with average family farm income (FFI) in 2011 of -€72/ha and -€82/ha for suckler beef cow and cattle finishing farms, respectively when non-market payments, such as the support payments from the EU Common Agricultural Policy (CAP), are excluded. Thus, family farm viability is heavily dependent on these non-market payments and therefore, the review of CAP is of significant importance. Nevertheless, from a market returns perspective, there are a number of factors which favour Irish livestock production. In particular, the capacity to grow high yields of highly digestible grass at low cost is a key competitive advantage of Irish livestock systems. The grass based nature of Irish cattle systems also underpins the sustainability of beef production. In Ireland, agriculturally derived greenhouse (GHG) emissions have shown steady reductions in recent years with emissions in 2010 8.3% lower than 1990 levels. A recent study showed that there is potential to further reduce GHG emissions per kg of beef on Irish suckler beef farms by 12% by adopting management practices prevailing on research farm experiments. Similarly, on dairy beef systems there is scope to reduce GHG emissions; however in this case the potential reduction is 18% per kg beef. These studies also indicated that these efficiency gains were consistent with improved farm economic performance. The objective of this paper is to outline current levels, and technical and biological drivers, of profitability and GHG emissions from Irish beef farms.

Session 48 Theatre 6

Some considerations for the future of grassland dairy farming in Austria

M. Kapfer, J. Kantelhardt and T. Moser
Institute of Agricultural and Forestry Economics, BOKU, Feistmantelstraße 4, 1180 Vienna, Austria; martin.kapfer@boku.ac.at

The conditions for agricultural production are changing. Special emphasis is placed on raising the importance of environmental and animal protection issues and the volatile markets. Furthermore, the future amount of transfer payments is unclear, but a decrease is expected. These uncertainties implicate a challenge for farms. While cash-crop farms can respond in a relatively flexible way by changing crop rotation and cultivation practices, the situation for grassland-based dairy and cattle farming is difficult. This is due to the high fixed costs, the long-term commitment of capital and the lack of alternatives. The present work attempts to answer the following question: what changes can be experienced in grassland areas from both rising producer prices in grassland farming and a decline of transfer payments (with special regard to the direct payments scheme). In our study we assume that land rent from grassland use depends directly on the site quality and the intensity of use. We consider only two intensity levels where the high level may stand for dairy farming and the low level can be interpreted as suckler-cow keeping. From an economic point of view intensive grassland use is to be preferred in high-yield regions. On marginal sites the extensive grassland use is more profitable. A decline of the (site-quality independent) transfer payments would affect both extensive grassland use on marginal sites as well as intensive grassland use on high-yield sites. However, a decline of state payments affects marginal sites relatively more due to a larger share of these payments on the farm income. Abandonment would be the consequence. At the same time the decline in state payments can be compensated in sites of average quality (transition zone between intensive and extensive production) by intensification. We expect that the environment-friendly extensive grassland use is threatened two-fold and therefore extensively used grassland will decline.

Session 48 Theatre 7

An empirical assessment of microeconomic effects from suckler cow farming in Austria

S. Kirchweger, M. Eder, M. Kapfer and J. Kantelhardt
Institute of Agricultural and Forestry Economics, University of Natural Resources and Life Sciences, Feistmantelstraße 4, 1180 Vienna, Austria; stefan.kirchweger@boku.ac.at

Many farms in Austria leave dairy farming and keep suckler cows to produce meat from grassland. This is encouraged through the possibility to reduce farm labour and engage in non-farm work as well as through coupled direct payments. However, it is very likely that these payments will be decoupled in the next period of the CAP. In combination with the change from the historical to the regional payment scheme in the first pillar of the CAP, this might lead to a decrease in direct payments. The objectives of this study is to analyse microeconomic effects from suckler cow farming compared to dairy farms and give insights in their economic performances under volatile markets. We use a panel of accounting data for 86 specialized suckler cow farms and 483 specialized dairy farms in the time period 2005 to 2010. We apply Propensity Score Matching (PSM) to balance variables between suckler and dairy cow farms. To do so, PSM basically tries to identify similar dairy farms for each suckler cow farm. First results indicate a negative statistical significant effect for farm income on suckler cow farms, but not for family income, which also includes non-farm income. This follows from the fact that suckler cow farmers work in non-farm activities. Furthermore, these farms use their labour alternatively in forestry, direct marketing and other services and can therefore keep a high level of family income even in years with low output prices. Especially when times of low output prices occur, suckler cow farming is more depending on public payments than dairy farms. With regard to reduced and decouple direct payments and volatile markets, this implicates that farmers have to increase productivity of their used labour and area. One possibility is to reduce their farm inputs, which are not different from dairy farms.

Session 49 Theatre 1

Growth performance and behaviour of entire male pigs under various management conditions

N. Quiniou, V. Courboulay, T. Goues, A. Le Roux and P. Chevillon
IFIP-Institut du Porc, BP 35104, 35651 Le Rheu cedex, France; nathalie.quiniou@ifip.asso.fr

Most literature data on growing entire males were obtained in *ad libitum* feeding conditions with diet supplied as pellets or mash (dry), and up to a slaughter body weight (sBW) below 120 kg. As a general ban on castration is expected in the EU by the year 2018, more knowledge was required in different management conditions, representative of those observed in the field. Six batches (b) were used to investigate the effect of six bi-modal factors on performance, behaviour (b 3, 4, 5, 6) and boar taint risk (not b 3): group (6/pen) vs. single housed pigs (b 1), *ad libitum* feeding with dry pellets or liquid feed (b 2), pigs fed *ad libitum* or restrictively with dry pellets (b 3) or liquid feed (b 4), standard or heavy sBW (119/134 kg, b 5). Boars from b 1 to 5 were obtained from Large White × Pietrain sires, those from b 6 were obtained either from Pietrain (PP) or crossbred Duroc × Pietrain (D×PP) sires. High growth rate and feed efficiency (G:F) are obtained under *ad libitum* fed pigs both in dry and liquid feeding systems, up to a heavy sBW, and with different types of crossbreeding. Feed efficiency was not influenced by a feed restriction of 5% (b 4), whereas it increased with a feed restriction of 9% (b 3). Neither feeding conditions nor sBW influenced sexual behaviour, which remained at a low level (<2.5% of the active behaviour). However, amount of negative social behaviour (aggression, biting, head knocking) was more important in restrictively fed pigs, both in dry and liquid feeding systems. Lesion score tended to be higher in restricted than in *ad libitum* fed males with the dry feeding system where only 1 or 2 pigs have a simultaneous access to the feeder. The PP offspring presented a higher lesion score than D×PP ones. Based on performance and behaviour observations, entire males should be fed *ad libitum*, with attention paid to the sire in order to limit aggression and subsequent lesions. However, when the boar taint risk is high, a small feed restriction could be pertinent.

Session 49 Theatre 2

Phosphorus requirements and retention in entire male and female pigs, a dose-response study

P. Bikker[1], A.W. Jongbloed[1], M.M. Van Krimpen[1], R.A. Dekker[1], J.T.M. Van Diepen[1] and S. Millet[2]
[1]Wageningen UR, Livestock Research, P.O. Box 65, 8200 AM Lelystad, the Netherlands, [2]Institute for Agricultural and Fisheries Research (ILVO), Animal Sciences Unit, Scheldeweg 68, 9090 Melle, Belgium; paul.bikker@wur.nl

Phosphorus (P) requirements of pigs are determined in dose-response studies with increasing dietary P content or by calculating the P-requirements for (near) maximum bone mineralisation using a factorial model. The use of the latter is widespread and based on an estimate of P requirements for maintenance processes (endogenous losses) and whole body P retention, combined with an assumption for P utilisation. Recent data on carcass and bone ash content are scarce, especially in entire male pigs, since much of the earlier work has been conducted with castrated male pigs. Therefore we conducted a comparative dose-response study with 118 growing male and female pigs from 25-50, 50-80 and 80-125 kg of body weight (BW). The pigs received diets with incremental P contents from 50 to 130% of current P recommendations in 5 equidistant steps by inclusion of increasing levels of mono calcium phosphate in the diet. Daily gain and feed conversion ratio (FCR) were determined. Groups of pigs were killed at 25, 50, 80 and 125 kg of BW and empty bodies were analysed for whole body ash, Ca and P content. Daily gain increased and FCR decreased curvilinearly ($P<0.05$) with increasing dietary P content in both male and female pigs between 25 and 80 kg BW. In each BW range, male pigs realised a higher daily gain and lower FCR ($P<0.001$) than female pigs. The mean growth rate was 50 g/d higher and FCR was 0.2 lower in the male pigs. We used a linear-plateau model to estimate the dietary digestible P-content required for maximum growth performance. The mean requirements for maximum gain and feed efficiency were approximately 1.8 g digestible P/kg of diet, and not significantly different for male and female pigs. Requirements based on body mineral content and bone mineralisation will be determined and may be higher.

Effect of chirurgical or immune castration on postprandial nutrient profiles in male pigs

N. Lefloch, A. Prunier and I. Louveau
INRA, UMR1348 PEGASE, 35590 Saint Gilles, France; nathalie.lefloch@rennes.inra.fr

Male pigs (EM) eat less and exhibit higher growth performance and feed efficiency than castrated pigs (CM). Despite large differences in feed efficiency between EM and CM, the mechanisms involved in these differences have been poorly investigated. The current study was undertaken to determine whether difference in feed efficiency between CM, immune castrated (IC) and EM pigs can be explained by difference in nutrient utilization after a meal. Male pigs, 6 CM, 6 EM and 6 IC, were fitted with a jugular catheter. Three postprandial plasma nutrient profiles were established on each pig at 16, 18 and 20 weeks of age. These periods (P) were chosen to surround the second injection of Improvac© to IC pigs: P1 being before the decrease in testicular hormones, P2 and P3 being during and after the decrease in testicular hormones in IC pigs. On each test day, serial blood samples were collected prior to a test meal (400 g), after pigs were fasted overnight, and for 4 h after the meal. Plasma concentrations of glucose, urea and amino acids (AA) were determined on each blood sample. Data were submitted to repeated measures ANOVA (MIXED procedure of SAS). Individual AA kinetics were analysed with the NLIN procedure (SAS) using the one-compartment model of Erlang. During P2 and P3, profiles of plasma glucose were similar in IC and CM pigs. Plasma urea profiles did not differ between CM, EM and IC pigs during P1 whereas CM pigs showed higher plasma urea concentrations than EM and IC pigs during P2 and P3. Lys, Leu, Ile and Val plasma concentrations were greater in CM than in EM and IC pigs during the 3 periods and only during P3 for Met, Thr, Tyr and Phe. Plasma hypro concentrations were lower in CM than in EM and IC during P1 whereas those of IC were intermediate during P2 and P3. Plasma glucose profiles were affected by immunocastration much faster and earlier than urea and AA profiles, suggesting that IC pigs kept the advantages of EM pigs in terms of protein metabolism during the experimental period.

Response of blood hormones and nutrients to stress in male pigs differing by their gonadal status

A. Prunier, C. Leclercq, N. Le Floc'h and E. Merlot
INRA, Phase, UMR 1348, 35590 Saint-Gilles, France; armelle.prunier@rennes.inra.fr

With the aim of stopping chirurgical castration of pigs by 2018, it is necessary to evaluate all the consequences of rearing entire or immunocastrated (vaccinated against GnRH to suppress testicular activity) male pigs, especially their response to stressors. We compared the endocrine and metabolic responses to a physical stressor (nose lasso applied during 3 min = NL) or to an iv injection of ACTH (5 µg/kg live weight) in male pigs left intact (E, n=9) and vaccinated against GnRH at 82 and 117 days (IC, n=8), submitted to neonatal castration (C, n=9) and vaccinated (SIC, n=8). Hormones and nutrients were measured in blood samples collected through a jugular catheter, before and after NL (3 to 95 min at 140 days) or ACTH (3 to 240 min at 142 days) applied 6 to 10 h after eating. Data were analyzed by ANOVA using SAS. Before ACTH or NL, plasma ACTH, cortisol and lactate were similar in all groups. Glucose and glutamic acid were lower whereas several essential AA (Leu, Lys, Thr, Val) were higher in surgical castrated males (SC + SIC vs. E+IC). FFA, Leu and Val were higher in vaccinated pigs (IC + SIC vs. E+SC). After NL, the increase in ACTH was lower in surgically castrated males whereas the increase in cortisol was similar. The increases in lactate and glucose were similar in all groups whereas the increase in FFA was slightly lower in vaccinated males. The decrease (most AA) or increase (Hypro, Glu) in AA were similar in all groups except for Hypro which increased only in E and IC males. After the ACTH injection, the increases in cortisol, FFA and glucose (120 to 180 min post ACTH), the decrease in glucose (30 to 60 min) were more marked in surgical castrated males whereas the increase in lactate was similar. For AA, the decrease at 60 min post ACTH (most AA) or the increase at 240 min post ACTH (Ala, Glu, Tyr) were similar in all groups. Present data suggest that surgical castrated males have a lower sensitivity of the pituitary gland but a higher sensitivity of the adrenal cortex to stress.

Session 49 — Theatre 5

Effect of dietary net energy content on performance and lipid deposition in immunocastrated pigs

N. Batorek[1,2,3,4], J. Noblet[2,3], M. Bonneau[2,3], M. Čandek-Potokar[1,4] and E. Labussiere[2,3]
[1]University of Maribor, Faculty of Agriculture and Life Sciences, Pivola 10, 2311 Hoče, Slovenia, [2]INRA, UMR 1348 Pegase, 35590 Saint-Gilles, France, [3]Agrocampus-Ouest, UMR 1348 Pegase, 35000 Rennes, France, [4]Agricultural Institute of Slovenia, Hacquetova ulica 17, 1000 Ljubljana, Slovenia; nina.batorek@kis.si

Immunocastration seems a suitable alternative to surgical castration of pigs, but results in increasing feed intake (FI) after second vaccination (V2), leading to increased fat deposition. The aim of the present study was to evaluate the effect of reduced dietary net energy (NE) content on performance, carcass and meat quality in 45 immunocastrated pigs (IM) after V2 (eight weeks prior slaughter at mean BW of 126 kg). At the age of 70 days, pigs were spread according to litter over three treatment groups (TG) differing in the dietary NE content (HNE: 10.6 MJ/kg DM, MNE: 11.1 MJ/kg DM and LNE: 10.5 MJ/kg DM) and fed *ad libitum*. In the period between V2 and slaughter, average daily gain and average daily feed intake were 1.14 kg/day and 2.94 kg of DM for the three TG, respectively, whereas gain to feed ratio (G:F) was lower for LNE pigs (0.34 vs. 0.37 and 0.38 for LNE, MNE and HNE; $P<0.05$). The NE intake was higher in HNE (2.46 vs. 2.32 and 2. 24 MJ/kg BW 0.60/day for HNE, LNE and MNE; $P<0.05$) and gain to NE intake ratio averaged 39.9 g/MJ NE ($P>0.05$). Body weight, carcass weight and dressing yield averaged 126 kg, 98 kg and 78%, respectively and did not differ among TG ($P>0.05$). Backfat thickness measured with ultrasound, neck intermusculat fat and fat over loin eye indicate that pigs in HNE deposited more fat after V2, than those in other two TG (for 0.03 mm/day, 3.4% and 3 cm^2, respectively; $P<0.05$). Loin eye area and intramuscular fat in loin did not differ among TG ($P>0.05$). Results suggest that NE reduction in growth-finishing diet fed *ad libitum* to IM limits fat deposition without affecting growth performance.

Session 49 — Theatre 6

Breeding against boar taint: selection strategies against boar taint in a Swiss pig sire line

A.M. Haberland[1], H. Luther[2], A. Hofer[2], P. Spring[3], E. Tholen[4], H. Simianer[1] and C.F. Baes[3,5]
[1]Georg-August University, Department of Animal Sciences, Albrecht-Thaer-Weg 3, 37075 Göttingen, Germany, [2]SUISAG, Breeding Division, Allmend 8, 6204 Sempach, Switzerland, [3]Bern University of Applied Sciences, School of Agriculture, Forest and Food Science, Langgasse 85, 3052 Zollikofen, Switzerland, [4]University of Bonn, Institute of Animal Science, Endenicher Allee 15, 53115 Bonn, Germany, [5]Swiss Federal Institute of Technology, Institute of Agricultural Sciences, Tannenstrasse 1, 8006 Zürich, Switzerland; baesc@ethz.ch

Surgical castration as a reliable means for producing meat free of boar taint is common practice in worldwide pig production. A ban on surgical castration, including that performed using anesthesia or analgesia, will likely be anchored in the legislation of many European countries in the foreseeable future; feasible alternatives are required as soon as possible. The aim of this study was to model a terminal sire line breeding program to assess the potential of selection against boar taint using different breeding goals (Human Nose Score HNS or chemical compounds) using selection index calculations. The Swiss terminal sire line PREMO® was used as an example for comparing the information sources: (1) biopsy-based performance testing (BPT) of live boars; (2) assessment of HNS on station; and (3) genomic selection. Due to high heritabilities, natural genetic gain was highest when breeding for the main chemical components responsible for boar taint (androstenone (AND), skatole (SKA) and indole (IND)) conducting BPT. Genomic information of young selection candidates adds little to natural genetic gain and is costly. Even if HNS is considered the target trait, the (correlated) natural genetic gain in HNS was highest when breeding for chemical compounds. Concerning BPT, the amount of AND (SKA, IND) could be halved within 9 (7, 11) years.

Maximising energy intake and the growth potential of entire male pigs

F.R. Dunshea
The University of Melbourne, Agriculture and Food Systems, Royal Parade, Parkville 3010, Australia; fdunshea@unimelb.edu.au

In pigs, entire males are generally more muscular, less fat and more metabolically efficient than gilts or barrows. Despite these advantages the traditional practice in most countries is to physically castrate male piglets early in life. The major reason for this is to control boar taint, which is an offensive smell present in the meat of many mature entire male pigs. A second reason is to control aggression and sexual behaviors in entire male pigs as they approach slaughter weight. While it is accepted that boars are leaner and more efficient than barrows, the growth performance of entire males housed in groups is generally less than that of individually-housed entire males, suggesting that the benefits under commercial conditions may not be as marked as assumed. This could be because of increased sexual and aggressive activities between entire males. Therefore, it is essential that producers optimise energy and protein (lysine) intake in order to make use of the growth potential of entire male pigs. The lysine requirements of entire male pigs are higher than that of barrows and gilts, particularly after the peri-pubertal period and so any reduction in feed intake at this time will constrain performance. Strategies to maintain energy intake or reduce energy expenditure in finisher entire male pigs include providing adequate access to feeders, appropriate stocking density, immunocastration and use of dietary supplements such as betaine and neuroleptics. For example, regression of energy intake versus energy deposition indicates that dietary betaine reduces energy expenditure and increases energy deposition by 0.5 MJ DE/d ($P<0.05$). Also, dietary neuroleptics such as bromide, tryptophan and magnesium can improve ($P<0.05$) growth performance in entire male pigs, particularly in heavy for age pigs. In conclusion, entire male pig production requires specific management and nutritional strategies to maximise their potential for lean tissue growth.

Body lesions in entire male pigs during growth and on the carcass

C. Larzul[1], N. Muller[2], V. Courboulay[3], L. Udin[2] and A. Prunier[4]
[1]INRA, UMR1313 GABI, Domaine de Vilvert, 78350 Jouy-en-Josas, France, [2]INRA, UETP, Domaine de la Motte, 35653 Le Rheu, France, [3]IFIP, Domaine de la Motte, 35653 Le Rheu, France, [4]INRA/Agrocampus, UMR1348 PEGASE, Domaine de la Prise, 35590 Saint-Gilles, France; catherine.larzul@jouy.inra.fr

The European pig industry is engaged in a voluntary abandonment of surgical castration of piglets by 2018. Entire males being more aggressive than castrated males, it is very important to take into account their behaviour for genetic selection in order to avoid problems linked to aggressiveness. Aggressiveness of pigs can be estimated from the number of skin lesions. Present work describes the number of skin lesions measured at two stages during fattening (2 days after transfer to the fattening unit, shortly before first pigs are sent to the slaughterhouse) as well as on carcasses. Three populations of Pietrain (PP) type boars were used to produce the males for the present experiment using PP or Large White dams. About 1,500 males, from nine batches, were followed. The total number of skin lesions was very high after transfer to the fattening unit and decreased at the end of fattening ($P<0.001$). Total numbers of skin lesions located on both sides of the pigs were linked (r around 0.7, $P<0.001$) regardless of the stage, but in some animals, lesions on one side could be twice the number of those on the other side. This made measurements on both sides necessary for a precise evaluation at individual level. Correlations between stages were relatively low even though. Significant differences existed between genetic types of pigs that were present at the three stages ($P<0.001$). Overall, skin lesions were less numerous in pigs of the Pietrain type than in cross-bred pigs. Genetic parameters (heritability values and genetic correlations) were also estimated for each trait to look for possibilities of selection on these criteria.

Session 49 Theatre 9

Effect of mixing entire males with females and slaughter strategy on behavior, growth and boar taint

V. Courboulay, C. Leroy, N. Quiniou and P. Chevillon
IFIP Institut du Porc, BP 35104, 35651 Le Rheu cedex, France; valerie.courboulay@ifip.asso.fr

Management may influence behavior and growth performance of entire male, as well as boar taint risk. Two batches of 12 groups of 10 pigs each were used in a 3×2 factorial design based on group composition (10 males-M, 10 females-F, 5 males and 5 females-MF) and slaughter strategy (1 (1D) or 2 (2D) departures per pen). In the 2D treatment, the four heaviest pigs (2 males and 2 females for MF treatment) were slaughtered 2 weeks before the others (22.5 weeks old). Pigs were 9 weeks old at the beginning of the experiment. Posture and behavior were recorded by scan sampling every 10 minutes during 2.5 h, every week from the fifth week before the first departure to the slaughterhouse (D1). Gender of the recipient and the performer of social interactions were considered. Lesions and lameness were scored individually 1 week before D1, 1 and 2 weeks later. Data were analyzed with a mixed model or with the Kruskal Wallis test. Analyses were performed for both periods, before and after D1. Social and sexual behavior did not differ significantly between M and MF pens and were significantly more frequent than in F pens. In MF pens, more aggressions were observed between males than in any other combination ($P<0.05$). Slaughter strategy had no impact on social behavior after D1, but sexual behavior was significantly less performed in 2D pens, whatever the group composition. Consequently, lesion score was lower in 2D than in 1D pens, with a significant difference only in MF pens. Both factors affected boar taint components. Androstenone level in backfat was lower in males mixed with females than in single sex pens ($P<0.05$). Skatole level in backfat was higher for the 2D males at the second departure than for 1D males ($P<0.01$). Our results indicate that mixing pigs reduces boar taint and improves animal welfare in both slaughter strategies.

Session 49 Theatre 10

Pre slaughter conditions influence skatole and androstenone in adipose tissue and blood of boars

I. Jungbluth, R. Wesoly, V. Stefanski and U. Weiler
Universität Hohenheim, FG Verhaltensphysiologie landwirtschaftlicher Nutztiere, Garbenstr. 17, 70593 Stuttgart, Germany; inajung@uni-hohenheim.de

The amount tainted boar carcasses may be influenced by management conditions but also by treatment before slaughter. The relevant environmental factors and the underlying physiological mechanisms are not fully known. Thus, in the present study 169 boars from three farms were slaughtered at two different slaughter plants. Skatole and androstenone were measured in fat and blood plasma and additional endocrine parameter were included. Each farm delivered boars at two consecutive days. Duration of transport ranged between 1 and 5 h, the time spent on the vehicle after arrival at the slaughter plant until unloading (pre-unloading time) ranged between 17 and 480 min. Blood, faeces, urine, and adipose tissue samples were collected from each animal along the slaughter line. Each carcass was classified according to the number of skin injuries ('none': 0 to '>25 per side': 3). Androstenone was measured in adipose tissue and blood plasma with an EIA after extraction, skatole was determined with UHPLC. Cortisol was measured in faeces and urine after extraction with RIA. Analysis of variance revealed a strong influence of transport time on androstenone in fat and plasma, whereas the pre-unloading time was without effect. In contrast, skatole in both substrates was mainly affected by the pre-unloading time. The influence of farm was significant for both, skatole and androstenone in plasma. The number of skin injuries was influenced by pre-unloading time, farm, and slaughter plant and was significantly related to skatole concentrations in fat. Similarly fecal cortisol levels correlated with skatole concentrations in fat. Thus we conclude, that androstenone and skatole respond differently to pre slaughter conditions. Androstenone was influenced by the duration of transport. Skatole in contrast was mainly affected by pre unloading time, possibly reflecting stressful fighting encounter before slaughter.

Different preslaughter fasting periods and their effects on boar taint compounds in finishing pigs

C. Knorr, H. Wahl, A.R. Sharifi and D. Mörlein
Department of Animal Sciences/Georg-August-University of Göttingen, Albrecht-Thaer-Weg 3, 37075 Göttingen, Germany; cknorr@gwdg.de

In Germany, fattening of boars is regarded as the most probable alternative to the surgical castration of male piglets. Castration is commonly practised to prevent boar taint mainly caused by androstenone, skatole and indole in fat tissues. Castration without anaesthesia will, however, be banned in the EU by 2018. The goal of this study was therefore to investigate the effect of different preslaughter fasting periods (24 h, 18 h, and 12 h) on skatole and indole levels in blood and backfat of finishing boars. Each group comprised of 30 Duroc-sired Danish crossbreds. The animals were raised on one farm and fed the same diet. Blood was sampled at slaughter during exsanguination. Backfat samples were excised from the neck region about 1 h post mortem. Skatole and indole were measured applying liquid chromatography. All boars were also genotyped for SNP g.2412 C>T (at -586 ATG) located in CYP2E1 as this has recently been reported to affect skatole levels in backfat. The effects of fasting group, CYP2E1 genotype and their interaction on skatole and indole were assessed on ln-transformed values. Correlations between backfat and blood levels for skatole were $r=0.69$ and $r=0.56$ for indole. Group effected indole levels in backfat and blood. Genotype affected all traits except blood skatole. Boars of the 24 h fasting group tended to have the highest skatole levels in blood and backfat, but no significant differences existed between the groups. By contrast, blood indole was significantly ($P\leq0.05$) higher after 24 h (8.8 ng/g) fasting compared to 18 h (5.4 ng/g) and 12 h (3.8 ng/g). Back transformed levels (skatole backfat, indole backfat, indole blood) of CC (147.05 ng/g, 64.47 ng/g, 7.47 ng/ml) boars differed significantly ($P\leq0.05$) from genotypes CT (91.45 ng/g, 41.55 ng/g, 5.06 ng/ml) or TT (93.76 ng/g, 36.50 ng/g, 4.24 ng/ml) indicating the importance of CYP2E1 on boar taint compounds in Duroc-sired populations.

Farmers' expectations and experiences with alternatives for surgical castration

M. Aluwé[1], F. Vanhonacker[2], W. Verbeke[2], S. Millet[1] and F.A.M. Tuyttens[1]
[1]ILVO (Institute for Agricultural and Fisheries Research), Animal Sciences Unit, Scheldeweg 68, 9090 Melle, Belgium, [2]Ghent University, Department of Agricultural Economics, Coupure Links 653, 9000 Gent, Belgium; marijke.aluwe@ilvo.vlaanderen.be

A ban on castration of piglets is planned in the EU for 2018. Pig farmers' expectations of and experiences with four alternative strategies for surgical castration were evaluated on 19 Flemish pig farms. Treatments were castration without anaesthesia (CN), castration with 100% CO_2-anaesthesia (CO_2), castration with analgesia (MET), immunocastration (IC), and raising entire males (EM). At each farm, 120 male pigs per treatment were included. The farmers completed the same questionnaire before (ex ante, year 2009) and after (ex post, year 2012) the experiment to register eventual shifts in attitudes and preferences. The five treatments were ranked with regard to overall preference, labour effort, animal welfare, effectiveness against boar taint, performances, profitability and expected consumer acceptance from most to least favourable (non-parametric Friedman test with chi-square statistics). There was an ex ante overall preference for the current practice of CN, followed by IC, EM and MET. CO_2 was least preferred in 2009. Ex post, average overall preference shifted in favour of EM, followed by CN and MET, IC and CO_2. These shifts were mainly due to a lower ranking of IC and an improved ranking of EM for both performances and profitability. Farmers' ranking of effectiveness against boar taint was not affected, but their ranking of consumer acceptance improved for EM, probably because assessed boar taint levels were lower than expected. Ranking results indicated that farmers' experience matched expectations for labour effort and perceived animal welfare for all strategies. Providing correct information as well as giving the opportunity to test the alternatives on-farm may stimulate farmers to ban surgical castration and shift towards EM or IC.

Session 50 — Theatre 1

Biomarkers of beef tenderness, moving towards analytical tools

B. Picard[1], D. Micol[1], I. Cassar-Malek[1], C. Denoyelle[2], G. Renand[3], J.F. Hocquette[1], C. Capel[2] and L. Journaux[2]
[1]INRA-VetAgro Sup, UMRH 1213, Theix, 63122, France, [2]Institut de l'Elevage, 149 rue de Bercy, 75012 Paris, France, [3]INRA, UMR 1313, 78352 Jouy en Josas, France; picard@clermont.inra.fr

Variation in beef tenderness is a major problem for the beef industry. Currently, there are no accurate commercial techniques available to evaluate the tenderness of carcasses, and this capacity is even poorer for live animals. The beef industry needs commercially viable tenderness measurements for progress to occur. For years, functional genomic programs have been conducted to search for biomarkers of this quality. The main results showed that, in the Longissimus thoracis of the tenderest animals, the muscle proteins associated with the fast glycolytic type (phosphoglucomutase, lactate dehydrogenase B, triophosphate isomerase, glyceraldehyde 3-phosphate dehydrogenase, isoforms of fast troponin T, β enolase ...) were less abundant. Several proteins involved in calcium metabolism such as parvalbumin were also identified as positive markers of tenderness, in agreement with the role of calcium in post-mortem ageing. Proteins of heat shock (Hsp) family have been revealed as markers of tenderness at mRNA and protein levels in different experiments (Hsp40, Hsp27, Hsp20, α-B crystallin, Hsp70). This is consistent with a theory that the anti-apoptotic activity of Hsp could slow down the process of cell death early post-mortem and consequently improve tenderness. Other proteins involved in oxidative stress, such as superoxide dismutase 1 or peroxiredoxin 6, have been related negatively with tenderness. Currently, the abundance of these proteins is being measured in different muscles within a representative sample of the French production systems. The final step is to provide a tool for the beef industry to enable it to measure the abundance of these tenderness markers. Its use will provide objective data on the potential tenderness of individuals pre or post slaughter allowing production to be tailored to maximise tenderness and/or decide the appropriate market.

Session 50 — Theatre 2

Need to conciliate beef quality, farm efficiency, environment preservation and animal welfare

R. Botreau, M. Doreau, V. Monteils, D. Durand, M. Eugène, M. L'herm, M. Benoit, S. Prache, J.F. Hocquette and C. Terlouw
INRA, VetAgro Sup, UMRH 1213, Theix, 63122 Saint-Genès Champanelle, France; jfhocquette@clermont.inra.fr

Meat quality is a complex concept which can be divided into intrinsic quality traits (which are the characteristics of the product itself) and extrinsic quality traits (animal welfare and health, environmental impacts, profitability and workload of the production system, and product image and price). Modelling methods have been developed to aggregate different quality traits (for instance, the Meat Standard Australia System aggregates tenderness, flavour, juiciness and overall liking), and thus to help improving simultaneously different meat quality aspects. Concerning environmental issues, research demonstrated that greenhouse gases emission, risk of eutrophication and energy consumption were lowest with a diet based on concentrate, hay and maize silage, respectively. Thus, each feeding system had its own advantages and disadvantages. However, other studies identified win-win strategies. Thus, the most economically efficient farms were those with the lowest carbon footprint, presenting a high variability in both parameters. Similarly, animals that were less stressed at slaughter (lower heart rates) were those producing tenderer meat. Supplying simultaneously linseed and antioxidants in the animal diet increased polyunsaturated fatty acid content in beef while providing protection against peroxidation thus ensuring stable colour traits. To sustain beef production, future studies need to use multicriteria approaches combining indices related to sensory and nutritional quality, social and environmental expectations and economic efficiency to help identifying the best compromises and to optimise strategies. Such research needs, in addition, to conduct studies at different levels (tissue, animal and system levels).

Mid-infrared prediction of cheese yield from milk and its genetic variability in first-parity cows

F.G. Colinet[1], T. Troch[1], S. Vanden Bossche[1], H. Soyeurt[1], O. Abbas[2], V. Baeten[2], F. Dehareng[2], E. Froidmont[2], G. Sinnaeve[2], P. Dardenne[2], M. Sindic[1] and N. Gengler[1]
[1]University of Liege, Gembloux Agro-Bio Tech, 5030 Gembloux, Belgium, [2]Walloon Agricultural Research Center, 5030 Gembloux, Belgium; frederic.colinet@ulg.ac.be

Economically, cheese yield is very important. It would be interesting to predict fresh Individual Laboratory Cheese Yield (ILCYf) without the need to estimate milk components and to use empirical or theoretical formulae. In order to study the genetic variability of this trait on a large scale, mid-infrared (MIR) chemometric methods were used to predict ILCYf. A total of 258 milk samples collected in the Walloon Region of Belgium from individual cows (Holstein, Red-Holstein, Dual Purpose Belgian Blue, and Montbeliarde) were analyzed using a MIR spectrometer and ILCYf was determined for each sample. An equation to predict ILCYf from milk MIR spectra was developed using PLS regression after a first derivative pre-treatment applied to the spectra to correct the baseline drift. During the calibration process, 22 outliers were detected and removed from the calibration set. The ILCYf mean of the final calibration set was 26.8 g coagulum/100 g milk (SD=6.5). The coefficient of determination (R^2) was 0.83 for the calibration with a standard error (SE) of 2.6. A cross-validation (cv) was performed (R^2_{cv}=0.81 with SE_{CV}=2.8). This equation was then applied on the spectral database generated during the Walloon routine milk recording. The variances components were estimated by REML using single-trait random regression animal test-day model. The dataset used includes 51,537 predicted records from 7,870 Holstein first-parity cows, the ILCYf mean was 24.2 (SD=4.5) and ILCYf ranged from 13.6 to 40.9. Estimated daily heritabilities ranged from 0.27 at 5th day in milk to 0.55 at 231th day in milk indicating potential of selection. Further research will study phenotypic and genetic correlations between ILCYf and milk production traits.

Predicting cheese yield and nutrient recoveries of individual milk using FTIR spectra

A. Ferragina, A. Cecchinato, C. Cipolat-Gotet and G. Bittante
University of Padova, Department of Agronomy, Food, Natural resources, Animals and Environment (DAFNAE), Viale dell'Università 16, 35020 Legnaro (PD), Italy; alessandro.ferragina.1@studenti.unipd.it

The aim of this study was to investigate the potential application of Fourier-transform infrared (FTIR) spectra for the prediction of: (1) cheese yield (CY) defined as CY_{CURD}, CY_{SOLIDS} and CY_{WATER} which were the ratios between, the weight of fresh curd, the total solids of the curd, and the water content of the curd, respectively, and the weight of the milk processed; (2) three measures of nutrient recovery (REC) defined as REC_{FAT}, $REC_{PROTEIN}$ and REC_{SOLIDS}, which represent the ratios between the weights of the fat, protein and total solids in the curd, respectively, and the corresponding nutrient in the milk; and (3) the energy recovery (REC_{ENERGY}) which represents the energy content of the cheese versus that in the milk. A total of 1,260 individual Brown Swiss cows, reared in 85 herds, were milk-sampled once. Two FTIR spectra for each sample were collected and averaged. A model cheese manufacturing process was applied to asses milk samples from all individual cows using 1.5 l of milk. Modified partial least square regression combined with some spectra pretreatments (first derivative, standard normal variate, combined spectral ranges of 925 to 1,582 cm^{-1}, 1,701 to 3,048 cm^{-1} and 3,673 to 5,011 cm^{-1}) was used for the development of calibrations equations. Means (SD) of the investigate traits were: CY_{CURD} 15.04% (1.89), CY_{SOLIDS} 7.22% (0.93), CY_{WATER} 7.80% (1.28), REC_{FAT} 89.87% (3.58), $REC_{PROTEIN}$ 78.08% (2.41), REC_{SOLIDS} 52.05% (3.58) and REC_{ENERGY} 67.31% (3.32). The best predictions were obtained for CY_{SOLIDS} and REC_{SOLIDS} (1-VR of 0.95 and 0.86, respectively). Results indicated that predictions of CY and REC were accurate enough to enable selection of individual animals based on milk FTIR spectra with the only exception for REC_{FAT}.

Selection of assessors for boar taint evaluation: effects of varying androstenone sensitivity

L. Meier-Dinkel[1], A.R. Sharifi[1], L. Frieden[2], E. Tholen[2], M. Wicke[1] and D. Mörlein[1]
[1]University of Göttingen, Department of Animal Sciences, Albrecht-Thaer-Weg 3, 37075 Göttingen, Germany, [2]University of Bonn, Institute of Animal Science, Endenicher Allee 15, 53115 Bonn, Germany; daniel.moerlein@agr.uni-goettingen.de

Raising boars is regarded as an alternative to surgical piglet castration, but requires the control of boar taint. This off odor is mainly due to increased levels of androstenone and skatole in fat tissues. Consumer acceptance thresholds are needed to be implemented in breeding programs. Recent studies showed that consumer tolerated boar loins although androstenone exceeded previously established values of 0.5 to 1.0 μg/g melted fat. It thus appears appropriate to re-evaluate loins by trained panelists. Olfactory acuity to androstenone, however, varies among individuals and little is known how that affects objective boar taint evaluation. This study investigated how assessors' olfactory acuity affects taint evaluation in boar loins. Olfactory acuity was determined after 10 weeks training using smell strips with varying levels of androstenone. SENS (n=7) and SENSHIGH (n=9) panelists then assessed boar and reference loins following descriptive sensory analysis. Smell strips were provided as references for scale use of androstenone odour intensity. SENSHIGH assessors scored low-fat boar loins with 1.5 to 2.0 μg of androstenone per gram of melted back fat being significantly different from castrate and gilt loins for androstenone odour and flavor. Less sensitive assessors with lower olfactory acuity, however, were less discriminating. To conclude, panelists' olfactory acuity should be considered for selection and training. The presented paper strip system is suggested for objective screening and training purposes and to be used as quantitative references in descriptive analysis. Acceptance thresholds for boar taint compounds need to be elucidated with respect to the fat content of the respective pork cuts as established backfat values appear ineligible for lean cuts such as, e.g. loins.

Session 50 Theatre 6

Coagulation properties and composition of milk of crossbred cows compared with Holstein cows

F. Malchiodi, C. Cipolat-Gotet, M. Penasa, A. Cecchinato and G. Bittante
University of Padova, Department of Agronomy, Food, Natural resources, Animals and Environment (DANFNAE), Viale dell'università 16, 35020 Legnaro (PD), Italy; francesca.malchiodi@studenti.unipd.it

Primiparous and multiparous crossbred cows were compared with pure Holstein cows for milk composition and milk coagulation properties (MCP). Milk samples of Swedish Red × Holstein (SR × HO) crossbred cows (n=92) Montbéliarde (MO) × (SR × HO) crossbred cows (n=66) and pure Holstein (HO) cows (n=102) were collected from two herds located in northern Italy. Individual milk samples were analyzed for lactose, fat, protein, and casein contents using MilkoScan FT6000 (Foss). Somatic cell count (SCC) was obtained from the Fossomatic FC counter. Milk rennet coagulation time (RCT), curd-firming time (k20) and curd firmness (a30) were assessed by Formagraph. All traits were analyzed through ANOVA considering the fixed effects of herd-test-day, parity, days in milk and breed. For milk composition, the crossbreed between SR × HO was significantly different ($P<0.05$) from pure HO for all traits (+0.12% of protein, +0.07% of casein, -0.15% of lactose +0.32% of fat). MO × (SR × HO) crossbred cows were significantly different to pure HO for protein and lactose content (+ 0.08% and -0.12%, respectively) while MCP and SCC were not significantly different for any crossbred compared with pure HO.

Effect of diets on bovine muscle composition and sensory quality characteristics

M. Gagaoua[1,2], D. Micol[2], J.-F. Hocquette[2], A.P. Moloney[3], K. Nuernberg[4], D. Bauchart[2], N.D. Scollan[5], R.I. Richardson[6], A. Boudjellal[1] and B. Picard[2]

[1]INATAA, Constantine 1 University, 25000, Algeria, [2]INRA, Amuvi UMR1213, 63122 St-Genès-Champanelle, France, [3]Teagasc, Grange Beef Research Centre, Grange, Co Meath, Dunsany, Ireland, [4]FBN, Leibniz Institute for Farm Animal Biology, W-Stahl-Allee 2, 18196 Dummerstorf, Germany, [5]Institute of Biological, Environmental and Rural Sciences, Aberystwyth University, Wales, SY23 3EB, United Kingdom, [6]Division of Farm Animal Science, University of Bristol, Langford, BS40 5DU, United Kingdom; brigitte.picard@clermont.inra.fr

An objective of the EU ProSafeBeef project was to determine effects of diets enriched with PUFAs and antioxidants on bovine muscle characteristics and meat sensory qualities. This study used 265 animals finished in 4 experimentations under different EU production systems. Animals were from 8 breeds including bulls (B), steers (S) and heifers (H), i.e. 25 Limousin (B), 25 Blond d'Aquitaine (B) and 24 Angus (B) from France; 47 Belgian-Blue × Friesian (H) and 47 Angus × Friesian (H) from Ireland; 25 Holstein (B) from Germany and 40 Belgian-Blue × Holstein (S) and 32 Charolais (S) from United Kingdom. The diets were aggregated in 4 classes consisting of silage (Si) or concentrates (C) supplemented with PUFAs (L) and/or antioxidants (AO). Statistical analyses were all performed using GLM procedure of SAS 9.2. Longissimus thoracis muscle of animals given the Si diet had a higher proportion of SO fibres and higher ICDH activity ($P<0.0001$), and a lower proportion of FG fibres and lower LDH activity ($P<0.0001$) associated with higher ultimate pH values ($P<0.05$). Muscles of animals of C and L groups had a higher lipid content than those of Si and AO groups ($P<0.0001$). Moreover, muscles of animals given C and L diets were more tender and juicy with a higher flavor intensity rating ($P<0.0001$) than those of S and AO groups. These results demonstrate that diets enriched with lipids (PUFAs) during the finishing period affect bovine muscle properties and meat sensory qualities.

Variations in milk production and composition in response to a lengthened milking interval

J. Guinard-Flament[1], H. Larroque[2] and S. Pochet[3]

[1]Agrocampus Ouest, UMR1348 PEGASE, 35000 Rennes, France, [2]INRA, UR631SAGA, 31326 Toulouse, France, [3]INRA, UR342 URTAL, 39801 Poligny, France; jocelyne.flament@agrocampus-ouest.fr

Milking intervals can be punctually and voluntary lengthened to reduce workload using once-daily milking or involuntary in the case of milking robot, when herd is oversized compared to the robot capacity. Longer milking interval such as once-daily milking when repeated consecutively induces changes in milk composition. However little is known about the early variations in milk biochemical composition observed on the first lengthened milking interval. The trial included 32 primi- and multiparous Holstein cows at 184+29 days in milk according to a continuous design with a 7-d control period (twice-daily-milking at 07:00 and 16:00) and a 1-d experimental period where morning milking was omitted (MO, milking omission). Milk fat, protein, lactose, and cell contents were analysed 3 consecutive days at each milking on control period, and on the MO day. Two days before MO and on MO day, additional milk samples were collected to determine lipolysis and contents in lactoferrin, plasmin, and plasminogen. In response to MO, milk yield decreased by 20%, from 34.1 to 27.3 kg/d ($P=0.001$ by GLM SAS procedure). Milk lipolysis and milk contents in protein and lactoferrin remained unchanged. Milk lactose content decreased by 2.1 g/kg ($P=0.0001$) whereas milk fat content increased by 5.5 g/kg ($P=0.001$). Milk plasmin and plasminogen activities increased from 2.06 to 2.35 and from 21.6 to 22.9 10-3 AU/min/ml (i.e. by 14.5 and 6%, respectively; $P=0.001$). Plasmin:plasminogen ratio increased from 0.098 to 0.106 units ($P=0.02$), as did milk SCS (from 1.99 to 3.18; $P=0.0001$). In conclusion, milk biochemical composition is modified from the first 24h-lengthened milking interval. Increased plasmin activity could modify the amount and profile of the soluble peptide fraction in milk and also have further possible implication in hard cooked cheese technology where plasmin activity is enhanced by high temperature cooking.

Session 50 Poster 9

Relationships between muscle characteristics and quality parameters of Arabian camel meat

A.N. Al-Owaimer[1], G.M. Suliman[1], A.S. Sami[2], B. Picard[3] and J.F. Hocquette[3]
[1]King Saud University, P.O. Box 2460, 11451 Riyadh, Saudi Arabia, [2]Cairo University, Gamaa Street, 12613, Giza, Egypt, [3]INRA-VetAgro Sup, UMRH 1213, Theix, 63122 Saint-Genes Champanelle, France; jfhocquette@clermont.inra.fr

Twenty-four Saudi male Arabian camels of four different breeds were used to evaluate characteristics of their muscles and quality parameters of their meat. The camels were on average six-month-old and weighed 133.83±2.83 kg. The results indicated the chemical composition of Longissimus thoracis (LT) muscle did not differ significantly between the four camel breeds. This was also true for shear force and fiber cross-sectional area. On the contrary, myofibrill fragmentation index (MFI) and sarcomere length were statistically different ($P<0.05$ respectively) between the camel breeds. The four breeds of Saudi Arabian camel also did not show any significant differences in cooking loss, pHu, and color values, except ($P<0.05$) for the redness (a*) color of their LT muscles. The LT muscles were all found to express similar proportions of only the Myosin Heavy Chain I (MyHC I) and IIa (MyHC IIa) isoforms. They also contained similar proportions of total, insoluble and soluable collagen. Moreover, the four breeds did not differ significantly in the activities of the metabolic enzymes involved in muscle glycolytic metabolism and muscle oxidative metabolism. Activities of the mitochondrial enzymes (CS, COX, ICDH) were positively correlated to each other ($r>0.49$, $P<0.05$). As were the glycolytic activities (PFK and LDH) ($r=0.16$, $P<0.01$). The activities of the mitochondrial enzymes were negatively correlated with the glycolytic activities. A strong positive correlation was observed between MyHC IIa proportions and LDH activity. As in beef, intramuscular fat content was shown to be positively associated with redness and muscle oxidative metabolism whereas toughness (shear foce) had a slight positive association with collagen content and muscle glycolytic metabolism and a negative association with muscle oxidative metabolism and muscle fibre area.

Session 50 Poster 10

Physicochemical and textural properties of white soft cheese from Greek buffalo milk during ripening

G. Dimitreli[1], S. Exarhopoulos[1], K.D. Antoniou[1], A. Zotos[1] and V.A. Bampidis[2]
[1]Alexander Technological Educational Institute, Department of Food Technology, P.O. Box 141, 57400 Thessaloniki, Greece, [2]Alexander Technological Educational Institute, Department of Animal Production, P.O. Box 141, 57400 Thessaloniki, Greece; bampidis@ap.teithe.gr

The physicochemical properties and the textural behavior of white soft cheese made from Greek buffalo (*Bubalus bubalis*) milk during ripening were investigated. Cheese samples using cow milk, for the purpose of comparing the above mentioned variables, were also prepared. The textural properties of the samples were studied using Texture Profile Analysis and organoleptic evaluation. The fresh cheese yield of buffalo cheese was higher (31.4%) when compared to the cow one (18.9%). Cheese samples made from buffalo milk had higher fat content (32%) and reduced moisture content (47.83%) than cheese samples made with cow milk (15.3% and 59.03%, respectively). As compared to cow milk cheese samples, buffalo milk cheese samples exhibited lower textural (fracturability, hardness and cohesiveness) and organoleptic properties (fracturability, hardness and cohesiveness). During the two months of ripening the water soluble nitrogen content for both of the samples was increased. Increasing the ripening time of cheese resulted in reduced textural and organoleptic properties. This research has been co-financed by the European Union (European Social Fund – ESF) and Greek national funds through the Operational Program 'Education and Lifelong Learning' of the National Strategic Reference Framework (NSRF) – Research Funding Program: ARCHIMEDES III. Investing in knowledge society through the European Social Fund.

Session 50 — Poster 11

Use of visible-near infrared spectroscopy to determine cheese properties

T. Troch[1], S. Vanden Bossche[1], C. De Bisschop[1], V. Baeten[2], F. Dehareng[2], P. Dardenne[2], F.G. Colinet[1], N. Gengler[1] and M. Sindic[1]

[1]University of Liege, Gembloux Agro-Bio Tech, Passage des déportés 2, 5030 Gembloux, Belgium, [2]Walloon Agricultural Research Center, Chaussée de Namur 24, 5030 Gembloux, Belgium; ttroch@ulg.ac.be

This study is aimed on the possibility from cheese visible-near infrared (VIS-NIR) spectra to determine cheeses characteristics. Analyzes (cheese yield, texture, color, dry matter, fat content, total nitrogenous matter) were performed on 34 cheeses made in duplicate from individual cow milks. Milk samples were selected to have the maximum level of cheese variability. Each cheese analyzes and the VIS-NIR spectra were carried out at 3 times (day 1, 28 and 42) during ripening. The principal component analysis of the VIS-NIR spectra allowed to distinguish cheeses at day 1 compared to the same cheeses at other times. Moreover, the distribution of the spectra on the first day of ripening could discriminate the different cheeses depending on the farm from which the cow comes. Using partial least square regression of laboratory values on VIS-NIR spectra and a leave-one-out cross-validation (cv), high determination coefficients (R^2) were obtained for the color and the texture of cheeses. For the color, R^2 of calibration (c) was 0.94 and R^2_{cv} was 0.93 whatever the time of measurement considered unlike the texture (R^2_c=0.79 and R^2_{cv}=0.77) for which R^2_c and R^2_{cv} were less accurate for cheeses at day 1 and increased with refining. These results indicated the possibility of using spectral analysis in the context of the analysis of color and texture. The use of VIS-NIR spectra could be a tool to determine directly cheese characteristics and allow characterization of the evolution of the properties of the cheeses during ripening. Although these results need to be confirmed with a larger number of samples, the analysis of these spectra has highlighted that there would be an effect related to the herd management.

Session 50 — Poster 12

Linseed in the maternal diet modifies fatty acid composition of brain and muscle tissues in lambs

A. Nudda, G. Battacone, R. Boe, M. Lovicu, N. Castanares, S.P.G. Rassu and G. Pulina

University of Sassari, Dipartimento di Agraria, Sezione Scienze Zootecniche, Viale Italia 39, 07100 Sassari, Italy; battacon@uniss.it

We investigated the effect of supplementation of the maternal diet with linseed, rich in 18:3-n3, on fatty acid composition in the muscle and brain of suckling lambs. Thirty-eight ewes were fed experimental diets from 8 weeks before lambing to 28 d of lactation. A control diet (CTL) and a C18:3-n3 enriched diet by adding linseed (LIN) were used in a 2×2 factorial design. During gestation, ewes were divided into two groups: one fed the CTL diet and the other the LIN diet. After lambing, both CTL and LIN groups were divided into two subgroups, fed with CTL and LIN diets. Thirty-eight lambs born from the experimental ewes were slaughtered at 28 d of age. Brain samples and the Longissimus dorsi (LD) muscle of lamb carcasses were collected to determine their fatty acid composition by gas-chromatography. Data were analyzed with a model that included fixed effects of ewe's gestation diet, ewe's lactation diet and their interaction. Feeding LIN to ewes during late-pregnancy increased significantly the C18:3-n3 (+27%) and C20:5-n3 (+54%) proportions in LIN lambs. Feeding LIN to ewes during lactation increased the C18:3-n3 (+111%), but did not affect the proportions of their longer-chain metabolites. The maternal LIN supplementation during gestation increase the concentration of C22:5-n3 (+45%) and C22:6-n3 (+19%) in brain tissue. The LIN supplementation during lactation increase the C22:5-n3 (+35%) and tended to increase the C22:6-n3 (+12%) in brain tissue. In conclusion, the maternal linseed supplementation to ewes diet increase the EPA and DHA in brain tissue but was ineffective to increase the C22:5-n3 and C22:6-n3 concentration in LD muscle of suckling lambs.
This research was supported by Cargill, Animal Nutrition Division, Milan, Italy.

Session 51a — Theatre 1

Training the future generations of animal breeders: example of EGS-ABG, a European joint PhD program

E. Verrier[1], E. Norberg[2], A. Lundén[3] and H. Komen[4]
[1]AgroParisTech, 16 rue Claude Bernard, 75231 Paris 05, France, [2]Aarhus University, P.O. Box 50, 8830 Tejle, Denmark, [3]Swedish University of Agricultural Sciences, P.O. Box 7023, 75007 Uppsala, Sweden, [4]Wageningen University, P.O. Box 338, 6700 AH Wageningen, the Netherlands; etienne.verrier@agroparistech.fr

Animal breeding is a knowledge intensive sector, with a high rate of innovation in tools (e.g. NGS), methods (e.g. genomic selection) or organization of breeders. It is also an international business, with EU playing a leading role. The European Graduate School in Animal Breeding and Genetics (EGS-ABG) is a joint PhD program, accredited with the Erasmus-Mundus label. The aim is to generate the future key international leaders of the animal breeding and genetics sector, whether for employment in academia or in industry. EGS-ABG is jointly organized by four European higher education institutions (Partners), with the input of five international Associated Partners. The program is based on four-year PhD projects, jointly supervised by two Partners, with a compulsory mobility between these two. There is also a course plan, with a minimum of 30 ECTS, which represents approximately half a year. Two groups of higly qualified and extremely motivated PhD candidates have been enrolled in EGS-ABG, comprising a total number of 17 candidates originating from Europe, Africa, Asia and South America. A third group of 9 candidates will be enrolled on September 2013. After more than two years of implementation, added values of such a program are clearly identified: (1) the definition of harmonized and stringent rules for selection, supervision and awarding, intended to warrant the high quality of both training and research; (2) the mobility, which allows candidates to experiment diverse ways of thinking and working, and to benefit from a wide range of expertise; and (3) a definitive multicultural touch with good networking opportunities.

Session 51a — Theatre 2

Essential skills for young professionals holding a master degree of animal science

L. Montagne and C. Disenhaus
Agrocampus Ouest, Animal sciences, 65 rue de saint-Brieuc, 35042 Rennes, France; montagne@agrocampus-ouest.fr

Higher education needs to provide students qualifications that allow them to be efficient in their professional integration in the uncertain and complex context of European animal productions. To prepare possible changes of an animal sciences master program, the scientific, technical and personal skills mobilised by postgraduate student during their first years of professional activities were analysed. In 2012, a semi-qualitative inquiry was submitted to young French professionals who graduated from 2006 to 2011 in Agrocampus Ouest. Gathered information mainly concerned professional occupation, daily activities and necessary skills. Analyses were performed using Chi2 tests and correspondence analysis from R software. A total of 53 persons were questionned, working either on public research institute (15/53), private companies (15/53), co-operative or associative structures (13/53) or advisory and transfer structures (10/53). Their main missions consisted in research & development studies (16/53), advice in animal production (16/53) or projects management (10/53). Scientific and technical knowledge in animal science across all fields were mentioned by all participants as essential skills. A majority of them (48/53) underlined that interpersonal skills and communication facilities are also of importance. Moreover, statistics skills (27/53), abilities to analyze and to gather different sources of information (24/53), as well as adaptative abilities (23/53) and English language (21/53) were quoted as essential. Such results indicate that the acquisition of transversal skills might be one of the main targets in animal science master course.

Session 51a — Theatre 3

The analysis of animal food chains, a tool for engineers education: procedure, interest, conditions

J. Lossouarn
AgroParisTech, 16 Rue Claude Bernard, 75231 Paris cedex 05, France; jean.lossouarn@agroparistech.fr

Often, teaching animal science, we refer simultaneously to farming systems and the market chains they are included in. These are two complex and managed systems, their evolutions require technical means suitable with the general objective of sustainability, i.e. economical, social and environmental criteria. We present and discuss a teaching procedure used for two decades at AgroParisTech, at the end of the 'Ingénieur agronome' course. We applied it to various animal food chains, working with milk, eggs, meats, or fishes. They included very small chains (the Basque swine chain), as well as big ones (the biggest French swine chain). The procedure includes: (1) a theoretical course on food chain analysis, focusing the concepts of fluxes, agents, functions and final demand (2) a field case study, running on a complete week, including talks with professional agents of the different links (farmers, feed manufacturers, breeders, managers of slaughterhouses or dairies, distributors…), and visits of farms and plants (3) the presentation of their analysis of the chain by the students at the end of the week, in front of the professional partners. They underline its forces and weaknesses, the opportunities or threats it can be faced with, related to its different steps. A debate with the professional audience follows. Finally, the students deliver their Pwpt and a short text (about 20 p.) supporting it. After 24 issues, we conclude that it is the best way to initiate into the analysis of animal food chain. The procedure is specially efficient to obtain a very high implication of the students, to provoke their creativeness, to bring them to manage collectively under a very short time. As conditions of success, we mention: (1) a team of teachers having a current practice of systemic approaches and a real knowledge of the professional circles; (2) a good partnership with the professionals involved in the procedure, who generally highly appreciate the external look provided by the students.

Session 51a — Theatre 4

MAN-IMAL: a new and innovative 'One Health' Master's degree

S. Couvreur[1], S. Malherbe[2], M. Krempf[3], M. Eveillard[4] and C. Magras[5]
[1]LUNAM Université, Groupe ESA, UR Systèmes d'Elevage, 55 rue Rabelais, 49007 Angers, France, [2]Oniris, La Chantrerie, 44307 Nantes, France, [3]Université Nantes INSERM CNRS, UMR1087/6291, 8 quai Moncousu, 44007 Nantes, France, [4]CHU Angers, Laboratoire Bactériologie-hygiène, 4 rue Larrey, 49100 Angers, France, [5]Oniris INRA, UMR1014 SECALIM, La chantrerie, 44307 Nantes, France; s.couvreur@groupe-esa.com

Today food production chains have become more international, industrial and complex: multiple professionals, regulations and trade routes are involved. The sanitary and nutritional crises no longer have any boundaries – it is more and more difficult to assure public health. Human diets meet new objectives which influence animal production chains: nutrition, reproduction and health management of livestock, animal products processing, conservation and composition (less salt and sugar, fatty acid profile…). These new aims have consequences on food safety and use-by dates. These changes have erased the borders between the fields of health care and food processing industries, meaning that new kind of professionals able to understand these circumstances are urgently needed. Our new training course, MAN-IMAL, based on the One health – concept, aims to reduce compartmentalisation within those fields training managers and technicians able to join forces to take on the worldwide challenges of ensuring sufficient and efficient food production chains, associated with the issues of public health and food safety. In this course, veterinarians, doctors, pharmacists, engineers and biologists learn to work together, to acquire the same shared culture as well as to share their specific expertise. The course is structured around 5 subject-related blocks, each of them jointly led by a trio of experts from complementary disciplines, supported by research units in these areas and closely linked to industry professionals. In plus, the programme distinguishes by the use of innovative educational methods. The Master's degree, first one of its kind in Europe, will begin in autumn 2013.

A 3D-serious game for teaching the environmental sustainability of pig farming systems

J.Y. Dourmad[1], K. Adji[2], A.L. Boulestreau-Boulay[3], L. Emeraud[4] and S. Espagnol[5]
[1]INRA Agrocampus Ouest, UMR Pegase, 35590 Saint-Gilles, France, [2]DRAFF, Bretagne, 35047 Rennes, France, [3]Chambre d'Agriculture, Pays de la Loire, 49105 Angers, France, [4]Lycée Agricole, Théodore Monod, 35650 Le Rheu, France, [5]IFIP, Institut du Porc, 35650 Le Rheu, France; jean-yves.dourmad@rennes.inra.fr

Models have been developed in the literature to evaluate the environmental sustainability of pig farming systems. These models require a precise description of the farm, including feed composition, animal performance, housing and manure management. Criteria are calculated, such as N or P load per ha or ammonia and greenhouse gas emissions. Environmental impacts can also be determined using life cycle assessment (LCA). However, these models are generally complex to understand and difficult to teach to students or farmers, mainly because of their lack of realism and attractiveness. The idea of this project was to develop a 3D serious game using a pig farm simulator, in order to facilitate this learning process. A farrow-to-finish farm was designed with a realistic 3D representation of the different farm components: pigs, buildings, equipments, manure storage, fields, crops... The user travels in his farm, may enter the different rooms and modify their characteristics, just by 'touching' a component, e.g. the sow for modifying reproductive performance, or the feeder for feed composition. A dashboard shows in real time the situation of the farm, as regard to different environmental indicators. Red lights are put on when N or P loads per ha exceed maximal allowed amounts. Overflow of manure storage is also detected as well as inconstancy between availability of manure and their spreading, as regard to the crop rotation. The user has access to a detailed report with different graphs and tables. It also is possible for the teacher to build scenarios in order to address a specific topic (e.g. LCA of pork) or to explore different strategies aiming at achieving a given objective.

Project of animal biology, the opportunity to sensitize master's students to research approaches

V. Lollivier[1,2], M. Boutinaud[1,2], F. Dessauge[1,2] and F. Lecerf[1,2]
[1]INRA UMR1348 PEGASE, Domaine de la Prise, 35590 St Gilles, France, [2]AGROCAMPUS UMR1348 PEGASE, 65 Rue de St Brieuc, 35000 Rennes, France; vanessa.lollivier@agrocampus-ouest.fr

The teaching unit 'Experimental Project of Animal Biology' offers an opportunity to master's students to discover a specific topic within the biological sciences, to sensitize them to the research approaches in a real-life working conditions and constitutes good prerequisites for students who want to start a scientific career. It is organized by teachers from Agrocampus Ouest, in collaboration with researchers from INRA and takes place mainly in the research lab of INRA. It allows students to understand the multi-scale (from *in vivo* experiments to the study of organs and cells in the lab) and multidisciplinary approaches that are necessary in physiological studies, including the combined use of computational methodologies, molecular and cell biology, immunology and microscopy. It is organized around a research problematic in the field of the regulation of cellular dynamics of the udder in dairy ruminants. During the 1rst day, the research problematic is introduced by supervisors. Students have to perform a literature research and should also consider the appropriate methodologies to carry out the research. After a validation by the supervisors, students are divided into 4 groups, each being in charge of a part of scientific and technological approach. Then, they realize the experiments (3-4 d), the critical analysis of their results and they compare the results obtained between the groups. At the end of the teaching unit, they present the results to the supervisors and colleagues from the research unit and propose perspectives. The evaluation focuses on the ability to have a scientific approach, to find the information needed to solve a problem, to propose solutions consistent with a set of constraint, to carry out the experiments, to work in groups and share the results with colleagues and supervisors and finally to clearly state the obtained results clearly (oral and written project reports).

Session 51a — Poster 7

Survey on expected farmers' skills for the future in the western part of France

A. Bedin[1], N. Cardoso-Leal[2], C. Disenhaus[2], A. Fischer[2], H. Fuchey[2], L. Grolleau[2], M. Perrot[2], J. Mayot[2], S. Pattier[2] and Y. Le Cozler[2]

[1]LEGTPA du Manoir, rte de Fougères, 50600 Saint Hilaire du Harcouët, France, [2]Agrocampus-Ouest, 65, rue de St-Brieuc, 35042 Rennes, France; yannick.lecozler@agrocampus-ouest.fr

Actual and future farmers sometimes disagree on what should be the required skills for the future and how it should be included in new teaching programs. Surveys were then conducted in the western part of France from 7-18 January 2013, to determine these probable skills in the coming 10 years. Two qualitative inquiries were submitted to farmers and teachers in agriculture, and a closed-questionnaire was sent to students in farming schools. A total of 75 students from 4 different farming high-schools were inquired, 2/3 of them being male of 17.9 years of age on average (17 to 20). If 72% of them had the strong willing to settle down in a farm, only 60% felt ready to do it right after school. According to them, a strong motivation, in close connection with passion (64/75) and being a hard-worker (43/75), were the main success factors for becoming a farmer. The 38 farmers inquired were mainly men (33/35), aged 47 years on average (23 to 63). The mean number of full-time worker was 2.3 per farm and varied from 1 to 5. Management, associated with basic and technical knowledge, were the main factors of becoming a successful farmer. The 14 teachers from 3 farming high-schools inquired had on average 20.3 years of teaching experience. They mainly taught animal sciences (9), economical management (3) or crops science (2). Management and the ability to learn new knowledge (9/14 in each case) were also identified as important skills. In all surveys, management appeared then to be the most important skill, before others such as technical knowledge. Such surveys conducted in a local part of France should be carried on other European territories to have a better idea of the opinion of the agricultural representatives, which could be useful to prepare some of expected changes in the future agricultural trainings.

Session 51b — Theatre 1

Current and future studies of the dog-human relationship and possible welfare implications

T. Rehn

Swedish University of Agricultural Sciences, Animal Environment and Health, Box 7068, 750 07 Uppsala, Sweden; therese.rehn@slu.se

Some types of dog-human relationships may be predisposed to conflicts and a better understanding of the quality and factors associated with different types of relationships could be used as a tool to increase the success in such relationships. It may help us e.g. tailoring training techniques, which is useful in areas where society benefits from the dog as well as improving the welfare of the dog. However, we need reliable methods to study this relationship. In this presentation, the aim is to reveal findings from previous research on the dog-human relationship, but also to suggest future directions in the scientific study of this relationship. When evaluating a widely used test of dogs' attachment to humans, the Strange Situation Procedure (SSP), inherent order effects in the test were found. The SSP is based on the attachment theory in humans where different types of attachment styles has been described, while the main focus in dog research has been on the secure type of attachment. It is intriguing to adapt the idea about different attachment styles also to the dog-human relationship. Since the SSP seems to be an inappropriate method to investigate attachment in dogs, other measures should be used, focusing on e.g. reunion behaviour, which is commonly used when identifying attachment styles in humans. Specific reunion behaviour studies have shown that dogs behave differently towards their owners after longer times of separation and that the behaviour of the owner affects the dog's behavioural as well as physiological response to the reunion. These findings can be useful when further investigating greeting behaviour in relation to the dogs' bond to their human caretaker in new setups. Increased understanding about the underlying mechanisms of the dog-human relationship is a crucial next step in dog welfare research and is beneficial also for the community as such, since e.g. behavioural problems caused by conflicts in the relationship pose a possible threat.

Breed differences in everyday behavior of Swedish dogs

H. Eken Asp, K. Grandinson, F. Fikse and E. Strandberg
Swedish University of Agricultural Sciences, Dept. of Animal Breeding and Genetics, P.O. Box 7023, 750 07 Uppsala, Sweden; helena.eken@slu.se

The objective of this study was to increase the knowledge of the genetic background of everyday behavior, which could be used for breeding purposes in the future. For the study ten breeds were selected based on their ancestry and population size. Breeds included in the study belonged to six of the FCI breed groups and include both working breeds and companion breeds. The study is performed in cooperation with the Swedish Kennelclub (SKC). A web-survey based on the canine behavior and research assessment questionnaire with some additional question has been open from June 2012 to February 2013. The survey was advertised both online and in papers, by the SKC as well as the breed clubs. When the data were collected there were 1,522 answers for the 10 breeds. The survey contained questions concerning general information about the dog and its owner, and the dogs' behavior reaction in different situations, in total 152 questions. The answers can be grouped into 16 behavioral traits. The dogs included in the analyses were born in the years 2000-2011. The average age, in years, of the dogs when the survey was answered was 4.0 (SD 2.7). Preliminary analyses of the ten breeds showed a breed differences in most of the behavior traits. In general the breeds that had higher average scores for Human-directed play interest and Stranger-directed interest showed less Stranger-directed fear. These breeds also had higher average scores for the trait Trainability. Breeds that had a higher average score for Dog-directed interest also had a lower Dog-directed fear. The working breeds had a higher average score for the traits Excitability, Trainability and Human-directed play interest compared to the companion breeds. They also had a lower score for Non-social fear compared to the companion breeds. The survey was able to show breed difference and could therefore be a useful tool in finding genetic differences within breeds.

Is the strange situation procedure a reliable method to study dog attachment?

T. Rehn, R.T.S. McGowan and L.J. Keeling
Swedish University of Agricultural Sciences, Department of Animal Environment and Health, Box 7068, 750 07 Uppsala, Sweden; therese.rehn@slu.se

The Strange Situation Procedure (SSP) originally developed for children is increasingly being used to study attachment between dogs and humans. In this experiment, 12 female beagle dogs were tested in two treatments to identify possible order effects in the test, a potential weakness in the SSP. In one treatment (FS), dogs participated with a 'familiar person' and a 'stranger'. In a control treatment (SS), the same dogs participated with two unfamiliar people, 'stranger A' and 'stranger B'. Comparisons were made between episodes within as well as between treatments. As predicted in FS, dogs explored more ($T=39$, $P<0.001$) in the presence of the familiar person (0.20 ± 0.03 (mean±SE)) than the stranger (0.10 ± 0.02). Importantly, they also explored more ($T=26.5$, $P=0.004$) in the presence of stranger A (who appeared in the same order as the familiar person and followed the same procedure) (0.16 ± 0.03) than stranger B (0.07 ± 0.02) in SS. Furthermore, there was no difference between treatments in how much the dogs explored in the presence of the familiar person in FS and stranger A in SS. In combination, these results indicate that the effect of a familiar person on dogs' exploratory behaviour, a key feature when assessing secure attachment styles, could not be tested reliably due to order effects in the SSP test. It is proposed that in the future only counterbalanced versions of the test are used. It is further proposed that an alternative approach might be to compare the response of the dog at reunion, since it was found that dogs reliably initiated more contact with the familiar person compared to the strangers. This could be achieved by focussing either on the behaviour of the dog in those episodes of the SSP when the person returns, or on reunion behaviour in other studies.

Inbreeding management in Dutch Golden Retrievers

J.J. Windig[1,2], S.J.F. Van De Kraats[2] and K. Oldenbroek[1]
[1]Wageningen UR, Centre for Genetic Resources the Netherlands, P.O. Box 16, 6700 AA Wageningen, the Netherlands, [2]Wageningen UR Livestock Research, Animal Breeding and Genomics Centre, P.O. Box 65, 8200 AB Lelystad, the Netherlands; jack.windig@wur.nl

Pedigreed dogs frequently suffer from genetic defects due to excessive inbreeding rates. The Golden Retriever is an example of such a breed, despite being one of the largest dog breeds. Analysis of the pedigree of the Dutch Golden Retriever population, which goes back to the early 1900s containing up to 34 generations, revealed inbreeding rates in excess of 1.3% / generation. Recently, inbreeding rates dropped to 0.2% / generation, but average kinship still increased with 3.7% / generation. Consequently, the population runs the risk of future excessive inbreeding rates. Several inbreeding restriction schemes were compared by computer simulation. Although initially effective, minimisation of kinship of mates was counterproductive in the long run. Restrictions on the number of matings per father or number of sons per father allowed to breed were only successful if the number of dogs used as a father increased at the same time. If not, the main effect was a faster turnover of the generations. Excluding dogs with a high average relatedness with the rest of the population proved to be the most efficient method to decrease inbreeding rates in the long run. In conclusion, genetic management can be successful in a closed population, but requires an active policy by the breed organisation and cooperation of the breeders. For the golden retriever increasing the number of breeding dogs, and excluding dogs with high average relatedness with the rest of the population are efficient strategies to reduce inbreeding rates.

Genetic diversity in 23 dog breeds in Belgium as revealed by pedigree analysis & molecular markers

K. Wijnrocx, S. Janssens and N. Buys
KU Leuven, Biosystems, Livestock Genetics, Kasteelpark Arenberg 30 bus 2456, 3001 Leuven, Belgium; katrien.wijnrocx@biw.kuleuven.be

Breeding for specific characteristics and behaviour in purebred dogs has resulted in a wide variety of phenotypic diversity between breeds, but also contributed to reduced genetic variation and high levels of inbreeding. Furthermore, several genetic disorders have been identified in purebred dogs, compromising their health and welfare. This study assessed the genetic diversity of 23 dog breeds in Belgium using both pedigree data and molecular marker genotypes. Genealogical analyses were performed on pedigree data from the Belgian studbook KMSH (Koninklijke Maatschappij Sint-Hubertus). A total of 298,296 dogs (varying from 207 to 63,408 registrations per breed) between 1950 and 2012 were analyzed. From these, a total of 7,570 dogs (ranging from 6 to 1,185 dogs per breed) were genotyped for 19 microsatellite markers of the ISAG Dog Panel within the framework of parentage control by the studbook. Pedigree analysis was performed on dogs born between 2000 and 2011, using PEDIG and own software routines. The genealogical analysis will be complemented by diversity measures (observed and expected heterozygosities, Wright's F-statistics, deviations from Hardy-Weinberg equilibrium,...) computed from molecular marker data and the relationship between the two sources of data will be compared. Preliminary results indicated that the average coefficient of inbreeding ranged from 1.2 to 29.7% and the effective population size from 5 to 202. According to this approach, it was concluded that genetic diversity and effective population size differ strongly between breeds, so that a breed-specific approach will be necessary to maintain genetic diversity while selecting against heritable disorders.

Session 51b — Theatre 6

Inbreeding effect on litter size and longevity in dogs

G. Leroy[1,2], B. Hedan[3], X. Rognon[1,2], F. Phocas[1], E. Verrier[1,2] and T. Mary-Huard[4,5]
[1]INRA, UMR1313 GABI, Domaine de Vilvert, 78352 Jouy-en-Josas, France, [2]AgroParisTech, UMR1313 GABI, 16 rue Claude Bernard, 75005 Paris 05, France, [3]CNRS, UMR 6061 IGDR, 2 avenue du Professeur Léon Bernard, 35043 Rennes, France, [4]INRA, UMR 518 MIA, 16 rue Claude Bernard, 75005 Paris 05, France, [5]AgroParisTech, UMR 518 MIA, 16 rue Claude Bernard, 75005 Paris 05, France; gregoire.leroy@agroparistech.fr

Inbreeding is a major concern for dog breeders and users, and is supposed to affect health, welfare and fitness on each stage of life. We analyzed the consequences of inbreeding on prenatal and postnatal survival in purebred dogs. Two traits were considered, litter size and longevity, based on birth and deaths declared by French breeders and owners over the 2007-2011 period. Six breeds, with various body size/shape, uses and population sizes, have been considered, namely Basset Hound, Epagneul Breton, German Shepherd, Leonberger, Cairn Terrier and West Highland White Terrier, the two later being investigated for their common history. According to breeds, 607 (Leonberger) to 8,989 (German Shepherd) litters and 811 (Basset Hound) to 12,091 (German Shepherd) dogs were investigated. According to the breed, the average litter size ranged from 3.5 (West Highland White Terrier) to 6.1 puppies (Leonberger), while the average longevity ranged from 8.2 (Leonberger) to 12 years (Cairn Terrier). Depending on the breed and the model used, inbreeding was found to have no significant effect or a significantly negative effect on traits. In order to investigate if phenomenon of inbreeding purge may have occurred in those breeds, we also compare effects of 'old' versus 'recent' inbreeding on those traits.

Session 52 — Theatre 1

Genetics of feed efficiency in dairy cattle

Y. De Haas[1], J.E. Pryce[2] and R.F. Veerkamp[1]
[1]Wageningen UR Livestock Research, Animal Breeding and Genomics Centre, P.O. Box 135, 6700 AC Wageningen, the Netherlands, [2]Department of Primary Industries Victoria, Biosciences Research Division, 5 Ring Road, Bundoora, VIC 3083, Australia; yvette.dehaas@wur.nl

Given that feed costs are a major proportion of the total cost of dairy farming, reducing feed costs for the same output will improve profitability. Intense selection for higher milk yields and little or negative pressure on body weight in recent decades has increased gross efficiency. However, exploiting genetic variation in maintenance requirements has proved difficult because measuring dry matter intake (DMI) is complicated and expensive. In this study we have examined the genetic basis of feed efficiency. Residual feed intake (RFI, in MJ/d) is calculated as the difference between net energy intake and calculated energy requirements for maintenance (based on metabolic live weight) and milk production. In an experimental dataset of 548 heifers, we demonstrated that RFI is heritable (40%), indicating that genetic selection is possible. The estimated phenotypic and genetic correlation between RFI and milk yield (-0.45 and -0.84, respectively) confirm that selecting on higher milk yield will lower the residual feed intake; i.e. will improve the feed efficiency. We have also demonstrated that genomic selection could be used to improve RFI. However, the extent of recording would need to be improved to increase the accuracy. To increase the size of the database, we have set up an international collaboration between Australia, UK and The Netherlands, and combined data on DMI recorded on 1,800 animals in these countries. We showed that the accuracy of genomic breeding values for DMI can be increased by combining datasets across countries, and by using a multi-trait approach. However, the number of high quality phenotypes still limits achieving acceptable accuracies for bull proofs. To address this, a major international collaboration to assemble DMI data on more than 6,000 cows with high quality phenotypes and genotypes has started.

Genetic parameters for energy efficiency traits in Nordic Red dairy cattle

A.-E. Liinamo[1], P. Mäntysaari[2], M. Lidauer[1] and E.A. Mäntysaari[1]
[1]MTT Agrifood Research Finland, Biotechnology and Food Research, Biometrical Genetics Group, 31600 Jokioinen, Finland, [2]MTT Agrifood Research Finland, Animal Production Research, 31600 Jokioinen, Finland; anna-elisa.liinamo@mtt.fi

The aim of this paper was to estimate genetic parameters for residual energy intake and energy conversion efficiency, and their relationships with production, intake, body weight, body condition and energy balance in Nordic Red dairy cattle. The data was collected at the MTT Agrifood Research Finland experimental farms between 1998 and 2012, and included lactation weeks 2-30 for 393 Nordic Red MOET nucleus heifers descending from 114 different sires. Residual energy intake (REI) was defined as the difference between total energy intake of each animal, and the energy required for milk, maintenance and body weight change. REI was expressed as weekly averages in ME MJ/d. Energy conversion efficiency (ECE) was defined as the ratio of weekly average ECM yield to the weekly average ME intake. The other studied traits included the weekly averages for energy corrected milk yield kg/d (ECM), dry matter intake kg/d (DMI), body weight kg (BW), body condition score 1-5 (BCS), and energy balance ME MJ/d (EB). The data were analyzed with multitrait and random regression models. The highest heritability estimates for REI and ECE were obtained in the beginning and the end of the 30 week lactation period and they were moderate (0.25-0.46). Modeling energy efficiency in the beginning of the lactation was challenging and would benefit from more detailed information on the body composition changes. However, the results suggest that the in early and mid-lactation periods REI and ECE described partially different fysiological traits. The genetic correlations of REI were high and positive for DMI and EB, and moderate and positive for BW and BSC. The genetic correlations of ECE were high and positive with ECM, and moderate to low and negative with DMI, BW and BCS.

Session 52 Theatre 3

Association of SNPs of NPY, LEP and IGF-1 genes with residual feed intake in grazing Angus cattle

A.I. Trujillo, A. Casal, V. Cal and P. Chilibroste
Facultad de Agronomia. Universidad de la Republica, Produccion Animal y Pasturas, Avda. Garzon 780, 12900 Montevideo, Uruguay; anatrujister@gmail.com

Single nucleotide polymorphisms (SNPs) that showed associations with residual feed intake (RFI) may be useful for marker-assisted selection. Neuropeptide Y (NPY), leptin (LEP) and insulin like growth factor-1 (IGF-1) are candidate genes due to their roles in the regulation of feed intake, growth and energy balance. Thus, our aim was to study the associations of SNPs previously identified, in NPY (A/G, intron 2), LEP (C/T, exon 2) and IGF-1 (C/T, promoter region) genes, to dry matter intake (DMI), metabolizable energy intake (MEI), average daily gain (ADG) and RFI. Twenty four Angus heifers carrying 3 putative favorable alleles simultaneously (V=validation group; n=12; 288±39 kg BW, 364±34 d old at the beginning of test (BT)) or carrying 3 putative unfavorable alleles (C=control group, n=12, 299.8±35, 375±17 d old at BT) were selected from thirty eight heifers previously evaluated for RFI with a high-energy diet in confinement and allocated in 8 paddocks. Heifers were continuously stocked at high quality, high herbage mass mixed pasture (fescue, birdsfoot trefoil and white clover). Herbage DMI was determined using the n-alkane technique. The RFI were calculated as the residual from regression of DMI on ADG and mid-test metabolic weight (MidMW) (RFIK) and as the residual from regression of metabolizable energy intake (MEI) on ADG and MidMW (RFIME). Data were analyzed as a completely randomized design using a mixed model. Herbage DMI (kg/day), ADG (kg), MEI (Mcal/day), RFIK, RFIME and FCR for V and C groups were 8.76 and 10.93 (0.71), 1.4 and 1.37 (0.05), 25.5 and 31.7 (2.09),-1.02 and 1.02 (0.41), -2.91 and 2.91 (1.20), 6.96 and 6.40 (0.46), respectively. DMI (P=0.005), MEI (P=0.006), RFIK (P=0.002) and RFIME (P=0.002) were greater in V than C heifers while ADG and FCR did not differ between groups. Our results demonstrated a strong effect between these SNPs and RFI when animals were grazing on a high quality, high availability pasture.

Session 52 Theatre 4

Investigating potential interactions between methane emission and rumen microbial profiles

J. Lassen[1], B. St-Pierre[2], R. Smith[2], A.D.G. Wright[2] and P. Løvendahl[1]
[1]Aarhus University, Department of Molecular Biology and Genetics, Blichers Alle, 8830, Denmark, [2]University of Vermont, Department of Animal Science, Burlington, VT 05405-0148, USA; jan.lassen@agrsci.dk

Enteric methane emission from ruminants contributes substantially to the greenhouse effect. Few studies have focused on the interaction between the genetic variation in enteric methane emission and the rumen microbial profile in dairy cattle. The objective of this study was to investigate this interaction in a Danish Holstein herd. A Fourier Transformed Infrared (FTIR) measuring unit was used to measure individual methane emission records on 109 cows. The cows were measured in 7 herds during their visits to automatic milking systems (AMS). The FTIR unit air inlet was mounted in the front part of an AMS close to the cows head for 7 days, recording continuously every 5 s. The phenotype analysed was the mean methane to carbon dioxide ratio across visits during the measuring period, as well as the total methane production in liters per day. From 85 cows 50 ml of rumen fluid was sampled using an oral probe. 16S rRNA genes sequences were amplified from purified rumen microbial DNA by PCR. Microbial population profiles were determined using a combination of the programs MOTHUR and RDP Classifier. Substantial variation between animals was found for methane emission both in terms of methane to carbon dioxide ratio (0.025-0.09) and liters of methane produced (137-680 l). This variation was affected by production level and weight. Rumen microbial diversity was substantial for Bacteroidetes, Firmicutes and Proteobacteria. Some interactions were found eg cows with high production of methane tended to have low occurrence of Bacteroidetes. The data available and the setup in this study show huge potential to investigate interactions between phenotypes and rumen microbial profiles.

Session 52 Theatre 5

Assessment of the selenium status in cattle herds in Wallonia

Y. Mehdi, O. Dotreppe, V. Robaye, L. Istasse, J.-L. Hornick and I. Dufrasne
University of Liege, Nutrition Unit, bd de colonster 20, 4000 Liege, Belgium; ymehdi@doct.ulg.ac.be

The selenium content is low in the soil of Wallonia, the southern part of Belgium. It is therefore logical that the selenium content in feedstuffs grown in these fields -grass, conserved forages, cereals and by-products- is also low. A survey was carried on 166 farms in Wallonia. The farms were located in 4 specific agronomy areas: (1) Pays de Herve with mainly pastures (4 farms); (2) Ardennes (a sub mountainous area and pastures, 24 farms); (3) Hesbaye with mainly arable lands (55 farms); and (4) Condroz with both pastures and arable (83 farms). There were herds with dairy cows, herds with beef cows and herds with both types of animals. The average age was 44±15 months for the cows and 18±6 months for the heifers. The blood samples were taken on five healthy animals of each herd, on the end of the winter just before the animals went to pastures. The Se status was assessed by the activity of the glutathione peroxydase. The Se status was extremely variable for the different farms of the 4 areas with a variation coefficient of 56%. It was in Pays de Herve and in Hesbaye that the average Se status was the highest at 53 and 56 μg Se/l. By contrast, it was in Ardenne and in Condroz that the status was the lowest, at 39 and 43 μg Se/l. The selenium status was, on the whole, higher in the dairy herds than in the beef herds owing to the higher selenium provision by the mineral mixtures and the compound feedstuffs added to the diet. Assuming that the normal selenium concentration in plasma being 70 μg/l, it appeared also that 81% of the individual animals tested were below the normal Se concentration. It can thus be concluded from the present survey that the cattle herds in Wallonia are deficient in Selenium.

Session 52 Theatre 6

Major components of feed efficiency in ruminants

D. Sauvant[1], P. Faverdin[2] and N. Friggens[1]
[1]AgroParisTech, UMR 791 MOSAR, 16 r C Bernard, 75005 Paris, France, [2]INRA, UMR 1348 PEGASE, La Prise, 35590 St-Gilles, France; daniel.sauvant@agroparistech.fr

In ruminants, there are new perspectives for increasing feed efficiency (FEff) afforded by the advent of genomic selection. FEff can be expressed as an index of efficiency, (i.e. MY or ADG/DMI). In a first approach it is relevant to split FEff into its digestive (DEff) and metabolic (MEff) components. This can be simply done by, for instance, introducing digestible organic matter intake (DOMI) such that: FEff = (DOMI/DMI) × (MY or ADG/DOMI). Thus FEff = DEff × MEff. This principle was applied to experimental databases pooled from literature, of various types of ruminant. We show that FEff is mainly influenced by MEff when MY, or ADG are low because FEff and MEff are also decreasing. For high levels of performance FEff and MEff are reaching a plateau, and then DEff can be more influential than MEff. For growing ruminants housed with very various dietary contexts, there was a positive relationship between DEff and ADG, and FEff, because higher performance is generally achieved by providing more digestible diets (less NDF). In contrast, the relationship between DEff and MY is negative in high yielding dairy cows, due to the digestive interactions which increase when DMI and dietary concentrate increase. Due to maintenance needs, global protein efficiency (PEff) is also improved when milk protein yield or ADG is increased. However, dietary factors can also be very influential on PEff. For instance, in lactating ruminants, Peff is systematically decreased when the supply of metabolisable protein beyond maintenance requirement level increased. Moreover, the level of NH_3 in the rumen, or urea content in plasma or milk, which are closely linked with urea N outflow through urine, is negatively linked with PEff. In conclusion, FEff and PEff are better in high performing ruminants as a consequence of the variations in MEff caused by the existence of maintenance requirements. DEff does not present the same relation, however its influence on FEff, or MEff, can be marked in some feeding situations.

Session 52 Theatre 7

Nitrogen partitioning into faeces, urine and milk of dairy cows according to feeding strategy

E. Cutullic[1], A. Bannink[2], J. Carli[3], L. Crompton[4], M. Doreau[5], N. Edouard[1], P. Faverdin[1], S. Jurjanz[6], A. Klop[2], J. Mills[4], J. Moorby[3], P. Noziere[5], C. Reynolds[4], A. Van Vuuren[2] and J.L. Peyraud[1]
[1]INRA, UMR1348, Rennes, France, [2]WUR, LR, Wageningen, the Netherlands, [3]IBERS, AU, Aberyswyth, United Kingdom, [4]UReading, SAPD, Earley Gate, United Kingdom, [5]INRA, UMRH1213, Theix, France, [6]INPL, INRA, Nancy, France; erwan.cutullic@rennes.inra.fr

Increasing nitrogen (N) efficiency thereby reducing N losses towards air or water and increasing N excretion in milk is a major challenge for European dairy production. Since N losses in faeces and in urine do not have the same fate at the farm level, a precise prediction of N partitioning into milk, faeces and urine at the cow level is required for plausible modelling of farm emissions and management recommendations. In that context, a database gathering more than 1,700 complete individual N balances (N intake, in milk, faeces and urine) of lactating dairy cows, of which at least two thirds have a known diet composition, was constructed within the REDNEX European project (FP7, KBB-2007-1). Data originate from trials conducted over the last 40 years in France, United Kingdom and the Netherlands. A wide range of feeding practices is covered with grass hay, grass silage, fresh grass and maize silage based diets, supplemented with 0 to 70% concentrate, and ranging from 120 to 220 g CP/kg DM (5th and 95th percentiles). A wide variation is also observed in performance, with milk yields varying from 12 to 39 kg/day, BW from 480 to 720 kg, and DM intakes from 12 to 24 kg/day. N outputs vary from 80 to 300 g/day in urine, 100 to 230 g/day in faeces and 65 to 190 g/day in milk. The urinary part of N excretion, which varies from 35 to 70%, is strongly reduced with a decrease in CP content of the diet as expected. It is also reduced with an increase in N use efficiency for milk production, defined as the ratio of N in milk to N intake, which varies from 16 to 37%. Further meta-analysis on this large database will precisely characterise N partitioning according to the diet and cow characteristics.

Session 52 Theatre 8

Nitrogen supplies and manure handling improve feed efficiency and reduce emissions in dairy cattle

N. Edouard[1], A. Charpiot[2], P. Robin[3], E. Lorinquer[2], J.-B. Dollé[2] and P. Faverdin[1]
[1]INRA-Agrocampus Ouest, UMR1348 PEGASE, 35590 Saint-Gilles, France, [2]Institut de l'élevage, 149 rue de Bercy, 75012 Paris, France, [3]INRA-Agrocampus Ouest, UMR1069 SAS, 35000 Rennes, France; nadege.edouard@rennes.inra.fr

Improving nitrogen (N) use efficiency in dairy cattle is a challenge in a context of increasing prices of high protein feedstuffs and environmental concerns. Decreasing the amount of protein in the diet will reduce N excretion, but sometimes at the cost of animal performances. Moreover, manure management as liquid or solid will influence polluting emission processes in relation with the amount of N excreted. The aim of this trial is to assess the consequences of an important reduction in degradable N in the diet for animal production and for ammonia and greenhouse gas emissions when considering two housing systems: free stalls with cubicles (CU) and straw deep litters (DL) producing respectively slurry and farm yard manure. Mixed rations were offered *ad libitum* either with a low (N-; CP: 12%DM) or a high (N+; CP: 18%DM) dietary crude protein level. Two groups of three dairy cattle were offered the four combinations (N-CU, N+CU, N-DL, N+DL) in a Latin-Square design during four periods of four weeks. Milk yields and dry matter intakes were individually recorded. Gas concentrations were measured continuously with an infrared photo-acoustic gas analyser. Milk yields and DM intakes were similar for all treatments. Ammonia emissions were divided by 2 on N- compared with N+ for CU and by 4 for DL, bringing NH_3 emissions at a very low and similar level for CU and DL on N- diet. Methane emissions were always higher on DL compared with CU due to emissions from the litter itself. However, CH_4 emissions were lower on N- and the difference between CU and DL was reduced. Feeding a low degradable CP diet reduced losses towards the environment without any significant effect on performances. This reduction was amplified for deep litters showing that these can be low emission systems when combined with adjusted nitrogen supplies.

Session 52 Theatre 9

Effect of maturity and conservation of grass/clover on digestibility and rumen pH in heifers

A.S. Koch[1], P. Nørgaard[1] and M.R. Weisbjerg[2]
[1]Dept. of Veterinary Clinical and Animal Science, University of Copenhagen, Grønnegårdsvej 3, Frederiksberg, Denmark, [2]Dept. of Animal Science, Aarhus University, Blichers Allé, Tjele, Denmark; anne-k@sund.ku.dk

The study aimed to evaluate effects of maturity and conservation of primary growth grass/clover on apparent digestibility and rumen pH. Two batches of mixed ryegrass, red and white clover harvested in 2009 on May 9 and 25 were conserved as either silage or hay. The forages early silage (ES) and hay (EH), and late silage (LS) and hay (LH) had DM contents of 45, 84, 25 and 83%, and NDF contents of 32, 44, 42 and 50% of DM, respectively. Forages were fed as sole feed to four Jersey heifers of 435±30 kg BW in a 4×4 Latin square experiment. Feeding level was 90% of individual *ad libitum* intake, divided in two daily meals at 08:00 and 15:30 h. Potentially digestible NDF (DNDF) was determined after 288 h in situ. Apparent digestibility of OM and NDF was estimated using Cr_3O_2 as marker. Rumen fluid pH in the medial and ventral rumen was measured with 1 h intervals from 07:30 to 15:30 h. Data was analysed by Proc MIXED in SAS 9.2 with period, conservation, harvest time, and conservation × harvest time as fixed effects and heifer as random. Early compared to late harvest increased DM intake (DMI) ($P<0.001$), whereas DMI was unaffected by conservation. Daily DMI of ES, EH, LS, and LH averaged 8.7, 9.4, 7.2, and 7.2 kg, respectively. In early compared to late harvest, digestibilities were: OM 82% vs. 78%; NDF 88% vs. 80% and DNDF 95% vs. 90% ($P<0.01$). Digestibility of OM was higher in silage (81%) than hay (80%, $P<0.02$), while NDF and DNDF digestibility were similar. Rumen pH ranged from 7.2 before feeding to 5.9 three h post feeding. Rumen pH was unaffected by type of conservation, but early compared to late harvest resulted in lower minimum and mean rumen pH ($P<0.05$). Results indicate that early harvest improves OM, NDF and DNDF digestibility and lowers rumen pH, and that ensiling improves OM digestibility without affecting rumen pH.

Session 52 — Theatre 10

Influence of prolonged vs. Instant use of Natuzyme® on *in vitro* fermentation of two ruminant's diets

M. Danesh Mesgaran, E. Parand, A. Faramarzi Garmroodi and A. Vakili
Ferdowsi University of Mashhad, Mashhad, Ferdowsi University of Mashhad, College of Agriculture, Departmen of Animal Science, 0098, Iran; danesh@um.ac.ir

The influence of prolonged pre-treatment vs. Instant administration of Natuzyme® (Bioproton Co.) on *in vitro* gas production (GP, ml/200 mg DM), dry matter disappearance (IVDMD) methane production (MET, ml/200 mg DM) and fermentation efficiency as mg IVDMD per ml MET (FE) of various ruminant diets containing wheat straw (D1: 6.2% wheat straw, 39.4% corn silage) or alfalfa hay (D2: 21.3% alfalfa, 37.2% corn silage) at half time (t1/2) of gas production. Approximately 200 mg (DM) of each diet was weighted into a 125 ml serum bottle at 0.0 or 24 h prior to incubation (T1 and T2, respectively). Each bottle received 2.52 g/kg DM of the enzyme in an aqueous suspension to maintain same moisture content maintain same moisture content (40%), run=3 and n=3. The gas production procedure was followed by pipetting buffered rumen fluid into the bottles and incubated at 38.6 °C for desired intervals. In a pre-trail, pressure of gas was recorded at 2, 4, 6, 8, 10, 12, 24, 48, 72 and 96 h of incubation. Pressure data was converted to volume using an experimental curve and was modeled to estimate t1/2. Main trail incubation was continued until t1/2 and volumes of GP and ME, and residual DM was measured. Data were analyzed as 2×2 factorial arrangement in a completely randomized design. Results revealed that D2 comparing to D1, had higher GP (43.95 vs. 43.18) and Met (12.10 vs. 8.62) but less IVDMD (30.90 vs. 33.90) and FE (5.35 vs. 8.24) ($P<0.05$). In addition T1 compared to T2 had higher GP (47.43 vs. 36.70) and less IVDMD (30.90 vs. 33.90) ($P<0.05$) and Met and FE did not differ for T1 and T2. Although T2 in D1 decreased FE (6.44 vs. 10.46) and increased Met (10.38 vs. 6.85) comparing to T1($P<0.05$), situation was different for D2 and FE was higher for T2 in D2 (6.64 vs. 4.10) ($P<0.05$) while Met was not affected. Therefore it seems optimal Pre-treatment duration for Natuzyme® widely depends on diet type and forage content.

Session 52 — Poster 11

Relationship between chemical composition and rumen degradation characteristics of maize silages

M. Ali[1,2], G. Van Duinkerken[1], J.W. Cone[2] and W.H. Hendriks[2,3]
[1]Wageningen UR Livestock Research, P.O. Box 65,8200 AB Lelystad, the Netherlands, [2]Animal Nutrition Group, Wageningen University, P.O. Box 338, 6700 AH, Wageningen, the Netherlands, [3]Faculty of Veterinary Medicine, Utrecht University, P.O. Box 80.163, 3508 TD Utrecht, the Netherlands; gert.vanduinkerken@wur.nl

Several in situ studies have been conducted on maize silages to determine the effect of individual factors such as maturity stage, chop length and ensiling of maize crop on the rumen degradation of maize silages but the information on the relationship between chemical composition and in situ rumen degradation characteristics of maize silages remains scarce. The objective of this study was to determine and describe relationship between the chemical composition and the rumen degradation characteristics of dry matter (DM), organic matter (OM), crude protein (CP), starch and neutral detergent fibre (NDF) of maize silages. Seventy-five maize silage samples were selected, with a broad range in chemical composition and quality parameters. The samples were incubated in the rumen of three cows for 2, 4, 8, 16, 32, 72 and 336 h, using the nylon bag technique under uniform experimental conditions and protocols. The new database with in situ rumen degradation characteristics of DM, OM, CP, starch and NDF of the maize silages was obtained under uniform experimental conditions; same cows, same incubation protocol and same chemical analysis procedures. Regression equations were developed with significant predictors ($P<0.05$) describing strong, moderate and weak relationships between the chemical composition of maize silages and the washout content, rumen undegradable content, potentially rumen degradable content, fractional degradation rate and effective rumen degradability of DM, OM, CP, starch and NDF. The developed regression equations can be used for the rapid assessment and accurate estimation of rumen degradation characteristics of maize silages used in practice.

Dose response effects of some Iranian native essential oils on *in vitro* ruminal methane production

H. Jahani-Azizabadi, M. Danesh Mesgaran and A.R. Vakili
Ferdowsi University of Mashhad, Dept. of Animal Science, Excellence Center for Animal Science, Mashhad, Iran, P.O. Box 91775-1163, Iran; danesh@um.ac.ir

The aim of the present study was to evaluated effect of some Iranian native essential oils at different doses on *in vitro* ruminal methane production of a high concentrate diet. Experimental diet that used for batch cultures was a mix of concentrate: alfalfa hay (80:20, based on DM) which was ground to pass from 1-mm screen. Approximately, 500 mg of the diet alone (as control) or plus cinnamon (Ci), Oregano (Or), Pistachio hull (PH), Coriander (Cor) and Thyme (Th) essential oils (35, 70, 140 and 280 µl/l of culture fluid) were placed into a 125 ml serum bottles (n=6) containing 50 ml of buffered rumen fluid (ratio of buffer to rumen fluid was 2:1), then bottles were placed in water bath for 24 h at 38.5 °C. Rumen fluid was obtained from three adult ruminally fistulated sheep, before the morning feeding. In the end of incubation gas production of each bottles was recorded and a sample of the gas was collected into a 10 ml evacuated tubes. Methane content of the produced gas was determined using gas chromatography procedure. Data were statistically analyzed using SAS (V. 9/1) and differences between each treatment and control were tested using Dunnett's test ($P<0.05$). The results indicate that all of the essential oils (except Cor) were effective ($P<0.05$) in decrease of methane produced compared with those of the control. Relative to the control increasing concentration of essential oil (except for Cor) resulted to decrease in methane production. Results of the present study concluded that TH, Ci and PH essential oils have a high potential to decrease ruminal methane production.

Tea leaves improve *in vitro* degradability but reduce rumen ammonia from rice straw-based diets

D. Ramdani[1,2], A.S. Chaudhry[2] and C.J. Seal[2]
[1]Universitas Padjadjaran, Faculty of Animal Husbandry, 45363 Jatinangor Sumedang, Indonesia, [2]Newcastle University, School of Agriculture, Food and Rural Development, NE1 7RU Newcastle Upon Tyne, United Kingdom; diky.ramdani@newcastle.ac.uk

Green (GTL) and black (BTL) tea leaves are good sources of crude protein (CP) and phenolic compounds including tannins. Therefore, their use as novel additives may be useful to improve the utilization of low quality rice straws (RS) based ruminant diets. Five iso-nitrogenous (152±7.6 g CP & 11±0.13 MJ ME/kg DM) diets were prepared by mixing (% DM basis) 70 concentrate with either 30 RS (control) or 25 RS + 5 GTL (GTL5) or 20 RS + 10 GTL (GTL10) or 25 RS + 5 BTL (BTL5) or 20 RS + 10 BTL (BTL10). An *in vitro* incubation experiment with a 5×5 factorial design, in triplicate, was used to compare these 5 diets at 5 incubation times (0, 6, 24, 48 and 72 h) for their *in vitro* dry matter degradability (IVDMD) and ammonia (NH_3) production using sheep rumen fluid under anaerobic conditions at 39 °C. The statistical comparison showed that the mean IVDMD (g/kg DM) of GTL5 (308.4) and GTL10 (311.8) were significantly higher ($P<0.001$) than the control (275.6) but not BTL5 (278.0) and BTL10 (280.2). Conversely, the mean NH_3 concentration (mg/l) for GTL5 (114.9), GTL10 (100.4) and BTL10 (117.1) were lower ($P<0.001$) than the control (126.3) but not for BTL5 (125.9) which was similar to the control. GTL have the more potential than BTL to improve IVDMD of RS based diets. Moreover, lower NH_3 production for tea leaves-containing diets was likely due to the ability of tea tannins to bind and protect plant proteins from rumen degradation but make these available as by-pass protein for their absorption in small intestine.

Session 52 Poster 14

Performance of crossbreed young bulls fed corn silage inoculated with microbial inoculants

C.H.S.R. Rabelo, A.L.S.V. Valente, R.B.P. Pavezzi, F.C.B. Basso, E.C.L. Lara and R.A.R. Reis
University of State of São Paulo, Animal Science, Via de Acesso Prof. Paulo Donato Castellane s/n, 14884-900 Jaboticabal, SP, Brazil; carlos.zoo@hotmail.com

The aim of this study was to evaluate the performance of crossbreed young bulls fed corn silage inoculated. Corn hybrid was harvested at 40.4% dry matter (DM), chopped and treated with 1×10^5cfu of *Lactobacillus plantarum* 'MA18/5U' combined with 1×10^5cfu of *L. buchneri* 'CNCM I-4323' (LPLB) or 1×10^5cfu of *B. subtilis* 'AT553098' (LPBS) per gram of wet forage, remaining a treatment uninoculated (control silage). Thirty six crossbred young bulls (average initial body weight of 316.3±33.9 kg) were used. Diet was composed by 40% of silage and 60% of concentrate (in DM basis) and provided to animals once daily (*ad libitum*, 6 am) in individual stalls. Orts were weighed daily and DM intake was measured. Animals were weighed after fasting (12 h) at the beginning and end of the trial to obtain the average daily gain (ADG) and feed conversion (FC). Animal ingestive behavior was evaluated for 3 consecutive days at intervals of 10 minutes between observations. Data were analyzed as completely randomized by ANOVA using MIXED procedure of SAS. Corn silages DM intake by crossbred young bulls was similar (P>0.05; Control=8.95; LPLB=8.76; LPBS=8.75 kg/day), even as the ADG (P>0.05; Control=1.45; LPLB=1.48; LPBS=1.53 kg/day) and FC (P>0.05; Control=5.82; LPLB=5.83; LPBS=5.58). Animals fed control silage showed lower time of feeding (P<0.05) in minutes (Control=129; LPLB=169; LPBS=176). In relation to rumination time, the animals fed with silage LPLB showed higher time (P<0.05; Control=286; LPLB=326; LPBS=257), while the control treatment showed higher time of idle in relation to LPLB silage (P<0.05; Control=1035; LPLB=943; LPBS=993). Inoculation in corn silage did not promote improvements on animal performance of crossbreed young bulls.

Session 52 Poster 15

Growth and carcass quality of crossbred Jersey bull and heifer calves

M. Vestergaard[1], P. Spleth[2], L. Kristensen[3] and M. Kargo[1,2]
[1]Aarhus University, Foulum, Tjele, Denmark, [2]Knowledge Centre Agriculture, Skejby, Aarhus, Denmark, [3]Cattle Reserch Centre, Foulum, Tjele, Denmark; mogens.vestergaard@agrsci.dk

Most newborn Jersey (JER) bull calves are killed at birth due to low growth rate, poor feed utilization and low carcass value. Use of sexed semen to produce heifer calves on the genetically best cows and the low culling rate in JER lead to excess of heifer calves. The objective was to investigate if use of crossbreeding by inseminating less superior cows with a beef breed will produce offspring that could be utilized in rosé veal calf production (<8 mo. of age). 12 purebred JER bull calves (PURE), 12 beef × JER bull (BULX) and 12 beef × JER heifer (HEIX) calves were purchased at 3-4 wk of age. Two sires were used; a Limousine and a Belgium Blue (6 calves per crossbred group) chosen not to cause calving difficulties in JER cows. Calves were fed milk replacer, hay and concentrate before weaning at 8 wk of age, and then gradually changed to a high energy TMR based on concentrate pellets, corn-cob silage (35% of DM in TMR), ground barley, sugar beet pulp, and soy bean and canola meal fed *ad libitum*. The concentrate pellets (1.4 kg/calf/d) were removed from the TMR at 200 kg BW. Individually feed intake was recorded from 4 to 8 mo. of age. At 4 mo, PURE, BULX and HEIX weighed 101, 116 and 112 kg, ADG from 4-8 mo. was 1.15, 1.48 and 1.21 kg/d, BW at slaughter was 256, 315 and 263 kg, and FCE was 4.4, 4.0 and 4.6 Scandinavian Feed Units/kg gain for PURE, BULX and HEIX, respectively (all P<0.05). Dressing increased from 48.1 in PURE to 53.2 and 51.7% in BULX and HEIX, respectively, leading to carcasses of 118, 162 and 129 kg with EUROP conformation of 2.8, 5.4 and 4.6, respectively (both P<0.05). Despite a 32% higher payment for HEIX compared with PURE carcasses this was not sufficient to make these two groups economically attractive whereas the 89% higher payment for BULX carcasses compared with PURE and improved feed conversion suggests a market potential for such <8 mo. old veal calves.

Session 52 Poster 16

Metabolic parameters of three levels of concentrate supplement for beef cattle in pasture

R.P. Barbero, A.A. Oliveira, R.A. Reis, U.B. Carneiro, M.V. Azenha and F.H.M. Souza
UNESP, Animal Science, Via de Acesso Prof. Paulo Donato Castellane s/n, 14884-900, Brazil; rondinelibarbero@zootecnista.com.br

The aim of this study was to evaluate the effects generated by three levels of supplementation, 1.0, 1.5 and 2.0% of body weight (BW), for finishing beef cattle in the dry season, raised on pastures of *Brachiaria brizantha* (HOCHST ex A. Rich) Stapf cv Marandu on the parameters: pH and ammonia nitrogen. The experimental area was 18 acres, divided into 18 paddocks of 1.3, 1.0 and 0.7 hectares. Six Nelore cattle with an average weight of 300 kg fitted with ruminal cannula were used. The experimental design was a Latin square 3×3, conducted simultaneously with three treatments and three experimental periods, lasting 18 days each. For evaluation of pH and ammonia concentration of ruminal fluid, sampling was done at 0, 3, 6, 12, 18 and 24 h after the provision of the concentrate. The average mass of leaves was 2,446.7±128.95 kg/ha, and the supply of leaves of 3.76±0.26% BW with 8.66±0.7% CP, 74.33±0.88% of NDF. It was noted that the minimum value of 6.08 was reached after 6 h from the start of ingestion of the supplement when the concentrate intake was 2% in diet 6.69 BW and 1% BW. When the supply was 1.5%, the minimum pH was observed after 12 h of delivery. The concentrations of ruminal ammonia-N (mg/dl) in the present study had higher rates after 3 h of supplementation the animals without significant effect among the three levels of concentrate (P>0.05). This is due to high rumen degradation of protein in the concentrate and urea is rapidly hydrolyzed and converted into ammonia. Although a decrease in ruminal pH was observed when increasing the supply of concentrate (P<0.01) from zero to 18 h, critical levels of this parameter were not observed in animals receiving concentrated supplementation concentrated in pastures up to 2% concentrate of body weight.

Session 52 Poster 17

Nutritive value and Kinetics of *in vitro* fermentation with rumen liquor of *Paulownia* sp leaves

P. Descals, A. Seradj, D. Villalba, G. Villorbina and J. Balcells
University of Lleida, Animal Production, Av. Alcalde Rovira Roure 191, 25198 Lleida, Spain; dvillalba@prodan.udl.cat

An *in vitro* trial was performed to define nutritive value and fermentation kinetics of *Paulownia* sp. leaves. Leaf samples were harvested in spring, summer, autumn and fall and dried at 50°C for 24 h. Content of dry matter, crude protein ether extract and fiber (NDF and ADF) were determined. The four leaves-time samples plus two roughages used as positive (alfalfa hay) and negative (wheat straw) control were incubated (600 mg) in bottles (120 ml) with incubation solution (80 ml) including (20%; 16 ml) of strained rumen fluid under anaerobic conditions, during 72 h at 39°C in triplicate. The gas pressure was measured at different times and converted to volume. A nonlinear model for rate determination with lag time [$y = a\ (1\text{-}e^{-b\ (t-c)})$] was applied to determine the potential cumulative GP, fractional rate and discrete lag time. Finally the metabolizable energy (ME : Mcal/kg) content was estimated from the equation [$ME=2.2 + 0.1357GP + 0.0057CP + 0.0002859EE^2$]. Senescent leaves (autumn / fall samples) showed similar CP concentrations (17% of DM) than alfalfa hay (15-18% of DM), while in spring and summer samples CP content of the leaves was much higher (27.1% and 22.6% of DM, respectively). As expected, the concentrations of CF, NDF and ADF were lower in young leaves and increased with the maturity. Hemicellulose content varied from 6.7 to 0.3 g/kgDM in young and senescent leaves, respectively. In relation to GP the highest value corresponds to spring leaves decreasing along the maturation process, thus ME (Mcal/kg) content for spring, summer, autumn and fall leaves were 2.2, 2.3, 1.9 and 1.7, respectively. Spring-summer harvested paulownia leaves constitute a high-quality roughage with a highest CP than alfalfa hay although senescent leaves had a poor nutritive value similar to cereal straw.

Session 52 Poster 18

Effect of a dietary escape microbial protein on the performance of dairy cows

S.Z. Orosz[1], S. Andrieu[2] and M. Agovino[2]
[1]Livestock Performance Testing, Dózsa Gy. u. 58, Gödöllő, Hungary, [2]Alltech Bioscience Centre, Dunboyne, Co Meath, Ireland; soneill@alltech.com

Excess dietary protein can be costly in a dairy cow diet, both financially and physiologically. Optimising quality of dietary protein can help to reduce ration crude protein (CP). The aim of this study was to investigate the effect of a dietary escape microbial protein (DEMP®, Alltech Inc., KY) on milk yield and milk and blood urea of primiparous (PP) and multiparous (MP) cows on a large-scale farm in Hungary. Trial duration was 60 d (May-July 2012). Primiparous (PP) and multiparous (MP) cows were allocated according to milk yield (l/d) and days in milk (DIM) to one of two treatments: control (C) – a basal diet of by-product/forage mix, alfalfa haylage, corn gluten, brewer's grains, maize silage, molasses and commercial concentrate (PP:n=30, milk yield 37.3, DIM=96; MP:n=33, milk yield 46.7, DIM=79) and treatment (D) – basal diet reformulated to include DEMP at 300 g/h/day reducing the dietary crude protein (CP) level from 174 to 166 g/kgDM (PP:n=30, milk yield 37.3, DIM=97; MP:n=33, milk yield 46.8, DIM=80). Milk yield, milk urea (MU) and blood urea nitrogen (BUN) were measured during the trial. Data were analysed by ANOVA (Levene's test, $P \leq 0.05$), paired t-test, Tukey post hoc test, Welch's test and Dunett post hoc test ($P \leq 0.05$) (SPSS). Day 60 milk yield was numerically higher for treatment versus control at +1.8 and +0.6 kg/d for PP and MP cows, respectively. Milk urea increased non-significantly by 0.1 mg/dl for control PP cows. Conversely, MU for treatment PP cows decreased non-significantly by 3.7 mg/dl. For treatment MP cows, there was a significant ($P<0.05$) reduction in MU (-8.1 mg/dl) versus a non-significant reduction of 2.0 mg/dl for control MP cows. Treatment diet appeared to lower MU in PP and MP cows, while at the same time increase milk yield numerically, resulting in increased revenue for the producer. These data demonstrate benefits of improving N use efficiency by reformulating rations to provide a more quality protein profile protein profile.

Session 52 Poster 19

Reduction of nitrogen excreted in dairy farms through the application of high N efficiency ration

L. Migliorati and L. Boselli
Consiglio per la Ricerca e Sperimentazione in Agricoltura, Centro di Ricerca per le produzioni foraggere e lattiero-casearie, Via Porcellasco 7, 26100 Cremona, Italy; luciano.migliorati@entecra.it

Aim of this study was to demonstrate that dairy cows N excretion can be reduced by adopting feeding techniques based on the reduction of diet crude protein (CP) content and on high diet N efficiency, while supporting high milk production. Two commercial farms placed in North Italy in Grana Padano (GP) and Parmigiano Reggiano (PR) areas were used. One hundred and twenty Italian Friesian cows in GP and 220 in PR were used to compare two diets with different CP content provided in two consecutive years. In the 1st year GP cows were fed a 16% CP diet on DM (GP1) and PR cows were fed a 14.75% CP on DM diet (PR1). In the 2nd year, CP was 5% reduced by 15.2 (GP2) and 14% on DM (PR2). Experimental design was a pretest-posttest, with one group and two consecutive trials of one year each. Data were subjected to statistical analysis with the paired student's t-test. Nutrient balance was estimated according to ERM/AB-DLO and EMR methodology using N input/output flows to determine dairy cow N excretion such difference between dairy cow N intake and milk N retention. N intake was estimated using dry matter intake (DMI) and CP diet. DMI was estimated by live weight and milk yield recorded once a month, while CP content was determined every 15 days. In GP farm, no difference was found between two years in milk yield, fat and protein contents. In PR farm, fat content was higher in PR2 than in PR1, while no difference was found in milk yield and protein content. Estimate N excretion was significantly reduced by 8% in GP2 in comparison to GP1 with 0.420 and 0.387 kg/cow/day respectively, and by 9% in PR2 compared to PR1 with a reduction from 0.330 to 0.300 kg/cow/day respectively. Balancing for cows RUP and RDP MP requirements will have positive effects, not only on profitability, but also on environment. This work was funded by European Commission area Environment (LIFE+ AQUA Project).

Session 52 Poster 20

Ingestive behavior of beef cattle confined in feedlots and fed with different concentrate levels

R.P. Barbero, A.A. Oliveira, R.A. Reis, R.C.G. França, M.V. Azenha and F.H.M. Souza
UNESP, Animal Science, Via de Acesso Prof. Paulo Donato Castellane s/n, 14884-900, Brazil;
rondinelibarbero@zootecnista.com.br

The aim of this study was to evaluate the ingestive behavior of beef cattle confined in feedlots fed with different concentrate levels. Thirty-six Nellore steers with initial weight of 336 kg and 24 months of age were used. The treatments consisted of three levels of concentrate based on the diet dry matter (40, 60 and 80%), corn silage as forage. The diet was provided once a day, and adjusted daily allowing surplus of 10%. Thus, an experiment was conducted in a completely randomized design with twelve replicates per treatment. The rumination time standing and lying, dry matter intake (kg/day and% PV) decreased ($P<0.05$) in animals fed with 80% of concentrate. Lying idle times were higher ($P<0.05$) in animals treated with high grain content, but the days of idleness and stay standing in the trough did not differ ($P>0.05$) among treatments. The results of this study showed that increasing the participation of concentrate in the diet resulted in reduced dry matter intake by animals, but without reducing the time spent at the trough.

Session 52 Poster 21

Effect of abomasal casein infusion on transport of urea-N from blood to gut in cows fed low N diet

B.A. Røjen, N.B. Kristensen, M. Vestergaard and M. Larsen
Aarhus University, Department of Animal Science, Blichers Allé 20, 8830 Tjele, Denmark;
betina.amdisenrojen@agrsci.dk

The aim of the study was to investigate ruminal and PDV extraction of arterial urea-N when supply of rumen undegradable protein is increased without changing supply of rumen degradable protein (RDP). Five Holstein cows fitted with ruminal cannulas and permanent indwelling catheters in major splanchnic blood vessels were abomasally infused with water (CON; 13.9% CP) and 800 g/d of Na-casein (CAS; 18.2% CP). Eight blood sample sets were taken from arterial, portal, and ruminal catheters on d 14 of each period. Data was analyzed using the mixed procedure of SAS with fixed effects of treatment, sampling time and the interaction; sampling time was considered as repeated measurement within cow × treatment. Period and treatment was confounded. DMI was unaffected by abomasal casein infusion (19.6±0.7 kg/d, $P=0.73$), whereas milk yield increased for CAS (+4.4±0.8 kg/d, $P<0.01$) compared with CON. Arterial urea-N concentration increased for CAS (+4.7±0.6 mM, $P<0.01$) compared with CON, while arterial ammonia concentration was unaffected by treatment ($P=0.60$). Ruminal vein – arterial concentration difference of urea-N was unaffected by treatment ($P=0.37$). Combined with increased arterial concentration of urea-N for CAS, this resulted in decreased ruminal and PDV extractions of arterial urea-N ($P\leq0.07$). This indicates that the rumen epithelia adapt to the increased protein supply with CAS by down-regulating the permeability. The PDV uptake of urea-N was however increased for CAS (+109±36 mmol/h, $P<0.01$) compared with CON and PDV release of NH_3 was increased equivalently with CAS (+120±35 mmol/h, $P<0.01$). Increased PDV release of NH_3 from ruminal degradation of dietary protein appears unlikely, as the supply of RDP was unaffected by treatment. The present results suggest that cows fed on low N diet are able to increase transfer of urea-N from blood to gut when N supply bypasses the rumen and blood urea-N concentration is kept at a higher level.

Session 52 Poster 22

Effects of NaOH treatment of corn and sunflower straws on composition and digestibility by sheep

E. Yosef, M. Nikbahat and J. Miron
Institute of Animal Science, Agricultural Research Organization (ARO), Department of Animal Science, P.O. Box 6 Bet Dagan, 50250, Israel; edithyos@agri.gov.il

The semi-arid climate of Israel interferes with the production of high quality forages and motivates the use of chemically treated lignocelluloses as potential forage replacers. Our previous study with serial of lignocelluloses demonstrated the potential of corn straw from mono-cotyledons and of sunflower straw from di-cotyledons plants as potential ruminant feeds. The aim of this study was to measure the effect 5% NaOH treatment of corn and sunflower straws on their composition and *in vivo* digestibility by sheep. Two experiments were conducted in ARO. 1. Two groups of 5 male sheep each held in individual metabolic cages were fed individually by rations containing either untreated or 5% NaOH treated corn straw plus soybean meal and mineral supplement to reach 12% protein in the diet. Data of intake and digestibility was analyzed according to the ANOVA model of SAS using each individual sheep as replicate. Voluntary dry matter intake by sheep fed the treated corn straw diet was 39.2% higher than that of the untreated corn straw diet ($P<0.05$). Dietary organic matter digestibility was 63.6% in the untreated corn straw and 67.8% in treated corn straw diets. NaOH treatment improved neutral detergent fiber (NDF) and cellulose digestibility of corn straw diets by 14.2% and 16.6%, respectively ($P<0.05$). 2. Two groups of 4 male sheep each held in individual metabolic cages were fed individually by rations containing either untreated or 5% NaOH treated sunflower straw plus soybean meal and mineral supplement. *In vivo* dry mater digestibility of the untreated and treated sunflower straw rations were 50.5% and 57.8%, respectively ($P<0.05$). Both trials show a significant advantage of using NaOH treatment of corn and sunflower straws for increasing voluntary consumption of digestible dry matter by sheep. However, the NaOH treatment was more effective in improving NDF digestibility in corn straw than in sunflower straw.

Session 52 Poster 23

Total mixed ration prepared by trailer mixer or self-propelled mixer on diet quality and cows' yield

U. Moallem
ARO, P.O. Box 6, Beit Dagan, 50250, Israel; uzim@volcani.agri.gov.il

The objectives were to evaluate the accuracy, and homogeneity of total mixed ration (TMR) prepared by trailer mixer (TM) or self-propelled mixer (SPM), and the effects on yields of high producing dairy cows. Cows (n=216) were divided into two groups by milk yield, days in milk (DIM), number of lactation and body weight. Later, calving cows at 21 DIM were randomly assigned into the treatment groups. The cows were fed for 10 wks the same formulated ration prepared either by TM or SPM. TMRs' samples were taken weekly from 5 spots along the manger to evaluate the accuracy and homogeneity of the rations' chemical composition. In another 7 times, 3 samples from each TMR were taken to evaluate the particle size distribution. TMRs' chemical composition and milk yields were analyzed as repeated measurements with the PROC MIXED procedure of SAS. The homogeneity of the TMRs was analyzed using the Homogeneity of Variance test – HOVTEST-BARLETT procedure of SAS. The deviation of TMRs from the formulated ration was greater in the TM than in the SPM in ADF and NDF, and numerically greater in protein content. The SPM TMR was more homogeneous in DM, ADF and NDF, and less homogeneous in ASH as compared to the TM TMR. The long particles size fraction (≥19 mm) was 2.9 percent unit greater and the short size (≤8 mm) was 3.3 percent unit lesser in the TM than in the SPM TMR, while the intermediate size (<8 >19) was similar. No differences were observed in particle size homogeneity between the TMRs. The milk production was 3.7% greater for all cows and 5.6% higher for cows from 2 and 3 lactations in the SPM than in the TM cows. For all cows, the milk-fat content was 0.18 percent unit lower in the SPM than in the TM cows. Also, the protein and lactose yields (kg/d) in cows from 2 and 3 lactations were 7.6% greater in the SPM than in the TM cows. In conclusion, SPM TMR was more accurate and homogenous than the TM TMR, which might explain the higher milk production observed in the SPM cows than in the TM cows.

Session 52 — Poster 24

Variation between individual cows in rumen degradation characteristics of maize and grass silages

M. Ali[1,2], J.W. Cone[2], G. Van Duinkerken[1] and W.H. Hendriks[2,3]

[1]Wageningen UR Livestock Research, P.O. Box 65, 8200 AB, Lelystad, the Netherlands, [2] Animal Nutrition Group, Wageningen University, P.O. Box 338,6700 AH, Wageningen, the Netherlands, [3]Faculty of Veterinary Medicine, Utrecht University, P.O. Box 80163, 3508 TD, Utrecht, the Netherlands; mubarak.ali@wur.nl

Factors affecting in situ rumen degradation characteristics of different feedstuffs have been extensively studied. Data on the variability in rumen degradation of maize and grass silages between individual animals remain scarce. The objective of this study was to determine the variation in in situ rumen degradation characteristics of dry matter (DM), organic matter (OM), crude protein (CP) and neutral detergent fibre (NDF) in maize and grass silages and starch in maize silages using three individual cows. Fifteen maize and 15 grass silage samples, with a broad range in chemical composition, were selected. The samples were incubated in the rumen for 2, 4, 8, 16, 32, 72 and 336 h, using the nylon bag technique. Three cows were used for nylon bag incubation of maize silages and three other cows for grass silages. The variation between individual cows was significant for degradation rate (k_d) of DM (P=0.02), OM (P=0.03) and CP (P=0.005), and the effective rumen degradability (ED) of DM (P=0.01) and CP (P<0.001) of maize silages whereas no significant (P>0.05) differences were found for the rumen undegradable fraction (U) and potentially rumen degradable fraction (D) of DM, OM and CP, and all the parameters of starch and NDF. The variation between individual cows was not significant (P>0.05) for U, D, k_d and ED of DM, OM, CP and NDF of grass silages. Pooling of rumen incubated residues of three cows in in situ degradation studies after each rumen incubation period is sufficient to obtain accurate estimates for degradability of dietary fractions or nutrients for grass silages. For maize silages, a number of estimates related to DM, OM and CP were significantly different between cows, necessitating the use of four or more cows.

Session 52 — Poster 25

Studying on effect of adding powdered thyme on digestibility, metabolizable energy and gas production

K.H. Ghazvinian[1], A. Mahdavi[1], B. Afsharhamidi[2], S. Heydaryan[3], A. Gharibi[3] and M.S. Ghodrati[3]

[1]University, Faculty of Vet Med, Semnan, Iran, [2]Agricultural and Natural Resource Research Center, West, Azarbijan, Iran, [3]university, Semnan, Iran; khosroghazvinian@yahoo.com

This study is based on the purpose of different additional values of powdered thyme on organic matter digestibility (OMD), available metabolizable energy (ME) and gas production of alfalfa hay, wheat straw and barley grain with gas test technique. That was done in a completely randomized design, with 4 treatments including different additional levels(0, 1.5, 3 and 6%)of thyme powder and 3 replicates. First of all analysis was done on feed stuffs;Gas production by Menke technique was measured in 24 h incubation and after it Produced gases was measured and recorded. Calculated gas production parameters were measured by using F-curve, and relative equation were used for estimating of ME and OMD. Data were analyzed by SAS. Results showed that gas production from rapid degradable part(a) increased by increasing in thyme levels in barley grain and we observed reduction in gas production from slow degradable part(b) in wheat straw. Rate of degradation increased by adding thyme in wheat straw and decreased in alfalfa hay. There was a significant reduction in produced methane gas in alfalfa hay and barley. Estimated ME and OMD showed an increase in alfalfa hay and a reduction in wheat straw. This increase may be due to an increase in soluble carbohydrates in wheat straw which leads to increasing in slope of gas production. Different values of produced methane gas are probably based on the values of produced volatile fatty acids especially Acetic acid and Changes in activity and population of microorganisms. Effective compounds of thyme cause an increase in growth and milk production and reduction in waste energy and improve rumen fermentation by rumen microorganisms. As the conclusion use of essential oils(thymol and carvacrol)may leads to better performance by reduction in fermentable gas production and improving ruminal fermentation.

Improvement of feed efficiency: lessons from residual feed intake studies in pigs: part 1

H. Gilbert[1] and J.C.M. Dekkers[2]
[1]INRA, UMR444 LGC, 31326 Castanet Tolosan, France, [2]Iowa State University, Department of Animal Science, 50011 Ames, IA, USA; helene.gilbert@toulouse.inra.fr

The ratio output/input (body weight gain/feed intake in growing animals) or its inverse have been used for decades as indicators of the efficiency of the conversion of feed to body growth to quantify feed cost in animal breeding. It corresponds to gross feed efficiency (GFE). Most of the improvement of GFE has been achieved, and is still ongoing, by increasing lean growth rate through management, nutrition and genetic strategies, ie targeting the component of GFE that is related to performance differences between animals. However, production and maintenance requirements explain only 65 to 80% of the variation in feed intake in the growing pig for example. Net feed efficiency (NFE) is used to target the remaining 20 to 35%. Residual feed intake (RFI), the difference between measured feed intake and predicted feed intake based on production and maintenance requirements, is a measure of NFE. In pigs, it is usually estimated as the residual of the multiple linear regression of feed intake on growth rate and backfat thickness (production requirements), and metabolic body weight (maintenance requirements). Issues remain to specifically improve this component in livestock. Evaluating RFI for large groups of animals bred in various environments is not straightforward, but integrating RFI per se in selection criteria that already use growth rate and body composition, and potentially feed intake and/or GFE is probably not necessary. Indicator traits or markers, as well as best management and nutritional practices to improve RFI, have yet to be defined. Finally, RFI has been proposed as a buffer compartment for animals to face stresses or changes in their environment. How large a reduction in RFI is acceptable and how to prevent loss of homeostatic balance is yet to be determined. We will illustrate these points by exploring results from two selection experiments conducted in pigs independently at INRA and ISU during the last 14 years.

Response to a high level fibers diet in pigs divergently selected for residual feed intake

L. Hauptli[1], A. Priet[2], H. Gilbert[3] and L. Montagne[4]
[1]Universidade Federal do Paraná, Ciências Agrárias, Centro Curitiba, 80060-000, Brazil, [2]INRA, UE 1372 GENESI, Surgères, 17000, France, [3]INRA, UMR 444 LGC, Castanet-Tolosan, 31326, France, [4]Agrocampus Ouest, INRA-UMR 1348 PEGASE, Rennes, 35042, France; montagne@agrocampus-ouest.fr

Residual feed intake (RFI) is the difference between observed and predicted daily feed intake (DFI) estimated from maintenance and production requirements. Genetic selection for lower RFI is considered to improve feed efficiency. A divergent selection experiment on RFI was conducted for 6 generations in a Large White population: the RFI+ line consumed more feed than predicted compared with the RFI- line. The study aims at measuring the performances of pigs fed a high fiber diet (244 g NDF – 8.36 MJ NE/kg) or a control diet (133 g NDF – 9.69 MJ NE/kg), to test the impact of the selection on the lines' ability to cope with a high fiber diet. Forty three castrated males were used for each line. Pigs were fed with an automatic feeder during the test. All pigs received the control diet on weeks 10 to 11 and 15 to 17. From weeks 12 to 14, half of the pigs in each line were submitted to a dietary challenge and received the high fiber diets. Body Weight (BW), average daily gain (ADG), DFI and feed/gain ratio were analysed using a linear model (GLM procedure, Minitab). BW at 10 weeks was used as a covariate. Before the challenge, DFI and ADG were 16% greater in RFI+ than RFI- pigs (for a similar adjusted BW of 25.3 kg). The first week of the challenge, pigs from both lines fed the high fiber diet had a 20% lower DFI than pigs receiving the control diet ($P_{Diet}<0.0001$). Consequences on ADG were more marked in RFI+ (970 and 541 g/d for control and high fiber diets respectively) than RFI- (934 and 696 g/d; $P_{Diet}<0.001$, $P_{Line\times Diet}=0.10$). At the end of the challenge, when compared with the control, the BW of pigs receiving the high fiber was 8% lower for the RFI+ line and 3% lower for the RFI- line ($P_{Diet}<0.01$). To conclude, pigs selected for lower RFI had a better ability to adapt to a high fiber diet.

Session 53 Theatre 3

Responses of adipose tissue to feed efficiency: effects of genetics, diet and feed restriction

F. Gondret[1], I. Louveau[1], A. Vincent[1], T. Le Naou[1], M. Jegou[1], J. Van Milgen[1] and H. Gilbert[2]
[1]INRA, UMR PEGASE, 35590 Saint-Gilles, France, [2]INRA, UMR LGC, 31326 Castanet-Tolosan, France; florence.gondret@rennes.inra.fr

How body fat content and energy metabolism may be associated with feed efficiency remain rather controversial. This study aims to decipher the responses of adipose tissue features to variations in feed efficiency by taking advantage of a divergent selection program on residual feed intake (RFI) in growing pigs. In two successive experiments, pigs (n=12-17 per line) divergently-selected for low (RFI-) or high (RFI+) RFI were fed *ad libitum* during the growing period. In the first experiment, an additional group of RFI+ pigs (n=14) were pair-fed to the daily feed intake of RFI- pigs. In the second experiment, pigs of both RFI lines (n=12 per group) were fed either a high-starch (LF) diet or a high-fat/high-fiber (HF) diet, both formulated for the same metabolizable energy content. In both experiments, RFI+ pigs had a lower gain to feed ratio (GF) than the RFI- pigs, whatever their diet. The RFI+ pigs were fatter than RFI- pigs when fed *ad libitum*, resulting in adipocyte hypertrophy in subcutaneous adipose tissue. Plasma concentrations in lipid-connected substances were different between lines. Feed restriction in RFI+ pigs did not affect GF, but reduced body fat content. Compared to diet LF, feeding diet HF tended to deteriorate GF to a similar extent in both RFI lines. It also reduced adiposity in both lines. This response might result from the lower daily feed intake of HF than LF pigs, and from the lower metabolic efficiency of using dietary fat for lipid deposition compared to carbohydrates. Transcriptomic analyses indicate various gene pathways associated with feed efficiency or feeding in adipose tissues. This study suggests that the relationship between adiposity and feed efficiency depends more on daily feed intake than on a strictly-related genetic reorientation of metabolic pathways in the less-efficient pigs. Financially-supported by ANR PIG_FEED and FatInteger projects.

Session 53 Theatre 4

Divergent selection on residual feed intake influences gene and protein expressions in pig muscle

I. Louveau[1], A. Vincent[1], F. Gondret[1], H. Gilbert[2] and L. Lefaucheur[1]
[1]INRA, UMR PEGASE, Domaine de la Prise, 35590 Saint-Gilles, France, [2]INRA, UMR LGC, Chemin de Borde Rouge, 31326 Castanet-Tolosan, France; isabelle.louveau@rennes.inra.fr

Residual feed intake (RFI) is an alternative criterion to feed-to-gain ratio to improve feed efficiency. Selection experiments on RFI have indicated that low RFI is associated with higher lean meat content in pigs. To gain insights into the molecular mechanisms underlying these differences, 24 Large White females from 2 lines divergently selected for RFI during more than 7 generations were examined. Pigs with low (RFI^-) and high RFI (RFI^+) were individually fed *ad libitum* from 27 kg to 115 kg body weight (n=8 per group). Additional RFI^+ pigs (RFI^{+R}, n=8) were pair-fed to the daily feed intake of RFI^- pigs to investigate the impact of feed intake independently of selection. Longissimus muscle samples were collected to evaluate gene and protein expressions. The analyses involved a 44K porcine commercial microarray and two-dimensional gel electrophoresis using colloidal blue staining to quantify abundance in each protein spot as the percentage of total spot volume within each gel. In this study, about one thousand probes were found to be differentially expressed ($P<0.01$) between RFI^- and RFI^+ pigs and 90% of those probes were not affected by feed restriction. Gene functional classification indicated a lower expression of genes involved in mRNA translation and a greater expression of genes associated with mitochondrial energy metabolism in RFI^+ compared with RFI^- pigs. At the protein level, 19 spots had a differential abundance ($P<0.05$) between RFI^- and RFI^+ pigs; a greater abundance in several mitochondrial proteins was observed in RFI^+ pigs. Only 20% of these spots were still differentially expressed between RFI^- and RFI^{+R} pigs. Altogether, both the transcriptomic and proteomic approaches suggest a higher mitochondrial energy metabolism in RFI^+ compared with RFI^- pigs.

Selection for residual feed intake in growing pigs: effect on sow performance in a tropical climate

D. Renaudeau[1], C. Anais[2], Y. Billon[3], J.L. Gourdine[1], J. Noblet[4] and H. Gilbert[5]
[1]INRA UR143, URZ, 97170 Petit Bourg, France, [2]INRA UE 1294, PTEA, 97170 Petit Bourg, France, [3]INRA UE 1372, GENESI, 17700 Surgères, France, [4]INRA UMR 1348, PEGASE, 35590 Rennes, France, [5]INRA UMR 144, LGC, 31326, Toulouse, France; david.renaudeau@antilles.inra.fr

Climatic factors are major limiting factors of sow performance in tropical areas. During lactation, impaired sow performance are related to a decline of their average daily feed intake (ADFI). The objective of this study was to evaluate the consequences of a divergent selection for residual feed intake (RFI) conducted in a temperate environment on Large White growing pigs on subsequent performance of lactating sows bred in a tropical environment. Twenty gilts from low (RFI-) and high RFI (RFI+) lines were imported to the tropical experimental facilities of INRA in the French West Indies. Data from 86 lactations (39 in warm season and 45 in hot season) were available. The daily temperature averaged 23.5 °C in warm and 25.7 °C in hot season. The ADFI was lower and maternal BW loss was higher in the RFI- than in the RFI+ line (4.06 vs. 4.67 kg/d and 1.25 vs. 0.78 kg/d, respectively; $P<0.05$). However, the high BW loss in RFI- line did not impair reproductive performance after weaning. The litter BW at weaning was higher in the RFI- than in RFI+ line (74.0 vs. 63.0 kg; $P<0.05$). Whatever the line, ADFI was reduced ($P<0.01$) and BW loss tended to increase ($P=0.07$) in the hot season (-0.94 kg/d and +0.25 kg/d, respectively). The litter BW gain was not affected by season ($P>0.05$). The season × line interaction was not significant for most of the traits. The sow feed efficiency in the RFI- line (sow ADFI/litter BW gain, kg/kg) did not change according to the season whereas it declined during the hot season in the RFI+ line. This is interpreted as an inability of the RFI+ sow to mobilize its body reserves for compensating their reduced ADFI during the hot season. In conclusion, selecting for low RFI during growth could help improving sow performance under tropical climate.

Impact of technological treatment of feed ingredients on feed efficiency in farm animals

P. Bikker, A.F.B. Van Der Poel, T.G. Hulshof, S. Salazar-Vilanea, E.M.A.M. Bruininx and G. Van Duinkerken
Wageningen UR, Livestock Research, P.O. Box 65, 8200 AM, the Netherlands; gert.vanduinkerken@wur.nl

The increasing demand for food, feed and (bio)fuel emphasises the need for efficient use of feed ingredients. An increasing proportion of feed ingredients is derived as by-product from the food industry and biofuel production. The technological processes are optimised for the production of the primary product, but these will also influence the (variation in) nutritive value of the by-products used for animal feed, either directly by an effect on nutrient availability or indirectly by an effect on endogenous losses in the intestinal tract. A precise estimate of the nutritive value of the feed ingredients need to be included in the feed optimisation to assure that the diets meets the animals requirements. However, present feed evaluation systems not always reflect possible influences on the nutritive value, e.g. protein quality. For example, we observed a large variation in the reactive (RL) to total lysine (TL) ratio in a survey of processed feed ingredients, whereas these ingredients are included on the basis of the same (ileal)digestible TL. The resulting imbalance in amino acids may cause a lower feed efficiency and/or a loss in other amino acids. Thus, new feed characteristics need to be developed and implemented to adequately describe the nutritive value of (processed) feed ingredients. Moreover, ingredient and diet processing should better take into account the effects on the nutritional value. At present, compound feed production is largely optimised to limit costs of production and improve handling properties. Limited information is available regarding the consequences for the nutritive value of ingredients and diets and results are conflicting because of the large variation in process conditions and diet composition. More insight in interactive effects of diet characteristics and feed processing, on digestive physiology and nutrient utilisation will add to a further improvement in nutrient utilisation.

Session 53 Theatre 7

Amino acid incorporation in feeds reduces the environmental impacts of pig production

F. Garcia-Launay[1], H.M.G. Van der Werf[2], T.T.H. Nguyen[2], L. Le Tutour[3] and J.Y. Dourmad[1]
[1]INRA-AgrocampusOuest, UMR1348 PEGASE, Domaine de la Prise, 35590 Saint-Gilles, France, [2]INRA-AgrocampusOuest, UMR1069 SAS, Rue de Saint-Brieuc, 35000 Rennes, France, [3]Ajinomoto Eurolysine SAS, 153 rue de Courcelles, 75817 Paris, France; florence.garcia-launay@rennes.inra.fr

Feed-use amino acids (AA) allow reducing, at constant performance, the protein content of pig feeds and nitrogen excretion by the animals. The aim was to assess the environmental impact of one kg of live pig produced in a conventional farm in Brittany by Life Cycle Assessment (LCA) according to several scenarios of AA incorporation. Two modalities of effluent management (slurry or solid manure) and two hypotheses of protein sources (soybean only, or soybean, rapeseed and pea) were considered. In three scenarios, feeds were least-cost formulated: no AA (NoAA), AA incorporation with protein content corresponding to current feeding practices (LowCP), and AA incorporation with free protein content (Min€). In the last scenario (MinCP), feeds were formulated to minimize protein content. Fattening pigs were fed either with only one feed (1P), two feeds (2P) or according to multiphase feeding (MP). The average protein content of pig feeds and the incorporation of soybean meal into feeds decreased with AA incorporation, down to 123 g/kg and 70 kg/t, respectively, in MP-MinCP. This reduction of soybean meal incorporation was concomitant with a decrease of feed cost. The incorporation of AA in low protein diets reduced the impacts of pig production on Climate Change (CC), Acidification (AC) and Eutrophication (EU). The lowest CC, AC and EU impacts were reached either with the Min€ or MinCP scenarios, for which tryptophane and valine were incorporated in pig feeds. The impacts on terrestrial ecotoxicity, cumulated energy demand and land occupation were barely affected by the studied scenarios. It was concluded that AA incorporation in least-cost formulated feeds substantially reduced the feeding cost and environmental impacts of pig production.

Session 53 Theatre 8

Impact of dietary energy content and feed level on the digestive efficiency in growing rabbit

C. Knudsen[1], S. Combes[1], C. Briens[2], J. Duperray[3], G. Rebours[4], J.-M. Salaun[5], A. Travel[6], D. Weissman[7] and T. Gidenne[1]
[1]INRA, UMR TANDEM, Castanet Tolosan, France, [2]CCPA, ZA du Bois de Teillay, Janzé, France, [3]In vivo NSA, Talhoüet, Vannes, France, [4]TECHNA, Route de St Etienne de Montluc, Coueron, France, [5]SANDERS, Centre d'affaires Odyssée, Bruz, France, [6]ITAVI, Centre INRA de Tours, Nouzilly, France, [7]INZO, Rue de l'Eglise, Chierry, France; christelle.knudsen@toulouse.inra.fr

Restricted feeding is commonly used in rabbit breeding to reduce mortality and morbidity but results in decreased growth and lower slaughter yield. A high energy (HE) diet offered at 75% of the *ad libitum* (AL) intake could prevent these detrimental effects. This work studies the consequences of such a feeding strategy on the digestive efficiency of growing rabbits. 48 animals were divided into four groups differing in dietary energy content (2,417 vs. 2,168 kcal DE/kg GM) and feeding level (AL or restricted at 75% (R)) according to a 2×2 experimental design. Animals were fed the experimental diets from weaning (35 d of age) to 74 d of age, and feed restriction was applied from 35 to 63 d. Digestive efficiency was assessed during feed restriction (42 to 46 d) and after one week of AL feeding (70 to 74 d). Feed restriction improved fecal digestibility of organic matter (OM, +3.8 pts), protein (+5.4 pts) and fiber (NDF +5.5 pts, ADF +5.9 pts; $P<0.001$). The digestibility of OM (+6.1 pts), protein (+4.7 pts) and fiber (NDF +7.2 pts, ADF +6.2 pts; $P<0.001$) was higher in HE animals. Interaction between DE content and feeding level was significant for OM and ADF digestibility: R animals fed the HE diet had better digestibility (+5.4 pts and +8.9 pts) compared to the HE AL fed animals. When returning to an AL feeding, no effect of the previous feeding level was observed while the effects of DE content on the digestibility of protein (+2.9 pts), OM (+4.3 pts, $P<0.001$) and NDF (+4.3 pts, $P<0.01$) were maintained. In conclusion restricted feeding of a high energy diet would be favorable to the digestive efficiency, thus growth and slaughter yield.

Using a mechanistic growth model to investigate energy efficiency and nitrogen excretion in pigs

M. Shirali[1,2], A. Doeschl-Wilson[3], P.W. Knap[4], C. Duthie[2], E. Kanis[1], J.A.M. Van Arendonk[1] and R. Roehe[2]
[1]Wageningen University, 6700 AH Wageningen, the Netherlands, [2]SRUC, Edinburgh, EH9 3JG, United Kingdom, [3]The Roslin Institute and R(D)SVS, University of Edinburgh, Roslin, EH25 9RG, United Kingdom, [4]PIC International Group, 24837 Schleswig, Germany; rainer.roehe@sruc.ac.uk

The aim of this study was to use a mechanistic growth model to investigate the effect of improving growth on residual energy intake (REI) and total nitrogen excretion (TNE) from 30 to 100 kg body weight (BW). An existing dynamic mechanistic pig growth model was extended to estimate individual's REI and TNE. The model was fitted, using a modified computational genetic algorithm, to longitudinal body protein and lipid measurements of 315 pigs. The fitting procedure produced distributions for the model input parameters that are representative of a current commercial pig population. Based on these input parameters, an in-silico population of 1000 pigs was generated and individual predictions for growth, body composition, REI and TNE at 100 kg were obtained. The effect of selection for average daily gain (ADG) at 100 kg BW was explored in this population by comparing model predictions for body composition, REI and TNE of pigs, whose ADG was at least 1 standard deviation below or above the average ADG at 100 kg BW. Comparison between the upper and lower ADG category showed significantly ($P<0.001$) greater average protein (0.159±0.001 vs. 0.108±0.001, kg/day) and lipid (0.244±0.002 vs. 0.167±0.002, kg/day) gain, and lower REI (473±15 vs. 706±15, MJ ME), TNE (3.06±0.04 vs. 3.86±0.04, kg), days of growth (71±1 vs. 105±1, days), Gompertz growth rate (0.0140±0.0001 vs. 0.0096±0.0001, kg/(kg × day)), and maximum protein deposition (169±1 vs. 116±1, kg/day). Thus, the mechanistic growth model indicates that a selection for ADG has a strong positive impact on energy and nitrogen efficiency. The growth model provided insight at what stages of growth energy and nitrogen efficiency changes most between fast and slow growing pigs.

Influence of synbiotic on broiler chicken performance and meat quality

I.H. Konosonoka, V. Krastina, A. Jemeljanovs, I.I. Vitina, A. Valdovska, V. Sterna and V. Strazdina
Research institute of Biotechnology and Veterinary Medicine Sigra, Instituta street No 1, 2150 Sigulda, Latvia; biolab.sigra@lis.lv

The objective of the present study was to evaluate the influence of the synbiotic (composition: prebiotic Jerusalem artichoke *Helianthus tuberosus* (JA) dry form 0.5% and probiotic *Lactobacillus reuteri* 10 mg 1.0×10^8 cfu/kg) on broiler growth performance, microflora of intestinal tract, carcass yield, and meat quality. 120 cross-breed ROSS 308 one-day-old broiler chickens were randomly allocated in the trial and divided in two groups of 60 birds each. The birds in the control group (CG) were fed with the commercial basic feed for cross ROSS 308 birds. Basic feed ration of the trial group (G1) chickens was supplemented with 0.5% JA in dry form and 10 mg *Lactobacillus reuteri* in concentration 1×10^8 cfu/kg. Records for live body weight and feed consumption were obtained weekly. At the 42nd day of the experiment, five chickens from each group were slaughtered, the pH of intestinal tract content was measured, content of ileum was bacteriologically tested for counts of lactic acid bacteria, and meat biochemical testing was performed. Results confirmed that average live body weight of G1 on the 42nd day of trial was significantly higher ($P<0.05$) than in control group. G1 group intestinal contents pH tended to decrease, but the number of lactic acid bacteria significantly ($P<0.05$) increased in comparison to the control group. Trial group broilers' meat quality was better: meat quality index was 21.35 compared to 20.70 for control group; the energy value of meat was 91.86 kcal/100 g compared to 101.90 kcal/100 g for control group; the cholesterol content 71.43 mg/100 g compared to 91.10 mg/100 g for control group broilers' meat. Results confirmed that broiler performance and meat quality is most favourably influenced by adding the basic feed synbiotic dried JA in 0.5% concentration in combination with 10 mg 1.0×10^8 cfu/kg *Lactobacillus reuteri*.

Effects of synbiotic on growth performance and gut health of piglets

A. Jemeljanovs[1], M. Pilmane[2], A. Valdovska[3], I. Zitare[1], I.H. Konosonoka[1] and I. Jansons[1]
[1]Research Institute of Biotechnology and Veterinary Medicine 'Sigra', Instituta street 1, 2150 Sigulda, Latvia, [2]Institute of Anatomy and Anthropology, Riga Stradins University, Dzirciema 16, 1007 Riga, Latvia, [3]Latvia University of Agriculture, Faculty of Veterinary Medicine, K. Helmana street 8, 3001 Jelgava, Latvia; anda.valdovska@llu.lv

The feeding trial was conducted to investigate the effect of dietary supplementation of synbiotic and probiotic Jerusalem artichoke on piglets production values, and gut microbiology and architecture. A total 60 weaning pigs (30 d of age; 9.09±0.12 kg of BW) were selected. The piglets in group 1 received a basal diet without any supplements, group 2 – basal diet with *Lactobacillus reuteri* (0.5 g/day/piglet), *Pediococcus pentosaceus* (0.5 g/day/piglet) and from the diet 3% of Jerusalem artichoke (JA) powder, but group 3 – basal diet only with 3% JA powder. Supplementation with probiotics and prebiotic improved average daily gain by 18.6% and feed conversion ratio by 12.6% compared with the control (basal) diet. Amount of microorganisms were reduced significantly compared to basal diet after 35 days supplementation with probiotic. In 2nd group, the amount of microorganisms of genera *Enterobacteriaceae* at the end of trial were by 6%, but *Escherichia coli* – by 9% decreased. Increase by 2% of *Enterobacteriaceae* and *E. coli* levels were seen only in group 1. Pathogenic microorganism *E. coli* O157 was found at the closing stage of the trial in group 3 piglets, but *S. enteritidis* – only in group 1. In groups 1 and 2 animals jejunum villi were slim and the small intestinal mucosa revealed no histopathological changes, but in group 3 – distinct degeneration process up to crypts, moderate inflammation process and plasmocytes were seen. Our results confirm that the synbiotic action of *L. reuteri* and *P. pentosaceus* with JA has the potential of inhibiting pathogenic microflora of gut and has positive effect on feed conversion ratio. Research was supported by ERAF project Nr. 2010/0226/2DP/2.1.1.1.0/10/APIA/ VIAA/099.

Technologies, resources and tools for the exploitation of sheep and goat genomes

B.P. Dalrymple[1], G. Tosser-Klopp[2], N. Cockett[3], A. Archibald[4], W. Zhang[5] and J. Kijas[1]
[1]CSIRO, 306 Carmody Road, St Lucia 4067, Australia, [2]INRA, 24 Chemin de Borde Rouge, Auzeville, 31326 Castanet Tolosan cedex, France, [3]Utah State University, Logan, Utah 84322, USA, [4]The Roslin Institute, Easter Bush, Midlothian EH25 9RG, United Kingdom, [5]Inner Mongolia Agricultural University, 306 Zhaowuda Rd, Hohhot 010018, China, P.R.; brian.dalrymple@csiro.au

For sheep, more than 32 million single nucleotide polymorphisms (SNPs) have been discovered and 7K, 50K and 600K SNP chips and parentage SNP sets are available. For goats ~14.5 million SNP have been discovered and a 50K SNP chip is available. Application of the different chips and targeted genome resequencing allows imputation of the genome sequence of all individuals in a breed or population. Good quality annotated draft reference genome assemblies are also available for both species. Surveys of gene expression across a wide range of tissues have contributed to the construction of high quality gene models facilitating detailed analysis of the genes underlying traits. However, the current genome assemblies are not adequate for future applications. The focus of SNP discovery has now turned to the identification of causative mutations underlying traits. The resequencing of key individuals from a diverse range of breeds of each species, and of their undomesticated relatives, is enabling the exploration of the full extent of diversity within each species. Much of this diversity will be in regions of copy number variation (CNV), however, the imputation of CNV status is unlikely to be as successful as SNP alleles. New tools and resources, including improved assemblies of the reference genomes and alternate assemblies in CNV regions, and a better understanding of the relationship between the genome and the expression of the genes across the different tissues, will provide the foundation for a much broader utilisation of the sheep and goat genomic resources by their respective industries in the future. This work was partially supported by the EU FP7 3SR project (no. 245140).

Session 54 Theatre 2

Exploring resistance to nematodes in 3SR sheep and goat populations

S.C. Bishop[1], V. Riggio[1], G. Sallé[2], N. Mandonnet[3], M.G. Usai[4], S. Casu[4], O. Keane[5], M.A. Poli[6], A. Carta[4] and C.R. Moreno[2]
[1]University of Edinburgh, The Roslin Institute, Midlothian, EH25 9RG, United Kingdom, [2]INRA, SAGA, 31326 Castanet-Tolosan, France, [3]INRA, URZ, 97170 Petit-Bourg, Guadeloupe, [4]AGRIS, SGB, Olmedo, 07040 Sassari, Italy, [5]Teagasc, Grange, Co. Meath, Ireland, [6]INTA, CICYyA CNIA, Buenos Aires, Argentina; stephen.bishop@roslin.ed.ac.uk

Gastrointestinal nematodes are a major constraint on sheep and goat production worldwide and, with the incidence of anthelmintic resistance growing, selection for nematode resistance is an increasingly attractive option. One of the aims of the EU-funded 3SR project was to mine genomic information relating to nematode resistance. Specifically, to genotype, with 50k SNP chips, sheep and goat populations phenotyped for nematode resistance, explore functional data describing host responses to nematode infections and determine SNPs that may be used to select for nematode resistance. Genome-wide association studies were performed in three sheep populations (Martinik Black-Belly × Romane (MBR), Sarda × Lacaune (SL) and Scottish Blackface (BF)) and are underway in Creole goat lines divergently selected for resistance. These studies have revealed a complex pattern of inheritance; each population contains a few QTL of moderate effect expressed against a largely polygenic background of resistance. Joint analyses, accounting for the differing population structures, revealed regions of OAR4, 12, 19 and 20 that were significant across the three populations. Results from the MBR and BF populations have been pursued through gene expression studies, including RNAseq, using tissues from deliberately infected lambs from both populations. Full genome sequence data are also available on key animals. Transferability of results across populations will be explored on several breeds (Manech Romane, Pampinta, Corriedale, Suffolk and Texel) using small chips with dense SNP coverage of QTL regions from the individual and joint population analyses. Results obtained with EC FP7 funding: Project 3SR-245140.

Session 54 Theatre 3

Genome-wide association analysis of resistance to gastro-intestinal parasites in dairy sheep

S. Casu[1], M.G. Usai[1], S. Sechi[1], M. Casula[1], G.B. Congiu[1], S. Miari[1], G. Mulas[1], S. Salaris[1], T. Sechi[1], A. Scala[2] and A. Carta[1]
[1]AGRIS-Sardegna, Settore Genetica e Biotecnologie, s.s. Sassari-Fertilia km 18.6, 07040 Olmedo, Italy, [2]Università di Sassari, Dipartimento di Medicina Veterinaria, Via Vienna 2, 07100 Sassari, Italy; scasu@agrisricerca.it

The aim of this study was to identify genomic regions affecting the resistance to gastro-intestinal parasites in dairy sheep by performing a genome-wide association analysis in an experimental population. The population consisted of 917 backcross Sarda × Lacaune ewes sired by 10 F1 rams and two generations of their descendants (1,497 ewes) that were sired by 33 Sarda rams. The whole population was genotyped with the Illumina 50K BeadChip. Ewe faeces were sampled between 1 and 3 times per year. Fecal egg count (FEC) was determined by floating the faeces in saturated salt solution in a McMaster slide and counting the eggs. FEC were log-transformed prior to further analysis (lnFEC = ln (Eggs Number + 14)). Yield deviations were analyzed by LD, LDLA and LA approaches. All models were based on a multiple regression of phenotypes on the probabilities of carrying the analyzed Sarda haplotypes. One location on OAR 20 was 5% genome-wise significant. Moreover, 1% chromosome–wise significant peaks were found on OAR 7 and 12. Further 10 regions were significant at the 5% chromosome-wise threshold. The most significant regions are being further investigated by whole genome re-sequencing trios of animals in which the QTLs were expected to be segregating. These results are obtained through the EC-funded FP7 Project 3SR-245140.

Session 54 Theatre 4

Loci underlying variation in nematode resistance in three sheep populations: a joint analysis

V. Riggio[1], R. Pong-Wong[1], G. Sallé[2], M.G. Usai[3], S. Casu[3], C. Moreno[2], O. Matika[1] and S.C. Bishop[1]
[1]The Roslin Institute and R(D)SVS, University of Edinburgh, Easter Bush, EH25 9RG Midlothian, United Kingdom, [2]INRA, SAGA, BP 27, 31326 Castanet-Tolosan, France, [3]SGB, AGRIS Sardegna, loc. Bonassai, 07040 Olmedo, Sassari, Italy; oswald.matika@roslin.ed.ac.uk

Gastrointestinal nematode infections are one of the main health/economic issues in the sheep industry. Indicator traits for resistance such as faecal egg count (FEC) are commonly used in genomic studies; however published results are inconsistent among breeds. Meta (or joint) analysis is a tool for aggregating information from multiple independent studies. The aim of this study was to identify loci underlying variation in FEC, as indicator of nematode resistance, in a joint analysis, using data from three populations (Scottish Blackface, Sarda × Lacaune, and Martinik Black-Belly × Romane), genotyped with the ovine 50k SNP chip. The trait analysed was the average animal effect for Strongyles FEC data (estimated by fitting a repeatability model across different time points for each population). Analyses were performed using two different software packages: Regional Heritability Mapping (RHM) and QTLMap, which make contrasting assumptions when identifying QTL. After editing, a total of 4,123 and 2,644 animals were available for RHM and QTLMap analyses respectively; after quality control 38,991 SNPs (common across breeds) were available for both analyses. RHM identified significant regions on OAR4, 19 and 20, with the latter being the most significant. The OAR20 region is close to the Major Histocompatibility Complex, which has often been proposed as a functional candidate for nematode resistance. This region was significant in only one of the three individual breeds. The most significant region identified from QTLMap was on OAR12. This result was confirmed by RHM, when using the reduced dataset edited for QTLMap. These regions will be investigated in other breeds to validate the results. These results are obtained through the EC-funded FP7 Project 3SR-245140.

Session 54 Theatre 5

Functional investigation of a QTL affecting resistance to *Haemonchus contortus* in sheep

G. Sallé[1,2], C.R. Moreno[2], J. Ruesche[2], M. Aletru[3], J.L. Weisbecker[3], F. Bouvier[4], F. Prévot[1,5], J.P. Bergeaud[1,5], C. Trumel[5], C. Grisez[1,5], D. François[2], A. Legarra[2], E. Liénard[1,5] and P. Jacquiet[1,5]
[1]INRA, Animal health, UMR1225 IHAP, 23 Ch. des Capelles, 31076, Toulouse Cedex 3, France, [2]INRA, Animal Genetics, UR631, SAGA, BP 27, 31326, Castanet-Tolosan, France, [3]INRA, Animal Genetics, Domaine de Langlade, 31450, Pompertuzat, France, [4]INRA, Animal Genetics, Domaine de La Sapinière, 18390, Osmoy, France, [5]Université de Toulouse, INP-ENVT, 23 Ch. des Capelles, 31076, Toulouse Cedex 3, France; guillaume.salle@tours.inra.fr

In previous studies using the ovine 50K SNP chip, a region of OAR12 was found to been associated with variation in fecal egg count (FEC) during *Haemonchus contortus* infection in both naive and primed back-cross (BC) lambs. Our study aimed at investigating the functional properties of this QTL region. Based on the linkage analysis results, BC sheep were selected for different haplotypes in a 20-Mbp-wide QTL region, and mated together to produce BC×BC progenies carrying two favorable or unfavorable QTL alleles. Subsequent to this marker-assisted mating, an association analysis pinpointed a 4-SNP haplotype. Based on the 4-SNP region, 61 BCxBC lambs were selected and experimentally infected with 10,000 *H. contortus* larvae. FEC were determined every three days from 18 days post-infection (dpi) and blood cell populations were determined at 0, 14 and 27 dpi. After one month, sheep were sacrificed to determine worm burden and worm fertility and to sample tissues for gene expression analysis. Significant differences in FEC and hematocrit drop were found. In addition, the female worms recovered from predicted resistant sheep were less fecund (250 eggs/female less). A gene expression analysis was performed between carriers of two 4-SNP alleles with opposite effects. A 4-fold over-expression was found for IL4 and IL13 genes in the abomasal mucosa of the sheep carrying the favorable QTL allele, indicating a higher Th2-biased cytokinic environment. Still, no functional candidate gene underlying the QTL region has been validated so far.

Genome-wide association analysis of resistance to paratuberculosis and mastitis in dairy sheep

S. Sechi[1], S. Casu[1], M. Casula[1], G.B. Congiu[1], S. Miari[1], G. Mulas[1], S. Salaris[1], T. Sechi[1], M.G. Usai[1], C. Ligios[2], G. Foucras[3] and A. Carta[1]
[1]AGRIS-Sardegna, Settore Genetica e Biotecnologie, SS Sassari-Fertilia km 18.6, 07040 Olmedo, Italy, [2]Istituto Zooprofilattico Sperimentale della Sardegna G. Pegreffi, Via Vienna 2, 07100 Sassari, Italy, [3]ENVT, UMR1225 IHAP INRA, 31076 Toulouse, France; acarta@agrisricerca.it

The aim of this study was to identify genomic regions that affect resistance to paratuberculosis (PTB) and mastitis in dairy sheep by using a genome-wide association analysis. The experimental population considered included 917 backcross Sarda × Lacaune ewes sired by 10 F1 rams and two generations of their descendants (1,497 ewes) that were sired by 33 Sarda rams. All animals were genotyped with the Illumina 50K BeadChip. Daily somatic cell count (SCC) was calculated as the arithmetic mean of a.m. and p.m. milking that were recorded bimonthly. Lactation Somatic Cell Score (LSCS) was calculated as the arithmetic mean of the log-transformed daily SCC. Blood samples of ewes were collected from 1 to 2 times per year and tested by ELISA for presence of antibodies against PTB. Two phenotypes were analyzed: the ewe's status (positive or negative) at 3 years of age, and the status when her entire productive life was considered. Yield deviations were analyzed by LD, LDLA and LA approaches by multiple regression of phenotypes on the probabilities of carrying the analyzed Sarda haplotypes. For LSCS, one region on OAR 4 and further 7 regions were 1% and 5% chromosome–wise significant respectively. For PTB, significant locations at 5% genome–wise threshold were found on OAR 14 and 20. Moreover, 1% chromosome–wise significant peaks were found on OAR 1, 6, 9, 10, 12, and 24. Further 8 regions were significant at the 5% chromosome-wise threshold. The most significant regions are being further investigated by whole genome re-sequencing of trios of animals in which the QTLs were expected to be segregating. These results are obtained through the EC-funded FP7 Project 3SR-245140.

Fine mapping of a QTL for mastitis resistance on OAR3 in Lacaune dairy sheep

R. Rupp[1], P. Senin[2], J. Sarry[3], O. Bouchez[4], G. Foucras[5] and G. Tosser-Klopp[3]
[1]INRA, U631, 31326, Castanet-Tolosan, France, [2]INRA, Sigenae, 31326, Castanet-Tolosan, France, [3]INRA-ENVT, UMR444, 31326 Castanet-Tolosan, France, [4]INRA, GeT-PlaGe, Genotoul, 31326 Castanet-Tolosan, France, [5]INRA-ENVT, UMR1225, 31326 Toulouse, Castanet-Tolosan, France; rachel.rupp@toulouse.inra.fr

A QTL controlling somatic cell score (SCS), as the trait pertaining to mastitis resistance, has been previously detected on OAR3 in an association study of 1013 AI rams from a Lacaune grand-daughter design using the Illumina Ovine SNP50 chip. Linkage and linkage disequilibrium analyses showed very close localization with narrow confidence intervals (<2 Mb) around 130 Mb. The QTL explained 5% of the variance of the analyzed trait (DYD for SCS). Here we report validation and fine mapping for this QTL. The QTL was confirmed in an independent population of 117 Lacaune rams. SCS EBVs of rams carrying the most favorable phases were significantly higher (+0.82 standard deviation) than for rams carrying the most unfavorable phases. For further fine mapping, full sequencing, with a coverage of 12X, was performed in one trio of individuals. The trio included a segregating sire (Qq), and two sons of extreme divergent phenotype suspected to be homozygous for alternative alleles of the QTL (QQ and qq). Among a total of 1543 SNPs found in a region of 0.5 MB, one SNP mapped to a coding region of a highly conserved functional candidate gene with a non-synonymous change in one amino acid. A KASPar™ test was implemented to genotype 614 Lacaune sheep in the discovery population for the potential causal mutation. The frequency of the mutation was 20.6% and highly significant correlation with SCS EBVs was confirmed. Altogether these results provided good evidence for the identification of a causal mutation controlling SCC in sheep milk. Further functional assays should reinforce the hypothesis and allow characterizing the effect of the mutation on the mastitis resistance trait. This work was funded through the EC-FP7 '3SR'project (no. 245140) with the support of the French 'Roquefort'in' project.

QTL detection for traits of interest for the dairy goat industry

C. Maroteau[1], I. Palhière[2], H. Larroque[2], V. Clément[3], G. Tosser-Klopp[4] and R. Rupp[2]
[1]UNCEIA, 149 rue de Bercy, 75012 Paris, France, [2]INRA, UR631, SAGA, 31326 Castanet-Tolosan, France, [3]Institut de l'élevage, 31326 Castanet-Tolosan, France, [4]INRA, UMR444, Génétique Cellulaire, 31326 Castanet-Tolosan, France; cyrielle.maroteau@unceia.fr

Since 2008, a program for mapping traits of interest in dairy goats has been carried out in France as part of the national 'PhénoFinlait' program and the EC FP7 funded '3SR' project (no. 245140). The project was based on a large daughter design of Alpine and Saanen bucks and data were gathered on production traits, mastitis resistance (SCC), type and fatty acid (FA) composition based on mid infrared spectra prediction. A total of 2,254 goats and 20 AI sires (11 Alpine, 9 Saanen) were genotyped with the 50K Illumina SNP goat beadchip. After classical quality control, a total of 49,647 out of 53,347 synthetized SNPs were validated for further analyses. Yield deviations were computed for 57 traits: milk, fat and protein contents and yields (FY, PY), SCC, 11 type traits and 38 FA-related traits. QTL detection based on linkage analyses (LA) and linkage disequilibrium (LD) was implemented using the QTLmap software. Bonferroni correction was applied to P-values in order to provide experience-wise significant thresholds. Fourteen regions controlling milk production traits, conformation and FAs were found with LA analyses on CHI 1, 6, 7, 8, 11, 14, 18, 19, 21, 29. LD analyses identified many more QTLs (480) and confirmed LA regions. There was evidence for QTLs with a major effect for PY and FY on CHI 6 (caseins region) and for FY and FAs on CHI 14 (DGAT1 region). Interestingly, a region of CHI 19 showed a QTL for SCC and also for milk and udder type traits. FA results seemed to be breed-specific as clusters of QTL were generally found on CHI 8, 11 and 14 in the Alpine but on CHI 18, 25 and 26 in the Saanen breed. This project receives financial support from Apis-Gène, UE 7th PCRDT, Ministry of Agriculture (CASDAR), FranceAgriMer and FGE.

Mapping a putative autosomal gene controlling ovulation rate and infertility in Cambridge sheep

O.M. Keane[1], J.P. Hanrahan[1], G. Tosser-Klopp[2], J. Sarry[2], S. Fabre[2], J. Demars[2] and L. Bodin[2]
[1]Teagasc, Animal & Bioscience, Grange, Dunsany, Co. Meath, Ireland, [2]INRA, SAGA, BP 52627, 31326 Castanet-Tolosan, France; orla.keane@teagasc.ie

Lambing percentage is a major driver of profitability in a sheep enterprise. The identification of a major gene controlling ovulation rate and hence litter size is a viable option to increase lambing percentage via genetic testing and selection. Cambridge sheep are characterised by having a high ovulation rate with extreme variation between individuals, consistent with segregation of major genes controlling ovulation rate. Ovarian hypoplasia with resultant female sterility is also found in the Cambridge breed, and polymorphisms in both GDF9 and BMP15 have been shown to be associated with increased ovulation rate in heterozygous carriers and sterility in homozygous carriers within breed. Recent data has provided evidence of a third major gene controlling ovulation rate in the Cambridge breed. The inheritance pattern suggests that this gene is autosomal and unlinked to GDF9 or BMP15. Mapping this gene was one of the goals of the EU-funded project 3SR, Sustainable Solutions for Small Ruminants. Sterile ewes with ovarian hypoplasia, where sterility cannot be explained by GDF9 or BMP15 polymorphisms, were identified along with their parents, and 26 animals were genotyped using the Illumina ovine 50 Beadchip. Homozygosity mapping was then used to identify regions of homozygosity in the unexplained sterile animals that were heterozygous in their carrier parents. Two regions of homozygosity were mapped to OAR 2 and OAR 8. The coding region of 2 candidate genes in the region of interest on OAR2, namely ACVR1 and ACVR1C were sequenced but no polymorphisms associated with sterility were identified. Whole genome sequencing of 5 animals is currently being performed in help refine the region of interest, and to fine map the major gene regulating ovulation rate. These results were obtained through the EC-funded FP7 Project 3SR-245140.

Novel BMP15 mutations responsible for an atypical hyperprolificacy phenotype in sheep

J. Demars[1], S. Fabre[1], J. Sarry[1], R. Rossetti[2], H. Gilbert[1], L. Persani[2], G. Tosser-Klopp[1], P. Mulsant[1], Z. Nowak[3], D. Drobik[3], E. Martyniuk[3] and L. Bodin[1]

[1]INRA, 24 Chemin de Borde Rouge, Auzeville CS 52627, 31326 Castanet-Tolosan, France, [2]University of Milan, Via Zucchi 18, 20100 Milano, Italy, [3]Faculty of Animal Science, 8, Ciszewskiego street, 02-786 Warsaw, Poland; julie.demars@toulouse.inra.fr

Major genes increasing litter size (LS) and ovulation rate (OR) were suspected in the French Grivette and the Polish Olkuska sheep populations, respectively. To identify the genetic variants responsible for the highly prolific phenotype in these two breeds, genome-wide association studies (GWAS) followed by complementary genetic and functional analyses were performed. Highly prolific ewes (cases) and normal prolific ewes (controls) from each breed were genotyped using the Illumina OvineSNP50 Genotyping Beadchip. In both populations, an X chromosome region, close to the BMP15 gene, harbored clusters of markers with suggestive evidence of association at significance levels between $1E^{-05}$ and $1E^{-07}$. The BMP15 candidate gene was then sequenced and 2 novel non-conservative mutations called $FecX^{Gr}$ and $FecX^{O}$ were identified in the Grivette and Olkuska breeds, respectively. The two mutations were associated with the highly prolific phenotype ($p_{FecX}{}^{Gr}=5.98E^{-06}$ and $p_{FecX}{}^{O}=2.55E^{-08}$). Homozygous ewes for the mutated allele showed a significantly increased prolificacy ($FecX^{Gr}/FecX^{Gr}$, LS=2.50±0.65 vs. $FecX^{+}/FecX^{Gr}$, LS=1.93±0.42, $P<1E^{-03}$ and $FecX^{O}/FecX^{O}$, OR=3.28±0.85 vs. $FecX^{+}/FecX^{O}$, OR=2.02±0.47, $P<1E^{-03}$). Both mutations are located in very well conserved protein coding motifs and would alter the *in vitro* BMP15 signaling activity. Thus, we have identified 2 novel mutations in the BMP15 gene associated with increased LS and OR. Noteworthy, homozygous $FecX^{Gr}/FecX^{Gr}$ Grivette and homozygous $FecX^{O}/FecX^{O}$ Olkuska ewes are hyperprolific in striking contrast with the sterility exhibited by all other known homozygous BMP15 mutations. These results were obtained through the EC fnded FP7 project 3SR-245140.

Genome re-sequencing in sheep used to detect signatures of selection

J. Kijas[1], Y. Jiang[1], K. Worley[2], H. Daetwyler[3], R. Brauning[4], J. McEwan[4], B. Dalrymple[1], Y. Li[1], M. Heaton[5] and International Sheep Genomics Consortium[6]

[1]CSIRO, Animal, Food and Health Science, QBP, St Lucia, 4067, Brisbane, Qld, Australia, [2]Baylor College of Medicine, Human Genome Sequencing Center, One Baylor Plaza, 77030, Houston, TX, USA, [3]Department of Primary Industries, 5 Ring Rd, 3083, Bundoora, Vic, Australia, [4]AgResearch, Invermay Research Center, Mosgiel, 9053, New Zealand, [5]USDA, ARS, Meat Animal Research Center, P.O. Box 166, State Spur 18D, 68933, Clay Center, NE, USA, [6]ISGC, www.sheephapmap.org, Australia; james.kijas@csiro.au

Patterns of genetic variation are highly informative for understanding the diversity and evolutionary history of domestic animal species. Within this context, the International Sheep Genomics Consortium sequenced 73 sheep to a depth of 10-fold coverage. Following variant calling to identify approximately 20 million high confidence SNP, three methods were applied to search for regions that have undergone accelerated change in response to domestication and/or the establishment of sheep breeds with divergent phenotypic characteristics. These were: (1) detection of runs of homozygosity within individual genomes; (2) searching for regions of heterozygote deficiency within a set of 68 domestic sheep genomes; and (3) a polymorphism-divergence test that compares within, to between, species variability rates (HKA test) using 5 out-group wild sheep genomes. Known pigmentation genes MC1R, MITF, and ASIP were identified as outliers using one or more of the tests, confirming that the experimental strategy successfully detected selection events. The spectrum of variability within each gene will be presented, along with predictions concerning their functional consequence. Evidence was obtained to suggest developmental genes were selected during the domestication process, particularly those involved in controlling aspects of skeletal morphology. These results advance our understanding of the genetic history of this important livestock species.

Session 54 Theatre 12

Do selective sweeps in sheep breeds indicate the genomic sites of breed characteristics?

G.E. Pollott
RVC, Royal College Street, London NW1 0TU, United Kingdom; gpollott@rvc.ac.uk

Most breeds are the result of recent selection for phenotypic characteristics which make that breed unique, often referred to as a selective sweep in natural populations. This process should result in long runs of homozygosity (ROH) around the genomic polymorphisms which determine a breed's key characteristics, particularly if their mode of inheritance is autosomal recessive. Since recombination in each generation has the potential to breakdown ROH, the length of a ROH is indicative of its age; more recent mutations being associated with longer runs. The 67 breeds in the Sheep HapMap dataset, with >20 genotyped animals, were analysed to identify the distribution of ROH within each breed. A ROH was identified when an individual animal had consecutive homozygous SNP genotypes of the majority type, within the breed, at each locus. Missing genotypes were scored as if they were homozygous. Thus a ROH score represents the length, in numbers of SNP, of a continuous run of homozygosity comprising the majority homozygous genotype for the breed at each SNP. The probability of any given score, within a breed, was calculated by 1000 permutations of the whole HapMap dataset drawing 20 animals at random at each permutation. Over 30 breeds were found to have long ROH (>50 SNP; P<0.05). For example, the various Texel subgroups were found to have a long ROH on OAR2 between bp positions 116,277,389 and 127,499,743. This contains the Myostatin gene and a mRNA previously identified as being a characteristic of the breed. Interestingly, the Soay breed, a primitive breed from Scotland, was shown to have a long ROH in the same position as that found on OAR2 in the Texel, but comprised a different haplotype. A range of other ROH of interest will be discussed and suggestive sites for breed characteristics presented. This method provides the possibility to identify genomic regions which determine breed characteristics when no candidate gene is suggested by previous studies.

Session 54 Theatre 13

Genomic selection in the multi-breed French dairy goat population

C. Carillier[1], H. Larroque[1], I. Palhière[1], V. Clément[2], R. Rupp[1] and C. Robert-Granié[1]
[1]Institut National de la Recherche Agronomique (INRA), Station d'amélioration génétique des animaux (UR631), CS 52627, 31326 Castanet-Tolosan Cedex, France, [2]Institut de l'élevage (Idele), CS 52627, 31326 Castanet Tolosan Cedex, France; celine.carillier@toulouse.inra.fr

In French dairy goats, 2,246 females and 872 males from Alpine and Saanen breeds were recently genotyped with the Illumina 50K SNP bead chip, as part of a large genomic project supported by the dairy goat industry stakeholders. The first goal of this study was to investigate linkage disequilibrium (LD) within the population and between the two breeds. The second objective was to examine the effect of adding males, females or males and females in the reference population on the ranking and accuracy of genomic breeding values for young bucks. The level of LD in the multi-breed population (0.14 for 50 kb) was lower than the one found in each breed (0.17) or in literature for cattle (0.18 to 0.30). In addition, the persistence of LD phases between the two breeds decreased rapidly with distance (0.56 for 50 kb). Conventional and genomic evaluations using GBLUP for milk production traits, somatic cell score and type traits were calculated in several multi-breed reference populations (from 67 males to 677 males and 1,985 females). The ranking of animals based on EBV and GEBV were close, with correlations between EBV and GEBV of up to 97%. Rankings were improved by adding animals, males or females only for some traits. Accuracies of genomic predictions were low (from -5% to 38%) because of the small size of the reference population analyzed. For young bucks, average difference between genomic or conventional breeding value accuracies were lower than those reported for other species. Altogether, this first genomic study in Alpine and Saanen goats suggest that the current data is not sufficienct to allow genomic selection to be performed. Other models such as a multiple-trait model, single step genomic BLUP model or models using haplotypes instead of SNP will also be examined in the future.

Detection of QTL influencing somatic cell score in Churra sheep employing the OvineSNP50 BeadChip

B. Gutiérrez-Gil, E. García-Gámez, A. Suárez-Vega and J.J. Arranz
Universidad de León, Dpto Producción Animal, Facultad de Veterinaria, Universidad de Leon, 24071 Leon, Spain; beatriz.gutierrez@unileon.es

Subclinical mastitis is a major problem for the dairy sheep industry. Somatic cell score (SCS) for milk is generally considered as a good indicator of this complex disease. A previous genome scan performed in a commercial population of Spanish Churra sheep based on the analysis of microsatellite markers, identified a single significant QTL influencing SCS on sheep chromosome (OAR) 20. In the present study we performed a higher density genome-wide analysis in a new commercial population of the same breed using the OvineSNP50 BeadChip. A total of 1696 animals belonging to 16 half-sib families were analysed in this study. Yield Deviations (YD) were considered as the dependent variables in the QTL detection analysis. YD's were calculated, from the raw phenotypic data, as deviations from the population mean and corrected for environmental effects. After a quality control of genotypes, QTL detection was performed using two approaches, based on Linkage Analysis (LA) and combined linkage and linkage disequilibrium analysis (LDLA). Significance thresholds were estimated through permutations and simulations for LA and LDLA respectively. The LA results showed two chromosome-wise significant QTL on ovine chromosomes OAR5 and 25, and one genome-wise significant QTL on OAR20. Segregating families for each of these QTL were identified based on the corresponding within-family analyses. Several chromosome-wise and nine genome-wide significant QTL, on OAR1, 2, 3, 13, 17, 18, 19, 20 and 25, were also identified through LDLA. A preliminary list of positional candidate genes located within the confidence intervals of the most promising QTL regions has been obtained. Additional analyses will be required to help better understand the genetic architecture of these genetic effects. These results were obtained through the EC-funded FP7 Project 3SR-245140 and the Spanish National Project AGL2009-07000.

Estimation of LD and haplotype block sizes in European sheep populations

M.G. Usai[1], S. Casu[1], C. Moreno[2], R. Rupp[2], G. Salle[2], E. Garcia-Gamez[3], J.J. Arranz[3], V. Riggio[4], S.C. Bishop[4], N. Cockett[5] and A. Carta[1]
[1]AGRIS Sardegna, Settore Genetica e Biotecnologie, 07040 Olmedo, Italy, [2]INRA, UR631, SAGA, BP 27, 31326 Castanet-Tolosan, France, [3]Universidad de Leon, Produccion Animal, 24071 Leon, Spain, [4]The Roslin Institute and R(D)SVS, University of Edinburgh, Midlothian EH25 9RG, United Kingdom, [5]Utah State University, ADVS, Logan, UT 84322-4815, USA; gmusai@agrisricerca.it

Animals from the Sarda (SAW); Lacaune (LAC); Churra (CHU); Scottish Blackface (SBF); Martinique Blackbelly (MBB) and Romane (RMN) breeds were genotyped with the Illumina 50K BeadChip. The resulting data were used: to evaluate the LD decay for increasing distances between SNP; to analyse the LD pattern along the genome and to identify haplotype blocks and their size. LD was measured by r and r^2 statistics for pairs of SNPs from 0 to 1 Mb apart. The average r^2 was calculated for distances between SNPs increasing by 10 Kb steps. The LD pattern along each chromosome was calculated by averaging r^2 in sliding windows of 1 Mb which overlapped 0.5 Mb. The correlations between r at common SNP pairs were calculated to study the level of haplotype sharing between SAW, LAC, SBF and CHU. Correlations between r were calculated both for the whole genome and sliding windows. Haplotype blocks (HB) were estimated using the |D'| based method. No relevant differences in LD decay were observed between breeds, with the highest difference being 0.06 between MBB and CHU. The LD pattern and HB analysis gave similar results. Several sites with an excess of LD were identified, with the most relevant excess LD found on OAR2 and OAR10. In these two sites high r correlations between the four major breeds (SAW, LAC, SBF and CHU) were observed. This result suggests that there is a strong similarity of conserved haplotypes among breeds. These results are obtained by EC-funded FP7 Project 3SR (no. 245140); French SNP data were funded by SHEEPSNPQTL ANR project.

Whole genome association study for reproductive seasonality trait in the Rasa Aragonesa sheep breed

A. Martinez-Royo[1], J.L. Alabart[1], J. Folch[1], B. Lahoz[1], E. Fantova[2] and J.H. Calvo[1,3]
[1]CITA, Unidad de Tecnología en Producción Animal, Avda Montañana 930, 50059 Spain, [2]Oviaragón, Grupo Pastores, Camino Cogullada s/n, 50014 Spain, [3]Fundación ARAID, C/ María de Luna 11, 1ª, 50018 Spain; amartinezroyo@aragon.es

Many sheep breeds from the Mediterranean area have seasonal patterns of oestrous behaviour and ovulation. Maximal reproductive activities occur from August to March. This reproductive seasonality induces great variation in lamb production and, therefore, in the market price of lamb meat. This spring ovulatory activity is under genetic control, yet to date only a small proportion of the total variation has been explained by genes identified through linkage analysis and candidate gene association studies. To address the lack of information at the genomic level we present a results from a preliminary whole genome association study that was performed using data on reproductive seasonality for 141 ewes of the Rasa Aragonesa Spanish sheep breed that were genotyped using the OvineSNP50K BeadChip platform of Illumina®. Biannual registration data of seasonality was measured in an experimental flock as total days of anestrus (TDA) based on weekly blood progesterone levels from January to August, and corrected for age and body condition score. TDA was defined as the sum of periods in which three or more consecutive samplings having plasmatic progesterone levels lower than 0.5 ng/ml. Results have pointed to new regions across the genome not previously described that are potentially associated with natural cyclicity in this breed. Use of a greater number of samples and next generation high density BeadChips could help to accurately identify loci that are involved in controling the seasonality of sheep. Financed by Ministerio de Economía y Competitividad and FEDER (INNPACT Project IPT-010000-2010-33) and INIA (B. Lahoz grant).

Session 54 Poster 17

Towards a genomic monitoring of intra-breed sheep variability

L. François[1,2,3,4], G. Leroy[1,5], G. Baloche[6], J. Raoul[2,3], C. Danchin-Burge[2,3] and D. Laloë[1,5]
[1]AgroParis Tech, UMR1313 GABI, 16 Rue Claude Bernard, 75231 Paris, France, [2]IDELE, Animal Genetics, BP 42118, 31321 Castanet-Tolosan, France, [3]IDELE, Animal Genetics, 149 Rue de Bercy, 75595 Paris, France, [4]BOKU, Institut für Nutztierwissenschaften, Gregor Mendel Straße 33, 1180 Vienna, Austria, [5]INRA, UMR1313 GABI, CRJ, 78352 Jouy-en-Josas, France, [6]INRA, SAGA, BP 52627, 31326 Castanet-Tolosan, France; francois.liesbeth@gmail.com

Limiting the number of breeding stock used can result in the loss of genetic variability in most breeds, and may restrict the capacity to improve and adapt breeds to future needs. The objective of the VARUME project (genetic Variability of RUMinants and Equine species) was to quantify the genetic variability in French ruminant and equine species, based both on pedigree and molecular data. The selection program of the four main dairy sheep breeds in France (Lacaune, Manech Tête rousse, Manech Tête noire and Basco-Béarnaise) is based on the selection of a few elite breeding animals. To assess the variability in genomic information, males were genotyped using the SNP50 chip. The genetic variability in these four breeds will be determined by calculating the effective population size based on linkage disequilibrium over multiple generations. As the effective population size may depend on parameters, data source and methods used, a comparison will be made with the effective population size based on pedigree information, the differences will be interpreted considering the size of the sample and structure of the population. This information can be used to observe the evolution of genetic variability in the breeds and in turn improve the breed's capability for adaptation.

Effects of genetic markers on milk traits in Romanian sheep breed

M.A. Gras, G. Pistol, C. Lazar, R. Pelmus, H. Grosu and E. Ghita
National Institute for Research and Development for Biology and Animal Nutrition, Laboratory of Animal Biology, Calea Bucuresti No. 1, Balotesti, 077015 Ilfov, Romania; gras_mihai@yahoo.com

Substantial attention was attributed to the genetic structure of a native sheep populations and to the possible relationships between genetic variants of milk protein genes and milk related traits. Amongst the main genes associated with milk production traits (milk, fat and protein yield and content), β-lactoglobulin (LGB), κ-casein (CSN3) and prolactin (PRL) plays important roles. Our study aims to find molecular markers linked to milk production, for increasing selection accuracy in Teleorman Black Head breed. After DNA extraction, PCR-RFLP was used to identify genetic polymorphisms for LGB, CSN3 and PRL genes. The allele frequency and genotype effects were estimated. After LGB gene amplification, three different genotypes, AA, AB and BB were detected. The allelic frequency was 68% for A allele and 32% for B allele. Genotyping of the CSN3 gene lead to identification of two PCR products with different lengths, T and C allele, respectively. Only a single CT heterozygote genotype was identified. Digestion of the PRL amplicon differentiated alleles A and B. Identified genotypes of PRL gene where AA, AB and BB. The frequency was 53% for allele A and 47% for B allele. Analyzing the marker effect on milk production traits, our study showed that effect of LGB is quite constant. Genotype AA perform better then genotypes BB. Casein shows a null effect for all individuals. On the other hand variation in PRL is assciated with small differences. For milk, fat and protein yield, the AA genotype for PRL had a smaller positive impact than variation in LGB. For fat and protein content, PRL variation show the reversed effect, being negative for AA genotype and positive for BB genotype. In conclusion, allelic variation in PRL and LGB genes showed a strong association with milk production, indicating that these polymorphisms could potentially be used as a DNA marker for milk yield in Teleorman Black Head population.

Analysis of pedigree and marker based inbreeding coefficients in German Fleckvieh

D. Hinrichs and G. Thaller
Institute of Animal Breeding and Husbandry, Christian-Albrechts-University, Olshausenstrasse 40, 24098 Kiel, Germany; dhinrichs@tierzucht.uni-kiel.de

In Germany commercial milk production is dominated by two dairy cattle breeds and one of these breeds is German Fleckvieh. The aim of the present study was the estimation of different pedigree based inbreeding coefficients, i.e. classical- and ancestral-inbreeding coefficients, and to analyze the relationship between these inbreeding measurements. The base year for the analysis of the inbreeding coefficients was 1950 and the pedigree file includes 19,184 animals descending from 4,314 different sires and 16,724 different dams. The next step was the estimation of three different SNP-marker based inbreeding coefficients. Therefore a data set was constructed including 3,323 bulls with a call rate above 90%. All bulls were geno-typed with the Illumina 50K chip and 35,957 SNPs with a minor allele frequency of at least 5% were used for the estimation of genomic inbreeding coefficients. Finally, the relationship between pedigree based inbreeding coefficients and SNP-marker-based inbreeding coefficients was analyzed. From the 19,184 animals in the pedigree 3,529 were inbreed with an average classical inbreeding coefficient of 0.02. Ancestral inbreeding was found in 1150 animals and the mean ancestral inbreeding coefficient was 0.01. The correlation between classical and ancestral inbreeding was 0.58. The mean of the three different genomic inbreeding coefficients was very similar (approximately -0.005) and the correlations between the genomic inbreeding coefficients vary between -0.68 and 0.57. Correlations between classical inbreeding measurements and genomic inbreeding measurements ranged from 0.09 to 0.45 and were lower for ancestral inbreeding and genomic inbreeding, where correlations were estimated in the interval between 0.03 and 0.26. All in all, this study showed that further research is needed for a better understanding of the relationship between different concepts of inbreeding.

High overlap of CNVs and selection signatures revealed by varLD analyses of taurine and zebu cattle

A.M. Pérez O'brien[1], G. Meszaros[1], Y.T. Utsunomiya[2], J.F. Garcia[2], C.P. Van Tassell[3], T.S. Sonstegard[3] and J. Sölkner[1]
[1]BOKU, Gregor-mendel-strasse 33, 1180 Vienna, Austria, [2]UNESP, Rua Clóvis Pestana 793, 16050-680 Aracatuba, Brazil, [3]ARS-USDA, 10300 Baltimore Av, 20705 Beltsville, USA; anita_op@students.boku.ac.at

Selection Signatures (SS) assessed through analysis of genomic data are being widely studied to discover population specific regions selected via artificial or natural selection. Different methodologies have been proposed for these analyses, each having specific limitations as to the age of the selection process aimed to discover and the genomic parameters behind the methods used, being able to assess population, breed or species-wise SS. On a different side of the work on genomic characterization of the genome, interest has re-emerged on the identification of Copy Number Variants (CNVs) and other types of variations, showing that a considerable amount of the variation in the genome can be explained by this type of genomic variants. We identified putative regions of divergent selection between Indicine and Taurine cattle, of Dairy and Beef type, through the varLD methodology, analyzing differential regional Linkage Disequilibrium (LD) variation across populations, compared to population specific background LD levels. Subsequently, the identified regions were compared to recently reported CNVs, to find possible candidate regions where the selection process has been acting on advantageous CNVs. Four cattle breeds, Angus, Brown Swiss, Nellore and Gyr (30 to 100 individuals per breed) were included in the analyses, genotyped with the Illumina Bovine HD Beadchip. Our results show a large overlap of regions discovered through the SS analyses and reported CNVs. Of the 164 varLD regions covering 0.35% of the autosomal genome, 30 (18.3%) overlapped with CNV regions reported in another study, covering 2.1% of the genome. Additionally, some of the regions found in common, have been previously identified by other types of studies (GWAS, SS) as having productive or adaptive importance in different bovine breeds.

Fine-mapping of a chromosomal region on BTA17 associated with milk-fat composition

S.I. Duchemin[1,2], M.H.P.W. Visker[1], J.A.M. Van Arendonk[1] and H. Bovenhuis[1]
[1]Wageningen University, Animal Breeding and Genomics Centre, P.O. 338, 6700 AH Wageningen, the Netherlands, [2]Swedish University of Agricultural Sciences, Animal Breeding and Genetics, P.O. 7023, 750 07 Uppsala, Sweden; sandrine.duchemin@wur.nl

It is well established that milk-fat composition in dairy cattle shows genetic variation and is influenced by genes such as DGAT1 located on BTA14 and SCD1 on BTA26. In addition, a genomic region on BTA17 has been found to be associated with milk fatty acids (FA), however, no candidate gene or causal variant has been identified so far. Based on the 50k SNP array, we previously identified 10 significantly associated SNPs distributed over 3.7 Mbp on BTA17. The aim of this study was to fine-map this region using the 777k SNP array. FA were determined based on winter and summer milk samples of 2,001 cows on 398 herds. Phenotypes were available on 14 FA (saturated C4:0 through C18:0, and unsaturated C10:1 through C18:1-cis9, trans11 (CLA)). 50k SNP genotypes were available on 1,813 daughters and 55 sires. 777k SNP genotypes were available on the same 55 sires. Daughters were imputed from 50k to 777k SNP genotypes using Beagle. Imputation was based on 777k genotypes of an independent set of 1,330 animals, in addition to the 777k genotypes available for the sires. After imputation the number of SNPs on BTA17 increased from 1,570 to 22,240. Single SNP analysis was done with an animal model in ASReml. Our results indicated significant association of C6:0, C8:0, C10:0 and C12:0 with 29 SNPs distributed over 2.8 Mbp. The minor allele frequency of the SNP with the strongest association was 0.44 and its association with C6:0, C8:0, C10:0, and C12:0 was in the same direction, both in winter and in summer milk. However, the effects in summer milk were up to twice as large as in winter milk. Adjusting for DGAT1 K232A and SCD1 A293V polymorphisms did not change the effects, which suggests that this gene on BTA17 acts independently of these previously identified genes affecting FA composition.

Session 55 Theatre 4

Imputation of non-genotyped individuals based on genotyped relatives: a real case scenario

A.C. Bouwman[1], J.M. Hickey[2], M.P.L. Calus[1] and R.F. Veerkamp[1]
[1]Animal Breeding and Genomics Centre, P.O. Box 135, 6700 AC Wageningen, the Netherlands, [2]International Maize and Wheat Improvement Center (CIMMYT), Apdo. 06600, Mexico D.F., Mexico; aniek.bouwman@wur.nl

Imputing genotypes for non-genotyped individuals is attractive because it enables inclusion of historic datasets with valuable phenotypes (e.g. feed intake) to a training set, and it might help to reduce genotyping cost of breeding programs. The objective of this study was to see if, and how accurate non-genotyped individuals can be imputed from genotyped relatives. This study was based on a real dataset for feed intake of dairy cows with 1,021 cows phenotyped and genotyped; 1,344 cows were only phenotyped and thus needed to be imputed; and 3,076 relatives with genotypes only. Genotypes were simulated for all individuals in the pedigree. Subsequently genotypes were set to missing in different scenarios: the real situation, adding sire and maternal grandsire information, and adding information from 1, 2 or 4 offspring. AlphaImpute was used to impute missing genotypes based on pedigree information. Accuracy of imputation was assessed per individual using correlations between true and imputed genotype dosage, both corrected for mean gene content. As expected, imputation accuracy increased when more close relatives were genotyped. Most interesting is the increase in accuracy without phasing from 0.59 (0 offspring) to 0.73, 0.82 and 0.92 by adding 1, 2, and 4 genotyped offspring, respectively. With genotyped offspring, imputation accuracy appeared to be higher than the expected accuracy based on selection index theory. This is because the imputation method can make use of correlations between markers due to linkage and linkage disequilibrium. In these situations a two-step approach, where imputed genotypes are used in further analyses, will therefore give better results than an one-step approach using for instance a H-matrix. In conclusion, imputation of non-genotyped individuals was possible with acceptable accuracy when multiple offspring were genotyped.

Session 55 Theatre 5

Required increase in training set to keep accuracy of genomic selection constant across generations

M. Pszczola[1,2,3], T. Strabel[1], R.F. Veerkamp[2,3], H.A. Mulder[2], J.A.M. Van Arendonk[2] and M.P.L. Calus[3]
[1]Poznan University of Life Sciences, Department of Genetics and Animal Breeding, Wolynska 33, 60-637 Poznan, Poland, [2]Wageningen University, Animal Breeding and Genomics Centre, Wageningen, 6700 AH, the Netherlands, [3]Wageningen UR Livestock Research, Animal Breeding and Genomics Centre, 8200 AB Lelystad, the Netherlands; mbee@jay.up.poznan.pl

The accuracy of genomic predictions, among other factors, depends on the relationship of selection candidates with the reference population (RP). Part of the observed drop in accuracy per generation comes from the decay of LD, but most likely the largest part is attributable to the decay of the relationships between selection candidates and the RP. As a consequence, the accuracy of genomic predictions can only be kept at a constant level across generations, by adding animals' phenotypes and genotypes to the RP every generation. The number of animals that have to be added to the RP per generation is, however, unknown. Therefore, the objective of this study was to investigate how many animals per generation need to be added to the RP to keep the accuracy at a constant level across generations. Considering decrease of relationships across generations, the drop in accuracy of direct genomic values, without adding animals from new generations to RP, is expected to be 1/(number of generations between selection candidate and RP)^2, as compared to the first generation. For instance, the accuracy in the second generation is only 25% of the accuracy in the first generation. Simulations will be used to determine the number of animals per generation that need to be added to the RP such that prediction accuracies are constant across generations.

GENIFER: fine mapping and effects of QTL affecting fertility in Holstein cattle

R. Lefebvre[1], S. Fritz[2], D. Ledoux[3,4], J. Gatien[5], L. Genestout[6], M.N. Rossignol[6], B. Grimard[3,4], D. Boichard[1], P. Humblot[7] and C. Ponsart[5]
[1]INRA, Domaine de Vilvert, 78350 Jouy en Josas, France, [2]UNCEIA, 149 rue de Bercy, 75595 Paris Cedex 12, France, [3]ENVA, 7 av du général de Gaulle, 94704 Maisons-Alfort Cedex, France, [4]INRA, Domaine de Vilvert, 78350 Jouy-en-Josas, France, [5]UNCEIA, 13 rue Jouet, 94704 Maisons-Alfort, France, [6]LABOGENA, Domaine de Vilvert, 78350 Jouy-en-Josas, France, [7]SLU, Department of Clinical Sciences, 750-07 Uppsala, Sweden; rachel.lefebvre@jouy.inra.fr

The GENIFER project was built to confirm some fertility QTL, fine mapping and precise effects thanks to monitoring events between 0 and 90 days after first insemination (AI). Phenotyping involved 4,559 Holstein cows born from 12 sires and located in 1,028 farms. Combining progesterone assays at day 0 and 21, pregnancy diagnosis at day 40 and 90, and subsequent calving information lead to determination of the time of pregnancy failure at first AI and to the following diagnostics: inappropriate time of AI, no fertilization or early embryonic mortality, late embryonic mortality, total embryonic mortality, fetal mortality, abortion or lack of calving. Genetic analysis included 2,669 females with clear phenotype and genotyped for 353 SNP chosen in 16 regions of 13 chromosomes (1-6, 9, 10, 14, 15, 18, 26, 27). These regions were selected based on previous QTL mapping results from a large granddaughter design. QTL were detected by association analysis with Fasta method and GenABEL package of R software. The main QTL targeted in this study, located on chromosome 3, was confirmed and its location was refined around 24 cM. Its maximum effect appeared on late embryonic mortality, whereas an effect on early mortality and abortions was not excluded. 2 to 8 QTL were confirmed (P<0.01) for each trait: 6 for calving rate, 3 for no fertilization or early mortality, 3 for late embryonic mortality, 8 for total embryonic mortality, 2 for fetal mortality and 3 for abortion.

Session 55 Theatre 7

Genome wide association study for calving performance in Irish cattle

D. Purfield[1,2], D.G. Bradley[2], J.F. Kearney[3] and D.P. Berry[1]
[1]Teagasc Animal & Grassland Research and Innovation Centre, Moorepark, Fermoy, Co. Cork, Ireland, [2]Smurfit Institute of Genetics, Trinity College Dublin, Dublin 2, Ireland, [3]Irish Cattle Breeding Federation, Bandon, Co. Cork, Ireland; purfield@tcd.ie

Dystocia (CD) and perinatal mortality (PM) are complex quantitative traits that are known to exhibit genetic variation. There is evidence from monitoring of genomic evaluations that there may be a major gene effect for calving dystocia segregating in the Irish population. The aim of this study was to conduct a genome wide association study to identify genomic regions associated with calving traits in Irish dairy cattle. Genotypic and phenotypic data was available on 1,970 Holstein-Friesian sires. Sires were genotyped using the Illumina BovineSNP 50K Beadchip comprising 54,001 Single Nucleotide Polymorphisms (SNPs). SNP edits were applied, including removing sires with genotype call rates <90%, SNPs with >0.5% mendelian inconsistencies, call rates <95%, monomorphic SNPs and SNPs that deviated from Hardy Weinberg equilibrium ($P<1\times10^{-9}$). After edits a total of 43,204 SNPs remained. Missing genotypes were imputed. Deregressed predicted transmitting abilities (PTAs) and their reliabilities were available for all sires. Bayesian genomic selection methods were applied on only sires with >40% adjusted reliability, where parental contribution was removed for each trait (1,970 sires for CD and 740 for PM). Bayes Cπ was used to for the estimation of π, the probability of a marker having zero effect. The Bayes B algorithm was run using the posterior mean π from the Bayes C analysis. The posterior mean π for CD was 0.98 and for PM 0.99. Several different chromosomal regions were associated with CD and PM. The Bayesian posterior probability was greatest for regions associated with CD, with SNPs on BTAs 12, 18, and 23 yielding the highest posterior probabilities. The probability of SNPs being associated with PM was much lower than that for CD; however BTAs 1, 2, and 25 displayed genomic regions of the highest posterior probabilities.

Genetic and non-genetics effects on female reproductive performance in seasonal-calving dairy herds

M.M. Kelleher[1,2], F. Buckley[2], R.D. Evans[3], D. Ryan[4], K. Pierce[1] and D.P. Berry[2]
[1]University College Dublin, School of Agriculture, Food Science & Veterinary Medicine, Belfield, Dublin 4, Ireland, [2]Teagasc, Animal & Grassland Research and Innovation Centre, Moorepark, Fermoy, Co. Cork, Ireland, [3]3Irish Cattle Breeding Federation, Bandon, Co. Cork, Ireland, [4]Reprodoc LTD, Fermoy, Co. Cork, Ireland; margaret.kelleher@teagasc.ie

Excellent reproductive performance is paramount to the profitability and efficiency of seasonal-calving dairy herds. However optimum reproductive targets are currently not being realised in Ireland. Furthermore, most genetic research on reproductive performance in dairy cattle has focused primarily on lactating cows and relatively few studies have attempted to quantify the genetic contribution to reproductive performance in nulliparous heifers. The objective of this study was to estimate additive genetic, non-additive genetic and permanent environmental components for a range of fertility traits in nulliparous, primiparous, and pluriparous seasonal calving dairy cattle. Fertility information including calving dates, as well as information on services, pregnancy diagnoses, slaughter and animal movements (including death) was available for 143,447 cows after editing, from the Irish Cattle Breeding Federation between the years 2006 to 2012 inclusive. Variance components were estimated using (repeatability where appropriate) animal models. Repeatability estimates for the fertility traits (0.05 to 0.11) suggest a significant contribution of permanent environmental effects to phenotypic differences among cows. Coupled with the significant heterosis effects of up to 6% of the phenotypic mean for some fertility traits, and up to 2% of the variation attributable to recombination loss effects, suggests that non-additive genetic and permanent environmental in addition to additive genetic effects, could be accounted for in the development of an index for ranking cows more aligned to their expected phenotypic performance.

Genetic parameters for pre-weaning traits in Charolais × Montbéliard and Charolais × Holstein calves

A. Vallée[1,2], J.A.M. Van Arendonk[2] and H. Bovenhuis[2]
[1]Gènes Diffusion, 3595 route de Tournai, 59500 Douai, France, [2]Wageningen University, Animal Breeding and Genomics Centre, P.O. Box 338, 6700 AH Wageningen, the Netherlands; amelie.vallee@wur.nl

Charolais sires can be mated to Montbéliard or Holstein dairy cows to produce crossbred calves sold for meat production. Heritabilities and correlations between traits can differ when they are calculated within Charolais × Montbéliard or within Charolais × Holstein. Moreover, a trait measured on Charolais × Montbéliard and on Charolais × Holstein is not necessarily genetically identical. First objective of this study was to estimate heritability and genetic correlations between traits within each population. Second objective was to investigate if traits were genetically identical between populations. Traits studied were calving difficulty, birth weight, height, bone thinness, and muscular development. Data included 22,852 Charolais × Montbéliard and 16,012 Charolais × Holstein calves from 391 Charolais sires. Heritabilities estimated separately within each population were similar. Stronger genetic correlations were observed in Charolais × Holstein compared with Charolais × Montbéliard between calving difficulty and height (0.67 vs. 0.54), calving difficulty and bone thinness (0.42 vs. 0.27), birth weight and bone thinness (0.52 vs. 0.20), and between birth weight and muscular development (0.41 vs. 0.18). Bivariate analysis considering observations on Charolais × Montbéliard and on Charolais × Holstein as different traits showed that genetic variances and heritabilities were similar for all traits except for height. Birth weight and muscular development were genetically identical traits in both populations, with genetic correlations of 0.96 and 0.99. Genetic correlation was 0.91 for calving difficulty, 0.70 for bone thinness and 0.80 for height and these genetic correlations were significantly different from 1. Results suggest different breeding values of Charolais sires for calving difficulty, bone thinness and height when mated to Montbéliard or Holstein cows.

Session 55 Theatre 10

Predicting lifespan of dairy cows: phenotypic and genetic change during life

M.L. Van Pelt[1,2], R.F. Veerkamp[2] and G. De Jong[1]
[1]CRV BV, P.O. Box 454, 6800 AL Arnhem, the Netherlands, [2]Wageningen UR Livestock Research, Animal Breeding and Genomics Centre, P.O. Box 65, 8200 AB Lelystad, the Netherlands; mathijs.vanpelt@wur.nl

Longevity of dairy cattle is an important trait from an economic and welfare perspective. Reducing involuntary culling will improve farm profit and animal welfare. Longevity or lifespan is the number of days between first calving and last milk recording date, but is only known after an animal is culled. Methods for genetic evaluations need to include censored data. Currently these methods assume that longevity is genetically the same trait during the total lifespan of a cow. However, the expectation is that culling rate differs during life and within lactation. The aim of this study was to investigate this expectation and estimate the genetic variation of the total lifespan of a cow. For this study the Dutch national dataset was used. Survival rates per parity and per month within parity grouped were estimated, and also the importance of fixed effects and the change over the past two decades. Genetic analyses were performed on subsets of the complete data with random regression or spline models to estimate genetic covariances over the total lifespan and for each month within parity. Results show that survival rates are phenotypically not equal during the total lifespan. In the first year after first calving monthly survival rate is around 99%, whereas five years after first calving monthly survival rate declined to 95% or lower. Also, in first parity monthly survival rate in the first eight months is constant, followed by a decline in later lactation. Survival rates have changed over years. For example, survival rates of 12, 24, 36, 48 and 60 months after first calving increased from 83 to 91, 65 to 77, 50 to 60, 34 to 43 and 21 to 28% respectively in the last two decades for Holstein cows. During these years the cumulative distribution of culling has changed more towards a Gaussian cumulative distribution.

Session 55 Theatre 11

Genetic parameters for major milk proteins in three French dairy cattle breeds

M. Brochard[1], M.P. Sanchez[2], A. Govignon-Gion[2], M. Ferrand[1], M. Gelé[1], D. Pourchet[3], G. Miranda[2], P. Martin[2] and D. Boichard[2]
[1]Idele, 149 rue de Bercy, 75012 Paris, France, [2]INRA, UMR1313 GABI, 78350 Jouy en Josas, France, [3]ECEL, Doubs et Territoire de Belfort, 25640 Roulans, France; mickael.brochard@idele.fr

Genetic parameters of the major milk protein contents were estimated in the three main French dairy cattle breeds i.e. Montbéliarde, Normande and Holstein in the framework of the PhénoFinLait programme. Protein composition was estimated from Mid-Infrared (MIR) spectrometry on 266,508 test-day milk samples from 57,477 cows in first lactation. Lactation means, expressed in percentage of milk or protein, were analyzed with an animal mixed model including fixed environmental effects (test-day × herd, month × year of calving and spectrometer) and a random genetic effect. Genetic parameter estimates were very consistent across breeds. In milk or in protein, heritability estimates (h^2) were moderate to high for αs1, αs2, β and κ-caseins and for α-lactalbumin ($0.21<h^2<0.58$). In each population, β-lactoglobulin was the most heritable trait ($0.57<h^2<0.75$). Genetic coefficients of variation ranged from 1 to 10%. Genetic correlations (r) were very sensitive to the expression unit. Protein fractions were generally in opposition when they were expressed in protein, except between whey proteins and αs2-casein ($0.39<r<0.60$) and between β and κ-caseins ($0.55<r<0.64$). In milk, r estimates were close to zero or positive, with highest r values found between different caseins ($0.45<r<0.92$). In the three populations, β-lactoglobulin was positively correlated with αs1-casein ($0.21<r<0.42$) and αs2-casein ($0.46<r<0.62$) whereas the correlations were close to zero with β and κ-casein ($-0.15<r<0.21$). These results, obtained from a large panel of cows, show that routinely collected MIR could be used to modify milk protein composition by selection. However, antagonisms between proteins will have to be considered. This program receives financial support from ANR, Apis-Gène, Ministry of Agriculture (CASDAR), CNIEL, FranceAgriMer and FGE.

Genetic correlations among fighting ability, fertility, and productive traits in Valdostana cattle

C. Sartori, S. Mazza, N. Guzzo and R. Mantovani
University of Padua, Dept. of Agronomy, Food, Natural Resources, Animals and Environment, Viale dell'Universitá 16, 35020 Legnaro (PD), Italy; roberto.mantovani@unipd.it

An empirical selection for fighting ability (FIGH) in traditional competitions was carried out for centuries in Valdostana cattle, giving cows a masculine phenotype. Moving from the recent addition of FIGH to dual-purpose selection, this study aimed at investigating implications of selecting for FIGH by analyzing genetic correlations (r) with milk yield (MY), morphology and fertility, and trends of breeding values (EBVs) in recent years. About 34,000 data of 10,700 cows (21,200 animals in pedigree) from 12 years of fights were joined to annual scoring of primiparous cows for type traits and whole lactation MY. Calving interval (CI) and parity to conception period (PC) were joined to data as indicators of fertility. Genetic parameters were estimated via EM-REML in bi-trait linear analyses. Genetic correlations of FIGH resulted moderate but positive with front muscularity (FM) and thorax depth (TD; average r=0.12), and negative with udder traits and thinness (r=-0.20), indicating that FIGH is related to an enhanced masculine aspect in cows. FIGH and MY traits showed a negative genetic correlation too (r=-0.21), as FIGH with CI and PC (r=-0.33) suggesting that selecting for masculine traits as fighting ability may increase dominance, but also depresses relevant female attributes as milk and fertility. EBVs for FIGH showed a moderate but positive trend, revealing that also without planned breeding schemes mean population for FIGH is changing. A slight but positive trend in FM and TD EBVs was observed too, as a reduction in overall udder score. However, a genetic selection mainly focused on MY led to an increment in milk, fat and protein EBVs, and had no negative effects on fertility. Therefore, a depression in female attitude due to the enhancement of masculine traits has not yet realized, but the measured negative genetic relationships warned about possible further negative implications of a selection for fighting ability.

Longevity and reasons of culling of German Holstein Friesian under Libyan conditions

S.A.M. Abdalah Bozrayda[1], F.H. Alshakmak[2] and R.S. Gargum[3]
[1]University of Benghazi Faculty of Agriculture, Animal Production, Benghazi sulug Libya, 21861, The State of Libya, [2]University of Benghazi faculty of Science and Arts, zoology Department, Sulug Libya, 21861, The State of Libya, [3]University of Benghazi faculty of Science, zoology Department, Benghazi Libya, 21861, The State of Libya; sbozrayda@yahoo.com

Factors affecting true herd life THL, productive life PL, number of lactations NL and 305-day milk yield of German Holstein Friesian cows were investigated using 2,196 first lactation records. Imported pregnant cows in 1986 were considered as first generation whereas subsequent generations born in Libya were identified through pedigrees. The mixed model included level of of milk production, origin of sire, generation, age at first calving, year and month of calving as fixed and sires as random effects. Days open was included in the model as a covariate. Level of production was the most important factor for survival of the cow. Daughters of Libyan sires as compared with those of North American and European sires had shorter true herd and productive life, less number of lactations and lower 305-day milk yield. After the 3rd generation, true herd life and productive life were decreased due to using the local sires. 305-day milk yield showed fluctuation through generation. Younger age at first calving cows showed lower longevity compared with medium age cows. Heritability estimates of THL, PL, NL and 305-day milk yield were low (0.064, 0.072, 0.056 and 0.055) respectively. Very high genetic and phenotypic correlations (0.941-0.993) were found between longevity traits. Genetically, 305-day milk yield was moderately correlated (0.350-0.420) with longevity traits. Reasons of culling were mainly due to low fertility (37%), Mastitis and udder injures (15%)and accidental defects (16%). This study showed that longevity traits depends on level of milk production but at expense of fertility traits. In addition longevity traits were negatively affected by inbreeding.

Session 55 Theatre 14

Effects of inbreeding on milk production, fertility, and somatic cell count in Norwegian Red

K. Hov Martinsen[1], E. Sehested[2] and B. Heringstad[1,2]

[1]Department of Animal and Aquacultural Sciences, Norwegian University of Life Sciences, P.O. Box 5003, 1432 Ås, Norway, [2]Geno Breeding and A.I. Association, IHA, Ås, Norway; bjorg.heringstad@umb.no

Unfavorable effects of inbreeding have been reported for many dairy populations, but have so far not been studied for Norwegian Red. The aim was to estimate the effects of inbreeding on milk production, fertility, and somatic cell count (SCC) in Norwegian Red. Data from relatively large herds with mainly Norwegian Red cows (>95%) and high use of AI (>95%) over the last 10 years were used in this study. This was to ensure complete pedigree many generations back for each animal. The average inbreeding coefficient was 2.5%, and there were very few animal with high inbreeding coefficients, about 7% had inbreeding coefficient >5. Effects of inbreeding were calculated for 305 days lactation yield for kg milk, kg fat, and kg protein, fat- and protein percentage, lactation mean somatic cell score (LSCS), non-return within 56 days after first insemination for heifers, first- and second/third lactation cows, and interval from calving to first insemination for first and second/third lactation cows. For all traits the effect of inbreeding was best explained by a linear regression model. Inbreeding had significant unfavorable effect on milk yield traits and a small favorable effect on SCC. The regression coefficients showed that the 305 day lactation yield will be reduced by 34.2 kg milk, 1.28 kg fat, and 1.15 kg protein per 1% increase in inbreeding. A small reduction in LSCS, equivalent to about 1000 cells lower lactation mean SCC per 1% increased inbreeding, was also found. These results imply that an offspring of half-sibs will produce 427.5 kg less milk, 16 kg less fat, and 14.5 kg less protein in a 305 day lactation and have 12,500 cells lower lactation mean SCC than a non-inbred cow, under otherwise equal conditions. No significant effects of inbreeding were found for any of the fertility traits or for fat- and protein percentage.

Session 56a Theatre 1

Detection of emerging vector-borne diseases in the Dutch surveillance system

G. Van Schaik, H. Brouwer and A. Veldhuis

Animal Health Service, R&D Epidemiology, Arnsbergstraat 7, 7400 EZ Deventer, the Netherlands; g.v.schaik@gddeventer.com

In the Netherlands, a surveillance system is in place with three major objectives: (1) early detection of known exotic diseases; (2) early detection of new or emerging diseases or syndromes; and (3) description of trends and developments in ruminant health. It is organised by a private organisation, GD Animal Health Service, and is financed by both public and private stakeholders. The Netherlands encountered two recent outbreaks of emerging vector-borne diseases in ruminants, i.e. Bluetongue serotype 8 (BTV8) in 2006 with a re-emergence in 2007 and Schmallenbergvirus in 2011. Several passive and active instruments can be used for detection of emerging diseases, such as syndromic surveillance, surveys, risk analysis, and monitoring of vectors. The aim of the presentation is to describe the relation between the epidemiological components developed to achieve the second objective, early detection of emerging (vector-borne) diseases. An important component of the Dutch system is a telephone service that provides free advice to veterinarians and farmers about animal health related problems that they encounter. The signals that are obtained from the field are discussed weekly in a multidisciplinary team of veterinarians, epidemiologists and pathologists. In addition, active syndromic surveillance on production data, and (sentinel) surveys are carried out. The epidemiological tools are compared on timeliness of detection, feasibility, efforts and costs. Timeliness of detection of emerging vector-borne diseases depends on the transmission rate and the impact of the infection, i.e. the quicker it spreads and the higher the impact on production, the quicker the infection is detected by active syndromic surveillance on production data. For re-emergence of BTV8, a risk-based survey seemed most sensitive to detect the infection. Finally, epidemiological tools that may assist in improving the sensitivity of the surveillance system for emerging vector-borne diseases in general are discussed.

Session 56a Theatre 2

Use of monthly collected milk yields for the early detection of vector-borne emerging diseases

A. Madouasse[1], A. Lehébel[1], A. Marceau[1], H. Brouwer[2], Y. Van Der Stede[3] and C. Fourichon[1]
[1]LUNAM Université, Oniris, INRA, UMR BioEpAR, CS 40706, 44307 Nantes, France, [2]GD Animal Health Service Ltd., P.O. Box 9, 7400 AA Deventer, the Netherlands, [3]Coda-Cerva, Groeselenberg, 99 1180 Brussels, Belgium; aurelien.madouasse@oniris-nantes.fr

In 2006 and 2011 two vector borne diseases have emerged in the European ruminant population: bluetongue (BTV) and Schmallenberg. Milk yields collected as part of milk recording represent an abundant source of information that could be used for emerging disease surveillance. The aim of this work was to evaluate monthly collected milk yields as an indicator of vector borne disease emergence. Milk recording data were used to detect the 2007 BTV emergence in France. Expected test-day milk productions per cow were predicted for 2006 and 2007 from linear mixed models, based on 3 years of herd milk records history. Clusters of deviation between predicted and observed milk productions were detected using a scan statistic as implemented in SaTScan™. Log likelihood ratios (LLR) were used to rank the clusters and to set a threshold for the definition of an alarm. The choice of a threshold was a trade-off between the number of alarms before the emergence (false alarms) and the timeliness of the detection once the disease had emerged. Using an LLR of 50 (100), there were an average of 1.7 (0.8) false alarms per week and the BTV emergence was detected 7 (9) weeks after the first notification. The first cluster with an LLR>100 located in the emergence area was further investigated. A difference between observed and predicted production of greater than 1 kg/cow/day was observed around the time of emergence. However, a difference of equal magnitude was observed during the year preceding the outbreak. We conclude that milk production predicted from herd history alone does not allow the detection of an emerging infectious disease that would have had an effect on milk production similar to BTV.

Session 56a Theatre 3

Use of milk production data to improve early detection of vector borne diseases

H. Brouwer[1], A. Veldhuis[1], A. Madouasse[2] and G. Van Schaik[1]
[1]Animal Health Service, Epidemiology, P.O. Box 9, 7400 AA, Deventer, the Netherlands, [2]LUNAM Université, Oniris, UMR BioEpAR, Epidemiologie, CS 40706, 44307 Nantes, France; g.v.schaik@gddeventer.com

Traditional surveillance for detection of disease outbreaks are triggered by specific symptoms, whereas syndromic surveillance focuses on deviations of non-specific disease indicators such as drop in milk production. Milk production data on herd and test-day level are continuously available for 80-85% of Dutch dairy herds. Real time analyses of milk production data may allow detection of disease outbreaks before any confirmation by laboratory diagnosis and therefore can accelerate the detection of emerging diseases. The objective of this study was to evaluate to what extent syndromic surveillance on milk production data can contribute to early detection of vector borne diseases in the Dutch cattle population. Data were used from January 1st 2003 to March 31st 2012 and evaluated for the detection of bluetongue (re) emergence in 2006 and 2007 (1st analysis) and the Schmallenberg outbreak in 2011 (2nd analysis). Therefore, the methodology described by Madouasse *et al.* was applied on Dutch data. For the first analysis, a multilevel linear regression model was fitted on 2005 (in which no epidemics occurred) and mean milk production was predicted for 2006-2007. For the second analysis, a multilevel linear regression model was fitted on 2009-2010 and mean milk production was predicted for 2011 and 2012. The differences between observed and predicted milk production for the two models were plotted and analyzed with SaTScan to detect spatio-temporal clusters of decreased milk production. The results showed that no significant drop in milk production was detected during the 2006 and 2007 bluetongue outbreak, whereas a significant drop in milk production was detected at the start of the Schmallenbergvirus outbreak in the eastern part of the Netherlands. This study shows that real time analysis of milk production data can trigger an early signal in a surveillance program.

Surveillance of emerging diseases in cattle based on reproduction data

C. Fourichon[1], A. Marceau[1], T. Lesuffleur[1], A. Madouasse[1], A. Lehebel[1], Y. Van Der Stede[2] and G. Van Schaik[3]
[1]INRA, Oniris, Lunam University, UMR1300 Biology, Epidemiology and Risk Analysis in animal health, Atlanpole La Chantrerie, CS 40706, 44307 Nantes Cedex 3, France, [2]CODA-CERVA, Brussels, Belgium, [3]Animal Health Service, Deventer, the Netherlands; christine.fourichon@oniris-nantes.fr

The recent emergence of Bluetongue virus serotype 8 and Schmallenberg disease in European cattle illustrates the increasing risks of disease emergence due to global changes. To early detect emergence of unforeseen diseases in cattle, syndromic surveillance based on data collected routinely, used as non specific indicators of health status is of interest. The objective of this study was to evaluate if and how reproduction data in dairy cattle can be used for surveillance. The bluetongue outbreak in France in 2007 was used as a case model. Five indicators targeting different effects of disease on reproduction were defined. The capacity of statistical models for temporal series to early detect bluetongue emergence was evaluated in districts with different levels of prevalence of bluetongue. The criteria for evaluation were: ability to detect a significant increase in the district during the bluetongue outbreak, precocity of the detection (in comparison to notification of clinical cases), number of false alarms (i.e. detection of an increase in the indicator in a time-period without bluetongue). Four indicators based on frequency of return-to-service and duration of gestation showed a detectable increase during the outbreak. The occurrence of early calvings was the most sensitive. The time between the detection based on reproduction data and the first case notification varied from -15 to 60 days depending on the threshold to trigger an alarm. The proposed surveillance method based on reproduction data was effective to early detect of bluetongue, with a limited number of false alarms and is of potential interest for surveillance of diseases with an effect on reproduction.

Risk-analysis for the reintroduction of BTV-8 through cattle imports

I. Santman[1], G. van Schaik[1] and J.D. Groot[2]
[1]GD Deventer, 7400 AA Deventer, the Netherlands, [2]VanDrie group, 3050 Ermelo, the Netherlands; i.santman@gddeventer.com

On 15 February 2012, Germany, the Netherlands and Belgium regained their BTV-8 free status. After this date, the question arose what the probability was of reintroduction of BTV-8 through cattle imports. An epidemiological tool that can be used to asses this probability is a risk analysis, in which the probability (and its uncertainty) of reintroduction of BTV-8 can be quantified. This was done for the Netherlands, using a stochastic Risk model in @Risk. The risk of reintroduction of BTV-8 was quantified for three different cattle industries: (1) white veal; (2) rose veal; and (3) other cattle. In the model the probability on re-introduction of BTV-8 through cattle imports per year was based on the probability that one imported cow was BTV-8 virus positive multiplied by the number of imported cattle per year. Where the probability of being BTV-8 virus positive at the moment of import depended on: (1) the BTV-8 status in the country of origin; (2) the probability that vertical transmission took place; (3) the probability that a calf received enough colostrum; (4) the age at import; (5) the probability that horizontal transmission took place; (6) the infectious period. To quantify each parameter available data and literature were used. When no data were available, an expert group made assumptions, which were evaluated using sensitivity analyses. Furthermore, 5 scenario's including expanding the risk towards all BTV types were added to the model. With the assumptions included, the probability to import one BTV-8 viremic calf in the white veal industry was once per 5.6 (2.6-25) year, was once per 4.5 (2.1-20) year in the rose veal industry and was 1.7 (0.3-3.7) cow per year in the other cattle industries. The scenario's showed that when a new BTV outbreak occurs, the risk of introduction through cattle imports may increase, depending on the country of emergence. Using a risk analyses provided us more insight on the probability for a reintroduction of BTV-8 through cattle imports.

Session 56b Theatre 1

Why using epidemiological models to evaluate control strategies for livestock infectious diseases?

P. Ezanno[1], E. Vergu[2], F. Beaudeau[1], A. Courcoul[3], C. Marcé[1], B.L. Dutta[1,2], N. Go[1,2] and C. Belloc[1]
[1]Oniris-INRA UMR BioEpAR, CS 40706, 44307 Nantes, France, [2]INRA MIA, Allée de Vilvert, 78350 Jouy en Josas, France, [3]ANSES, Av Général Leclerc, 94701 Maisons Alfort, France; pauline.ezanno@oniris-nantes.fr

Modelling is a pertinent approach: (1) to better understand and to predict pathogen spread in host populations according to the biological system characteristics under various management scenarios; and (2) to evaluate the epidemiological and the economic effectiveness of control strategies. To end with useful models, a back-and-forth between models and biological data is needed. First, building epidemiological models consists in proposing from all of the up-to-date available knowledge an integrated conceptual view of the system. This highlights which processes are well known vs. which are still of the biological assumption type. Second, observed data can be used to estimate observable parameters (such as disease-related mortality rates and production losses), whereas epidemiological models can be used to estimate unobservable parameters (such as transmission rates). Sensitivity analysis is a powerful tool to identify parameters with a major influence on model outputs, these parameters need to be precisely informed. Third, data can be used to evaluate / validate models, which in turn can help to identify potential control points of the biological system, to compare scenarios and test biological assumptions, and even (when the model has been evaluated) to predict future states of the system according to past (known) states. We illustrate such interactions between observations and models in the context of livestock infectious disease spread and control, with examples as Q fever, paratuberculosis, bovine viral diarrhea in cattle, and *Salmonella* carriage and the PRRS in pigs. Focus is made on the multi-scale modeling (from the within-host immune response to the infection dynamics at a regional scale), and the coupling of epidemiological and economic models to account for farmer decisions in evaluating collective control options.

Session 56b Theatre 2

Development of a simple model for the control of gastrointestinal strongylosis in cattle herds

N. Ravinet[1,2], R. Vermesse[3] and A. Chauvin[1,4]
[1]LUNAM Université, Oniris, UMR1300BioEpAR, CS 40706, 44307 Nantes, France, [2]IDELE, CS 40706, 44307 Nantes, France, [3]GDS Bretagne, CS 110, 56003 Vannes, France, [4]INRA, UMR1300BioEpAR, CS 40706, 44307 Nantes, France; alain.chauvin@oniris-nantes.fr

Control of gastrointestinal strongylosis in heifers is often based on systematic treatments at standard periods, without any adaptation to the specific parasitic risk for the animals in each herd. This parasitic risk depends on the immunological status against gastrointestinal nematodes (GIN), and on the pasture infectivity level (PIL) (GIN larval challenge). A model based on these 2 key factors has been developed to evaluate in field conditions parasitic risk periods. The development of immunity is modeled using data regarding heifers' grazing and treatment history: the time of effective contact (TEC, in months) with GIN larvae is calculated, heifers being considered as non-immune when TEC<8 months. The increasing PIL during the grazing season results from the succession of parasite cycles. In the model, this increase is evaluated calculating the numbers parasitic cycles realized since turn out, the duration of one cycle being the addition of the prepatent period and the development time from egg to infective larvae. As this development time depends on temperature, it is calculated using a previously model developed by Smith *et al.* based on daily average temperature. By taking into account the grazing management (one or several pastures, time spent on each pasture), a parasitic risk is estimated by the model when non-immune heifers are grazing a pasture where the PIL generated by the number of parasitic cycles is considered to be high enough to represent a danger (e.g. to cause weight gain losses or clinical signs). As long as such heifers remain on such a pasture, the risk period continues, and different therapeutic or agronomic control strategies can then be tested and recommended. This model is a useful tool on the field to optimize the use of anthelmintics in the control of GIN through targeted treatment.

Modeling the spread of BVDV in a beef cow-calf herd to evaluate the efficiency of vaccine strategy

A. Damman[1,2], A.-F. Viet[1,2], M.-C. Guerrier-Chatellet[3], E. Petit[3] and P. Ezanno[1,2]
[1]L'Université Nantes Angers Le Mans, Oniris, UMR BioEpAR, 44307 Nantes, France, [2]INRA, UMR1300 BioEpAR, BP 40706, 44307 Nantes, France, [3]FRGDS Bourgogne, 21000 Dijon, France; alix.damman@oniris-nantes.fr

The bovine viral diarrhea virus (BVDV) causes losses for farmers that may be reduced using vaccination. However, vaccination efficiency is barely known and depends on the herd and the region characteristics. Bourgogne is one of the main beef production regions in France. A BVDV alert program has been set up in this region since 2006, vaccination strategy being the privileged solution for controlling BVDV spread. Based on the characteristics of beef cow-calf herds present in this region, a stochastic model was developed to evaluate the efficiency of vaccination as a control strategy of BVDV spread. The herd was structured into subgroups. The within-herd virus dynamics includes both horizontal and vertical transmission. First, BVDV spread without any control strategy has been investigated. The virus was introduced either by inserting a persistently infected (PI) animal before, after, and in the middle of the breeding period, or via contamination due to neighbouring contacts on pasture. Second, vaccination of the breeding females before and after the virus introduction has been tested. Bred heifers only or bred heifers and cows were vaccinated, either every year or every two years. Simulation tests were done for three herd sizes. We have shown that vaccination impacts not only BVDV persistence, but also the number of PI animals, abortions, and deviations in sales and purchases. Finally, this model was a suitable tool to predict the consequences of BVDV introduction into a naive herd and to evaluate vaccination programs. It should prove to be a useful tool to help cow–calf producers in controlling the spread of BVDV in their herds.

Identifying factors underlying heritable variation and response to selection in R0: simulation study

M. Anche, P. Bijma and M. De Jong
Wageningen University, De Elst 1, 6708 WD, the Netherlands; mahlet.anche@wur.nl

The reproduction ratio of an infectious disease, R_0, is the average number of secondary cases a typical infectious individual produces during its infectious life time. Since an epidemic can occur only when $R_0>1$, selective breeding could aim at reducing R_0 to a value ≤ 1. This requires knowledge of heritable variation in R_0 and of efficient selection schemes. Here we investigated heritable variation and response to selection in R_0, using a Susceptible-Infected-Recovered (SIR) model for the disease, and two bi-allelic loci segregating in the host population; one locus affecting susceptibility, and the other infectivity. Though R_0 is a characteristics of a population, rather than of a single individual, we can still define individual breeding values and heritable variation in R_0, using direct-indirect genetic effect models commonly applied to socially-affected traits. Susceptibility affects an individual's own disease status (0/1) and is therefore a direct genetic effect. Infectivity, in contrast, affects the disease status of an individual's social partners, and is therefore an indirect genetic effect. Based on the Total Breeding Value concept, our theoretical findings show that the heritable variation in R_0 depends on both the average of, and the genetic variation in, susceptibility and infectivity. When interacting individuals are unrelated, selection for individual disease status (0/1) results in response in susceptibility only. In contrast, when interacting individuals are related, selection yields response also in infectivity, and furthermore an additional response in susceptibility. This shows that susceptibility has an indirect genetic component as well. As a result, response in R_0 was substantially higher when data were collected on groups consisting of related individuals.

Spatiotemporal evolution of cattle movement network in France

B.L. Dutta[1,2,3], P. Ezanno[2,3] and E. Vergu[1]
[1]INRA UR341 MIA, 78352 Jouy en Josas, France, [2]LUNAM Université, Oniris, 44307 Nantes, France, [3]INRA UMR1300 BioEpAR, 44307 Nantes, France; bhagat-lal.dutta@oniris-nantes.fr

A network representation of the exchange of animals among the holdings can act as the backbone of the spatial dependence analysis of disease spreading. Therefore, the analysis of this network at different scales of time and space is an important part of livestock epidemiological studies. Here, the objective is three fold: (1) to characterize the network; (2) to investigate its evolution over time; and (3) to identify the modifications in its structure due to restrictions. Based on the French National Cattle Database (BDNI) from 2005-2011, we construct networks at different spatiotemporal scales, analyze them with methods of graph theory (indicators such as, clustering, reciprocity, assortativity, distributions of degree and strength, centrality, activity, entropy, disparity, preference index). Comparisons of means and distributions of these indicators (e.g. Kolmogorov-Smirnov test) lead to identification of patterns in cattle movement in terms of seasonality, demographics and dependence on geographical and administrative units. The structure of the network at different spatiotemporal scales and its response to interventions (restrictions imposed due to bluetongue outbreak in France in 2007) reveals the scales (e.g. communes as nodes and monthly time window), where interventions affect network descriptors, which may be used in redesigning control policies. We observed that although most of the average indicators do not show statistically significant variations, centrality distributions related to capacity of bridging among different nodes show significant differences for the period of intervention (2007). We also study the network in terms of measures used for competition in market (e.g. preference index) and note that with interventions imposed, the network reorganized itself (through changes, for a given holding, in the composition of subnetworks of its neighbors and/or the intensity of corresponding trade) to sustain its structure.

Control of classical swine fever focusing on emergency vaccination and rapid PCR testing

I. Traulsen, J. Brosig and J. Krieter
Institute of Animal Breeding and Husbandry, Olshausenstr. 40, 24098 Kiel, Germany; itaulsen@tierzucht.uni-kiel.de

Following legislation of the European Union control of classical swine fever outbreaks is based upon culling of swine in infected farms, movement restrictions in protection and surveillance zones, contact tracing as well as preventive culling as additional measure. Currently discussed alternatives to preventive culling are emergency vaccination and rapid PCR testing. The aim of the present study was to evaluate control measures mentioned above under varying outbreak conditions. In an individual-based, spatial and temporal Monte-Carlo simulation model control strategies 'Traditional Control', 'Emergency Vaccination', 'Test To Slaughter', 'Test To Control' and 'Vaccination in conjunction with Rapid Testing' are projected. Varying outbreak conditions were described by different farm densities (0.8 and 3.0 farms/km^2), compliance with movement restrictions (80, 90, 100%) and delay in establishment of an emergency vaccination (3/15 days for primary, 15/ 6 days for secondary outbreaks). Results showed that all factors had a significant influence on the number of infected and culled farms. In the low-density region, the basic measures are sufficient to control an epidemic, provided strict compliance with movement restrictions (100%) is adhered to. In the high-density region additional measures are necessary for a rapid eradication of classical swine fever. These measures can even compensate non-strict compliance with movement restrictions to a certain extent. In the high-density region, 'Emergency Vaccination' and 'Vaccination in conjunction with Rapid Testing' reached the same level of infected farms as 'Traditional Control', independent of the level of compliance with movement restrictions. However, in the case of an emergency vaccination, an early start of the vaccination campaign is essential for a successful disease control. In conclusion, additional control measures improve a rapid control of classical swine fever in high-density areas can overcome shortcomings in other control options.

Session 56b Theatre 7

Modelling framework to coordinate disease control decisions: example of the PRRS

A.-F. Viet[1], S. Krebs[1], O. Rat-Aspert[2], L. Jeanpierre[3], P. Ezanno[1] and C. Belloc[1]
[1]INRA, Oniris, UMR 1300 BioEpAR, Oniris, site de la chantrerie; CS 40706, 44307 Nantes Cedex 3, France, [2]AgroSup Dijon-INRA, UMR 1041 CESAER, 26, bd Docteur Petitjean; BP 87999, 21079 Dijon Cedex, France, [3]Université Caen Basse-Normandie, UMR 6072 GREYC, Campus Côte de Nacre; Boulevard du Maréchal Juin; CS 14032, 14032 Caen cedex 5, France; catherine.belloc@oniris-nantes.fr

For non-regulated diseases within a livestock population, the farmer decides whether to control them or not on a voluntary basis. Nevertheless, individual decisions have an impact on the risk for other farmers to be infected. Since some farmers are grouped in associations/geographical areas, it is relevant to investigate how a group of farmers can coordinate individual decisions, implementing incentives for individual disease management. This issue is applied to a major viral disease in swine production: the porcine respiratory and reproductive syndrome (PRRS). A Markov sequential decision model is defined including stochastic compartmental models representing PRRS virus spread within a group of herds among which some are PRRS virus positive. We describe an approach to propose control strategies which are adaptive to the evolution of the epidemiological situation over time. We assumed that the collective decision-maker should at each time-step select the incentive to optimise a criterion, for example the minimisation of the total cost at the group level (incentive, control and disease costs). The decision-maker can choose among many incentives levels, ranging from cheap no-incentive to costly incentives. We compute a policy corresponding to a guideline indicating the action to use according to the observed epidemiological situation. By simulation, we illustrate that different levels of incentive are used over time inducing an average total cost at the group level lower than if we systematically used each incentive level. While optimising the total cost, the model can be extended to consider also an objective in terms of prevalence decrease.

Session 56b Theatre 8

An economical tool for the assessment of *Salmonella* control strategies in the pork supply chain

S. Krebs, M. Leblanc-Maridor and C. Belloc
INRA-Oniris, UMR1300 BioEpAR, Oniris, Site de la Chantrerie, CS 40706, 44307 Nantes Cedex 3, France; stephane.krebs@oniris-nantes.fr

The aim of this study is to develop an useful tool for decision making process regarding food pathogen control in the pork supply chain. The different levels considered are farm, transport-lairage and slaughtering process. A mathematical model has been developed in order to determine the influence of control measures implemented at one or several steps along the pork supply chain on food pathogen's prevalence on carcasses at the end of the slaughtering process. The model is parameterized using the outcomes of epidemiological models as well as expert knowledge and observational data. Mathematical simulations (Monte Carlo simulations) are then performed using economic parameters to determine which measures implemented at which step(s) are the most cost-effective for food pathogen control. The performed cost-effectiveness explicitly includes a target prevalence, which has to be achieved at the end of the slaughtering process. To illustrate this new approach, a numerical application concerning *Salmonella* control is given. Simulation results enable us to highlight the influence of the heterogeneity of *Salmonella* prevalence between slaughter pigs' batches on the choice of an intervention strategy. For each considered strategy, the probability to overcome the target prevalence can also be assessed. This study enables us to develop a flexible tool, which can be parameterized to take into account the diversity of field situations (levels of *Salmonella* infection, slaughter processes). It can also be adapted to specific stakeholders' needs such as the ex ante assessment of incentive systems.

Phenotyping for optimized decision making on cow and herd level

K.L. Ingvartsen
Aarhus University, Department of Animal Science, Blichers Allé 20, P.O. Box 50, 8830 Tjele, Denmark; kli@agrsci.dk

Major changes have occurred in the dairy industry during the last couple of decades. In particular milk yield has increased substantially and changes in price relations and technological development have caused a structural development in the dairy industry resulting in a rapid increase in the average herd size in many countries, and e.g. in Denmark it has doubled over the last decade while the number of herds has been halved. In these larger units the farmer or farm staff has to look after an increasing number of animals concerning disease, reproduction, production and welfare and therefore risk management and optimization becomes important issues. This, together with consumer and society demands on product quality, animal welfare, and concerns about the environmental and climate effects of livestock production, calls for phenotyping for optimized decision making on cow and herd level. Phenotyping is also a major limiting factor in genomic selection. The phenotyping of cows has so far primarily been focused on performance data such as milk yield and composition, reproductive data, veterinary records, etc. The future challenges in optimization and proactive risk management call for new more detailed phenotyping based on large scale collection of physical, behavioural and physiological data. One of the challenges is to combat subclinical states causing increased risk of disease and suboptimal performance and reproduction. This calls for e.g. physiologically based measures from easily accessible samples, e.g. milk, that can be collected and analysed automatically in-line and used real-time. Challenges for future disease prevention and management of individual dairy cows are believed to include monitoring of physiological imbalance and understanding how e.g. nutrition and management of the individual cow should be changed to bring the cow in balance and thereby reduce risk of disease and suboptimal performance and improve welfare.

Session 57 Theatre 2

Standardisation of milk MIR spectra: a first step to create new tools of dairy farm management

C. Grelet, J.A. Fernandez Pierna, F. Dehareng and P. Dardenne
Walloon Agricultural Research Center, Valorisation of Agricultural Products, 24 Chaussée de Namur, 5030 Gembloux, Belgium; c.grelet@cra.wallonie.be

This work is performed in the framework of the OptiMIR EU project that aims to use mid infrared (MIR) spectrum of milk as the mirror of the cow status. Spectra from different countries are matched with physiological data in a common database to create calibrations predicting cow fertility, health, environmental and feeding indicators. In order to reach this goal, a procedure to standardize dairy milk MIR spectra from different apparatus from several brands inside this European dairy network is necessary to create a common and transnational spectral database. The method used is Piecewise Direct Standardization (PDS), which match slave apparatus spectra on those obtained with a 'master', wavelength by wavelength. The procedure is validated using different prediction models applied on the master and slaves spectra before and after its application. In the case of fat prediction, for instance, a decrease after the application of PDS from 0.3781 to 0 and from 0.4609 to 0.0156 for bias and RMSE respectively has been obtained. Stability of these results in time has been proved by the application of the PDS coefficients on slave spectra collected one month later. Bias and RMSE also decreased, respectively from 0.4118 to 0.0350 and from 0.4458 to 0.0393. These results have shown that the PDS method reduces the inherent spectra variability between apparatus, and then allowing the grouping of spectra in a common database. This should lead to the creation of universal prediction equations to be used by all apparatus of the OptiMIR network, giving new types of farm management indicators to the dairy sector.

EOL: a new ontology for livestock system and rearing conditions

L. Joret[1], J. Bugeon[1], J. Aubin[2], J.P. Blancheton[3], M. Hassouna[2], C. Hurtaud[4], S. Kaushik[5], F. Médale[5], M.C. Meunier-Salaün[4], J. Vernet[6], A. Wilfart[2], J.Y. Dourmad[4] and P.Y. Le Bail[1]
[1]INRA, UR 1037 LPGP, 35000 Rennes, France, [2]INRA-AgroCampus, UMR1069 SAS, 35042 Rennes, France, [3]IFREMER, Laboratoire d'Aquaculture, 34520 Palavas les flots, France, [4]INRA-AgroCampus, UMR 1348 PEGASE, 35590 Saint-Gilles, France, [5]INRA, UR1067 NUMEA, 64310 St Pée sur Nivelle, France, [6]INRA, UMR 1213 UMRH, 63122 Saint-Genes-Champanelle, France; marie-christine.salaun@rennes.inra.fr

The development of ontologies is prerequisite for better organizing and exploiting the knowledge coming from the large quantity of data available nowadays in biology. ATOL (http://www.atol-ontology.com) for 'Animal Trait Ontology for Livestock' has been devoted to the definition and organisation of the phenotypic characters of farming animals (fish, poultry, mammals). Taking into account that a phenotype arises from the action of both the genotype and the environment (plus eventually epigenetics mechanism), a precise description of the animal rearing environments is critical. EOL for 'Environmental Ontology for Livestock' has been created to describe, in a generic manner, the livestock systems and the rearing conditions. In February 2013, EOL contains about 600 concepts distributed in 4 main branches: (1) the livestock farming system describing shortly how and for what reasons the system was conceived; (2) the farming structure for the physical environment of animal; (3) the farming environment for the biological, chemical and physical rearing conditions; (4) the feed nature, quality and distribution conditions. The combination of animal's traits concepts (ATOL) and the rearing conditions concepts (EOL) will allow a powerful annotation of animal phenotypic databases with explicit metadata. Such ontologies pave the way to future predictive breeding programs. ATOL and EOL are also relevant tools to develop semantic analysis in order to retrieve precise information.

A graph database to store and manage phenotypic, pedigree and genotypic data of livestock

F. Biscarini, M. Picciolini, D. Iamartino, A. Stella, F. Strozzi and E. Nicolazzi
PTP (Parco Tecnologico Padano), Bioinformatics, Via Einstein, Loc. Cascina Codazza, 26900 Lodi, Italy; filippo.biscarini@tecnoparco.org

Thanks to recent developments in sequencing and genotyping technologies, increasing amounts of data are being generated for the livestock industry and research. This is urging operators to find solutions for the storage, manipulation, visualization and analysis of large data sets. Livestock examples are pedigree records, phenotypes, SNP genotypes, and whole-genome sequences. Binary or ASCII files are often used to archive and transfer such data, but have shortcomings: are cumbersome to manipulate, need to be read and parsed each time, and are far from having standard formats. Databases offer a better way to handle big data in terms of space, data safety and stability, ease of access and format standardization. In relational databases data are represented in tables related through unique keys. Graph databases are composed of nodes (the elements) and edges (the connections among the elements). For instance, nodes may represent animals and markers, and the relationships between nodes model the animal genotypes at marker loci. We have implemented a graph database using the Neo4j technology. Data on 529 Italian Mediterranean buffaloes were used as case study. Pedigree, lactation records and Affymetrix 90k SNP-chip genotypes were available. We modelled a base node for the entire database and a category node for each of the three layers of elements: animal nodes (ID, sex, age), SNP nodes (name, chromosome, position) and trait nodes (name, class, value range). Connections between nodes modelled SNP genotypes, phenotypic observations and pedigree relationships. Cypher and Neo4j native methods were used as query languages to extract, add, edit and remove records. Graph databases are well suited to represent interconnected data, may have billions of nodes and relationships, are traversed and queried much faster than relational databases, and allow for the ready implementation of algorithms (e.g. minimum spanning tree).

Locomotor activity of dairy cows in relation to season and lactation

A. Brzozowska[1], M. Łukaszewicz[1], G. Sender[1], D. Choromańska[2] and J. Oprządek[1]
[1]Institute of Genetics and Animal Breeding of the Polish Academy of Sciences, Jastrzębiec, Postepu 36A, 05-552 Magdalenka, Poland, [2]Warsaw University of Life Sciences, Ciszewskiego 8, 02-786 Warsaw, Poland; a.brzozowska@ighz.pl

All systems recording activity of cattle measure it independently of environmental and animal factors. Such recordings may hide differences in activity caused by extraneous interferences or may be inaccurately interpreted. The information produced by pedometers could be more accurate, if behaviour pattern is put against other factors affecting activity. The objective of the study was to verify if information obtained from activity sensors can be improved, by including additional factors such as parity, stage of lactation and season of year. Activity of 132 Polish Holstein-Friesian cows was continuously monitored between April 2011 and August 2012 using IceQube Sensors. Generalized linear models were used to analyse activity parameters of dairy cows. The results show that parity, stage of lactation and season have a significant impact ($P<0.01$) on cows activity. Total lying time decreases with consecutive lactations, however older cows prefer longer and less frequent lying bouts. Decreased lying time, higher number of steps and less number of lying bouts during the first 30 days after calving were observed. Including the fact of restlessness behaviour after calving in data from pedometers would prevent false conclusions. In winter cows spend more time lying down than in other seasons. Therefore lying time increased over 650 minutes in summer provides different information (may be sign of lameness) than in winter (common phenomenon). The results of the study represent the activity of cows from commercial dairy herd. Activity parameters are highly affected by stage of lactation, parity and season of year, therefore data obtained from pedometers should be put against these factors. It is highly recommended to include such factors in software analysing activity parameters, to improve the information obtained from sensors.

Dynamic monitoring of mortality rate for sows and piglets

C. Bono[1], C. Cornou[1], S. Lundbye-Christensen[2] and A.R. Kristensen[1]
[1]University of Copenhagen, Department of Large Animal Sciences, Health and Production, Grønnegårdsvej 2, 1870, Frederiksberg C, Denmark, [2]Aarhus University Hospital, Department of Cardiology, Cardiovascular Research Center, Sdr. Skovvej 15, 9000, Aalborg, Denmark; clbo@life.ku.dk

Management and monitoring systems may enable the farmers to enhance production results and reduce labor time. The aim of this paper is to develop a dynamic monitoring system for mortality rate of sows and piglets. For this purpose a mortality rate model is developed using a Dynamic Generalized Linear Model. Variance components are pre-estimated using an Expectation-Maximization algorithm applied on a dataset consisting of data from 15 herds in a period ranging from 3 to 9 years. Data are registrations of events for insemination, farrowing (including stillborn and live born), number of weaned piglets and death of sows. The model provides reliable forecasting on weekly basis. Detection of impaired mortality rate is performed by statistical control tools that give warnings when the mortality (rate) has sudden or gradual changes. For each herd, mortality rate profile, analysis of variance components over time and detection of alarms are computed for two categories i.e. sow and piglet. The combination of this model with the previous two on litter size and farrowing rate, represents a significant step into the creation of a new, dynamic, management tool.

Multiple births limit the advantage of using high growth sires

K.R. Kelman[1,2], C.L. Alston[3], D.W. Pethick[1,2] and G.E. Gardner[1,2]
[1]Division of Veterinary Biology and Biomedical Science, Murdoch University, WA, 6150, Australia, [2]Australian Cooperative Research Centre for Sheep Industry Innovation, Armidale, NSW, 2351, Australia, [3]School of Mathematical Sciences, Queensland University of Technology, Brisbane, QLD, 4001, Australia; k.kelman@murdoch.edu.au

The Australian lamb industry uses breeding values to select for progeny with increased post-weaning weight at 150 days (PWWT). Accurate weight prediction is essential to provide age estimates for lambs to reach target weights, to underpin breeding values, and for assessing the influence of growth rate on factors such as intramuscular fat and myoglobin concentration of lamb muscle. As growth curves can be biased when predicting weights at the edge of the available weight data, the key aim was to develop a population based random regression model to predict lamb PWWT. This fit was compared to an individual based Brody curve fit with comparable results confirming the rigour of the model. The PWWT results were then used to assess the impact of factors such as lamb birth-type rear-type and sire PWWT breeding value on lamb weight. Multiple births were hypothesised to limit the progeny of high PWWT sires from reaching their full weight due to nutritional restriction pre-weaning. Weight data totalling 164,797 observations was collected from 17,525 lambs across eight sites and five years of the Sheep Cooperative Research Centre Information Nucleus Flock. A Bayesian linear mixed model was fitted to the live weight data with fixed effects for site, year of birth, gender, birth type-rear type, age of dam, sire type, dam breed within sire type, sire PWWT as a covariate and random terms for sire, dam by drop and individual. Lamb PWWT was then analysed in a linear mixed model of similar structure. In line with our hypothesis, the weight of singles, twins and triplets at 150 days increased by 9.43, 6.67 and 3.68 kg across the 23 kg PWWT range ($P<0.05$) confirming that multiple births limit the full expression of weight potential.

Practical integration of genomic selection into dairy cattle breeding schemes

A. Bouquet[1,2] and J. Juga[2]
[1]IFIP, French Institute for Pig and Pork Industry, BP 35104, 35651 Le Rheu Cedex, France, [2]University of Helsinki, Department of Agricultural Sciences, P.O. Box 27, 00014 Helsinki, Finland; alban.bouquet@ifip.asso.fr

Genomic selection (GS) consists of integrating information at thousands of genetic markers in the statistical models used for the prediction of breeding values of animals. The basic principle is to take advantage of a reference population, with both phenotypes and genotypes, to build prediction equations of genetic merit that can be applied to candidates having only genotypes. The use of GS was shown to greatly increase the technical and economical efficiency of dairy cattle breeding schemes. With this technology, progeny-testing is no longer necessary to select artificial insemination (AI) sires with reasonable accuracy. The use of young AI sires allows boosting annual genetic gain (ΔG), up to double, due to the shortening of generation intervals. However, the number of AI sires in service must be sufficiently large to avoid large increases in inbreeding rates. It is also recommended for breeders to use 'teams' of young sires to limit the risk of using bulls with poor merit. When females are genotyped, a substantial part of ΔG is achieved on the bull dam selection path due to increased selection accuracy. The benefits are even larger when GS is coupled with embryo transfer technologies. Thus, strategies using SNP chips of different densities and imputation techniques were proposed to reduce genotyping costs and, hence, widen the use of genotyping to females. Many factors affect the success of GS. The size and constitution of the reference population were shown to influence the accuracy of genomic predictions and, hence, achieved ΔG. As gains in ΔG are the largest for low heritability traits, GS has the potential to improve the balance of ΔG achieved between production and functional traits in the total merit index. In practice, diverse strategies were applied to integrate GS in the breeding schemes of a few large dairy cattle populations of national and international extent.

Genotyping cows for the reference increases reliability of genomic predictions in a small breed

J.R. Thomasen[1,2], A.C. Sørensen[1], M.S. Lund[1] and B. Guldbrandtsen[1]
[1]Aarhus University, Department of Molecular Biology and Genetics, Center for Quantitative Genetics and Genomics, P.O. Box 50, 8830 Tjele, Denmark, [2]VikingGenetics, Ebeltoftvej 16, 8960 Randers SØ, Denmark; jotho@vikinggenetics.com

We hypothesized that adding cows to the reference population in a breed with a small number of reference bulls would increase reliabilities of genomic breeding values and genetic gain. We tested this premise by comparing two strategies for maintaining the reference population for genetic gain, inbreeding and reliabilities of genomic predictions: (1) adding 60 progeny tested bulls each year (B); and (2) in addition to 60 progeny tested bulls, adding 2,000 genotyped cows per year (C). Two breeding schemes were tested: (1) a turbo scheme (T) with only genotyped young bulls used intensively; and (2) a hybrid breeding scheme (H) with use of both genotyped young bulls and progeny tested bulls. The genomic selection schemes were simulated over 15 years, and the reference population of the first year of genomic selection consisted of approximately 1000 progeny tested bulls. T-B yielded 8.6% higher ΔG compared to the H-B, at the same level of ΔF. T-C yielded 15% higher ΔG compared to T-B. Changing the breeding scheme from H-B to H-C increased ΔG by 5.5%. The lowest ΔF was observed with genotyping of cows. Reliabilities of GEBV in the C schemes showed a steep increase in reliability during the first four years, from 0.2 to just above 0.4. With B schemes the reliability increased from 0.19 to 0.27. Genotyping cows for the reference population increases genetic gain, particularly in breeding schemes with intense use of young bulls.

Large scale genomic testing within herd does not affect contribution margin

L. Hjortø[1], J.F. Ettema[2], A.C. Sørensen[3] and M. Kargo[3]
[1]Knowledge centre for Agriculture, Agro Food park 15, 8200, Denmark, [2]SimHerd A/S, Niels Pedersens alle 2, 8830, Denmark, [3]Aarhus University, QGG, P.O. Box 50, 8830, Denmark; morten.kargo@agrsci.dk

There are three overall reasons for performing genomic tests among females in dairy cattle production: (1) tests for improvement of the reference population; (2) tests for finding the best females for MOET (bull dams); and (3) tests for management purposes. The first two have an effect at population level, while (3) and to some degree (2) have an effect at herd level. In this study the effect of genomic tests for management purposes was investigated using the simulation models SimHerd, a computer program to simulate the effect of management decisions and ADAM, a computer program to simulate selective breeding schemes for animals. We studied 5 scenarios of performing genomic tests in combination with 8 scenarios using sexed semen. In addition to a baseline strategy of not using genomic testing, four strategies were investigated: all heifer calves, the top 50% of all heifer calves or an intermediate group of 25% or 50% of all heifer calves. Selection for genomic tests was based on pre-sorting by traditional parent averages or predicted transmitting abilities. The genetically best females were inseminated with sexed dairy semen and all others were inseminated with conventional semen or semen from beef sires. All scenarios were compared to the economic result without genomic tests and sexed semen. The scenarios were combinations of: (1) breeding 40-80% of the heifers with sexed semen; (2) breeding 20-40% of 1st lactation cows with sexed semen; and (3) breeding the lowest yielding proportion of the herd with beef semen. This proportion was chosen so to expect neither surplus nor shortage of replacement heifers. The study demonstrated that, under Nordic circumstances and a price of $130 per test, the use of genomic testing can improve the farmers' economy in some cases, where sexed semen was used intensively. However, it was never economic to genotype all heifers.

Session 57 Theatre 11

SNPchiMp: a database to disentangle the SNPchip jungle

E.L. Nicolazzi[1], M. Picciolini[1], F. Strozzi[1], R.D. Schnabel[2], C. Lawley[3], A. Eggen[3], A. Pirani[4], F. Brew[5] and A. Stella[1]
[1]Parco Tecnologico Padano, Via Einstein, 26900 Lodi, Italy, [2]University of Missouri, Columbia, MO 65203, USA, [3]Illumina, Inc., 5200 Illumina way, San Diego, CA 92121, USA, [4]Affymetrix, Inc., 3420 Central Expressway, Santa Clara, CA 95051, USA, [5]Affymetrix UK Ltd., Mercury Park, HP10 0HH, Wycombe Lane, High Wycombe, United Kingdom; ezequiel.nicolazzi@tecnoparco.org

Currently, there are six commercial SNP chips available for cattle, produced by two different genotyping platforms. Technical issues need to be addressed before managing data from different platforms, or even different versions of the same assay: (1) higher density chips do not always include all the SNPs present in the lower density chips; (2) SNP names may not be consistent across chips and platforms; (3) genome coordinates for these chips may refer to different genome assemblies and reference genome sequences are updated over time; (4) producer SNP names are not searchable terms in any public database; and (5) SNPs can be coded using different formats. Most researchers and breeding associations need to manage these SNP data in real-time and there is need for a set of tools that provide best practice to address issues in a straightforward, consistent and user-friendly manner. Here we present SNPchiMp, a MySQL database linked to an open-access web-based interface. Features of this interface include, but are not limited to, the following functionality: (1) download chip(s) map information updated to the latest assembly; (2) extract information contained in dbSNP for SNPs in any commercial chip; and (3) list SNPs in common -or not-between two or more chips. In addition, you can retrieve the above information on a subset of SNPs (e.g. the output of a GWAS), accessing such data either by physical position of any supported assembly, or by a list of SNP names, rs or ss IDs. This tool is currently available for cattle data only. However, we include plans to cover SNP chip data for livestock, fishery as well as companion animals.

Session 57 Theatre 12

Generating large scale on-farm methane measurements in exhaled air of individual cows

Y. De Haas[1], J. Van Riel[2], N. Ogink[3] and R.F. Veerkamp[1]
[1]Wageningen UR Livestock Research, Animal Breeding and Genomics Centre, P.O. Box 135, 6700 AC Wageningen, the Netherlands, [2]Wageningen UR Livestock Research, Systems Group, P.O. Box 65, 8200 AB Lelystad, the Netherlands, [3]Wageningen UR Livestock Research, Environment Dept., P.O. Box 135, 6700 AC Wageningen, the Netherlands; yvette.dehaas@wur.nl

Methane (CH_4) is a greenhouse gas that contributes to climate change. Preliminary international data suggest that genetic selection to reduce CH_4 emissions is possible. However, successful breeding programs require large datasets of individual measurements which cannot be generated through respiration chambers. The aim of this preliminary study is to show whether realistic values for and individual differences in enteric CH_4 emission could be measured during milking, so that a large scale data collection can be set up for genetic evaluation of CH_4 production in dairy cattle. Data was collected between October and December 2012. Breath air was sampled and analysed directly for CH_4 using a portable Fourier Transformed Infrared (FTIR) gas analyser. The equipment was installed in an automatic milking system (AMS) and the air inlet was placed in front of the cow's head in the AMS. The total number of unique cows that visited the AMS was 78, and a total of 92,360 measurements of 1 minute each were recorded. Mean CH_4 concentration was 351 ppm, with a range of individual cow means between 90 and 630 ppm. Daily means of the background concentration for CH_4 was 33 ppm. The high concentrations measured in the AMS indicate that the sampled air included a high portion of exhaled air of the milked cow. Individual variation is shown in the mean enteric CH_4 concentration, and between day repeatability is 86%. This preliminary study has shown that using a portable FTIR measuring unit in an AMS to measure individual cow CH_4 emissions gave realistic values and ranges. The FTIR instrument combined with AMS may therefore be useful in the future to generate large scale data for genetic evaluation of CH_4 production in dairy cattle.

Session 57 — Poster 13

Comparison between bivariate and multivariate joint analyses on the selection loss

V.B. Pedrosa[1], E. Groeneveld[2], J.P. Eler[3] and J.B.S. Ferraz[3]
[1]Ponta Grossa State University, Department of Animal Science, Av. Carlos Cavalcanti, 4748, 84030-900 Ponta Grossa, PR, Brazil, [2]Friedrich Loeffler Institute, Institute of Farm Animal Genetics, Department of Animal Breeding and Genetic Resources, Höltystrasse 10, 31535 Neustadt, Germany, [3]University of Sao Paulo, Av Duque de Caxias Norte, 225, 13635-900 Pirassununga, SP, Brazil; vbpedrosa@uepg.br

For genetic evaluation of beef cattle, univariate or bivariate analyses are often performed as an alternative to decrease the complexity of matrices and mathematical models when compared to multivariate analysis that consider a larger number of joint traits. Utilization of bivariate methods to calculate genetic predictors may cause bias in the estimation of breeding values and, as a consequence, reclassification in the rank of top selected sires, resulting in a loss of genetic gain in future generations. Thus, the objective of this study was to compare the bivariate to multivariate joint methods of genetic evaluation, verifying selection loss and the reclassification in the ranking of the best animals in different selection intensities. Records of 431,224 Nellore animals were evaluated for birth weight, weaning weight, post-weaning gain, muscle score, scrotal circumference and selection index. The pedigree file consisted of 505,848 animals, including 218,727 males and 287,121 females. The predicted breeding values were obtained using the program PEST 2, and the complete pedigree analysis was performed by PopReport software. The results demonstrated that for the four different selection intensities considered, TOP 10, 1%, 10% and 30%, selection loss and reclassification of animals in ranking were detected for all evaluated traits when the two methodologies of analyses were compared. From the previous results could be concluded that multivariate joint analyses is the more appropriated method of genetic evaluation, especially for large databases. Funded by FAPESP and Fundação Araucária.

Session 57 — Poster 14

Dynamic monitoring of farrowing rate at herd level

C. Bono[1], C. Cornou[1], S. Lundbye-Christensen[2] and A.R. Kristensen[1]
[1]University of Copenhagen, Department of Large Animal Sciences, Health and Production, Grønnegårdsvej 2, 1870 Frederiksberg C, Denmark, [2]Aarhus University Hospital, Department of Cardiology, Cardiovascular Research Center, Sdr. Skovvej 15, 9000 Aalborg, Denmark; clbo@life.ku.dk

Good management in animal production systems is becoming of paramount importance. The aim of this paper is to develop a dynamic monitoring system for farrowing rate. A farrowing rate model is implemented using a Dynamic Generalized Linear Model (DGLM). Variance components are pre-estimated using an Expectation-Maximization (EM) algorithm applied on a dataset containing data from 15 herds, each of them including insemination and farrowing observations over a period ranging from 150 to 800 weeks. The model includes a set of parameters describing the parity-specific farrowing rate and the re-insemination effect. It also provides reliable forecasting on weekly basis. Statistical control tools are used to give warnings in case of impaired farrowing rate. For each herd, farrowing rate profile, analysis of model components over time and detection of alarms are computed. Together with a previous model for litter size data and a planned similar model for mortality rate this model will be an important basis for developing a new, dynamic, management tool.

Session 57 Poster 15

Genetic parameters across lactation for dry matter intake and milk yield in Holstein cattle

C.I.V. Manzanilla P[1,2], R.F. Veerkamp[1,2], J.E. Pryce[3], M.P.L. Calus[1] and Y. De Haas[1]
[1]Wageningen UR Livestock Research, Animal Breeding and Genomics Centre, P.O. Box 65, 8200 AB Lelystad, the Netherlands, [2]Wageningen University, Animal Breeding and Genomics Centre, P.O. Box 338, 6700 AH Wageningen, the Netherlands, [3]Biosciences Research Division, Department of Primary Industries, 1 Park Drive Bundoora, VIC 3083, Australia; coralia.manzanillapech@wur.nl

Breeding for lower feed intake will reduce the feed costs for farmers. However, measuring individual feed intake is expensive; therefore there is an interest to select on indicator traits instead (e.g. milk yield). In this study, genetic parameters across lactation for dry matter intake (DMI) and fat and protein corrected milk (FPCM) were estimated using random regression test-day models. A dataset was available with 30,500 weekly DMI records on 2,278 Dutch Holstein Friesian first parity cows between 1989 and 2011, and 49,953 test-day records of FPCM. The pedigree included 8,860 animals. The single trait model included random effects for experiment, year-month of recording, direct additive, permanent environmental and residual. The included fixed effects were herd, year-season of calving, age at calving and days in milk, modelled with a 3rd and 4th order polynomial, respectively. The additive genetic and permanent environmental covariance functions were estimated using Legendre orthogonal polynomials of 3rd order. Bivariate analysis between DMI and FPCM were also carried out, with the same fixed and random effects. Heritability estimates for DMI ranged from 0.20 to 0.50 at different stages of lactation, being highest at 150 d. The genetic correlation for DMI at different stages ranged between 0.47 and 1.00 being lowest between early (50 d) and late lactation (250 d). Genetic correlations between DMI and FPCM across lactation ranged between 0.56 and 1.00. Genetic parameters for DMI vary according to stage of lactation, and this should be considered when breeding for a lower DMI. Also, due to its high correlation, FPCM could be used as an indicator trait for DMI.

Session 57 Poster 16

Computational efficiency of software for the analysis of Next Generation Sequencing data

M. Mielczarek[1], J. Szyda[1], A. Gurgul[2], K. Żukowski[2] and M. Bugno-Poniewierska[2]
[1]Wrocław University of Environmental and Life Sciences, Department of Animal Genetics, Biostatistics Group, Kożuchowska 7, 51-631 Wrocław, Poland, [2]National Research Institute of Animal Production, Krakowska 1, 31-047 Kraków, Poland; joanna.szyda@up.wroc.pl

Based on the data set of 44,926,270 reads of DNA sequence of a single dog, generated by the sequencing-by-synthesis Illumina next generation sequencing (NGS) technology we compared the quality and computational efficiency of various software tools publicly available for handling NGS data. The data set is in FASTQ format and contains paired-end reads of 100 bp length. The particular steps of the analysis comprised the comparison of: (1) tools for sequence quality visualisation (e.g. FASTQ and HTQC); (2) tools for sequence alignment to the reference genome (e.g. BFAST, SOAPaligner, Bowtie); and (3) variant calling programmes (e.g. SAMTools, PolyBayes). Software quality was assessed based on ease of installation, flexibility towards data formats, user friendliness, reliability expressed by trouble-free computations and first of all by their computational efficiency expressed by CPU time as well as by the quality of results related to alignment as single or paired end reads and variant detection.

Associations between gestation length, stillbirth, calving difficulty and calf size in Norwegian Red

H. Hopen Amundal[1], M. Svendsen[2] and B. Heringstad[1,2]
[1]Department of Animal and Aquacultural Sciences, Norwegian University of Life Sciences, P.O. Box 5003, 1432 Ås, Norway, [2]Geno Breeding and A.I. Association, IHA, 1430 Ås, Norway; bjorg.heringstad@umb.no

In this first genetic analysis of gestation length (GL) in Norwegian Red cows the aims were to infer heritability of and genetic correlations among GL, stillbirth (SB), calving difficulty (CD), and calf size (CS), and to examine genetic change, range and distribution of EBV for GL in Norwegian Red AI sires. Records on GL, SB, CD, and CS at first calving for 631,510 cows were analyzed with a multivariate linear sire model including direct and maternal genetic effects. The mean GL was 279 days. Stillbirth was recorded as a binary trait, born alive (97%) or dead at birth or within 24 h (3%). Calving difficulty had 3 categories, easy calving (90%), slight problems (7%), and difficult calving (3%). Calf size was scored as small (13%), medium (76%), or large (11%). Heritability was 0.37 for direct GL, 0.06 for maternal GL, 0.12 for direct CS, and $\leq$0.05 for the other traits. Maternal GL showed moderate genetic correlation to maternal CS (0.63) and maternal CD (0.26). Moderate genetic correlations were also estimated between direct GL and direct CS (0.29), direct CD (0.26), and direct SB (0.17), respectively. All genetic correlations between direct and maternal effects were close to zero within as well as between traits. EBV for direct GL varied between -10.6 and +9.6 days and EBV for maternal GL varied from -4 to +3 days. Mean EBV for direct GL decreased slightly over time (-2 days from 1995 to 2009), while mean EBV for maternal GL showed variation between years but no indications of genetic change. Too short as well as too long GL cause problems, so this is not a trait we aim to change genetically. However, GL may contribute correlated information in genetic evaluation of calving traits, and breeding values for gestation length may be used to predict expected calving date more accurately, and thereby provide useful information for herd management purposes.

Genetic analysis of claw disorders in Norwegian Red

C. Ødegård[1,2], M. Svendsen[1] and B. Heringstad[1,2]
[1]Geno Breeding and A.I. Association, P.O. Box 5003, 1432 Ås, Norway, [2]Department of Animal and Aquacultural Sciences, Norwegian University of Life Sciences, P.O. Box 5003, 1432 Ås, Norway; cecilie.odegard@umb.no

The aim was a first genetic analysis of claw disorders in Norwegian Red cows. Heritabilities and genetic correlations were estimated using claw health data from the Norwegian Dairy Herd Recording System. Recording of claw health at claw trimming started in 2004 and nine claw disorders were defined: corkscrew claw (CSC), heel horn erosion (HH), dermatitis (DE), sole ulcer (SU), white line disorder (WLD), haemorrhage of sole and white line (HSW), interdigital phlegmon (IDP), lameness (LAME) and acute trauma (AT). Cows with normal claws were recorded as healthy. Claw disorders were also grouped and analyzed as: (1) infectious claw disorders (INFEC), including HH, DE and IDP; (2) laminitis related claw disorders (LAMIN), including SU, WLD and HSW; and (3) overall claw disorder (ALL), including all nine claw disorders. Data from 2004 to 2011 contained 243,158 claw health records (including normal claws and disorders) from 141,659 cows, 1,904 sires, and 6,156 herds. Cows were defined as either affected (1) or healthy (0) for each trait and lactation. Univariate threshold sire models were used to analyze the nine disorders and the three groups. Multivariate analyses were performed for the five most frequent disorders; CSC, HH, DE, SU and WLD, and for CSC, together with the groups INFEC and LAMIN. Posterior mean of heritability of liability ranged from 0.04 (LAME and AT) to 0.23 (CSC), with standard deviations from 0.01 to 0.03, except for IDP (0.06). Heritability of liability for groups were 0.11 (INFEC and LAMIN) and 0.15 (ALL). Posterior mean of genetic correlations ranged from 0.02 to 0.79, where the highest genetic correlations were between HH and DE (0.65) and between SU and WLD (0.79). The results show that claw disorders are heritable. Claw health status recorded at claw trimming is valuable information that can be used for genetic evaluation of claw health for Norwegian Red.

Cow udder cistern storage capacity affects lactation performance in a NZ pasture based dairy system

A. Molenaar[1], S. Leath[1], G. Caja[2], H. Henderson[1], C. Cameron[1], K. Taukiri[1], T. Chikazhe[1], S. Kaumoana[1], A. Dorleac[1], A. Guy[1], C. Gavin[1] and K. Singh[1]
[1]AgResearch, Ruakura, 3240 Hamilton, New Zealand, [2]Ruminant Research Group, Universitat Autonoma de Barcelona, 08193 Barcelona, Spain; adrian.molenaar@agresearch.co.nz

Ultrasound technology has been developed to measure the cow udder cisternal volumes to establish the relationship with lactation performance in a feedlot dairying system non-invasively. The aim of this study was to determine if ultrasonography could be used in the New Zealand pasture fed dairying system to understand the relationship between cistern size and lactation performance. Daily milk production records from the previous season of 204 Friesian, Jersey and New Zealand crossbreed cows, between 2 and 7 years of age from a commercial research herd (AgResearch Tokanui) were examined. From these, groups of cows (n=20 each) with 4 combinations of high/low milk yield and lactation persistency were chosen. Additionally, 30 post-partum heifers (mixed breeds) and 19 cows from a preliminary trial were included. Their cistern capacities were measured by collecting milk volumes and by ultrasound imaging, during their afternoon milking. All cow udders were scanned 1-3 days after calving and again near peak production. Soon after peak production half the cows from each group were further split into balanced groups to evaluate the effects of passing from twice (2×) to once (1×) daily milking and scanned 10 days later. In the 2nd and 3rd rounds, milk let down was initially inhibited by using the oxytocin receptor blocker atosiban before scanning, then induced by oxytocin and rescanned. Preliminary results show that cistern size does influence yield and that almost all cows in the trial left the milking platform with between 5 and 15% of their milk remaining in the udder. Scanning of udders could allow farmers to select cows (perhaps as early as post partum heifers) for enhanced lifetime yield, persistency, ability to milk out fully and the ability to tolerate once daily milking without lowering yields.

Compensatory growth at pasture in weaned suckler bulls offered contrasting winter feeding levels

D. Marren[1,2], M. McGee[2], A.P. Moloney[2], A. Kelly[1] and E.G. O'Riordan[2]
[1]UCD, School of Agriculture and Food Science, Belfield, Dublin 4, Ireland, [2]AGRIC Teagasc, Grange, Dunsany, Co. Meath, Ireland; declan.marren@teagasc.ie

Optimal growth rate during the first indoor winter for weaned suckler bulls to exploit subsequent compensatory growth at pasture, prior to finishing indoors on a high concentrate diet, was evaluated. A total of 120 spring-born, Charolais and Limousin sired bulls, (372 kg) were used in a 3 (first winter growth rate – W1GR) × 2 (carcass weights – CW, 380 and 420 kg) factorial arrangement of treatments. Feeding regimes to achieve the 3 W1GR were: (1) grass silage *ad libitum* (GS) + 2 kg of barley-based concentrate (C) daily (GS2); (2) GS + 4 kg C (GS4); and (3) GS + 6 kg C (GS6). Duration of W1 was 123 d. Subsequently, bulls were turned out to pasture and rotationally grazed for 99 d and then re-housed for a finishing period on *ad libitum* C plus GS. Bulls were slaughtered on reaching the treatment mean live weight (LW) to achieve the target CW. Data were analysed using the GLM procedure of SAS. The model contained effects for PS, SW, their interactions and sire breed. Initial LW was included as a covariate. There were no W1GR × CW interactions. Increasing C level decreased silage DMI ($P<0.05$). Feed conversion ratios were 9.4, 7.9 and 7.1 kg DM for GS2, 4 and 6, respectively, ($P<0.001$). At the end of W1 GS4 and GS6 were 30 and 63 kg heavier, respectively, than GS2 ($P<0.001$). Corresponding values at the end of the grazing period were 3 kg ($P>0.05$) and 21 kg ($P<0.001$). Increasing C level increased ADG during W1 ($P<0.001$) but decreased ADG at pasture ($P<0.001$) for GS4 and GS6 compared to GS2 ($P<0.05$). Slaughter and carcass weight, kill out proportion, and carcass grades did not differ ($P>0.05$) between W1GR. Increasing CW increased all these factors ($P<0.001$). Supplementing GS with more than 2 kg of C daily (equivalent to >0.67 kg ADG) during the first winter had short-term effects on ADG but, due to compensatory growth, had no effect on carcass traits of suckler bred bulls.

Genetic parameters for carcass conformation in 5 beef cattle breeds

A. Kause[1], L. Mikkola[1], I. Stranden[1] and K. Sirkko[2]
[1]MTT Agrifood Research Finland, Biometrical Genetics, 31600 Jokioinen, Finland, [2]Faba Co, Urheilutie 6, 01301 Vantaa, Finland; antti.kause@mtt.fi

Profitability of beef production and quality of carcasses can be increased by genetically improving carcass traits. To construct breeding value evaluations for carcass traits, breed-specific genetic parameters were estimated for carcass weight, fat and carcass muscularity conformation in 5 most common breeds used in Finland (Hereford, Aberdeen Angus, Charolais, Limousin, Simmental). Fat and muscularity conformation were visually scored using the EUROP scale. A total of 4,020-12,708 pure-bred animals per breed were phenotyped. Animal model heritabilities varied between 0.31-0.51 (s.e.=0.03-0.05) for carcass weight, 0.30-0.50 (s.e.=0.03-0.05) for fat score, and 0.25-0.55 (s.e.=0.03-0.05) for muscularity score. The genetic correlations between carcass weight and muscularity score were highly favourable (r_G from 0.65±0.04 to 0.71±0.04), heavy carcasses being muscular. Genetic correlations between carcass weight and fat score were close to zero or weakly favourable (r_G from -0.02±0.06 to -0.27±0.06). Genetic correlations between fat and muscularity score were moderately favourable in Limousin, Charolais and Simmental, muscularity increasing with decreasing fat (r_G from -0.27±0.08 to -0.55±0.05), and close to zero in Hereford (r_G=-0.06±0.07) and Angus (r_G=-0.08±0.08). These results indicate ample genetic variation for the carcass traits and mostly favourable genetic correlation structure for simultaneous genetic improvement of carcass weight and carcass conformation.

Crossbreeding in French Holstein farms

C. Dezetter[1,2,3], C. Côrtes[3], C. Lechartier[3], P. Le Mezec[4], S. Mattalia[4] and H. Seegers[5]
[1]COOPEX Montbeliarde, 4 rue des Epicéas, 25640 Roulans, France, [2]ONIRIS, UMR BIOEPAR, La Chantrerie, 44307 Nantes, France, [3]ESA Angers, 55 Rue Rabelais, 49000 Angers, France, [4]Institut de l'élevage, 149 Rue de Bercy, 75595 Paris, France, [5]INRA, UMR1300, La Chantrerie, 44307 Nantes, France; c.dezetter@groupe-esa.com

Fertility of Holstein cows has decreased while crossbreeding is known to be a way to improve fertility. Previous sources have reported an increasing interest of French Holstein farmers for inter-breed crossing. This study aimed at describing crossbreeding practices in French Holstein herds between July 2002 and July 2012, from artificial insemination (AI) records. Among 20,078 Holstein farms gathering >30 cows in average over the study period, 3,893 had at least one crossbreeding AI (CAI) during campaign 2002-2003 or 2003-2004. Five patterns were identified: (1) low use of CAI (<10% of cows with CAI) only for return-AI and no crossbred F1 offspring kept in herd (n=1,471); (2) low use of CAI only for return-AI and some F1 kept (n=2,161); (3) low use of CAI at 2 first campaigns followed by increased use over the latter ones with F1 at least partially kept (n=55); (4) intermediate use (10 to 50% of cows with CAI) on first AI or return AI and some F1 kept (n=188); and (5) frequent use of CAI (>50% of cows with CAI, all on the first AI) and F1 kept (n=18). Out of these farms, only 1,646 were implementing true dairy-purpose crossbreeding, based on two criteria (>10% cows with CAI on one campaign; and existence of milk records for crossbred F1). In these farms, 27% CAI occurred at the first AI and 47% after the third one. Farmers tended to choose multiparous cows with lower genetic merit for milk or primiparous cows with higher genetic merit for protein content to inseminate them with a different breed. Holstein cows targeted for CAI were 2 to 3 times less frequently pregnant at 110 days in milk than cows without CAI. In conclusion, true dairy-purpose crossbreeding was little used and diversity of crossbreeding practices confirmed a need of further study.

Gascon cattle breed in Spain: a model of introduction and establishment of a foreign cattle breed

A. Guerrero[1], P. Sans[2], C. Sañudo[1], J.P. Gajan[3], A. Tranier[4], J.A. Mateos[1] and P. Santolaria[1]
[1]University of Zaragoza, Animal Production, Miguel Servet 177, 50013 Zaragoza, Spain, [2]ENVT, Animal Production, 23 Chemin des Capelles, 31076 Toulouse, France, [3]Groupe Gascon, Villeneuve du Paréage, 09100, France, [4] Institut de l'Elevage, Castanet-Tolosan Cedex, 31321, France; p.sans@envt.fr

Gascon is one of the main French cattle rustic breeds, which origin is placed in the South of France near Pyrenees. Its long tradition and its presence in border areas, as well as it meat quality characteristics, favour its progressive presence in the neighbour country of Spain. That presence started in the 19th century, but the most important growth was in 1980-1990´s, until obtaining its current census of 5,666 heads and presence in more than 552 Spanish farms. Being Cataluña, Aragón and Cantabria the regions which concentrate the 69.2% of the national census. After personal interviews to Spanish Gascon breeders, the main reasons to initiate and rear with this breed were: rusticity, adaptability and easy calving, which combined with its meat production potential, make the real profitability of the breed. However, one of main problems to the breed is the lack of official recognition in Spain and the scarce information about their situation and productivity under Spanish conditions. So, last years several studies have been performed in collaboration of both Spanish and French breeders associations and research centres, under a European Project POCTEFA-OTRAC, which increased the information available about the breed and its characteristics. The Spanish Association of Gascon Breeders was created and it was started to develop the requirements (R.D.2129/2008) for the official registration. Also, the control of performances and the basis for Spanish herd-book are in process of development. This paper will illustrate how and why a breed increases its presence in a foreign country, as well as what is the common evolution of a breed and steps until it became completely established in a foreign country.

Building of biomimetic structures in order to reproduce the outer membrane of bull spermatozoa

J. Le Guillou[1], M.H. Ropers[2], D. Bencharif[1], L. Amirat-Briand[1], S. Desherces[3], E. Schmitt[3], M. Anton[2] and D. Tainturier[1]
[1]Oniris, 44, Site de la Chantrerie, route de Gachet, 44300 Nantes, France, [2]INRA, 44, Rue de la Geraudière, 44300 Nantes, France, [3]IMV Technologies, 61, ZI no. 1 Est, 61300 L'Aigle, France; chalawak@yahoo.fr

The knowledge about the mechanisms of action of protective extenders used to preserve bull semen for frozen or chilled is essentially empirical. The purpose of this study was to create biomimetics structures which can reproduce these mechanisms. To achieve this goal, a structure which reproduces the outer membrane of spermatozoa has been built. The biomimetics structure chosen for experiments is a lipid monolayers at the air-water interface formed on a Langmuir balance. Composition of subphase is controlled during experiments and can be changed to test interactions between biomolecules and biomimetic membrane. The purpose of these tests is to highlight biomolecules implicated in the protective effect on the semen. First, a lipids mix deposit was done, then the barriers positions was modulated to get the desired molecular compression with a controlled pressure. Then, molecules with protective effect were introduced in the subphase. The monolayer changes were monitored. Each experiment was replicated twice. Miscibility studies at 34 °C and 8 °C shows the formation of homogeneous domains of sphingomyelin and cholesterol, located in fluids domains composed of phosphatidylcholine. Complex biomolecules extracted from egg yolk like Low Density Lipoprotein was incorporated into the monolayer, contrary to other purified molecules like egg phospholipids. Contact tests with the monolayer were conducted with bull seminal plasma. Seminal plasma was injected either alone or associated with protective molecules which seems to inhibit the effect of seminal plasma on the monolayer. This model could be an opportunity for further studies where monolayer composition can vary as lipids composition depending the species studied. Other protective molecules, as well as the composition of the subphase can be also tested.

Different forage allowances in the pre- and postpartum period and the production of beef cows

G. Quintans, A. Scarsi, J.I. Velazco and G. Banchero
National Institute for Agricultural Research, Beef and Wool, Ruta 8, km 281, 33000, Uruguay; gquintans@tyt.inia.org.uy

The experiment was a 2×2 factorial design, in which the factors were forage allowance (FA) of native pastures in the prepartum (Low (PreP-L) and High (PreP-H)) and postpartum period (Low (PP-L) and High (PP-H)). It involved 48 multiparous AAxHH cows (472 kg BW and 4.4 u BCS -1-to-8scale). The L level of FA was 5 kg of DM/100 kg BW and H level, 15 kg of DM/100 kg BW. The prepartum treatments were applied from Day -56 to calving (Day 0) and the postpartum treatments from calving to Day 56. BCS was recorded biweekly from Day -56 to 168 (weaning). Cows were blood sampled weekly from Day -56 to 119. The breeding season started at Day 60 and lasted 60 d. BCS, NEFA, progesterone and postpartum anoestrous period (PPA) were analyzed by repeated measures (PROC MIXED procedure). The probability of pregnant cows were analyzed with a generalized linear model (PROC GENMOD procedure). Preliminary results revealed no significant interaction between the main treatments effects. There was a prepartum and postpartum treatments by time interaction ($P<0.01$) in BCS. Cows in PreP-L lost 0.6u of BCS ($P<0.01$) from Day -56 to calving while cows in PreP-H lost ($P<0.01$) 0.1u. After calving cows in PP-H increased ($P<0.01$) BCS (0.45u) until Day 84 respect to cows in PP-L that maintained their BCS. Also there was a postpartum treatment × time interaction ($P<0.01$) in NEFA concentrations. In PP-H it decreased sharply from 0.49 to 0.23±0.04 mmol/l from Day 7 to 14 and in PP-L it decrease gradually from 0.61 to 0.35±0.04 mmol/l between Day 7 and 49. After Day 49 NEFA concentrations were similar in both group. The PPA was shorter ($P<0.05$) in PP-H than in PP-L cows (102 vs. 113±3 d). Also, the probability of cows to get pregnant was greater ($P<0.05$) in PP-H than in PP-L cows (63 vs. 25%). Under the conditions of this experiment, high FA only in the postpartum period improved reproductive performance, probably explained by higher BCS and lower NEFA concentrations.

Author index

A

C

E

G

H

L

N

O

P

Q

S

T

W

Printed in the United States
by Baker & Taylor Publisher Services